油气田生产常用化学剂
（第二版）

［挪威］马尔科姆·A. 凯兰（Malcolm A.Kelland） 著

李琼玮 陆红军 刘广胜 刘爱华 周志平 等译

石油工业出版社

内容提要

本书介绍了油气田堵水调剖、结垢和沥青质、腐蚀、细菌、水合物、结蜡防治等18类生产中常用化学剂的最新进展、技术实践等内容，特别是近年来国外油气田化学剂配方的最新研究和基于生产实践的一些新认识、环保性要求。

本书适合油田工程技术领域和从事油田化学技术开发的科研人员、高等院校相关专业师生参考使用。

图书在版编目（CIP）数据

油气田生产常用化学剂：第二版 /（挪威）马尔科姆·A.凯兰（Malcolm A.Kelland）著；李琼玮等译.
北京：石油工业出版社，2024.7. -- ISBN 978-7-5183-6624-8

Ⅰ.TE39
中国国家版本馆 CIP 数据核字第 2024GX9414 号

Production Chemicals for the Oil and Gas Industry, 2nd Edition
Malcolm A. Kelland
ISBN: 9781439873793

© 2014 by Taylor & Francis Group, LLC
CRC Press is an imprint of Taylor & Francis Group, an Informa business
Authorized translation from English language edition published by CRC Press, part of Taylor & Francis Group LLC.
All Rights Reserved.
Petroleum Industry Press is authorized to publish and distribute exclusively the Chinese (Simplified Characters) language edition. This edition is authorized for sale throughout Mainland of China. No part of the publication may be reproduced or distributed by any means, or stored in a database or retrieval system, without the prior written permission of the publisher.
本书经 Taylor & Francis Group, LLC 授权石油工业出版社独家翻译出版并仅在中国大陆地区销售，简体中文版权归石油工业出版社所有，未经出版者书面许可，不得以任何方式复制或发行本书的任何部分。
Copies of this book sold without a Taylor & Francis sticker on the cover are unauthorized and illegal. 本书封面贴有 Taylor & Francis 公司防伪标签，无标签者不得销售。
北京市版权局著作权合同登记号：01-2023-3639

出版发行：石油工业出版社
（北京安定门外安华里2区1号　100011）
网　　址：www.petropub.com
编辑部：（010）64523825　图书营销中心：（010）64523633
经　　销：全国新华书店
印　　刷：北京中石油彩色印刷有限责任公司

2024 年 7 月第 1 版　2024 年 7 月第 1 次印刷
787×1092 毫米　开本：1/16　印张：27.75
字数：650 千字

定价：160.00 元
（如出现印装质量问题，我社图书营销中心负责调换）
版权所有，翻印必究

《油气田生产常用化学剂（第二版）》
翻 译 组

组　长： 李琼玮　陆红军
副组长： 刘广胜　刘爱华　周志平
成　员： 杨会丽　李成龙　尹太恒　刘晓庆　苑慧莹
　　　　　郭　钢　毕卫宇　杨　乐　贺华镭　赵　静
　　　　　连宇博　姬忠文　戚建晶　王　燕　苟利鹏
　　　　　李炘泽　李　慧　陈　伟　吴春生　吕　伟
　　　　　李　楷　刘　宁　刘　伟　田宇欣　杨　鹏

译者前言

2018年，我们接触本书英文版，其内容和形式不同于国内石油院校相关教材或著作，在油气工业生产常用化学剂技术发展脉络和专利体系分析等方面有其独特之处，工作中翻阅相关章节时，经常有所收获或启发。2019年底新冠疫情暴发期间，与Ping Chen、Qiwei Wang等专家的线上交流中，他们也推荐此书，认为内容丰富值得国内同行参考，我们萌生了将此书翻译为中文的想法。自2020年下半年起的3年多时间里，我们的团队依托油气田开发工艺和油田化学的生产实践，以推动国内相关油田生产化学剂技术进步为目标，以准确、流畅的高质量翻译为标准，克服诸多困难，最终得以成书。

本书共19章。第1章、第6章至第19章及附录由李琼玮、陆红军、刘广胜、刘爱华等翻译，第2章至第5章由刘爱华、李琼玮、尹太恒、刘晓庆等翻译。杨会丽、毕卫宇、苑慧莹、杨乐、贺华镭等参与了部分校对工作。

在本书翻译过程中，我们得到了Nalco公司Ping Chen、Baker Hughes公司Peng Jin、Saudi Aramco公司Qiwei Wang、中国石油大学（北京）杨子浩、华中科技大学陈振宇、陕西师范大学俞斌勋、西南石油大学李年银和贾虎及西安石油大学李善建等教授和专家的支持与帮助，还得到了中国石油长庆油田公司胡建国、慕立俊、张矿生、张箭啸、张钊、杨海恩、李宪文等专家的支持，在此一并表示感谢！

由于译者水平有限，书中难免存在疏漏和不当之处，敬请读者批评指正！

原书前言

首先感谢购买了 *Production Chemicals for the Oil and Gas Industry*（First Edition）（简称第一版）著作的读者，希望该书对大家有所裨益。同时也感谢那些给我反馈建议的读者，几乎所有的反馈都很正面，鼓励我对第一版进行了修订。

Production Chemicals for the Oil and Gas Industry（Second Edition）（简称第二版）的参考文献更新到2013年秋季。尽管第一版和第二版之间只相距4.5年，但这期间技术工作者又开展了大量的研究和现场应用。例如，第二版的参考文献数量增加了近50%，几乎没有早于2009年以前的。另外，我撰写的报告反映了在油田生产化学品的发展中观察到的其他变化和趋势（M.A. Kelland，*Production Chemicals and Their Future*，*Chemistry in the Oil Industry* 第13期新前沿栏目，英国曼彻斯特，2013年11月）。

为了便于参考文献中新文章的定位，所有新文献都以数字顺序方式添加在第一版文献之后。例如，第1章中的参考文献［1-54］来自第一版，而参考文献［55-96］则是第二版中新增内容。除此之外，第二版还有什么新内容呢？首先，我将排水采气用泡排剂从气井排水采气章节中拆分，并独立成章（第19章）。应许多读者的要求，还增加了有关静水压测试用化学剂的新章节（第18章）。此外，在第1章中还增加了关于化学加注方式的内容。以上新增内容都不甚详尽，希望读者多提意见和建议，还有读者希望增加如表面活性剂和（或）聚合物驱提高采收率、稠油集输、压裂液和示踪剂等章节，由于出版物的篇幅限制和个人时间有限，第二版中没有添加这些主题。

自第一版出版以来，出现了一些有参考价值的新网站。在Oilfieldwiki网站可以查找一些油田化学品的基本信息，同时还可以学习相关油田术语，其链接如www.oilfieldwiki.com/w/Emulsion 或 www.oilfieldwiki.com/w/Scale_Inhibitor。在www.LinkedIn.com，还有如Oilfield production chemistry、Oilfield chemicals、Flow assurance、Gas hydrates、Multiphase（no chemistry here）等相关论坛。

感谢在本书部分主题的编写上给予我特别帮助的Alan Hunton（Humber技术服

务公司），Henry Craddock（HC 油田和化学咨询公司），Ian Gilbert（M-I Swaco，斯伦贝谢公司），Niall Fleming 和 Lars Ystanes（Statoil 公司），Øystein Bache（康菲公司），Paul Barnes（科莱恩公司）和 Kolbjørn Johansen（bp 挪威）。此外，还要感谢 Amor Nanas 对书稿的校对。再次感谢我的妻子 Evy 在第二版写作过程中的耐心支持。

我的电子邮件地址是 malcolm.kelland@uis.no。

Malcolm A. Kelland

斯塔万格大学

目录

1　简介和环境问题 ... 1
1.1　生产化学概述 ... 1
1.2　影响化学品选择的因素 ... 4
1.3　环境和生态毒理学条例 ... 8
1.4　绿色化学品设计 ... 11
1.5　汞和砷的生产 ... 15
参考文献 ... 16

2　堵水调剖及气体封窜 ... 22
2.1　概述 ... 22
2.2　树脂和弹性体 ... 23
2.3　无机凝胶 ... 23
2.4　用于永久封堵的交联有机聚合物凝胶 ... 24
2.5　黏弹性表面活性剂凝胶 ... 31
2.6　不等比例渗透率降低剂（DPR）或相对渗透率调节剂（RPM） ... 31
2.7　使用微粒控水 ... 36
2.8　热敏水溶性聚合物 ... 37
2.9　水溶胀聚合物 ... 38
2.10　控制产气 ... 38
参考文献 ... 39

3　结垢防治 ... 52
3.1　概述 ... 52
3.2　垢的类型 ... 52

  3.3 非化学防垢 ··· 55
  3.4 元素周期表的ⅡA族元素碳酸盐和硫酸盐防垢剂 ············· 57
  3.5 硫化物防垢剂 ··· 68
  3.6 岩盐防垢剂 ··· 69
  3.7 防垢剂应用方法 ··· 70
  3.8 防垢剂性能测试 ··· 78
  3.9 化学除垢 ·· 80
  参考文献 ··· 85

4 沥青质的控制 ··· **115**
  4.1 概述 ·· 115
  4.2 沥青质分散剂和沥青质抑制剂 ··································· 118
  4.3 沥青质溶解剂 ··· 135
  参考文献 ··· 138

5 酸化增产 ·· **150**
  5.1 概述 ·· 150
  5.2 碳酸盐岩地层压裂酸化 ··· 150
  5.3 基质酸化 ·· 151
  5.4 酸化中的酸 ··· 151
  5.5 酸化对储层的潜在伤害 ··· 153
  5.6 酸化添加剂 ··· 153
  5.7 井筒轴向布酸酸化 ·· 160
  5.8 井筒径向布酸酸化 ·· 168
  参考文献 ··· 171

6 油井防砂 ·· **185**
  6.1 概述 ·· 185
  6.2 化学防砂 ·· 185
  参考文献 ··· 188

7 环烷酸盐和其他羧酸盐垢的控制 ... 190
7.1 概述 ... 190
7.2 使用酸控制环烷酸盐沉积 ... 191
7.3 低剂量的环烷酸盐抑制剂 ... 191
参考文献 ... 193

8 生产过程中的腐蚀控制 ... 196
8.1 概述 ... 196
8.2 腐蚀控制的方法 ... 197
8.3 缓蚀剂类型 ... 198
8.4 成膜型缓蚀剂 ... 200
参考文献 ... 214

9 天然气水合物控制 ... 225
9.1 概述 ... 225
9.2 水合物堵塞的化学预防 ... 227
9.3 水合物堵塞物的清除 ... 249
参考文献 ... 250

10 蜡（石蜡）控制 ... 270
10.1 概述 ... 270
10.2 蜡质控制策略 ... 272
10.3 化学除蜡 ... 274
10.4 化学防蜡 ... 276
参考文献 ... 287

11 破乳剂 ... 299
11.1 概述 ... 299
11.2 破乳方法 ... 300
11.3 油包水破乳剂 ... 301
参考文献 ... 312

12 泡沫控制 320
12.1 概述 320
12.2 消泡剂和抑泡剂 320
参考文献 323

13 絮凝剂 325
13.1 概述 325
13.2 絮凝理论 326
13.3 絮凝剂的种类及性能 326
参考文献 334

14 杀菌剂 339
14.1 概述 339
14.2 控制细菌的化学剂 340
14.3 杀菌剂的种类及性能 342
14.4 生物抑制剂（"控制性生物杀伤剂"或代谢抑制剂） 352
参考文献 355

15 硫化氢清除剂 365
15.1 概述 365
15.2 非再生型脱硫剂 367
参考文献 376

16 除氧剂 383
16.1 概述 383
16.2 除氧剂的种类 383
参考文献 387

17 减阻剂 390
17.1 概述 390
17.2 减阻剂机制 391

17.3 油溶性减阻剂 ·· 393

17.4 水溶性减阻剂 ·· 397

参考文献 ··· 404

18 静水压测试用化学剂 ·· 412

18.1 概述 ·· 412

18.2 静水压测试用化学剂配方 ·· 413

参考文献 ··· 416

19 排水采气用泡排剂 ·· 418

19.1 概述 ·· 418

19.2 泡排剂的特性和类别 ·· 418

参考文献 ··· 419

附录 油田化学品的 OSPAR 环境条例 ·· 421

附录1 英国和荷兰的北海生态毒理学条例 ·· 423

附录2 挪威海上生态毒理学条例 ·· 424

参考文献 ··· 425

1 简介和环境问题

1.1 生产化学概述

油气流体从油藏到井底、井口、地面集输系统等过程中,其化学和物理特性的变化会导致生产化学问题。液态烃(石油或凝析油)、气态烃(天然气)和伴生水等混合物组成的井下流体从储层,经过井下管柱和井口,进入集输系统到达处理厂后,被分离成不同相态。分离过程因压力大幅度下降、温度和流态变化,流体会出现可预测或不可预测的状态变化,从而影响整个生产运行的效率。在处理厂的下游,原油被输送到炼油厂,气体进行加工处理,水进行除杂质处理,这些过程导致生产问题更加复杂。

一般来说,生产化学问题有以下四种类型:

(1)污垢造成的问题。包括垢、腐蚀产物、蜡(石蜡)、沥青质、生物垢和天然气水合物等任何有害物质在系统中的沉积。

(2)流体物理性质变化引起的诸如泡沫、乳液和黏性流等问题。

(3)影响设施结构完整性和员工安全的问题,主要与腐蚀有关。

(4)环境或经济问题。含油污水排放破坏环境,H_2S 等硫化物造成环境和经济问题。

上述问题可以通过应用非化学技术和选择适当的化学添加剂来解决。常用的非化学技术包括以下几种:

(1)绝热保温(保留热量以延缓结蜡或产生天然气水合物);

(2)加热出油管路(防控结蜡和天然气水合物);

(3)在分离器中加热(解决乳化液问题);

(4)降低压力(延缓生成天然气水合物);

(5)保持高压(延缓沥青质絮凝、碳酸盐结垢、环烷酸盐沉积);

(6)使用耐腐蚀材料和涂层(尽量降低腐蚀);

(7)增加流量/湍流(减少沥青质、蜡、微生物膜);

(8)降低流速(最大限度减少起泡、乳化液);

(9)增加分离器尺寸(提高油水分离效果);

(10)利用电场(增大油水乳状液中水滴的聚并);

(11)离心(分离油水乳状液);

(12)膜和精细过滤器(去除机械杂质、悬浮物和特定离子);

(13)管道清管(防止管道内固体积聚);

(14)刮削工具(清除沉积物,特别是井下沉积物);

（15）磨铣或钻孔/划眼（清除井下沉积垢）；

（16）真空脱气（从水中去除气体）；

（17）筛管和桥塞（用于井下堵水）；

（18）筛管和砾石充填（用于防砂）。

石油行业中，无机垢造成的损失（不包括腐蚀）估算每年约20亿美元，与腐蚀相关的损失成本更高。目前针对蜡质、天然气水合物、沥青质和环烷酸盐污垢带来的成本损失没有准确估算，但普遍认为这些问题比无机垢更严重。

合理的设备设计和正确的材料选择虽然可以减少油田生命周期内生产用化学品的问题，但原油生产过程中产量经常波动，且产出流体的性质变化不可预测，化学技术人员可以采用一系列化学添加剂来解决上述无法彻底解决的生产问题。最新加工工艺、提升不同品质原油的需求及环境要求，也都需要化学品来解决。

为解决或尽量减少上述化学问题的影响，油田化学品从功能上可分为以下几类：

（1）最大限度减少污垢的抑制剂，以及清除各类沉积物的溶剂。

（2）有助于提高气液分离和油水分离工艺效果的化学品。

（3）提高完整性管理的缓蚀剂。

（4）为环保合规等其他生产目的而添加的化学品。

许多化学污垢问题都涉及流动保障，该术语出现在20世纪90年代，用于描述产出流体从油井到加工设备的流动过程所涉及的相关保障问题。流动保障的化学问题通常与管线中蜡（石蜡）、沥青质、垢、环烷酸盐和天然气水合物等固体沉积问题有关。防止这些沉积的两种常用方法如下：一是使用分散剂，允许固体颗粒产生，将其分散在生产流体中而不沉积；二是使用抑制剂来控制形成固体。近几年，新开发的大型油田都面临海上水深不断增加和（或）更寒冷环境。此外，小型的海上油田通常存在现有平台或其他需要水下多相流长距离管道等基础设施并行的情况。高压、海底低温和液体滞留时间过长等极端条件对流体的流动保障带来了更大的挑战，特别是在控制天然气水合物和蜡沉积方面。与其他化学问题一样，在油田设计阶段必须制定防止天然气水合物沉积的方案，可以使用热力学水合物抑制剂（THIs）或低剂量水合物抑制剂（LDHIs）。当使用防蜡剂不能完全控制蜡沉积时，可以配套采用定期机械清管措施。

在井口上游，主要可能出现的化学沉积问题是结垢和沥青质沉积，如果井筒的上段温度较低，也会出现蜡沉积。在海上或寒冷的陆上环境，天然气水合物会出现在井口上游或停输的管道内，通常采用水合物抑制剂溶解或加热来解决海上油井上游的水合物堵塞问题，大多数海底管道水合物堵塞则通过减压等技术来解决。通过机械或化学的堵水处理方法来阻断和减少产出水，降低处理设备需要处理的水量，也能缓解结垢问题。通常情况下，在井下或井口投加防垢剂控制结垢，还能用不同的化学溶解剂来清除沥青质、蜡和无机垢。可以通过优选酸化液体系或使用低剂量环烷酸盐抑制剂来减缓环烷酸盐及其相关的乳化问题。以提高油气产量为目的的酸化（压裂酸化或基质酸化）通常用于溶蚀部分天然岩层（砂岩或碳酸盐岩），也可去除沉积的碳酸盐和硫化物垢。酸化过程中的腐蚀问题解决需要在强酸性条件下有良好性能的特殊缓蚀剂。其他井下化学处理包括堵水、防气窜和固结砂。

分离的油、气和水必须满足生产规定的最低杂质要求。破乳剂和消泡剂等辅助剂用以满足原油生产技术要求。但由于原油中存在沥青质、胶质、环烷酸盐和其他天然表面活性剂，以及井口的高剪切和运输过程中的混合，部分或全部产出水与液态烃相发生乳化，需要在地面处理设施中解决乳化问题。这些处理设施的有效运行使原油达到外输质量要求。油水分离过程中通常将破乳剂与热或电处理等非化学技术联合使用。分离后的采出水通常含有很多未溶解和分散的烃类（水包油型乳液）及悬浮物颗粒，不允许排放环境中，因此还要加入絮凝剂（也称除油剂、净水剂或反相破乳剂）处理这类采出水，分离絮凝的杂质（或絮体），排放相对更洁净的采出水。还有其他技术从水中分离游离的和溶解的（水溶性）烃类，甚至某些生产化学品。对陆上油田采出水的处理，有包括限制盐浓度等指标在内的更高环保要求，此时需要回注采出水，这种方法在北海等油田也有应用。在油气分离器处理设备中，还需要加入消泡剂和抑泡剂以解决泡沫问题。

金属与水接触后，会发生影响设施和管道内外表面的电化学腐蚀。腐蚀速率与如 CO_2 和 H_2S 等水溶性酸性气体的浓度成正比，也与水的矿化度相关。在高温井中，该问题的影响更严重，此时特殊的耐蚀合金更具有经济优势。使用 H_2S 清除剂、杀菌剂/生物抑制剂、硝酸盐/亚硝酸盐等，可降低 H_2S 的浓度以缓解腐蚀问题，在第 14 章和第 15 章中有相关介绍。通常油田生产时，间歇或连续加注缓蚀剂，将腐蚀控制在油田可接受的使用寿命范围内。

在油气上游工业中，还应用了其他许多化学品：H_2S 清除剂既可以减缓腐蚀问题，也可以避免炼油厂的原油加工问题或环境问题（减少毒性）；加入絮凝剂可减少分离水中的有毒污染物，满足环境要求；减阻剂（DRAs）不会影响地层胶结、腐蚀，或改变乳液、泡沫，其作用是提高管道或注入井中流体的流量，所以不归为常规化学品。

化学品可以在井下、井口或井口与处理设施（分离系统）之间加注。原油输送管道可加注缓蚀剂、防蜡剂、防垢剂和杀菌剂等化学剂。缓蚀剂也可用于输气管道。

化学剂可以通过毛细管柱或气举系统进行井下注入。间歇式处理通常用于井下位置，其中包括将化学药剂挤入储层的技术。挤注处理是在储层内放置大量的化合物，在数月内随产出流体缓慢释放。挤注处理将在 3.7.2 中详述。酸化、堵水、防垢挤注或溶垢处理等井下化学剂作业可采用高压方式，即通过井口或生产管线从平台、船或泵车泵入井中。可以采用多种转向方法，获得适宜的地层挤注处理部位。如果挤注存在风险（需要精确的位置，或者处理液可能会造成伤害和产量损失），则可以使用连续油管等油井干预技术，将化学剂注入所需的区域，但该技术应用于海上修井作业时成本很高。压裂作业期间，化学品也可以注入在近井区域。

除了挤注处理，还开发了在井中控制释放化学品的技术。例如，将固体颗粒放置在井底（鼠洞）及上部等位置，如果颗粒非常小，还可以挤注到近井区域，当储层流体流过时，固体颗粒会缓慢释放所需的化学物质。这些技术主要用于控制结垢，更多细节和参考文献可见第 3 章。

化学品服务公司从化学品原料供应商、特种化学品供应商处采购或利用自有的专用生产设施等渠道制备化学品。这些产品可以是单一成分（如甲醇），也可以是在溶剂中含有多种活性成分的复杂配方。

化学品服务公司还销售用于注水井、管道静水压测试、其他维护和公用系统的化学品。用于注水系统的化学品包括降低腐蚀的除氧剂、减少微生物和硫化氢腐蚀的杀菌剂、提高注水速度的水基减阻剂、缓蚀防垢剂和消泡剂。为了进一步提高采收率，还可以注入聚合物或表面活性剂。油基减阻剂通常用于提高原油管道的输送能力或减少对增压站的需求。

油田生产中化学问题由现场作业人员管理。在许多生产地区，化学品服务公司负责优化和实施化学处理，为整个生产过程制订全面的化学处理方案，包括选择适宜的化学品及其剂量、考虑药剂相容性（见 1.2 节）和置放、现场生命周期的需求等。油田化学工作者也关注非化学技术的贡献，以及其对化学添加剂需求的影响。实验室推荐的化学品用量往往与实际现场条件的需要相差很大，预测最佳现场用量也因此变得较为困难。在油田开发阶段，生产人员需提前考虑，与服务公司合作确定最佳解决方案，尽量减少或避免油田开发中后期出现的化学处理问题。对修井或措施成本较高的海上油田来说，这项工作尤为重要。

1.2　影响化学品选择的因素

影响化学品选择的因素包括性能，价格，稳定性，操作、储存的健康和安全，环境限制和兼容性等。

一般来说，油田运营商希望产品的性能令人满意，价格合理。化学品的总体性能基于多项测试，如用于挤注处理的防垢剂可能防垢性能好，但其在岩石表面的吸附能力较差，导致挤注有效期较短（详见第 3 章）。如果缓蚀剂能更好地吸附于岩石，则可以选择缓蚀性能稍差的产品。对于防垢、蜡、沥青质、腐蚀和低剂量水合物抑制剂等化学品，油田运营商会要求几家服务公司提交选定的产品，自主或委托第三方公司对产品进行性能测试和排序。最终选择的不一定是性能最优的产品，而是性价比综合最优的产品。低价格产品可能短期内比较经济，但如果其性能明显比稍贵的产品差很多，并导致更多的生产问题、更频繁的修井或管道维护和生产损失，则从长期生产成本考虑，应选择价格稍贵的产品。

生产用化学剂在加注前的运输和储存期间，必须保持性能稳定。在寒冷环境中，产品不能过于黏稠或结冰。相反，在高温的储存环境，需要产品不会出现过快降解或相态变化。在对化学品生物降解性等环保要求较高的地区，可能出现两难情况。油田运营商既希望产品在海水中快速降解，又要求现场储存时不会降解。

根据化学品可能对使用者的安全和健康带来的危害和风险，很多国家都制定了化学品分类条例。化学品的材料安全数据表中，必须依法提供所有潜在危险产品的必要信息。例如，蜡和沥青质抑制剂配方或破乳剂中的某些非极性芳香溶剂等挥发性有毒溶剂，使用者吸入后，会因剂量和接触时间的不同，导致健康问题。许多化学品供应商和服务公司尽量采用更安全、毒性更小、挥发性更低、闪点更高的替代溶剂。适用于油田化学品的绿色有机溶剂包括丙二醇甲醚乙酸酯、二丙二醇甲醚乙酸酯、二异丁基酮和甲基异丁基酮。

使用化学品时，与兼容性相关的操作问题主要包括：

（1）化学品是否会导致或加剧其他化学品问题？是否与其他化学品有协同作用？

（2）是否与生产流程的所有材料兼容？

（3）是否会造成下游生产的问题？

（4）是否存在注入问题——黏度、浊点、起泡？

（5）是否与同时使用的其他化学品兼容？两种不同的化学品彼此影响性能？能否与其他化学品共注入？

第一类中有一些为人熟知的问题。例如，一些成膜型缓蚀剂会加剧分离器中的乳液和泡沫问题。使用热力学水合物抑制剂（如甲醇和乙二醇）会加速垢沉积。三嗪类脱硫剂增加采出水中的pH值，可能加剧碳酸盐结垢。井下酸化作业中使用酸会导致沥青质析出，但在配方中添加的某些化合物也可以减少这些情况。季铵天然气水合物防聚剂可以起到缓蚀作用，有时甚至不需要单独使用缓蚀剂。一些供应商提供防垢剂和缓蚀剂复合配方，其优点是只使用一套储罐、泵和注入管线。这种复合配方既有几种产品的简单混合，也有一些多功能的单组分产品可供选择。在某些条件下，减阻剂也可以提高缓蚀性能。本书的相关章节将进一步讨论这些实例。

油田生产中使用新化学品（可能是纯/浓缩的，或被产出液稀释的）时，要检查其与生产中所用材料的兼容性（如弹性密封件）。

一些化学品会进入油相或气相，造成炼油厂中的催化剂污染、过多的水合物抑制剂甲醇降低油品价值等问题。运营商还需要注意蜡、沥青质抑制剂和某些天然气水合物抗凝结剂等油溶性化学品，是否会造成裂化器结垢等下游问题。

化学品注入方式包括：油井井口或井下连续或间歇注入；通过井下挤注处理；通过缓释产品。

在第3章中详细讨论挤注处理。在化学挤注领域，目前应用范围最广的是挤注防垢剂，也可用沥青质抑制剂等其他化学物质挤注处理，以及放置井底鼠洞具有缓释性能的防垢剂等多种加注方法。

化学品可连续或间歇注入。井下注入防垢剂需要仔细设计。典型的化学注入系统如图1.1至图1.3所示，有文章讨论了化学品连续注入系统的相关经验和后果。

化学品还可以在井口（陆上或海底）和运行设备处加注。海底井口注入化学品时，通过一个或多个小型脐带缆从平台注入井口。脐带缆的长度可达50km，直径0.6~2.5cm。要保证在海底温度环境下注入的化学剂黏度足够低，能够沿管线泵送。在低温环境中应用化学产品，会限制某些化学品的浓度和溶剂，需要对产品进行防冻处理。当不同的化学品使用同一条脐带缆输送时，偶尔会发生化学品间的不相容。例如，将缓蚀剂注入之前加注

图1.1　德国 Bran & Luebbe 多头计量泵（现为 SPX）

图1.2　Williams 的气泵（现为 Milton Roy）

图1.3　配套分配阀和泵的化学剂储存系统（Milton Roy）

过防蜡剂或防垢剂的管线中，如果不使用溶剂或水充分冲洗，可能会造成管线堵塞。在长期停产期间，应该用溶剂置换脐带缆中的化学品，确保管道完整性。通常为避免不同化学品立刻混合和可能的不相容，注入点之间的距离应大于3倍管径。有报道称在注入压力高达70MPa和海底温度条件下，改进了注入化学品的测试方法。

受到化学品黏度和泵的局限性，很多化学物质溶解在水、醇、乙二醇和烃类（煤油、二甲苯等）溶剂中。应用甲醇、乙醇和乙二醇等热力学水合物抑制剂时，要考虑随着油田生产期内的产出水量增加，加注能力是否足够。

在注入化学品时，必须考虑相关因素。下面简要总结最重要的因素：

（1）注入药剂的最佳部位在哪里，才能产生最好的处理效果？

所有的抑制剂应该在潜在危险点的上游加注。例如，防蜡剂在蜡析出点之前加入，沥青质抑制剂在沥青质絮凝之前加入，水合物抑制剂在水合物平衡温度之前加入，防垢剂在结垢过饱和点之前加入。通常最好在流体进入混合装置（如节流器和阀门）之前注入化学剂，使其均匀分散于流体（聚合物减阻剂为例外情况，可能会因此降解并失去减阻能力）。这样井下加注或挤注等措施可以保护井口以外的出油管道，防止结垢或沥青沉积。但是海底井口注入水合物抑制剂后，关井期间仍会在井口形成水合物堵塞。

（2）化学品在管线中是否有保证正常起效的足够的停留时间？

例如，一些 H_2S 清除剂在混合不足的情况下，由于其反应动力学相当缓慢，如果注入延迟，会降低清除剂效率。破乳剂在井口注入与在地面集输设施注入相比，其黏度降低，使油水提早分离，增加管道 6 点钟部位的腐蚀。另一个与化学停留时间有关的实例是除氧剂的使用，除非水中存在铁离子或添加过渡金属离子的催化剂，否则其除氧反应过程比较缓慢。

（3）化学品在注入部位的温度范围内是否稳定？

一些动力学水合物抑制剂在热产出水中易沉积，尤其在高矿化度环境中。如果溶剂系统蒸发过快，药剂通过井下气举管线注入，会导致化学沉积。防垢剂应用中，也会出现类似现象。

（4）化学品与其注入的流体是否相容？

化学品必须与流体的矿化度和 pH 值相容。如前文所述，一些动力学水合物抑制剂与高矿化度盐水不相容（特别是高温环境），如果同时注入采出水和海水，会出现硫酸盐结垢。因此在流体混合之前，必须在管汇上游注入防垢剂。

（5）化学品在注入、生产或设备中是否会产生副作用？

操作人员需要在整个注入或生产过程中检查化学品与材料的长期相容性。缓蚀剂等表面活性高的化学品可能会影响处理设备的破乳效率，也可能影响排放的水质。例如，在注水管线中使用季铵盐杀菌剂，优点是其高表面活性可以去除生物膜，但在生物膜进入储层的 1~2h 内，会导致药剂注入性暂时下降。在乙二醇再生回收塔中，气流中微量的缓蚀剂会引起泡沫和其他负面影响。

关于化学品与材料的相容性，如氨基缓蚀剂与常用的密封材料氟化橡胶不相容，在使用过程中会造成密封材料老化问题。甲醇与尼龙 66 不相容，甲醇通过尼龙脐带缆输送到相邻脐带缆的防垢剂中，会导致防垢剂沉淀。如果是金属脐带缆，就需特别重视腐蚀问题。

（6）化学品会影响其他化学品的性能吗？

某些化学品会影响其他化学品的性能。任何化学品的性能测试必须在可能同时存在其他化学品的条件下进行。例如，很多成膜缓蚀剂可以降低多数除垢剂和破乳剂的性能，有些甚至使动力学水合物抑制剂失效。有些碱性的 H_2S 清除剂（如三嗪）提高产出液的 pH 值，导致结碳酸盐垢或加剧已有的结垢问题。注入热力学水合物抑制剂也可能会影响结垢，使垢盐更难水溶。

如果生产中使用的化学品数量很大，与相容性有关的问题可能会变得极其复杂，甚至导致生产中断。现场的油田化学技术人员要根据自身经验，结合实际分析具体生产问题，保障正常生产。

（7）注入管线中黏度和泵送问题。

使用纯或高浓度化学品可降低其储存、运输、现场使用等成本。为满足油田生产中化学品注入和泵送能力，就需权衡化学品的活性浓度与最大黏度之间的关系。在冷注入管道和长输管道（数十千米）中，很难控制注入速度，需要添加特殊的防冻配方，有时只是将产品稀释或使用特殊溶剂。如减阻剂（用于油或水的减阻剂）、动力学水合物抑制剂和蜡抑制剂—倾点抑制剂等高浓度的黏性聚合物溶液，特别是高分子量的聚合物溶液可能难以注入。减阻剂是超高分子量的聚合物，在溶液中具有极强的黏性。可以通过溶剂将高黏度化学品分散成颗粒或使用乳剂来降低其黏度等新方法，克服这类产品难注入的问题，详见第17章。

一些化学品是水溶性的，而另一些是油溶性的，这就意味着这两种产品不能通过同一条管线注入，但是实际生产中经常会遇到一条管线需要同时注入两种化学剂，如井口同时注入油溶性蜡或沥青质抑制剂和水溶性防垢剂，此时就需探索合适的互溶剂或开发乳化剂，以克服该问题。

1.3 环境和生态毒理学条例

世界各地越来越重视油田化学品在储存、应用中带来的危险，并限制油田采出水中化学品排放浓度。目前难题是一方面高性能的"更绿色"化学品开发难度大，另一方面许多新油田开发条件更恶劣，如更冷或更深的水、储层温度更高等，需要特殊的高性能化学品。

世界上大多数地区都有关于使用和排放油田化学品的环境条例。化学品注入井中，进入水、烃类化合物或气相。如甲醇等低沸点溶剂会部分进入气相，污染原始天然气，降低其经济价值。但多数化学品会在采出水或液态烃相或在两者之间分布。如果其在烃相中，一般不会被排放到环境中（除非发生泄漏或其他事故）。在正常生产过程中，对环境构成最大风险的是含油和化学品采出水的排放。采出水残余油中含有多环芳香烃（PAHs）等有毒、致癌物质，不同地区对排放浓度要求不同，如北海当前的排放标准是30mg/L，一些地区要求5~10mg/L。陆上或滩海油田对采出水水质有非常严格的要求，通常将采出水进行回注以避免排放带来的环境问题。回注采出水也有利于海上油田补充储层压力、提高采收率。有文献综述了降低采出水中脂肪烃、苯、多环芳香烃、烷基酚和一些其他化学品的处理技术。例如，在陆上浅层气藏中，流体可能会污染浅水层或水井，这就需要限制油井作业用化学品。

全球多个监管机构通过了化学品统一分类和标签制度（GHS），达成了国际共识以规范对危害的标准和定义。在欧盟，该制度与REACH法规正在实施。美国职业安全与健康管理局提出GHS标准，作为修订危险化学品管理制度的基础，这种标准化确保更易获得相关化学品危害和毒性信息，以加强化学品在处理、运输和使用过程中对人类健康和环境的保护。

1.3.1 OSPAR 油田化学品环境法规

含有化学物质的采出水的排放规定因地而异。2001 年,《保护东北大西洋海洋环境公约》（OSPAR）颁布了对东北大西洋油田化学品中所有化学成分生态毒理学测试的强制管制指南。无论化学品最后是否存在于产出水、气或油相中，生产使用前都要进行测试。将化学、物理和生态毒理学测试数据提交给相关的国家污染主管部门，由其对数据进行评估。海上化学品统一格式（HOCNF）申请表格，包含用于规范北海 OSPAR 海上勘探和生产行业使用的化学品信息。评估结果是根据海上使用的适宜性对化学品进行排序。最终用户（现场操作者）必须向当局申请在一定期限内使用该化学品的许可证。此外，当地主管部门规定了每个平台下游的石油和化学品排放范围。挪威要求运营商定期分析采出水中苯和酚等有毒成分，以及化学品的含量。运营商有责任将化学品的排放控制在环保主管部门规定的范围内。

OSPAR 要求对生产用化学品进行三类生态毒理学测试：急性中毒、生物积累性、海水中生物降解性。本书附录中列出了这些测试。

必须确定配方中每种成分（而不是成品）的环境特性。一些具有生态毒理学测试能力的服务公司，将实验室测得的油田化学品生态毒理学数据提供给地方管理部门。由于其他环境原因，北海近海禁止使用一些化学品，如烷基酚等内分泌干扰物会带来海洋生物的慢性健康问题。在 OSPAR 的 PLONOR 清单上列出的化学物质可以在北海使用，而无须特别批准（PLONOR 的意思是对环境"几乎或没有风险"）。

北海国家要求的 OSPAR 生态毒理学测试都相同，在解释和实现数据的方式上有一些细微差异。北海国家的法规还在不断修订，读者可以查询相关国家污染管理部门最新制度。挪威使用的评估制度与英国、丹麦和荷兰等其他北海石油生产国使用的评估制度略有差异（见附录）。加拿大可能将采用 OSPAR 规则。自 20 世纪 90 年代后期以来，人们更加关注油田化学物质的生物降解，特别是易于生物积累的化学物质。因此，依据 OECD 306 测试方法，28 天生物降解能力小于 20% 的新化学品不允许在海上油田使用，已经在用的应优先选择替代品。

北海地区对某些化学品的生物降解问题可能过于重视。例如，二乙烯三胺甲基膦酸盐（DETPMP）等氨基膦酸盐防垢剂因生物降解性差，在北海油田作业中已被逐步淘汰，而海水中生物降解性与 DETPMP 相当的聚乙烯磺酸盐，却仍在使用。膦酸盐主要以钙镁复合物的形式存在于天然水体中，因此不影响金属形态的变化和迁移。

化学品也会发生光降解，而且在某些情况下会很严重。例如，氨基膦酸酯防垢剂螯合铁（Ⅲ）的配合物迅速光降解。但排放到海里时，随着海水深度增加，阳光会明显地减少。另一种用于环境的降解机制是空气中或溶解在水中氧的氧化降解。环境中存在锰（Ⅱ）（海水中）和氧会将氨基多膦酸盐迅速氧化降解。油田化学品的生态毒理学规定没有考虑光降解和氧化降解作用。

1.3.2 欧盟 REACH 法规

REACH 是欧盟关于化学品及其安全使用的法规（EC1907/2006），其处理的问题可以缩写为 REACH：化学品的注册、检验、批准和限制。这是迄今为止适用于化学工业的最广范围的一项立法。2007 年发表了关于上游石油工业化学品新法规的三篇论文。推荐一篇最近的论文，报告了 REACH 对北海海上在用化学品监管的影响。对于年使用量超过 100t 的特殊化学品，可能要求石油和天然气行业对其进行注册，REACH 注册截止日期为 2013 年 6 月。

新法规于 2007 年 6 月 1 日生效，所有 REACH 条款在 11 年内逐步实施，对欧盟内部化学品的制造、进口和出口实行管制。

到 2018 年 6 月，对于每年供应超过 1t 的化学品，其中已预注册的物质必须符合正式注册。要求进入欧洲自由贸易区且每年供应超过 1t 的任何物质必须完全注册。因此，对供应商来说，最大的问题是大量化学品和环境测试的成本。聚合物不受此限制，但是所用的单体受此限制。化学制造商在决定继续生产哪些化学品和引进哪些新化学品时，需要考虑这些成本问题。在欧洲使用的任何新化学品也都需要根据这些新规定进行检查。

为实现 REACH 注册，开发了针对海洋化学品使用的通用暴露情景模拟方法。有文献概述了 REACH 的影响，列出了各国在应用 OSPAR HMCS 各方面的差异。英国皇家化学学会的石油工业化学等一系列会议通常于 11 月在英国曼彻斯特举行，发表许多关于 REACH 条例和一般油田化学品环境要求的论文。建议读者从这些会议和其他会议中获得相关信息。欧洲化学品部门的网页上有 REACH 和 CLP 实施指南（CLP 代表分类、标签和包装）。CLP 于 2009 年 1 月生效，根据《全球化学品统一分类和标签制度》（GHS），规定了化学品分类及标签方法。

REACH 法规已经对油田化学品的选择产生了影响。例如，由于 REACH 新法规的成本问题（有关化学成分的详细信息，可参考 9.2.3.2），荷兰已停止生产季铵盐表面活性低剂量水合物抑制剂（LDHI）。

1.3.3 美国环境法规

美国的环境法规与北海不同，在美国境内的不同地点也各有不同。美国环境法规由美国环境保护署（EPA）实施。《有毒物质控制法案》（TSCA）赋予美国环境保护署有权要求化学品的制造商和加工者检测其化学品对健康和环境的影响。《有毒物质控制法案》清单是化学品供应商和操作人员需要检查的主要清单，列入这个清单的物质（除非是特别有毒或危险的物质）才允许在美国不受太多限制地使用。美国加利福尼亚州等会有更严格的环保排放法规要求。

在美国墨西哥湾海岸外大陆架（以及墨西哥湾部分地区，简称 GOM）作业的海上石油和天然气平台或生产设施的排放，受国家污染物排放消除系统（NPDES）总许可证的监管。这些 NPDES 排放许可证的制定标准是依据排出流体的毒性，而非每种特定化学品的危险评估。GOM 运营商必须就其全部废水向美国环境保护署提交下列毒性数据：

（1）采出水：7 天最高无可观察效应浓度（NOEC）。

① 糠虾。

② 银边鲦鱼。

（2）添加了化学品的海水或淡水的多项排放：48h 最高无可观察效应浓度。

① 糠虾。

② 银边鲦鱼。

每个设备的采出水排放都指定了一个临界稀释系数（CDF）。必须根据许可证获得水样品，在指定期间内，产出水在 CDF 或其以下不得对海洋生物（银边鲦鱼和糠虾）有毒。而墨西哥湾采出水排放的毒性测试建议全部修改，包括符合亚致死效应要求。

1.3.4　其他地方的环境法规

北海地区根据毒性、生物降解和单个物质的生物积累来制定法规，通常被认为是油田化学品环境规则最复杂的地区。但是，在环境法规不太明确的地区，管理机构或石油公司可以要求服务公司提供符合北海分类要求的产品，一般认为，如果化学品足够环保可以用于北海，也可能被允许在北海之外的其他地方使用。一些管理机构或石油公司在采出水排放法规没有明确规定的地方，采用北海标准作为其全球部分或全部标准。在某些情况下，地方当局会调整生态毒理学测试，并引入一种或多种本地物种。

1.4　绿色化学品设计

已提出的绿色化学品的 12 个原则，可以分为两大类：首先，指高效化学品的绿色制造过程，并且少产或不产生废料，最好使用可再生原料；其次，参考化学品对环境产生的实际影响。虽然不应忽视制造过程对环境的影响，但是石油行业最关心的还是第二种含义。

有关低环境影响化学品设计的文章很多，但不是针对油田化工行业。提升化学品风险状况的策略已有报道。如果以 OSPAR 规则作为标准，可以从生物降解率、较低的生物积累潜力（$\lg P_{ow}$）和较低的毒性三个方面改进，设计更为绿色的化学品。

1.4.1　生物积累

人们基本上无法改变许多种化学品的生物积累潜力。例如，水溶性防垢剂必须在水相中起作用，这样生物积累量低。如果产品被设计成在油相中工作（如蜡和沥青质抑制剂），将具有很高的 $\lg P_{ow}$ 值。如缓蚀剂和水合物抗凝结剂等化学品，溶于水也溶于油。传统的脂肪酸咪唑啉表面活性剂等成膜缓蚀剂大多分散于油相。如第 8 章所述，人们围绕更好水溶性和更低毒性的缓蚀剂开展了很多工作，如通过增加表面活性剂端基的亲水性或减少疏水尾部的长度。一些不含有疏水尾部的缓蚀剂是水溶性的，详见第 8 章。生物积累潜力的考虑仅限于分子量低于 700 的有机物（不同的国家，这个值有所不同）。确定表面活性剂

的 $\lg P_{ow}$ 值非常困难,虽然已经开发了一些软件程序来估算该值,但是软件的准确性还存在一些争议。

1.4.2 降低毒性

聚合物一般毒性较低,但某些阳离子聚合物具有杀菌性能,可能对其他海洋生物有毒(见第 14 章)。对于表面活性剂,阴离子表面活性剂和非离子表面活性剂的毒性一般小于阳离子表面活性剂。后者中最常见的是季铵盐表面活性剂,其中包括吡啶类和喹啉类表面活性剂,以及在酸性产出液中能部分季铵化(离子化)的长链叔胺和咪唑啉。有文献研究了生物杀菌剂和缓蚀剂配方中常用的季铵盐表面活性剂,特别是苯扎氯铵、苯扎溴铵表面活性剂等的毒性。

降低阳离子表面活性剂的毒性有以下方法:

(1)在表面活性剂的阳离子中心附近连接阴离子基团,以中和其端基的电荷,使表面活性剂两性化,如通过添加丙烯酸衍生的羧酸基团改性的咪唑啉类缓蚀剂(见第 8 章)。

(2)将阳离子表面活性剂与阴离子分子(以离子对的形式)混合——已被用于气体水合物抗凝结剂(见第 9 章)。

(3)将表面活性剂的疏水性尾部长度减少到 8 个碳或以下——已经用于成膜缓蚀剂,这样做可能会降低缓蚀性能(见第 8 章)。

生物表面活性剂的生物可降解性强、毒性低,可作为某些油田化学品使用。但目前大多数这类产品应用于油田的成本高(如果这类化学品在石油行业之外的领域得到大规模应用,价格就会有大幅下降)。由于很难获得低毒性的表面活性剂,快速降解或可裂解的表面活性剂是去除表面活性剂固有毒性的首选。

1.4.3 实现生物降解性

一种化学物质很难既具有良好的生物降解性,又保持长期的储存稳定性和高温稳定性。如果化学物质分子量太大,就不能穿过细胞膜而不能生物积累,这就是北海 OSPAR 规定将分子量设定为 700 的原因,超过 700 就认为不会有生物积累。但是这个分子量值并没有考虑到化学品的体积或横截面大小,而且不同种类海洋生物的细胞膜大小差别很大。

淡水中,化学品的生物降解数据较多,而海水中此类降解的系统性数据鲜见报道。海水与河水中的化学品浓度、细菌种类截然不同,且海水中的细菌不会像河水中的细菌那样能充分暴露于特定化学品组分和浓度环境。关于生物降解、修复的书中,介绍了测量非水溶性化合物的生物降解难点及化学结构对生物降解率的影响等。

在题为《生物降解性小分子的设计》(不包括聚合物)综述中,作者给出了一个结构因素的简短列表。这些结构因素通常降低了有氧生物的降解性(但总有例外)。

(1)卤素(默认与碳成键),特别是氟或氯,尤其是分子中有 3 个以上的卤原子。

(2)多分支碳链——主要是季碳。

（3）叔胺，默认还有季铵盐。
（4）硝基、亚硝基、偶氮和芳胺。
（5）多环残基，如多环芳香烃。
（6）一些杂环残留物，如咪唑。
（7）脂肪醚键，乙氧基醚除外。

用烷基芳基表面活性剂为例说明上述第 2 个因素。如果烷基是高度支化的"四丙烯"，很难被生物降解，直链烷基则降解快得多。以聚丙氧基化物为例说明因素 2，与聚乙氧基化物相比，多出的甲基分支使化学剂生物降解性降低。因此，某些表面活性剂和破乳剂中用聚丙氧基封端聚乙氧基化物分子，实际上会降低其生物降解性。天然化合物维生素 A、胆固醇、泛酸和合成的季戊四醇 [$C(CH_2OH)_4$] 等含有季碳的小分子化合物可生物降解。

如溶剂 N-乙基吡咯烷酮（NEP）等杂环残留物可生物降解。相比之下，聚乙烯吡啶烷酮和聚乙烯己内酰胺（用作 KHIs）具有与 NEP 相同的环结构，但是其生物降解性非常差，这可能是受空间拥挤的影响，酶对聚合物环的攻击非常有限。

同一篇综述还列出了一些能提高小分子生物降解率的特征：
（1）易被酶水解的基团——主要是酯（包括磷酸盐酯）和酰胺，但有不确定性。
（2）羟基、醛或羧酸基形式的氧原子，可能还有酮，但不包括醚，除非是乙氧基。
（3）未取代的直链烷基（特别是 4 个碳以上的链）和苯环——这些是加氧酶"攻击"的好位置，如果分子中没有氧，这是最好的结构因素。

第 9 章讨论的双尾季铵型表面活性抗凝结剂很好地说明了上述的经验法则。第一代双尾抗凝结剂的结构类似于图 1.4，其中一个季铵盐基团与两个丁基（对性能来说是关键结构特征）结合，另外两个是长烃链的尾基。与许多季铵盐表面活性剂类似，该分子具有毒性，生物降解性差。为提高生物降解性，第二代双尾抗凝结剂将酯基放在长尾中（图 1.5），这种分子在酯基部位降解，留下两个长链脂肪羧酸和一个表面活性较低、毒性较弱的带有两个羟基的小分子季铵盐化合物。形成的线型脂肪羧酸降解得非常快，但是小分子季铵盐化合物降解非常慢，甚至长于标准要求 OECD 306 的 28 天测试。这些数据说明了双尾季型表面活性抗凝结剂不允许在挪威近海使用的部分原因，尽管该类化学物质在荷兰近海使用。如果脂肪羧酸尾部的支化度高，其生物降解速率会较低。第三代双尾抗凝结剂在酯基附近增加一个甲基（图 1.6），这种分支可能会引起空间位阻，从而影响酯基的酶解速率。

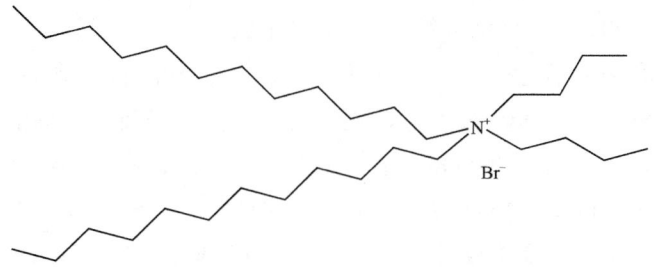

图 1.4 第一代双尾抗凝结剂

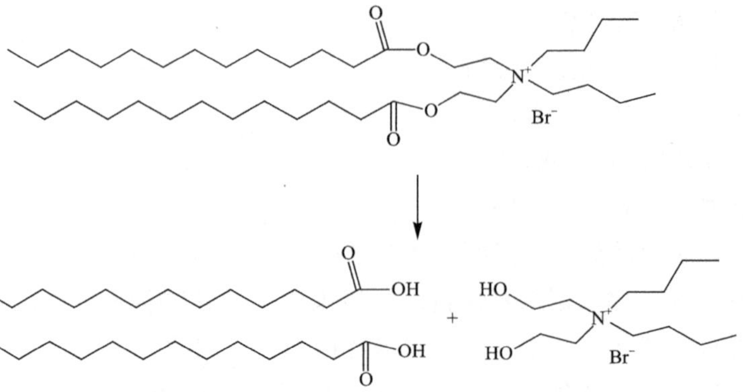

图 1.5 第二代双尾抗凝结剂和生物降解产物

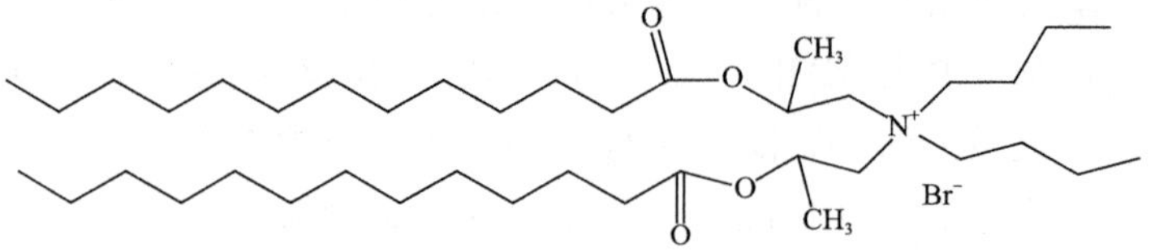

图 1.6 第三代双尾抗凝结剂

对于聚合物，即使是在低分子量的情况下，以聚乙烯醇为骨架的许多聚合物也很难被生物降解，尽管这也取决于侧基的种类和大小，如聚丙烯酸及其酯类（用于防垢剂、蜡和沥青质抑制剂）、聚乙烯己内酰胺（一种动力学水合物抑制剂）、聚丙烯酰胺（用于堵水处理和减阻剂）和聚二甲基二烯丙基氯化铵（絮凝剂）。使用易受酶解影响的基团（主要是酯和酰胺）和乙氧基有助于提高聚合物的生物降解性能，如多糖和多氨基酸等许多天然聚合物表现良好的生物降解性能，但木质素是例外。因此开发了基于多氨基酸（如聚天冬氨酸）和多糖（如菊粉的衍生物）的生物降解防垢剂。然而，有一些非正式的报告指出，有些乙烯基聚合物防垢剂（如基于马来酸的化学剂）也可以很好地快速生物降解（见第 3 章）。此外，一系列可生物降解的破乳剂以各种易于生物可降解的聚合物结构为主，然后被乙氧基化和丙氧基化（见 11.3.3.9）。

可生物降解的聚合物作为塑料的课题研究相当多。其中大多具有聚酯骨架，如聚乳酸和聚 3-羟基丁酸酯。但是这些聚合物的衍生物，无论是在水基还是油基应用方面，都没有开发作为可生物降解的油田化学品。其他以聚酯为主干的接枝聚合物已用于破乳剂和动力学水合物抑制剂（KHIs）（分别见第 9 章和第 11 章）。

表面活性剂具有表面活性，有一定毒性（EC_{50} 小于 10mg/L，通常小于 1mg/L）。对用于生产化学品的阳离子表面活性剂尤为如此。文献对作为表面活性剂的脂肪胺衍生物生物降解性进行了相关研究。有毒性的表面活性剂会杀死细菌，影响生物降解结果（生物降解试验浓度通常为 2~40mg/L，但毒性试验为 μg/L 级别）。一般来说，烷基尾部越长，表面活性剂的毒性就越大，8~10 个碳原子以下的单一尾基［取决于亲水—亲油平衡（HLB）值］，毒性显著降低。

由于毒性，要求排放的表面活性剂具有更高生物降解性、更低生物积累，以减少对环境的影响（水溶性更好）。有专业人员出版了关于表面活性剂生物降解的书籍。该书涵盖了生物降解性能测试、策略、法律要求及阴离子表面活性剂、阳离子表面活性剂、非离子表面活性剂和两性表面活性剂的生物降解性。还有另一本稍早出版的关于表面活性剂生物降解的专著。

在表面活性剂中可加入弱的、可断裂的键，增加其降解性。有文献综述了单分子型和双子型可断裂表面活性剂的研究进展。

减少表面活性剂生物积累潜力的一种方法是增加化学物质的尺寸（分子量），超过一定的分子量（约700），化学物质穿过细胞膜杀死生物的可能性就大大降低，这就解释了多数聚合物具有低毒性的原因，但有些化合物通过表面机理而无须跨膜，具有毒性。另一种更绿色的方式是使用生物表面活性剂。一些生物表面活性剂可以从天然成分中分离出来，如植物表面活性剂或由胆汁酸制成的脱氧胆酸钠。可以通过修饰天然产物（如糖苷或多糖苷、月桂酰乳酸钠等）制成更绿色的表面活性剂。其他生物表面活性剂也可以在生物技术过程中制成，但当前其成本都过高，难以满足油田大规模使用的需求，例如：（1）糖脂类，包括海藻糖脂、鼠李糖脂和槐糖脂。（2）肽类抗生素表面活性剂（如表面蛋白、聚混合素、短杆菌肽）。（3）多聚杂糖蛋白质。

1.5 汞和砷的生产

本书仅在此处讨论汞和砷等有毒金属和准金属的生产化学问题。

最近出版了一份关于汞污染的资料。在环境法规严格的挪威，发布了一项关于汞暴露的指导方针，列出了从石油中去除汞的七个原因，以及报道了一项商业实施方案。汞来源于油气藏下的火山岩，并与烃类产生有天然联系。烃类中溶解的汞包括二甲基汞（$HgMe_2$）和其他二烷基汞，甚至是汞金属和无机汞盐。对很多原油的测试表明：原油中的汞大部分存在于沥青质中。有机金属汞化合物的毒性特别大，比无机汞或金属毒性更大，这些物质有挥发性，以 μg/L 级别的浓度存在于液体，特别是气体烃中。在生产过程中，汞及其金属有机化合物会与硫化物发生反应，导致硫化汞沉积于石油加工设备。在铝合金换热器中沉积的金属汞会导致设备腐蚀和灾难性故障。在原油或凝析油的处理过程中，汞会导致催化剂中毒、材料降解、产品质量下降，特别是使石化生产原料的精制产品质量下降，对炼油产生不利影响。

随着更深入地了解汞对生产和处理系统的影响，原油、冷凝析油和原始天然气中的汞浓度测试日益重要。安装除汞设备用于降低汞浓度，该设备安装在气体预处理或液体产品处理或采出水排水的前端。已经开发了多种去除烃类中汞化合物的技术，最常用的方法是使用吸附剂。例如，吸附于以氧化铝为载体的银或硫化铜、吸附于亚铜或锡金属盐、活性炭吸附剂、金属氧化物或硫化物的吸附剂、含活性硫醇基团的离子交换树脂和含硫醇基团的聚合物。油溶性硫化物可以首先加到烃流体中，然后在吸附剂上"捕获"汞—硫化合物。但是硫化汞等悬浮态（胶体）物质很难被一些吸附剂床"捕获"。碱性聚硫化物和油溶性有机二硫代氨基甲酸盐或硫化异丁烯用于去除原油、天然气凝析物和其他液态烃类中的汞。最近有人研究了硫化钠、2,4,6-三巯基硫嗪三钠盐、二硫代氨基甲酸钠三种药剂对模拟脱硫溶液中 Hg^{2+} 的沉淀效果，其中二硫代氨基甲酸钠效果最好。所有的沉淀剂都可有效抑制 Hg^{2+} 的还原和烟气脱硫液中氧化汞的释放。有研究报道了以层状金属硫化物捕获水中包括汞在内的重金属。

一些生产排放水中含有少量的无机汞阳离子，会对环境造成危害，因为汞（特别是转化为烷基汞时）会在食物链中生物积累，最终进入鱼类和人类（如果食用鱼类）体内。针对上述问题，可以采用化学沉淀法、混凝法和活性炭吸附法，或者将排出液回注深干井或已枯竭生产井。去除水溶性汞和砷的最佳化学沉淀技术是添加含硫化合物。Hg^+ 和 Hg^{2+} 是软酸，与硫化物、硫醇、硫代氨基甲酸盐和硫代碳酸盐等含硫软碱的结合力强。自2003年以来，在东南亚的两个油田成功实现了连续除砂、氧化剂（NaOCl）杀菌、Fe^{3+} 助凝、硫醇除汞、絮凝等水处理工艺。支链聚合物硫代氨基甲酸酯可用于结合和沉淀汞离子，同时具有絮凝作用，使水变得澄清。硫代氨基甲酸聚合物是由二氯乙烷和氨气合成，然后将其衍生化至含有至少5%（摩尔分数）的硫代氨基甲酸盐基团。现场应用中，也可以加入阴离子絮凝剂来帮助沉淀。硫醇和硫代氨基甲酸盐等硫化物也会与 Ag^+、Pb^{2+}、Cu^{2+} 和 Cd^{2+} 等金属离子形成不溶性配合物。

页岩油中砷含量最高达100mg/L，石油和凝析油中砷含量通常不超过5～10mg/L。砷也可能以三氢化砷（AsH_3，剧毒气体）或三甲基胂［$As(CH_3)_3$］等砷氢化物的形式存在于天然气体中。有毒的砷化合物也可能伴随在油田产出水中。为符合环保要求，去除天然气和水中砷化合物的方法包括热解、吸附和吸收。许多用于去除汞的吸附剂法也适用于从烃类流体中去除砷。在化学方面，采用与氧化剂反应的方法，从天然气中去除砷化合物。如果水中主要含有 As^{3+}（如亚砷酸盐）和少量 As^{5+}（砷酸盐），其危害比汞更大。As^{3+} 与汞一样，可通过硫化物（如硫醇）沉淀去除，随后絮凝。过去用于酸化增注配方中的含砷缓蚀剂，已被更环保的有机缓蚀剂和增强剂所取代，详见第5章。

参 考 文 献

[1] T. A. Hjelmas, S. Bakke, T. Hilde, E. A. Vik, and H. Grüner, "Produced Water Reinjection: Experiences from Performance Measurements on Ula in the North Sea," SPE 35874 (paper presented at the Third International Conference on Health, Safety & Environment in Oil & Gas Exploration & Production, New Orleans, LA, 9-12 June 1996).

[2] S. Szymczak, G. Brock, J. M. Brown, D. Daulton, and B. Ward, "Beyond the Frac: Using the Fracture Process as the Delivery System for Production Chemicals Designed to Perform for Prolonged Periods of Time," SPE 107707(paper presented at the Rocky Mountain Oil & Gas Technology Symposium, Denver, CO, 16-18 April 2007).

[3] C. Berkland, M. Cordova, J.-T. Liang, and G. P. Willhite, International Patent Application WO/2008/030758.

[4] S. E. Campbell, U.S. Patent 7135440, 2006.

[5] M. W. Kendig, M. M. Hon, and L. F. Warren, U.S. Patent Application 20050201890.

[6] R. Alapati, L. Sanford, T. Williams, and S. Gao, "New Test Method to Qualify Production Chemicals for Subsea Injection," SPE 114924 (paper presented at the SPE Annual Technical Conference and Exhibition, Denver, CO, 21−24 September 2008).

[7] J. Yang and V. Jovancicevic, U.S. Patent Application 20060166835.

[8] R. J. Dyer, U.S. Patent Application 20080047712.

[9] B. L. Knudsen, M. Hjelsvold, T. K. Frost, M. B. E. Svarstad, P. G. Grini, C. F. Willumsen, and H. Torvik, "Meeting the Zero-Discharge Challenge for Produced Water," SPE 86671 (paper presented at the SPE International Conference on Health, Safety, and Environment in Oil and Gas Exploration and Production, Calgary, Alberta, Canada, 29−31 March 2004).

[10] OSPAR Guidelines for Completing the Harmonised Offshore Chemical Notification Format (2010-05) (http://www.ospar.org).

[11] S. Glover and I. Still, "HMCS (Harmonised Mandatory Control Scheme) and the Issue of Substitution," Ninth Annual International Petroleum Environmental Conference, 22−25 October 2002.

[12] M. Thatcher and G. Payne, "Impact of the OSPAR Decision on the Harmonised Mandatory Control System on the Offshore Chemical Supply Industry," Proceedings of the Chemistry in the Oil Industry VII Symposium, RSC, Manchester, UK, 13−14 November 2001.

[13] G. M. Rand, *Fundamentals of Aquatic Toxicology: Effects, Environmental Fate and Risk Assessment*, 2nd ed., Philadelphia, PA: Taylor & Francis, 1995.

[14] G.-G. Ying, "Fate, Behavior and Effects of Surfactants and Their Degradation Products in the Environment," *Environment International* 32 (2006): 417.

[15] J. Beyer, A. Skadsheim, M. A. Kelland, K. Alfsnes, and S. Sanni, "Ecotoxicology of Oilfield Chemicals: The Relevance of Evaluating Low-Dose and Long-Term Impact on Fish and Invertebrates in Marine Recipients," SPE 65039 (paper presented at the SPE International Symposium on Oilfield Chemistry, Houston, TX, 13−16 February 2001).

[16] J. M. Getliff and S. G. James, "The Replacement of Alkylphenol Ethoxylates to Improve the Environmental Acceptability of Drilling Fluid Additives," SPE 35982 (paper presented at the International Conference on Health, Safety & Environment, New Orleans, LA, June 1996).

[17] OSPAR List of Substances/Preparations Used and Discharged Offshore Which Are Considered to Pose Little or No Risk to the Environment (PLONOR). Reference number: 2013-06 (http://www.ospar.org).

[18] http://ec.europa.eu/environment/chemicals/reach/publications_en.htm.

[19] http://europa.eu.int/comm/enterprise/reach/index.htm or http://ecb.jrc.it/REACH/.

[20] Council Directive 67/548/EEC in order to Adapt it to Regulation (Ec) of the European Parliament and of the Council Concerning the Registration, Evaluation, Authorisation and Restriction of Chemicals (http://www.reach-compliance.eu/).

[21] G. Dougherty, "REACH—The Big Picture" (paper presented at the RSC/EOSCA Chemistry in the Oil Industry X Symposium, November 2007).

[22] S. W. Longworth, "REACH—From a Supplier's Perspective" (paper presented at the RSC/EOSCA Chemistry in the Oil Industry X Symposium, November 2007).

[23] J. Hislop and D. Knight, "REACH—Intelligent Safety Evaluation and Prediction of Properties" (paper presented at the RSC/EOSCA Chemistry in the Oil Industry X Symposium, November 2007).

[24] Final NPDES Permit for the Western Portion of the GoM OCS (GMG290000) (http://www.epa.gov/region6/water/npdes/genpermit/index.htm) 2012.

[25] R. S. Boethling, E. Sommer, and D. DiFiore, "Designing Small Molecules for Biodegradability," *Chemical Reviews* 107 (2007): 2207.

[26] D. K. Platt, *Biodegradable Polymers*, *Rapra Market Report*, Cambridge, UK: Woodhead Publishing, 2006.

[27] G. T. Howard, "Biodegradation of Polyurethane: A Review," *International Biodeterioration & Biodegradation* 49 (2002): 245.

[28] NIIR Board, *The Complete Book on Biodegradable Plastics and Polymers (Recent Developments, Properties, Analysis, Materials & Processes)*, Delhi, India: Asia Pacific Business Press, 2006.

[29] E. S. Stevens, *Green Plastics*, Princeton, NJ: Princeton University Press, 2001.

[30] C. E. Cooke, U.S. Patent Application 20080015120.

[31] S. M. Wilhelm, L. Liang, and D. Kirchgessner, "Identification and Properties of Mercury Species in Crude Oil," *Energy & Fuels* 20 (2006): 180−186.

[32] S. M. Wilhelm and N. Bloom, "Mercury in Petroleum," *Fuel Processing Technology* 63 (2000): 1−27.

[33] M. Abu El Ela, I. S. Mahgoub, M. H. Nabawi, and M. Abdel Azim, "Mercury Monitoring and Removal at Gas-Processing Facilities: Case Study of Salam Gas Plant," SPE 106900, *SPE Projects, Facilities & Construction* 3 (1)(2008): 1.

[34] K. J. Grice, L. D. Van Orman, L. A. Young, and C. R. Manning, "Evaluation of Portable Mercury Vapor Monitors and Their Response to a Range of Simulated Oil Processing Environments," SPE 111837 (paper presented at the SPE International Conference on Health, Safety, and Environment in Oil and Gas Exploration and Production, Nice, France, 15−17 April 2008).

[35] M. Abu El Ela, M. H. Nabawi, and M. Abdel Azim, "Behavior of the Mercury Removal Absorbents at Egyptian Gas Plant," SPE 114521 (paper presented at the CIPC/SPE Gas Technology Symposium 2008 Joint Conference, Calgary, AB, 16−19 June 2008).

[36] M. R. Sainal, T. Mat, A. Shafawi, and A. J. Mohamed, "Mercury Removal Project: Issues and Challenges in Managing and Executing a Technology Project," SPE 110118 (paper presented at the SPE Annual Technical Conference and Exhibition, Anaheim, CA, 11−14 November 2007).

[37] T. Y. Yan, U.S. Patent 4909926, 1990.

[38] T. Torihata and E. Kawashima, U.S. Patent 4946582, 1990.

[39] J. D. McNamara, U.S. Patent 5336835, 1994.

[40] M. Roussel, P. Courty, J.-P. Boitiaux, and J. Cosyns, U.S. Patent 4911825, 1990.

[41] H. A. M. Duisters, and P. C. Van Geem, U.S. Patent 4950408, 1990.

[42] F. R. Van Buren, L. Deij, G. Merz, and H. P. Schneider, U.S. Patent 553844, 1994.

[43] T. F. Degnan and S. M. LeCours, U.S. Patent 6350372, 2002.

[44] T. Y. Yan, U.S. Patent 4915818, 1990.

[45] T. C. Frankiewicz and J. Gerlach, U.S. Patent 6685824, 2004.

[46] P. Dhanasin, W. Utoomprurkporn, M. Hungspreugs, and T. Soponkanabhorn, "Monitoring of Mercury Content of Fishes Near the Bongkot Platform, Gulf of Thailand," SPE 96492 (paper presented at the SPE Asia Pacific Health, Safety and Environment Conference and Exhibition, Kuala Lumpur, Malaysia, 19−21 September 2005).

[47] M. Rangponsumrit, S. Athichanagorn, and T. Chaianansutcharit, "Optimal Strategy of Disposing Mercury Contaminated Waste," SPE 103843 (paper presented at the International Oil & Gas Conference and

Exhibition, China, Beijing, China, 5-7 December 2006).

[48] G. S. Elfline, U.S. Patent 4612125, 1986.

[49] D. L. Gallup and J. B. Strong, "Removal of Mercury and Arsenic from Produced Water" (paper presented at the 13th Annual International Petroleum Environmental Conference, San Antonio, TX, 17-20 October 2006).

[50] T. C. Frankiewicz and J. Gerlach, U.S. Patent 6117333, 2000.

[51] K. S. Siefert, P. L. Choo, J. W. Sparapany, and J. H. Collins, U.S. Patent 5346627, 1994.

[52] M. L. Braden, "Mercury: Real Problems…Not Roman Mythology" (paper presented at 13th Annual International Petroleum Environmental Conference, San Antonio, TX, 17-20 October 2006).

[53] A. Block-Bolten and L. Glowacki, "Natural Gas Separation from Arsenic Compounds," SPE 19078 (paper presented at the SPE Gas Technology Symposium, Dallas, TX, 7-9 June 1989).

[54] T. J. Pinnavaia, J. I. Dulebohn, and E. J. McKimmy, U.S. Patent 7404901, 2008.

[55] W. W. Frenier and M. Ziauddin, "A Multifaceted Approach for Controlling Complex Deposits in Oil and Gas Production," SPE 132707 (paper presented at the SPE Annual Technical Conference and Exhibition, Florence, Tuscany, Italy, 20-22 September 2010).

[56] D. J. Poelker, U.S. Patent Application 20130046048.

[57] C. Berkland, M. Cordova, J.-T. Liang, and G. P. Willhite, U.S. Patent Application 20100056399.

[58] N. J. Goodwin, O. G. Svela, J. H. Olsen, T. Tjomsland, B. M. Hustad, and G. M. Graham, "Qualification Procedure for Continuous Injection of Chemicals in the Well—Method Development," SPE 154934 (paper presented at the SPE International Conference on Oilfield Scale, Aberdeen, UK, 30-31 May 2012).

[59] B. M. Hustad, O. G. Svela, J. H. Olsen, K. Ramstad, and T. Tjomsland, "Downhole Chemical Injection Lines—Why Do They Fail? Experiences, Challenges and Application of New Test Methods," SPE 154967 (paper presented at the SPE International Conference on Oilfield Scale, Aberdeen, UK, 30-31 May 2012).

[60] J. H. Olsen, "Experiences and Consequences Related to Continuous Chemical Injection," SPE Paper 146625 (paper presented at the SPE Annual Technical Conference and Exhibition, Denver, CO, 30 October-2 November 2011).

[61] J. G. Willmon, and M. A. Edwards, "Precommissioning to Startup: Getting Chemical Injection Right," Paper SPE 96144 (paper presented at the SPE Annual Technical Conference and Exhibition, Dallas, TX, 9-12 October 2005).

[62] D. Bowering, M. McCall, and G. Graham, "Measuring the Impact of Evaporation Effects in Chemical Injection Lines and Gas Lift Systems on the Risk of Corrosion," Oilfield Chemical Symposium, Geilo, Norway, 17-20 March 2013.

[63] A. Hunton, private communication.

[64] D. J. Daulton, A. K. Jordan, J. Cobb, and T. G. Grumbles, "Quantitative Rankings Measure Oilfield Chemicals' Environmental Impacts," SPE 135517 ATC 2010 (SPE Annual Technical Conference and Exhibition, Florence, Italy, 19-22 September 2010).

[65] B. Nowack, "Environmental Chemistry of Phosphonates," *Water Research* 37 (2003): 2533.

[66] G. Payne, I. Still, and N. Robinson, "The Development of the EOSCA Generic Exposure Scenario Tool (EGEST)—Why We Need It? The Development of a Method to Conduct a Generic Exposure Scenario for Offshore Chemical Use to Enable REACH Registration" (paper presented at the RSC Chemicals in the Oil Industry XI, Manchester, UK, 2-4 November 2009).

[67] I. Still, "The Impact of REACH on the Regulation of Chemicals Used Offshore in the OSPAR Convention

Area—Another Chance for Harmonisation？" (paper presented at the RSC Chemistry in Oil Industry XI, Manchester, UK, 2-4 November 2009).

[68] C. E. Kirchof, H. Romijn, and S. C. Danican, "Lessons Learned by an Oil and Gas Service Company in the Development of a Chemical Management System Towards Compliance with REACH," SPE 26727 (paper presented at the SPE International Conference on Health, Safety and Environment in Oil and Gas Exploration and Production, Rio De Janeiro, Brazil, 12-14 April 2010).

[69] J. Galvan and C. Smith, "The Impact of REACH on the E&P Industry: A Service Company's Perspective," SPE 126995 (paper presented at the SPE International Conference on Health, Safety and Environment in Oil and Gas Exploration and Production, Rio De Janeiro, Brazil, 12-14 April 2010).

[70] J. Sanders, D. A. Tuck, and R. J. Sherman, "Are Your Chemical Products Green？: A Chemical Hazard Scoring System," SPE 126451 (paper presented at the SPE International Conference on Health, Safety and Environment in Oil and Gas Exploration and Production, Rio De Janeiro, Brazil, 12-14 April 2010).

[71] http://echa.europa.eu/web/guest/support/guidance-on-reach-and-clp-implementation.

[72] http://www.epa.gov/oppt/chemtest/.

[73] P. Anastas and J. Warner, *Green Chemistry: Theory and Practice*, New York: Oxford University Press, 1998.

[74] T. McLean, E. Dalrymple, M. Muellner, and S. Garcia-Swofford, "A Method for Improving Chemical Product Risk Profiles as Part of Product Development," SPE 159355 (paper presented at the SPE Annual Technical Conference and Exhibition, San Antonio, TX, 8-10 October 2012).

[75] J. Wharfe, "Environmental Toxicity Testing," in *Sheffield Analytical Chemistry Series*, eds. K. C. Thompson, K. Wadhia, and A. P. Loibner, Oxford, UK: Blackwell Publishing, 2005.

[76] M. T. Garcia, I. Ribosa, T. Guindulain, J. Sanchez-Leal, and J. Vives-Rego, *Environmental Pollution* 111 (2001): 169.

[77] G. Jawecki, E. Grabinska-Sota, and P. Narkiewicz, *Ecotoxicology and Environmental Safety* 54 (2003): 87.

[78] F. Ferk, M. Mišikl, C. Hoelzl, M. Uhl, M. Fuerhacker, B. Grillitsch, W. Parzefall, A. Nersesyan, K. Mičietal, T. Grummt, V. Ehrlich, and S. Knasmuller, *Mutagenesis* (2007): 1-8.

[79] D. K. F. Santos, R. D. Rufino, J. M. Luna, V. A. Santos, A. A. Salgueiro, and L. A. Sarubbo, *Journal of Petroleum Science and Engineering* 106 (2013): 43-50.

[80] P.-E. Hellberg, K. Bergström, and K. Holmberg, *Journal of Surfactants and Detergents* 3 (1)(2000).

[81] M. Alexander, *Biodegradation and Bioremediation*. San Diego, CA: Academic Press, 1999.

[82] C. G. van Ginkel, C. A. Stroo, and A. G. M. Kroon, "Biodegradability of Ethoxylated Fatty Amines and Amides and the Non-toxicity of Their Biodegradation Products," *Tenside Surfactants Detergents* 30 (1993): 3.

[83] C. G. van Ginkel, A. Louwerse, and B. van der Togt, "Substrate Specificity of a Long Chain Alkylamine Degrading *Pseudomonas* sp. Isolated from Active Sludge," *Biodegradation* 19 (2008): 129-136.

[84] C. G. van Ginkel, C. Gancet, M. Hirschen, M. Galobardes, P. Lemaire, and J. Rosenblom, "Improving Ready Biodegradability Testing of Fatty Amine Derivatives," *Chemosphere* 73 (2008): 506-510.

[85] D. R. Karsa and M. R. Porter, eds., *Biodegradability of Surfactants*. Blackie Academic and Professional, Glasgow, Scotland: Springer, 1994.

[86] R. D. Swisher, *Surfactant Biodegradation*, *Surfactant Science Series*, vol. 18, New York: Marcel Dekker, 1987.

[87] D. Shukla and V. K. Tyagi, *Tenside Surfactants Detergents* 1 (2010): 7-12.

[88] S. L. Zuber and M. C. Newman, eds., *Mercury Pollution: A Transdisciplinary Treatment*. Boca Raton, FL: CRC Press, 2011.

[89] J. Naerheim, "Mercury Guideline for the Norwegian Oil and Gas Industry," SPE 164950 (paper presented at the SPE European HSE Conference and Exhibition—Health, Safety, Environment and Social Responsibility in the Oil & Gas Exploration, London, UK, 16–18 April 2013).

[90] I. Still, "The Further Impact of Reach on the Regulation of Chemicals Used Offshore in the North Sea New Challenges for 2013" (paper presented at the RSC Chemistry in the Oil Industry XII: Innovative Chemistry—Value, Risks and Rewards, Manchester, UK, 7–9 November 2011).

[91] C. A. Salva and D. L. Gallup, "Mercury Removal Process Is Applied to Crude Oil of Southern Argentina," SPE, 138333 (paper presented at the SPE Latin American and Caribbean Petroleum Engineering Conference, Lima, Peru, 1–3 December 2010).

[92] T. Tang, J. Xu, R. Lu, J. Wo, and X. Xu, *Fuel* 89 (2010): 3613.

[93] M. J. Manos and M. G. Kanatzidis, *Chemistry—A European Journal* 15 (2009): 4779.

[94] R. Sen, *Biosurfactants (Advances in Experimental Medicine and Biology)*, Austin, TX: Springer, 2010.

[95] N. G. K. Karanth, P. G. Deo, and N. K. Veenanadig, "Microbial Production of Biosurfactants and Their Importance," *Current Science* 77 (1)(1999): 116–126.

[96] H. J. Oschmann, M. C. Huijgen, and H. F. Grondman, "Production Chemicals Based on Active Dispersions—Alternatives to Conventional Solvent Based Products Reduces Viscosity Also for Low T" (paper presented at the RSC Chemistry in the Oil Industry XII: Innovative Chemistry—Value, Risks and Rewards, Manchester, UK, 7–9 November 2011).

[97] G. Payne and N. Robinson, "Harmonised REACH Exposure Scenario Mapping for the Oilfield Chemicals Industry" (Chemistry in the Oil Industry XIII: Oilfield Chemistry—New Frontiers', Manchester Conference Centre, UK, 5–6 November 2013).

[98] B. Rowles, R. Gioia, L. Hughes, L. Jones, M. La Vedrine, C. Moran, S. Kroeger, C. Phillips and S. Supple, "The Harmonisation of REACH and HMCS: Evolution of the Regulatory Framework for Offshore Chemicals" (Chemistry in the Oil Industry XIII: Oilfield Chemistry—New Frontiers', Manchester Conference Centre, UK, 5–6 November 2013).

[99] R. Gioia, B. Rowles, S. Kröger, C. Phillips, L. Jones, S. Supple, L. Hughes, C. Moran, M. La Vedrine, and L. R. Henriquez, "Proposed Guidance on the Assessment of Polymers under the OSPAR HMCS" (Chemistry in the Oil Industry XIII: Oilfield Chemistry—New Frontiers', Manchester Conference Centre, UK, 5–6 November 2013).

2 堵水调剖及气体封窜

2.1 概述

本章将讨论油气井生产中控制气和水产出的化学方法，其中部分技术可用于堵水或气体封窜。有文章回顾了解过去20年成功应用的技术。

随着油田进入开发中、后期，产水量越来越大。处理如此大量的水成本高昂，甚至会成为地面处理设施的瓶颈。2007年发布的一份报告以北海50口井的某典型油田为例说明了油田产出水的处理成本，如果水处理成本是0.50美元/bbl❶，每口井产水以5000bbl/d计算，则油田每天的水处理成本达12.5万美元（每年4560万美元）。产出水不仅与水处理成本有关，还会导致结垢、微粒运移、砂面塌陷和油管腐蚀，甚至因流体静液柱压力过高而导致油井被压井停产，因此应尽可能长时间地抑制地层水的产出。

封堵产水层最常见的技术是对见水区域进行机械封隔、水泥挤注和聚合物凝胶处理，这些技术要求分区封隔。不等比例渗透率降低剂（DPR）或相对渗透率调节剂（RPM）等可以从井口加注，是分区封隔处理的一种替代方案。

生产井最初是在靠近产层底部射孔。当底部出水开始占产出流体的主要比例时，通过水泥挤注封堵或封隔器、桥塞来封隔这些射孔段，然后在封堵段上方重新射孔，恢复原油生产。这个过程一直持续到整个产层都出水为止。只用水泥进行挤注往往效果不足，成功率仅约为30%，其原因是标准的水泥颗粒尺寸限制其渗透进较小的孔道、裂缝和高渗透区域。在这些情况下，无水水泥浆的挤注已证明是解决出水的有效方法。该方法应用于墨西哥南部的几个油田，成功降低了采出液含水率，堵水成功率接近100%。使用聚合物凝胶可以提供良好的渗透性，并且一次处理后，有效期可以维持数年。而组合处理（水泥/聚合物凝胶）可能更加有效，首先将凝胶挤入储层所需的径向深度，然后用尾追的水泥封堵近井通道。水泥有助于将凝胶或聚合物锁定在地层中，并防止残留聚合物返排。

另一种堵水方法是在井中放置由吸水膨胀、吸油不膨胀材料制成的膜。当油井产水时，膨胀材料将封堵出水层。但这种方法无法渗透到深部地层。

本章将讨论油井化学堵水或控水的系列技术。有文献回顾了截至2001年的可用市售产品。为了获得更好的注水定向剖面，堵水处理也已成功应用于注水井。化学堵水技术包括水泥浆，树脂和弹性体，无机凝胶，交联有机聚合物凝胶（单体基、聚合物基），黏弹性表面活性剂（VESs）；DPR或RPM（包括乳化凝胶），微粒和胶体，水溶胀和超吸水聚合物。

凝胶处理也适用于气体封窜。在2.8节中，将对一种采用稳定泡沫封堵气窜（不能堵水）的技术进行回顾。

---

❶ 1bbl=158.987dm³。

2.2 树脂和弹性体

基于酚醛树脂、环氧树脂或糠醇等的树脂和弹性体可用于封堵产水区，有足够的物理强度来封堵裂缝、孔道和射孔段，但其价格相对较高，因此使用通常局限于井筒径向 1ft[1]范围内。此类材料在较低浓度时，还可用于固砂。酚醛树脂是基于苯酚和甲醛的树脂。一项降低产水副作用的专利提出了一种堵水剂及其多种施工工艺，该堵水剂为酚醛与亚硫酸氢钠、焦亚硫酸氢钠或其混合物中的至少一种的组合物。亚硫酸盐发挥了重要作用，确保水溶性聚合物有足够时间以低黏度流动到目标位置。也可使用由盐水激活的苯乙烯-丁二烯胶乳。环氧树脂是环氧氯丙烷和双酚A的反应产物，固化剂常采用二乙烯三胺。还使用了三氯甲苯和吡啶与糠醇的受控催化剂体系。另一项专利提出一种用于减少微粒和出水的固结液，包括可固化的树脂成分、液体固化剂、硅烷偶联剂和表面活性剂。其中典型的树脂有双酚A-环氧氯丙烷树脂、聚环氧树脂、酚醛清漆树脂、聚酯树脂、酚醛树脂、脲醛树脂、呋喃树脂、聚氨酯树脂和缩水甘油醚。

2.3 无机凝胶

基于硅酸盐的无机凝胶已广泛用于堵水。基于硅酸盐的不同技术已在匈牙利、塞尔维亚、挪威、美国、阿曼等国家应用了数百井次。常规的方法是使用水溶性的硅酸钠溶液与胶凝剂反应，生成凝胶。硅酸盐溶液通常与地层水不相容（硅酸钠与 Ca^{2+} 会迅速反应，生成凝胶）。在此方法中，两种溶液可以任意顺序注入，要采用惰性的水性隔离段塞。常用的胶凝剂是硫酸铵。但这种技术无法产生堵塞多孔介质的均匀凝胶，也难以将凝胶挤入地层深部。在使用这些技术时，需要进行分段的多级泵送。有文章介绍了挪威海上油田为提高采收率，采用此项技术的深部水转向评估和建模。

一种常见的无机凝胶封堵剂是基于内部活化的硅酸钠（IAS）。由硅酸盐和活化剂组成的IAS体系通常制备为低黏度的淡水溶液，注入目标位置后，可在指定时间触发硅酸盐凝胶化。IAS体系的成胶时间由pH值和温度控制。当硅酸钠被酸化到pH值低于10.6后，就会形成三维网状结构的硅酸聚合物凝胶。pH值稍微降低就会大大缩短成胶时间，因此很难通过在地面添加酸来控制成胶时间。但加入缓慢释放酸的材料，可以控制在地层中达到目标pH值。加入醇也会使高pH值的硅酸钠溶液凝胶化。通常将与水反应产生醇和（或）酸的有机材料（活化剂）与硅酸盐溶液混合以调节胶凝时间。

在地层水和高温下，释放酸或醇的活化剂包括如三氯乙酸及其盐类和酯类等卤代有机化合物，会反应释放盐酸（HCl）。另一种产生酸的方法是使用酯、内酯或酰胺的原位水解，如以微乳液形式溶解的二羧酸二酯（如琥珀酸二烷基酯）。产生醇的活化剂包括多羟基化合物（如甘油、甘露醇或糖，或乳糖、木糖等还原性糖）。

---

[1] 1ft=30.48cm。

另一种使用尿素/甲醛树脂的硅酸盐凝胶活化剂体系会在地层中分解释放尿素，从而使硅酸盐凝胶化。加入 EDTA 等螯合剂可大幅增加硅酸盐凝胶系统的二价离子耐受性，从而可以使硅酸盐被外加的阳离子胶凝。硅酸盐凝胶稳定温度可达到 200℃，可用于高温井及气体封窜作业（见2.8节）。

硅酸盐/聚合物法已用于极高渗透储层的处理。含聚合物的硅酸盐溶液在处理液置放和原位分散方面有相应的缺点，可以通过纳米材料代替聚合物以消除这些缺点。详细的实验室研究侧重于无聚合物但含 SiO_2 和 Al_2O_3 纳米颗粒的硅酸盐溶液。2013年开展了先导试验。

有机硅酸盐（或硅氧烷）封堵凝胶可以通过油溶性有机硅与水接触时的原位水解生成。在低浓度下，对固砂很有用；在高浓度下，可用于堵水。典型的有机硅烷有 3-氨基丙基三乙氧基硅烷和双（三乙氧基硅基丙基）胺，或其混合物。

与此相关的是开发一种环境友好型基于胶体二氧化硅的堵剂体系。该体系是含有液态胶体二氧化硅和活化剂的双组分系统，有效温度范围是 30~150℃。优选的活化剂是氯化钾。

鉴于有机硅药剂价格昂贵，研究了成本相对更低的基于聚乙二醇/表面活性剂混合物等的油包水乳液，在特定储层条件下可以乳液反转。该体系已用于气井堵水作业。

现场已使用的另一种无机凝胶体系是基于三价阳离子溶液的原位水解。最易于采用的是酸性溶液中稳定的 Fe^{3+}，地层中自发老化导致 pH 值上升，容易水解形成 $Fe(OH)_3$ 沉淀，这些颗粒絮凝后起到堵水作用。这种新型封堵材料在现场条件下的稳定性很好，即使挤注失败，也可以进行简单的补救。此外，该方法具有的自调控化学机制，即使在低渗透和致密地层应用时，也不会出现难注入的问题。

还有专利介绍了一种相关的体系，包括将氢氧化钠溶液和含有多价阳离子（如 Mg^{2+}）的水溶液，由烃类段塞隔开后，依次注入近井筒地层。隔离段塞可以确保两种水溶液在近井地层中混合，并生成 $Mg(OH)_2$ 等不溶沉淀物，以优先降低近井筒相对高渗透区域的渗透率，从而改善随后的地层流体或产出流体的一致性和流动剖面。在北海油田通过连续油管的堵水作业，使用注入氯氧化镁水浆的化学相关技术。该组合物在快速相变后，形成基本是固体的团块，相变发生所需的时间与其温度之间存在近乎线性的关系，从而可精确地确定设定的时间。

部分水解氯化铝也用作无机凝胶前体。当活化剂（氰酸钠与尿素或六亚甲基四胺）对温度做出响应，并将体系 pH 值提高到一定值以上时，Al^{3+} 和 OH^- 反应，会析出氢氧化铝凝胶，形成一个无定形、不规则的三维不渗透网络。高浓度的 SO_4^{2-} 和 CO_3^{2-} 会破坏系统的稳定性。铝酸钠 $[NaAl(OH)_4]$ 已用于气井的控水作业，其与地层水接触后，pH 值升高会形成氢氧化铝凝胶。

2.4 用于永久封堵的交联有机聚合物凝胶

强交联聚合物凝胶是首个应用于现场的化学堵水处理方法。封隔 100% 产水层后，将含有交联剂的聚合物水溶液泵注入近井地层。随着溶液温度升高，交联形成凝胶，阻断孔隙。这些凝胶在堵水的同时也会堵塞油流通道，因此需要仔细设计隔离不同区域，以免堵塞油层。行业内现在已经很好地掌握了凝胶堵水处理技术，整体成功率约为80%，在油、气井都有应用。针对碳酸盐岩储层聚合物凝胶堵水处理，可参见相关综述文章。

聚合物溶液（含交联剂）可以提前配制并泵入地层，或者将单体溶液与聚合引发剂一起泵入，关井过程中在近井筒处高温交联形成聚合物。聚合物凝胶处理可以与固砂处理或表面活性剂吞吐增产技术结合起来。吞吐是一种增产过程，首先将表面活性剂溶液（通常为阴离子或非离子）注入生产井（吞），然后关井若干天，让表面活性剂浸泡，最后将该井重新投入生产（吐）。注入的表面活性剂可以冲刷井筒附近与其接触的残余油。此外，注入碳酸盐岩储层的表面活性剂可以浸入基质中，使基质从油润湿变为水润湿，从而使水相渗透到富集油的基质中，并驱出原先被绕流过的原油。聚合物凝胶处理中，还可结合使用类似水泥的刚性凝固材料或非水泥颗粒，以更好地控制解决高含水井的堵水调剖和封窜等问题。该体系可以从井口注入井内，颗粒凝胶体系的温度范围为21～177℃，而且不同于水泥必须钻塞，该体系也很容易从井内循环返出。有研究分析了碳酸盐岩层中开展聚合物凝胶和酸化增产组合处理的技术难点。

2.4.1 聚合物注入

有多种水溶性聚合物可通过交联作用形成固体凝胶。已经发表了关于交联聚合物凝胶提高稠油采收率的综述文章。西非Dalia深海油田的注聚合物项目是世界首次在地面和井下方面开展的相关试验。

目前实施的大多数交联聚合物凝胶堵水处理都使用现成的聚合物，在高温地层中交联和凝胶化。常见的凝胶体系如下：

（1）部分水解聚丙烯酰胺（PHPA，或丙烯酸酯/丙烯酰胺共聚物）。

① 与金属离子交联。

② 与有机化合物交联。

（2）阳离子或磺化聚丙烯酰胺（PAMs）。

① 与金属离子交联。

② 与有机化合物交联。

（3）PAM。

与有机化合物交联。

（4）丙烯酰胺/丙烯酸特丁酯共聚物。

与聚乙烯亚胺（PEI）交联。

（5）生物聚合物。

与金属离子交联。

（6）聚乙烯醇或聚乙烯胺。

与有机化合物交联。

高pH值环境下，与硼酸盐离子和（或）金属氧化物交联剂交联的聚丙烯酰胺或生物聚合物，可作为堵水处理的胶凝转向剂。胶凝时间可以通过改变交联剂浓度或注入预冲洗液冷却地层来控制。一般来说，聚合物的分子量非常高（10^6）。PAM具有中性的酰胺侧基，很难交联。部分水解聚丙烯酰胺（PHPA）含有以丙烯酸酯单体形式存在的羧酸盐基团（0~60%不等），这些羧基可以用各种有机化合物及三价或四价金属离子在高温下交联。

2.4.1.1 含羧基的丙烯酰胺和生物聚合物的金属离子交联

多价金属离子会与PHPA、丙烯酸酯共聚物或某些天然聚合物中的羧基交联。铬（Ⅵ，如重铬酸盐离子）可用作交联剂前体，被还原剂还原为Cr^{3+}交联剂。堵水处理早期多用该体系。但该体系对凝胶化时间的控制不够充分。储层中的H_2S也可能过早地还原出Cr^{3+}并随后发生交联。

已研究的最常见的金属离子交联剂是Cr^{3+}、Ti^{4+}、Al^{3+}和Zr^{4+}等。Al^{3+}交联剂的反应无法像其他体系那样可控或延迟，所以很少使用。例如有报道称，采用柠檬酸铝的现场试验需要对近井筒进行预冷以避免过早胶结。有关Al^{3+}交联聚丙烯酰胺凝胶的研究表明，有两种类型的交联反应：一种是发生在同一聚合物分子不同链上的分子内交联反应；另一种是发生在不同聚合物分子之间的分子间交联反应。温度越高，分子内交联的聚合物越容易形成。矿物质影响交联聚合物的分子构型，其影响程度由大到小的顺序是Ca^{2+}>Mg^{2+}>Na^+。

现场Cr^{3+}已大量用于交联PHPA。尽管Cr^{3+}在环境中被氧化成Cr^{4+}的可能性很小，但因Cr^{6+}化合物致癌，在北海地区已禁止使用。

为控制凝胶化时间，并使凝胶渗透距离更远，使用Cr^{3+}的羧酸盐。早期和现在使用较多的体系是PHPA和Cr^{3+}乙酸盐（图2.1）。该系统对pH值和矿化度相对不敏感，能在高达124℃下形成堵水凝胶。高分子量（约$5×10^6$）和低分子量的PHPA分别用于封堵裂缝和封堵基质孔隙。已有相关的研究者对高矿化度下Cr^{3+}交联聚合物溶液的作用机制和性能进行了研究。

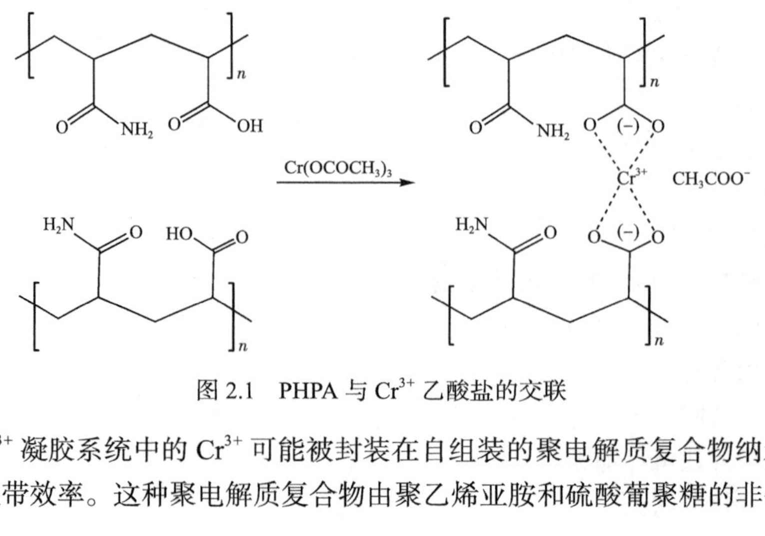

图2.1 PHPA与Cr^{3+}乙酸盐的交联

PHPA/Cr^{3+}凝胶系统中的Cr^{3+}可能被封装在自组装的聚电解质复合物纳米粒子内，并具有很高的夹带效率。这种聚电解质复合物由聚乙烯亚胺和硫酸葡聚糖的非化学计量混合形成。还研究了海水或地层水中的盐类影响，发现存在二价离子时，无论是单独还是在现场盐水或海水中，凝胶化时间都进一步延长到40℃时的35d和80℃时的35h。

另一个市售体系使用丙酸铬（Ⅲ）交联的PHPA，据称比PHPA/乙酸铬（Ⅲ）体系更耐高温。在90～135℃范围内，PHPA/Cr^{3+}乳酸盐或丙二酸盐具有很好的凝胶时间。由于与PHPA交联的金属离子受制于金属离子与保护性阴离子的相互作用强度，可以调节pH值或在金属/聚合物溶液中加入螯合剂，控制金属离子的释放。NaF也可以作为PHPA/Cr^{3+}胶凝溶液的稳定性添加剂。有室内研究表明，高、低分子量聚合物复配能够提高堵水聚合物凝胶对裂缝的封堵性能。这些凝胶主要用于与油井直接接触的裂缝或其他高渗透率的异常区域。

除Cr^{3+}外，Zr^{4+}是最常用的金属阳离子交联剂，如已经使用的Zr^{4+}乳酸盐。聚合物可以是PHPA或弱水解的磺化PAM。用于高温的金属离子交联凝胶，可以使用含有额外数量开链的磺化聚合物，特别是如N-乙烯基吡咯烷酮或N-乙烯基己内酰胺等环状乙烯基酰胺来制备。这些聚合物与二价阳离子也有良好的兼容性。加入少量的膦酸基团可以改善其吸附特性。通过使用有机交联剂也可以提高聚合物凝胶的热稳定性（见下文）。

2.4.1.2 使用天然聚合物的凝胶

基于金属交联天然聚合物的凝胶已研究并应用。但此类凝胶强度低，易受细菌作用而快速降解，效果一般比合成聚合物差，最好在低温下应用。例如，纤维素基凝胶和黄胞胶凝胶在90℃下的半衰期很短，取决于浓度和聚合物类型，从不到1d（完全脱水收缩）到45d不等。使用硬葡聚糖时，观察到中等的稳定性，而用木质素磺酸盐和苯甲醛材料在90℃观察到了最稳定的凝胶，其半衰期超过1年。硬葡聚糖的三重螺旋结构可能是其稳定性优于黄胞胶的原因。与Cr^{3+}交联的黄胞胶已成功地用于许多低温油田，以堵塞见水区，并将水转向到未被波及的含油区。还有报道由产碱菌（*Alcaligenes*）产生的胞外多糖与Cr^{3+}盐交联，形成高温下长期稳定的高强度凝胶。如戊二醛等醛类可作为多糖（如硬葡聚糖）的交联剂，用于温度高达130℃的储层。六亚甲基四胺可以用作醛的前体❶。

2.4.1.3 有机交联

羧酸盐聚合物（如PHPA）的金属交联不适用温度非常高的储层，会发生过度的聚合物凝胶水解。而且与二价阳离子（如Mg^{2+}和Ca^{2+}）进行有害的交联，会发生脱水。在软水中，PAM凝胶更为稳定。

部分水解的聚丙烯酰胺可以与酚类化合物和醛的混合物交联，适用于高温应用。最简单的交联剂体系是苯酚和甲醛（图2.2）。为避免使用有毒的苯酚或甲醛，可以使用在储层中产生苯酚或甲醛的前体。六亚甲基四胺可以生成甲醛，间苯二酚、水杨醇、苯酯（如乙酸苯酯、糠醇、氨基苯甲酸和水杨酸）则是更环保的苯酚前体。未水解的PAM也可以与

---

❶ 邻位有羟基的聚合物（如半乳甘露聚糖和聚乙烯醇），与含硼的化合物交联形成的凝胶，在高温下会失去黏性。但对Ti^{4+}和Zr^{4+}交联剂未观察到此现象。

六亚甲基四胺、对苯二甲醛和己醛（与增溶表面活性剂）等醛类及戊二酸发生交联。通过添加对苯二酚、二羟基萘或没食子酸等二级交联剂，可以使凝胶更加稳定。事实上，酚类化合物可以单独与甲醛形成热固性凝胶，而不需要丙烯酰胺聚合物。一种基于间苯二酚和甲醛的体系已经在现场试用。与乙二醛交联的阳离子PAM也有应用。还开发了对苯二酚/六亚甲基四胺交联的聚丙烯酰胺，是一种环保友好的有机交联聚合物体系。

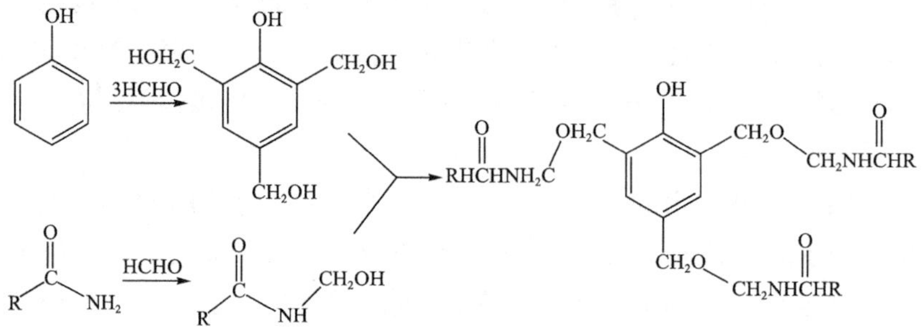

图2.2 丙烯酰胺与苯酚和甲醛的交联机理

延迟凝胶体系的另一种方法是使用丙烯酰胺/丙烯酸酯共聚物。研究和应用最多的共聚物是丙烯酰胺/丙烯酸叔丁酯共聚物（PAtBA）。现场围绕该体系开展了很多工作，使用碳酸盐缓速剂甚至能在177℃下应用。丙烯酰胺和叔丁酯基团都会在高温储层中水解，产生丙烯酸酯基团，叔丁酯更容易水解。产生的丙烯酸酯共聚物可以与金属离子交联，但这些凝胶在高温下不能长期稳定。更好的交联剂是如聚乙烯亚胺（PEI）等聚胺，形成酰胺交联（图2.3）。在相同的PHPA浓度下，用PHPA和PEI制备的凝胶明显强于用Cr^{3+}交联剂制备的凝胶。由PEI和PAtBA形成的凝胶在室内156℃下可稳定若干个月，另一项研究发现其在177℃下有良好的稳定性。

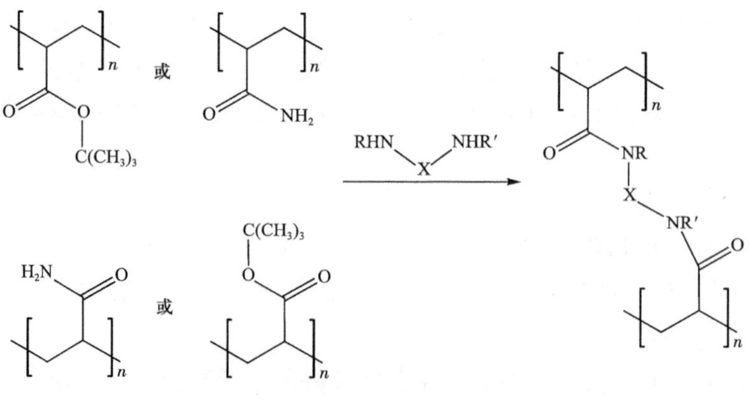

图2.3 丙烯酰胺/丙烯酸叔丁酯共聚物与PEI的水解和交联
（X是PEI的支链骨架，N是一级或二级胺）

在同等条件下，这些凝胶剂的渗透深度是铬基凝胶剂的8倍。使用丙烯酸叔丁酯是因其与PEI的反应速率比丙烯酸甲酯等分子量较小的酯类更慢，从而避免了过早凝胶化，可渗透更深。已经用这种体系进行了许多现场应用。

关于 PAtBA/PEI 体系的交联反应，PAtBA 聚合物上的酯基提供了掩蔽交联位点。这些基团会根据 pH 值和温度发生水解或热解：在低 pH 值和低温条件下，共聚物会水解并形成 PHPA 和叔丁醇；高温下的共聚物会热解，产生 PHPA 和异丁烯。酯基断裂后，形成羧基。在小于 100℃和小于 20h 条件下，PEI 中的亚胺氮亲核进攻与酯相连接的羰基碳，从而形成共价键。在高温下另一种机制可能是通过氨基攻击丙烯酰胺中的羰基来实现。

已提出多种凝胶缓凝剂或可交联的聚合物，如带有 HPEI 交联剂的聚丙烯酰胺/丙烯酸叔丁酯。这些包括弱的有机酸，最好是带有酚基（如间苯三酚和木质素磺酸盐）、弱布朗斯特碱和布朗斯特酸的盐（如弱酸的铵盐）、氨基醇、低聚或多聚胺，以及小分子季铵盐（如四甲基氯化铵）。

据报道有方法可以改进目前的丙烯酰胺和氨基聚合物配方，保持凝胶性能的同时，大幅降低聚合物的浓度。具体方法是加入有机活化剂（包括烷醇胺和季铵盐）以大幅提高多胺的交联反应性，从而减少在特定温度下实现设定凝胶时间所必需的聚合物负载。聚合物是聚丙烯酰胺或 PAtBA，也具有双峰分子量分布。

在北海的挪威地区，作为丙烯酰胺或丙烯酸酯聚合物交联剂的 PEI 因为环保问题被列为淘汰产品。有研究表明，壳聚糖可作为更环保的替代性交联剂，在 120℃的高温下形成稳定的封堵凝胶。壳聚糖是天然聚合物甲壳素的水解产物，含有伯氨基和羟基，主要是氨基交联形成的酰氨基（图 2.4）。

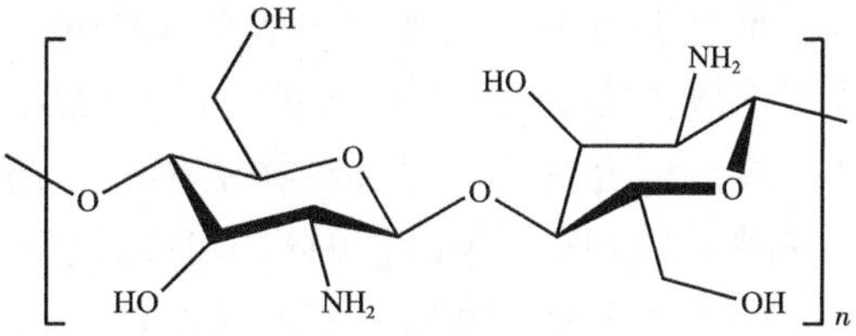

图 2.4 壳聚糖结构

壳聚糖交联剂只是凝胶组合物的一个次要成分，由于合成的基础聚合物生物降解性差，整个系统仍然是不可生物降解为主。为避免这个问题，有人主张使用淀粉等多糖基聚合物和氧化剂，该氧化剂至少能够部分氧化多糖基聚合物，使其可交联。

通过有机交联剂交联成型的聚合物微球（5～30μm），在现场应用时，同丙烯酰胺聚合物水溶液以悬浮液的形式注入。这些微球在 79.4℃、20000mg/L 氯化钠溶液中很稳定。微球在去离子水中可以膨胀到原始尺寸的 5 倍，并显示出良好弹性。在聚合物溶液中加入微球后，阻力系数增加，超过了单独使用聚合物溶液所获得的阻力系数。现场数据显示，注入曲线得到明显改善，提高了原油产量。

2.4.1.4　聚乙烯醇或聚乙烯胺凝胶

聚乙烯醇（PVA）或聚乙烯胺（PVAm）可以与酚类/萘酚类和醛类（甲醛）交联，形成非常稳定的凝胶，但现场应用很少有报道。工业上 PVA 是通过水解聚乙酸乙烯酯制成，PVAm 则由水解聚乙烯甲酰胺制成。PVA 加部分甲基化的三聚氰胺-甲醛树脂也有报道。PVA 也可以只使用二醛（如乙二醛或戊二醛）交联，反应通过半缩醛进行。据报道，唯一能够交联 PVA 的金属离子是 Ti^{4+}。

2.4.1.5 与聚合物凝胶堵水处理有关的问题

聚合物凝胶应用中的潜在问题有脱水收缩、沉淀、化学降解、机械降解（剪切降解）。

脱水收缩指的是聚合物凝胶结构的塌缩，在许多聚合物体系中都有发现，其特征是失去黏附力，体积缩小，并排出水分。可能的原因是交联剂过量、聚合物水解和有 Ca^{2+} 等二价离子。当存在过多的交联剂时，在凝胶化点之后会持续交联，导致聚合物凝胶的体积收缩，并在此过程中排出水分。根据不同的组成成分，脱水收缩的凝胶可能只占初始溶液体积的 5%。丙烯酰胺聚合物可以在热储层中水解成丙烯酸酯基团，这就为交联和可能的脱水收缩提供了更多的位点。地层中高浓度的二价阳离子可以提供比注入的金属交联剂所预期的更多的交联。在岩石基质孔隙中，即使聚合物凝胶的脱水收缩率非常高，其带来的渗透率降低，在技术上仍有一定效果。而裂缝性储层中，脱水收缩会对堵水效果下降造成很大的影响。

聚合物沉淀是二价阳离子和聚合物水解后的羧基之间相互作用的结果。随着羧基含量增加，二价阳离子会降低其溶解度，发生沉淀，最终聚合物无法形成凝胶。

聚合物可能会在地层中成功凝胶化，然后由于化学键断裂而变稀，导致现场措施有效期很短。化学键断裂的可能原因是储层温度过高、氧气污染或过氧化物污染，产生自由基，导致聚合物分子量和结构的损失。通过添加如丙烯酰胺/2-丙烯酰氨基-2-甲基丙烷磺酸（AMPS）等磺化单体，可以改善凝胶的稳定性。

许多堵水聚合物的分子量非常高，成为长链线型聚合物。当聚合物通过泵、在管道和油管的湍流中、流经射孔段及地层孔隙时，可能会发生剪切降解，使聚合物分子量降低，凝胶效果变差。

2.4.1.6 交联聚合物凝胶的其他改进措施

阳离子 PAM 可使聚合物更好地吸附于碳酸盐岩，并防止过早产生回流。通过添加一定量的胶体二氧化硅等惰性胶体颗粒材料，可以获得更高的凝胶强度。使用四甲基铵盐、聚乙烯吡咯烷酮、聚氧化乙烯、PEI 及非离子型、两性型、阴离子型和阳离子型表面活性剂化合物预冲洗，可使聚合物流入油藏深部，在更深的位置布放凝胶。这些化合物附着在地层孔隙表面的吸附位点上，当另一种流体流经处理的地层时，会缓慢带走化合物。此后，将阻止水流动的聚合物引入地层，使其在先前引入的化合物被带走之前深入到地层的孔隙中，并在化合物被带走时附着在吸附位点上。

自交联水溶性聚合物的一个实例是由单体丙烯酰胺、丙烯醛和4-乙烯基苯酚组成（图2.5）。泡沫凝胶也是一种堵水技术。泡沫凝胶由可交联的含羧基聚合物、含有反应性过

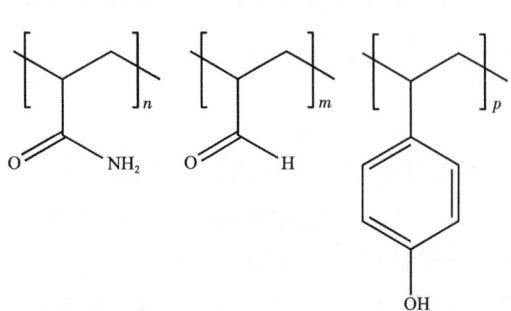

图 2.5　丙烯酰胺/丙烯醛/4-乙烯基苯酚三元共聚物

渡金属阳离子的交联剂、聚乙烯醇、水溶剂和添加的气体形成。

2.4.2 原位单体聚合

不使用预制的聚合物，而是将单体溶液和交联剂与聚合引发剂一起泵入，高温关井期间在近井筒内形成聚合物。原位单体聚合法的使用明显少于预制聚合物法。其中一个原因是这些聚合物中的主要成分丙烯酰胺单体是致癌物。另一个原因是偶氮或过氧化合物的聚合引发剂价格昂贵。此外，用这些体系很难获得足够长的聚合延迟时间。通过使用乳液已克服了后一个问题。典型的乳液体系使用 N- 羟甲基丙烯酰胺或 N- 羟甲基甲基丙烯酰胺单体、2，2- 双（叔丁基过氧）丁烷等双过氧化物聚合引发剂、聚氧乙烯失水山梨糖醇酯等乳化剂、对叔丁基邻苯二酚等聚合抑制剂及脂肪烃。

2.5 黏弹性表面活性剂凝胶

基于黏弹性表面活性剂的流体体系形成的黏性段塞，已用于水和（或）气体封窜。该流体体系可能包括盐水、增黏剂和 VES，以及可选的高温稳定剂。黏弹性表面活性剂包括非离子、阳离子、两性和两性离子表面活性剂（示例见 5.7.4）。优选的体系有氧化胺类 VES，其可能提供更高的单位重量胶凝力，比同类的其他液体更具成本优势。优选的稳定剂是氧化镁、氧化钛（Ⅳ）和氧化铝。增黏剂可包括热释电颗粒、压电颗粒及其混合物。

一种在压裂作业期间封堵产水区的化学方法是使用特定的黏弹性阴离子表面活性剂（VASs），是一类选择性渗透性封堵剂。黏弹性阴离子表面活性剂（VASs）存在阳离子时，会产生剪切变稀凝胶，容易泵送，可以渗透到多孔、可渗透的岩石中。VES 的实例是烷基肌氨酸盐类（图 2.6），一旦进入地层孔隙，其黏度可增加 100 倍之多，从而限制了流体的流动。与烃类化合物接触会破坏凝胶，大大降低黏度。这只释放了具有残余烃饱和度的孔隙，使其变得清洁并具有强烈的水湿性。相反，具有高含水饱和度的孔隙仍然被凝胶堵塞。

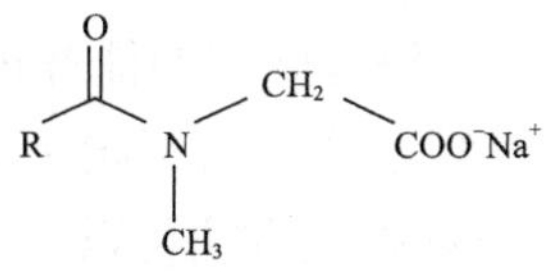

图 2.6 烷基肌氨酸盐

2.6 不等比例渗透率降低剂（DPR）或相对渗透率调节剂（RPM）

一些研究人员将 DPR 与 RPM 区分开来，尽管二者在不明显影响石油渗透率、降低水的渗透率方面效果相同，DPRs 有时被描述为堵塞孔隙的化学剂，但在烃类存在情况下不会像在水中那样沉淀、膨胀或粘连。其净效果是水相相对渗透率的降低系数远大于油相相对渗透率。DPR 和 RPM 处理可以从井口注入，是聚合物凝胶处理的一种替代方法，后者需要分段隔离。DPR 的实例是松香木衍生物在有水情况下的凝胶化。还可以使用油溶性硅化合物，与地层中的水反应形成阻水的硅酸盐凝胶，这样就不会明显地影响含油区。硅化合物的实例包括正硅酸甲酯或硅酸乙酯等烷基硅酸酯，以及卤代硅烷。这些硅化合物可与聚合物 RPM（见下文）结合使用，以提高堵水效果。DPR 还可以与防垢剂挤注处理相结合。

2.6.1 乳化凝胶作为 DPRs

为了优化 DPR 效果，重要的是将凝胶置于原油饱和度高于残余油饱和度的地方。基于此，开发了一种新的 DPR 技术，该技术使用含胶凝剂配方的热敏油包水乳剂。已有现场成功应用的报道，其使用的凝胶体系是丙烯酰胺/丙烯酸叔丁酯共聚物，PEI 作为交联剂。为了显著降低油的渗透性，凝胶应只能占据一小部分处理的体积。乳液的一个重要特点是，在形成凝胶之前，会自发地破裂。在关井后的静态条件下，水相分离并凝胶化，留下流动的油相。乳化体系比以前评估的油和胶凝剂的共同注入更容易处理和预测。在水润湿的岩心中，含油饱和度为 25% 时，胶凝剂对渗透率的降低要大得多，因为水润湿介质中的水基胶凝剂会阻塞孔隙喉道。使用乳化胶凝剂，有可能获得可衡量的渗透率降低，而不是完全堵塞。原因是油（在乳液中）有助于保持部分喉道畅通，这样油就有可能流过岩心，而不必首先机械地破坏凝胶。

2.6.2 亲水聚合物作为 RPMs

相对渗透率调节剂是一种水溶性的亲水聚合物，可以吸附在地层的岩石表面，已经有相关综述文章发表。

相对渗透率调节剂所要的效果是针对水润湿孔隙，利用吸附的水合聚合物层大幅减少孔喉通道尺寸。因此，重要的是最大限度地提高水合聚合物的体积。其效果是水相相对渗透率比油相相对渗透率降低系数更大。如果产能指数的损失可以通过更高的压降来补偿，那么经过处理的油井也可以增产更多的石油或天然气。如果 RPM 主要进入裂缝，对水相相对渗透率的影响不大，因为裂缝通道比孔喉宽很多。20 世纪 90 年代中期以后，因为 RPM 处理方法可以井口注入，不需要连续油管和分段隔离，其研究和开发得到迅速发展。与需要分段隔离的聚合物凝胶处理相比，RPM 更便宜。但如果操作正确，聚合物凝胶堵水处理可以持续数年，而 RPM 处理的有效期通常只有一年或更短。

现场使用 RPM 的处理成功率仍然没有聚合物凝胶处理高，这可能不是因为产品失败，而是因为选井不当。评估 RPM 和其他基质应用的选井方法已经有文章发表，以及使用 RPM 来控制气井产水的可行性。如果在已完成的区域中存在不同的含水饱和度和不连续的生产，RPM 选择性堵水将更加有效。如果不是这种情况，那么任何水量的降低都可能对产油产生不利影响。适合开展 RPM 处理的油井包括页岩层分隔产油区和产水区的直井，或产水量不均的长水平井。有研究报道，在注入 RPM 后再进行聚合物胶凝处理，可以阻断气井的产水。

尽管对 RPM 的机制仍有争议，但有人提出了一个基本的机制，即 RPM 受孔隙层级的油水分离或优先流动的制约。水基 RPM 流体分布于优先水流的孔隙部分，几乎不影响油流。膨胀的吸附聚合物层是亲水疏油性，当烃类流经孔隙时，聚合物层将脱水和收缩，因此不会损害油气生产。聚合物层还可以对流动的水产生阻力作用，但对流动的原油影响很小。该机制的作者还提出，RRF（定义为处理前后的渗透率之比）可由多孔介质中的凝胶饱和度确定。

有效的 RPM 必须进入完成区域并强力吸附于岩石。因此，清除近井筒的原油沉积物至关重要，通常是通过表面活性剂预冲洗来实现。更好的方式是将表面活性剂添加到聚合物 RPM 配方中。RPM 材料正常起效，对产油区的影响就很小。但是如果产层受到伤害，也可以添加如过硫酸盐等聚合物破胶剂清理处理。已经证明，利用相对渗透率调节剂的控水效果，在很大程度上取决于油井的流速，因为渗透率的降低幅度会随着流速的增加而减小。

除众所周知的使用亲水聚合物以外，很少有关于使用非 VES 来降低产水的报告。有一项专利涉及使用咪唑啉化合物（咪唑的脂肪酸、氧乙基衍生物），以控制砂岩储层的产水。

2.6.2.1 聚合物 RPM 的类型

已经研究了一系列的亲水聚合物作为 RPM，可以分为合成乙烯基聚合物和天然聚合物两类。已经应用的无须交联的 RPM 聚合物实例包括：

（1）PAM、PHPA、黄胞胶、硬葡聚糖、黏弹性聚合物。

（2）含有甲基丙烯酰胺丙基三甲基氯化铵、二甲基二烯丙基氯化铵或甲基丙烯酸二甲氨基乙酯（DMAEMA）单体的阳离子丙烯酰胺聚合物。

（3）阳离子改性天然聚合物，如阳离子淀粉。

（4）丙烯酸、丙烯酸盐和二甲基二烯丙基氯化铵单体的两性三元共聚物。

（5）丙烯酰胺 /AMPS 共聚物。

（6）乙烯基磺酸盐 / 丙烯酰胺共聚物和乙烯基磺酸盐 / 乙烯基甲酰胺 / 丙烯酰胺三元共聚物。

（7）丙烯酰胺 /N− 乙烯基吡咯烷酮共聚物。

（8）AMPS/ 乙烯基甲酰胺三元共聚物（图 2.7）。

（9）由聚合物骨架与聚氧化乙烯接枝组成的梳形聚合物，如聚［二烷基氨基丙烯酸酯 − 共 − 丙烯酸酯 −g− 聚氧化乙烯］。

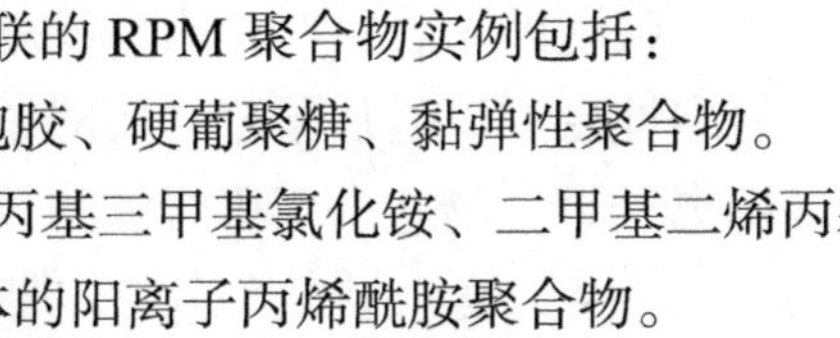

图 2.7　丙烯酰胺 /AMPS/ 乙烯基甲酰胺三元共聚物 RPMs

早期的 RPMs 使用简单的 PAM 或 PHPA 或天然聚合物如黄胞胶或硬葡聚糖等。但这些聚合物的回流或降解速度太快，有效期相当短。加入甲醛作为杀菌剂，使硬葡聚糖的热稳定性更好。但由于聚合物溶液的高黏度和絮凝的倾向，生物聚合物有注入性问题。阳离子丙烯酰胺或两性丙烯酰胺聚合物已用于提高岩石吸附、抗剪切敏感性及对温度和盐的耐受性。

聚丙烯酰胺由于热稳定性有限，效果较差，且有效时间缩短；通过引入阳离子基团改性的PAM提升功效，并显示出良好的岩石吸附和耐盐性。通过使用甲基丙烯酰胺丙基三甲基氯化铵单体，而非丙烯酰胺丙基三甲基氯化铵来制备共聚物，可以改善其热稳定性。有数百项工作针对中等分子量的阴离子丙烯酰胺/AMPS/乙烯基甲酰胺三元共聚物，大都应用于低产油井。这在很大程度上取代了两性共聚物的使用。乙烯基甲酰胺的功能是锚定单体。最初的压裂处理现场试验成功地使用这种RPM作为有机硼酸盐交联的前置液的添加剂。含有磺化单体的乙烯基聚合物甚至比非磺化聚合物（如PAM和PHPA）的稳定性更强，其与二价金属离子的相容性也比PAM和PHPA好，后者在储层温度下，容易进一步水解成丙烯酸酯基团。有人提出，相对于原油，聚合物可以选择性地降低水相相对渗透率，而且这种特性并不取决于聚合物的类型（只要其是亲水性的）或岩石类型。

在一项RPM处理的改进中，水合聚合物占据的体积可以通过改变水溶液的矿化度调节。因此，例如溶解在高矿化度盐水中的PHPA由于静电屏蔽而导致其收缩，当其被泵入井中时，低矿化度的地层水接触到吸附的聚合物时，聚合物就会膨胀。还有人声称，将两种相反电荷的聚合物依次注入砂岩地层，以增加聚合物的滞留。这样，带负电的地层首先与阴离子聚合物接触，然后是阳离子聚合物。由于聚合物的接触，阴离子和阳离子聚合物之间发生了凝聚（相分离），这就减少了被地层中产出液体带走的阴离子聚合物数量。先注入阳离子聚合物，再注入阴离子聚合物，对带正电的碳酸盐岩储层更有用。

如前所述，最大限度地增加水合聚合物的体积对控水很重要。为此设计了由聚合物骨架与聚氧化乙烯接枝组成的刷状（或梳状）聚合物，现场应用的成功率较高。刷状侧链在与烃类接触时将会塌缩，但在与水接触时会完全水合，占据更大的体积。

相对渗透性调节剂可用于控制水力压裂等几种不同综合处理中的产水。一些聚合物防垢剂和聚合物RPM之间的相似性并未被忽视，已经进行了RPM和防垢剂的联合挤注处理。还介绍了一种防垢和控水的联合处理方法，其需要的步骤比单独每种处理程序的总和要少。RPM控水处理也可以与固砂处理或酸化处理结合。例如，一种固砂/RPM组合处理包括以下步骤：向地层注入预冲洗液，注入表面活性剂水溶液（含RPM），注入低黏度固结液（含表面活性剂），注入后冲洗液。

2.6.2.2 疏水改性的合成聚合物作为RPMs

砂岩岩心的实验室结果表明，PAM的疏水改性修饰大大增加了矿物覆盖率，降低表面粗糙度，从而使RPM比全亲水聚合物效果更好。疏水改性并没有像亲水聚合物那样达到平衡的吸附效果，而是随着聚合物浓度的增加而产生持续增长的吸附效果。这种行为归因于聚合物链在先前吸附的聚合物层上的缔合吸附。其效果是通过改善油的流动性和减少水的流动性来提高采收率。所声称的疏水改性RPM的实例包括烷基卤化物部分季铵化的 DMAEMA 聚合物，其中烷基卤化物的烷基链长度为 6~22 个碳（图 2.8）。其他实例有丙烯酰胺/十八烷基二甲基胺甲基丙烯酸酯溴化物共聚物、甲基丙烯酸二甲氨基乙酯/乙烯基吡咯烷酮/十六烷基二甲基甲基丙烯酸酯溴化物三元共聚物，以及丙烯酰胺/2-丙烯酰氨基-2-甲基丙烷磺酸/2-乙基己基甲基丙烯酸酯三元共聚物。

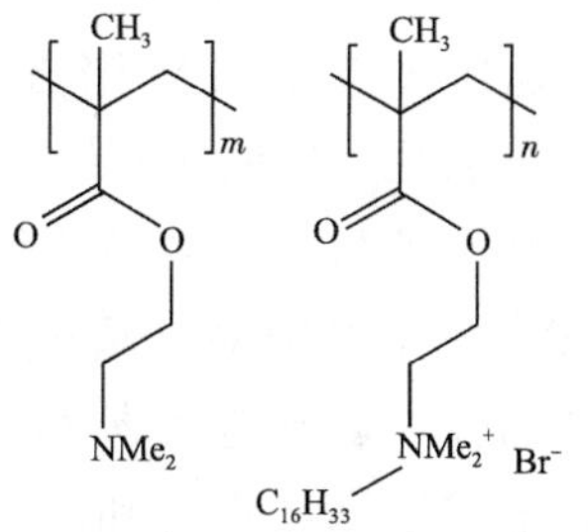

图 2.8 以 $C_{16}H_{33}Br$ 部分季铵化的疏水改性 RPM/DMAEMA 共聚物
（$m>10n$ 以增加水溶性）

已经证明，当注入水和油通道的凝胶量相等时，疏水缔合的水溶性聚合物凝胶系统表现出显著的不等比例的渗透率降低现象。聚合物凝胶的吸附性保留可能是导致不等比例的渗透率降低的主要因素。已声称有一种方法是将排水率增强剂引入地层，以便在进行 RPM 处理后提高天然气产量，从而减少产水量。RPM 是一种有 12~18 个碳的氨基甲基丙烯酸酯/烷基氨基甲基丙烯酸酯共聚物。排水率增强剂包括以下化合物中的至少一种：一种两性离子表面活性剂、一种包含至少两个季铵基团的阳离子聚有机硅氧烷和一种特定的烷氧基化非离子表面活性剂。

据称使用可交联的、疏水改性的水溶性聚合物与表面活性剂作为 RPM 处理剂。聚合物可以与 VES 一起使用，在产油和产水岩层中形成稳定的物理结构。这种结构在烃类流经时会被分解，使烃类顺利流动，而在产水区不会被分解。聚合物可以与交联剂一起配制，在地层中凝胶化，形成更持久的产水屏障。具体实例为在 0.5mol/L 的氯化钠溶液中，用 3%（摩尔分数）的丙烯酸正壬酯疏水改性的 PAM 和 VES［N-芥基-N，N-双（2-羟乙基）-N-甲基氯化铵］（图 2.9）反应。加入磷酸脲（一种氢键改性剂）后，表面活性剂黏弹性的增长过程被延迟，加入乙醛后，疏水改性的 PAM 发生交联。疏水改性的丙烯酰胺/丙烯酸酯共聚物可以与 Cr^{3+} 或 Zr^{4+} 等金属离子交联。第二个实例使用油酸钾在氯化钾电解质溶液中形成的 VES 溶液，与 2，2′-偶氮（双脒基丙烷）二盐酸盐等乙烯基聚合引发剂形成聚合表面活性剂凝胶。所形成的聚合表面活性剂溶液的黏弹性比原来的单体溶液略低，但观察到的黏弹性对与烃类的接触不敏感。聚合表面活性剂形成的凝胶在与水长期接触后仍能保持其黏弹性。

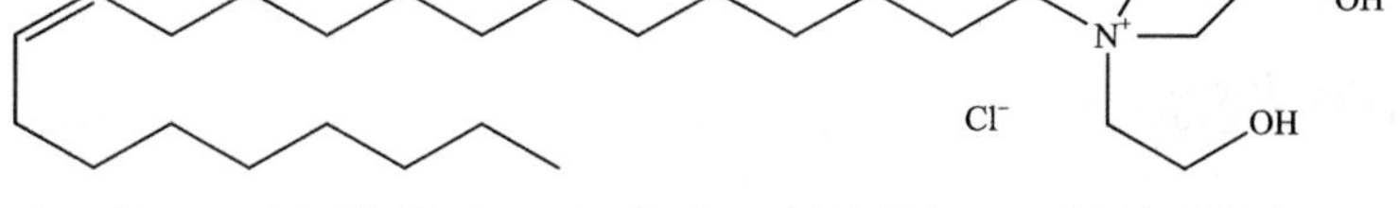

图 2.9 N-芥基-N，N-双（2-羟乙基）-N-甲基氯化铵

2.6.2.3 交联聚合物 RPMs

交联亲水聚合物不仅适用于前面所讨论的堵水凝胶处理，也适用于高渗透地层中的

RPM 处理。高度交联可产生全阻隔凝胶，部分交联则产生三维聚合物网络，连接孔隙空间。这比普通的 RPM 聚合物具有更强效果，后者只改变了地层的表面特性。

早期在该领域使用的实例是与乙二醛交联的阳离子 PAM。与醛交联的多糖也有报道。最近开发了与金属离子交联的磺化聚合物，其交联剂是 Zr^{4+}，典型的聚合物是丙烯酰胺/AMPS/乙烯基膦酸三元共聚物，含有少量乙烯基膦酸单体以增强岩石吸附（图2.10）。聚合物的分子量不能太高（$>1\times10^7$），否则难以处理，分子量也不能太低（$<3\times10^5$），否则会形成堵塞产水层和产油层的凝胶。相反，聚合物是中等分子量[$(1\sim3)\times10^6$]，交联形成一个开放、水合的柔性网络。当采出水与聚合物网络接触时，其就会膨胀并变硬，对水的流动产生阻力，并使水的渗透率降低。如果烃类通过聚合物网络，就会因水的置换而塌缩，对烃类流动阻力最小。该体系已成功应用于降低高硫酸钡垢油井的产水量，这种 RPM 处理方法对高温的高渗透地层很有用。对于渗透率较低的地层，可能没有必要将聚合物交联成三维网络。但在有 Fe^{2+} 而非 Fe^{3+} 的现场试验中，RPM 效果不佳。

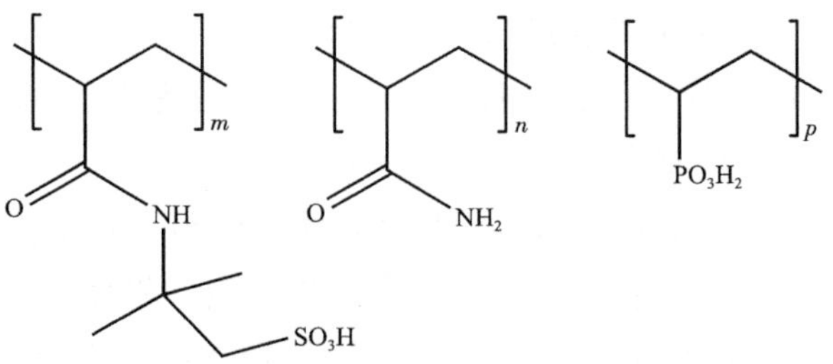

图 2.10　高渗透储层用金属离子交联的聚合物 RPM：丙烯酰胺/AMPS/乙烯基膦酸三元共聚物

据报道，含中性有机共价交联剂的丙烯酰氨基凝胶的胶体颗粒，可作为控制产水的 RPM。与其他交联 RPM 的不同之处是该技术在配方注入前预先交联。该微凝胶的尺寸（0.3～2.0μm）、一致性和化学性质都可变。由于其尺寸大，不容易渗透到低渗透地层中，从而适应了堵水技术的挑战之一。这种微凝胶具有良好的抗剪切、耐温和化学稳定性。首个现场应用是在储气的储层，在北海油田也有成功应用。

2.6.2.4　黏弹性 RPMs

对一种新型无固体黏弹性 RPM 的性能已有评估。这种产品的设计是为改进普通黏弹性表面活性剂和普通聚合物 RPM 有时遇到的孔隙堵塞问题。

2.7　使用微粒控水

使用固体颗粒堵水的想法长期存在，但现场应用不多（前面已经讨论过聚合物溶液与固体微粒的共同注入问题）。例如，有专利声称使用由二乙烯基苯和苯乙烯等单体的水包油乳液和聚合引发剂，原位形成不溶于水的聚合物固体球体。这种类似塑料的固体球体会渗入渗透率更高的地层并阻止流体流动。在二氧化碳驱之前的调剖过程，也可以利用这种乳液。还有相关专利称，乳液应含有不可聚合的液体载体和少量可聚合的单体（如丙烯酸酯或苯乙烯）。这样当乳液的非连续相与基质中的环境流体混溶时，由于乳液滴被稀释，形成的非理想颗粒将降到最低。油井防垢剂等也可通过乳液非连续相聚合制成颗粒态。

与 RPMs 和 DPRs 使用有关的是水溶胀交联聚合物颗粒。例如，由乙烯基酰胺（如丙烯酰胺或 N-乙烯基吡咯烷酮）与阳离子单体制成的聚合物颗粒分散体，已用于堵水处理。这些颗粒是通过聚合物乳液反相聚合制成，其颗粒尺寸比地层孔隙的尺寸小得多。在注入时，颗粒滞留于地层中，当水回流时，颗粒会膨胀并吸附在地层上，形成一层薄膜，限制流体的进一步流动。也有人提出了能够在中性 pH 值或弱酸性 pH 值下水解的交联可膨胀聚合物微粒。

另一项专利是关于涂覆颗粒。该方法包括用有机聚合物涂覆颗粒固体的步骤，该聚合物与水接触后会反应膨胀。这就通过堆积聚合物涂覆颗粒固体，减少了水的流动。该有机聚合物可以是乙烯基硅烷（如乙烯基三甲氧基硅烷或甲基丙烯酸酯三甲氧基硅烷）和一种或多种水溶性有机单体［如丙烯酸 2-羟乙酯、（甲基）丙烯酰胺和 N-乙烯基吡咯烷酮）的共聚物］（图 2.11），或聚二烷基二烯丙基聚合物或聚 DMAEMA 共聚物的季铵盐阳离子水溶性聚合物。颗粒尺寸可以小到足以穿透孔喉，也可以是压裂作业的支撑剂尺寸。

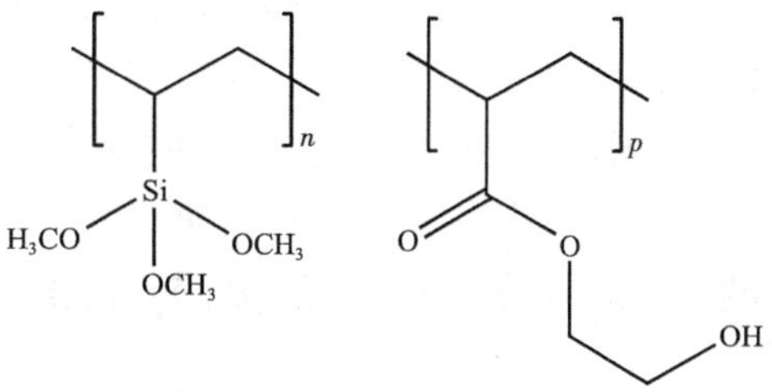

图 2.11 使用乙烯基三甲氧基硅烷和丙烯酸 2-羟乙酯涂覆颗粒的膨胀聚合物实例

另一种方法使用高岭石颗粒堵水。还测试评估了将相对高浓度的高岭石与固定剂一起注入的堵水技术潜力。这一概念是基于高岭石造成地层破坏的能力，加上高岭石固定剂保持稳定黏土的能力。到目前为止，已经证明使用这种技术可以承受 15MPa 的压差，而不会对密封剂的性能造成任何伤害。

用一种基于聚硅烷的特殊疏水粉末处理含水油井，可以显著提高油井产量。这种效果被解释为是由于在近井筒的储层基质中引入了疏水但亲油的颗粒，使储层中水和油的相对渗透率发生了变化，从而使油的流动途径增多，而水的流动途径则相应减少。典型的配方是由用二氯二甲基硅烷处理的二氧化硅（如云母海绿砂岩），加上膨润土或磷石膏等黏土，在煤油等烃类溶剂中制成。

2.8 热敏水溶性聚合物

在石油工业之外的生物医学等领域，热敏水溶性聚合物应用较为广泛。水敏性指当温度上升到浊点（也称低临界溶液温度）以上时，聚合物由于与水溶剂的氢键断裂，能够自我塌缩并与溶液相分离。因聚（N-异丙基丙烯酰胺）的低临界溶液温度接近人体温度，是研究最多的聚合物。热敏性聚合物的凝胶系统也有研究。例如，EO-PO-EO 三元共聚物凝胶在堵水和改善水驱中的潜在应用研究表明，水中添加 3%～6% 的 EO-PO-EO 聚合物制成的凝胶，无法适用于油井措施，但可以适用于注水井的深部调驱。这种凝胶在连续重新加热时的可重复胶凝特性，以及凝胶强度随温度变化率的变化，表明通过注入水缓慢冷却固化的凝胶，可以实现凝胶的移动和再固化。已经制备出如由丙烯酰胺和3-磺丙基丙烯酸钾盐制成的水凝胶，兼具强度、热敏性和超强吸水性。

2.9 水溶胀聚合物

水溶胀聚合物具有形成凝胶体的能力，可以阻止不需要的流体流过地层。水化后产生的凝胶体如果大小合适，很容易渗入地层裂缝中。水合前干燥的聚合物颗粒大小决定了凝胶体的大小。这些特性使凝胶聚合物颗粒能够挤入地层孔隙，形成密封。水溶胀聚合物可以与盐水一同泵送，通过增加盐的浓度可以延迟溶胀。水溶胀聚合物也可以在油载体中泵送，以防止过早膨胀，但如果其要在地层中膨胀，就必须保证与水接触。这一点往往很难保证。

一种改进的方法包括一种具有酸性水相的载体。水溶胀聚合物在酸性水相中基本上不会膨胀。酸在井下消耗后，聚合物可在地层中膨胀达到效果。也可以使用内相为酸性水相的油包水乳液。最理想的是随着内部水相所含的酸在井下消耗，油包水乳液破裂。

2.10 控制产气

部分用于控制产水的技术也可用于控制产气。例如，本章前面讨论的有机聚合物凝胶及无机凝胶（如硅酸钠）已成功应用。泡沫水泥或水泥颗粒凝胶也有应用。

另一种成功使用的方法是用泡沫控制高气油比（GOR）油井的气体。现场试验表明，泡沫可以有效地控制气体锥进、注入井到采出井间气体在高渗透层的突破及至少在某些情况下的气体舌进。限制在岩石孔隙网络内的泡沫由横跨孔隙的薄液膜组成，使气相不连续。这大大降低了气体的流动性，基本上不影响液体的相对渗透率。泡沫是由气体（通常是氮气）置换适宜的表面活性剂发泡剂溶液所产生。以阻止气体进入生产井为目的的泡沫，最好是在可能发生气窜的地方就地形成，然后保持停滞状态，在最长时间内实现控制气体流动性的最佳效果。经过一系列的现场应用，得出的结论是高渗透储层需要气体/泡沫剂的共同注入，以改善泡沫在储层中的位置。在渗透性较差的储层中，通过表面活性剂与气体的交替技术在原位产生/置入泡沫，就足以取得良好效果。

泡沫的一个基本特性是在原油存在时稳定。现场应用中，在非水溶液（如煤油）中产生的泡沫被流经泡沫屏障下方的原油冲走。基于α-烯烃磺酸盐（AOS）表面活性剂的水性泡沫被证明在原油存在时的泡沫阻断效果很差。AOS与氟化表面活性剂或AOS与如PAM等聚合物结合使用得到更好的泡沫稳定性。使用这些混合起泡剂防气窜处理取得了良好的效果。据报道，使用增稠的合成聚合物和生物聚合物可以改善泡沫稳定性。典型的混合物是不同比例的丙烯酰胺/丙烯酸钠共聚物、迪特胶（diutan gum）和椰油酰胺丙基甜菜碱表面活性剂。凝胶或交联泡沫也可以产生刚性更强的泡沫。但有实验表明，与单独使用 AOS 表面活性剂相比，添加聚合物并没有带来任何益处，如果使用浓度过高，可能会造成地层伤害。有研究评估了一种含钙源的表面活性剂体系用于碳酸盐岩油层的气体封窜。

参 考 文 献

[1] O. Vazquez, M. Singleton, K. S. Sorbie, and R. Weare, "Sensitivity Study on the Main Factors Affecting a Polymeric RPM Treatment in the Near-Wellbore Region of a Mature Oil-Producing Well," SPE 106012 (paper presented at the SPE International Symposium on Oilfield Chemistry, Houston, TX, 28 February-2 March 2007).

[2] R. S. Seright, R. H. Lane, and R. D. Sydansk, "A Strategy for Attacking Excess Water Production," SPE 84966, *SPE Production & Facilities* 18（3）(2003): 158.

[3] J. Hibbeler and P. Rae, "The Environmental Benefits of Reducing Unwanted Water Production," SPE 96582 (paper presented at the SPE Asia Pacific Health, Safety and Environment Conference and Exhibition, Kuala Lumpur, Malaysia, 19-21 September 2005).

[4] D. W. Ross, U.S. Patent Application, 20060175065, 2006.

[5] A. H. Kabir, "Chemical Water & Gas Shutoff Technology—An Overview," SPE 72119 (paper presented at the SPE Asia Pacific Improved Oil Recovery Conference, Kuala Lumpur, Malaysia, 6-9 October 2001).

[6] G. E. Anderson and W. J. Heaven, International Patent Application WO/2008/083468.

[7] D. D. Sparlin and R. W. Hagen Jr., "Controlling Water in Production Operations," *World Oil* July (1984): 1.

[8] P. D. Nguyen and D. L. Brown, U.S. Patent 7028774, 2006.

[9] B. B. Sandiford, U.S. Patent 4004639, 1977.

[10] G. D. Herring, J. T. Milloway, and W. N. Wilson, "Selective Gas Shut-Off Using Sodium Silicate in the Prudhoe Bay Field, AK," SPE 12473 (paper presented at the SPE Formation Damage Control Symposium, Bakersfield, CA, 13-14 February 1984).

[11] T. A. T. Lund, H. I. Berge, S. Espedal, R. Kristensen, T. A. Rolfsvaag, and G. Stromsvik, "The Technical Performance and Interpretation of Results from a Large Scale Na-Silicate Gel Treatment of a Production Well on the Gullfaks Field," Proceedings of the 8th European Symposium on Increased Oil Recovery, vol. 2, Vienna, 1995.

[12] R. S. Seright and J. Liang, "A Survey of Field Applications of Gel Treatments for Water Shut-Off," SPE 26991 (paper presented at the SPE Permian Basin Oil and Gas Recovery Conference, Midland, TX, 16-18 March 1994).

[13] J. J. Jurinak and L. E. Summers, "Oilfield Applications of Colloidal Silica Gel," *SPE Production Engineering* 6（4）(1991): 406.

[14] M. A. Hardy, D. W. van Batenburg, and C. W. Botermans, "Use of Temperature Simulations in Water-Control Design," SPE 60896, *SPE Production & Facilities* 15（1）(2000): 14.

[15] R. Boreng and O. B. Svendsen, "A Successful Water Shut Off: A Case Study from the Statfjord Field," SPE 37466 (paper presented at the SPE Production Operations Symposium, Oklahoma City, OK, 9-11 March 1997).

[16] M. R. Islam and S. M. Farouq Ali, "Use of Silica Gel for Improving Waterflooding Performance of Bottom-Water Reservoirs," *Journal of Petroleum Science and Engineering* 8 (1993): 303.

[17] B. Vinot, R. S. Schechter, and L. W. Lake, "Formation of Water-Soluble Silicate Gels by the Hydrolysis

of a Diester of Dicarboxylic Acid Solubilized as Microemulsions," SPE 14236, *SPE Reservoir Engineering* 4 (3)(1989): 291.

[18] B. Vinot, G. Berrod, and J.-L. Brun, U.S. Patent 4799549, 1989.

[19] T. Huang and P. M. McElfresh, U.S. Patent Application 20040031611, 2004.

[20] (a) W. H. Smith and E. F. Vinson, U.S. Patent 4640361, 1987. (b) T. M. Vickers and L. J. Powers, U.S. Patent 4384894, 1983.

[21] E. A. Elphingstone, H. C. McLaughlin, and C. W. Smith, U.S. Patent 4293440, 1981.

[22] M. A. H. Laramay, U.S. Patent 5320171, 1994.

[23] H. A. Nasr-El-Din and K. C. Taylor, *Journal of Petroleum Science and Engineering* 48 (2005): 141-160.

[24] K. C. Taylor and H. A. Nasr-El-Din, U.S. Patent 6660694, 2003.

[25] H. K. Kotlar, International Patent Application WO/2005/124099.

[26] I. Lakatos, J. Lakatos-Szabo, B. Kosztin, Gy. Palasthy, and P. Kristof, "Application of Iron-Hydroxide-Based Well Treatment Techniques at the Hungarian Oil Fields," SPE 59321 (paper presented at the SPE/DOE Improved Oil Recovery Symposium, Tulsa, OK, 3-5 April 2000).

[27] B. Kosztin, Gy. Palasthy, F. Udvari, L. Benedek, I. Lakatos, and J. Lakatos-Szabo, "Field Evaluation of Iron Hydroxide Gel Treatments," SPE 78351 (paper presented at the SPE European Petroleum Conference, Aberdeen, UK, 29-31 October 2002).

[28] R. D. Sydansk, U.S. Patent 4304301, 1981.

[29] B. H. Tomlinson, U.S. Patent 7044222, 2006.

[30] D. Barclay, K. Lawson, B. Mullins, and B. Cardno, "Stand Alone Coiled Tubing Water Shutoff Operations Reinstate Well on a Normally Unattended Installation," SPE 100132 (paper presented at the SPE/ICoTA Coiled Tubing and Well Intervention Conference and Exhibition, Woodlands, TX, 4-5 April 2006).

[31] A. Parker and C. Davidson, U.S. Patent 4889563, 1989.

[32] A. Parker, European Patent EP0266808, 1988.

[33] S. M. Lahiliah, *Journal of Applied Polymer Science* 106 (2007): 2076.

[34] Y. Wang, B. Bai, and H. Gao, "Enhanced Oil Production Through a Combined Application of Gel Treatment and Surfactant Huff'n' Puff Technology," SPE 112495 (paper presented at the SPE International Symposium and Exhibition on Formation Damage Control, Lafayette, LA, 13-15 February 2008).

[35] C. Deolarte, J. Vasquez, E. Soriano, and A. Santillan, "Successful Combination of an Organically Crosslinked Polymer System and a Rigid-Setting Material for Conformance Control in Mexico," SPE 112411 (paper presented at the SPE International Symposium and Exhibition on Formation Damage Control, Lafayette, LA, 13-15 February 2008).

[36] J. Vasquez, L. Eoff, D. Dalrymple, and J. van Eijden, "Shallow Penetration Particle-Gel System for Water and Gas Shutoff Applications," SPE 114885 (paper presented at the SPE Annual Technical Conference and Exhibition, Denver, CO, 21-24 September 2008).

[37] A. Moradi-Araghi, "A Review of Thermally Stable Gels for Fluid Diversion in Petroleum Production," *Journal of Petroleum Science and Engineering* 26 (2000): 1.

[38] (a) G. P. Karmakar and C. Chakraborty, "Improved Oil Recovery Using Polymeric Gelants: A Review," *Indian Journal of Chemical Technology* 13 (2006): 162. (b) J. K. Fink, *Oilfield Chemicals*, Burlington, MA: Gulf Professional, Elsevier, 2003.

[39] C. Harrison, M. Luyster, L. Moore, B. B. Prasek, and R. D. Ravitz, International Patent Application WO/2009/026021.

[40] S. M. Vargas-Vasquez and L. B. Romero-Zerón, "A Review of the Partly Hydrolyzed Polyacrylamide Cr (III) Acetate Polymer Gels," *Petroleum Science and Technology* 26 (2008): 481.

[41] A. Stavland and H. C. Jonsbråten, "New Insight into Aluminium Citrate/Polyacrylamide Gels for Fluid Control," SPE 35381 (paper presented at the SPE/DOE 10th Symposium on Improved Oil Recovery, Tulsa, OK, 21-24 April 1996).

[42] R. D. Sydansk and G. P. Southwell, "More Than 12 Years of Experience with a Successful Conformance-Control Polymer Gel Technology," SPE 49315 (paper presented at the SPE Annual Technical Conference and Exhibition, New Orleans, LA, 27-30 September 1998).

[43] P. Shu, "Gelation Mechanism of Chromium (III), Oilfield Chemistry-Enhanced Recovery and Production Stimulation," in *ACS Symposium Series*, eds. J. K. Borchardt and T. F. Yen, Washington, DC: American Chemical Society, 1989, 137.

[44] A. Sabhapondit and A. Borthakur, "A Comparative Study of Gelation Behaviour of Some Acrylamide Copolymers with Cr (III) Crosslinker," *Journal of Polymer Materials* 20 (3)(2003): 309.

[45] R. D. Sydansk, U.S. Patent 5421411, 1995.

[46] R. D. Sydansk, "Acrylamide-Polymer/Chromium (III) -Carboxylate Gels for Near Wellbore Matrix Treatments," SPE 20214, *SPE Advanced Technology Series* 1 (1)(1993): 146.

[47] P. D. Moffitt, "Long-Term Production Results of Polymer Treatments in Producing Wells in Western Kansas," *Journal of Petroleum Technology* 45 (4)(1993): 356.

[48] T. P. Lockhart and P. Albonico, "New Chemistry for the Placement of Chromium (III)/Polymer Gels in High-Temperature Reservoirs," SPE 24194, *SPE Production & Facilities* 9 (4)(1994): 273.

[49] M. Bartosek, A. Mennella, T. P. Lockhart, C. Emilio, R. Elio, and C. Passucci, "Polymer Gels for Conformance Treatments: Propagation of Cr (III) Crosslinking Complexes in Porous Media," SPE 27828 (paper presented at the SPE/DOE Improved Oil Recovery Symposium, Tulsa, OK, 17-20 April 1994).

[50] J. P. Feraud, S. Karlstad, and L. Aasberg, Water Control through Placement of Polyacrylamide Gel in Propped Fracture in the Gullfaks Field," Paper 23 (paper presented at the 8th International Oil Field Chemical Symposium, Geilo, Norway, 2-5 March 1997).

[51] R. D. Sydansk, U.S. Patent 6189615, 2001.

[52] R. D. Sydansk, A. M. Al-Dhafeeri, Y. Xiong, and R. S. Seright, "Polymer Gels Formulated with a Combination of High- and Low-Molecular-Weight Polymers Provide Improved Performance for Water-Shutoff Treatments of Fractured Production Wells," SPE 89402, *SPE Production & Facilities* 19 (4) (2004): 229.

[53] P. D. Moffitt, A. Moradi-Araghi, I. Ahmed, and V. R. Janway, "Development and Field Testing of a New Low Toxicity Polymer Crosslinking System," SPE 35173 (paper presented at the SPE Permian Basin Oil and Gas Recovery Conference, Midland, TX, 27-29 March 1996).

[54] G. Chauveteau, R. Tabary, M. Renard, and A. Omari, "Controlling In-Situ Gelation of Polyacrylamides by Zirconium for Water Shutoff," SPE 50752 (paper presented at the SPE International Symposium on Oilfield Chemistry, Houston, TX, 16-19 February 1999).

[55] C. Kayser, G. Botthof, K. H. Heier, A. Tardi, M. Krull, and M. Schaefer, U.S. Patent Application 20060019835, 2006.

[56] M. Zettlitzer, W. Schuhbauer, and N. Kohler, "Laboratory Evaluation of a New, Selective Water Control Treatment and Its Implementation in a North Sea Well," *Petroleum Geosciences* 2 (4)(1996): 325.

[57] S. S. Nagra, J. P. Batycky, R. E. Nieman, and J. B. Bodeux, "Stability of Waterflood Diverting Agents at Elevated Temperatures in Reservoir Brines," SPE 15548 (paper presented at the SPE Annual Technical Conference and Exhibition, New Orleans, LA, 5–8 October 1998).

[58] B. Kalpakci, Y. T. Jeans, N. F. Magri, and J. P. Padolewski, "Thermal Stability of Scleroglucan at Realistic Reservoir Conditions," SPE 20237 (paper presented at the SPE/DOE Enhanced Oil Recovery Symposium, Tulsa, OK, 22–25 April 1990).

[59] M. R. Avery, L. A. Burkholder, and M. A. Gruenenfelder, "Use of Crosslinked Xanthan Gels in Actual Profile Modification Field Projects," SPE 14114 (paper presented at the SPE International Meeting on Petroleum Engineering, Beijing, China, 17–20 March 1986).

[60] E. T. Strom, J. M. Paul, C. H. Phelps, and K. Sampath, "A New Biopolymer for High–Temperature Profile Control: Part 1—Laboratory Testing," SPE 1963, *SPE Reservoir Engineering* 6 (3)(1991): 360.

[61] K. Sampath, L. G. Jones, E. T. Strom, C. H. Phelps, and C. S. Chiou, "A New Biopolymer for High–Temperature Profile Control: Part II–Field Results," SPE 19867 (paper presented at the SPE Annual Technical Conference and Exhibition, San Antonio, TX, 8–11 October 1989).

[62] N. Kohler and A. Zaitoun, U.S. Patent 5082577, 1992.

[63] B. B. Sandiford, H. T. Dovan, and R. D. Hutchins, U.S. Patent 5486312, 1996.

[64] R. D. Hutchins, H. T. Dovan, and B. B. Sandiford, "Field Applications of High Temperature Organic Gels for Water Control," SPE 35444 (paper presented at the SPE/DOE Improved Oil Recovery Symposium, Tulsa, OK, 21–24 April 1996).

[65] R. G. Ryles, "Chemical Stability Limits of Water–Soluble Polymers Used in Oil Recovery Processes," SPE 13585, *SPE Reservoir Engineering*, February 3 (1)(1988): 23.

[66] A. Moradi–Aragahi and P. H. Doe, "Hydrolysis and Precipitation of Polyacrylamides in Hard Brines at Elevated Temperatures," *SPE Reservoir Engineering*, May 2 (2)(1987): 189.

[67] P. M. DiGiacomo and C. M. Schramm, "Mechanism of Polyacrylamide Gel Syneresis Determined by C–13 NMR," SPE 11787 (paper presented at the SPE Oilfield and Geothermal Chemistry Symposium, Denver, CO, 1–3 June 1983).

[68] D. O. Falk, U.S. Patent 4485875, 1985.

[69] B. L. Swanson, 4440228, 1984.

[70] R. Banerjee, B. Ghosh, and K. C. Khilar, *Oil Gas-European Magazine* 32 (4)(2006): 184.

[71] A. Moradi–Araghi, G. Bjornson, and P. H. Doe, "Thermally Stable Gels for Near–Wellbore Permeability Contrast Modifications," SPE 18500, *Advanced Technical Series* 1 (1)(1993): 140.

[72] R. Hutchins and M. Parris, "New Crosslinked Gel Technology," (paper presented at the PNEC 4th International Conference on Reservoir Conformance, Profile Control, Water and Gas Shut–Off, Houston, TX, 10–12 August 1998).

[73] R. L. Clampitt, H. M. Al–Rikabi, and M. K. Dabbous, "A Hostile Environment Gelled Polymer System for Well Treatment and Profile Control," SPE 25629 (paper presented at the SPE Middle East Oil Show, Bahrain, 3–6 April 1993).

[74] A. Moradi–Araghi, U.S. Patent 4994194, 1991.

[75] A. Moradi–Araghi, U.S. Patent 5179136, 1993.

[76] M. D. Parris and R. D. Hutchins, U.S. Patent 6011075, 2000.

[77] H. T. Dovan and R. D. Hutchins, "Delaying Gelation of Aqueous Polymers at Elevated Temperatures Using Novel Organic Crosslinkers," SPE 37246 (paper presented at the SPE International Symposium on Oilfield

Chemistry, Houston, TX, 18-21 February 1997).

[78] P. W. Chang, I. M. Goldman, and K. J. Stingley, "Laboratory Studies and Field Evaluation of a New Gelant for High-Temperature Profile Modification," SPE 14235 (paper presented at the SPE Annual Technical Conference and Exhibition, Las Vegas, NV, 22-26 September 1985).

[79] G. A. Al-Muntasheri, I. A. Hussein, and H. A. Nasr-El-Din, *Journal of Petroleum Science and Engineering* 55 (2007): 56.

[80] J. C. Morgan, P. L. Smith, and D. G. Stevens, "Chemical Adaptation and Deployment Strategies for Water and Gas Shut-off Gel Systems," RSC 6th International Symposium on Chemistry in the Oil Industry, Ambleside, UK, 14-17 April 1997.

[81] G. A. Al-Muntasheri, H. A. Nasr-El-Din, K. R. Al-Noaimi, and P. L. J. Zitha, "A Study of Polyacrylamide-Based Gels Crosslinked with Polyethyleneimine," SPE 105925 (paper presented at the SPE International Symposium on Oilfield Chemistry, Houston, TX, 28 February-2 March 2007).

[82] G. A. Al-Muntasheri, P. L. J. Zitha, and H. A. Nasr-El-Din, "Evaluation of a New Cost-Effective Organic Gel System for High Temperature Water Control," IPTC-11080 (paper presented at the International Petroleum Technology Conference, Dubai, UAE, 4-6 December 2007).

[83] J. Vasquez, E. D. Dalrymple, L. Eoff, B. R. Reddy, and F. Civan, "Development and Evaluation of High Temperature Conformance Polymer Systems," SPE 93156 (paper presented at the SPE International Symposium on Oilfield Chemistry, Houston, TX, 2-4 February 2005).

[84] M. Hardy, W. Botermans, A. Hamouda, J. Valdal, and J. Warren, "The First Carbonate Field Application of a New Organically Crosslinked Water Shutoff Polymer System," SPER 50738 (paper presented at the SPE International Symposium on Oilfield Chemistry, Houston, TX, 16-19 February 1999).

[85] M. B. Hardy, C. W. Botermans, and P. Smith, "New Organically Cross-Linked Polymer System Provides Competent Propagation at High Temperatures in Conformance Treatments," SPE 39690 (paper presented at the SPE/DOE Symposium on Improved Oil Recovery, Tulsa, OK, 1998).

[86] B. R. Reddy, L. Eoff, E. D. Dalrymple, K. Black, D. Brown, and M. Rietjens, "A Natural Polymer-Based Cross-Linker System for Conformance Gel Systems," SPE 84937, *SPE Journal*, June 8 (2) (2003): 99.

[87] B. R. Reddy, L. S. Eoff, and E. D. Dalrymple, U.S. Patent 7007752, 2006.

[88] S. M. Lahalih and E. F. Ghloum, "Rheological Properties of New Polymer Compositions for Sand Consolidation and Water Shutoff in Oil Wells," *Journal of Applied Polymer Science* 104 (2007): 2076.

[89] S. M. Skjæveland, A. Skauge, L. Hinderaker, and C. D. Sisk eds., *RUTH Program Summary*, Stavanger, Norway: NDP (Norwegian Petroleum Directorate), 1996 (http://www.npd.no/engelsk/projects/ruth/ruth.htm).

[90] G. Di Lullo and P. Rae, "New Insights into Water Control—A Review of the State of the Art," SPE 77963 (paper presented at the SPE Asia Pacific Oil and Gas Conference and Exhibition, Melbourne, Australia, 8-10 October 2002).

[91] M. Mahajan, N. Rauf, T. Gilmore, and A. Maylana, "Water Control and Fracturing: A Reality," SPE 101019 (paper presented at the SPE Asia Pacific Oil & Gas Conference and Exhibition, Adelaide, Australia, 11-13 September 2006).

[92] D. H. Hoskin and P. Shu, U.S. Patent 4896723, 1990.

[93] P. Shu, U.S. Patent 4964463, 1990.

[94] C. Victorius, U.S. Patent 5061387, 1991.

[95] M. L. Marrocco, U.S. Patent 4664194, 1987.

[96] M. L. Marrocco, U.S. Patent 4498540, 1985.

[97] P. Shu, U.S. Patent 4678032, 1987.

[98] M. R. Avery, T. A. Wells, P. W. Chang, and J. P. Millican, "Field Evaluation of a New Gelant for Water Control in Production Wells," SPE 18201 (paper presented at the SPE Annual Technical Conference and Exhibition, Houston, TX, 2-5 October 1988).

[99] E. D. Dalrymple, L. S. Eoff, B. R. Reddy, D. L. Brown, and P. S. Brown, U.S. Patent 6364016, 2002.

[100] S. N. Davies, T. G. J. Jones, S. Olthoff, and G. J. Tustin, U.S. Patent 6920928, 2005.

[101] R. D. Sydansk, U.S. Patent 5834406, 1998.

[102] P. D. Nguyen, L. Sierra, E. D. Dalrymple, and L. S. Eoff, U.S. Patent Application 20070012445, 2007.

[103] K. A. Rodrigues, U.S. Patent 5335726, 1994.

[104] K. A. Rodrigues, U.S. Patent 5358051, 1994.

[105] H. Lane, "Design, Placement and Field Quality Control of Polymer Gel Water & Gas Shut-Off Treatments," PNEC 97#3 (paper presented at the 3rd International Conference on Reservoir Conformance, Profile Control, Water and Gas Shut-Off, Houston, TX, 6-8 August 1997).

[106] D. Dalrymple, J. T. Tarkington, and H. James, "A Gelation System for Conformance Technology," SPE 28503 (paper presented at the SPE Annual Technical Conference and Exhibition, New Orleans, LA, 25-28 September 1994).

[107] P. Woods, K. Schramko, D. Turner, D. Dalrymple, and E. Vinson, "In-Situ Polymerization Controls CO2/Water Channelling at Lick Creek," SPE 14958 (paper presented at the SPE/DOE Fifth Symposium on Enhanced Oil Recovery, Tulsa, OK, 20-23 April 1986).

[108] M.-C. P. Leblanc, J. A. Durrieu, J.-P. P. Binon, G. G. Provin, and J.-J. Fery, U.S. Patent 4975483, 1990.

[109] T. Huang, J. B. Crews, and J. R. Willingham, U.S. Patent Application 20080248978.

[110] G. Di Lullo, A. Ahmad, P. Rae, L. Anaya, and R. Ariel Meli, "Toward Zero Damage: New Fluid Points the Way," SPE 69453 (paper presented at the SPE Latin American and Caribbean Petroleum Engineering Conference, Buenos Aires, Argentina, 25-28 March 2001).

[111] G. F. DiLullo, P. J. Rae, and A. J. K. Ahmad, U.S. Patent 6767869, 2004.

[112] G. P. Karmakar, C. A. Grattoni, and R. W. Zimmerman, "Relative Permeability Modification Using an Oil-Soluble Gelant to Control Water Production," SPE 77414 (paper presented at the SPE Annual Technical Conference and Exhibition, San Antonio, TX, 29 September-2 October 2002).

[113] L. J. Kalfayan and J. C. Dawson, British Patent GB 2399364.

[114] J. C. Dawson, L. J. Kalfayan, P. H. Javora, M. Vorderbruggen, J. W. Kirk, and Q. Qu, U.S. Patent Application 20060065396, 2006.

[115] A. Stavland and S. Nilsson, International Patent Application WO/2001/021726.

[116] A. Stavland, K. I. Andersen, B. Sandøy, T. Tjomsland, and A. A. Mebratu, "How To Apply a Blocking Gel System for Bullhead Selective Water Shutoff: From Laboratory to Field," SPE 99729 (paper presented at the SPE/DOE Symposium on Improved Oil Recovery, Tulsa, OK, 22-26 April 2006).

[117] S. Nilsson, A. Stavland, and H. C. Jonsbraten, "Mechanistic Study of Disproportionate Permeability Reduction," SPE 39635 (paper presented at the SPE/DOE Improved Oil Recovery Symposium, Tulsa, OK, 19-22 April 1998).

[118] D. Dalrymple, L. Eoff, B. R. Reddy, and C. W. Botermans, "Relative Permeability Modifiers for Improved Oil Recovery: A Literature Review" (paper presented at the PNEC 5th International Conference on Reservoir Conformance, Profile Control, Water and Gas Shut-Off, Houston, TX, 8-10 November

1999）．

[119] J. T. Liang, R. L. Lee, and R. S. Seright, "Gel Placement in Production Wells," SPE 20211, *SPE Production & Facilities* 8（1993）: 276.

[120] A. Zaitoun, N. Kohler, D. Bossie-Codreanu, and K. Denys, "Water Shutoff by Relative Permeability Modifiers: Lessons from Several Field Applications," SPE 56740（paper presented at the SPE Annual Technical Conference and Exhibition, Houston, TX, 3-6 October, 1999）．

[121] J. Novotny, "Matrix Flow Evaluation Technique for Water Control Applications," SPE 30094（paper presented at the SPE European Formation Damage Conference, The Hague, Netherlands, 15-16 May 1995）．

[122] D. J. Ligthelm, "Water Shut Off in Gas Wells: Is There Scope for a Chemical Treatment？," SPE 68978（paper presented at the SPE European Formation Damage Conference, The Hague, Netherlands, 21-22 May 2001）．

[123] K. Munday, U.S. Patent 6516885, 2003.

[124] A. Stavland and S. Nilsson, "Segregated Flow Is the Governing Mechanism of Disproportionate Permeability Reduction in Water and Gas Shutoff," SPE 71510（paper presented at the SPE Annual Technical Conference and Exhibition, New Orleans, LA, 30 September-3 October 2001）．

[125] C. W. Botermans, D. W. van Batenburg, and J. Bruining, "Relative Permeability Modifiers: Myth or Reality？," SPE 68973（paper presented at the SPE European Formation Damage Conference, The Hague, Netherlands, 21-22 May 2001）．

[126] R. D. Sydansk and R. S. Seright, "When and Where Relative Permeability Modification Water-Shutoff Treatments Can Be Successfully Applied," SPE 99371（paper presented at the SPE/DOE Symposium on Improved Oil Recovery, Tulsa, OK, 22-26 April 2006）．

[127] D. Dalrymple, U.S. Patent 5146986, 1992.

[128] A. Zaitoun, N. Kohler, U.S. Patent 4842071, 1989.

[129] C. Tielong, Z. Yong, P. Kezong, and P. Wanfeng, "A Relative Permeability Modifier for Water Control of Gas Wells in a Low-Permeability Reservoir," *SPE Reservoir Engineering* 11（3）（1996）: 168.

[130] N. Kholer, R. Tabary, A. Zaitoun U.S. Patent 4718491, 1988.

[131] A. Zaitoun, N. Kohler, B. K. Maitin, and M. Zettlitzer, "Preparation of a Water Control Polymer Treatment at Conditions of High Temperature and Salinity," *Journal of Petroleum Science and Engineering* 7（1992）: 67.

[132] K. Denys, C. Fichen, and A. Zaitoun, "Bridging Adsorption Of Cationic Polyacrylamides In Porous Media," SPE 64984（paper presented at the SPE International Symposium on Oilfield Chemistry, Houston, TX, 13-16 February 2001）．

[133] L. Chiappa, M. Andrei, T. P. Lockhart, G. Burrafato, and G. Maddinelli, International Patent Application WO/2002/097236.

[134] P. D. Nguyen, S. Ingram, L. Sierra, E. D. Dalrymple, and L. S. Eoff, U.S. Patent Application 20070029087, 2007.

[135] L. Chiappa, M. Andrei, T. P. Lockhart, G. Burrafato, and G. Maddinelli, U.S. Patent 7188673, 2007.

[136] J. L. Boles and G. Mancillas, U.S. Patent 4476931, 1984.

[137] F. O. Stanley, M. E. Hardianto, and P. S. Tanggu, "Improving Hydrocarbon/Water Ratios in Producing Wells—An Indonesian Case History Study," SPE 36615（paper presented at the SPE Annual Technical Conference and Exhibition, Denver, CO, 6-9 October 1996）．

[138] D. D. Dunlap, J. L. Boles, and R. J. Novotny, "Method for Improving Hydrocarbon/Water Ratios in

Producing Wells," SPE 14822 (paper presented at the SPE Formation Damage Control Symposium, Lafayette, LA, 26-27 February 1986).

[139] G. Push, "Practical Experience with Water Control in Gas Wells by Polymer Treatments," *EAPG Improved Oil Recovery European Symposium*, Vienna, Austria 2 (1995): 48-56.

[140] M. Ranjbar, P. Czolbe, and N. Kohler, "Comparative Laboratory Selection and Field Testing of Polymers for Selective Control of Water Production in Gas Wells," SPE 28984 (paper presented at the SPE International Symposium on Oilfield Chemistry, San Antonio, TX, 14-17 February 1995).

[141] D. Coehlo, "Development and Application of Selective Polymer Injection to Control Water," (paper presented at the 5th Latin American and Caribbean Petroleum Engineering Conference and Exhibition, Rio de Janeiro, Brazil, 30 August-3 September 1997).

[142] J. C. Dawson, H. V. Le, and S. Kesavan, U.S. Patent 6228812, 2001.

[143] L. Eoff, E. D. Dalrymple, B. R. Reddy, and D. Everett, "Structure and Process Optimization for the Use of a Polymeric Relative-Permeability Modifier in Conformance Control," SPE 64985 (paper presented at the SPE International Symposium on Oilfield Chemistry, Houston, TX, 13-16 February 2001).

[144] S. G. Nelson, L. J. Kalfayan, and W. M. Rittenberry, "The Application of a New and Unique Relative Permeability Modifier in Selectively Reducing Water Production," SPE 84511 (paper presented at the SPE Annual Technical Conference and Exhibition, Denver, CO, 5-8 October 2003).

[145] P. Barriau, H. Bertin, and A. Zaitoun, "Water Control in Producing Wells: Influence of Adsorbed Polymer Layer on Relative Permeabilities and Capillary Pressure," SPE 35447 (paper presented at the SPE/DOE Symposium on Improved Oil Recovery, Tulsa, OK, 21-24 April 1996).

[146] A. Zaitoun and N. Kohler, "Improved Polyacrylamide Treatments for Water Control in Producing Wells," SPE 18501 (paper presented at the SPE International Symposium on Oilfield Chemistry, Houston, TX, 8-10 February 1989).

[147] D. Dalrymple and E. Vinson, U.S. Patent 4617132, 1986.

[148] R. Castano, J. Villamizar, O. Diaz, S. A. Hocol, M. Avila, S. Gonzalez, E. D. Dalrymple, S. Milson, and D. Everett, "Relative Permeability Modifier and Scale Inhibitor Combination in Fracturing Process at San Francisco Field in Colombia, South America," SPE 77412 (paper presented at the SPE Annual Technical Conference and Exhibition, San Antonio, TX, 29 September-2 October 2002).

[149] P. Powell, M. A. Singleton, and K. S. Sorbie, US 6913081, 2005.

[150] E. D. Dalrymple, L. Eoff, B. R. Reddy, and J. Venditto, U.S. Patent 7182136, 2007.

[151] M. Cordova, J. L. Mogollon, H. Molero, and M. Navas, "Sorbed Polyacrylamides: Selective Permeability Parameters Using Surface Techniques," SPE 75210 (paper presented at the SPE/DOE Improved Oil Recovery Symposium, Tulsa, OK, 13-17 April 2002).

[152] E. Volpert, J. Selb, F. Candau, N. Green, J. F. Argillier, and A. Audibert, *Langmuir* 14 (1998): 1870.

[153] L. S. Eoff, B. R. Reddy, and E. D. Dalrymple, U.S. Patent 6476169, 2002.

[154] L. S. Eoff, E. D. Dalrympel, B. R. Reddy, and F. Zamora, U.S. Patent 7114568.

[155] T. G. J. Jones and G. J. Tustin, U.S. Patent 6194356, 2001.

[156] J. C. Dawson, L. J. Kalfayan, and G. Brock, U.S. Patent Application 20040177957.

[157] M. J. Faber, G. J. P. Joosten, K. A. Hashmi, and M. Gruenenfelder, "Water Shut-off Field Experience with a Relative Permeability Modification System in the Marmul Field (Oman)," SPE 39633 (paper presented at the SPE/DOE Improved Oil Recovery Symposium, Tulsa, OK, 19-22 April 1998).

[158] K. H. Heier, C. Kayser, A. Tardi, M. Schaefer, R. Morschhaeuser, J. C. Morgan, and A. M. Gunn, U.S. Patent 7150319, 2006.

[159] J. J. Wylde, G. D. Williams, and C. Shields, "Field Experiences in Application of a Novel Relative Permeability Modifier Gel in North Sea Operations," SPE 97643 (paper presented at the SPE International Improved Oil Recovery Conference in Asia Pacific, Kuala Lumpur, Malaysia, 5-6 December 2005).

[160] J. Morgan, A. Gunn, G. Fitch, H. Frampton, R. Harvey, D. Thrasher, R. Lane, R. McClure, K. H. Heier, and C. Kayser, "Development and Deployment of a 'Bullheadable' Chemical System for Selective Water Shut Off Leaving Oil/Gas Production Unharmed," SPE 78540 (paper presented at the Abu Dhabi International Petroleum Exhibition and Conference, Abu Dhabi, UAE, 13-16 October 2002).

[161] Y. Feng, R. Tabary, M. Renard, C. Le Bon, A. Omari, and G. Chauveteau, "Characteristics of Microgels Designed for Water Shutoff and Profile Control," SPE 80203 (paper presented at the SPE International Symposium on Oilfield Chemistry, Houston, TX, 5-7 February 2003).

[162] G. Chauveteau, R. Tabary, N. Blin, M. Renard, D. Rousseau, and R. Faber, "Disproportionate Permeability Reduction by Soft Preformed Microgels," SPE 89390 (paper presented at the SPE/DOE Symposium on Improved Oil Recovery, Tulsa, OK, 17-21 April 2004).

[163] D. Rousseau, G. Chauveteau, M. Renard, R. Tabary, A. Zaitoun, P. Mallo, O. Braun, and A. Omari, "Rheology and Transport in Porous Media of New Water Shutoff/ Conformance Control Microgels," SPE 93254 (paper presented at the SPE International Symposium on Oilfield Chemistry, The Woodlands, TX, 2-4 February 2005).

[164] C. Cozic, D. Rousseau, and R. Tabary, "Broadening the Application Range of Water Shutoff/ Conformance Control Microgels: An Investigation of Their Chemical Robustness," SPE 115974 (paper presented at the SPE Annual Technical Conference and Exhibition, Denver, CO, 21-24 September 2008).

[165] A. Zaitoun, R. Tabary, D. Rousseau, T. Pichery, S. Nouyoux, P. Mallo, and O. Braun, "Using Microgels to Shut Off Water in a Gas Storage Well," SPE 106042 (paper presented at the SPE International Symposium on Oilfield Chemistry, Houston, TX, 28 February-2 March 2007).

[166] C. H. Phelps, E. T. Strom, and M. L. Hoefner, U.S. Patent 5048607, 1991.

[167] H. K. Kotlar, B. Schilling, and J. Sjoblom, U.S. Patent 7270184, 2007.

[168] J. C. Dawson, H. V. Le, and S. Kesavan, U.S. Patent 5735349, 1998.

[169] P. D. Nguyen and B. T. Dewprashad, U.S. Patent 6109350, 2000.

[170] N. Fleming, K. Ramstad, A.-M. Mathisen, Alex Nelson, and S. Kidd, "Innovative Use of Kaolinite in Downhole Scale Management: Squeeze Life Enhancement and Water Shutoff," SPE 113656 (paper presented at the SPE 9th International Conference on Oilfield Scale, Aberdeen, UK, 28-29 May 2008).

[171] J. F. Hurtado, A. Milne, and E. Olivares, "Shallow Gas Shut-Off Using Rigid Polymer Gel in Lake Maracaibo, Venezuela," SPE 94532 (paper presented at the SPE Latin American and Caribbean Petroleum Engineering Conference, Rio de Janeiro, Brazil, 20-23 June 2005).

[172] M. A. Llamedo, F. V. Mejias, E. R. González, J. Espinoza, E. M. Valero, and N. Calis, SPE 96696 (paper presented at the Offshore Europe, Aberdeen, UK, 6-9 September 2005).

[173] L. Perdomo, H. Rodríguez, M. Llamedo, L. Oliveros, E. González, O. Molina, and C. Giovingo, "Successful Experiences for Water and Gas Shutoff Treatments in North Monagas, Venezuela," SPE 106564 (paper presented at the Latin American & Caribbean Petroleum Engineering Conference, Buenos Aires, Argentina, 15-18 April 2007).

[174] G. Burrafato, E. Pitoni, G. Vietina, L. Mauri, and L. Chiappa, "Rigless WSO Treatments in Gas Fields. Bullheading Gels and Polymers in Shaly Sand: Italian Case Histories," SPE 54747 (paper presented at

the SPE European Formation Damage Conference, The Hague, Netherlands, 31 May–1 June 1999).

[175] E. Ali, F. E. Bergren, P. DeMestre, E. Biezen, and J. van Eijden, "Effective Gas Shutoff Treatments in a Fractured Carbonate Field in Oman," 102244 (paper presented at the SPE Annual Technical Conference and Exhibition, San Antonio, TX, 24–27 September 2006).

[176] J. E. Hanssen and S. Ekrann, U.S. Patent 4903771, 1990.

[177] J. E. Hanssen and M. Dalland, "Foams for Effective Gas Blockage in the Presence of Crude Oil," SPE 20193 (paper presented at the SPE/DOE Enhanced Oil Recovery Symposium, Tulsa, OK, 22–25 April 1990).

[178] M. N. Bouts and M. Dalland, "Foam Treatments Against Unwanted Gas Production in Oil Producers," *Chemistry in the Oil Industry VI*, Manchester, UK: Royal Society of Chemistry, 1998, 132.

[179] V. O. Chukwueke, M. N. Bouts, and C. E. van Dijkum, "Gas Shut–Off Foam Treatments," SPE 39650 (paper presented at the SPE/DOE Improved Oil Recovery Symposium, Tulsa, OK, 19–22 April 1998).

[180] M. N. Bouts, A. S. de Vries, M. Dalland, and J. E. Hanssen, "Design of Near Well Bore Foam Treatments for High GOR Producers," SPE 35399 (paper presented at the SPE/DOE Improved Oil Recovery Symposium, Tulsa, OK, 21–24 April 1996).

[181] M. Dalland, J. E. Hanssen, and T. S. Kristiansen, "Oil Interaction with Foams under Static and Flowing Conditions in Porous Media," *Colloids and Surfaces A: Physicochemical and Engineering Aspects* 82 (1994): 129.

[182] S. Thach, K. C. Miller, Q. J. Lai, G. S. Sanders, J. W. Styler, and R. H. Lane, "Matrix Gas Shut–Off in Hydraulically Fractured Wells Using Polymer Foams," SPE 36616 (paper presented at the SPE Eastern Regional Meeting, Columbus, OH, 23–25 October 1996).

[183] J. Van Houwelingen, "Chemical Gas Shut–Off Treatments in Brunei," SPR 57268 (paper presented at the SPE Asia Pacific Improved Oil Recovery Conference, Kuala Lumpur, Malaysia, 25–26 October 1999).

[184] R. D. Sydansk, "Polymer–Enhanced Foams: Laboratory Development and Evaluation," SPE 25168 (paper presented at the SPE International Symposium on Oilfield Chemistry, New Orleans, LA, 1993).

[185] J. E. Hanssen and M. Dalland, "Increased Oil Tolerance of Polymer–Enhanced Foams: Deep Chemistry or Just 'Simple' Displacement Effects?," SPE 59282 (paper presented at the SPE/DOE Improved Oil Recovery Symposium, Tulsa, OK, 3–5 April 2000).

[186] A. M. Al–Dhafeeri, H. A. Nasr–El–Din, H. K. Al–Mubarak, and J. Al–Ghamdi, "Gas Shut–Off Treatment in Oil Carbonate Reservoirs in Saudi Arabia," SPE 114323 (paper presented at the SPE Annual Technical Conference and Exhibition held in Denver, CO, 21–24 September 2008).

[187] O. Jaripatke and D. Dalrymple, "Water–Control Management Technologies: A Review of Successful Chemical Technologies in the Last Two Decades," 127806 (paper presented at the SPE International Symposium and Exhibition on Formation Damage Control, Lafayette, LA, 10–12 February 2010).

[188] A. Sourget, A. Milne, L. Diaz, E. Lian, H. Larios, P. Flores, and M. Macip, "Waterless Cement Slurry Controls Water Production in Southern Mexico Naturally Fractured Oil Wells," SPE 151646 (paper presented at the SPE International Symposium and Exhibition on Formation Damage Control, Lafayette, LA, 15–17 February 2012).

[189] I. Lakatos and J. Lakatos–Szabó, "Reservoir Conformance Control in Oilfields Using of Silicates: State–of–the–Arts and Perspectives," SPE 159640 (paper presented atv the SPE Annual Technical Conference and Exhibition, San Antonio, TX, 8–10 October 2012).

[190] K. Skrettingland, N. H. Giske, J.–H. Johnsen, and A. Stavland, "Snorre In–depth Water Diversion

Using Sodium Silicate—Single Well Injection Pilot," SPE 154004 (paper presented at the SPE Improved Oil Recovery Symposium, Tulsa, OK, 14-18 April 2012).

[191] A. Stavland, H. C. Jonsbrøten, O. Vikane, K. Skrettingland, and H. Fischer, "In-Depth Water Diversion Using Sodium Silicate on Snorre—Factors Controlling In-Depth Placement," SPE 143836 (paper presented at the SPE 9th European Formation Damage Conference, Noordwijk, Netherlands, 7-10 June 2011).

[192] P. L. Zitha, D. C. Standnes, and R. Huseynov, "Modeling and Experimental Study Deep Diversion for IOR Using Silicate Gels," SPE 144233 (paper presented at the SPE 9th European Formation Damage Conference, Noordwijk, Netherlands, 7-10 June 2011).

[193] I. J. Lakatos, J. Lakatos-Szabo, B. Kosztin, H. H. Al-Sharji, E. Ali, R. A. Al-Mujaini, and N. Al-Alawi, "Application of Silicate/Polymer Water Shutoff Treatment in Faulted Reservoirs with Extreme High Permeability," SPE 144112 (paper presented at the SPE 9th European Formation Damage Conference, Noordwijk, Netherlands, 7-10 June 2011).

[194] I. Lakatos, J. Lakatos-Szabo, G. Szentes, and A. Vaga, "Improvement of Silicate Well Treatment Methods by Nanoparticle Fillers," SPE 155550-MS (paper presented at the SPE International Oilfield Nanotechnology Conference, Noordwijk, Netherlands, 12-14 June 2012).

[195] B. Boye, M. Mulrooney, A. Rygg, C. Jodal, and I. Klungland, "Development and Evaluation of a New Environmentally Friendly Conformance Sealant," SPE 142417 (paper presented at the SPE 9th European Formation Damage Conference, Noordwijk, Netherlands, 7-10 June 2011).

[196] G. Lende and S. Olaussen, International Patent Application WO/2009/034287.

[197] I. J. Lakatos, J. Lakatos-Szabo, G. Szentes, M. Vadaszi, University of Miskolc; and A. Vago, "Novel Water Shutoff Treatments in Gas Wells Using Petroleum External Solutions and Microemulsions," SPE 165175 (paper presented at the SPE 10th International Conference and Exhibition on European Formation Damage, Noordwijk, Netherlands, 5-7 June 2013).

[198] S. Canbolat and M. Parlaktuna, "Well Selection Criteria for Water Shut-Off Polymer Gel Injection in Carbonates," SPE 158059 (paper presented at the Abu Dhabi International Petroleum Conference and Exhibition, Abu Dhabi, UAE, 11-14 November 2012).

[199] M. S. Al-Anazi, S. Caliskan, and I. S. Al-Yami, "Coreflooding Experiments on the Combination of Water Shutoff and Acid Treatments in Carbonate Formations," SPE 151760 (paper presented at the SPE International Symposium and Exhibition on Formation Damage Control, Lafayette, LA, 15-17 February 2012).

[200] N. Topguder, "A Review on Utilization of Crosslinked Polymer Gels for Improving Heavy Oil Recovery in Turkey," SPE 131267 (paper presented at the SPE EUROPEC/EAGE Annual Conference and Exhibition, Barcelona, Spain, 14-17 June 2010).

[201] D. Morel, M. Vert, S. Jouenne, R. Gauchet, and Y. Bouger, "First Polymer Injection in Deep Offshore Field Angola: Recent Advances in the Dalia/Camelia Field Case," SPE 13573, *SPE Oil and Gas Facilities* 1 (2)(2012): 43.

[202] W. Wang and X. Lu, *Petroleum Science and Technology* 27 (7)(2009): 699-711.

[203] M. Simjoo, A. Dadvand Koohi, M. Vafaie-Sefti, and P. L. J. Zitha, "Water Shut-Off in a Fractured System Using a Robust Polymer Gel," SPE 122280 (paper presented at the 8th European Formation Damage Conference, Scheveningen, Netherlands, 27-29 May 2009).

[204] X. Lu, J. Liu, R. Wang, Y. Liu, and S. Zhang, *Petroleum Science* 9 (1)(2012): 75-81.

[205] S. Johnson, J. Trejo, M. Veisi, G. P. Willhite, J.-T. Liang, and C. Berkland, *Journal of Applied Polymer*

Science 115 (2)(2009): 1008.

[206] M. S. Al-Anazi, S. H. Al-Mutairi, M. Alkhaldi, A. A. Al-Zahrani, and M. N. Gurmen, "Laboratory Evaluation of Organic Water Shut-Off Gelling System for Carbonate Formations," SPE 144082 (paper presented at the SPE European Formation Damage Conference, Noordwijk, Netherlands, 7-10 June 2011).

[207] B. Sengupta, V. P. Sharma, and G. Udayabhanu, "Development and Performance of an Eco-Friendly Cross-Linked Polymer System for Water Shut-Off Technique," SPE 14381 (paper presented at the International Petroleum Technology Conference, Bangkok, Thailand, 7-9 February 2012).

[208] M. Mercado, J. C. Acuna, J. Vasquez, C. Caballero, and E. Soriano, "Successful Field Application of a High-Temperature Conformance Polymer in Mexico," SPE 121143 (paper presented at the 8th European Formation Damage Conference, Scheveningen, Netherlands, 27-29 May 2009).

[209] S. Jayakumar and R. H. Lane, "Delayed Crosslink Polymer Flowing Gel System for Water Shutoff in Conventional and Unconventional Oil and Gas Reservoirs," SPE 151699 (paper presented at the SPE International Symposium and Exhibition on Formation Damage Control, Lafayette, LA, 15-17 February 2012).

[210] B. R. Reddy, L. S. Eoff, J. E. Vasquez, and E. D. Dalrymple, U.S. Patent Application 20110214865.

[211] B. R. Reddy, L. S. Eoff, J. E. Vasquez, and E. D. Dalrymple, U.S. Patent Application 20110214866.

[212] B. R. Reddy and L. S. Eoff, U.S. Patent Application 20130000905.

[213] B. R. Reddy and L. S. Eoff, U.S. Patent Application 20130000911.

[214] B. R. Reddy, L. S. Eoff, F. Crespo, and C. A. Lewis, "Recent Advances in Organically Crosslinked Conformance Polymer Systems," SPE 164115 (paper presented at the SPE International Symposium on Oilfield Chemistry, The Woodlands, TX, 8-10 April 2013).

[215] G. Lei, L. Li, and H. A. Nasr-El-Din, "New Gel Aggregates for Water Shut-Off Treatments," SPE 129960, *SPE Reservoir Evaluation & Engineering* 14 (1)(2011): 120.

[216] J. Vasquez and C. Miranda, "Profile Modification in Injection Wells through Relative Permeability Modifiers: Laboratory Validation and Reservoir Simulation," SPE 135107 (paper presented at the SPE Production and Operations Conference and Exhibition, Tunis, Tunisia, 8-10 June 2010).

[217] A. Stavland, "How to Apply the Flow Velocity as a Design Criteria in RPM Treatments," *SPE Production & Operations* 25 (2)(2010): 223.

[218] J. Wang, X. Zhu, H. Guo, X. Gong, and J. de Hu, *Journal of Petroleum Science and Engineering* 80(1) (2011): 69-74.

[219] M. Lastre and A. Milne, "A Solids-Free Permeability Modifier for Application in Producing and Injection Wells," SPE 151836 (paper presented at the SPE International Symposium and Exhibition on Formation Damage Control, Lafayette, LA, 15-17 February 2012).

[220] Z. Ye, E. He, S. Xie, L. Han, H. Chen, P. Luo, Z. Shu, L. Shi, and N. Lai, *Journal of Petroleum Science and Engineering* 72(2010): 64.

[221] L. E. Eoff, B. R. Reddy, E. D. Dalrymple, D. M. Everett, M. Gutierrez, and D. Zhang, International Patent Application WO/2009/063161.

[222] C. Cozic, D. Rousseau, and R. Tabary, "Novel Insights into Microgel Systems for Water Control," SPE 115974, *SPE Production & Operations* 24 (4)(2009): 590-601.

[223] R. J. R. Cairns, "Modification of Relative Permeability in Wet Wells in Sandstone Reservoirs: Multiple Well Treatments and Core Experiments," RSC Chemistry in the Oil Industry XI, Manchester, UK, 2-4 November 2009.

[224] S. Poletaev and A. Demakhin, U.S. Patent 7316991, 2008.

[225] H. Frampton and S. Taylor, "Gel Systems with Properties Suitable for Remediation of Water Injection Wells with Performance Impaired by Fractures and Super-Permeability Zones," RSC Chemistry in the Oil Industry XI, Manchester, UK, 2-4 November 2009.

[226] S. Scognamillo, V. Alzari, D. Nuvoli, and A. Mariani, *Journal of Polymer Science Part A: Polymer Chemistry* 48 (2010): 2486.

[227] R. A. Kalgaonkar, A. Sabhapondit, and A. S. Aldridge, U.S. Patent Application 20130000900.

[228] C. Okeke, S. Mofunlewi, A. Bankole, and G. U. Agbogu, "Polymeric Solutions to High Gas Influx in an Oil-Producing Well: A Field Experience," 151895-MS (paper presented at the North Africa Technical Conference and Exhibition, Cairo, Egypt, 20-22 February 2012).

[229] E. Miquilena and A. Milne, International Patent Application WO/2011/135553.

[230] P. Kurian, M. Wei, and K.-T. Chang, International Patent Application WO2013112664.

3 结垢防治

3.1 概述

油田垢的形成主要源自水溶液中难溶无机盐的沉积，以及金属离子与羧酸或环烷酸阴离子形成的盐垢（见第 7 章）。垢可在几乎任何类型的表面沉积，且垢层形成后如不做处理，会持续变厚（图 3.1）。垢会堵塞近井地带的储层岩心孔喉或井筒，导致储层伤害和产能损失。垢可沉积在电潜泵、侧通滑套等井下设备表面，导致其故障失效；沉积于地面生产管道内的垢还会缩小管径并阻碍流动；在后续的处理设备中也会成垢。与水有关的三个最大生产问题是垢、腐蚀和天然气水合物，需要提前预测，以确定最佳防治方案。某些油田的垢防治是运营成本中占比最高的部分，相关资料可以参考油服公司从运营角度的系列出版物和其他公开文献。

图 3.1 管道中的垢

3.2 垢的类型

石油工业中的垢，按照常见程度从高到低的排序如下：
（1）碳酸钙（方解石和文石）。
（2）硫酸盐：硫酸钙（石膏）、硫酸锶（天青石）和硫酸钡（重晶石），这些垢的晶格中，特别是硫酸钡中可能共存镭元素。
（3）硫化物垢：常见的有硫化亚铁、硫化锌和硫化铅。
（4）氯化钠（岩盐）。

多数矿物垢的溶解度会随着温度降低而降低（碳酸钙是例外情况，后续会讨论）。因此，海上油田的产出液在长距离回输过程中，冷却到海床温度后，会加剧管道结垢问题。此外，生产系统加注的甲醇或乙二醇等有机热力学水合物抑制剂（THI）会加剧无机垢沉积，其中甲醇的影响最严重。因为这些溶剂的极性比水低，会降低盐垢的溶解度。随着深海、低温地区的油藏开发越来越普遍，为应对水合物的水合物抑制剂使用量增加，结垢问题的频次和严重程度也将上升。深海油田所面临的结垢防治挑战可参见相应论文。

油田生产中，有时还会发现碳酸亚铁（菱铁矿，主要源自腐蚀）、氟化钙（氟石，HF 酸化的副产品）、硅酸盐和天然碱 [$Na_3H(CO_3)_2 \cdot 2H_2O$] 等特殊少见的垢型。蒸汽驱作业中有时会发现二氧化硅/硅酸盐垢，防控方法包括：用淡水稀释；降低水的pH值；采用还原剂、络合剂和螯合剂；通过石灰软化去除水中的二氧化硅，用金属离子或阳离子表面活性剂沉淀水中的二氧化硅；最后可以用地热硅垢抑制剂/分散剂处理水质。油田条件下二氧化硅垢影响因素有许多研究资料。还有研究报道了使用一系列环保型、小分子和聚合物等化合物抑制二氧化硅垢。对硅酸镁防垢剂的静态和动态测试也有文章报道。二氧化硅防垢剂现场应用的加注浓度最高达500mg/L。还有研究将多酚类化合物作为一类生物可再生的二氧化硅聚合物防垢剂。其他二氧化硅垢抑制剂也有报道。某些生产环境下，大多是气井，会发现放射性铅、钋金属和盐垢，还发现井下有单质硫的沉积。

3.2.1 碳酸钙垢

碳酸氢钙易溶于水，但碳酸钙难溶。地层水通常含有HCO_3^-和Ca^{2+}，当如下的反应平衡向右移动时，就沉淀出碳酸钙：

$$2HCO_3^- \rightleftharpoons CO_3^{2-} + H_2O + CO_2(g)$$

因此，根据勒夏特列原理（化学平衡移动原理），压力下降时，上述平衡向右移动以生成CO_2气体来增加压力。这样会形成较多的CO_3^{2-}，pH值上升。超过特定临界值后，高浓度的CO_3^{2-}就会形成碳酸钙沉淀。

$$Ca^{2+} + CO_3^{2-} \longrightarrow CaCO_3(s)$$

成垢的临界压降可能出现在如射孔段、节流阀或生产油管、井底到井口等生产系统的任何位置。油井产出水流中二氧化碳含量足够高，从而pH值足够低，井内不会沉积出碳酸钙，但在井口节流阀下游压力大幅下降。此外，碳酸钙垢可能在油田生产数年后发生，因为此后管线内的压力才会降至碳酸盐结垢的水平。随着压力进一步下降，结垢会向上游移动（即进一步深入生产井）。如果该段管线的结垢概率高，可能造成井下安全阀的卡堵问题。有多种商业软件可用于预测碳酸钙的热力学结垢，也开发出了预测何时、何处碳酸钙结垢的动力学模型。Mg^{2+}对碳酸盐垢生长的影响也有相关文献报道。

另外，基质增产处理后的酸溶液消耗，会导致碳酸盐垢的再沉淀而形成储层伤害。碳酸钙的溶垢剂和螯合剂可以打破这种再沉淀循环。

在含高浓度Fe^{2+}的盐水中，可同时存在碳酸亚铁和碳酸钙沉积物。乙二醇再生系统中因为高浓度Fe^{2+}和pH值升高（≥8.5），会出现铁垢。有实验室研究表明，膦基聚羧酸（PPCA）和双六亚甲基三胺-五（亚甲基膦）酸等常见碳酸钙防垢剂，对防止碳酸亚铁垢沉积的效果不明显，而柠檬酸盐的防垢效果较好。其他碳酸钙防垢剂对碳酸亚铁也有效。防腐处理也有助于减少碳酸亚铁结垢。

其他改变生产流体pH值的化学剂将影响碳酸钙成垢的可能性。例如，碱性的三嗪类H_2S清除剂会增加结碳酸盐垢的可能性，现场已经观察到此现象。

3.2.2 硫酸盐垢

除镁以外，元素周期表的ⅡA族元素离子均可与硫酸盐离子混合生成难溶垢，反应如下：

$$M^{2+} + SO_4^{2-} \longrightarrow MSO_4(s)$$

ⅡA族元素自上而下的硫酸盐溶解度逐渐降低，而硫酸钡最难溶也最难防控。硫酸盐垢通常在地层水和注入的海水混合时形成。当这种情况发生在生产井的近井地带时，硫酸盐垢沉积会造成地层伤害。此外，两种不同的非结垢井液可能会在平台顶部集输管线中混合，造成平台顶部的硫酸盐垢问题。海水中高浓度的硫酸盐离子（约2800mg/L）与地层水中的二价金属离子混合，导致硫酸盐结垢。注海水驱的油藏中常出现硫酸盐垢，同时海上油井修井过程中使用海水，也会出现这种问题，改用清水作业可以防止该问题。有越来越多的证据表明，水驱过程中，储层中的反应改变了采出水成分，使其不同于注入水和地层水的简单混合物。这些反应包括硫酸盐和碳酸盐的溶蚀/沉积及离子交换，对注入水、注入水和地层水的混合物都有影响。结垢管理计划和决策通常假设采出水是注入水和地层水的简单混合物。而当反应发生时，该假设是不正确的，结垢预测、防垢剂测试和挤注处理性能评价（将在后面讨论）等都会受到影响。

硫酸钙（硬石膏或石膏）微溶于水，可溶于许多螯合型溶垢剂，是最容易处理的硫酸盐垢。另一种形式的硫酸钙称为半水石膏，是石膏的一种稳定的前体相，可能对硫酸钙防垢剂的设计有影响。中东部分油田有纯硫酸锶垢，但通常其是与钡盐共存。硫酸钡非常坚硬、最难处理，并且只有在最好的溶垢剂中才能以合理的速度溶解。由于其极难溶（极低的溶解度），地层水中往往无须高浓度的Ba^{2+}就可以沉积硫酸钡垢（重晶石）。产出水也可能被硫酸钡饱和，如果在生产过程中温度下降，甚至会析出更多的硫酸钡垢。Sr^{2+}和Ba^{2+}会与SO_4^{2-}共沉淀，形成混合硫酸盐垢。一些地层水还含有浓度较低的放射性镭，其在硫酸钡垢和硫酸锶垢的晶格中共同沉淀。因此，清除自然产生的放射性物质（NORM垢）是必须处理的环境问题。与碳酸盐垢一样，可以用计算机软件来确定硫酸盐垢结垢趋势。此外，为便于确定生产井何时处理硫酸盐垢，对注海水井的海水突破进行预测和监测十分重要。对储层地球化学特性的了解有助于确定结垢控制策略。据称，通过在储层内形成结垢（大多容易形成硫酸盐垢），可以提高储层的油气采收率。硫酸盐垢的严重程度取决于地层水与海水比例的突破和过饱和程度。在油田开发的后期，由于采出水主要是海水，就可能只有很少量或没有硫酸盐垢。然而，当海水首次突破这个比例时，硫酸盐结垢量可能达到惊人的严重程度。英国北海油田的一口油井就曾出现24h内产量从3×10^4bbl/d降至0。

3.2.3 硫化物垢

硫化物垢不像碳酸盐垢和硫酸盐垢那么常见，但如不加以控制，仍然会造成严重的问题。许多地层水中天然存在H_2S，但对油井来说，多数H_2S源自硫酸盐还原菌（SRB）对注入海水中硫酸盐离子的作用。SRB 将硫酸根离子还原为 H_2S，而 H_2S 与 HS^- 和 S^{2-} 处于平衡态：

$$H_2S + H_2O \rightleftharpoons H_3O^+ + HS^-$$
$$HS^- + H_2O \rightleftharpoons H_3O^+ + S^{2-}$$

生产井或注水井的钢腐蚀产生的 Fe^{2+}，与 S^{2-} 反应形成硫化亚铁垢。

$$Fe^{2+} + S^{2-} \longrightarrow FeS(s)$$

类似地，如果地层水含有 Zn^{2+}、Pb^{2+} 等金属离子并与 S^{2-} 混合，形成硫化锌（ZnS，锌混合物）和硫化铅（PbS，方铅矿）等更难溶硫化物垢。在北海的高压高温井（HPHT）中，硫化锌/硫化铅的混合垢很常见。使用锌基盐水完井液时，也可形成硫化物垢。在含 H_2S 气井中，硫化铅可能是由于 S^{2-} 与采出水中的 Pb^{2+} 发生反应，并在局部过饱和（由快速的温度或压力降低引起）而沉积。在某些地层水中，Pb^{2+} 的浓度高达 150mg/L。

3.2.4 氯化钠垢（岩盐）

氯化钠易溶于水，而且溶解性随着温度的升高而增加。某些地层水特别是高温高压地层水中的氯化钠含量非常高。有时地层水的氯化钠可能是过饱和的。随着采出水温度降低，氯化钠可能会析出。这个动力学过程非常快，如果不及时处理，管道会很快堵塞。即使是含水率很低（<0.5%）的油井，可能一夜之间迅速盐化。随着生产过程的压力降低，水闪蒸为气相，导致氯化钠溶液浓缩和岩盐沉淀。可以参见几起含岩盐地层的案例，其中有清除和控制策略。

3.2.5 混合垢

垢通常是层状的并且是混合垢质组成，如在特定的现场条件下同时含有碳酸盐和硫酸盐的垢。垢可能呈现油性，甚至含有如沥青质或蜡（"软垢"）等其他沉积物，此时的化学溶垢处理更为复杂。如果沥青质混合过多的无机垢，会使水性溶垢剂失效。

3.3 非化学防垢

有效的垢管理意味着在油田开发初期做出决策，并在油田全生命周期中持续地回顾这些决策。防垢首先主要针对生产设备，但注海水或采出水回注也可能需要防垢。在注水前有时还要除去 Fe^{2+}。

缓解结垢形成的基本方法有三种：
（1）注入脱硫酸根的海水。
（2）抑制垢/防垢。
（3）待垢形成后，再用物理或化学方法除垢。

方法1只能防止硫酸盐和硫化物垢，但是地层水中可能会自然存在一些H_2S，会导致硫化物结垢（在石油生产行业中，通过去除Ca^{2+}来软化注入水从而避免结碳酸钙垢的方法仅有很少实践案例）。采用纳滤膜法对注入海水进行脱硫。有的采出水含有较低硫酸根离子也可用作注入水。使用脱硫酸根设施需要的投资相当高，但对预期硫酸盐结垢严重的大油田来说是最佳选择。脱硫酸根技术并不能去除所有的硫酸盐离子（可将海水中硫酸根含量从约2700mg/L减少到40～100mg/L），但可以显著减少硫酸盐和硫化物（来自SRB活性）的结垢问题。此外，脱硫大幅降低了储层的次生H_2S，并可能减少微生物腐蚀的程度。值得一提的是，在碳酸钙储层中，由于硫酸钙的沉淀而自然减少硫酸根。通过这种方法可除去最高95%的硫酸盐，这样生产井中再不产生硫酸钡垢。有资料提出了向注水井注入垢微粒的新方法。垢微粒作为结垢的晶种/成核位点使结垢离子到达生产井前结垢，从而避免了这些井的垢沉积。当晶种的平均粒径小于2.5μm时，控制垢尤其有效。另一种在实验室和现场都成功测试的方法是将随机脉冲高频电信号引入管道系统，使垢晶形成在生产流体中，而不是在井下或平台顶部设备表面沉积。电场不能防止沉积，只能改变沉积发生的物理位置。通过在管径上诱导电场，离子在溶液中聚成簇。

将离子溶解度降到低于饱和水平，也会使已有的垢沉积物重新溶解，从而清除管道中垢，特别是碳酸钙。在水量少且温度、流量、硬度和压力变化不大的地方，硬垢可能要缓慢地分解。在这些条件下，设备最好是新的或刚化学清洗后。

也有利用声波来控制结垢的方法。一种方法是在管道内安装产生声波的液体口哨，并使部分流体流经液体口哨产生声波。由于是靠流经管道的流体驱动产生声波，因而无须外部电源。也有使用超声波辐照防垢的报道。

有文献分析了结垢管理中表面的重要性。不同的涂层能防止或延缓垢沉积，有时也抑制腐蚀，如环氧树脂、含氟聚合物、聚硅氧烷和聚硅烷。这些可能对保护生产系统中的关键部件等有用（如井下安全阀、智能井系统的关键组件）。

但如果涂层表面开始形成垢，表面状态就会变化，容易沉积更多的垢。光滑的涂层表面也会被湍流流体中的携砂和其他颗粒而磨蚀，这可能会加速垢沉积。

利用磁场控制结垢已经取得了一些成功，特别是在石油行业以外的如换热器等领域，也有许多失败的报告。可以使用永磁体或电子产生的磁场，这种设备已用于石油工业特别是碳酸钙垢的控制，如荷兰的海上油田。但因缺乏对其作用机理的有说服力阐述，人们普遍持怀疑态度。迄今为止的研究认为至少有三种可能的磁处理机理。首先是磁流体力学效应；其次是铁磁和超顺磁微粒的团聚；最后可能的机制是管壁发生微小pH值变化及晶体形态的变化。磁防垢可能涉及这些机制的组合，也可能涉及其他现象。管道材质可能也会影响磁场的效果。电场防垢据称也能抑制结垢。磁场或者电子设备在低过饱和度的情况下，可能对控制碳酸钙垢有效。

本章将讨论化学防垢剂和除垢。第5章的酸化增产部分应与化学除垢一起阅读。控制采出水量也能降低结垢的严重程度（见第2章）。

3.4 元素周期表的ⅡA族元素碳酸盐和硫酸盐防垢剂

防垢剂是一种可溶于水的化学物质，可以防止或阻碍无机垢成核和（或）结晶生长。已有文献对早期化学防垢技术进行了简要评述。简而言之，聚合物是一种很好的成核抑制剂和分散剂。当结垢结晶发生时，其吸附在晶体表面并融入晶格中，当测试低于其阈值水平时，减缓垢晶生长。小分子量、非聚合物抑制剂，如众所周知的氨基膦酸盐，通过阻断活性生长点，能很好地阻止晶体生长，但如果测试低于其阈值水平，则不太可能阻止成核。因此，增加聚合物的剂量将确保垢晶停止生长，类似地，增加氨基膦酸盐的浓度将确保垢晶停止成核。实际上，防垢剂的关键作用是防止垢的沉积，因此防垢剂的作用机理可以是抑制成核、晶体生长及分散垢。研究表明，只有3%~5%的碳酸盐或硫酸盐垢晶体的表面需要覆盖一些聚合物抑制剂，以实现完全的抑制垢晶生长。对作用于重晶石垢的氨基膦酸盐防垢剂，其成核抑制的覆盖率需为16%。例如，对于硫酸钡垢，已建立了防垢剂的防垢效果模型和防垢机理。对于如小分子氨基膦酸盐等硫酸盐垢，其防垢机理更为复杂。钙防垢复合物首先与晶格中的钙反应，干扰晶格并抑制进一步的生长。有文献证明氨基膦酸盐在低Ca^{2+}浓度下作为硫酸盐垢的防垢剂时，效果有限。

文献研究了氨基三亚甲基膦酸（NTMP）、氨基三乙酸（NTAA）和两种介于两者之间的混合磷酸盐-羧酸盐对硫酸钡的抑制作用。研究者认为：电荷匹配更恰当地解释了这些添加剂的行为，特别是在离子强度提高的情况下。这并不是说，晶格匹配在硫酸钡体系中不发生或不重要，因为晶格匹配可能有助于添加剂更有效地进行"电荷匹配"。另一方面，聚合物聚羧酸盐抑制剂在Ca^{2+}浓度非常低的情况下仍然有效，表明聚合物和晶体表面之间形成的多重键，带来了更强的吸附和抑制作用。

油田应用防垢剂的方式很多，将在后续详细讨论。首先，了解一些已经应用于不同垢型的各种类型防垢剂。需要注意的是，大多数油田垢，如碳酸盐垢和硫酸盐垢都是由二价阴离子（CO_3^{2-}和SO_4^{2-}）及二价金属阳离子组成。对于亚临界核抑制或晶体生长抑制，防垢剂要与结垢离子结合，就要与产出水中的阴离子或阳离子相互作用。通常需要发生几种相互作用才能使防垢剂在表面紧密结合，因此需要几个类似的官能团，并且这些官能团之间有适当的间距，这样才能与晶体表面的晶格离子相互作用。

为了能很好地与产出水中的阴离子结合，防垢剂需要带相反电荷的阳离子。通常唯一简单的方法是通过季铵、磷或磺基，将几个阳离子接入同一个分子中。但聚季铵盐[如聚二烯丙基二甲基氯化铵和聚（丙烯酰胺丙基）三甲基氯化铵]是较差的防垢剂，可能是由于在垢层中，季铵盐基与阳离子（如钙）之间的尺寸不匹配。然而，在挤注处理过程中，在阴离子防垢聚合物中加入季铵盐基团，有利于其在地层岩石的吸附。这将在本章后面的防垢剂应用技术中讨论。

有机分子上有一些阴离子基团能很好地与垢晶格表面上的ⅡA族阳离子相互作用。其中，常用的阴离子基团是磷酸盐离子（—OPO_3H^-）、有机膦酸盐离子（—PO_3H^-）、亚膦酸盐离子（—PO_2H^-）、羧酸盐离子（—COO^-）和磺酸盐离子（—SO_3^-）。

防垢剂分子结构中含有两种或两种以上的上述阴离子或这些离子的混合物，可以成为油田垢的良好抑制剂。这些分子可以酸的形式制备（如羧酸、膦酸），但常以钠、钾或铵盐形式成为最活跃的防垢剂。除聚磷酸盐外，阴离子基团都通过碳原子与分子的主链结合。$d-$葡萄糖酰胺是一种环保型 $BaSO_4$ 防垢剂，但是其防垢效果表现一般。单宁酸可作为挤注处理的防垢剂，室内实验评价表明经单宁酸处理后岩心的渗透率略有提高，注入 250PV 盐水时未发生结垢。槟榔膏提取物（含有儿茶素）也被证明可以防止碳酸盐结垢。相关文献报道了环保型防垢剂——抗坏血酸，也有报道实验室研究了叶酸作为环保型碳酸钙防垢剂。

含有上述离子或酸的最常见的防垢剂类别是聚磷酸，小分子、非聚合膦酸和氨基膦酸，聚膦酸，聚羧酸，膦基聚合物和聚次膦酸，聚磺酸。

含羧酸、膦酸和（或）磺酸基团的各种共聚物和三元共聚物也是很好的防垢剂，本文也将对其进行讨论。由于官能团之间的相似性，许多防垢剂能够抑制几种类型的垢。一些商业膦酸盐防垢剂以钾盐的形式出售，表明该阳离子可能比其他类型阳离子防垢性能更强。聚氨基羧酸盐螯合溶垢剂也类似（见 3.9.2）。

丙烯酸酯基聚合物或氨基膦酸等许多传统防垢剂的生物降解性差。此外，排放含 N 或 P 防垢剂的采出水会使环境中营养物丰富，从而导致环境失衡（即富营养化）。自 20 世纪 90 年代初以来，人们越来越重视寻找对环境更友好的防垢剂，其中一些现在已经在商业上应用并在下面讨论。

以下对防垢剂的分类之间会有一些重叠，如同时含有羧基和膦基或磺酸基的聚合物可以归入两类。在这种情况下，优先考虑的是膦酸基团，其次是磺酸基团，再次是羧酸基团。此外，还有许多专利详细说明了例如小分子氨基膦酸盐和多羧酸不同类防垢剂之间的协同作用，也有文献综述了各种防垢剂浓度检测方法。

3.4.1 聚磷酸

如三聚磷酸钠或六偏磷酸钠中的聚磷酸阴离子，长期以来一直被认为是碳酸钙防垢剂（图 3.2）（六偏磷酸也是一种很好的 $BaSO_4$ 防垢剂）。但其主要用于低钙离子浓度的锅炉水处理，作为油田防垢剂产品需要具有更强的热稳定性和兼容性。聚磷酸也表现出一定的缓蚀特性。柠檬酸磷酸盐虽然还没有商业化，但其是哺乳动物软组织中的一种天然晶体生长抑制剂，具有低毒性，作为磷酸基和羧基的组合是良好的防垢剂。植酸（肌醇六磷酸）已用于防垢剂配方，但其本身形成不溶性的钙、镁、铁盐。

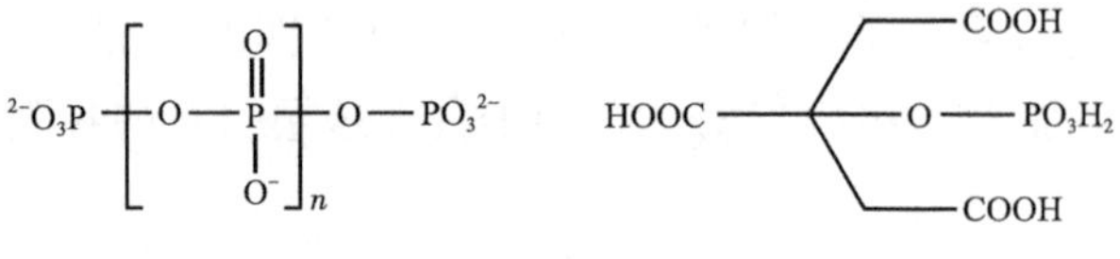

(a) 聚磷酸阴离子　　　(b) 柠檬酸磷酸酯

图 3.2　聚磷酸阴离子和柠檬酸磷酸酯

3.4.2 磷酸酯

磷酸酯（ROPO$_3$H$_2$）是众所周知的环境友好型防垢剂，尤其适用于碳酸钙和硫酸钙垢，如果应用的水溶液环境不是偏酸性，也适用于硫酸钡垢，但其通常不是最有效的防垢剂。制备过程由磷酸和醇反应生成，通过改变醇中烷基链的长度，可制成水溶性或油溶性磷酸酯。磷酸酯比聚磷酸盐的耐酸性更好，与高钙盐水的相容性更好。磷酸酯的热稳定性有限，文献报道控制平台顶部硫酸盐垢时的最高耐温可达 95℃，用于控制碳酸盐垢时的耐温可达 110℃。磷酸酯也可以在温度高达 100℃ 的情况下作为挤注化学剂使用。至少早在 20 世纪 80 年代已开始使用具有良好生物降解性的三乙醇胺磷酸单酯，但由于水解不稳定，只能在 80℃ 以内的温度使用（图 3.3）。有文献报道磷酸酯也表现出一定的缓蚀性能。

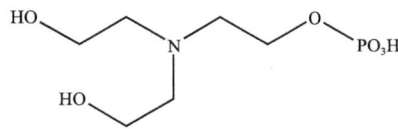

图 3.3　三乙醇胺磷酸单酯

通过采用单体甲基丙烯酸乙二醇酯磷酸酯可以将磷酸基团引入防垢聚合物。例如，这种单体也可以与乙烯基羧基单体或乙烯基磺酸单体共聚。磷酸的功能性使聚合物防垢剂在挤注处理过程中具有良好的吸附/解吸附特性，使聚合物滞留在油藏中，延长了挤注处理寿命。聚氧乙烯烷基醚磷酸酯与羧酸或磺酸单体的共聚物已用于油田防垢剂。

磷酸化蛋白质可以控制磷灰石成核、晶体生长和抑制作用，其还能抑制方解石的形成。例如，牙本质胶原蛋白就含有共价结合的磷酸盐。

3.4.3 非聚合型膦酸和氨基膦酸

只含有羧酸或磺酸基的非聚合型分子是较差的防垢剂，而含有膦酸的防垢剂并非如此。与许多聚合物防垢剂相比，膦酸有一个较低的"临界"温度，低于这个温度，效果低很多。很多防垢剂含有一个膦酸基团和几个羧酸基团，最常见的例子是 2- 膦酸基 -1，2，4- 三羧酸丁烷（PBTCA）（图 3.4），其变体如膦酸琥珀酸和 1- 膦酸基 -2，3- 二羧酸丙烷也是有用的防垢剂。PBTCA 主要用于碳酸钙防垢剂，PBTCA 盐也可用于硫酸盐垢的溶垢剂（见 3.9 节）。

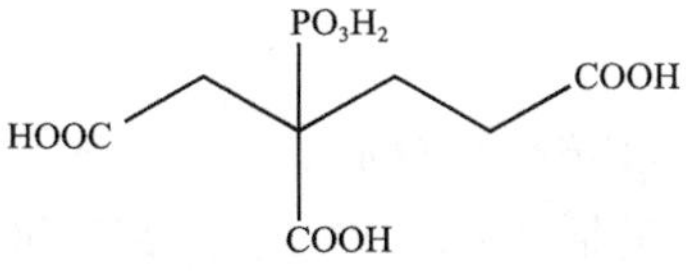

图 3.4　2- 膦酸基 -1，2，4- 三羧酸丁烷

改进 PBTCA 及其相关分子如膦酸琥珀酸的合成路线，又生成了其他膦酸酯防垢剂。膦酸琥珀酸的防垢性能较差，但与该分子的低聚物混合后，对碳酸盐垢和硫酸盐垢表现出良好的防垢性能（图 3.5）。这些分子还可以转化为油溶性膦酸羧酸酯，也可以作为沥青质分散剂。

另一种常见的膦酸防垢剂是羟基亚乙基二膦酸（HEDP）（图 3.6）。

通过 Irani–Moedritzer 改进的卡巴奇尼克 - 菲尔茨反应（Kabachnik-Fields）反应，在膦酸分子中引入氨基，从而获得—NH$_2$—CH$_2$—PO(OH)$_2$ 基团，通过氨和膦酸的相互作用，提高了分子的金属结合能力。有一系列市售的氨基膦酸防垢剂，主要用于碳酸盐和硫酸盐防垢。文献报道了几种氨基膦酸在方解石上的吸附性能，其中最重要的分子结构如图 3.7 所示。相关文献对许多氨基膦酸的耐钙性也进行了研究。

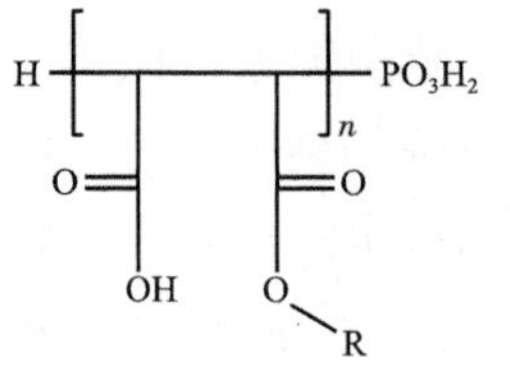

图 3.5　膦酸琥珀酸低聚物（R=H）和酯类
（R= 烷基、烷基芳香基或烯基）（$n=2\sim10$）

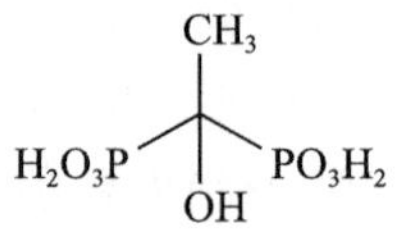

图 3.6　羟基亚乙基二膦酸

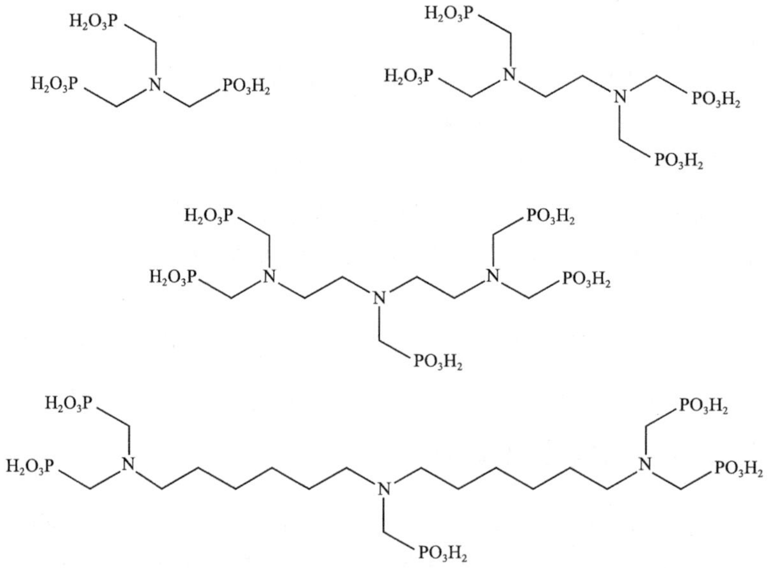

图 3.7　氨基三亚甲基膦酸，1,2-二氨基乙烷四亚甲基膦酸，二乙烯三胺五亚甲基膦酸，二六亚甲基三胺五亚甲基膦酸

乙醇胺 -N,N- 双（亚甲基膦酸）（EBMP）是最小的氨基膦酸防垢剂之一，是由乙醇胺、甲醛和磷酸反应形成。实际上，该反应产生的混合物包括大约 50% 的 EBMP 和 50% 的 EBMP 环酯。该酯本身对防垢没有贡献，但是可以在碱中水解成 EBMP，有效地将原混合物的防垢性能提高一倍，对高温井尤其有效果。

EMBP 和其胺氧化物都是很好的油田防垢剂，甚至也可用于二氧化硅垢。EBMP 具有很好的钙、铁耐受性。

氨基三亚甲基膦酸（ATMP）是一种简单、便宜但不是最有效的抑制剂，其通过三个膦酸基团和叔氮原子上的孤对电子与金属离子结合。增大分子量，就可以得到 1,2-二氨基乙烷四亚甲基膦酸（EDTMP），是一种良好的全方位膦酸防垢剂。下一个更大的分子是二乙烯三胺五亚甲基膦酸（DTPMP）。DTPMP 是一种优良的碳酸盐和硫酸盐防垢剂，可能是石油工业中使用最多的膦酸盐防垢剂。金属阳离子和 pH 值会影响其性能。另一种氨基膦酸是二六亚甲基三胺五亚甲基膦酸，可提高钙耐受性（即在钙浓度高的情况下不沉淀），并适用于 120~140℃以上的高温条件。而另一种膦酸防垢剂是六亚甲基二胺四亚甲基膦酸，其膦酸基团之间的距离与 DTPMP 不同。N, N'- 双（3- 氨基丙基）乙烯二胺或 1，7- 双（3- 氨基丙基）乙烯二胺的膦酸衍生物已被认为是高钡盐水的优良防垢剂。显然，氨基甲基膦酸基团之间的距离对防垢有显著影响。3- 氯丙基亚氨基双（亚甲基膦酸）与如聚乙烯亚胺或巯基乙酸等多种分子反应形成防垢剂。

大多数氨基膦酸生物降解性低，但毒性和生物积累低。在环境法规严格的地区，已经开始用更环保的替代品取代氨基膦酸，有些国家的规定并不是强制执行。但在北海油田低分子量的膦酸仍是环境可接受。其他生物可降解膦酸防垢剂已被报道。而将膦酸排放到环境中，会导致湖泊和沿海地区不必要的富营养化。膦酸铵盐可能比ⅠA族阳离子盐表现出更高的降解率，因为铵离子中的氮可以作为降解抑制剂细菌的营养来源。膦酸盐 Fe（Ⅲ）配合物可快速光降解。氨基多膦酸在 Mn^{2+} 存在下也会迅速氧化，已在废水中检测到氧和稳定的分解产物。在油田化学品的生态毒理法规中，似乎没有考虑光降解或氧化降解。

由小分子多乙二醇二胺（如三乙二醇二胺）衍生的氨基亚甲基膦酸是良好的碳酸盐和硫酸盐防垢剂，具有更好的相容性。在 Fe^{2+} 存在下，具有羟基或羧酸基团的氨基膦酸的性能，明显优于传统单独使用或与聚羧酸防垢剂共混使用的氨基膦酸（图 3.7）。

氨基酸烷基膦酸由氨基酸与甲醛和磷酸反应制成，是碳酸盐和硫酸盐防垢剂，如 D, L- 亮氨酸双亚甲基膦酸、L- 苯丙氨酸双亚甲基膦酸和 L- 赖氨酸四亚甲基膦酸。后者也含有部分膦酸化的赖氨酸（图 3.8）。氨基酸基团的膦酸防垢剂已经在油田应用多年，包括挤注处理。可以通过控制烷基膦酸化水平，得到部分烷基膦酸化的氨基酸产品。与完全取代的产品相比，其在防垢和环境性能（生物降解）方面都表现出更好的效果。

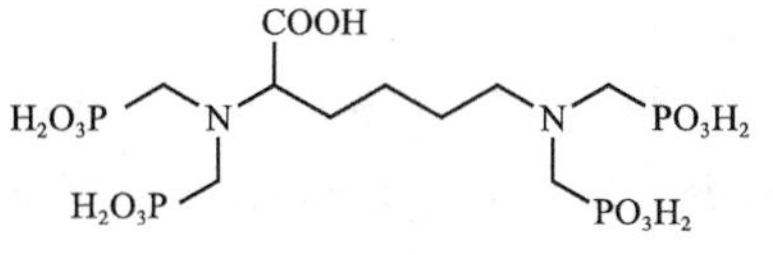

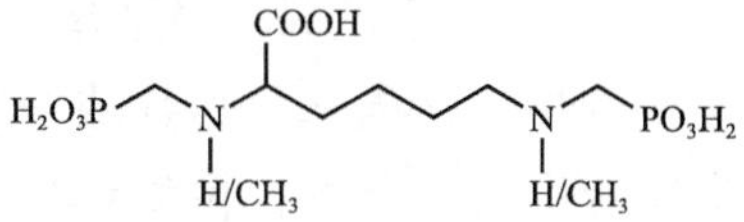

图 3.8　赖氨酸四亚甲基膦酸和部分膦酸化赖氨酸

另一类膦酸防垢剂是基于乙烯基膦酸（VPA）或亚乙烯 -1，1- 二膦酸（VDPA）与小的聚醇反应，结构如图 3.9 所示。如三甘醇和 VDPA 的 1∶1 混合物的反应产物，或者这类抑制剂也可以由氧化亚烯烃（如环氧乙烷或环氧丙烯）与从 VPA 或 VDPA 衍生的羟基膦酸（或其盐或酯）反应制成。这类防垢剂具有部分生物降解、缓蚀功能。

3.4.4　聚膦酸

聚膦酸主要有以多胺为主链和以聚乙烯为主链的两类。由小的聚烯基胺如三亚乙基四胺与环氧氯丙烷反应，然后氨基与甲醛和磷酸反应得到具有多胺主链的防垢剂，如 N- 膦酸甲基氨基-2-羟丙烯聚合物，分子量在300～5000之间（图3.10）。这类化合物特别适用于油田钡垢和挤注处理。

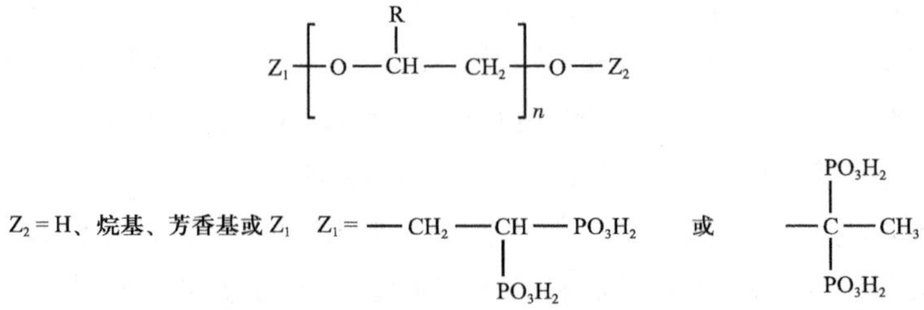

图3.9 聚醇膦酸（R=H 或 CH₃ 或其混合物）

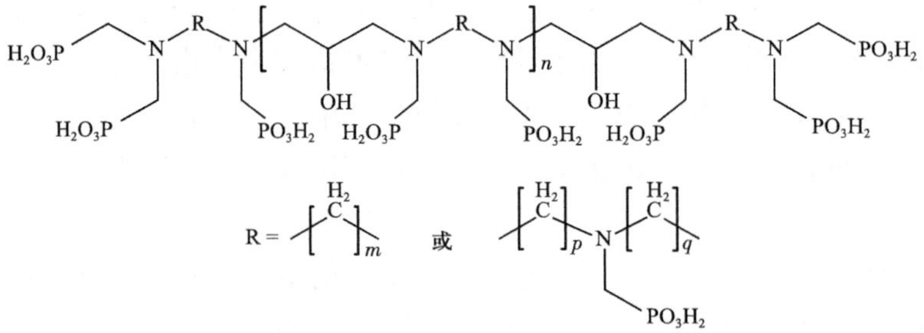

图3.10 N-膦酸甲基氨基-2-羟丙烯聚合物

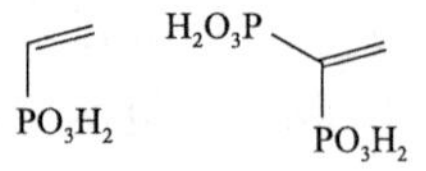

图3.11 乙烯基膦酸和乙烯基二膦酸单体

使用VPA或VDPA单体与任何数量的如丙烯酸、顺丁烯酸和乙烯基磺酸等其他单体，可以制备具有聚乙烯主链的聚膦酸（图3.11）。VPA与不饱和二羧酸酐（在水中开环成羧酸）的共聚物是Ba/Sr防垢剂，例如，VPA/异丁烯/马来酸酐共聚物水解得到羧酸基团。VPA和VDPA成本较高，因此其主要用于制备以膦酸封端的膦基聚合物（见3.4.5）。

其他膦酸聚合物包括羰基化合物或亚胺和次磷酸的反应产物，再与乙烯基单体（如丙烯酸）进一步反应生成聚合物，再氧化得到膦酸聚合物。带有膦酸端基和氨基膦酸的聚丙烯酸也被认为是防垢剂。

除了VPA和VDPA，其他含膦酸的单体包括异丙烯基膦酸（图3.12）、异丙烯基膦酸酐、烯丙基膦酸、亚乙基二膦酸、乙烯苄基膦酸、2-（甲基）丙烯酰胺-2-甲基丙基膦酸、3-（甲基）丙烯酰胺-2-羟基膦酸、2-甲基丙烯酰胺乙基膦酸（图3.12）、苄基膦酸酯和3-烯丙氧基-2-羟丙基膦酸，其中一些也用作牙科材料。在更高温度下（>250℃），含磷烷基（甲基）丙烯酰胺聚合物的酰胺键比磷烷基（甲基）丙烯酸酯更稳定。含磷酰胺单体与丙烯酰胺（甲基）丙磺酸（AMPS）的共聚物可作为高温防垢剂。

据报道有新型烯不饱和膦酸化合物，其将磷融入防垢剂中，便于检测。具体由乙烯不饱和取代的氧烷与氨基（或羟基）官能的膦酸或其前体反应制成。膦酸官能团对Ca^{2+}的亲和力，以及其在钙垢上的吸附行为，随着其与聚合物主链距离的增加而降低。

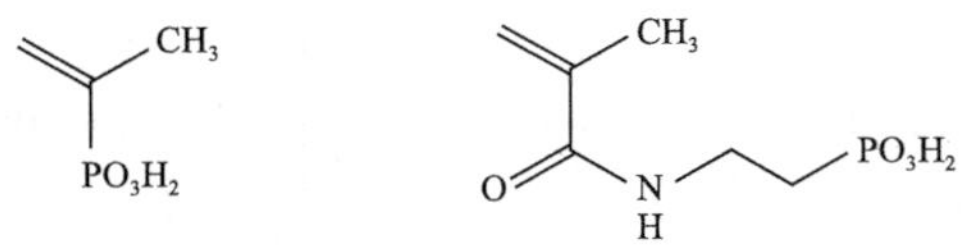

(a) 异丙烯基膦酸　　(b) N-（乙基-2-膦酸）-2-甲基丙烯酰胺

图 3.12　异丙烯基膦酸和 N-（乙基-2-膦酸）-2-甲基丙烯酰胺

3.4.5　膦基聚合物

石油工业中最常用的膦基聚合物是 PPCA，其包含一个膦酸基连接在两个聚丙烯酸或聚马来酸链上（图3.13）。例如，次亚磷酸离子与马来酸反应生成膦二十琥珀酸低聚物与各种更小的分子，产物的摩尔比可变。这些混合物，特别是膦二十琥珀酸低聚物，可作为碳酸盐垢挤注防垢剂。磷原子的存在，使 PPCA 聚合物比多羧酸更容易检测分析，具有更好的效果（特别是防硫酸钡垢）、钙相容性和岩石吸附的性能（延长油田挤注寿命）。但次膦酸基团与岩石的结合能力不如膦酸基团。在 60℃ 的温度下，PPCA 与五膦酸（DETPMP）和其他膦酸基防垢剂相比，对 $BaSO_4$ 有良好的防垢效果。PPCA 在沉淀挤注处理中的相行为也有报道。

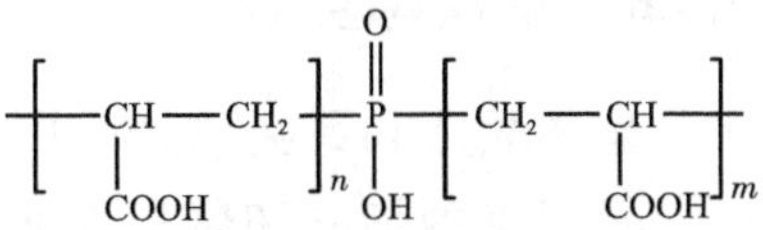

图 3.13　聚膦酸羧酸

利用次磷酸和炔烃化学反应，可以将若干个次膦酸基团引入低聚物中。使用乙炔，可以得到乙烷-1,2-二次膦酸和二乙烯基三次膦酸的混合物，这是一种有用的防垢剂（图3.14）。这些低聚物（或短链聚合物）可以与乙烯基单体反应形成性能更好的多膦聚合物（图3.15）。用单羧酸和二羧酸（如丙烯酸和马来酸）可以制成主链中含有大量磷（多数是次膦酸，但也有膦酸）的膦酸基羧酸。

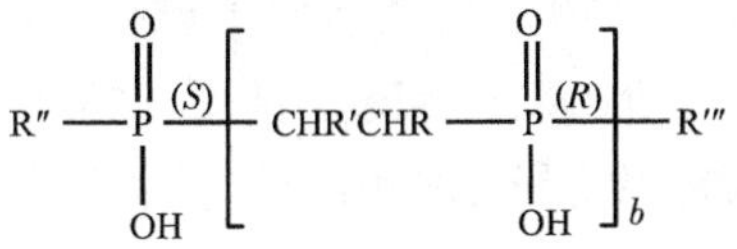

图 3.14　膦基低聚物（$b=1\sim2$）

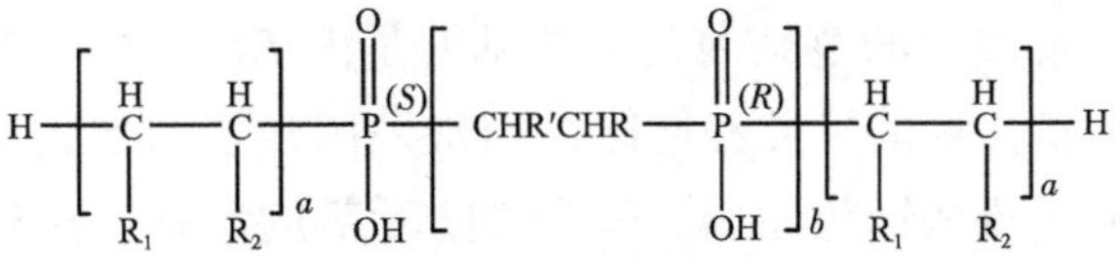

图 3.15　聚次膦酸（在大多数首选实例中，R_1 和 R_2 中的一个是羧基、磺酸或膦酸基，或者 R_1 和 R_2 都是羧酸基）

氨基次膦酸聚合物也可以使用膦基短链聚合物来制备（图3.16）。膦酸基封端的膦基聚合物对 $BaSO_4$ 垢特别有用（图3.17）。与常见的含等量磷的随机共聚高分子防垢剂相比，该聚合物具有较好的岩石吸附性能和热稳定性。OECD 306 的 28 天生物降解试验表明这些封端聚合物具有大于 20% 的生物降解性。封端的 VDPA 聚合物已经实现商业化。

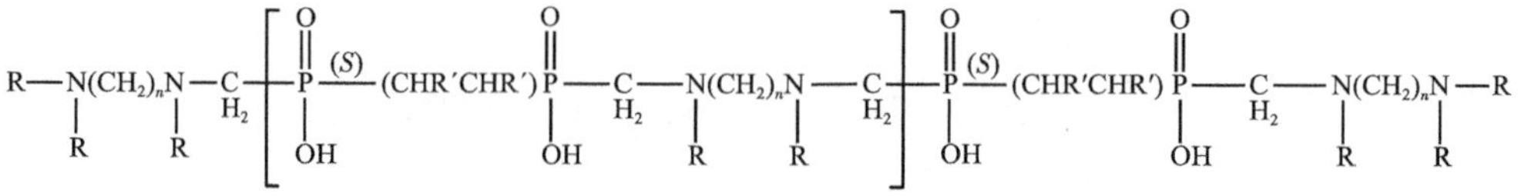

图 3.16 氨基次膦酸聚合物

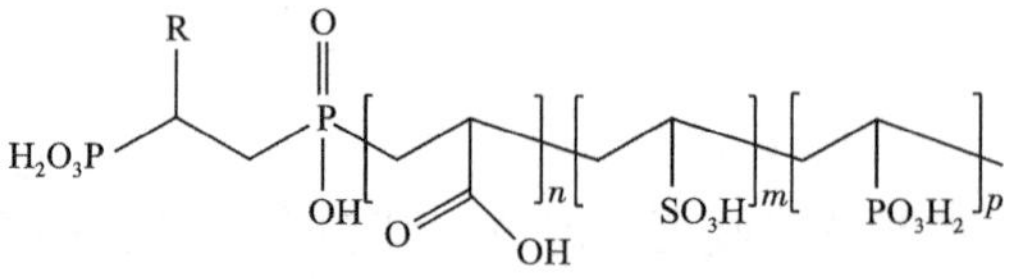

图 3.17 膦酸基封端聚合物（R=H 或最好是 PO_3H_2，n、m 和 p 可以为 0 或任何数字）

3.4.6 聚羧酸盐

聚羧酸盐和聚膦酸开始用作防垢剂的时间相近。聚羧酸最常见的种类是基于聚丙烯酸、聚甲基丙烯酸和聚马来酸（图 3.18）。所有这些线型聚合物都有碳主链。一般来说，聚丙烯酸的生物降解性较差，但有些聚马来酸的生物降解性相当好。一种由 OECD 306 核准的生物降解率大于 60% 的聚羧酸防垢剂最近已被批准在北海使用。与大多数聚合物防垢剂一样，聚合物中活性重复单元的数量需要至少 15~20 个才能达到最佳防垢效果，否则就没有足够的活性基团结合到垢晶体表面使之固定。对于聚丙烯酸的分子量，至少应在1000~1500 之间。大多数聚合物防垢剂的分子量在 1000~30000 之间，在高分子量时性能会下降。众所周知，对于一般生物降解性较差的含丙烯酸聚合物，当重均分子量在 700以下，可大大提高其生物降解性，但这通常是以牺牲性能为代价。对于羧基化和磺化的聚合物，支链或部分交联聚合物的性能优于线型聚合物。而丙烯酸/异戊二烯及相关共聚物认为比聚丙烯酸具有更好的生物降解性，尽管疏水单体的加入可能会降低聚合物与采出水的相容性。

丙烯酸和马来酸单体是成本相对较低的原料，与磺酸或膦酸基团可制成各种共聚物和三元共聚物防垢剂。在聚羧酸的范畴内，最常见的是丙烯酸/马来酸共聚物。马来酸本身不容易聚合，为了提高聚合物的尺寸和性能，通常采用共聚的方法。其与疏水单体共聚物被认为是良好的硫酸钡、硫酸锶防垢剂。聚合物防垢剂的制备方法对防垢剂的最终性能极为重要。因此，一种好的防垢剂产品配方和工艺一旦确定，制造商都保持相同的配方和工艺。其他不饱和羧酸单体包括甲基丙烯酸、巴豆酸、衣康酸、谷氨酸、替格酸和天使酸；但这些单体通常更昂贵（除甲基丙烯酸外），较少用于制备防垢剂。

据称在聚羧酸中加入一定比例的酰胺或羟基可以改善其防垢性能，并增加其钙耐受性。例如，马来酸与一些马来酰胺或马来酰亚胺基团的共聚物，或丙烯酸聚合物与丙烯酰胺、甲基丙烯酰胺、N,N-二甲基丙烯酰胺或羟丙基丙烯酸单体的聚合物。引入磺化基团或不饱和聚乙二醇单体也可以提高钙耐受性。

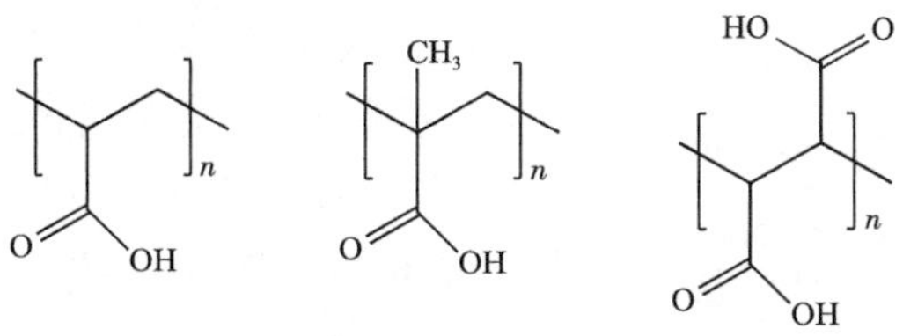

图 3.18 聚丙烯酸、聚甲基丙烯酸和聚马来酸

丙烯酸与阳离子单体（如甲基丙烯氧乙基三甲基氯化铵）的共聚物除具有防垢性能外，还具有杀菌和防腐性能。将季胺单体加入羧酸基聚合物中也可以增强岩石的吸附，延长挤注寿命。

除在 3.4.6 中提到的一种高度可生物降解的聚乙烯基多羧酸外，另一类多羧酸是多氨基酸或多肽，如聚天冬氨酸盐（图 3.19）。聚天冬氨酸是由 L- 天冬氨酸或马来酸酐和氨通过聚琥珀酰亚胺在碱中水解得到。聚天冬氨酸的生物降解性好，对碳酸盐和硫酸盐垢具有良好的防垢性能，也有一定的缓蚀特性。其与碳酸钙相互作用的分子动力学模拟研究也有开展。聚天冬氨酸在化学反应过程中由于存在某种程度的支链，结构比较复杂，包含 α- 和 β- 组，α- 组含有的羧酸侧基离肽主链的距离比 β- 组多一个碳原子。聚天冬氨酸目前在石油工业中主要用于防碳酸盐和硫酸盐垢，特别是在环境法规通常要求大于 20% 生物降解性的地区（如北海盆地）。其也用于最高 85℃挤注处理，但一种改进的聚天冬氨酸可用于 120℃挤注处理并且已经商业化。聚天冬氨酸还可以减少挤注处理后的微粒迁移。一项专利声称通过注入基因工程改造的嗜热微生物和适当的营养物质，在井下生物合成聚天冬氨酸或类似的聚氨基酸防垢剂。文献也综述了聚氨基酸和其他分子在自然界中形成和溶解碳酸钙的研究进展。聚天冬氨酸的衍生物也可以作为防垢剂。据称，如果单体单元含有 N-2- 羟基烷基天冬氨酸酰胺结构的含氧烃酰胺和其他氨基酸，这样的聚天冬氨酸是具有良好的钙相容性和生物降解性的防垢剂。具有二羟基的聚天冬氨酸衍生物可用于碳酸钙防垢剂。胶原蛋白也作为防垢剂。

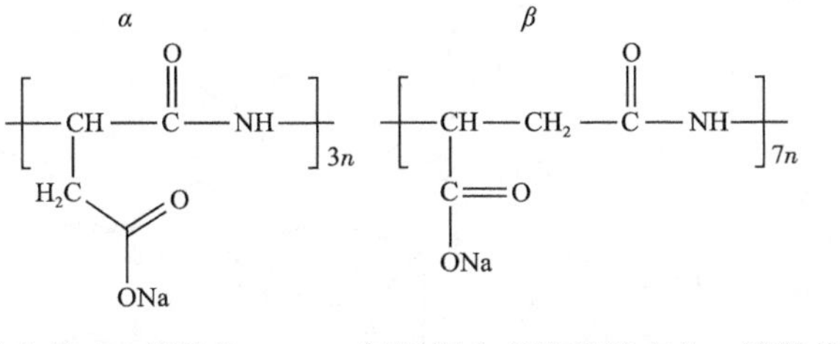

图 3.19 聚天冬氨酸钠（此处 α/β 值约为 3∶7，但根据生产流程的变化，该数值和支链程度也会变化）

为了更好地监测和分析，合成了咔唑和羟基标记的聚天冬氨酸。使用谷氨酸等其他氨基酸的聚天冬氨酸的异变体，也可作为防垢剂，但由于其他氨基酸的成本较高，还没有商业化。聚谷氨酸可通过生物技术获得。聚天冬氨酸也是一种低毒缓蚀剂，并作为油田防垢和防腐蚀的组合缓蚀剂。含羟肟酸基团的多氨基酸是一种具有生物可降解的防垢缓蚀剂。

聚乙醇酸是一种主链中具有杂原子的羧酸聚合物，这可能使其具有一定的生物降解性，但是未见报道（图3.20）。带有聚乙二醇侧基的均聚物或共聚物作为碳酸盐防垢剂，可提高钙的相容性。聚环氧琥珀酸（也称聚酒石酸、PESA）是另一种比聚丙烯酸等聚乙烯醇羧酸有更好生物降解性的聚合物。PESA是低分子量聚合物，很容易生物降解，已用作防垢和缓蚀剂，或与更传统的缓蚀剂混合使用。改性PESA（环氧琥珀酸-2，3-环氧丙基磺酸共聚物）也有报道。实验表明，PESA在$CaCO_3$和$SrSO_4$垢上的性能优于聚天冬氨酸，而对$CaSO_4 \cdot 2H_2O$和$BaSO_4$垢的防垢性能不佳。羧基化低聚物（如由顺丁烯二酸和环氧琥珀酸等聚醚羧基化的低聚物）被用作油田防垢剂。使用胺或胺盐的抑制剂可以解决现场高浓度钠盐沉淀问题。以顺丁烯二酸酐、苯乙烯磺酸钠、丙烯酰胺和壳聚糖为原料制备的壳聚糖接枝共聚物对碳酸钙表现出良好的防垢性能。生物可降解的接枝聚合物混合物，以褐煤、多酚或天然聚酰胺（如酪蛋白、明胶和胶原蛋白）为接枝基，可作为油田化学品，但不是专门用作防垢剂，这些聚合物可以接枝丙烯酸、顺丁烯二酸酐或磺化单体。

羧甲基菊粉是一种寡糖的羧基化衍生物，作为可生物降解的油田防垢剂已商业化，尤其适用于环境敏感地区。其对碳酸盐防垢性能好，对硫酸盐防垢较差。一项挤注处理防垢剂的可行性研究表明，该聚合物很有前景，但还需要优化。所有寡糖和多糖中的羟基都可以用碱和氯乙酸衍生成羧酸基团（如羧甲基纤维素）。但此类聚羧酸溶液黏度太高，不能用于挤注处理。羧甲基菊粉可能是因为分子量相当低，是个特例（图3.21）。但多糖可以被选择性氧化以得到羧酸基团，然后通过加热多糖控制降解，形成具有羧基和醛基官能团

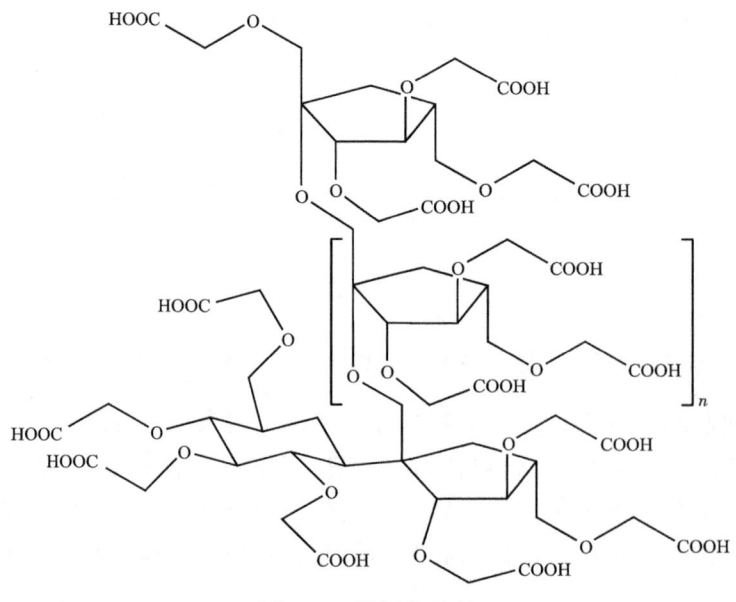

图3.21 羧甲基菊粉

的较低分子量的改性多糖。这类聚合物也可以作为防垢剂。例如，过氧化物解聚羧烷基多糖（最好是从瓜尔豆胶、刺槐豆胶等中天然发现的多半乳甘露聚糖）是每个糖单位中COOH进行了0.5～3.0的取代度形成。含有羧基和醇基的芦荟多糖衍生物也可以用作防垢剂。接枝到麦芽糊精或玉米淀粉多糖上的烯不饱和羧酸单体也是有效的防垢剂。相关文献还报道了其他生物可降解的防垢剂。

丙烯酸、马来酸和AMPS等是天然衍生的羟基化合物和乙烯不饱和单体的混合共聚物。可以引入二级链转移剂二水次亚磷酸钠将磷包含到聚合物中。用氧化淀粉改性得到共聚物也有报道。

一些可生物降解的聚羧酸可以通过生物培养形成。例如，使用绒泡菌（Physarum）培养的无分支的β-聚（1-苹果酸），数均分子量高达100000。芦荟衍生的多糖也可以起到防垢作用。

3.4.7 磺化聚合物

大多数市售的磺化聚合物防垢剂都有聚乙烯主链，最常用的乙烯基单体是乙烯基磺酸、AMPS、烯丙氧基-2-羟丙基磺酸和苯乙烯磺酸（SSA）（图3.22）。烯丙基磺酸/马来酸共聚物也作为硫酸盐防垢剂。在使用最多的单体中，乙烯基磺酸聚合物似乎对$BaSO_4$垢的抑制性能最好。许多类型的含有这些单体之一的共聚物已经商业化应用，如乙烯基磺酸和丙烯酸的共聚物，或AMPS和丙烯酸共聚物。甲基（丙烯酸）或马来酸、AMPS、阳离子单体如二烯丙基二甲基氯化铵的三元共聚物，以及乙烯基磺酸盐和聚亚烷基二醇单或二甲基丙烯酸酯的共聚物也有报道。与膦酸或羧基防垢剂相比，聚乙烯磺酸（PVS）的pK_a值较低，与Ca或Mg的稳定常数低，因此通过成核抑制作用，可以在较低的pH值下作为防垢剂。首选的聚合物是含少量乙烯基膦酸和酰胺单体，如乙烯基甲酰胺或丙烯酰胺或乙烯基吡咯烷酮的AMPS共聚物。

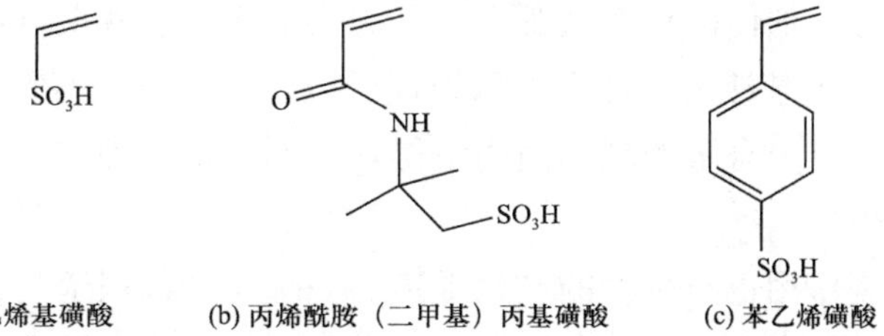

(a) 乙烯基磺酸　　(b) 丙烯酰胺（二甲基）丙基磺酸　　(c) 苯乙烯磺酸

图3.22　乙烯基磺酸、丙烯酰胺（二甲基）丙基磺酸、苯乙烯磺酸

聚磺酸由于其高的热稳定性和钙耐受性，特别用于硫酸盐垢的高温挤注处理。PVS和乙烯基磺酸共聚物及乙烯基磺酸/SSA共聚物的热稳定性优于AMPS基聚合物。但是，曾经被认为在高温溶液中不稳定的其他几类防垢剂，一旦吸附到地层岩石上，稳定性就会提高，也可以适用于挤注处理。例如，聚磺酸对岩石的吸附不像磷酸那样强，其挤注寿命会更短。PVS在低温下（4～5℃）效果很好，优于许多其他类别的防垢剂。

3.5 硫化物防垢剂

最常见的硫化物垢是 Fe^{2+}、Zn^{2+} 和 Pb^{2+} 盐。与碳酸盐或硫酸盐防垢相比，硫化物防垢的研究在 2000 年后得到了显著的发展。通过使用杀菌剂或注入硝酸盐/亚硝酸盐来避免生物成因 H_2S，可以抑制硫化物结垢（见第 14 章）。但地层中自然存在的 H_2S 会导致硫化物结垢。另外，H_2S 清除剂也可用在生产中去除 H_2S（见第 15 章），但常用的三嗪基 H_2S 清除剂会提高体系的 pH 值，可能加剧硫化物和碳酸盐垢的形成。

在注水井和生产井中，会形成硫化铁垢。中等温度的含硫气井中，注入氨基羧酸类螯合物（如 NTAA）可有效螯合铁，防止硫化亚铁结垢。室内研究表明，温度高于 149℃，NTAA 在 5h 内分解，因而不适用于高温环境。另一份报告阐述了一种氨基羧酸（没有给出明确的结构）通过螯合来抑制硫化亚铁垢。有机酸螯合铁的处理，要使用等于（或通常大于）溶解铁所对应的化学计量的量。如此高剂量的酸会对生产设备产生很强的腐蚀性。

三羟甲基膦（THP）和四羟甲基膦盐的共混物（如 THPS）和足够的螯合物（氨基羧酸盐或氨基膦酸盐）也可以抑制和溶解硫化亚铁垢。为避免 THP 或 THP 盐在低 pH 值下发生聚合副反应，可以将这些膦化物与氨或小分子量伯胺（如甲胺）混合使用（第 14 章有 THPS 相关参考文献）。

不用螯合 Fe^{2+}，也能直接抑制硫化亚铁垢，但只有少量文献报道。例如，常用的防垢剂 DETPMP 和 PPCA 已证实能抑制硫化亚铁垢。在使用 DETPMP 类膦酸盐时，发现形成硫化亚铁后变澄清的现象。首先形成黑色的硫化亚铁，24h 后溶液又澄清，这表明了防垢剂滞后"抑制"，可能是防垢剂直接螯合而不是通过低剂量（低于化学计量）的防垢机理作用。

防止硫化亚铁沉积的另一种方法是分散垢微粒而不是抑制其形成。市面上某些硫化亚铁分散剂通常将垢转移到油相中，有可能形成复杂的乳状液，造成额外的生产操作问题。又开发了一些新的分散剂使硫化亚铁既亲水又失去活性。这些聚合物含有阳离子单体，如 3-（丙烯酰）丙基三甲基铵盐。证据显示处理后微粒尺寸小，不太可能在生产系统中聚集和沉积。这一概念在一个有限的现场使用中得到证明，该处理方法成功地防止了一级分离器和加热炉内的沉积。

关于硫化锌抑制作用也有少量的文献报道。一个早期的报道是在油井生产过程中使用羟乙基丙烯酸酯/丙烯酸共聚物作为硫化锌阈值抑制剂。最近发表的文献阐述小分子的氨基膦酸盐（DTPMP、HEDP、ATMP）在浓度 50~100mg/L 能有效抑制硫化锌。研究者表示目前还不清楚这种抑制是由于防垢剂降低了 pH 值，还是发生了阈值防垢。用聚合物防垢剂对硫化物可获得更好的效果。膦酸/马来酸共聚物、AMPS/马来酸共聚物、AMPS/丙烯酸共聚物均优于聚丙烯酸共聚物。一般而言，硫化物防垢剂的用量是硫酸盐防垢剂的 10 倍。其他文献报道了聚合物防垢剂相对于小分子膦酸有明显的改善效果。螯合物如 Na_4EDTA 和 Na_5EDTA 对硫化物也具有良好的抑制效果，但其仅在化学计量时有效，因此在金属硫化物浓度高的情况下使用，成本可能会很高。相关文献报道了 PbS/ZnS 聚合物防垢剂对高温高压井的成功挤注处理。

3.6 岩盐防垢剂

针对油田岩盐（氯化钠）的沉积，可以通过不断稀释岩盐中过饱和的采出水起到控制作用。例如，可以通过小直径油管用淡水稀释井下。但此时需要大量与产出水配伍的低矿化度水，并进行脱氧，以限制腐蚀。用海水稀释时，还需加入一定量的硫酸盐防垢剂，预防硫酸盐结垢。

岩盐只含有一价阴离子，因此前面讨论的二价阴离子垢（碳酸盐、硫酸盐等）的防垢剂对岩盐垢不起作用。已知有两类化学物质可以抑制岩盐垢（即改变其晶体形态）：

（1）亚铁氰化物盐，$M_4Fe(CN)_6$，M=Na 或 K。
（2）氮三烷酰胺和季铵盐，$N[(CH_2)_nCONH_2]_3$，$n=1$ 或 2。

与硫酸盐和碳酸盐防垢剂相比，岩盐用抑制剂的加注浓度通常很高。有实验室研究表明，新开发的产品能在较低浓度下取得良好效果。有文献介绍了两种无机物产品，其中一种是基于有机氮，但化学成分的细节尚未被报道。有机类产品在储层条件下表现出良好的环境友好和吸附特性。pH 值测试的结果表明，溶液的 pH 值对吸附量有很大影响，这为挤注配方的开发优化提供了重要信息。

亚铁氰化物长期以来一直被认为是岩盐晶体的改性剂。以钠离子为中心的八面体氯离子可以被 $Fe(CN)_6^{4-}$ 取代。在 $2.48×10^{-4}$～$2.85×10^{-3}$mol/L 范围内，亚铁氰化物能增加溶液的临界过饱和度（高达 8%），结晶抑制作用显著。亚铁氰化物已成功应用于现场，大大减少了气体压缩设备中岩盐垢的沉积，其使用浓度通常在 250mg/L 左右。亚铁氰化物在实验室性能测试效果优于现场，主要是因为铁离子的不配伍。亚铁氰化物可以改变岩盐晶体，通过定期用低矿化度的水冲刷，很容易清除沉积物，目前还没有亚铁氰化物在井下应用的报告。亚铁氰化物与产出水中的 Fe（Ⅲ）盐反应生成普鲁士蓝。为了防止这种情况发生，可以添加诸如 NTAA 三钠盐和碱金属柠檬酸盐等螯合剂。

在氮三烷酰胺类中，氨三乙酰胺和氨三丙酰胺都有抑制岩盐沉淀的作用（图 3.23）。也可使用盐酸或硫酸氢盐抑制岩盐沉淀。晶体改性效果与添加剂的分子结构与晶体晶格的拟合有关。氨三乙酰胺盐在减缓油田地面、井下岩盐结垢方面已经有成功案例。市售产品中通常含有氯化铵，这是合成氨三乙酰胺的副产物。而氯离子是岩盐常见的离子，最好从产品中去除，但这又将极大地提高产品价格。氨三乙酰胺在高温下稳定和具有可生物降解特性。

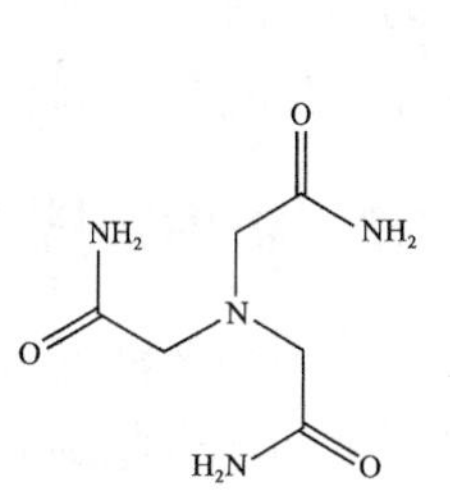

图 3.23　氨三乙酰胺

通过计算机模拟，证明了 NTAA 是 NaCl 的媒晶剂。同一研究组发现柠檬酸可以将氯化钠晶体结构由立方改性成八面体，但仅是通过模型模拟。

据报道，有一种含一个或多个部分水溶性的ⅢA到ⅦA主族金属盐的岩盐抑制剂。首选的金属是镉或铅，也可以是如脲类、硫脲类、甲酰胺类和乙酰胺类等的小分子酰胺类（图3.24）。用酸（如盐酸）或碱（如氨水溶液）对井筒进行预处理可以增强防盐剂的吸附。

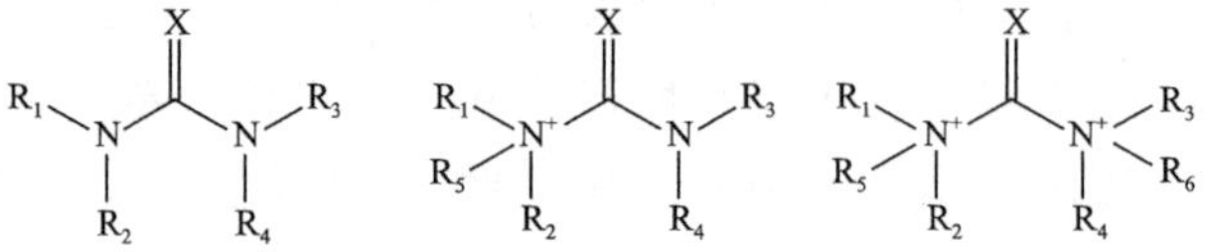

图3.24 尿素或硫脲基盐抑制剂

3.7 防垢剂应用方法

近年来，在高效防垢剂设计方面进展甚微，更多的是在开发可生物降解产品方面。防垢剂的应用仍是防垢控制中最能取得进展的领域。在油田中防垢剂应用的方法有多种，主要方法有连续加注、挤注处理、使用非水基或固体防垢剂。

3.7.1 连续加注

防垢剂可以通过化学注入管线在平台顶部或井下连续注入。对于注水井特别是在采出水回注井或生产井流体，可能需要在井口或井下连续注入。为了防止生产井结垢，也要在注水井中连续注入防垢剂。通常在井口上部将防垢剂连续注入产出水中，缓蚀剂等其他生产化学品也可能一起注入。事实上，许多防垢剂与某些缓蚀剂不相容。如果有毛细管管柱或气举注入系统，也可以将防垢剂注入井下。在气举注入过程中，在水基防垢剂溶液中加入乙二醇等低蒸气压溶剂（蒸气压抑制剂，VPD）非常重要，可避免溶剂过度蒸发导致的防垢剂"黏稠"。此外，可能需要加入乙二醇或其他水合物抑制剂来抑制形成天然气水合物。有现场报告称，使用VPD并不能防止黏稠。此时将固体防垢剂溶解在非常高沸点的溶剂中，成功实施了气举注入。此外，在气举中也加入与防垢剂混合的溶垢剂。有时在井下注入某些防垢剂也会增加井下设备的腐蚀。

3.7.2 防垢剂挤注处理

防垢剂挤注处理的基本思路是预防、减缓井下发生结垢和储层伤害。防垢剂还可对井口以上和管道起到防垢作用，但地面设施可能需要更高剂量的防垢剂。在挤注处理中，防垢剂溶液以高于地层压力注入井中，推入近井地层岩石孔隙；然后关井数小时，使防垢剂通过各种机制留在岩石基质中；当油井重新投入生产时，产出水流经留存化学物质的孔隙，溶解部分化学物质（图3.25）。这样采出水应含有足够的防垢剂，以防止结垢。当防垢剂浓度降至MIC（防止结垢的最小抑制剂浓度）以下时，应对井进行重新挤注。在

图 3.26 中，防垢剂浓度在 92 天后降至 MIC 以下（约 2mg/L）。当然，长挤注寿命将使整个井下结垢处理成本降至最低。有商用的挤注建模程序可帮助防垢剂挤注处理的设计。挤注设计中还应考虑岩石和防垢剂相互作用。挤注处理也可以与沥青质抑制剂等生产化学品一起进行（见第 4 章）。

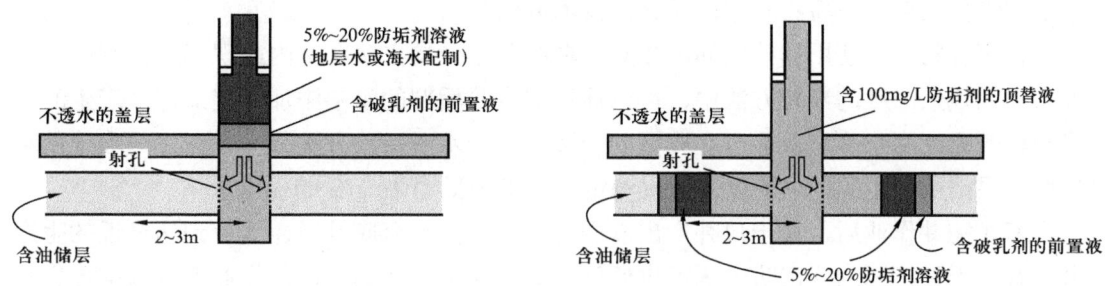

图 3.25　防垢剂挤注处理

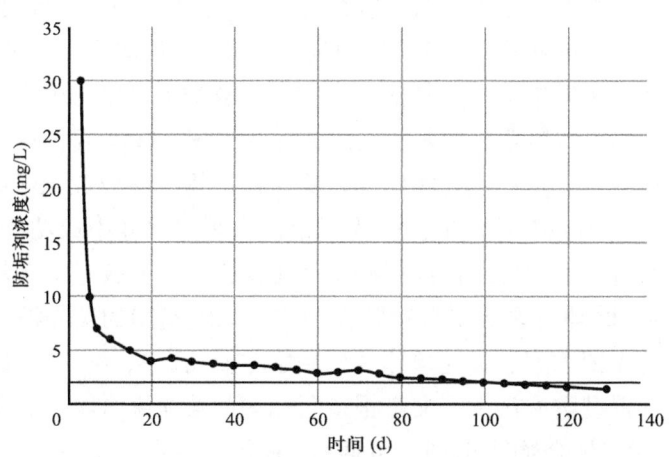

图 3.26　挤注后防垢剂浓度随时间的变化

在一些油田生产水中很难准确分析残余防垢剂或更重要的活性抑制成分浓度。含磷量高的分子容易检测。在其他油田化学品存在的情况下，检测低浓度防垢剂的现代技术包括：凝胶渗透色谱/体积排阻色谱（GPC/SEC）、高效液相色谱（HPLC）、电感耦合等离子体发射光谱（ICP-OES）和液相色谱质谱法（LC-MS）。

作为更环保的磷酸盐替代品，聚合物在油田防垢领域的应用日益增多。对于不含磷原子的聚合物防垢剂有许多可行的检测方法，但检测结果不是很精确。作为补充，更直接的分析方法有扫描电子显微镜、压力测试、在线监测。近年来，在需要高热稳定性的井下挤注应用中，磺化聚合物或其共聚物的应用有所增加。因为从盐水相分离防垢剂的操作往往耗时费力，防垢剂残余浓度分析的难度很大。为克服这些困难，近年来市场上推出了含标记物的药剂。

由于磷比硫或碳更容易检测和测量，一些羧基化或磺化聚合物通过添加少量含磷单体进行标记。这种方法的局限性是任何一个聚合物链中的磷数量都很难控制。在对不同磷含量的聚合物进行挤注处理后，岩石的解吸速率不同，通过磷含量的测试可能误导挤注寿命的预测结果。背景磷也会影响检测的准确性。相关文献对一系列新型膦酸官能化聚合物防垢剂进行了详细的表征研究，从采出水中可以更精准地检测出这些防垢剂。文献报道了油田盐水中磺化聚合物的高效液相色谱分析技术进展。

混合聚合物可以使用目前已知的相对简单的化学分析方法在室内或现场进行测试。在一种防垢剂残余浓度的专有方法中，可以通过向含防垢剂的水中添加特定试剂，产生的显色络合物的吸光度来确定，甚至在典型的合成油田盐水中，该方法都具有良好的准确性和重现性。根据所使用的试剂浓度，可以测定出溶液中低至 1mg/L 的残留量。

在海底完井作业后，若干口井产出液中的防垢剂流入海底管道，然后到达最近的生产平台，有若干种不同的标记防垢剂可供使用。通过对标记物的分析，作业者可以推算出每口井产出液中防垢剂的浓度，从而确定每口井需要何时进行重新挤注。标记防垢剂的另一个作用就是能够从混合水中区分防垢剂，这在海底多相流生产中是个难题。

有文献介绍了具有不同吸附特性的防垢剂类别及其在油田的应用，优选通过紫外/可见光谱分析来检测。活性防垢单体（和其他化合物如次磷酸）可以与少量含芳香基团的乙烯基单体或含共轭双键的单体聚合，如苯乙烯磺酸和含噻唑、咪唑或吡啶基团的乙烯单体。

标记聚合物的最新方法是在聚合物中加入荧光标记物。接枝荧光标记物的聚合物防垢剂最低检出限可以达到 1mg/L 或更低。荧光光谱法在采出水样中成功检测到荧光防垢剂，凝胶渗透色谱法对其进行了验证。最终目标是使用便携式荧光仪进行现场检测，实时监测产出流体。N-乙烯基咪唑可融入聚合物防垢剂中，咪唑基团的荧光波长约为 424nm，可用于检测水溶液中防垢剂浓度。荧光单体 8-烯丙氧基-1,3,6-芘三磺酸三钠盐也被引入碳酸盐聚合物和环保型防垢剂中。文献报道了标记的荧光聚合物，其中荧光团包括奎宁或奎尼丁。据称硼标记聚合物防垢剂，可通过 ^{11}B 核磁共振、ICP-MS 或 SEC-ICP-MS 分析。

免疫测定试剂可以在不受其他处理成分或污染物干扰的情况下识别特定浓度的磷酸羧酸等聚合物。这类聚合物包含一个可检测的端基。相关的防垢剂聚合物和免疫分析试剂盒已经商业化。

在氯化钾或海水中，传统的"吸附"防垢挤注处理采用含有 5%～20% 活性溶液的水基防垢剂。在关井期间，防垢剂被吸附在地层岩石。前置液（如在氯化钾或海水中加入 0.1% 的防垢剂，可选择与破乳剂配合使用）可用于射孔段附近的清洁和准备。后置液将防垢剂推入地层深处。这种处理方法适用于含水大于 10% 的油井。在含水为 0～10% 的区块，需要评估更多的非常规处理方法。膦酸基团通常比羧酸基团更容易吸附，羧酸基团比磺酸盐基团更容易吸附。因此，磺化防垢剂（如 PVS）在岩石基质中滞留率低，挤注寿命可能比具有多膦酸基团的防垢剂要低。但吸附性也不能太强，否则会导致返排液中防垢剂浓度低于 MIC。在吸附挤注处理中，通常在油井投产后占总量 25%～35% 的防垢剂会立即随着采出液采出，因防垢剂的早期大量流失，会降低挤注寿命。比较理想的挤注效果是从第 1 天开始，采出液中防垢剂浓度始终保持在恒定的 MIC 水平。注入的浓缩防垢剂 pH 值不能是高酸性，否则会造成腐蚀。例如，文献报道了一种 pH 值 0～1 的酸性防垢剂导致井下毛细管柱出现点蚀。

目前，有很多关于提高防垢剂在岩石上的滞留率、延长挤注处理寿命的技术，主要如下：

（1）沉淀挤注处理。
（2）使用一些过渡金属离子和 Zn^{2+}。
（3）井下原位提高 pH 值。
（4）两相溶剂改变岩石润湿性。
（5）与阳离子聚合物共混。
（6）在防垢聚合物结构中加入阳离子单体。
（7）交联防垢剂。
（8）使用高岭石或其他黏土来增强防垢剂的吸附。
（9）防垢剂微颗粒。

提高防垢剂在井筒近井地带滞留率的常用方法是进行沉淀挤注处理。在高温和油藏 pH 值条件下，许多防垢剂和较高含量的钙或镁不相容。可以通过将这些阳离子或 Fe^{2+} 注入防垢剂（或在储层矿物中将这些离子溶出），防垢-阳离子复合物可以在井筒近井地带沉淀，比单独使用防垢剂具有更好的滞留效果。这些离子从储层矿物中原位溶解，也可以提高抑制剂的滞留率。

值得注意的是，不要在射孔段形成沉淀复合体，以免影响油井产量，还要使用合适的注入泵。阳离子或聚阳离子的固体来源（如碱性阴离子交换树脂的微粒），可以先被挤注到地层，然后再挤入防垢剂，增加防垢剂的保留率。高磺化度聚合物（如 PVS）不容易与钙盐形成沉淀，磺化聚合物往往很快解吸，挤注寿命短，但是可以采取吸附挤注等其他技术解决这个问题。一种最新的挤注方法包括在地层中加入一种金属螯合剂、一种防垢剂和二价金属阳离子的组合物，其中金属螯合剂-金属阳离子螯合物在环境温度下的稳定常数不小于由金属阳离子和防垢剂形成的螯合物的稳定常数，金属阳离子与防垢剂形成的螯合物的溶解度随温度升高而降低。螯合剂可防止形成 Ca/Mg-Si 络合物的伤害。

实验室研究表明，挤注配方中的 Zn^{2+} 显著增加了抑制剂的保留率。此外，Zn^{2+} 对某些 $BaSO_4$ 垢防垢剂具有显著的协同作用。Zn^{2+} 与膦酸同用也表现出协同缓蚀作用。常见防垢剂与 Zn^{2+} 形成的溶液络合物比与碱土金属离子（如 Ca^{2+}、Mg^{2+} 和 Ba^{2+}）形成的溶液络合物更强。这种与 Zn^{2+} 形成的更强溶液络合物可能是提高防垢效率的原因。

沉淀挤注工艺中另一种增加防垢剂滞留时间、延长挤注寿命的方法是就地提高近井地层中防垢剂溶液的 pH 值。通过这种方式，防垢剂分子中的酸基变成阴离子，更容易与阳离子络合形成钙/镁络合物沉淀。例如，防垢剂溶液可以与尿素或烷基尿素衍生物混合。在地层温度升高时，尿素分解释放出碱性气体氨或烷基胺，从而提高防垢剂溶液的 pH 值。也可以使用如乙酰胺或二甲基甲酰胺等其他产碱酰胺。

尿素只能在85℃以上分解，不能单独用于低温井。针对这种情况，可采取两种方法：一种方法是可以进行一种新的多级沉淀挤注处理。低pH值的磷酸防垢剂和钙离子混合（实际上是作为温度敏感的反相乳液），将其与高pH值、含低剂量碳酸钠防垢剂的碳酸钠溶液交替泵送。水相的混合导致钙抑制剂复合物在岩石上沉淀。另一种方法是采用一种可以在40℃低温下分解尿素的酶。两种方法均成功应用于油田生产。

还有增加挤注寿命的方法是使用互溶剂（小分子非离子两亲体，NIA），如小分子烷基二醇（如三甘醇丁基醚）。互溶剂增强岩石的湿润性，提高防垢剂的滞留率，也有助于提高井的清洁速度。其还可以清除全水相挤注处理造成的滞留水（水封），从而提高油井产量。因此，通常在挤注的前置液中加入一些互溶剂。使用相同的互溶剂，沉淀挤注法的挤注寿命延长远远优于使用吸附性防垢剂挤注法。互溶剂增强与尿素pH值改性结合的技术在油田已成功应用。

一种据称可以增加挤注寿命的方法是，在岩石表面预先添加如聚二烯丙基二甲基氯化铵等阳离子聚合物，该聚合物最初用作黏土稳定/防砂添加剂。这能避免常规沉淀挤注可能带来的储层渗透性降低。低毒性非聚合物二季胺化合物也可以延长挤注寿命。相关文献对其他阳离子聚合物也进行了研究。带正电的表面能够更好地吸附带负电的防垢剂离子。研究发现，在防垢聚合物（如PPCA）中加入阳离子单体也会产生一种产物，在大于MIC时，产物在岩石上，有更长的停留时间。人们认为，通过将聚合物"固定"到地层表面（例如，通过带正电的单体和带负电的砂岩之间的静电吸引），可以增强抑制剂的吸附。在OECD 306的28天海水测试中，丙烯酸和季铵盐单体组成的聚合物生物降解率高达60%（在聚乙烯聚合物中不常见）。优化带正电荷的季铵盐单体数量后，吸附的抑制剂可以释放到地层盐水中。带有亚甲基膦酸侧基的聚二烯丙基胺（图3.27）已有报道，但作为防垢剂没有给出评价。尽管阳离子聚合物可以提高防垢剂的挤注寿命，但有证据表明，当存在这些聚合物时，防垢剂会因拮抗效应而性能降低。

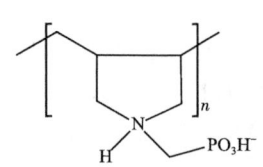

图3.27 聚二烯丙氨基亚甲基膦酸盐

另一种提高挤注寿命的方法是使用交联防垢剂。室内实验已经证明，使用羧基聚合物防垢剂可延长挤注寿命一倍，但尚未进行现场试验。该技术是垢挤注处理和交联聚合物堵水处理的组合。而通过在交联聚合物基体中加入防垢剂，来实现防垢的堵水处理技术已得到证实。

另一种提高防垢剂滞留率并可用于堵水的方法是使用高岭石。高岭石是砂岩储层中常见的黏土类型，可以增大防垢剂吸附表面并延长挤注寿命。对于高渗透率、低黏土含量的砂岩油藏，因其中普遍存在的石英和长石颗粒表面，防垢剂吸附较低，经常遇到的问题是挤注寿命相对较短。岩心驱油实验证明，注入高岭石可以改变井眼附近的矿物学特性，增加防垢剂的吸附性，从而可能显著延长高渗透油藏的挤注寿命。在类似储层内，注入低浓度的高岭石不会对高渗透储层造成严重伤害和影响油井产能。其他类型的黏土或材料可能更有效。

有机硅烷是一种高岭石固化剂和表面改性剂，也可以提高防垢剂的吸附性能。有文献研究了自组装有机硅烷薄膜的直接防止结垢。利用石英晶体微天平分析了石英晶体上形成的薄膜，表明使用相同的有机硅烷分子可以产生不同的膜结构。用涂覆有机硅烷的单晶石英片进行的盐水试验表明，碳酸钙垢沉积可减少66%。高岭石技术已经扩展到包括$CaCO_3$颗粒与高岭石和有机硅烷（KaCaO）。

已研发出交联的AlOOH-磺化聚羧酸（AlOOH-SPCA）纳米抑制剂，通过将液相聚合物防垢剂转化为黏性凝胶，来增加SPCA在地层中的保留。相关文献讨论了在有/无Ca^{2+}辅助下，不同pH值时SPCA在AlOOH纳米颗粒上的吸附。

在压裂液中加入防垢剂，可以预防注水突破的早期阶段因结垢导致的储层伤害。钻井液的设计中可以加入防垢剂，这样通过钻井液滤液进入井筒近井地带，就能在油井产水前吸附。研究表明，向水基压裂液中添加水溶性防垢剂并不那么简单。水力压裂地层中的挤注处理并不总是有效的。防垢剂与配套压裂液一起使用，可以更有效地将防垢剂放置在压裂裂缝的有效位置。在压裂作业期间或者压裂后，将防垢剂嵌入可渗透支撑剂的技术已经成功应用于油田。同时压裂和防垢挤注处理技术也可以用于气田。防垢剂嵌入砾石技术也已应用于油田。

由于酸、酸液添加剂、防垢剂和含较高阳离子盐的残酸之间存在相容性问题，HCl酸驱油和防垢剂的结合也不是一个简单的过程。这些技术已在油田开展试验。据称已经开发出与氟离子相容的防垢剂（用于砂岩酸性增产液）。某些防垢剂不仅与HCl相容，而且在强酸性条件下仍能吸附于储层岩石。这样防垢剂可能直接应用于酸体系，而不需要单独进行防垢处理。另一种方法是采用防垢剂和酸前驱体（如酯或聚酯）的混合物，在储层温度下水解形成酸。酸化除垢和防垢联合处理也有相关报道。但有研究表明，作为硫酸盐垢溶垢剂的螯合物严重影响如膦酸、多羧酸和磷酸酯等一系列防垢剂的性能，只有高度磺化的聚合物（如PVS）似乎受影响不大。

还有相关文献报道了相对渗透率调节剂（堵水用）和防垢剂的联合成功治理案例。防垢剂和堵水技术的联合处理方法，比单独进行的每个处理步骤之和，措施步骤更少。现场无须进行两次单独的作业，而是在一个步骤中完成了堵水和防垢剂挤注处理。这样将羧酸聚合物防垢剂和交联剂注入地层，在近井地带交联形成堵水凝胶。当凝胶分解后，就会释放防垢剂控制结垢。除防垢剂外，其他生产化学品也可保留在地层的聚合物凝胶中。

3.7.3 用于挤注处理的非水基或固体防垢剂

井下水基挤注处理的问题是防垢剂的水溶液可能会改变岩石润湿性。一旦岩石成为水润湿，其水相渗透率就会发生变化，有时是永久性的，因此水通道可能最终被打开，导致所谓的水锥效应，使井被不可逆地破坏。这样的井无法恢复产能，需要新的射孔作业才能经济地开采原油。相对渗透率的变化也可能发生在水基挤注处理中。因此在低含水井或水敏井中应避免使用水基挤注产品，对于避免原油产量损失非常重要。此外，有气举问题的高含水井也宜采用低密度非水基挤注处理。

为了克服这一问题，人们开发了多种非水基防垢剂，主要有油溶性防垢剂，有机溶剂混合液中完全不含水材料，反相乳液、微乳液或纳米乳液，胶囊产品。

3.7.3.1 油溶性防垢剂

烃类溶剂中的防垢剂通常又被称为油溶性防垢剂，该类型产品避免了注入水溶性或钙敏感防垢剂时造成的不利影响。其成分是一种烷基胺与酸形式的防垢剂，形成油溶性的电子对。这种混合物可以用柴油或煤油等低成本烃类溶剂稀释，如叔烷基伯胺与氨基膦酸（如 DTPMP）的混合物。市售的酸性防垢剂往往含水，所以最终的油溶性防垢剂混合物实际上含有少量水（5%～10%）。2-乙基己胺的环境特性更好，是一种更好的胺。油溶性防垢剂在注入时，会分散到地层水中并吸附于岩石。如柴油等烃类可以用于挤注处理的后置液阶段，让更多的防垢剂深入地层。油溶性防垢剂已成功应用于水敏性油井。

油溶性防垢剂也可用于连续注入。此时其可以与甲醇、乙二醇等热力学水合物抑制剂（与某些防垢剂不相容）一起注入。此外，还可以将其他油溶性化学品（如蜡和沥青质抑制剂）与油溶性防垢剂混合使用，从而避免在海上油田使用多个脐带缆。为了油溶性防垢剂在储层中更好停留，还开展了油溶性增黏剂的研究。

3.7.3.2 有机溶剂混合液中完全不含水材料

磷标记的聚合物型磺化防垢剂可以溶解在有机溶剂的混合物中，完全不含水。其工作原理与油溶性防垢剂类似，通过挤注将其分散到地层水中，然后沉淀到岩层上。此外，防垢剂还可以通过与载钙的水性溶剂（也可以含有机添加剂）发生乳化，进一步帮助和增强其在储层中的沉淀作用。这种产品虽然引入了水，但反相乳液提供了连续的油相，与标准油溶性和乳化防垢配方相比，对水敏性地层的伤害要小得多。

3.7.3.3 乳液防垢剂

反相乳液是油包水型大粒径乳液。可以调节其在储层温度下破裂，从而释放所含的水溶性防垢剂或其他水相生产化学品。在粗乳液后注入浊点为 40～60℃的破乳剂，会引起粗乳液破裂和释放防垢剂。微乳液挤注处理已有相关报道。在乳化沉淀防垢剂的挤注处理过程中，使用尿素加强防垢剂沉淀已有成功案例。室内实验中，测试了对温度敏感且含有 3% 防垢剂的纳米乳液。这种乳液用可生物降解的非离子表面活性剂制备，且用量低于制造反相乳液或微乳液的用量，但尚未有现场试验的报道。与其他乳液相比，纳米乳液在室温下的稳定性明显更优。

低渗透油藏需要油包水乳液能长时间稳定滞留于孔隙中，因此开发了液-液微胶囊、缓释乳液产品。这种液滴包裹技术适用于杀菌剂、H_2S 清除剂、缓蚀剂等许多化学品的挤注处理，并已有现场应用的报道。对于不会发生滞留的高渗透油藏，可以采用反相乳液沉淀挤注来解决。

3.7.3.4 固体防垢剂（用于挤注和其他处理）

在挤注处理中避免用水的技术是用一种包覆防垢剂的微固体颗粒。这些微囊化的防垢剂颗粒悬浮于液态烃，挤注后滞留于地层孔隙。当恢复生产后，缓释防垢剂进入产出水。该技术已成功地用于硫酸盐垢的防治。

生产含防垢剂颗粒的方法有很多种，包括100%固体产品、胶囊产品和可吸附防垢剂的高孔隙材料。一种含防垢剂颗粒的改进制备方法是将合适的单体在含防垢剂的反相乳液或悬浮液中，进行热自由基引发的聚合。通过这种方法制备的颗粒特性（如溶胀性、孔隙度、降解性、粒径、分子量、交联程度）可控制防垢剂的释放速率、前体或生成剂，这些特性与井下环境特性（如温度、pH值、矿化度）相结合，就能控制浸出或释放速率。有相关文献介绍了粉煤灰和膦酸固化剂的混合物形成防垢颗粒的研究。

悬浮液防垢颗粒也可以用于注水井。悬浮颗粒通过地层渗透到生产井中，并在生产井近井区域可控地释放防垢剂。这种控释防垢剂的颗粒通过酯交联剂与多元醇交联制得。

除挤注处理以外，含防垢剂或其他生产化学品的固体颗粒也用于其他领域。例如，在前面讨论了压裂作业。防垢支撑剂也用于砾石充填完井。悬浮在盐水中的加重固体防垢剂胶囊从井口泵入，在重力作用下落入井底口袋。在到达井底口袋后，防垢剂从胶囊中扩散形成浓度梯度，在整个处理周期内提供接近恒定浓度的防垢剂。防垢剂颗粒也可以放置在砾石充填的筛管后面，当与产出水接触时，会逐渐释放防垢剂，或者在压裂作业中作为支撑剂。防垢剂颗粒也可以放置在滑套后面，当检测到水时滑套会打开。防垢剂可以放置在分配器中随产出流体连续释放。

通过PPCA辅助化学沉淀法，直接制备出亚微米级Ca-DTPMP防垢剂，岩心驱替实验结果表明，其促进了膦酸防垢剂的滞留性能。PPCA吸附在Ca-DTPMP颗粒上，增加了颗粒的表面负电荷，减少了颗粒在多孔介质中的沉积。同时对合成纳米防垢剂的有效参数进行了评价。与比表面积很大的碳基纳米颗粒共同注入，可提高防垢剂在岩石上的滞留性能。通过超声波处理，将晶体相的磷酸钙纳米颗粒分散到表面活性剂溶液中，形成纳米材料悬浮液，用于将膦酸防垢剂输送到地层岩心中，以控制结垢。煅烧过的多孔吸附剂可以吸附多种类型的化学品，在井下缓慢释放。

3.7.4 挤注处理中防垢剂的位置

为了对所有产水层进行处理，挤注过程中防垢剂挤注位置非常重要。在大斜度井或水平井中，不希望所有的防垢剂都挤入水平段跟部，而是渗透到趾部。如果待处理层段存在显著的渗透率或压力变化，处理液将进入渗透率较高、压力较低的层段，剩下很少的流体用于处理其他层段，这些层段可能是产水层。尤其在渗透率和压力变化明显的长水平井中挑战更大。在防垢剂配方中使用示踪剂是一种评估防垢剂位置的方法，可以替代PLT（生产测井工具）。

为了实现更均匀的流体覆盖，通常需要改变层间的初始注入流体分布。用来改变这种情况的"转向"方法，目的是将流体从区间的一部分转移到另一部分。第 5 章酸化增产中有一节有化学转向剂技术相关内容。黏弹性表面活性剂已用于防垢剂的挤注处理。

在防垢剂挤注应用中，主要采用的两种转向方法是蜡转向剂和增黏水溶液。近几年出现了泡沫防垢剂处理相关专利。泡沫防垢剂在注入后会破碎成液体。

蜡转向剂是挤注过程中注入蜡颗粒，堵塞井跟部的射孔段，将流体转向井的趾部。当达到地层的温度时，小蜡球会熔化，避免对地层造成伤害。

另一种调节防垢剂在储层中分布的方法是使用一种临时、自转向的稠化（或胶化）防垢剂溶液。在防垢剂溶液中使用黄胞胶或琥珀聚糖等聚合物，增加黏度，形成自导流，有助于在水平井和复杂井中更好地调节防垢剂位置。室内研究中，基于黄胞胶的轻度黏性剪切变稀流体，已成功用于高达 170℃ 的环境。在海水中，琥珀聚糖在 70℃ 以上不可逆分解，对储层渗透率没有影响，其仅限于低温度井应用。与非黏性处理相比，稠化防垢剂溶液还能延长挤注寿命。黄胞胶是一种剪切变稀的多糖，当加热到 115℃ 以上时，根据矿化度的不同会被热解。矿化度越高，分解需要的温度越高。对于黄胞胶，通常需要使用氧化破胶剂进行热聚合物降解，以彻底"打破"黏性凝胶，使流体黏度恢复到接近于传统的水基挤注处理的水平。理论建模表明，为了使防垢剂在多裂缝区域的最佳位置注入，增黏防垢剂溶液应该是约 20mPa·s 的轻微剪切变稀液体，并以低流速 1～10bbl/min 注入。在北海某个油田，黏稠防垢剂挤注处理仅对部分斜井成功。防垢剂在储层的位置看似正常，然而该井在注入过程中因堵塞而受损。据称有一种使用烃类化合物胶凝剂作为自转向防垢剂的方法。关于黏性挤注和非黏性挤注的比较见相关文献。

乳化防垢剂可获得较高的黏度（见 3.7.3.3）。该方法已成功用于油田酸化增产处理。与黄胞胶相比，瓜尔胶聚合物被认为是更好的增黏剂，有效浓度只需要 0.3%（质量分数）。一种聚阳离子瓜尔胶表现出特别好的效果，也可能增强抑制剂在储层中滞留性能。有关其他转向技术的更多信息，请参阅第 5 章。

3.8 防垢剂性能测试

对新型防垢剂应用最常见的室内测试包括：静态测试、毛细管测试、配伍性测试、热稳定性测试（在溶液中或岩石上）、静态吸附测试、动态吸附测试（岩心驱替，渗透率变化）。

静态测试是一种防垢剂的粗筛方法，用于硫酸盐防垢剂的效果排序，最常用于硫酸钡垢（另一种评价防垢剂效率的方法是在给定的水质条件下，在一定的防垢剂用量或不加防垢剂的情况下，测定任一成垢化合物的饱和度）。

最近的报告表明，与使用搅拌或流动的测试槽相比，静态测试低估了可能形成的结垢量。一种流动实验系统已经开发出来，实验证明膨胀引起的管内回流增加了碳酸钙结垢速率。一种可能的解释是垢晶核停留时间的增加，加强了垢晶的沉淀并减少垢沉积物的流失。因此，建议在进行现场应用验证时，对防垢剂在湍流条件下的性能进行系统测试。与垢沉淀相比，预防垢沉积需要的碳酸钙防垢剂浓度更高。相关文献也报道一些其他评价方法，如一种研究钢表面碳酸钙生长的新方法、一种快速评价防垢剂的碳酸钙沉淀法、动力学浊度法快速有效筛选硫酸盐防垢剂、测试岩盐防垢剂的新方法等。

对于油田现场连续注入防垢剂的评价，先是通过静态测试评价出防垢性能最好的防垢剂，然后将最佳防垢剂进行动态毛细管测试。毛细管测试是碳酸盐防垢剂测试的第一步，测试中逐步降低防垢剂浓度，直到管路上出现压降。通常在管内预沉积垢以增加结果的可重现性。通过这种技术，可以确定完全抑制垢所必需的 MIC，进而比较防垢剂的效果。MIC 是现场使用的指导性数值，使用后其效果需要监测。防垢剂性能测试很重要的是在产出水的 pH 值环境和存在主要采出阳离子的条件下进行。通常采出水的 pH 值为 5~6，而不是浓缩防垢剂溶液的 pH 值 2~3。

许多防垢剂的性能也受采出水中阳离子的影响，二价阳离子尤其如此。例如，Fe^{3+} 和 Fe^{2+}（后者存在于厌氧系统中）都会影响许多防垢剂的性能。有研究表明，Fe^{2+} 实际上改善了膦酸基防垢剂的性能。随后的研究表明像 DETPMP 这样的小分子氨基膦酸对 Fe^{2+} 最不耐受；聚羧酸似乎受影响居中；磺化盐受影响最小。在 200mg/L Fe^{2+} 存在时，两种不含—OH 或—COOOH 基团的膦酸防垢剂对碳酸盐垢表现良好。

低浓度 Zn^{2+} 对碳酸盐的结垢可能性影响不大，但会影响防垢剂性能。锌-防垢剂络合物可以改善聚丙烯酸和 DTPMP 的性能，但会降低 ATP 的性能。Cu^{2+} 也能影响方解石的生长。用于硫酸钡垢的膦酸基防垢剂，其防垢效率受溶液中硫酸钡饱和度（SR）和二价离子 Ca^{2+} 和 Mg^{2+} 的影响。这些因素（SR 和 Ca^{2+}/Mg^{2+} 值）的相互作用对不同防垢剂的差别可能很大。

海底低温对防垢剂性能的影响很大，这不仅仅是因为方解石或重晶石过饱和的变化。小分子的氨基膦酸（如 DTPMP）在 5℃时比在 95℃时在重晶石上的吸附量低很多，因而效果差很多；而 PVS 则有很大的改善；PPCA 仅受轻微影响。这是由于在低温条件下，硫酸钡过饱和度的改变比任何防垢剂防垢机理的变化都要大。

盐水配伍性对连续注入的防垢剂也很重要。极端情况下，防垢剂与产出流体直接混合处的下游，可能立即形成一种包含防垢剂的 Ca/Mg 盐垢。一般来说，在高 Ca/Mg 浓度下，膦酸的相容性最差，羧酸居中，磺酸的相容性最好。在相容性测试中，溶液 pH 值对结果有很大的影响。

对于挤注处理，还需要进一步进行三项研究。首先进行配伍性测试，以确定防垢剂与地层水混合时不会发生沉淀、不会对地层产生伤害。如果在正确的地方沉淀，可以有利于提高防垢剂在地层中的保留。其次是对于高温油藏，需要进行热老化测试，以确保防垢剂在预期挤注寿命期间和油藏温度下保持稳定。防垢剂溶液在静态瓶中进行老化，并将其性能与未老化的样品比对。长期以来，人们基于静态测试、热老化试验和随后的动态毛细管路试验，认为许多类型的防垢剂在高温油藏不具备热稳定性。例如，乙烯基磺酸聚合物在高温下比氨基膦酸更具有热稳定性。但实际应用中，防垢剂沉积在地层岩石上，可以减缓热降解过程。因此，需要对吸附在岩石样品上的防垢剂进行室内热老化研究。一项随后的研究表明，200℃时，氨基膦酸的热稳定性低于几类聚合羧酸和磺酸防垢剂，而且腐蚀产物中的金属离子会使情况变得更差。但是 NTMP 和 DETPMP 在 125℃的热老化影响可以忽略不计。相关文献用 $^{31}P/^{1}H$ NMR 分析方法研究了防垢剂热降解的热力学和动力学过程。

评价挤注处理的第二种方法是静态吸附测试，以确定防垢剂在破碎地层岩石的吸附能力（不是沉淀）。第三种方法是最重要的动态岩心驱替试验。在储层条件下，将防垢剂注入岩心样品，然后利用地层水进行驱替。监测采出水中的防垢剂浓度，直到其降至 MIC 以下。根据这些数据可以绘制吸附等温线并与其他产品比较，确定预期挤注寿命。渗透率变化也可以通过动态岩心驱替试验进行监测。在长时间驱替过程中，要特别注意保持岩心厌氧条件，以避免地层环境的人为破坏。

有研究表明，采用岩心驱替测试方法可能会出现严重误导的结论，如油田预期的防垢剂挤注回流曲线及根据实验室岩心数据选择防垢剂。现场实例报告称，即使防垢剂浓度高于实验室确定的 MIC，但是在井下和安全阀附近的油管中还是观察到硫酸盐垢。有人认为，在岩心驱油评价中获得的最佳防垢剂（浓度降低到 MIC 前挤注寿命最长）不一定是现场使用的最佳防垢剂。对低渗透白垩地层的研究证明了这一点。据推测，防垢剂被推入含圈闭水的水润湿油层，然后通过水润湿孔隙中的连通水向裂缝扩散，并以更快的速度进入井筒。如果防垢剂对岩石的吸附太强，就不能足够快地扩散到裂缝中，那么产出水中就不能获得足够的防垢剂。因此，长挤注寿命的主要机制可能与扩散有关，而与防垢剂在地层岩石上的吸附无关。

防垢剂与相同井或管道中加入的其他生产化学品之间的配伍性也很重要。影响采出水 pH 值的化学物质（如三嗪基 H_2S 清除剂）将影响碳酸盐结垢。部分缓蚀剂、低剂量水合物抑制剂与某些防垢剂不相容。

有些研究专注于金属表面形成的沉积物。一项研究表明体积沉淀和表面沉积对过饱和指数有不同的依赖性。因此，为了完全了解一个工业防垢系统，应该研究这两个过程。

3.9 化学除垢

本节回顾去除碳酸盐垢、硫酸盐垢、硫化物垢和铅垢的化学物质。软的岩盐（NaCl）垢可以简单地用低矿化度水冲洗或喷射除垢。如果使用海水，可能需要添加硫酸盐防垢剂。综述了机械除垢方法。磨料喷射（水中含有锋利边缘的沙粒）比单独水喷射更好，但会损坏管柱。圆形的珠子可以尽量减少对管柱的伤害。因为珠子也是酸溶性的，清除工序更简单。用黄胞胶等聚合物使水增黏，还可以减少摩擦。

3.9.1 碳酸盐垢除垢

第 5 章是关于酸化增产内容，也介绍了碳酸盐岩储层增产的方法。清除碳酸盐垢的化学溶液与此非常相似，也是增产的一种形式。以下只对碳酸盐垢进行简要讨论。

碳酸钙（方解石）是迄今为止最常见的碳酸盐垢，其次是碳酸亚铁（菱铁矿）。碳酸盐可被各种酸溶解，最容易和最便宜的是盐酸（HCl）。

$$2HCl + CO_3^{2-} \longrightarrow H_2O + CO_2 + 2Cl^-$$

该反应的放热和 CO_2 气体释放都能加快反应速率。随着酸的消耗，溶液中金属离子（如钙离子）的浓度增加，延缓反应。同时必须在酸溶液配方中添加缓蚀剂。另外两种常见的添加剂是铁离子抑制剂和水润湿活性剂。

这些添加剂和其他添加剂，如应对淤积的添加剂，将在第5章进行综述。有机酸也可以用作碳酸盐垢的溶垢剂，如乙酸（CH_3COOH）会减慢与碳酸盐的反应，其腐蚀性比盐酸小，长链有机酸的使用也有相关报道。还原酸、甲酸和含甲酸离子的缓冲混合物可以有效地缓解高温下去除碳酸盐垢过程中的腐蚀问题。在裸眼水平井中，采用胶凝有机酸以提高除碳酸钙的效果。柠檬酸去除碳酸盐和氧化铁垢也有相关研究。甲基磺酸是碳酸钙的溶垢剂和除锈剂。

在去除碳酸盐垢的过程中，避免腐蚀和淤积问题的方法是使用螯合物溶垢剂而不是酸，如 EDTA 盐（Na_2EDTA）和其他氨基羧酸螯合剂已经用于去除碳酸盐垢的地层堵塞，特别是在高温环境（图 3.28）。还研究了各种螯合物溶垢剂的热稳定性。除此之外，可能是因为盐酸和有机酸更便宜，商业上很少使用螯合剂溶碳酸盐垢。如 2- 膦酸基 -1，2，4- 三羧酸丁烷等含羧酸基团的膦酸防垢剂，也可用于去除碳酸盐垢。

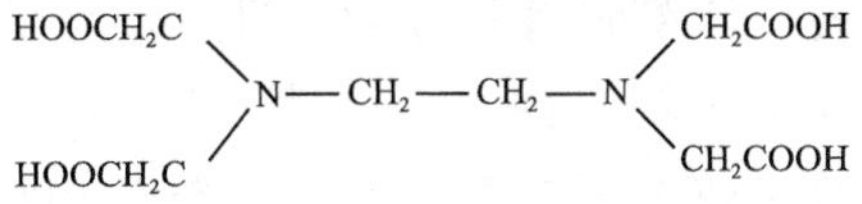

图 3.28 乙二胺四乙酸（EDTA）

有文献研究了其他可生物降解的螯合剂，包括羟乙基亚胺二乙酸盐（HEIDA）和羟乙基二胺三乙酸盐（HEDTA）和谷氨酸 N，$N-$ 二乙酸盐（GLDA）（图 3.29）。GLDA 或甲基甘氨酸 N，$N-$ 二乙酸（MGDA）的铵盐是碳酸盐垢、硫酸盐垢或硫化物垢的溶垢剂。一种 HEIDA 的溶解能力与 7.5% HCl 相当，其可生物降解，而 EDTA 则不能。但是，HEIDA 对石膏垢的溶解能力不如 HEDTA。这些羟烷基螯合物与 EDTA 相比，在酸溶液中更容易溶解，有助于酸化增产中的控铁。另一种市售可生物降解的螯合溶垢剂是乙二氨基二丁二酸三钠（图 3.29）。

(a) 羟乙基亚胺二乙酸

(b) 羟乙基二胺三乙酸

(c) 乙二氨基二丁二酸三钠

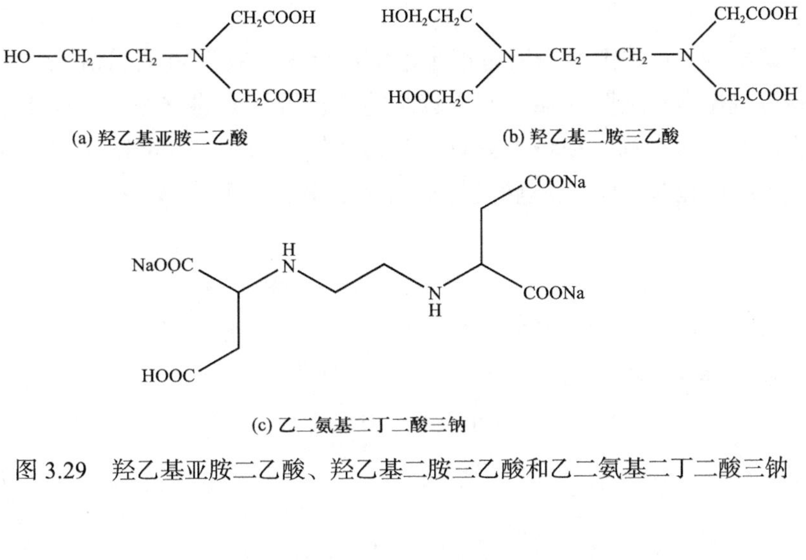

图 3.29 羟乙基亚胺二乙酸、羟乙基二胺三乙酸和乙二氨基二丁二酸三钠

黏弹性表面活性剂［如 N- 芥酸基 -N，N- 二（2- 羟乙基）-N- 甲基氯化铵］可以使溶垢剂到达地层更深的位置。地层的烃类化合物作用于表面活性剂，降低流体黏度，从而使流体选择性地进入含油气层。

碳酸亚铁（菱铁矿）垢可以用上述常用的螯合物除去。氧化铁很难用高 pH 值的螯合剂去除。在 pH 值不小于 7 的条件下，氨基酸/羟基芳香螯合物和磺化羟基芳香螯合物对氧化铁的溶解效果更好。在 pH 值较低的条件下，生物可降解的柠檬酸可以溶解碳酸亚铁垢。膦酸对碳酸亚铁垢也具有良好的溶解性能。

3.9.2 硫酸盐垢除垢

硫酸钙（石膏或硬石膏，$CaSO_4 \cdot H_2O$）是最容易化学除去的硫酸盐垢，硫酸钡（重晶石，$BaSO_4$）最难去除。硫酸钙不溶于酸，但高 pH 值的螯合剂可以溶解，即上面讨论的碳酸盐除垢剂。多年来，EDTA 钠盐是标准的硫酸盐除垢产品；而像北海对环境要求严格的地区，人们正在寻求更容易生物降解的产品。据称还有其他生物可降解的螯合溶垢剂，如碳酸钙垢的溶垢剂，以及谷氨酸 N，N- 二乙酸、2-（1，2- 二羧乙基亚胺）-3- 羟基丁烷二酸、3- 羟基 -2，2′- 亚氨基二琥珀酸四钠（HIDS）、2-（1，2- 二羧基乙基亚胺）丁二酸、乙二胺单琥珀酸（EDMS）［也称 N-（2- 氨基乙基）天冬氨酸）］及（S）天冬氨酸 -N- 乙酸。螯合剂中加入有机硅烷可以稳定微粒的迁移。在 EDTA 配方中加入螯合氨基膦酸（也是防垢剂）可以增强对硫酸钙垢的溶垢能力。一种据称比 EDTA 更能促进石膏垢溶解的产品是羟肟酸和（或）其某些盐。螯合溶垢剂还可以去除如氧化铁、钙和铝硅酸盐和溴化物等其他垢。生物可降解的谷氨酸 N，N- 二乙酸（GLDA）铵盐或甲基甘氨酸 N，N- 二乙酸（MGDA）铵盐对硫化亚铁或硫酸钙垢的效果优于其钠盐。

另一种去除石膏垢（甚至可能是重晶石垢）的方法是通过泵入 pH 值小于 7 的 CO_2/碳酸氢盐水溶液，将硫酸钙转化为碳酸钙。这种酸性溶液就可以溶解碳酸盐。使用纯碱（碳酸钠）将石膏"转化"为碳酸钙，然后使用盐酸溶液进行冲洗溶解，这是一种低成本的两步过程。

硫酸钡或硫酸钡/硫酸锶混合垢是很难溶解的硬垢。在严重结垢情况下，可能最好采用磨铣或喷砂等机械方法清除硬垢。有文献综述了机械除垢的方法。一种使用等离子体通道钻孔的方法已经获得专利。现场中只有一类持续使用的化学剂能以较好的速率溶解硫酸钡垢，即 pH 值大于 12 时的二乙三胺五乙酸盐（DTPA）（图 3.30）。

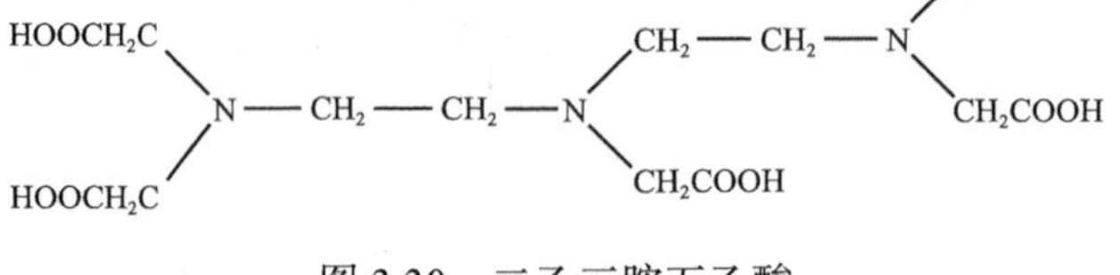

图 3.30 二乙三胺五乙酸

几乎所有效果好的市售硫酸钡溶垢剂至少含有部分 DTPA 和增效剂（协同剂）。但据称有一种新型重晶石溶垢剂，其性能与 DTPA 溶出剂相似，其结构信息很少，只知道它是一种二酯，适用于 pH 值 5～6。这类二酯溶垢剂也易生物降解，因此特别适用于环境敏感地区。在 OECD 306 测试中，一种重晶石溶垢剂 28 天生物降解率为 41%，但没有报道其结构。20 世纪 70 年代末，冠醚和笼状多醚作为潜在的硫酸钡溶垢剂曾有研究，但是其成本依然太高。

回到熟悉的 DTPA 基溶垢剂，DTPA 盐是十八酸螯合配体，通过五个羧酸基团上的氧原子和三个氮原子上的孤对电子与 Ba^{2+} 结合，形成可溶的 $[Ba(DTPA)]^{3-}$。通过非接触式原子力显微镜研究表明，初始反应是一个 DTPA 分子的活性位点与暴露在重晶石表面的两个或三个 Ba^{2+} 结合。更昂贵的螯合剂（如三乙基四胺六乙酸和 DIOCTA）也是重晶石垢的良好溶垢剂。提出了一种从溶垢溶液中再生螯合物（EDTA、DTPA 等）的方法。另一类完全不同的重晶石溶垢剂是离子液体（如七氯化三甲胺二铝），其溶解性能比 EDTA 盐更好，但是目前其成本对现场应用来说太高。

硫酸钡与 DTPA 基溶垢剂在室温下反应相当缓慢，在井下温度和搅拌时，硫酸钡的溶解速度会加快。实验室测试通常是 10g 固体溶于 100mL 溶垢剂，这并不能反映现场真实的固液比，这也就解释了为什么现场应用效果低于室内评价。DTPA 的钾盐（K_5DTPA）比钠盐效果更好。如果溶垢剂在井下与 CO_2、H_2S 或有机酸发生反应，降低了 pH 值，溶垢剂的效果变差，添加少量过量的碱（如氢氧化钠）可以提升溶垢效果。实验室研究表明，含有 I A 族无机碱（K 和 Na）的溶液作为 pH 值控制剂，会使人工固结砂岩岩心的渗透性变差，这与其是否含有螯合物无关。使用碱性乙胺也有同样的效果，但使用甲胺则没有。

有几种增效剂可以用以催化螯合剂与重晶石垢的反应速率。例如，常用的碳酸根和甲酸根离子及过去使用的草酸根离子。柠檬酸根、硫代硫酸根、次氮基乙酸根、巯基乙酸根、羟基乙酸根或氨基乙酸根离子也作为增效剂。据称，葡萄糖庚酸钠是 EDTA 或 DTPA 基溶垢剂的增效剂，也是 2-膦酸基-1，2，4-三羧酸丁烷的增效剂。

根据 OECD 306 测试实验，效果最好的重晶石螯合剂（DTPA 和 EDTA）在海水中不能生物降解。但是在现实的海洋条件下，通过光降解和随后的光降解产物降解，可以实现生物降解。

硫酸锶可以用上面提到的如 EDTA 或 HEIDA 盐等各种螯合物来溶解。硫酸锶经常与硫酸钡共同沉淀，因此在配方中需要一些 DTPA。方解石和重晶石的混合垢需要至少含有部分 DTPA 的螯合剂来溶解。

加入少量 NIA（如三甘油单丁醚），也可以加速垢的溶解。NIA 的工作原理是清除重晶石表面的烃类化合物，以便可与溶垢剂发生反应，改变其润湿性。硫酸钡和硫酸锶垢在晶格中含有少量的放射性镭离子，称为 NORM（自然发生的放射性物质）或 LSA（低比活度）垢。虽然放射性水平很低，但这种垢的固体或溶液需要特定的处理步骤和排放规定。

3.9.3 硫化物除垢

去除硫化亚铁垢的化学物质包括盐酸和有机酸、丙烯醛（也是生物杀菌剂）、四羟甲基磷盐（也是生物杀菌剂）、螯合剂（多价螯合剂）。

硫化亚铁是最常见的硫化物垢，一般溶于盐酸和有机酸（如甲酸）。一般来说，硫含量低的硫化亚铁沉淀在酸中具有较高的溶解度。但当硫化亚铁样品长时间放置，其成分会发生一些变化，不容易溶于酸。研究发现黄铁矿和镁铁矿是酸不溶的，磁黄铁矿在酸中溶解度低，而四方硫铁矿的溶解度较高。据称马来酸是一种硫化亚铁的溶垢剂，产生很少的 H_2S。

盐酸也可以去除硫化锌垢，但不能除去硫化铅。盐酸具有腐蚀性，因此需要在配方中添加缓蚀剂。但研究表明成膜缓蚀剂降低了硫化亚铁的溶解速度。相同的研究表明表面活性剂能提高溶蚀速率，而互溶剂不能。但互溶剂（如乙二醇单丁醚）有利于从硫化亚铁沉积物表面去除烃类化合物。还要添加 H_2S 清除剂，以清除 HCl 与硫化亚铁反应产生的剧毒 H_2S 气体：

$$FeS + 2HCl \longrightarrow Fe^{2+} + 2Cl^- + H_2S$$

但醛基硫化物清除剂对铁的溶解效果低，这可能是由于垢表面形成了聚合物沉积。可以用三嗪清除剂代替。随着盐酸的消耗和 pH 值增加，可能会形成不溶性的铁盐。因此可以添加铁离子抑制剂（如柠檬酸）以防止这种情况。然而在盐酸中加入柠檬酸会降低对硫化亚铁的溶解量。也可以使用如 EDTA 等其他螯合物，但可能产生同样的效果。据称氨基羧酸螯合物本身在 pH 值为 8~10 时可以去除硫化亚铁，但反应缓慢。

在高温下，可以使用比盐酸腐蚀性低的酸（如甲酸、巯基乙酸、乙醛酸和顺丁烯二酸）去除硫化物垢。

丙烯醛（2-丙烯醛）除了是一种 H_2S 清除剂和杀菌剂外，还能溶解硫化亚铁垢。但丙烯醛的毒性极高，是疑似致癌物，需要谨慎处理。

四羟甲基磷盐如硫酸盐（THPS）与小分子烷基胺或优选铵盐共注入时，可以去除井下和平台顶部的硫化亚铁垢（见第 14 章）。该配方也可以与强酸（如盐酸）溶液混合。铁最终与氮磷配位体螯合，使水呈现红色。在油田应用案例中，使用表面活性剂和 THPS 杀死 SRB，去除硫化亚铁垢，几口井的产量惊人地提高了 300%。在另一个油田应用案例中，先用 THPS 去除硫化亚铁和控制细菌的生长。然后在第二步中加入甲酸去除近井区域残留的硫化亚铁和一些聚合物残留。亚氨基二琥珀酸和 THPS 混合物能协同地除硫化亚铁垢，比单独使用 THPS 的效果更好。含铵盐的 THPS 配方可能具有腐蚀性，尤其在高温环境下。添加缓蚀剂（如乙炔醇、丙炔醇）或巯基化合物（如巯基乙酸）可以缓解这一问题。

还可用如亚氯酸盐/二氧化氯或高锰酸盐等氧化剂将 Fe^{2+} 转化为可溶的 Fe^{3+} 以去除硫化亚铁。这些药剂的氧化性很强，有腐蚀性。二氧化氯用于去除注水系统和酸化增产处理中的硫化亚铁泥。

3.9.4 清除铅垢

铅垢只溶于极热的高浓度氧化性酸（如硝酸）。这种酸的腐蚀性很强，风险太高。醋酸（CH_3COOH）和过氧化氢（H_2O_2）的混合物可以除去铅垢。原位反应产生活性的氧化性更强的过氧乙酸（CH_3COOOH），这可能是活性物质。与之类似，用高锰酸钾（$KMnO_4$）为氧化剂时也可以溶解铅垢，但会导致 MnO_2 沉淀。可以用柠檬酸或其他螯合物溶液去除 MnO_2 沉淀。

参 考 文 献

[1] (a) W. Frenier and M. Ziauddin, *Formation, Removal, and Inhibition of Scale in the Oilfield Environment*, eds. N. Wolf and R. L. Hartman, Society of Petroleum Engineers, 2008. (b) J. C. Cowan and D. J. Weintritt, *Water-Formed Scale Deposits*, Gulf Publishing, 1976. (c) M. Davies and P. J. B. Scott, *Oilfield Water Technology*, Houston, TX: National Association of Corrosion Engineers (NACE), 2006.

[2] I. R. Collins, "A New Model for Mineral Scale Adhesion," SPE 74655 (paper presented at the SPE International Symposium on Oilfield Scale, Aberdeen, UK, 30–31 January 2002).

[3] M. M. Jordan, K. Sjursaether, I. R. Collins, N. D. Feasey, and D. Emmons, "Life Cycle Management of Scale Control within Subsea Fields and Its Impact on Flow Assurance, Gulf of Mexico and North Sea Basin," *Chemistry in the Oil Industry VII*, Manchester, UK: Royal Society of Chemistry, 2002, 223.

[4] (a) M. B. Tomson, A. T. Kan, and G. Fu, "Inhibition of Barite Scale in the Presence of Hydrate Inhibitors," SPE 87437, *SPE Journal* 10 (3) (2005): 256. (b) R. Masoudi, B. Tohidi, A. Danesh, A. C. Todd, and J. Yang, "Measurement and Prediction of Salt Solubility in the Presence of Hydrate Organic Inhibitors," SPE 87468, *SPE Production & Operations* 21 (2) (2006): 182.

[5] (a) M. M. Jordan and E. J. Mackay, "Scale Control in Deepwater Fields," *World Oil* 226 (9) (2005): 75-80. (b) G. M. Graham, E. J. Mackay, S. J. Dyer, and H. M. Bourne, "The Challenges for Scale Control in Deepwater Production Systems—Chemical Inhibition and Placement Challenges," Paper No. 02316 (paper presented at the Annual Spring Meeting of NACE International, CORROSION/2002, Denver, CO, 7–14 April 2002).

[6] K. Davis, G. Wooodward, Y. Mottot, and D. Joubert, International Patent Application WO/2006/103203, 2006.

[7] C. Sitz, J. Shumway, and C. Miller, "An Unconventional Scale from an Unconventional Reservoir," SPE 87434 (paper presented at the SPE International Symposium on Oilfield Scale, Aberdeen, UK, 26–27 May 2004).

[8] D. L. Gallup and C. J. Hinrichsen, "Control of Silicate Scales in Steam Flood Operations," SPE 114042 (paper presented at the SPE International Oilfield Scale Conference, Aberdeen, UK, 28–29 May 2008).

[9] G. Atkinson and M. Mecik, "The Chemistry of Scale Prediction," *Journal of Petroleum Science & Engineering* 17 (1997): 113.

[10] G. Rousseau, C. Hurtevent, M. Azaroual, C. Kervevan, and M.-V. Durance, "Application of a Thermo-Kinetic Model to the Prediction of Scale in Angola Block 3 Field," SPE 80387 (paper presented at the International Symposium on Oilfield Scale, Aberdeen, UK, 29–30 January 2003).

[11] H. K. Kotlar, S. Jacobsen, and E. Vollen, "An Integrated Approach for Evaluating Matrix Stimulation Effectiveness and Improving Future Design in the Gullfaks Field," SPE 50616 (paper presented at the SPE

European Petroleum, Conference, The Hague, The Netherlands, 20–22 October 1998).

[12] (a) G. M. Graham, R. Stalker, and R. McIntosh, "The Impact of Dissolved Iron on the Performance of Scale Inhibitors under Carbonate Scaling Conditions," SPE 80254 (paper presented at the SPE International Symposium on Oilfield Scale, Houston, TX, 5–8 February 2003). (b) S. Yean, H. Al Saiari, A. T. Kan, and M. B. Tomson, "Ferrous Carbonate Nucleation and Inhibition," SPE 114124 (paper presented at the SPE International Oilfield Scale Conference, Aberdeen, UK, 28–29 May 2008). (c) K. Chokshi, W. Sun, and S. Nesic, "Iron Carbonate Scale Growth and the Effect of Inhibition in CO_2 Corrosion of Mild Steel," Paper 05285 (paper presented at the NACE International Corrosion Conference, Houston, TX, 2005).

[13] I. R. Collins, "Predicting the Location of Barium Sulfate Scale Formation in Production Systems," SPE 94366 (paper presented at the SPE International Symposium on Oilfield Scale, Aberdeen, UK, 11–12 May 2005).

[14] (a) P. J. Webb and O. Kuhn, "Enhanced Scale Management through the Application of Inorganic Geochemistry and Statistics," SPE 87458 (paper presented at the SPE International Symposium on Oilfield Scale, Aberdeen, UK, 26–27 May 2004). (b) J. J. Tyrie, International Patent Application WO/2007/144562.

[15] M. Brown, "Full Scale Attack, Review," *BP Technology Magazine*, October–December 1998, 30–32.

[16] I. R. Collins and M. M. Jordan, "Occurrence, Prediction and Prevention of Zinc Sulfide Scale within Gulf Coast and North Sea High Temperature/High Salinity Production Wells," SPE 68317 (paper presented at the SPE International Symposium on Oilfield Scale, Aberdeen, UK, 30–31 January 2001).

[17] M. M. Jordan, K. Sjursaether, M. C. Edgerton, and R. Bruce, "Inhibition of Lead and Zinc Sulfide Scale Deposits Formed during Production from High Temperature Oil and Condensate Reservoirs," SPE 64427 (paper presented at the SPE Asia Pacific Oil and Gas Conference and Exhibition, Brisbane, Australia, 16–18 October 2000).

[18] F. A. Hartog, G. Jonkers, A. P. Schmidt, and R. D. Schuiling, "Lead Deposits in Dutch Natural Gas Systems," *SPE Production & Facilities* 17 (2002): 122.

[19] R. Jasinski and D. Frigo, "The Modelling and Prediction of Halite Scale," Proceedings IBC International Conference, *Advances in Solving Oilfield Scaling*, Aberdeen, UK, 22–23 January 1996.

[20] J. K. Smith and J. L. Przybylinski, "The Effect of Common Brine Constituents on the Effficacy of Halite Precipitation Inhibitors," Paper No. 06837 (paper presented at the NACE CORROSION Conference, 2006).

[21] V. Khoi Vu, C. Hurtevent, and R. A. Davis, "Eliminating the Need for Scale Inhibition Treatments for Elf Exploration Angola's Girassol Field," SPE 60220 (paper presented at the SPE International Symposium on Oilfield Scale, Aberdeen, UK, 26–27 January 2000).

[22] E. Mackay and M. M. Jordan, "Natural Sulfate Ion Stripping during Seawater Flooding in Chalk Reservoirs," *Chemistry in the Oil Industry VIII*, Manchester, UK: Royal Society of Chemistry, 2003, 133.

[23] E. Mackay, K. Sorbie, V. Kavle, E. Sørhaug, K. Melvin, K. Sjursæther, and M. M. Jordan, "Impact of In-Situ Sulfate Stripping on Scale Management in the Gyda Field," 100516-MS (paper presented at the SPE International Oilfield Scale Symposium, Aberdeen, UK, 31 May–1 June 2006).

[24] I. R. Collins and P. A. Sermon, International Patent Application WO/2006/008506, 2006.

[25] E. Acton and G. J. Morris, International Patent Application WO/2000/079095, 2006.

[26] (a) L. Rzeznik, M. Juenke, D. Stefanini, M. Clark, and P. Lauretti, "Two Year Results of a Breakthrough

Physical Water Treating System for the Control of Scale in Oilfield Applications," SPE 114072 (paper presented at the SPE International Oilfield Scale Conference, Aberdeen, UK, 28–29 May 2008). (b) D. Stefanini, International Patent Application WO/2008/017849.

[27] (a) J. Groenenboom, S.-W. Wong, and G. Nitters, U.S. Patent Application 20040195187. (b) B. Wang and C. Wang, International Patent Application WO/2009/000177.

[28] (a) O. G. Maxson and G. D. Achenbach, U.S. Patent 4115606, 1978. (b) S. Brand, A. Dierdorf, H. Liebe, F. Osterod, G. Motz, and M. Günthner, International Patent Application WO/2007/096070.

[29] (a) I. R. Collins, "A New Model for Mineral Scale Adhesion," SPE 74655 (paper presented at the International Symposium on Oilfield Scale, Aberdeen, UK, 30–31 January 2002). (b) W. C. Cheong, A. Neville, P. H. Gaskell, and S. Abbott, "Using Nature to Provide Solutions to Calcareous Scale Deposition," SPE 114082 (paper presented at the SPE International Oilfield Scale Conference, Aberdeen, UK, 28–29 May 2008).

[30] (a) M. G. Mwaba, M. R. Golriz, and J. Gu, *International Journal of Heat Exchangers VI* (2005): 235. (b) R. E. Herzog, Q. Shi, J. N. Patil, and J. L. Katz, "Magnetic Water Treatment: The Effect of Iron on Calcium Carbonate Nucleation and Growth," *Langmuir* 5 (1989): 861.

[31] (a) E. Chibowski, L. Holysz, A. Szczes, and M. Chibowski, "Precipitation of Calcium Carbonate from Magnetically Treated Sodium Carbonate Solution," *Colloids Surface A: Physicochemical and Engineering Aspects* 225 (2003): 63. (b) F. F. Farshad and S. M. Vargas, "Scale Prevention, a Magnetic Treatment Approach," SPE 77850 (paper presented at the SPE Asia Pacific Oil and Gas Conference and Exhibition, Melbourne, Australia, 8–10 October 2002).

[32] J. D. Donaldson and S. M. Grimes, "Control of Scale in Sea Water Applications by Magnetic Treatment of Fluids," SPE 16540 (paper presented at the SPE Offshore Europe, Aberdeen, UK, 8–11 September 1987).

[33] J. S. Baker and S. A. Parsons, "Anti-Scale Magnetic Treatment," *Water and Waste Treatment* 39 (1996): 36–38.

[34] S. A. Parsons, B. L. Wang, S. J. Judd, and T. Stephenson, "Magnetic Treatment of Calcium Carbonate Scale—Effect of pH Control," *Water Research* 31 (1997): 339.

[35] G. N. Jefferson, U.S. Patent 5738766, 1998.

[36] D. Walker, U.S. Patent 6145542, 2000.

[37] J. S. Gill, "Development of Scale Inhibitors," Paper 229 (paper presented at the NACE CORROSION Conference, 1996).

[38] G. M. Graham and A. J. B. Hennessey, "Scale Inhibitor Surface Interactions Using Synchroton Radiation Techniques" (paper presented at the *Chemistry in the Oil Industry VIII*, Royal Society of Chemistry, Manchester, UK, November 2003).

[39] G. M. Graham, L. S. Boak, and K. S. Sorbie, "The Influence of Formation Calcium and Magnesium on the Effectiveness of Generically Different Barium Sulphate Oilfield Scale Inhibitors," SPE 81825, *SPE Production & Facilities* 18 (1) (2003): 28.

[40] M. D. Yuan, E. Jamieson, and P. Hammonds, "Investigation of Scaling and Inhibition Mechanisms and the Influencing Factors in Static and Dynamic Inhibition Tests," Paper 98067 (paper presented at the NACE CORROSION Conference, San Diego, 22–27 March 1998).

[41] W. H. Leung and G. H. Nancollas, "Nitrolotri (Methylenephosphonic Acid) Adsorption on Barium Sulfate Crystals and Its Influence on Crystal Growth," *Journal of Crystal Growth* 44 (1978): 163.

[42] M. B. Tomson, G. Fu, M. A. Watson, and A. T. Kan, "Mechanisms of Mineral Scale Inhibition," *SPE*

Production & Facilities 18（3）（2003）：192.
- ［43］S. He, A. T. Kan, and M. B. Tomson, "Mathematical Inhibitor Model for Barium Sulfate Scale Control," *Langmuir* 12（1996）：1901.
- ［44］K. S. Sorbie and N. Laing, "How Scale Inhibitors Work: Mechanisms of Selected Barium Sulfate Scale inhibitors Across a Wide Temperature Range," SPE 87470（paper presented at the SPE International Symposium on Oilfield Scale, Aberdeen, UK, 26−27 May 2004）.
- ［45］P. Wilkie, S. Heath, and C. Strachan, "An Overview of Scale Inhibitor Detection Techniques and Some Recent Advances in Detection Methods for Mixtures of Polymeric Scale Inhibitors in Produced Brines that Enable Improved Scale Management in Subsea and Deep Water Fields"（paper presented at the 19th Oilfield Chemical Symposium, Geilo, Norway, 9−12 March 2008）.
- ［46］J. D. Sallis, W. Juckes, and M. E. Anderson, "Phosphocitrate: Potential to Influence Deposition of Scaling Salts and Corrosion," in *Mineral Scale Formation and Inhibition*, ed. Z. Amjad, New York: Plenum Press, 1995, 87.
- ［47］H. A. El Dahan and H. S. Hegazy, "Gypsum Scale Control by Phosphate Ester," *Desalination* 127（2000）：111.
- ［48］（a）M. M. Jordan, N. Feasey, C. Johnston, D. Marlow, and M. Elrick, "Biodegradable Scale Inhibitors. Laboratory and Field Evaluation of a 'Green' Carbonate and Sulfate Scale Inhibitor with Deployment Histories in the North Sea," *RSC Chemistry in the Oil Industry X*, Manchester, UK, 5−7 November 2007.（b）A. F. Miles, S. H. Bodnar, H. C. Fisher, S. Sidoe, and C. D. Sitz, "Progress Towards Biodegradable Phosphonate Scale Inhibitors," International Oilfield Chemistry Symposium, Geilo, Norway, 23−25 March 2009.
- ［49］W. R. Hollingshad, U.S. Patent 3932303, 1976.
- ［50］M. Crossman and S. P. R. Holt, U.S. Patent 6995120, 2006.
- ［51］C. Holzner, W. Ohlendorf, H.−D. Block, H. Bertram, R. Kleinstuck, and H.−H. Moretto, U.S. Patent 5639909, 1997.
- ［52］J. L. Hen, U.S. Patent 5059333, 1991.
- ［53］G. Woodward, C. R. Jones, and K. P. Davis, International Patent Application, WO/2004/002994.
- ［54］K. P. Davis, G. F. Docherty, and G. Woodward, International Patent Application WO/2000/018695.
- ［55］K. P. Davis, G. P. Otter, and G. Woodward, International Patent Application WO/2004/078662.
- ［56］N. J. Stewart and P. A. M. Walker, U.S. Patent 6527983, 2003.
- ［57］J. L. Przybylinski, G. T. Rivers, and T. H. Lopez, U.S. Patent Application 20060113505, 2006.
- ［58］T. Johnson, C. Roggelin, C. Simpson, and R. Stalker, "Phosphonate Based Scale Inhibitors for High Iron and High Salinity Environments," Proceedings of Chemistry in the Oil Industry IX, Manchester, UK: Royal Society of Chemistry, 2005, 368.
- ［59］F. A. Devaux, J. H. Van Bree, T. N. Johnson, and P. P. Notte, International Patent Application WO/2008/017338.
- ［60］P. P. Notte and F. A. Devaux, International Patent Application WO/2008/017339.
- ［61］G. Woodward, G. P. Otter, K. P. Davis, and R. E. Talbot, International Patent Application WO/2001/085616.
- ［62］D. Redmore, B. Dhawan, and J. L. Przybylinski, U.S. Patent 4857205, 1989.
- ［63］G. E. Jackson, K. Mclaughlin, N. Poynton, and J. L. Przybylinski, International Patent Application WO 97/21905.
- ［64］（a）G. E. Jackson, G. Salters, P. R. Stead, B. Dahwan, and J. Przybylinski, "Using Statistical

Experimental Design to Optimise the Performance and Secondary Properties of Scale Inhibitors for Downhole Application," *Recent Advances in Oilfield Chemistry V*, Manchester, UK: Royal Society of Chemistry, 1994, 164. (b) M. A. Singleton, J. A. Collins, N. Poynton, and H. J. Formston, "Developments in PhosphonoMethylated PolyAmine (PMPA) Scale Inhibitor Chemistry for Severe $BaSO_4$ Scaling Conditions," SPE 60216 (paper presented at the SPE International Symposium on Oilfield Scale, Aberdeen, UK, 26−27 January 2000).

[65] T. L. Herrera, M. Guzmann, K. Neubecker, and A. Göthlich, International Patent Application WO/2008/095945.

[66] E. A. Kerr and J. Rideout, U.S. Patent 5604291, 1997.

[67] (a) D. H. Emmons, D. W. Fong, and M. A. Kinsella, U.S. Patent 5213691, 1993. (b) G. B. Clubley and J. Rideout, European Patent EP0479465, 1994.

[68] J. F. Pardue and J. F. Kneller, U.S. Patent 5018577, 1991.

[69] M. J. Smith, P. Miles, N. Richardson, and M. A. Finan, U.K. Patent Application GB 1458235, 1976.

[70] K. P. Davis, G. Otter, and G. Woodward, International Patent Application WO/2001/057050.

[71] K. Davis, G. Woodward, J. Hardy, K. Carmichael, and G. Otter, International Patent Application WO/2005/023904.

[72] G. Woodward, G. Otter, S. Zafar, D. Bendejaco, and O. Anthony, International Patent Application WO/2007/017647.

[73] G. Otter, S. Zafar, and G. Woodward, International Patent Application WO/2006/032896.

[74] G. Woodward, G. P. Otter, K. P. Davis, and K. Huan, International Patent Application WO/2004/056886.

[75] (a) K. P. Davis, D. R. E. Walker, G. Woodward, and A. C. Smith, European Patent Application EP0861846, 1998. (b) K. P. Davis, S. D. Fidoe, G. P. Otter, R. E. Talbot, and M. A. Veale, "Novel Scale Inhibitor Polymers with Enhanced Adsorption Properties," SPE 80381 (paper presented at the SPE International Symposium on Oilfield Scale, Aberdeen, UK, 29−30 January 2003).

[76] Available at http://www.wateradditives.com.

[77] J. E. Losasso, U.S. Patent 6322708, 2001.

[78] C. Gancet, R. Pirri, B. Boutevin, C. Loubat, and J. Lepetit, U.S. Patent 6900171, 2005.

[79] M. Guzmann, J. Rieger, T. L. Herrera, and K.−H. Buechner, International Patent Application WO2007000398, 2007.

[80] M. Guzmann, Y. Liu, R. Konrad, and D. Franz, International Patent Application WO/2007/125073.

[81] C. A. Costello and G. F. Matz, U.S. Patent 4,460,477, 1984.

[82] H. Trabitzsch, J. Frieser, A. Koschik, and H. Plainer, U.S. Patent 4271058, 1981.

[83] P. Chen, T. Hagen, H. Montgomerie, R. Matheson, T. Haaland, B. Juliussen, and R. Benvie, "A Scale Inhibitor Chemistry Developed for Downhole Squeeze Treatments in a Water Sensitive and HTHP Reservoir" (paper presented at the 19th Oilfield Chemical Symposium, Geilo, Norway, 9−12 March 2008).

[84] P. Chen, T. Hagen, H. Montgomerie, R. Wat, and O. M. Selle, International Patent Application WO/2008/020220.

[85] (a) N. Kohler, G. Courbinm, C. Estievenart, and F. Ropital, "Polyaspartates: Biodegradable Alternatives to Polyacrylates or Noteworthy Multifunctional Inhibitors?" Paper 02411 (paper presented at the NACE CORROSION Conference, 2002). (b) Z. Quan, Y. Chen, X. Wang, C. Shi, Y. Liu, and C. Ma, "Experimental Study on Scale Inhibition Performance of a Green Scale Inhibitor Polyaspartic Acid," *Science in China Series B: Chemistry* 51 (7) (2008): 695−699. (c) L. P. Kosan and K. C. Low, U.S.

Patent 511651, 1992.

[86] (a) R. J. Ross, K. C. Low, and J. E. Shannon, "Polyaspartate Scale Inhibitors: Biodegradable Alternatives to Polyacrylates," *Materials Performance* 36 (1997): 53. (b) A. P. Wheeler and L. P. Koskan, "Large Scale Thermally Synthesized Polyaspartate as a Substitute in Polymer Applications," *Materials Research Society Symposium Proceedings* 292 (1993): 277. (c) K. C. Low, A. P. Wheeler, and L. P. Koskan, "Commercial Poly (aspartic Acid) and Its Uses," *Advances in Chemistry Series* 248, Washington D.C.: American in Chemical Society, 1996.

[87] I. R. Collins, B. Hedges, L. M. Harris, J. C. Fan, and L. D. G. Fan, "The Development of a Novel Environmentally Friendly Dual Function Corrosion and Scale Inhibitor," SPE 65005 (paper presented at the SPE International Symposium on Oilfield Chemistry, Houston, TX, 13-16 February 2001).

[88] R. Kleinstuck, H. Sicius, T. Groth, and W. Joentgen, U.S. Patent 5525257, 1996.

[89] P. Chen, T. Hagen, A. Maclean, O. M. Selle, K. Stene, and R. Wat, "Meeting the Challenges of Downhole Scale Inhibitor Selection for an Environmentally Sensitive Part of the Norwegian North Sea," SPE 76452 (paper presented at the SPE Oilfield Scale Symposium, Aberdeen, UK, 30-31 January 2002).

[90] H. T. R. Montgomerie, P. Chen, T. Hagen, R. Wat, O. M. Selle, and H. K. Kotlar, International Patent Application WO/2004/011771, 2004.

[91] H. K. Kotlar and J. A. Haugen, International Patent Application WO/2002/095187.

[92] (a) J. Tang and R. V. Davis, U.S. Patent 5776875, 1998. (b) P. Y. Zhang, L. P. Zheng, and Y. J. Chai, *Polymer Materials* 25 (2008): 259. (c) Z. H. Quan, Y. C. Chen, and X. R. Wang, *Science in China Series B Chemistry* 51 (7)(2008): 695.

[93] D. I. Bain, G. Fan, J. Fan, and R. J. Ross, "Scale and Corrosion Inhibition by Thermal Polyaspartates," Paper No. 120 (paper presented at the NACE CORROSION Conference, 1999).

[94] D. I. Bain, G. Fan, J. Fan, H. Brugman, and K. Enoch, "Laboratory and Field Development of a Novel Environmentally Acceptable Scale and Corrosion Inhibitor," Paper No. 2230 (paper presented at the NACE CORROSION Conference, 2002).

[95] J. Tang, R. T. Cunningham, and B. Yang, U.S. Patent 5750070, 1998.

[96] C. G. Carter, L.-D. G. Fan, J. C. Fan, R. P. Kreh, and V. Jovancicevic, U.S. Patent 5344590, 1994.

[97] (a) T. Saeki, A. Kanzaki, J. Nakamura, G. Fukii, and S. Yyamaguchi, Japanese Patent JP2001163941, 2001. (b) T. Saeki, H. Nishibayashi, T. Hirata, and S. Yamaguchi, U.S. Patent 5856288, 1999.

[98] (a) J. M. Brown and G. F. Brock, U.S. Patent 5409062, 1995. (b) R. C. Xiong, Q. Zhou, and G. Wei, "Corrosion Inhibition of a Green Scale Inhibitor Polyepoxysuccinic Acid," *Chinese Chemical Letters* 14(9) (2003): 955-957.

[99] D. L. Verraest, J. P. Peters, H. van Bekkum, and G. M. van Rosmalen, "Carboxymethyl Inulin: A New Inhibitor for Calcium Carbonate Precipitation," *Journal of the American Oil Chemists Society* 73 (1) (1996): 55.

[100] H. C. Kuzee and H. W. C. Raaijmakers, U.S. Patent 6613899, 2003.

[101] D. L. Verraest, J. G. Batelaan, J. A. Peters, and H. van Bekkum, U.S. Patent 5777090, 1998.

[102] (a) B. Bazin, N. Kohler, A. Zaitoun, T. Johnson, and H. Raaijmakers, "A New Class of Green Mineral Scale Inhibitors for Squeeze Treatments," SPE 87453 (paper presented at the SPE International Symposium on Oilfield Scale, Aberdeen, UK, 26-27 May 2004). (b) S. Baraka-Lokmane, K. S. Sorbie, and N. Poisson, "The Use of Green Scale Inhibitors for Squeeze Treatments, Carbonate Coreflooding Experiments," *Geophysical Research Abstracts* 9 (2007): 02444. (c) S. Baraka-

Lokmane, K. Sorbie, N. Poisson, and N. Kohler, *Petroleum Science and Technology* 27(2009): 427.

[103] (a) K. A. Rodrigues, J. S. Thomaides, A. L. Cimecioglu, and M. Crossman, U.S. Patent Application 20070015678, 2007. (b) F. Decampo, S. Kesavan, and G. Woodward, International Patent Application WO/2008/140729. (c) S. Baraka-Lokmane, K. Sorbie, N. Poisson, and N. Kohler, *Petroleum Science and Technology* 27(2009): 427.

[104] (a) A. Viloria, L. Castillo, J. A. Garcia, and J. Biomorgi, U.S. Patent Application 20070281866, 2007. (b) S. P. Holt, J. Sanders, K. A. Rodrigues, and M. Vanderhoof, "Biodegradable Alternatives for Scale Control in Oil Field Applications," SPE 121723 (paper presented at the SPE International Symposium on Oilfield Chemistry, The Woodlands, TX, 20-22 April 2009).

[105] M. H. Salimi, K. C. Petty, and C. L. Emmett, U.S. Patent 5263539, 1993.

[106] L. J. Persinski, P. H. Ralston, and R. C. Gordon Jr., U.S. Patent 3928196, 1975.

[107] G. F. Matz, U.S. Patent 4536292, 1985.

[108] D. O. Falk, U.S. Patent 5360065, 1994.

[109] R. Pirri, C. Hurtevent, and P. Leconte, "New Scale Inhibitor for Harsh Field Conditions," SPE 60218 (paper presented at the International Symposium on Oilfield Scale, Aberdeen, UK, 26-27 January 2000).

[110] R. Wat, L.-E. Hauge, K. Solbakken, K. E. Wennberg, L. M. Sivertsen, and B. Gjersvold, "Squeeze Chemical for HT Applications—Have We Discarded Promising Products by Performing Unrepresentative Thermal Aging Tests?" SPE 105505 (paper presented at the SPE International Symposium on Oilfield Chemistry, Houston, TX, 28 February-2 March 2007).

[111] M. M. Jordan, K. Mackin, C. J. Johnston, and N. D. Feasey, "Control of Hydrogen Sulphide Scavenger Induced Scale and the Associated Challenge of Sulphide Scale Formation within a North Sea High Temperature/High Salinity Field Production Wells: Laboratory Evaluation to Field Application," SPE 87433 (paper presented at the SPE 6th International Symposium on Oilfield Scale, Aberdeen, UK, 26-27 May 2004).

[112] K. C. Taylor, H. A. Nasr-El-Din, and J. A. Saleem, "Laboratory Evaluation of Iron-Control Chemicals for High-Temperature Sour-Gas Wells," SPE 65010 (paper presented at the SPE International Symposium on Oilfield Chemistry, Houston, TX, 13-16 February 2001).

[113] J. A. Billman, "Antibiofoulants: A Practical Methodology for Control of Corrosion Caused by Sulfate-Reducing Bacteria," *Materials Performance* 36(1997): 43.

[114] S. D. Fidoe, R. E. Talbot, C. R. Jones, and R. Gabriel, U.S. Patent 6926836, 2005.

[115] M. A. Mattox and E. J. Valente, U.S. Patent 6986358, 2006.

[116] (a) M. Ke and Q. Qu, U.S. Patent 7159655, 2007. (b) T. Chen, H. Montgomerie, P. Chen, T. H. Hagen, and S. Kegg, "Development of Environmentally Friendly Iron Sulfide Inhibitors and Field Application," (paper presented at the SPE International Symposium on Oil Field Chemistry, The Woodlands, TX, 20-22 April 2009).

[117] C. Okocha, K. S. Sorbie, and L. S. Boak, "Inhibition Mechanisms for Sulfide Scales," SPE 112538 (paper presented at the SPE International Symposium and Exhibition on Formation Damage Control, Lafayette, LA, 13-15 February 2008).

[118] M. Lehmann and F. Firouzkouhi, "A New Chemical Treatment to Inhibit Iron Sulfide Deposition," SPE 114065 (paper presented at the SPE International Oilfield Scale Conference, Aberdeen, UK, 28-29 May 2008).

[119] D. Emmons and G. R. Chesnut, U.S. Patent 4762626, 1988.

[120] I. R. Collins and M. M. Jordan, "Occurrence, Prediction, and Prevention of Zinc Sulfide Scale within Gulf Coast and North Sea High-Temperature and High-Salinity Fields," SPE 84963, *SPE Production & Facilities* 18(3)(2003): 200–209.

[121] T. H. Lopez, M. Yuan, D. A. Williamson, and J. L. Przybylinski, "Comparing Efficacy of Scale Inhibitors for Inhibition of Zinc Sulfide and Lead Sulfide Scales," SPE 95097 (paper presented at the SPE International Symposium on Oilfield Scale, Aberdeen, UK, 11–12 May 2005).

[122] J. L. Przybylinski, "Iron Sulfide Scale Deposit Formation and Prevention under Anaerobic Conditions Typically Found in the Oil Field," SPE 65030 (paper presented at the SPE International Symposium on Oilfield Chemistry, Houston, TX, 13–16 February 2001).

[123] S. Dyer, K. Orski, C. Menezes, S. Heath, C. MacPherson, C. Simpson, and G. Graham, "Development of Appropriate Test Methodologies for the Selection and Application of Lead and Zinc Sulfide Inhibitors for the Elgin/Franklin Field," SPE 100627 (paper presented at the SPE International Oilfield Scale Symposium, Aberdeen, UK, 31 May–1 June 2006).

[124] W. Kleinitz, M. Koehler, and G. Dietzsch, "The Precipitation of Salt in Gas Producing Wells," SPE 68953 (paper presented at the SPE European Formation Damage Conference, The Hague, Netherlands, 21–22 May 2001).

[125] (a) H. Guan, R. Keatch, C. Benson, N. Grainger, and L. Morris, "Mechanistic Study of Chemicals Providing Improved Halite Inhibition," SPE 114058 (paper presented at the SPE 9th International Conference on Oilfield Scale, Aberdeen, UK, 28–29 May 2008). (b) H. Guan, R. Keatch, N. Grainger, and C. Benson, "Development of a Novel Salt Inhibitor" (paper presented at the 19th Oil Field Chemistry Symposium, Geilo, Norway, 9–12 March 2008).

[126] (a) T. Chen, H. Montgomerie, P. Chen, O. Vikane, and T. Jackson, "Development of Halite Test Methodology, Inhibitors, and Field Application," International Oilfield Chemistry Symposium, Geilo, Norway, 23–25 March 2009. (b) T. Chen, H. Montgomerie, P. Chen, O. Vikane, and T. Jackson, "Understanding the Mechanisms of Halite Inhibition and Evaluation of Halite Scale Inhibitor by Static and Dynamic Tests," SPE 121458 (poster presented at SPE International Symposium on Oilfield Chemistry, The Woodlands, TX, 20–22 April 2009).

[127] M. A. Damme-van Weele, PhD thesis, Twente University of Technology, Twente, Netherlands, 1965.

[128] M. C. Van der Leeden, and G. M. van Rosmalen, *Chemicals in the Oil Industry*, Manchester, UK: Royal Society of Chemistry, 1988, 68.

[129] C. Rodriguez-Navarro, L. Linares-Fernandez, E. Doehne, and E. Sebastian, "Effects of Ferrocyanide Ions on NaCl Crystallization in Porous Stone," *Journal of Crystal Growth* 243(2002): 503.

[130] D. M. Frigo, L. A. Jackson, S. M. Dora, and R. A. Trompert, SPE 60191 (paper presented at the SPE Second International Symposium on Oilfield Scale, Aberdeen, UK, 26–27 January 2000).

[131] N. H. Zaid, U.S. Patent 5396958, 1995.

[132] D. A. Smith, S. Sucheck, S. Cramer, and D. Baker, "Nitrilotriacetamide: Synthesis in Concentrated Sulfuric Acid and Stability in Water," *Synthetic Communications* 25(24)(1995): 4123.

[133] J. W. Kirk, U.S. Patent 7028776, 2006.

[134] P. H. Ralston and L. J. Persinski, U.S. Patent 3367416, 1968.

[135] S. Sarig, A. Glasner, and J. A. Epstein, "Crystal Habit Modifiers," *Journal of Crystal Growth* 28(3)(1975): 295.

[136] S. Sarig and F. Tartakovsky, "Crystal Habit Modifiers: II. The Effect of Supersaturation on Dendritic Growth," *Journal of Crystal Growth* 28(3)(1975): 300.

[137] J. W. Kirk and J. B. Dobbs, "A Protocol to Inhibit the Formation of Natrium Chloride Salt Blocks," SPE 74662 (paper presented at the SPE Oilfield Scale Symposium, Aberdeen UK, 30–31 January 2002).

[138] S. Szymczak, R. Perkins, M. McBryde, and M. El-Sedaway, "Salt Free: A Case History of a Chemical Application to Inhibit Salt Formation in a North African Field," SPE 102627 (paper presented at the SPE International Symposium on Oilfield Chemistry, Houston, TX, 28 February–2 March 2007).

[139] S. M. Heath, A. R. Thornton, E. K. McAra, M. Sim, A. Arefjord, E. Samuelsen, and R. Frederiksen, "Downhole Scale Control on Amerada Hess South Arne through Continuous Injection of Scale Inhibitor in the Water Injection System" (paper presented at the International Oilfield Chemistry Symposium, Geilo, Norway, 23–25 March 2009).

[140] (a) G. Poggesi, C. Hurtevent, and D. Buchart, "Multifunctional Chemicals for West African Deep Offshore Fields," SPE 74649 (paper presented at the International Symposium on Oilfield Scale, Aberdeen, UK, 30–31 January 2002). (b) N. Fleming, K. Ramstad, S. H. Eriksen, E. Moldrheim, and T. R. Johansen, "Development and Implementation of a Scale-Management Strategy for Oseberg Soer," SPE 100371 (paper presented at the SPE International Oilfield Scale Symposium, Aberdeen, UK, 31 May–1 June 2006).

[141] I. Fjelde, "Scale Inhibitor Injection in Gas-Lift Systems" (paper presented at the 11th International Oil Field Chemicals Symposium, Fagernes, Norway, 19–22 March 2000).

[142] N. Fleming, J. A. Stokkan, A. M. Mathisen, K. Ramstad, and T. Tydal, "Maintaining Well Productivity through Deployment of a Gas Lift Scale Inhibitor: Laboratory and Field Challenges," SPE 80374 (paper presented at the International Symposium on Oilfield Scale, Aberdeen, UK, 29–30 January 2003).

[143] G. Poggesi, C. Hurtevent, and J. L. Brazy, "Scale Inhibitor Injection via the Gas Lift System in High Temperature Block 3 Fields in Angola," SPE 68301 (paper presented at the SPE International Symposium on Oilfield Scale, Aberdeen, UK, 30–31 January 2001).

[144] M. N. Kelly, J. S. James, D. M. Frigo, D. W. Driessen, and A. D. Waldie, "Application of Scale Dissolver and Inhibitor Squeeze through the Gas Lift Line in a Sub-sea Field," SPE 95100 (paper presented at the SPE International Symposium on Oilfield Scale, Aberdeen, UK, 11–12 May 2005).

[145] A. Daminov, V. Ragulin, and A. Voloshin, "Mechanism Formations of Corrosion Damage of Inter Equipment in Wells by Continuous Scale-Inhibitor Dosing Utilizing Surface Dosing Systems: Testing Scale and Corrosion Inhibitors," SPE 100476 (paper presented at the SPE International Oilfield Corrosion Symposium, Aberdeen, UK, 30 May 2006).

[146] K. S. Sorbie and R. D. Gdanski, "A Complete Theory of Scale-Inhibitor Transport and Adsorption/Desorption in Squeeze Treatments," SPE 95088 (paper presented at the SPE International Symposium on Oilfield Scale, Aberdeen, UK, 11–12 May 2005).

[147] E. J. Mackay and K. S. Sorbie, "An Evaluation of Simulation Techniques for Modelling Squeeze Treatments," SPE 56775 (paper presented at the SPE Annual Technical Conference and Exhibition, Houston, TX, 3–6 October 1999).

[148] A. T. Kan, G. Fu, M. Al-Thubaiti, J. Xiao, and M. B. Tomson, "A New Approach to Inhibitor Squeeze Design," SPE 80230 (paper presented at the SPE International Symposium on Oilfield Chemistry, Houston, TX, 5–7 February 2003).

[149] M. B. Tomson, A. T. Kan, and G. Fu, "Control of Inhibitor Squeeze through Mechanistic Understanding of Inhibitor Chemistry," *SPE Journal* 11 (3)(2006): 283–293.

[150] (a) E. Hills, S. Touzet, and B. Langlois, International Patent Application WO/2005/001241. (b) K. Du, Y. M. Zhou, and L. Y. Dai, *International Journal of Polymer Materials* 57 (2008): 785.

[151] J. E. Pardue, "A New Inhibitor for Scale Squeeze Applications," SPE 21023 (paper presented at the SPE International Symposium on Oilfield Chemistry, Anaheim, CA, 20-22 February 1991).

[152] P. A. Read and T. Schmidt, U.S. Patent 5090479, 1992.

[153] M. M. Jordan, K. S. Sorbie, G. M. Graham, K. Taylor, K. E. Hourston, and S. Hennessey, "The Correct Selection and Application Methods for Adsorption and Precipitation Scale Inhibitors for Squeeze Treatments in North Sea Oilfields," SPE 31125 (paper presented at the SPE Formation Damage Control Symposium, Lafayette, LA, 14-15 February 1996).

[154] M. M. Jordan, K. S. Sorbie, P. Chen, P. Armitage, P. Hammond, and K. Taylor, "The Design of Polymer and Phosphonate Scale Inhibitor Precipitation Treatments and the Importance of Precipitate Solubility in Extending Squeeze Lifetime," SPE 37275 (paper presented at the SPE International Symposium on Oilfield Chemistry, Houston, TX, 18-21 February 1997).

[155] D. C. Berkshire, J. B. Lawson, and E. A. Richardson, U.S. Patent 4357248, 1982.

[156] (a) M. R. Rabaioli and T. P. Lockhart, *Journal of Petroleum Science and Engineering* 15 (1996): 1156. (b) M. B. Tomson, A. T. Kan, G. Fu, D. Shen, H. A. Nasr-El-Din, H. Al-Saiari, and M. Al-Thubaiti, "Mechanistic Understanding of the Rock/Phosphonate Interactions and the Effect of Metal Ions on Inhibitor Retention," *SPE Journal* 13 (2008): 325.

[157] E. S. Snavely Jr. and J. Hen, U.S. Patent 4787455, 1988.

[158] A. T. Kan, G. Fu, D. Shen, M. B. Tomson, and H. A. AlSaiari, "Enhanced Inhibitor Treatments with the Aid of Transition Metal Ions," SPE 114060 (paper presented at the SPE 9th International Conference on Oilfield Scale, Aberdeen, UK, 28-29 May 2008).

[159] F. F. D. Rosario, C. N. Khalil, M. C. Bezerra, and S. B. Rondinini, U.S. Patent 5840658, 1998.

[160] J. Hen, International Patent Application WO94/03706.

[161] J. Hen, A. Brunger, B. K. Peterson, M. D. Yuan, and J. P. Renwick, "A Novel Scale Inhibitor Chemistry for Downhole Squeeze Application in High Water Producing North Sea Wells," SPE 30410 (paper presented at the SPE Offshore Europe, Aberdeen, UK, 5-8 September 1995).

[162] H. M. Bourne, G. D. M. Williams, J. Ray, and A. Morgan, "Extending Squeeze Lifetime through In-Situ pH Modification—Laboratory and Field Experience" (paper presented at the 8th International Oil Field Chemical Symposium, Geilo, Norway, 2-5 March 1997).

[163] R. J. Faircloth and J. B. Lawson, U.S. Patent 5211237, 1993.

[164] J. D. Lynn and H. A. Nasr-El-Din, "A Novel Low-Temperature, Forced Precipitation Phosphonate Squeeze for Water Sensitive, Non-carbonate Bearing Formations," SPE 84404 (paper presented at the SPE Annual Technical Conference and Exhibition, Denver, CO, 5-8 October 2003).

[165] J. A. McRae, S. M. Heath, C. Strachan, L. Matthews, and R. Harris, "Development of an Enzyme Activated, Low Temperature, Scale Inhibitor Precipitation Squeeze System," SPE 87441 (paper presented at the SPE International Symposium on Oilfield Scale, Aberdeen, UK, 26-27 May 2004).

[166] I. R. Collins, International Patent Application WO98/30783, 1998.

[167] R. G. Chapman, I. R. Collins, S. P. Goodwin, A. R. Lucy, and N. J. Stewart, U.S. Patent 5690174, 1997.

[168] I. R. Collins, L. G. Cowie, M. Nicol, and N. J. Stewart, "The Field Application of a Scale Inhibitor Squeeze Enhancing Additive," SPE 38765 (paper presented at the SPE Annual Technical Conference and Exhibition, San Antonio, TX, 5-8 October 1997).

[169] M. M. Jordan, C. J. Graff, and K. N. Cooper, "Development and Deployment of a Scale Squeeze Enhancer and Oil-Soluble Scale Inhibitor to Avoid Deferred Oil Production Losses During Squeezing

Low-Water Cut Wells, North Slope, Alaska," SPE 58725 (paper presented at the SPE International Symposium on Formation Damage Control, Lafayette, LA, 23-24 February 2000).

[170] I. R. Collins, L. G. Cowie, M. Nicol, and N. J. Stewart, "Field Application of a Scale Inhibitor Squeeze Enhancer Additive," SPE 54525 (paper presented at the SPE Annual Technical Conference, San Antonio, TX, 5-8 October 1997).

[171] I. R. Collins, S. P. Goodwin, J. C. Morgan, and N. J. Stewart, U.S. Patent 6225263, 2001.

[172] H. M. Bourne, S. L. Booth, and A. Brunger, "Combining Innovative Technologies to Maximize Scale Squeeze Cost Reduction," SPE 50718 (paper presented at the SPE International Symposium on Oilfield Chemistry, Houston, TX, 16-19 February 1999).

[173] H. T. R. Montgomerie, P. Chen, T. Hagen, R. Wat, O. M. Selle, and H. K. Kotlar, International Patent Application WO/2004/011772.

[174] O. M. Selle, R. M. S. Wat, O. Vikane, H. Nasvik, P. Chen, T. Hagen, H. Montgomerie, and H. Bourne, "A Way Beyond Scale Inhibitors, Extending Scale Inhibitor Squeeze Life through Bridging," SPE 80377 (paper presented at the SPE 5th International Symposium on Oilfield Scale, Aberdeen, Scotland, 29-30 May 2003).

[175] P. Chen, X. Yan, T. Hagen, and H. T. R. Montgomerie, International Patent Application WO/2007/015090, 2007.

[176] H. Montgomerie, P. Chen, T. Hagen, O. Vikane, R. Matheson, V. Leirvik, C. Froytlog, and J. O. Saeten, "Development of a New Polymer Inhibitor Chemistry for Downhole Squeeze Applications," SPE 113926 (paper presented at the SPE International Oilfield Scale Conference, Aberdeen, UK, 28-29 May 2008).

[177] J. Hen, U.S. Patent 5089150, 1992.

[178] T.-Y. Yan, International Patent Application WO93/05270, 1993.

[179] O. Bache and S. Nilsson, "Ester Cross-Linking of Polycarboxylic Acid Scale Inhibitors as a Possible Means to Increase Inhibitor Squeeze Lifetime," SPE 60190 (paper presented at the SPE International Symposium on Oilfield Scale, Aberdeen, UK, 26-27 January 2000).

[180] N. Fleming, K. Ramstad, A.-M. Mathisen, A. Nelson, and S. Kidd, "Innovative Use of Kaolinite in Downhole Scale Management: Squeeze Life Enhancement and Water Shutoff," SPE 113656 (paper presented at the SPE 9th International Conference on Oilfield Scale, Aberdeen, UK, 28-29 May 2008).

[181] (a) D. R. Watkins, J. J. Clemens, J. C. Smith, S. N. Sharma, and H. G. Edwards, U.S. Patent 5224543, 1993. (b) M. Garcia-Lopez de Victoria, J. W. Still, and T. Bui, U.S. Patent Application 20090025933.

[182] K. Cheremisov, D. Oussoltsev, K. K. Butula, A. Gaifullin, I. Faizullin, and D. Senchenko, "First Application of Scale Inhibitor during Hydraulic Fracturing Treatments in Western Siberia," SPE 114255 (paper presented at the SPE International Oilfield Scale Conference, Aberdeen, UK, 28-29 May 2008).

[183] A. M. Fitzgerald and L. G. Cowie, "A History of Frac Pack Scale Inhibitor Deployment," SPE 112474 (paper presented at the SPE International Symposium and Exhibition on Formation Damage Control, Lafayette, LA, 13-15 February 2008).

[184] O. J. Vetter, S. Lankford, and T. Nilssen, "Well Stimulations and Scale Inhibitors," SPE 17284 (paper presented at the SPE Permian Basin Oil and Gas Recovery Conference, Midland, TX, 10-11 March 1988).

[185] M. Norris, D. Perez, H. M. Bourne, and S. M. Heath, "Maintaining Fracture Performance through Active Scale Control," SPE 68300 (paper presented at the SPE International Symposium on Oilfield Scale, Aberdeen, UK, 30-31 January 2001).

[186] S. Szymczak, G. Brock, J. M. Brown, D. Daulton, and B. Ward, "Beyond the Frac: Using the Fracture Process as the Delivery System for Production Chemicals Designed to Perform for Prolonged Periods of Time," SPE 107707 (paper presented at the SPE Rocky Mountain Oil & Gas Technology Symposium, Denver, CO, 16–18 April 2007).

[187] S. Szymczak, J. M. Brown, S. Noe, and G. Gallup, "Long-Term Scale Inhibition Using a Solid Scale Inhibitor in a Fracture Fluid," SPE 102720 (paper presented at the SPE Annual Technical Conference and Exhibition, San Antonio, TX, 24–27 September 2006).

[188] R. J. Powell, R. D. Gdanski, M. A. McCabe, and D. C. Buster, "Controlled-Release Scale Inhibitor for Use in Fracturing Treatments," SPE 28999 (paper presented at the SPE International Symposium on Oilfield Chemistry, San Antonio, TX, 14–17 February 1995).

[189] D. V. S. Gupta, J. M. Brown, and S. Szymczak, "Multi-year Scale Inhibition from a Solid Inhibitor Applied during Stimulation," SPE 115655 (paper presented at the SPE Annual Technical Conference and Exhibition, Denver, CO, 21–24 September 2008).

[190] S. M. Heath and H. M. Bourne, U.S. Patent 7196040, 2007.

[191] M. Tyndall, L. Maschio, B. C. O. Bustos, and B. Lungwitz, "Simultaneous Fracturing and Inhibitor Squeeze Treatments in Gas Reservoirs: A New Approach to Tailor Chemistry to Reservoir Conditions," *Chemistry in the Oil Industry X: Oilfield Chemistry*, RSC, 5–7 November 2007.

[192] P. S. Smith, C. C. Clement Jr., and A. Mendoza Rojas, "Combined Scale Removal and Scale Inhibition Treatments," SPE 60222 (paper presented at the SPE 2nd International Symposium on Oilfield Scale, Aberdeen, UK, 26–27 January 2000).

[193] P. S. Smith, L. G. Cowie, H. M. Bourne, M. Grainger, and S. M. Heath, "Field Experiences with a Combined Acid Stimulation and Scale Inhibition Treatment," SPE 68312 (paper presented at the SPE International Symposium on Oilfield Scale, Aberdeen, UK, 30–31 January 2001).

[194] R. T. Barthorpe, "The Impairment of Scale Inhibitor Function by Commonly Used Organic Anions," SPE 25158 (paper presented at the SPE International Symposium on Oilfield Chemistry, New Orleans, LA, 2–5 March 1993).

[195] R. Castano, J. Villamizar, O. Diaz, S. A. Hocol, M. Avila, S. Gonzalez, E. D. Dalrymple, S. Milson, and D. Everett, "Relative Permeability Modifier and Scale Inhibitor Combination in Fracturing Process at San Francisco Field in Colombia, South America," SPE 77412 (paper presented at the SPE Annual Technical Conference and Exhibition, San Antonio, TX, 29 September–2 October, 2002).

[196] P. Powell, M. A. Singleton, and K. S. Sorbie, U.S. Patent 6913081, 2005.

[197] I. R. Collins, T. Jones, and C. G. Osborne, International Patent Application WO/2004/016906, 2004.

[198] A. F. Miles, O. Vikane, D. S. Healey, I. R. Collins, J. Saeten, H. M. Bourne, and R. G. Smith, "Field Experiences Using 'Oil Soluble' Non-Aqueous Scale Inhibitor Delivery Systems," SPE 87431 (paper presented at the SPE International Symposium on Oilfield Scale, Aberdeen, UK, 26–27 May 2004).

[199] M. M. Jordan, I. R. Collins, A. Gyani, and G. M. Graham, "Coreflood Studies Examine New Technologies that Minimize Intervention throughout Well Life Cycle," SPE 74666, *SPE Production & Operations* 21 (2)(2006): 161–173.

[200] H. Guan, K. S. Sorbie, and E. J. Mackay, "The Comparison of Non-aqueous and Aqueous Scale-Inhibitor Treatments: Experimental and Modelling Studies," SPE 87445, *SPE Production & Operations* 21 (4)(2006): 419.

[201] J. M. Reizer, M. G. Rudel, C. D. Sitz, R. M. S. Wat, and H. Montgomerie, U.S. Patent 6379612, 2002.

[202] J. M. Reizer, M. G. Rudel, C. D. Sitz, R. M. S. Wat, H. Montgomerie, and A. F. Miles, U.S. Patent

Application, 20020150499, 2002.

[203] R. Wat, H. Montgomerie, T. Hagen, R. Boereng, H. K. Kotlar, and O. Vikane, "Development of an Oil- Soluble Scale Inhibitor for a Subsea Satellite Field," SPE 50706 (paper presented at the SPE International Symposium on Oilfield Chemistry, Houston, TX, 16-19 February 1999).

[204] N. J. Jenvey, A. F. MacLean, A. F. Miles, and H. T. R. Montgomerie, "The Application of Oil Soluble Scale Inhibitors into the Texaco Galley Reservoir: A Comparison with Traditional Squeeze Techniques to Avoid Problems Associated with Wettability Modification in Low Water-Cut Wells," SPE 60197 (paper presented at the SPE International Symposium on Oilfield Scale, Aberdeen, UK, 26-27 January 2000).

[205] S. M. Heath, J. J. Wylde, M. Archibald, M. Sim, and I. R. Collins, "Development of Oil Soluble Precipitation Squeeze Technology for Application in Low and High Water Cut Wells," SPE 87451 (paper presented at the SPE International Symposium on Oilfield Scale, Aberdeen, UK, 26-27 May 2004).

[206] J. J. Wylde and E. K. McAra, "Optimization of an Oil Soluble Scale Inhibitor for Minimizing Formation Damage: Laboratory and Field Studies," SPE 86477 (paper presented at the SPE International Symposium and Exhibition on Formation Damage Control, Lafayette, LA, 18-20 February 2004).

[207] T. A. Lawless and R. N. Smith, "New Technology, Invert Emulsion Scale Inhibitor Squeeze Design," SPE 50705 (paper presented at the SPE International Symposium on Oilfield Chemistry, Houston, TX, 16-19 February 1999).

[208] S. Beare, "The Development and Laboratory Evaluation of an Emulsified Scale Inhibitor Squeeze Treatment" (paper presented at the NIF 11th International Oilfield Chemical Symposium, Fagernes, Norway, March 2000).

[209] C. Romero, B. Bazin, A. Zaitoun, and F. Leal-Calderon, "Behavior of a Scale Inhibitor Water-in-Oil Emulsion in Porous Media," SPE 98275 (paper presented at the SPE International Symposium and Exhibition on Formation Damage Control, Lafayette, LA, 15-17 February 2006).

[210] M. M. Jordan, F. Murray, A. Kelly, and K. Stevens, "Deployment of Emulsified Scale—Inhibitor Squeeze to Control Sulfate/Carbonate Scales within Subsea Facilities in the North Sea Basin," SPE 80249 (paper presented at the International Symposium on Oilfield Chemistry, Houston, TX, 5-7 February 2003).

[211] I. R. Collins, International Patent Application WO/2001/046553.

[212] A. F. Miles, H. M. Bourne, R. G. Smith, and I. R. Collins, "Development of a Novel Water in Oil Microemulsion Based Scale Inhibitor Delivery System," 80390 (paper presented at the International Symposium on Oilfield Scale, Aberdeen, UK, 29-30 January 2003).

[213] R. Collins and I. Vervoort, U.S. Patent 6581687, 2003.

[214] L. Del Gaudio, R. Bortolo, and T. P. Lockhart, "Nanoemulsions: A New Vehicle for Chemical Additive Delivery," SPE 106016 (paper presented at the International Symposium on Oilfield Chemistry, Houston, TX, 28 February-2 March 2007).

[215] I. R. Collins, M. M. Jordan, and S. E. Taylor, "The Development and Application of a Novel Scale Inhibitor Deployment System," SPE 80286, *SPE Production & Facilities* 17 (4)(2002): 221.

[216] I. R. Collins, M. M. Jordan, and S. E. Taylor, "The Development and Application of a Novel Scale Inhibitor for Deployment in Low Water Cut, Water Sensitive or Low Pressure Reservoirs," SPE 60192 (paper presented at the SPE International Symposium on Oilfield Scale, Aberdeen, UK, 26-27 January 2000).

[217] I. R. Collins, M. M. Jordan, N. Feasey, and G. D. Williams, "The Development of Emulsion-Based Production Chemical Deployment Systems," SPE 65026 (paper presented at the SPE International

Symposium on Oilfield Chemistry, Houston, TX, 13-16 February 2001).

[218] H. A. Nasr-El-Din, J. D. Lynn, M. K. Hashem, and G. Bitar, "Field Application of a Novel Emulsified Scale Inhibitor System to Mitigate Calcium Carbonate Scale in a Low Temperature, Low Pressure Sandstone Reservoir in Saudi Arabia," SPE 77768 (paper presented at the SPE Annual Technical Conference and Exhibition, San Antonio, TX, 29 September-2 October 2002).

[219] I. R. Collins, P. D. Ravenscroft, and C. I. Bates, International Patent Application WO97/45625, 1997.

[220] I. R. Collins, S. D. Duncum, M. M. Jordan, and N. D. Feasey, "The Development of a Revolutionary Scale-Control Product for the Control of Near-Well Bore Sulfate Scale within Production Wells by the Treatment of Injection Seawater," SPE 100357 (paper presented at the SPE International Oilfield Scale Symposium, Aberdeen, UK, 31 May-1 June 2006).

[221] H. K. Kotlar, O. M. Selle, O. A. Aune, L. Kilaas, and A. D. Dyrli, International Patent Application WO/2002/040827.

[222] H. K. Kotlar, O. M. Selle, O. A. Aune, L. Kilaas, and A. D. Dyrli, International Patent Application WO/2002/040826.

[223] C. R. Clark, D. L. Whitfill, D. P. Cords, E. F. McBride, and H. E. Bellis, U.S. Patent 4986353, 1991.

[224] L. A. McDougall, J. C. Newlove, and J. A. Haslegrave, U.S. Patent 4670166, 1987.

[225] H. K. Kotlar, O. M. Selle, O. A. Aune, L. Kilaas, and A. D. Dyrli, International Patent Application WO/2002/040828.

[226] I. R. Collins and S. N. Duncum, International Patent Application WO/2003/106810.

[227] P. J. C. Webb, T. A. Nistad, B. Knapstad, P. D. Ravenscroft, and I. R. Collins, "Economic and Technical Features of a Revolutionary Chemical Scale Inhibitor Delivery Method for Fractured and Gravel Packed Wells: Comparative Analysis of Onshore and Offshore Subsea Applications," SPE 39451 (paper presented at the SPE Formation Damage Control Conference, Lafayette, LA, 18-19 February 1998).

[228] H. M. Bourne, S. M. Heath, S. McKay, J. Fraser, L. Stott, and S. Muller, "Effective Treatment of Subsea Wells with a Solid Scale Inhibitor System," SPE 60207 (paper presented at the International Symposium on Oilfield Scale, Aberdeen, UK, 26-27 January 2000).

[229] P. A. Read, European Patent EP656459, 1995.

[230] H. M. Bourne and P. A. Read, International Patent Application WO/1996/027070.

[231] H. K. Kotlar, O. M. Selle, O. A. Aune, L. Kilaas, and A. D. Dyrli, U.S. Patent Application 20040060702.

[232] D. Shen, P. Zhang, A. T. Kan, G. Fu, J. Farrell, and M. B. Tomson, "Control Placement of Scale Inhibitors in the Formation with Stable Ca-DTPMP Nanoparticle Suspension and Its Transport Porous Media," SPE 114063 (paper presented at the SPE International Oilfield Scale Conference, Aberdeen, UK, 28-29 May 2008).

[233] E. J. Mackay and K. S. Sorbie, "Modelling Scale Inhibitor Squeeze Treatments in High Crossflow Horizontal Wells," SPE 50418 (paper presented at the SPE 3rd International Conference on Horizontal Well Technology, Calgary, Alberta, Canada, 1-4 November 1998).

[234] M. M. Jordan, M. C. Edgerton, J. Cole-Hamilton, and K. Mackin, "The Application of Novel Wax Diverter Technology to Allow Successful Scale Inhibitor Squeeze Treatment into a Subsea Horizontal Well, North Sea Basin," SPE 49196 (paper presented at the SPE Annual Technical Conference and Exhibition, New Orleans, LA, 27-30 September 1998).

[235] M. M. Jordan, C. J. Tomlinson, A. R. P. Pritchard, and M. Lewis, "Laboratory Testing and Field Implementation of Scale Inhibitor Squeeze Treatments to Subsea and Platform Horizontal Wells, North

Sea Basin," SPE 50436 (paper presented at the SPE International Conference on Horizontal Well Technology, Calgary, Alberta, Canada, 1–4 November 1998).

[236] M. M. Jordan, K. Sjursaether, and I. R. Collins, "Scale Control within the North Sea Chalk/Limestone Reservoirs—The Challenge of Understanding and Optimizing Chemical–Placement Methods and Retention Mechanisms: Laboratory to Field," SPE 86476, *SPE Production & Facilities* 20 (4)(2005): 262.

[237] J. S. James, D. M. Frigo, M. M. Townsend, G. M. Graham, F. Wahid, and S. M. Heath, "Application of a Fully Viscosified Scale Squeeze for Improved Placement in Horizontal Wells," SPE 94593 (paper presented at the SPE International Symposium on Oilfield Scale, Aberdeen, UK, 11–12 May 2005).

[238] R. Stalker, G. M. Graham, D. Oliphant, and M. Smillie, "Potential Application of Viscosified Treatments for Improved Bullhead Scale Inhibitor Placement in Long Horizontal Wells—A Theoretical and Laboratory Examination," SPE 87439 (paper presented at the SPE International Symposium on Oilfield Scale, Aberdeen, UK, 26–27 May 2004).

[239] R. Stalker, K. Butler, G. Graham, R. Wat, L.–E. Hauge, and K. E. Wennberg, "Evaluation of Polymer Gel Diverters for a High–Temperature Field with Special Focus on the Formation Shape Factor—An Important Parameter for Enhancing Matrix Placement of Stimulation Chemicals," SPE 107806 (paper presented at the European Formation Damage Conference, Scheveningen, Netherlands, 30 May–1 June 2007).

[240] S. M. Heath, M. Sim, M. Archibald, and R. Stalker, "Development of a Viscosified Scale Inhibitor for Improving Placement in Acid Fractured Limestone Reservoirs," SPE 114046 (paper presented at the SPE International Oilfield Scale Conference, Aberdeen, UK, 28–29 May 2008).

[241] O. M. Selle, M. Springer, I. H. Auflem, P. Chen, R. Matheson, A. A. Mebratu, and G. Glasbergen, "Gelled Scale Inhibitor Treatment for Improved Placement in Long Horizontal Wells at Norne and Heidrun Fields," SPE 112464 (paper presented at the SPE International Symposium and Exhibition on Formation Damage Control, Lafayette, LA, 13–15 February 2008).

[242] N. D. Feasey, M. M. Jordan, E. J. Mackay, and I. R. Collins, "The Challenge that Completion Types Present to Scale Inhibitor Squeeze Chemical Placement: A Novel Solution Using a Self–Diverting Scale Inhibitor Squeeze Process," SPE 86478 (paper presented at the SPE International Symposium and Exhibition on Formation Damage Control, Lafayette, LA, 18–20 February 2004).

[243] F. De Campo, A. Colaco, and S. Kesavan, International Patent Application WO/2008/066918.

[244] I. Drela, P. Falewicz, and S. Kuczkowska, "New Rapid Test for Evaluation of Scale Inhibitors," *Water Research* 32 (10)(1998): 3188.

[245] (a) S. J. Dyer and G. M. Graham, *Journal of Petroleum Science and Engineering* 35 (2002): 95.
(b) G. M. Graham, I. R. Collins, R. Stalker, and I. J. Littlehales, "The Importance of Appropriate Laboratory Procedures for the Determination of Scale Inhibitor Performance," SPE 74679 (paper presented at the International Symposium on Oilfield Scale, Aberdeen, UK, 30–31 January 2002).
(c) E. N. Halvorsen, A. K. Halvorsen, K. Reiersølmeon, T. R. Andersen, and C. Bjornstad, "New Method for Scale Inhibitor Testing," SPE 121663 (paper presented at the SPE International Symposium on Oilfield Chemistry, The Woodlands, TX, 20–22 April 2009).

[246] A. M. Pritchard, L. Cowie, J. R. Goulding, G. Graham, A. C. Greig, B. M. Hamblin, A. Hunton, and S. Terry, *Chemicals in the Oil Industry III*. Manchester, UK: Royal Society of Chemistry, 1988, 140.

[247] M. D. Yuan, K. S. Sorbie, P. Jiang, P. Chen, M. M. Jordan, A. C. Todd, K. E. Hourston, and K. Ramstad, "Phosphonate Scale Inhibitor Adsorption on Outcrop and Reservoir Rock Substrates—The

'Static' and 'Dynamic' Adsorption Isotherms," *Recent Advances in Oilfield Chemistry V*, Manchester, UK: Royal Society of Chemistry, 1994, 164.

[248] T. Chen, A. Neville, and M. Yuan, "Calcium Carbonate Scale Formation—Assessing the Initial Stages of Precipitation and Deposition," *Journal of Petroleum Science and Engineering* 46 (2005): 185.

[249] M. Crabtree, D. Eslinger, P. Fletcher, A. Johnson, and G. King, "Fighting Scale—Removal and Prevention," *Oilfield Review* (*Schlumberger*) 11 (1999): 30.

[250] A. Johnson, D. Eslinger, and H. Larsen, "An Abrasive Jetting Removal System," SPE 46026 (paper presented at the SPE/IcoTA Coiled Tubing Roundtable, Houston, TX, 15–16 April 1998).

[251] R. J. Tailby, C. B. Amor, and A. McDonough, "Scale Removal from the Recesses of Side-Pocket Mandrels," SPE 54477 (paper presented at the SPE/IcoTA Coiled Tubing Roundtable, Houston, TX, 25–26 May 1999).

[252] M. S. Mirza and V. Prasad, "Scale Removal in Khuff Gas Wells," SPE 53345 (paper presented at the Middle East Oil Show and Conference, Bahrain, 20–23 February 1999).

[253] T. Huang, P. M. McElfresh, and A. D. Gabrysch, "Acid Removal of Scale and Fines at High Temperatures," SPE 74678 (paper presented at the SPE Oilfield Scale Symposium, Aberdeen, UK, 30–31 January 2002).

[254] S. M. Proctor, "Scale Dissolver Development and Testing for HP/HT Systems," SPE 60221 (paper presented at the SPE International Symposium on Oilfield Scale, Aberdeen, UK, 26–27 January 2000).

[255] H. Williams, R. Wat, P. Chen, T. Hagen, K. Wenneberg, V. Viken, and G. M. Graham, "Scale Dissolver Application under HPHT Conditions—Use of an HPHT 'Stirred Reactor' for In-Situ Scale Dissolver Evaluations," SPE 95127 (paper presented at the SPE International Symposium on Oilfield Scale, Aberdeen, UK, 11–12 May 2005).

[256] O. M. Selle, R. M. S. Wat, H. Nasvik, and A. Mebratu, "Gelled Organic Acid System for Improved $CaCO_3$ Removal in Horizontal Openhole Wells at the Heidrun Field," SPE 90359 (paper presented at the SPE Annual Technical Conference and Exhibition, Houston, TX, 26–29 September 2004).

[257] M. Al-Khaldi, H. A. Nasr-El-Din, S. Metha, and A. Aamri, "Reaction of Citric Acid with Calcite," *Chemical Engineering Science* 62 (2007): 5880.

[258] C. M. Shaughnessy and W. E. Kline, "EDTA Removes Formation Damage at Prudhoe Bay," *Journal of Petroleum Technology* 35 (1983): 1783.

[259] J. S. Rhudy, "Removal of Mineral Scale from Reservoir Core by Scale Dissolver," SPE 25161 (paper presented at the SPE International Symposium on Oilfield Chemistry, New Orleans, LA, 2–5 March 1993).

[260] K. D. Demadis, P. Lykoudis, R. G. Raptis, and G. Mezei, "Phosphonopolycarboxylates as Chemical Additives for Calcites Scale Dissolution and Metallic Corrosion Inhibition Based on a Calcium-Phosphonotricarboxylate Organic–Inorganic Hybrid," *Crystal Growth & Design* 6 (5)(2006): 1064.

[261] (*a*) R. S. Boethling, E. Sommer, and D. DiFiore, "Designing Small Molecules for Biodegradability," *Chemical Reviews* 107 (2007): 2207. (*b*) T. O. Boonstra, M. Heus, A. Carstens, and J. LePage, International Patent Application WO/2009/024518. (*c*) J. LePage, C. DeWolf, J. Bemelaar, and H. A. Nasr-El-Din, "An Environmentally Friendly Stimulation Fluid for High-Temperature Applications," SPE 121709 (paper presented at the SPE International Symposium on Oilfield Chemistry, The Woodlands, TX, 20–22 April 2009).

[262] W. W. Frenier, "Novel Scale Removers are Developed for Dissolving Alkaline Earth Deposits," SPE 65027 (paper presented at the SPE International Symposium on Oilfield Chemistry, Houston, TX, 13–

16 February 2001).
[263] W. W. Frenier and D. Wilson, "Use of Highly Acid-Soluble Chelating Agents in Well Stimulation Services," SPE 63242 (paper presented at the SPE Annual Technical Conference and Exhibition, Dallas, 1–4 October 2000).
[264] K. B. Charkhutian, B. L. Libutti, and M. A. Murphy, International Patent Application WO/2006/028917.
[265] T. G. J. Jones, G. J. Tustin, P. Fletcher, and C.-W. Lee, U.S. Patent Application 20070119593.
[266] R. P. Kreh, W. H. Henry, J. Richardson, and V. R. Kuhn, "The Use of Chelants and Dispersants for Prevention and Removal of Rust Scale," in *Mineral Scale Formation and Inhibition*, ed. Z. Amjad. New York: Plenum Press, 1995.
[267] A. F. Clemmit, D. C. Balance, and A. G. Hunton, "The Dissolution of Scales in Oilfield Systems," SPE 140190, Offshore Europe, Aberdeen, UK, 10–13 September 1985.
[268] M. Asakawa, Y. Sumida, M. Shimomura, S. Okuno, T. Morimoto, M. Morita, and H. Suenaga, U.S. Patent 6103686, 2000.
[269] H. Yamamoto, Y. Takayanagi, K. Takahashi, and T. Nakahama, U.S. Patent 6221834, 2001.
[270] K. B. Charkhutian, B. L. Libutti, F. L. M. De Cordt, and J. S. Ruffini, U.S. Patent 6797177, 2004.
[271] L. J. Kalfayan, D. R. Watkins, and G. S. Hewgill, U.S. Patent 4992182, 1991.
[272] G. H. Zaid and B. A. Wolf, U.S. Patent 6494218, 2002.
[273] M. B. Lawson, U.S. Patent 4096869, 1978.
[274] J. M. Paul, U.S. Patent 5146988, 1992.
[275] S. J. Mac Gregor and S. M. Turnbull, International Patent Application WO/2003/069110.
[276] J. Hen, U.S. Patent 5068042, 1991.
[277] A. Putnis, C. C. Putnis, and J. Paul, "The Efficiency of a DTPA-Based Solvent in the Dissolution of Barium Sulfate Scale Deposits," SPE 29094 (paper presented at the SPE International Symposium on Oilfield Chemistry, San Antonio, TX, 14–17 February 1995).
[278] M. M. Jordan, G. M. Graham, K. S. Sorbie, A. Matharu, R. Tomlins, and J. Bunney, "Scale Dissolver Application: Production Enhancement and Formation Damage Potential," SPE 66565, *SPE Production & Facilities* 14 (4)(2000): 288.
[279] J. Rebeschini, C. Jones, G. Collins, S. Edmunds, and A. Archer, "The Development and Performance Testing of a Novel Biodegradable Barium Sulfate Scale Dissolver" (paper presented at the 19th Oil Field Chemistry Symposium, Geilo, Norway, 9–12 March 2008).
[280] R. Boereng, P. Chen, T. Hagen, C. Sitz, R. Thoraval, and H. K. Kotlar, "Creating Value with Green Barium Sulfate Scale Dissolvers—Development and Field Deployment on Statfjord Unit," SPE 87438, (paper presented at the SPE International Symposium on Oilfield Scale, Aberdeen, UK, 26–27 May 2004).
[281] F. De Jong, D. N. Reinhoudt, and G. Torny-Schutte, U.S. Patents 4215000 and 4190462, 1980.
[282] A. Van Zon, F. De Jong, and G. Torny-Schutte, U.S. Patent 4288333, 1981.
[283] H. A. Nasr-El-Din, S. H. Mutairim, H. H. Al-Hajji, and J. D. Lynn, "Evaluation of a New Barite Dissolver: Lab Studies," SPE 86501 (paper presented at the SPE International Symposium and Exhibition on Formation Damage Control, Lafayette, LA, 18–20 February 2004).
[284] K.-S. Wang, R. Resch, K. Dunn, P. Shuler, Y. Tang, B. E. Koel, and T. H. Yen, "Dissolution of the Barite (001) Surface by the Chelating Agent DTPA as Studied with Non-contact Atomic Force Microscopy," *Colloids Surfaces A, Physical and Engineering Aspects* 160 (3)(1999): 217.
[285] I. Lakatos and J. Lakatos-Szabo, "Potential of Different Polyamino Carboxylic Acids as Barium and

Sulfate Dissolvers," SPE 94633 (paper presented at the SPE European Formation Damage Conference, Scheveningen, Netherlands, 25−27 May 2005).

[286] I. Lakatos, J. Lakatos-Szabo, and B. Kosztin, "Optimization of Barite Dissolvers by Organic Acids and pH Regulation" (paper presented at the SPE Oilfield Scale Symposium, Aberdeen, UK, 30−31 January 2002).

[287] I. Lakatos, J. Lakatos-Szabo, and B. Kosztin, "Comparative Study of Different Barite Dissolvers: Technical and Economic Aspects," SPE 73719-MS (paper presented at the SPE International Symposium and Exhibition on Formation Damage Control, Lafayette, LA, 20−21 February 2002).

[288] R. Keatch, International Patent Application WO/2007/109798.

[289] B. J. Palmer, D. Fu, R. Card, and M. J. Miller, U.S. Patent 6924253, 2005.

[290] R. Boereng, K. O. Bakken, O. Vikane, and A. Angelsen, "A Stimulation Treatment of a Subsea Well Using a Scale Dissolver," SPE 50736 (paper presented at the SPE International Symposium on Oilfield Chemistry, Houston, TX, 16−19 February 1999).

[291] H. K. Kotlar, O. M. Selle, and F. Haavind, "A 'Standardized' Method for Ranking of Scale Dissolver Efficiency: A Case Study from the Heidrun Field," SPE 74668 (paper presented at the SPE Oilfield Scale Symposium, Aberdeen, UK, 30−31 January 2002).

[292] I. J. Lakatos, J. Lakatos-Szabo, J. Toth, and T. Bodi, "Improvement of Placement Efficiency of $BaSO_4$ and $SrSO_4$ Dissolvers Using Organic Alkalis as pH Controlling Agents," SPE 106015 (paper presented at the European Formation Damage Conference, Scheveningen, Netherlands, 30 May−1 June 2007).

[293] F. J. Quattrini, U.S. Patent 3660287, 1972.

[294] (a) R. L. Morris and J. M. Paul, U.S. Patent 4980077, 1990. (b) J. M. Paul and R. L. Morris, U.S. Patent 5282995, 1994.

[295] (a) T. F. D' Muhala, U.S. Patent 4708805. (b) R. L. Morris and J. M. Paul, U.S. Patent 5049297, 1991. (c) R. L. Morris and J. M. Paul, U.S. Patent 5084105, 1991.

[296] R. D. Tate, U.S. Patent 5685918, 1997.

[297] C. A. de Wolf, I. C. M. Huybens, K. van Ginkel, and R. Geerts, "Chemistry Biodegradable Solvents for Barium Sulfate Dissolution in Offshore Oilfields" (paper presented at the Chemistry in the Oil Industry X: Oilfield Chemistry, Royal Society of Chemistry, Manchester, UK, 5−7 November 2007).

[298] J. Al-Ashhab, H. Al-Matar, and S. Mokhtar, "Techniques Used to Monitor and Remove Strontium Sulfate Scale in UZ Producing Wells," SPE 101401 (paper presented at the SPE Abu Dhabi International Petroleum Exhibition and Conference, Abu Dhabi, UAE, 5−8 November 2006).

[299] M. M. Jordan, D. Marlow, and C. Johnston, "The Evaluation of Enhanced (Carbonate/Sulfate) Scale Dissolver Treatments for Near-Wellbore Stimulation in Subsea Production Wells, Gulf of Mexico," SPE 100356 (paper presented at the SPE Oilfield Scale Symposium, Aberdeen, UK, 31 May−1 June 2006).

[300] G. D. Williams and I. R. Collins, "Enhancing Mineral Scale Dissolution in the Near-Wellbore Region," SPE 56774 (paper presented at the SPE Annual Technical Conference and Exhibition, Houston, TX, 3−6 October 1999).

[301] H. A. Nasr-El-Din and A. Y. Al-Humaidan, "Iron Sulfide Scale: Formation, Removal and Prevention," SPE 68315 (paper presented at the SPE International Symposium on Oilfield Scale, Aberdeen, UK, 30−31 January 2001).

[302] W. G. F. Ford, M. L. Walker, M. P. Halterman, D. L. Parker, D. G. Brawley, and R. G. Fulton, "Removing a Typical Iron Sulfide Scale: The Scientific Approach," SPE 24327 (paper presented at the SPE Rocky Mountain Regional Meeting, Casper, WY, 18−21 May 1992).

[303] J. Leal, J. R. Solares, H. A. Nasr-El-Din, C. Franco, F. Garzon, H. M. Marri, S. A. Aqeel, and G. Izquierdo, "A Systematic Approach to Remove Iron Sulfide Scale: A Case History," SPE 105607 (paper presented at the SPE Middle East Oil and Gas Show and Conference, Kingdom of Bahrain, 11-14 March 2007).

[304] M. B. Lawson, U.S. Patent 4351673, 1982.

[305] H. A. Nasr-El-Din, B. A. Fadhel, A. Y. Al-Humaidan, W. W. Frenier, and D. Hill, "An Experimental Study of Removing Iron Sulfide Scale from Well Tubulars," SPE 60205 (paper presented at the SPE International Symposium on Oilfield Scale, Aberdeen, UK, 26-27 January 2000).

[306] A. R. Miller, U.S. Patent 6887840, 2005.

[307] W. W. Frenier, M. D. Coffey, J. D. Huffines, and D. C. Smith, U.S. Patent 4220550, 1980.

[308] W. W. Frenier, U.S. Patent 4310435, 1982.

[309] L. D. Martin, U.S. Patent 4276185, 1981.

[310] G. R. Buske, U.S. Patent 4289639, 1981.

[311] G. H. Zaid and B. A. Wolf, U.S. Patent 6774090, 2004.

[312] R. M. Jorda, "Aqualin Biocide in Injection Waters," SPE 280 (paper presented at the SPE Production Research Symposium, Tulsa, OK, 12-13 April 1962).

[313] C. Reed, J. Foshee, J. E. Penkala, and M. Roberson, "Acrolein Application to Mitigate Biogenic Sulfides and Remediate Injection Well Damage in a Gas Plant Water Disposal System," SPE 93602 (paper presented at the SPE International Symposium on Oilfield Chemistry, The Woodlands, TX, 2-4 February 2005).

[314] J. Penkala, M. D. Law, D. D. Horaska, and A. L. Dickinson, "Acrolein 2-Propenal: A Versatile Microbiocide for Control of Bacteria in Oilfield Systems," Paper 04749 (paper presented at the NACE CORROSION Conference, 2004).

[315] T. Salma, "Cost Effective Removal of Iron Sulfide and Hydrogen Sulfide from Water Using Acrolein," SPE 59708 (paper presented at the SPE Permian Basin Oil and Gas Recovery Conference, Midland, TX, 21-23 March 2000).

[316] M. A. Mattox, U.S. Patent 6866048, 2005.

[317] P. D. Gilbert, J. M. Grech, R. E. Talbot, M. A. Veale, and K. A. Hernandez, "Tetrakishydroxymethylphosphonium sulfate (THPS), for Dissolving Iron Sulfides Downhole and Topside—A Study of the Chemistry Influencing Dissolution," Paper 02030 (paper presented at the NACE CORROSION Conference, 2002).

[318] H. A. Nasr-El-Din, A. M. Al-Mohammad, M. A. Al-Hajri, and J. B. Chesson, "A New Chemical Treatment to Remove Multiple Damages in a Water Supply Well," SPE 95001 (paper presented at the SPE European Formation Damage Conference, Scheveningen, Netherlands, 25-27 May 2005).

[319] R. E. Talbot, C. R. Jones, and J. M. Grech, International Patent Application WO/2005/026065.

[320] P. R. Rincon, J. P. McKee, C. E. Tarazon, L. A. Guevara, and B. Vinccler, "Biocide Stimulation in Oilwells for Downhole Corrosion Control and Increasing Production," SPE 87562 (paper presented at the SPE International Symposium on Oilfield Corrosion, Aberdeen, UK, 28 May 2004).

[321] R. E. Talbot and J. M. Grech, International Patent Application WO/2005/040050.

[322] C. R. Jones and J. M. Grech, International Patent Application WO/2004/083131.

[323] J. Romaine, T. G. Strawser, and M. L. Knippers, "Application of Chlorine Dioxide as an Oilfield-Facilities-Treatment Fluid," SPE 29017, *SPE Production & Facilities* 11 (1)(1996): 18.

[324] J. F. McCafferty, E. W. Tate, and D. A. Williams, "Field Performance in the Practical Application of

Chlorine Dioxide as a Stimulation Enhancement Fluid," SPE 20626, *SPE Production & Facilities* 8（1）（1993）: 9.

［325］J. G. R. Eylander, D. M. Frigo, F. A. Hartog, and G. Jonkers, "A Novel Methodology for In-Situ Removal of NORM from E & P Production Facilities," SPE 46791（paper presented at the SPE International Conference on Health, Safety and Environment in Oil and Gas Exploration and Production, Caracas, Venezuela, 7-10 June 1998）.

［326］O. M. Selle, A. Mebratu, H. Montgomerie, P. Chen, and T. Hagen, International Patent Application WO/2008/152419.

［327］Z. Amjad, ed., *The Science & Technology of Industrial Water Treatment*. Boca Raton, FL: CRC Press, 2010.

［328］E. M. Flaten, M. Seiersten, and J.-P. Andreassen, *Chemical Engineering Research & Design* 88（2010）: 1659-1668.

［329］E. M. Flaten, M. Seiersten, and J.-P. Andreassen, *Journal of Crystal Growth* 311（2009）: 3533.

［330］C. F. Fan, A. T. Kan, P. Zhang, and M. B. Tomson, *SPE Journal* 16（2）（2011）: 440-450.

［331］M. Yuan, D. A. Williamson, J. K. Smith, and T. H. Lopez, "Effective Control of Exotic Minerals Scales Under Harsh System Conditions," SPE80234（paper presented at the SPE International Symposium on Oilfield Chemistry, Houston TX, 5-7 February 2003）.

［332］R. Wright, K. Melvin, R. Stalker, S. Dyer, and G. Graham, "Laboratory Examination of the Factors Influencing the Formation of Silicate Scales Under Oil Field Operating Conditions," RSC Chemistry in the Oil Industry XI, Manchester, UK, 2-4 November 2009.

［333］K. D. Demadis, E. Mavredaki, and M. Somara, *Industrial & Engineering Chemistry Research* 50（24）（2011）: 13866-13876.

［334］J. Arensdorf, D. Hoster, D. McDougall, and M. Yuan, "Static and Dynamic Testing of Silicate Scale Inhibitors," SPE 132212（paper presented at the International Oil and Gas Conference and Exhibition in China, Beijing, China, 8-10 June 2010）.

［335］J. Arensdorf, S. Kerr, K. Miner, and T. Ellis-Toddington, "Mitigating Silicate Scale in Production Wells in an Oilfield in Alberta," SPE 141422（paper presented at the SPE International Symposium on Oilfield Chemistry, The Woodlands, TX, 11-13 April 2011）.

［336］E. Reyes-Garcia, "Applications and Laboratory Studies of a New Class of Silica Polymerization Inhibitors of Biorenewable Origin," SPE 143086, Brasil Offshore, Macaé, Brazil, 14-17 June 2011.

［337］R. W. Zuhl and Z. Amjad, International Patent Application WO/2010/005889.

［338］Y. Tang, J. Voelker, C. Keskin, Z. Xu, B. Hu, C. Jia, "A Flow Assurance Study on Elemental Sulfur Deposition in Sour Gas Wells," SPE 147244（paper presented at the SPE Annual Technical Conference and Exhibition, Denver, CO, 30 October-2 November 2011）.

［339］N. Fleming, K. Ramstad, S. H. Eriksen, E. Moldrheim, and T. R. Johansen, "Development and Implementation of a Scale-Management Strategy for Oseberg Sør," *SPE Production & Operations* 22（2007）: 307-317.

［340］Y.-P. Lin and P. C. Singer, *Journal of Crystal Growth* 312（2009）: 136.

［341］H. Guan, G. Cole, and P. Clark, SPE 114059, *SPE Production & Operations* 24（4）（2009）: 543-549.

［342］G. M. Graham, N. Goodwin, J. Davidson, and H. Williams, "Laboratory Simulation of Scaling in the Presence of Hydrogen Sulphide Scavengers," RSC Chemicals in the Oil Industry, Manchester, UK, November 2011.

[343] M. Sumestry and H. Tedjawidjaja, "Case Study—Calcium-Carbonate-Scale Inhibitor Performance Degradation Because of H2S-Scavenger Injection in Semoga Field," *SPE Oil and Gas Facilities* 2 (1) (2013): 2224.

[344] A. E. S. Van Driessche, L. G. Benning, J. D. Rodriguez-Blanco, M. Ossorio, P. Bots, and J. M. García-Ruiz, *Science* 336 (6077)(2012): 69-72.

[345] C. Okocha and K. Sorbie, "Scale Prediction for Iron, Zinc, and Lead Sulfides and Its Relation to Scale Test Design," SPE 164111 (paper presented at the SPE International Symposium on Oilfield Chemistry, The Woodlands, TX, 8-10 April 2013).

[346] J. J. Wylde and J. L. Slayer, "Halite Scale Formation Mechanisms, Removal and Control: A Global Overview of Mechanical, Process, and Chemical Strategies," SPE 164081 (paper presented at the SPE International Symposium on Oilfield Chemistry, The Woodlands, TX, 8-10 April 2013).

[347] J. Kamisetty, "Sulphate Removal from Seawater Injection Processes: Technical Benefits & Economics," RSC Chemistry in the Oil Industry XI, Manchester, UK, 2-4 November 2009.

[348] X. Li, J. Zhang, and D. Yang, *Industrial & Engineering Chemistry Research* 51 (2012): 9266.

[349] A. Neville, *Energy & Fuels* 26 (7)(2012): 4158.

[350] B. Stuyven, G. Vanbutsele, J. Nuyens, J. Vermant, and J. A. Martens, *Chemical Engineering Science* 64 (2009): 1904.

[351] F. Alimi, M. M. Tlili, M. Ben Amor, G. Maurin, and C. Gaberielli, *Chemical Engineering and Processing* 48 (2009): 1327-1332.

[352] W. L. Parker, U.S. Patent Application 20110240131.

[353] M. C. van der Leeden and G. M. van Rosmalen, *Journal of Colloid and Interface Science* 171 (1995): 142.

[354] L. S. Boak, "A Review of Factors that Impact Scale Inhibitor Mechanisms," Oilfield Chemical Symposium, Geilo, Norway, 17-20 March 2013.

[355] E. Mavredaki, A. Neville, and K. S. Sorbie, *Crystal Growth & Design* 11 (11)(2010): 4751-4758.

[356] G. M. Parkinson and A. L. Rohl, *Crystal Engineering Communications* 7 (52)(2005): 320-323.

[357] M. I. P. Reis, A. D. Goncalves, F. D. da Silva, A. K. Jordao, R. J. Alves, S. F. de Andrade, J. A. L. C. Resende, A. A. Rocha, and V. F. Ferreira, *Carbohydrate Research* 353 (2012): 6-12.

[358] B. Ghosh, S. Kundu, and B. Senthilmurugan, *Petroleum Science and Technology* 30 (4)(2011): 402-441.

[359] B. Suharso, S. Bahri, and T. Endaryanto, *Desalination* 265 (1-3)(2011): 102-106.

[360] S. Kundua, B. Ghosha, S. Balasubramaniana, and M. Haroun, *Petroleum Science and Technology* 29 (2011): 1512.

[361] T. Kumar, S. Vishwanatham, and S. S. Kundu, *Journal of Petroleum Science and Engineering* 1-2 (2010): 71.

[362] Z. Amjad, U.S. Patent 4652377, 1987.

[363] H. Adam, D. Labarre, J. Wilson, J.-F. Argillier, B. Bazin, P. Gateau, and B. Herzhaft, International Patent Application WO/2010/081944.

[364] A. George and A. Veis, *Chemical Reviews* 108 (2008): 4670.

[365] C. Y. Chen, W. Lei, M. Z. Xia, F. Y. Wang, and X. D. Gong, *Desalination* 309 (2013): 208-212.

[366] B. R. Zhang, L. Zhang, F. T. Li, W. Hu, and P. M. Hannam, *Corrosion Science* 52 (12)(2010): 3883.

[367] G. Collins, B. Downward, and C. Jones, International Patent Application WO/2009/080498.

[368] J. Guo and S. J. Severtson, *Industrial & Engineering Chemistry Research* 43 (2004): 5411.

[369] L. A. Bromley, D. Cottier, R. J. Davey, B. Dobbs, S. Smith, and B. R. Heywood, *Langmuir* 9 (1993): 3594.

[370] P. P. Notte, J. H. J. Van Bree, and A. Devaux, International Patent Application 2008/071692.

[371] H. C. Fisher, A. F. Miles, S. H. Bodnar, S. D. Fidoe, and C. D. Sitz, "Progress Towards Biodegradable Phosphonate Scale Inhibitors," RSC Chemistry in the Oil Industry XI, Manchester, UK, 2–4 November 2009.

[372] B. Nowack, "Environmental Chemistry of Phosphonates," *Water Research* 37 (2003): 2533.

[373] S. H. Bodnar, H. C. Fisher, A. F. Miles, and C. D. Sitz, International Patent Application WO/2010/002738.

[374] A. M. Dabdoub, U.S. Patent 8101700, 2012.

[375] E. Greyson, J. Manna, and S. C. Mehta, U.S. Patent Application 20110046023.

[376] K. L. Opper, D. Markova, M. Klapper, K. Mullen, and K. B. Wagener, *Macromolecules* 43(8)(2010): 3690–3698.

[377] B. Akgun and D. Avci, *Journal of Polymer Science Part A: Polymer Chemistry* 23 (2012): 4854.

[378] W. Guo, W.-T. Leu, S.-H. Hsiao, and G.-S. Liou, *Polymer Degradation and Stability* 91 (2006): 21.

[379] S. Heath, G. Moir, M. Archibald, and J. Goulding, International Patent Application WO/2010/046026.

[380] G. Adusei, S. Deb, J. W. Nicholson, L. Mou, and G. Singh, *Journal of Applied Polymer Science* 88 (2003): 565.

[381] J. Plank, F. Dugonji-Bili, and N. Recalde Lummer, *Journal of Applied Polymer Science* 115(3)(2009): 1758.

[382] M. Laubender, E. Heintz, C. Seidl, B. Urtel, and A. Berger, U.S. Patent Application 20120004383.

[383] A. Benbakhti and T. Bachir-Bey, *Journal of Applied Polymer Science* 116 (5)(2010): 3095.

[384] A. B. BinMerdhah, *Journal of Petroleum Science and Engineering* 90–91 (2012): 124.

[385] N. M. Farooqui and K. S. Sorbie, "Phase Behaviour of Poly-Phosphino Carboxylic Acid (PPCA) Scale Inhibitor for Application in Precipitation Squeeze Treatments," Chemistry in the Oil Industry XIII—New Frontiers, Manchester, UK, November 2013.

[386] C. Hogan, "Improved Scale Control, Hydrothermal & Environmental Properties with New Maleic Polymer Chemistry Suitable for Downhole Squeeze & Topside Applications," Chemistry in the Oil Industry XIII—New Frontiers, Manchester, UK, November 2013.

[387] B. Senthilmurugan, B. Ghosh, G. M. Graham, and S. S. Kundu, "The Influence of Maleic Acid Copolymers on the Growth and Microstructure of Calcite Scale," SPE 131132 (paper presented at the SPE International Conference on Oilfield Scale, Aberdeen, UK, 26–27 May 2010).

[388] A. Jada, R. Ait Akbour, C. Jacquemet, J. M. Suau, and O. Guerret, *Journal of Crystal Growth* 306 (2007): 373.

[389] K. A. Rodrigues, U.S. Patent Application 20100012885.

[390] J. W. Morse, R. S. Arvidson, and A. Lüttge, *Chemical Reviews* 107 (2007): 342.

[391] L. J. Fu, Y. Z. Zhao, H. H. Ge, and Y. S. Li, "Study on Scale Inhibition Performance of Polyaspartic Acid Derivatives with Dihydroxyl Group on $CaCO_3$," in Advanced Materials Research Series, eds. Q. J. Xu, H. H. Ge, and J. X. Zhang, *Natural Resources and Sustainable Development*, Pts. 1–3, 361–363, 1982–1986.

[392] X. H. Qiang, Z. H. Sheng, and H. Zhang, *Desalination* 309 (2013): 237–242.

[393] L. J. Gao, J. Y. Feng, B. Jin, Q. N. Zhang, T. Q. Liu, Y. Q. Lun, and Z. J. Wu, *Chemistry Letters* 40(12) (2011): 1392–1394.

[394] G. Ho, T. Ho, K. Hsieh, Y. Su, P. Lin, J. Yang, and S. Yang, *Journal of the Chinese Chemical Society* 53 (6)(2006): 1363-1384.

[395] D. Liu, W. Dong, F. Li, F. Hui, and J. Lédion, *Desalination* 304(2012): 1-10.

[396] M. K. Stephen and H. Fairless, U.S. Patent 256332, 1993.

[397] J. M. Brown, J. F. McDowell, and K. T. Chang, U.S. Patent 5062962, 1991.

[398] H. Li, W. Liu, and X. J. Qi, *Desalination* 214(2007): 193.

[399] W. P. Zeng, F. H. Wang, C. Zhou, and X. D. Gong, *Chinese Journal of Chemical Physics*, 25 (2): 219-225.

[400] X. Hu, Y. Liu, and H. Zhu, *Journal of Chemistry*(Chinese)22(2005): 111-113.

[401] T. Gu, P. Su, X. Liu, J. Zou, X. Zhang, and Y. Hu, *Journal of Petroleum Science and Engineering* 102 (2013): 41-46.

[402] L. E. I. Wu, W. Fengyun, X. I. A. Mingzhu, and W. Fenghe, *Journal of Chemical Industry and Engineering*(Chinese)57(2006): 2207-2213.

[403] J. Sanders, S. P. R. Holt, and K. A. Rodrigues, International Patent Application WO/2009/062924.

[404] X. R. Guo, F. X. Qiu, K. Dong, X. Zhou, J. Qi, Y. Zhou, and D. Y. Yang, *Journal of Industrial and Engineering Chemistry* 18 (6)(2012): 2177.

[405] A. Assmann and R. Reichenbach-Klinke, International Patent Application WO/2009/019050.

[406] S. Kirboga and M. Öner, *Crystal Engineering Communications* 15(2013): 3678-3686.

[407] A. P. Narrainena and P. A. Lovell, *Polymer* 51 (26)(2010): 6115.

[408] K. A. Rodrigues and J. Sanders, U.S. Patent Application 20110046025.

[409] K. A. Rodrigues, M. M. Vanderhoof, and J. Ozbek, U.S. Patent Application 20130137799.

[410] K. A. Rodrigues, M. Vanderhoof, A. M. Carrier, and J. Sanders, U.S. Patent Application 20120128608.

[411] X. R. Guo, F. X. Qiu, K. Dong, K. C. He, X. S. Rong, and D. Y. Yang, *Polymer-Plastics Technology and Engineering* 52 (3)(2013): 261-267.

[412] E. Holler, U.S. Patent Application 20100216199.

[413] A. Viloria, L. Castillo, J. A. Garcia, M. A. C. Ordaz, and E. V. Torin, U.S. Patent Application 20100075870.

[414] W. Mattmann, W. Loth, B. Urtel, E. Guetlich-Hauk, C. H. Weidl, and A. Daisse, U.S. Patent Application 20110054071.

[415] M. Dietzsch, M. Barz, T. Schueler, S. Klassen, M. Schreiber, M. Susewind, N. Loges, M. Lang, N. Hellmann, M. Fritz, K. Fischer, P. Theato, A. Kuhnle, M. Schmidt, R. Zentel, and W. Tremel, *Langmuir* 29 (9)(2013): 3080-3088.

[416] G. Graham, H. Williams, and N. Goodwin, "Sulphide Scavengers, Performance, Deployment and Impact on Carbonate Scaling," Oilfield Chemical Symposium, Geilo, Norway, 17-20 March 2013.

[417] M. Sumestry and H. Tedjawidjaja, "Case Study—Calcium-Carbonate-Scale Inhibitor Performance Degradation Because of H2S-Scavenger Injection in Semoga Field," SPE 150705, *Oil & Gas Facilities* 2 (1)(2013): 40.

[418] M. N. Lehmann, F. F. Firouzkouhi, M. P. Squicciarini, and T. Salma, U.S. Patent Application 20090143252.

[419] J. K. Smith, D. Wahus, G. Boyd, Q. Fu, G. Boyce, F. Firouzkouhi, and D. Deichert, "Control and Removal of Downhole Interfacial Solids," SPE 130296 (paper presented at the SPE International Conference on Oilfield Scale, Aberdeen, UK, 26-27 May 2010).

[420] B. Wang, T. Chen, P. Chen, H. Montgomerie, T. Hagen, X. Liu, and X. Yang, "Development of Test

Method and Environmentally Acceptable Inhibitors for Zinc Sulfide Deposited in Oil and Gas Fields," SPE 156005 (paper presented at the SPE International Conference on Oilfield Scale, Aberdeen, UK, 30–31 May 2012).

[421] L. Phoenix, *British Chemical Engineering* 11 (1966): 34.

[422] M. A. S. Khan, A. Singh, S. Haldar, and B. Ganguly, *Crystal Growth & Design* 11 (2011): 1675–1682.

[423] M. A. S. Khan, A. Sen, and B. Ganguly, *Crystal Engineering Communications* 11 (2009): 2660–2667.

[424] R. Keatch and H. Guan, U.S. Patent Application 20110024366.

[425] R. Keatch and H. Guan, International Patent Application WO/2009/050561.

[426] A. Thompson, B. Juliussen, S. Heath, T. E. Gundersen, and A. T. Andresen, "Improvements in Analytical Techniques and Analysis of Mixed Polymeric Scale Inhibitors," Tekna 23rd International Oil Field Chemistry Symposium, Geilo, Norway, 18–21 March 2012.

[427] K. Marshall, S. McMahon, M. Smillie, and G. Graham, "Challenges Facing Residual Chemical (Polymer) Scale Inhibitor Assay in Oilfield Waters," RSC Chemistry in the Oil Industry XI, Manchester, UK, 2–4 November 2009.

[428] M. J. Todd, M. Archibald, L. S. Boak, and J. Goulding, "Development of the Next Generation of Phosphorus Tagged Polymeric Scale Inhibitors," SPE 130733 (paper presented at the SPE International Conference on Oilfield Scale, Aberdeen, UK, 26–27 May 2010).

[429] J. R. Kerr, J. Goulding, and K. S. Sorbie, "The Development and Application of Techniques for the Detailed Characterization of a Novel Series of P-Functionalized Polymeric Scale Inhibitors," SPE 164124 (paper presented at the SPE International Symposium on Oilfield Chemistry, The Woodlands, TX, 8–10 April 2013).

[430] K. Marshall, M. Kyle, and G. Graham, "Analysis of Sulphonated Polymers in Oilfield Brines," SPE 141575 (paper presented at the SPE International Symposium on Oilfield Chemistry, The Woodlands, TX, 11–13 April 2011).

[431] S. Holt and J. Sanders, "A Technology Platform for Designing High Performance, Environmentally Benign Scale Inhibitors for a Range of Application Needs," RSC Chemistry in the Oil Industry XI, Manchester, UK, 2–4 November 2009.

[432] B. Smith, S. Hurst, N. Williamson, C. Jones, J. Wilson, S. Edmunds, D. Labarre, and R. Jones, "The Development of Tagged Scale Inhibitors for Cost-Effective Scale Management in Deepwater, Subsea Fields," *Chemistry in the Oil Industry XIII: Oilfield Chemistry—New Frontiers*, Manchester Conference Centre, Manchester, UK, 4–6 November 2013.

[433] C. Jones, D. Labarre, C. Rouault, and J. Wilson, International Patent Application WO/2012/098186.

[434] A.-M. Fuller, F. Mackay, C. Mackenzie, C. Rowley-Williams, E. Perfect, and V. Magdalenic, "Applying Biochemistry Concepts to Oilfield Produced Fluid Monitoring: A Focus on Latently Detectable Tags and Tracers," RSC Chemistry in the Oil Industry XI, Manchester, UK, 2–4 November 2009.

[435] (a) A.-M. Fuller, F. Mackay, C. Mackenzie, C. Rowley-Williams, and E. "Perfect Development of New Chemical Additive Detection Methods Inspired by the Life Sciences," SPE 141242 (paper presented at the SPE International Symposium on Oilfield Chemistry, The Woodlands, TX, 11–13 April 2011).

[436] N. Poynton, A. Molliet, A. Leontieff, S. Cook, S. Toivonen, and R. Griffin, "Development of a New Tagged Polymeric Scale Inhibitor with Accurate Low-Level Residual Inhibitor Detection, for Squeeze Applications," SPE 155187 (paper presented at the SPE International Conference on Oilfield Scale, Aberdeen, UK, 30–31 May 2012).

[437] L. Moore and L. Clapp, U.S. Patent Application 20120032093.

[438] G. Q. Liu, J. Y. Huang, Y. M. Zhou, Q. Z. Yao, L. Ling, P. X. Zhang, H. C. Wang, K. Cao, Y. H. Liu, W. D. Wu, W. Sun, and Z. J. Hu, *Tenside Surfactants Detergents* 49(2012): 404.

[439] T. E. McNeel, M. S. Whittemore, R. A. Clark, and J. J. Grabowicz, U.S. Patent Application 20130043194.

[440] K. Zhou and L. Chen, International Patent Application WO/2010/128322.

[441] P. Weatherby and W. H. Stimson, U.S. Patent 6911534, 2002.

[442] L. H. Keller and B. Weinstein, U.S. Patent 6197522, 2001.

[443] Available at http://msdssearch.dow.com/PublishedLiteratureDOWCOM/dh_0516/0901b803805167df.pdf? filepath=oilandgas/pdfs/noreg/812-00023.pdf&fromPage=GetDoc.

[444] B. M. Hustad, O. G. Svela, J. H. Olsen, K. Ramstad, and T. Tjomsland, "Statoil ASA, Downhole Chemical Injection Lines—Why Do They Fail? Experiences, Challenges and Application of New Test Methods," SPE 154967 (paper presented at the SPE International Conference on Oilfield Scale, Aberdeen, UK, 30-31 May 2012).

[445] J. H. Olsen, "Statoil Experiences and Consequences Related to Continuous Chemical Injection," Paper SPE 146625 (paper presented at the SPE Annual Technical Conference and Exhibition, Denver, CO, 30 October-2 November 2011).

[446] H. Guan and S. Davis, "Precipitation Squeeze and the Use of Modeling to Maintain Production and Extend Squeeze Life Time for Marginal Wells," SPR 158555 (paper presented at the SPE Annual Technical Conference and Exhibition, San Antonio, TX, 8-10 October 2012).

[447] S. S. Shaw and K. S. Sorbie, "Structure, Stoichiometry, and Modelling of Mixed Calcium-Magnesium-Phosphonate Scale Inhibitor Complexes for Application in Precipitation Squeeze Processes," *Chemistry in the Oil Industry XIII—New Frontiers*, Manchester, UK, November 2013.

[448] S. S. Shaw and K. S. Sorbie, "Structure, Stoichiometry, and Modelling of Calcium Phosphonate Scale Inhibitor Complexes for Application in Precipitation Squeeze Processes," SPE 164051 (paper presented at the SPE International Symposium on Oilfield Chemistry, The Woodlands, TX, 8-10 April 2013).

[449] S. Baraka-Lokmane and K. S. Sorbie, *Journal of Petroleum Science and Engineering* 70(2010): 10.

[450] S. Heath and M. Todd, U.S. Patent Application 20130023449.

[451] A. T. Kan, G. M. Fu, H. Al-Saiari, M. B. Tomson, and D. Shen, "Enhanced Scale-Inhibitor Treatments with the Addition of Zinc," *SPE Journal* 14(4)(2009): 617-626.

[452] O. Vazquez, E. Mackay, K. Sorbie, and M. Jordan, "Impact of Mutual Solvent Preflushes on Scale Squeeze Treatments: Extended Squeeze Lifetime and Improved Well Clean-Up Time," SPE 121857 (paper presented at the SPE 8th Eur. Formation Damage Conference, Scheveningen, Netherlands, 27-29 May 2009).

[453] T. D. Welton and P. D. Nguyen, U.S. Patent Application 20090143258.

[454] T. Chen, R. Benvie, S. M. Heath, P. Chen, T. Hagen, and H. Montgomerie, "Development of an Environmentally Friendly Polymer Scale Inhibitor for Tight Carbonate Reservoir Squeeze Treatment," SPE 23352 (paper presented at the Offshore Technology Conference, Houston, TX, 30 April-3 May 2012).

[455] C. Sitz, J. Hardy, J. Merryman, and D. Varner, "A New Class of Non-damaging, Environmentally Responsible Scale Squeeze Enhancement Aids," SPE 155254 (paper presented at the SPE International Conference on Oilfield Scale, Aberdeen, UK, 30-31 May 2012).

[456] J. Lawson, M. Todd, P. Wilkie, A. Gibb, and M. Burgoyne, "Squeeze Enhancer for an Environmentally

Sensitive, Low Temperature, Clean Sandstone Reservoir," Tekna 23rd International Oil Field Chemistry Symposium, Geilo, Norway, 18−21 March 2012.

[457] H. Montgomerie, P. Chen, T. Hagen, and L. Zheng, U.S. Patent Application 20090170732.

[458] K. Riedelsberger and W. Jeger, *Designed Monomers & Polymers* 1 (1998): 387.

[459] O. C. S. Al-Hamouz and S. A. Ali, *Journal of Polymer Science Part A: Polymer Chemistry* 50 (17) (2012): 3580.

[460] Z. Amjad, *Desalination & Water Treatment* 36 (1−3)(2011): 270−279.

[461] N. Kazemi, M. Wilson, N. Kapur, N. Fleming, and A. Neville, "Preventing Adhesion of Scale on Rock by Nanoscale Modification of the Surface," SPE 156955 (paper presented at the SPE International Oilfield Nanotechnology Conference, Noordwijk, Netherlands, 12−14 June 2012).

[462] N. Fleming, K. Ramstad, L. W. Hoeth, and S. Kidd, "Mechanical Alteration of Near Wellbore Mineralogy for Improved Squeeze Performance," SPE 130147 (paper presented at the SPE International Conference on Oilfield Scale held in Aberdeen, UK, 26−27 May 2010).

[463] O. Vazquez, P. Thanasutives, C. Eliasson, N. Fleming, and E. Mackay, SPE 141384, *SPE Production & Operations* 26 (3)(2011): 270.

[464] C. Yan, A. T. Kan, W. Wang, F. Yan, L. Wang, and M. B. Tomson, "Synthesis and Sorption Study of AlOOH Nanoparticle Cross-Linked Polymeric Scale Inhibitors and Their Squeeze Performance in Porous Media," SPE 164086 (paper presented at the SPE International Symposium on Oilfield Chemistry, The Woodlands, TX, 8−10 April 2013).

[465] C. Yan, A. T. Kan, W. Wang, F. Yan, L. Wang, M. B. Tomson, H. Zhu, and R. Tomson, "Boehmite Based Sulphonated Polymer Nanoparticles with Improved Squeeze Performance for Deepwater Scale Control," SPE 24252 (paper presented at the Offshore Technology Conference, Houston, TX, 6−9 May 2013).

[466] M. Marquez, L. A. Schafer, and W. D. Norman, "Preemptive Scale Management: Treating with Scale Inhibitor while Frac Packing a Well," SPE 147006 (paper presented at the SPE Annual Technical Conference and Exhibition, Denver, CO, 30 October−2 November 2011).

[467] S. Szymczak, D. V. Gupta, and J. M. Brown, "A 5-Year Survey of Applications and Results of Placing Solid Chemical Inhibitors in the Formation via Hydraulic Fracturing," SPE 134414 (paper presented at the SPE Annual Technical Conference and Exhibition, Florence, Italy, 19−22 September 2010).

[468] R. B. Watson, P. Viste, N. Kaageson-Loe, N. Fleming, A. M. Mathisen, and K. Ramstad, "Smart Mud Filtrate: An Engineered Solution to Minimize Near-Wellbore Formation Damage Due to Kaolinite Mobilization: Laboratory and Field Experience, Oseberg South," SPE 112455 (paper presented at the SPE International Symposium and Exhibition on Formation Damage Control, Lafayette, LA, 13−15 February 2008).

[469] S. Yakimov, M. Mukhametshin, O. Sosenko, A. Sadykov, O. Levanyuk, M. Oparin, D. Gromakovsky, K. Mullen, B. Lungwitz, D. Fu, and K. Mauth, "A Combined Laboratory and Field Approach to Optimize Scale Inhibition Placement via Fracturing in Krasnoleninskoe Field, Russia," SPE 127874 (paper presented at the SPE International Symposium and Exhibition on Formation Damage Control, Lafayette, LA, 10−12 February 2010).

[470] J. M. Brown and S. J. Szymczak, "Long Term Scale Prevention with the Placement of Solid Scale Inhibitor in the Formation via Hydraulic Fracturing," Paper 07063, NACE CORROSION, 2007.

[471] O. Selle, F. Haavind, M. H. Haukland, A. Moen, C. F. Hals, K. Oien, C. J. Strachan, and G. Clark, "Downhole Scale Control on Heidrun Field Using Scale Inhibitor Impregnated Gravel," SPE 130788

(paper presented at the SPE International Conference on Oilfield Scale, Aberdeen, UK, 26-27 May 2010).

[472] M. Marquez and R. J. Wetzel, "Acid/Scale Inhibitor Stimulation Treatment for High Temperature CHFP Subsea Wells: Fluid Qualification and Formation Damage Assessment," SPE 127768 (paper presented at the SPE International Symposium and Exhibition on Formation Damage Control, Lafayette, LA, 10-12 February 2010).

[473] D. Patterson, M. Kendrick, W. Williams, and M. Jordan, "'Squimulation'—Simultaneous Well-Stimulation and Scale-Squeeze Treatments in Deepwater West Africa," *SPE Production & Operations* 28(1)(2013): 55-66.

[474] M. M. Jordan, "Simultaneous Well Stimulation and Scale Squeeze Treatments in Carbonate Reservoirs," SPE 164097 (paper presented at the SPE International Symposium on Oilfield Chemistry, The Woodlands, TX, 8-10 April 2013).

[475] Z. Xiao, M. Garcia-Lopez de Victoria, and F. Tuedor, U.S. Patent Application 20100200238.

[476] M. Garcia-Lopez de Victoria, J. W. Still, and T. Bui, U.S. Patent Application 20110127039.

[477] R. Stalker and G. Graham, "Oil-Based Placement Fluids for Sensitive Reservoirs," Tekna 23rd International Oil Field Chemistry Symposium, Geilo, Norway, 18-21 March 2012.

[478] M. Luo, H. Sun, Z. Jia, Q. Wen, and L. Liao, "Preparation and Performance of Environment-Friendly Nanoemulsion with Antiscaling," SPE 152688 (paper presented at the SPE International Oilfield Nanotechnology Conference, Noordwijk, Netherlands, 12-14 June 2012).

[479] T. D. Welton, U.S. Patent Application 20110105368.

[480] D. R. Wilson, U.S. Patent Application 20110162841.

[481] A. Haghtalab and Z. Kiaei, *Journal of Nanoparticle Research* 14(2012): 1210.

[482] N. Ghorbani, M. Wilson, N. Kapur, N. Fleming, and A. Neville, "Using Nanoscale Dispersed Particles to Assist in the Retention of Polyphosphinocarboxylic Acid (PPCA) Scale Inhibitor on Rock," SPE 156200 (paper presented at the SPE International Oilfield Nanotechnology Conference, Noordwijk, Netherlands, 12-14 June 2012).

[483] P. Zhang, C. Fan, H. Lu, A. T. Kan, and M. B. Tomson, *Industrial & Engineering Chemistry Research* 50 (4)(2011): 1819-1830.

[484] D. V. S. Gupta, International Patent Application WO/2012/148819.

[485] O. Vazquez, E. Mackay, T. Tjomsland, O. Nygard, and E. Storas, "Use of Tracers to Evaluate and Optimize Scale Squeeze Treatment Design in the Norne Field," SPE 164114 (paper presented at the SPE International Symposium on Oilfield Chemistry, The Woodlands, TX, 8-10 April 2013).

[486] F. De Campo, A. Colaco, B. Langlois, C. Jones, and G. Woodward, U.S. Patent Application 20110143972.

[487] L. Morris, U.S. Patent Application 20080257551.

[488] O. Rauf, S. Gerdes, and N. Recalde Lummer, "Foam and Scale Inhibitor Squeeze with Natural Gas: A Lifeline for Water Flooded and Damaged Gas Wells," SPE 165098 (paper presented at the SPE European Formation Damage Conference and Exhibition, Noordwijk, Netherlands, 5-7 June 2013).

[489] T. D. Welton, U.S. Patent Application 20110100629.

[490] N. Fleming, K. Ramstad, S. Kidd, and L. W. Hoeth, "Impact of Successive Squeezes on Treatment Lifetime and Well Productivity: Comparative Assessment of Viscosified and Nonviscosified Treatments," SPE 120969, *SPE Production & Operations* 25 (1)(2010): 99.

[491] L. Sutherland, C. J. Johnston, and W. Taylor, "The Influence of Turbulence (or Hydrodynamic Effects)

on Barium Sulphate Scale Formation and Inhibitor Performance," SPE 164070 (paper presented at the SPE International Symposium on Oilfield Chemistry, The Woodlands, TX, 8-10 April 2013).

[492] L. O. Jøsang, M. N. Psarrou, T. Østvold, T. Tjomsland, K. Ramstad, and B. M. Hustad, "Scale under Turbulent Flow," Tekna 23rd International Oil Field Chemistry Symposium, Geilo, Norway, 18-21 March 2012.

[493] F.-A. Sett and A. Neville, *Desalination* 281 (2011): 340.

[494] P. Kjellin, K. Holmberg, and M. Nyden, *Colloids and Surfaces A: Physicochemical and Engineering Aspects* 194 (2001): 49.

[495] X. N. Zhang, W. L. Wu, D. M. Li, and G. J. Zhao, "A New Evaluation Method of Scale Inhibitors for Controlling $CaCO_3$ Scale in Reverse Osmosis System Based on pH Measurement," in *Book Series: Advanced Materials Research, Progress in Environmental Science and Engineering* (ICEESD2011), eds. H. Li, Q. J. Xu, and D. Zhang, 356-360 (1-5)(2012): 2146-2152.

[496] T. D. Baugh, J.-Y. Lee, K. Winters, J. Waters, and J. Wilcher, "A Fast and Information-Rich Test Method for Scale Inhibitor Performance," SPE 23150 (paper presented at the Offshore Technology Conference, Houston, TX, 30 April-3 May 2012).

[497] M. A. Kelland, *Industrial & Engineering Chemistry Research* 50 (2011): 5852-5861.

[498] G. M. Graham, R. Stalker, and R. McIntosh, "The Impact of Dissolved Iron on the Performance of Scale Inhibitors under Carbonate Scaling Conditions," SPE 80254 (paper presented at the SPE International Symposium on Oilfield Scale, Houston, TX, 5-8 February 2003).

[499] M. D. Yuan and L. S. Stoppelenburg, "The Performance of Barium Sulphate Inhibitors in Iron Containing Waters in both Aerated and Anaerobic Systems," Paper 114, NACE International CORROSION 2000 Conference and Exhibition, March 2000.

[500] D. Shen, D. Shcolnik, R. Perkins, G. Taylor, and J. M. Brown, "Evaluation of Scale Inhibitors in Marcellus Waters Containing High Levels of Dissolved Iron," SPE 141145 (paper presented at the SPE International Symposium on Oilfield Chemistry, The Woodlands, TX, 11-13 April 2011).

[501] Q. W. Wang, F. Al-Dawood, and H. Al Saiari, *Materials Performance* 51 (11)(2012): 60-64.

[502] K. I. Parsiegla and J. L. Katz, *Journal of Crystal Growth* 213 (3-4)(2000): 368-380.

[503] S. S. Shaw, K. S. Sorbie, and L. Boak, "The Effects of Barium Sulfate Saturation Ratio, Calcium, and Magnesium on the Inhibition Efficiency—Part II: Polymeric Scale Inhibitors," *SPE Production & Operations* 27 (3)(2012): 390.

[504] S. S. Shaw, K. S. Sorbie, and L. Boak, "The Effects of Barium Sulfate Saturation Ratio, Calcium, and Magnesium on the Inhibition Efficiency—Part II: Polymeric Scale Inhibitors," *SPE Production & Operations* 27 (3)(2012): 306.

[505] S. S. Shaw, K. S. Sorbie, and L. S. Boak, "The Effects of Barium Sulphate Supersaturation, Calcium and Magnesium on the Inhibition Efficiency: 1. Phosphonate Scale Inhibitors," SPE 130373 (paper presented at the SPE International Conference on Oilfield Scale, Aberdeen, UK, 26-27 May 2010).

[506] N. Laing, G. M. Graham, and S. Dyer, "Barium Sulphate Inhibition in Subsea Systems—The Impact of Cold Subsea Temperatures on the Performance of Generically Different Scale Inhibitor Species," SPE 80229 (paper presented at the SPE International Symposium on Oilfield Chemistry, Houston, TX, 5-7 February 2003).

[507] H. Guan and P. Farmer, "Evaluation of Scale Inhibitors Suitable for Deepwater Fields," SPE 13110 (paper presented at the SPE International Conference on Oilfield Scale, Aberdeen, UK, 26-27 May 2010).

[508] W. Wang, A. T. Kan, and M. B. Tomson, "A Novel and Comprehensive Study of Polymeric and

Traditional Phosphonate Inhibitors for High Temperature Scale Control," SPE 155108 (paper presented at the SPE International Conference on Oilfield Scale, Aberdeen, UK, 30–31 May 2012).

[509] W. Wang, A. Kan, C. Yan, and M. Tomson, "The Use of Inhibition Kinetics and NMR Spectroscopy in Thermal Stability Study of Scale Inhibitors," SPE 164047 (paper presented at the SPE International Symposium on Oilfield Chemistry, The Woodlands, TX, 8–10 April 2013).

[510] S. Kidd, L. W. Hoeth, G. Graham, N. Fleming, and K. Ramstad, "Challenges for Long Term "Field Life" Core Testing," The 20th International Oil Field Chemistry Symposium, Geilo, Norway, 22–25 March 2009.

[511] L. W. Hoeth, E. Sorhaug, G. M. Graham, R. Stalker, S. L. Kidd, and R. Wright, "Effect of Coreflood Test Methodology on Appropriate Simulation of Field Treatments," Tekna 23rd International Oil Field Chemistry Symposium, Geilo, Norway, 18–21 March 2012.

[512] G. M. Graham, S. Kidd, R. Stalker, and R. Wright, "Effect of Coreflood Test Methodology on Appropriate Simulation of Field Treatments," 151779 (paper presented at the SPE International Symposium and Exhibition on Formation Damage Control, Lafayette, LA, 15–17 February 2012).

[513] K. Ramstad, H. C. Rohde, T. Tydal, and D. Christensen, "Scale Squeeze Evaluation through Improved Sample Preservation, Inhibitor Detection, and Minimum Inhibitor Concentration Monitoring," SPE 114085, *SPE Production & Operations* 24 (4)(2009): 530.

[514] M. Rondon Gonzalez, N. Passade-Boupat, P. Bascoul, S. Baraka-Lokmane, and C. Hurtevent, "Selection of Anti-scale and Anti-corrosion Products: How to Avoid Interactions?" SPE 155235 (paper presented at the SPE International Conference on Oilfield Scale, Aberdeen, UK, 30–31 May 2012).

[515] V. González Dávila, International Patent Application WO/2012/158009.

[516] J. Laffitte and B. Monguillon, International Patent Application WO/2010/061146.

[517] K. Sokhanvarian, H. A. Nasr-El-Din, G. Wang, and C. A. De-Wolf, "Thermal Stability of Various Chelates That Are Used in the Oilfield and Potential Damage Due to their Decomposition Products," SPE 157426 (paper presented at the SPE International Production and Operations Conference & Exhibition, Doha, Qatar, 14–16 May 2012).

[518] M. Asakawa, Y. Sumida, and Y. Kita, U.S. Patent 6063302.

[519] A. D. Strickland, D. A. Wilson, D. K. Crump, and B. Burkholder, International Patent Application WO/2000/058263.

[520] C. A. De Wolf, J. N. Lepage, H. Nasr-El-Din, M. A. Nasr-El-Din Mahmoud, E. R. A. Bang, and N. T. George, International Patent Application WO/2012/080299.

[521] E. H. Riad, "Soda Ash as Gypsum Dissolver in 8 Inch Production Line Case Study Gemsa Field, Gulf of Suez, Egypt," SPE 142457 (paper presented at the SPE European Formation Damage Conference, Noordwijk, Netherlands, 7–10 June 2011).

[522] C. V. Putnis, M. Kowacz, and A. Putnis, "The Mechanism and Kinetics of DTPA-Promoted Dissolution of Barite," *Applied Geochemistry* 23 (9)(2008): 2778–2788.

[523] L. Sutherland, M. M. Jordan, E. Sorhaug, and D. Marlow, "New Insights on the Impact of Mass to Volume and Volume to Volume as it Relates to Calcium Sulphate Dissolver Performance," Tekna 23rd International Oil Field Chemistry Symposium, Geilo, Norway, 18–21 March 2012.

[524] M. M. Jordan, E. O. Ajayi, and M. Archibald, "New Insights on the Impact of Surface Area to Fluid Volume as it Relates to Sulphate Dissolver Performance," SPE 153098 (paper presented at the SPE International Conference on Oilfield Scale, Aberdeen, UK, 30–31 May 2012).

[525] S. L. Berry, J. L. Boles, A. Singh, and I. Hashim, "Enhancing Production by Removing Zinc Sulfide

Scale from an Offshore Well: A Case History," SPE 140751, *SPE Production & Operations* 27（3）（2012）: 318.

[526] D. O. Trahan, U.S. Patent Application 20100099596.

[527] N. J. Goodwin, O. G. Svela, J. H. Olsen, T. Tjomsland, B. M. Hustad, and G. M., Graham, "Qualification Procedure for Continuous Injection of Chemicals in the Well—Method Development," SPE 154934（paper presented at the SPE International Conference on Oilfield Scale, Aberdeen, UK, 30-31 May 2012）.

[528] J. Y. Huang, G. Q. Liu, Y. M. Zhou, Q. Z. Yao, Y. Yang, H. C. Wang, L. Ling, K. Cao, Y. H. Liu, W. D. Wu, and W. Sun, *Polymer Engineering & Science* 53（6）（2013）: 1306-1313.

[529] M. Laubender, E. Heintz, C. Seidl, B. Urtel, and A. Berger, Alexsandro, U.S. Patent 8497318, 2013.

[530] C. Hogan, H. Davies, L. Robins, M. Toole, and K. Harris, "Improved Scale Control, Hydrothermal and Environmental Properties with New Maleic Polymer Chemistry Suitable for Downhole Squeeze and Topside Applications," *Chemistry in the Oil Industry XIII: Oilfield Chemistry—New Frontiers*, Manchester Conference Centre, UK, 4-6 November 2013.

[531] J. P. Zeng, F. H. Wang, and X. D. Gong, *Molecular Simulation* 39（3）（2013）: 169-175.

[532] M. Todd, C. Strachan, G. Moir, and J. Goulding, International Patent Application WO2013152832.

[533] G. Q. Liu, J. Y. Huang, Y. M. Zhou, Q. Z. Yao, H. C. Wang, L. Ling, P. X. Zhang, K. Cao, Y. H. Liu, W. D. Wu, and W. Sun, *International Journal of Polymeric Materials and Polymeric Biomaterials* 62（13）（2013）: 678-685.

[534] C. Ping, H. Montgomerie, S. Heath, O. Vikane, B. Juliussen, and T. Hagen, International Patent Application WO/2013/045906.

[535] C. Yan, A. T. Kan, W. Wang, L. Wang, and M. B. Tomson, "Synthesis and Size Control of Monodispersed Al-Sulfonated Polycarboxylic Acid Nanoparticles and Their Transport in Porous Media," *SPE Journal* 18（4）（2013）: 610-619.

[536] C. A. De Wolf, J. N. Lepage, H. Nasr-El-Din, M. A. Nasr-El-Din Mahmoud, E. R. A. Bang, and N. T. George, U.S. Patent Application 20130281329.

[537] M. M. Jordan and L. Sutherland, "How to Correctly Evaluate Scale Dissolvers for Tubing and Near Wellbore Stimulation—Conventional and Novel Methods," *Chemistry in the Oil Industry XIII: Oilfield Chemistry—New Frontiers*, Manchester Conference Centre, UK, 4-6 November 2013.

4 沥青质的控制

4.1 概述

沥青质沉积是石油工业上、下游面临的一个重大问题。沥青质如同原油中的"胆固醇",能堵塞近井地带的储层孔隙,并沉积在井下油管、下游管道及设施中。许多含沥青质的油藏在生产过程中,往往是原油的稳定性受到干扰或破坏后,才会出现沥青质沉积问题。气窜(压降)、凝析处理、气体或气液注入(CO_2气驱或天然气凝析液驱)、酸化增产(见第 5 章)、低 pH 值防垢剂的挤注处理、多种原油混合及潜在的高剪切或流动等因素都能导致沥青质不稳定。如果这种不稳定发生在有带电矿物的地层中,地层就会吸附沥青质,并导致润湿性和渗透率的改变。原油中沥青质含量较高不一定必然导致沥青质沉积,沥青质含量高达 20% 的原油可能不出现沉积问题,而含量仅 0.2%(质量分数)的原油却可能出现沥青质沉积。

稠油中沥青质含量高,在其采出、地面集输和加工过程中存在特殊问题。最近有文献综述了稠油和超稠油的管道集输技术,还探讨了原油高黏、沥青质和石蜡沉积、地层水含量、含盐量、腐蚀等问题。

沥青质没有明确的亲水端和疏水链,因此没有两亲性和典型的表面活性剂特征。但沥青质的分层聚集可分离邻近的水滴,在薄油膜中形成稳定的网状结构有助于稳定油包水乳液。也有资料表明,水可以与芳香环形成 π—氢键。

沥青质属于原油中最重质的组分之一,通常被定义为不溶于轻脂肪烃(如戊烷和庚烷),但溶于芳香烃类溶剂(如甲苯)的原油组分。但某些原油在实验室内加入过量的庚烷后,虽然没有发现沉积的沥青质,但在现场仍可能出现沥青质沉积问题。与沥青质相关的是较低分子量的胶质或简单"树脂",它们也具有多环极性基团,但具有更多的脂肪族侧链,并可溶于庚烷。尽管胶质不会形成破坏性的沉淀物,但一般认为其有助于稳定溶液中的沥青质。易溶的沥青质(含胶质的"过渡材料")对难溶的沥青质有胶溶作用。

沥青质是由具有脂肪链的稠环芳香烃结构、杂原子(如硫、氮、氧)和金属元素(如镍、钒和铁)组成的固体有机物。金属形成络合物并产生电荷,电荷反过来可能影响沥青质沉积。不同油品的沥青质组成元素往往含量各异,均值(质量分数)为碳(76%~86%)、氢(7.3%~8.5%)、硫(5.0%~9.0%)、氧(0.7%~1.2%)、氮(1.3%~1.4%)和微量的金属元素(主要为 Ni、V 和 Fe,含量为 0.1%~0.2%)。沥青质中还有不同的官能团。硫主要以硫化物和噻吩基团的形式存在,少量以亚砜的形式存在;沥青质氮元素主要存在于几乎所有的吡咯基,其次是吡啶基等芳香基中,极偶尔情况下以叔胺的形式存在；氧主要存在于羰基和羟基/酚基中，这包括酮和羧酸。

为掌握沥青质分散剂（AD）、抑制剂（AI）工作机制，有必要搞清楚沥青质的结构、分子量和聚集机理。原油中沥青质单体的结构和芳香环体系的结构大小都是广泛讨论的话题。文献中提出了两种结构模型。一是"大陆"或"岛"模型，其沥青质单体的分子量为500~1000，平均约750，分子结构由6~7个融合芳环组成的核和围绕核的若干个带杂原子的脂肪族基团组成（图4.1）。另一种模型是"群岛"或"念珠型"模型，其沥青质单体由5~7个芳香环的多缩合基团组成，芳香环间由短脂肪侧链连接，也可能由极性杂原子桥接（图4.2）。沥青质的结构范围很大，为说明其基本区别，图4.1和图4.2中绘制了两种模型的典型结构。

图4.1 沥青质分子的"大陆"典型结构

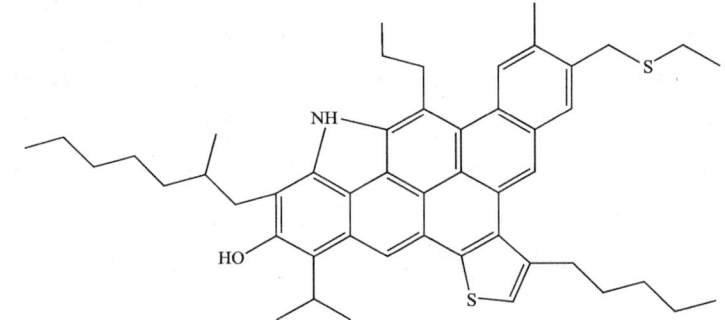

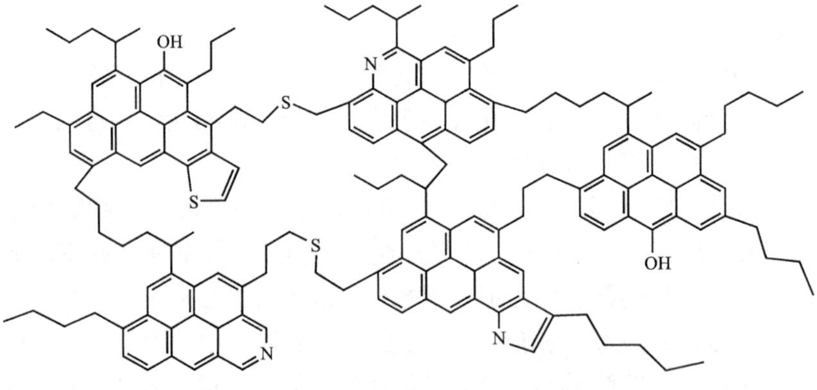

图4.2 沥青质分子的"群岛"典型结构

2008年发表的一篇文章认为，早期通过测量荧光衰减和去极化动力学时间得出的沥青质具有"大陆"模型的结论是错误的，沥青质不具备单一的稠环内核结构。2007年发表的另一篇文章认为，沥青质的分子量是双峰分布，其中一种组分的分子量大致在100万范围内，第二种组分的分子量在5000范围内。后续有文章对双峰分布的认识提出了异议，认为沥青质分子量是750~1000的单峰分布，与"大陆"模型吻合。该文章全面汇总了全球各研究团队采用4种分子扩散技术和7种质谱技术所获得的认识。例如，用双步激光质谱法（L^{2}MS）分析沥青质质谱，解吸和电离在空间和时间上相互独立，避免产生等离子体碎片。有资料分析认为沥青质存在"群岛"结构的可能不大。另一项激光解吸电离—质谱研究表明，与溶解度较高的沥青质样品相比，溶解度低的沥青质平均分子量更高。针对沥青质纳米团聚体及其团簇的纳滤研究表明，其直径尺寸都小于纳滤膜的标称孔径30nm。

高压储层的原油中，沥青质以单分子或者至多以纳米团簇的形式存在。近70年来，人们一直认为具有表面活性剂特性的极性胶质可以使溶液中的沥青质保持稳定。然而，通过沥青质重力梯度、高Q超声波和核磁共振扩散测量的数据表明，极性胶质与储层中的沥青质完全没有关联。而控制原油中沥青质胶束稳定性的两个关键参数是芳香烃与饱和烃的比例、胶质与沥青质的比例。当上述比例降低时，体系中的沥青质单体或聚集物将絮凝，并形成较大的聚集物而沉积。因此，某些单体表面活性剂类的沥青抑制剂（后面将讨论）就起到人造胶质或胶质增强剂的作用，而分散沥青质。在胶质的作用下，沥青质小团簇或聚集物不会沉积到储层中，只有在聚集量较高时才会从溶液中沉积。通过将简单的阴离子或阳离子表面活性剂的浓度从0.01mmol/L提高到1mmol/L，就可扭转和控制沥青质的Zeta电位，这表明在沥青质表面存在静电和疏水相互作用。

在采油过程中，当压力下降但仍保持在泡点压力以上时，原油的密度下降，轻质组分（C_{6-}）所占体积就会增加，且比不易压缩的更重组分（含芳香烃的C_{7+}）增加得更快。这样原油的极性降低，沥青质可能开始缔合并最终絮凝。这种情况通常发生在井下射孔段，也可能发生在全系统包括处理设备在内的任何地方。因此，生产过程中的压降最小化是控制沥青质沉积的一种方法。最近有关资料讨论了原油生产和精炼过程中，石油沥青质溶解度随温度、压力和组成的变化。

沥青质自聚集是一个广泛讨论的话题。石油沥青质最初的聚集体（或胶束）大小约为20Å❶或8～10个沥青质分子。另一些研究将初始沥青质聚集体描述为多分散的扁柱状物或松散、非球形的碎颗粒。聚集体的大小取决于溶剂的极性。然而在与胶质或合成沥青分散剂和沥青抑制剂的混合物中，这些聚集体不一定会团聚或沉积在管壁上。事实上，即使存在肉眼可见的沥青质颗粒，也可能不会发生沉积。虽然沥青质沉积一般认为是不可逆的，但也有相反情况的证据。沥青质絮凝是可逆的，其中加入胶质或分散剂可以防止其沉积。也可以通过化学处理剂将沉积的沥青质恢复为稳定态。这为控制沥青质沉积提供了有潜力的解决方案，特别是对于平台顶部的原油生产设施。还有研究人员提出从原油中分离活性胶质，利用其来优化分散沥青质的方法。

研究表明，聚集态的沥青质分子结构主要受控于稠合芳环的π键堆积能力（降低溶解度）和烷基空间位阻的破坏（增加溶解度）之间的平衡。沥青质中部分极性分子间也会发生如酸碱（电子供体—受体）和氢键作用等相互作用。有研究表明，在沥青质沉积体周围存在烷基层，这更符合沥青质单体的"大陆"模型。基于对许多实验结果的解释，提出了沥青质的胶束模型。该模型认为存在环绕芳香内核的烷基组分。还有研究认为，沥青质中

---

❶ $1Å=10^{-10}m$。

最难溶的组分会聚集形成高孔隙态的不溶颗粒。通过分子动力学模拟方法评价了侧链长度对沥青质聚集模式的影响。发现随着烷基侧链长度的增加，聚集作用减弱。

控制沥青质沉积的方法已有综述。沥青质控制的非化学法技术包括：

（1）避免某些类型原油的混合。原油混合是沥青质沉积的常见原因。不含沥青质的轻质原油可能是重质原油的沉淀剂。

（2）在沥青质形成区间（AFE）外生产。通过控制温度、压力或流速，最大限度地减少沥青质沉积条件的发生，提高油井生产和设备的运行效率。

（3）井和地面设备的机械清除：包括电缆清刮和定期清管、打开容器（如分离器）清理。

（4）高速流体冲洗沉积物。

有报道称超声波可辅助清洗井筒沥青质。还有研究采用陶瓷膜分离原油中的沥青质。

本章将详细讨论生产作业中控制沥青质的两种化学方法。

（1）用沥青质分散剂和沥青质抑制剂预防（连续或间歇注入、挤注处理、包覆药剂的压裂支撑剂）。

（2）用沥青质溶解剂补救处理（溶剂或脱沥青油）。

还有使用相对渗透率调节剂（RPM）的化学方法阻碍随后的沥青质沉积的方法（有关渗透率调节剂的更多信息，可参阅第2章）。有报道使用抑制储层沥青质伤害的纳米颗粒。在典型的储层压力和温度条件下，研究了纳米颗粒在多孔介质中的运移。结果表明，使用纳米颗粒可使体系顺利流动，抑制沥青质的沉淀和沉积，并增强持久性效果。

4.2 沥青质分散剂和沥青质抑制剂

沥青质分散剂和沥青质抑制剂可以防止沥青质沉积。沥青质抑制剂真正起抑制作用，阻止了沥青质分子的聚集，可以改变沥青质絮凝的起始压力值。这样能将井筒内的沥青质沉淀和随后的沉积，转移到生产系统中更容易处理的其他部位。例如，可以利用光纤传感器的光传输测定沥青质絮凝点，并以此对沥青抑制剂的性能进行排序。沥青质分散剂不影响沥青质絮凝点，但会减小絮凝后的沥青质粒径，使其在油中保持悬浮状态，还同时起到降低原油黏度的作用。最近有研究评价了某些分子结构已知和专有化学剂混合物的沥青质化学添加剂，表明其都不能抑制相分离，即都不具有真正的沥青质抑制剂性质。但一些沥青质分散剂可以有效减缓或停止沥青质的絮凝和生长。沥青质分散剂分散了已经形成的沥青质絮凝。很多沥青质抑制剂具有分散剂的功能，但沥青质分散剂则通常都不具有抑制剂的功能。

本章中已提及许多不同类别的沥青质分散剂和沥青质抑制剂。但只有少数在现场实际应用，并且还有部分是专利配方。例如，烷基芳香基磺酸、丙烯酸酯或马来酸酯的聚合物、酚醛树脂、聚异丁烯丁二酸酐衍生物、脂肪酸缩合物或脂肪环氧化物与（聚）胺或多元醇的加成物、脂肪醇或脂肪胺与羧酸的缩合物。

沥青质抑制剂和沥青质分散剂对原油具有选择性。例如，具有极性头和脂肪尾的聚合物沥青质抑制剂可以防止原油 A 中的沥青质沉积，但却不能在原油 B 中起效。相反，非聚合物胺在原油 B 中表现良好，但在原油 A 中却无效。这些现象可以从酸碱化学的角度来解释：碱性的非聚合胺会与原油 A 中的有机酸优先反应，而非沥青质分子；而质子化聚合物抑制剂与沥青质发生作用，不会与原油 A 中的酸性物质反应；原油 B 的沥青质中有较多的碱性氮元素，有利于与质子化聚合物抑制剂的相互作用。

一般来说，沥青质抑制剂是聚合物（或树脂），而沥青质分散剂通常是非聚合物的表面活性剂，但许多聚合物表面活性剂的沥青质抑制剂也具有沥青质分散剂的功能。为防止沥青质分子的聚集，沥青质抑制剂需要聚合物的若干分子相互作用点，来产生良好的抑制作用，因此需要聚合物形式。此外，如果沥青质抑制剂包含烷基长链，就有助于分散已形成的沥青质聚集体。这被认为也发生在非聚合沥青质分散剂中。沥青质分散剂的表面活性剂极性或芳香基头端与聚集的沥青质相互作用，而沥青质聚集体周边的长烷基链有助于改变聚集体外部的极性，在相似相溶效应下更好分散到原油中。一些研究表明，增加沥青质分散剂的用量后，可能因表面活性剂分子的自缔合作用，反而加剧沥青质的聚集。

对于沥青质抑制剂，一项关于沥青质在庚烷中沉淀的研究表明，在沥青质抑制剂的临界浓度以下没有任何效果；但达到临界浓度及以上，就会有阻止沥青质絮凝的明显效果。相同的研究中，沥青质分散剂则没有临界浓度效应，其作用效果几乎与浓度成正比。对一些市售沥青质抑制剂的研究表明，某些产品并没有降低沥青质絮凝的起始压力或平均粒径。而相对于未处理的样本，沥青质抑制剂确实减少了累计粒子数量。还提出一种关于抑制沥青质沉淀的热力学模型，该模型将沥青质按胶束处理。沥青质与两亲分子（天然胶质或合成添加剂）之间的吸附相互作用，被认为是原油中沥青质胶束稳定性的最重要参数。研究者还进一步提出，吸附焓可以作为寻找高效两亲化合物的最重要标准。

至少有两个研究小组已经表明，与未处理的沥青质体系相比，一些聚合物沥青质抑制剂实际上可以增加絮凝沥青质的含量。在另一项研究中，针对沥青质为中等质量分数的墨西哥湾原油，聚合物沥青质抑制剂在低浓度（100mg/L）时，对沥青质分散剂测试有负面影响，而在高浓度（500mg/L）时，该产品非常有效。

有研究者认为，这种效应是由于沥青质抑制剂分子的自组装，并被溶剂的相对亲水性或疏水性相互作用引发。还建立了一个沥青质沉积模型，也可以解释这种影响。

由于沥青质抑制剂能防止沥青质絮凝，因此最好在上游高泡点压力的井筒内应用，沥青质分散剂可以在下游使用，这样沥青质抑制剂处理不够充分时，可以有效分散絮凝沥青质，防止沉积。因为沥青质抑制剂本身通常有其特定的分散特性，油田一般只使用两类药剂中的一种，而不是同时使用。沥青质抑制剂通过毛细管柱或气举系统连续或间歇注入井内，也可以从井口注入下游。优质的商品沥青质抑制剂通常加量范围为 20～100mg/L。建模工作表明每个系统都有抑制剂的最佳使用浓度。有些报道公开了近井地带的沥青质抑制剂挤注处理，这种技术研究越来越普遍。技术的关键是使油溶性沥青质抑制剂对地层岩石有良好的吸附作用，而不是快速被采出。否则挤注处理的有效期短，不经济。有室内研究强调，如果对岩石的吸附能力较差，抑制性能最好的沥青质抑制剂也不一定是挤注的最佳选择。依据沥青质问题的严重程度，当前挤注处理的有效期为2～6个月。为辅助沥青质抑制剂吸附到岩石上，可以在聚合物中引入羧基等极性基团。膦酸基比羧酸基吸附力更强，也可以引入沥青质抑制剂聚合物中。

生产化学品的实验室测试仅是对实际现场条件的粗略近似，沥青质抑制剂和沥青质分散剂的研究方面尤为如此。有研究者开发了利用长不锈钢毛细管的新型沉积装置，以研究各种油类产生沉积物的趋势。通过模拟油流生产的现场原油试验，得到的沥青质沉积物极性比庚烷沉淀沥青质的更强，在组成上也有所不同。作者推断，现场沥青质沉积趋势与实验室筛选试验的相关性，对沥青质研究的发展具有重要意义。还有研究用石英晶体谐振器测量了沥青质的起始絮凝过程。

在检测沥青质抑制剂和沥青质分散剂时，使用新鲜原油和老化原油会产生不同的结果。粗略筛选试验不同类别的沥青质分散剂和沥青质抑制剂时，可以选用老化原油；而在压力系统或压降回路中，建议采用高压新鲜原油进行沉积试验。

有研究根据沥青质组分的一些相关溶解特性，提出了一种简易的新方法筛选沥青质的不稳定性。该方法利用新鲜原油的PVT特性和沥青质表征的简单流程方法，将后续精细研究聚焦于存在实际沉淀风险的流体。也可以采用老化原油样品表征沥青质。

一些室内研究使用了在芳香烃溶剂中重新溶解的沉淀沥青质。其中缺乏天然胶质，这就意味着依靠天然胶质才体现最佳性能的沥青质抑制剂和沥青质分散剂，不会取得良好的效果。有研究认为，沥青质稳定试验使用未减压的新鲜原油时，所需的沥青质抑制剂浓度会低于使用老化原油的标准实验室沉淀方法。

沥青质抑制剂和沥青质分散剂测试方法的新进展是使用更具有实际作用的CO_2，而不是烷烃来沉淀沥青质。添加剂在测试浓度约为500 mg/L时，效果达到稳定，也接近于现场使用浓度。

在控制沥青质的添加剂综述中，首先讨论常用作沥青质分散剂的单体表面活性剂。随后讨论聚合物类沥青质抑制剂（其中的一部分也同时是沥青质分散剂）。市售的聚合物通常既用作沥青质抑制剂，也用作沥青质分散剂，但有些单体表面活性剂在现场也有成功的应用。有一类市售的聚合物沥青质抑制剂产品在井下和地面系统都有应用，与破乳剂可以协同作用，去除油水界面的沥青质，实现油水分离。这种聚合物除了提及只含有碳、氢和氧外，没有其他结构细节。应用中可与生物降解溶剂配制，环保特性更佳。

4.2.1 低分子量、非聚合物沥青质分散剂

各类低分子量、单体沥青质分散剂可归类为极低极性的烷基芳香烃化合物、烷基芳香基磺酸、膦酸酯和膦羧酸、肌氨酸盐、两性表面活性剂、醚羧酸、氨基炔羧酸、烷基酚及其乙氧基化合物、咪唑啉和烷基酰胺咪唑啉、烷基琥珀酰亚胺、烷基吡咯烷酮、脂肪酸酰胺及其乙氧基酯、多元醇的脂肪酯、亚胺和有机酸的离子对盐、离子液体。

多年来，上游油田和炼油厂已在应用中将烷基芳香基磺酸及其碱盐（如十二烷基苯磺

酸及其镁盐）作为沥青质分散剂。现场也有使用如烷基酚（内分泌干扰物）和一些酰胺/亚胺产品等的其他表面活性剂。碱性或酸性沥青质分散剂的选择，取决于原油的类型及其极性和化合物含量。

4.2.1.1 低极性芳香族两亲化合物

在所有类型的沥青质控制添加剂中，先讨论极性最低的沥青质抑制剂。十六烷基萘和十六烷基萘氧化物被视为沥青质抑制剂，而非分散剂（图4.3）。据称，这些化学剂主要是阻止了沥青质的沉淀，而不是沉淀物的分散。

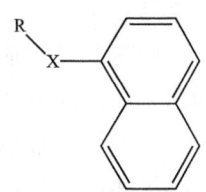

图4.3 基于萘的沥青抑制剂（R=长烷基链；X=醚、酯或酰氨基等间隔基团）

这些低极性分子通过萘的芳香环和沥青质单体之间产生π—π相互作用，也通过极性基团与沥青质相互作用。这种相互作用防止沥青质单体的堆积和聚集，而脂肪族尾部将与烃类溶剂起到互溶作用。与单个苯环相比，双环的萘基有更好的π—π相互作用。但比萘基极性更强的端基，可能与沥青质单体的相互作用更强。例如，有研究者测量了沥青质聚集物在几种溶剂中的中子和X射线散射强度，发现吡啶和四氢呋喃等强极性溶剂中的聚集体尺寸比在苯中的小2~4倍。另一组研究者发现芳香烃溶剂（如甲苯、二甲苯）和喹啉或烷基喹啉的混合物，在溶解沥青质能力方面，比芳香烃溶剂更强。二甲基甲酰胺和N-甲基吡咯烷酮（NMP）也是比芳香烃更好的沥青质溶剂。这意味着相比于苯基或萘基等弱极性端基，含有如吡啶基、喹啉基、四氢呋喃基和二甲氨基等强极性端基的沥青质抑制剂，能更好地与沥青质相互作用。分子模拟研究结果也证实了该认识，模拟显示喹啉中的沥青质团聚体，某些堆积相互作用可能被扰乱，而在1-甲基萘没有观察到类似效果。因此，十六烷基萘和十六烷基萘氧化物可能没有实现与沥青质最佳相互作用的强极性端基。

4.2.1.2 磺酸基非聚合表面活性剂沥青质分散剂

一种提升沥青质分散剂（或抑制剂）芳香端基极性的方法是添加磺酸基。该类中最常见的沥青质分散剂是十二烷基苯磺酸（DDBSA），价格便宜并且已经有成熟产品的成功应用（图4.4）。长链烷基芳香基磺酸也获得了专利。十二烷基苯磺酸的高碱性盐在油田也有应用。

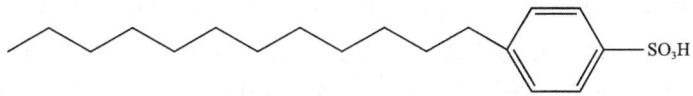

图4.4 十二烷基苯磺酸

有研究团队研究了十二烷基苯磺酸的沥青质分散和溶解能力。例如，发现十二烷基苯磺酸在低浓度时是絮凝剂，在高浓度时是分散剂。另一个团队研究了一些单体添加剂的性能，通过较强的酸碱作用，增强端基极性可以使沥青质更稳定。因此，十二烷基苯磺酸的沥青质分散剂性能优于壬基酚（NP），后者又优于壬基苯。傅里叶变换红外（FTIR）光谱也显示十二烷基苯磺酸通过氢键与沥青质相互作用。除酸碱作用外，该小组还提出磺酸基团可以将其质子提供给沥青质中的C=C键。但他们也承认过多的极性基团或一个极性过强的基团都可能降低表面活性剂在原油中的溶解度，使其对沥青质分散无效。为了支持该说法，另一个团队发现有一个强极性离子端基的十二烷基苯磺酸钠，沥青质分散性能相当差。其他研究人员也发现了类似的结果。例如，十二烷基间苯二酚、十二烷基苯磺酸、壬基酚和甲苯的沥青质分散性能依次下降。他们提出表面活性剂的作用是由于分子的酸性端基和沥青质之间的相互作用，用胶束模型解释了这种抑制机理。

另有研究结果显示，在芳香基端基连接一个极性更强的基团会具有更好的沥青质分散性能。结果表明，按沥青质分散性能排序，十二烷基苯磺酸＞壬基酚＞十二烷基酚双乙氧基酯＞壬基苯。唯一的顺序异常是，尽管十二烷基酚双乙氧基酯具有更强极性的端基［较高的亲水－亲油平衡值（HLB）］，但其表现不如壬基酚。这可能因为简单的醇类或醇聚氧乙烯酯与沥青质中极性基团的相互作用较差，不是很好的沥青质分散剂的端基。专利文献似乎证明了这一点，因为没有发现任何种类的聚氧基酯表面活性剂可以作为沥青质分散剂。原因是磺酸基中的强酸性质子比只有一个羟基质子能更好地与沥青质中的胺残基产生氢键作用。该团队还确定上述两亲物质通过两步吸附机制（LS型或S型）吸附到沥青质颗粒。第一步是两亲物质吸附到沥青质上；第二步是两亲物质吸附到被吸附的两亲物质上，形成双层。

十二烷基苯磺酸沥青质分散性能的相关改进，似乎已经走向了两个方向。在一项专利中，移除了十二烷基苯磺酸的芳香基端基，只留下脂肪族烷基链直接与磺酸盐结合。发明人声称仅使用链长为8~22个碳的仲烷烃磺酸，最好是11~18个碳的沥青分散剂（图4.5）。这类烷烃磺酸优选配制为溶液或微乳液，并可进一步包含烷基甲醛树脂、氧烷基化胺或蜡分散剂。烷烃磺酸降低了沉淀形成速度和沉淀量，形成的沉淀更细小，降低了沉淀物在表面上沉积的趋势。唯一的仲烷烃磺酸测试例子没有与任何其他沥青质分散剂进行比较。发明人称选用仲烷烃磺酸而不是伯烷烃磺酸的原因，可能是两个烷基尾链与烃溶剂的相互作用更好。

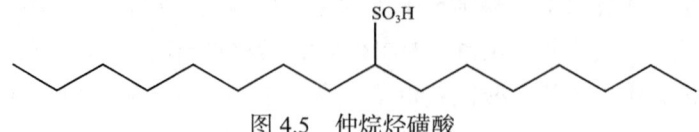

图4.5 仲烷烃磺酸

在十二烷基苯磺酸的第二个改进方向中，有专利使用支链烷基聚芳香类磺酸。优选的结构是尾链至少有一个分支的磺化烷基萘，最好尾链有30个以上碳。图4.6给出了该类中一种性能良好的沥青质分散剂（正构十五碳和异构十五碳）。磺酸基被认为优于羧酸、羟基和氨基，是最有效的连接芳香烃类结构的端基。直链烷烃尾的碳链数大于16个碳后无效，这是由于其与其他沥青质分散剂尾链和油中蜡的结晶，而降低了在油中的溶解度。此外，如果沥青质分散剂的芳香基部分是一个环或两个连接环（联苯），则性能不如两个融合环体系（如萘）。此外，研究人员发现，如十二烷基苯磺酸等的N-烷基芳香基磺酸，随着时间的推移会失去其分散沥青质的能力。这两个问题都通过使用两个长度不同的分支尾链来解决。结果表明，分散剂的有效性随尾链总长增加而增加，远高于30个碳，并长期保持有效。这些磺化烷基萘似乎是研究中的最佳磺酸基单聚体表面活性剂。

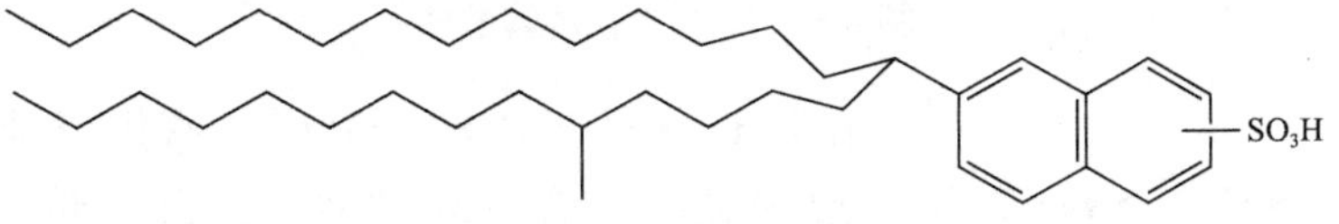

图 4.6　支化烷基（正构十五碳和异构十五碳）萘磺酸

4.2.1.3　其他具有酸性端基的非聚合表面活性剂沥青质分散剂

有几项关于非磺酸基酸性端基的表面活性剂类沥青质分散剂研究。一个研究团队研究了小分子烷基羧酸两亲物和烷基胺。他们提出两亲物质稳定沥青质的效果取决于其吸附在沥青质表面的能力。例如，他们发现己基苯甲酸比己胺更容易吸附在沥青质表面。而这两种添加剂的吸附效果都不如双十二烷基苯磺酸和十二烷基苯磺酸。与磺酸或羧酸等酸性基团相比，碱性烷基胺的效果很差，这表明两亲分子中的酸性基团与沥青质分子中的碱性位点（如胺、羟基）形成的氢键，比两亲分子的氨基与沥青质分子的酸性基团间的结合更紧密。这也反映在沥青质中含氧基团（羟基、羧基）的量要低于含氮基团（胺）。

在专利文献中提到较多的是在沥青质分散剂两亲体的端基使用如磺酸或羧酸等高酸性基团。虽然普通脂肪酸尚未在专利中提出，但已经提出将醚羧酸和膦酸酯作为沥青质分散剂（图 4.7）。两者都含有高酸性质子，这些质子可以与沥青质中的胺或羟基形成氢键，或者与金属离子相互作用，破坏沥青质的聚集过程。目前尚不清楚为什么醚羧酸中引入聚醚（聚烷氧酸酯）链比简单脂肪酸具有优势。聚醚链使两亲体的端基更具极性，并可能通过氧原子与酸性质子或金属离子相互作用。

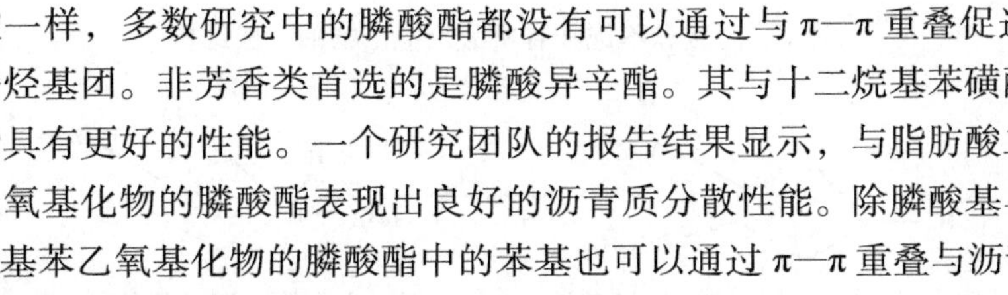

图 4.7　醚羧酸和膦酸单酯的结构（R= 烷基或烷基芳香基，R_1 和 R_2 为 H 或 Me）

与醚羧酸一样，多数研究中的膦酸酯都没有可以通过与 π—π 重叠促进与沥青质相互作用的芳香烃基团。非芳香类首选的是膦酸异辛酯。其与十二烷基苯磺酸的混合物比单独的膦酸酯具有更好的性能。一个研究团队的报告结果显示，与脂肪酸二乙醇酰胺共混的烷基苯乙氧基化物的膦酸酯表现出良好的沥青质分散性能。除膦酸基与沥青质之间的氢键外，烷基苯乙氧基化物的膦酸酯中的苯基也可以通过 π—π 重叠与沥青质相互作用（图 4.8）。膦酸酯与羧酸或羧酸衍生物的协同混合物也获得了专利，其性能明显优于后面讨论的壬基酚甲醛树脂。另一项专利中将膦酸（如膦丁二酸）与 C_6—C_{25} 烷基、烷基芳香基或烯基的醇类，酯化制成膦羧酸酯的沥青分散剂。

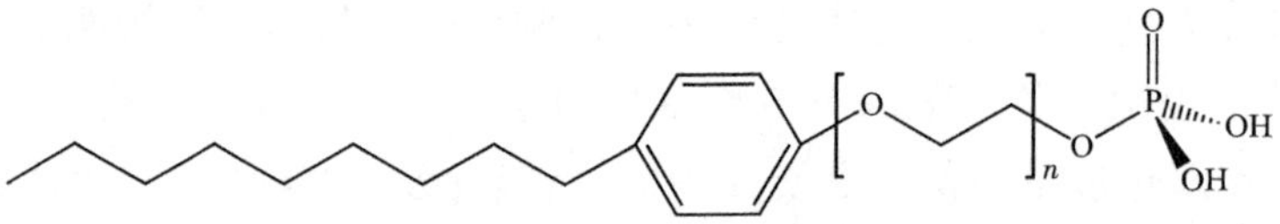

图 4.8 烷基苯氧基膦酸酯

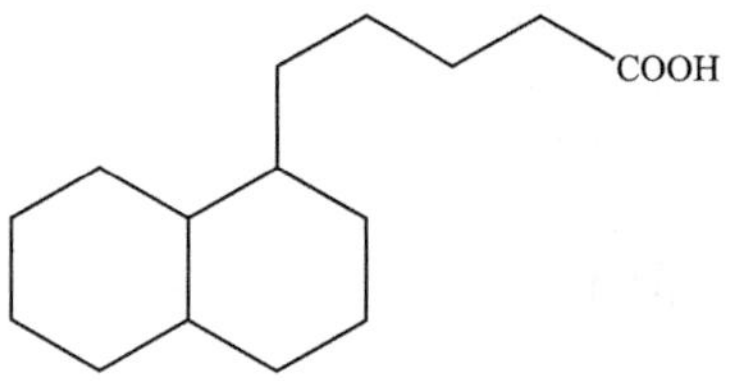

使用近红外光谱法研究了具有沥青质分散能力的天然环烷酸和一些合成环烷酸。结果表明，环烷酸能吸附并分散沥青质，但其效果低于一种市售脂肪酸和氨基的沥青质分散剂。合成的最佳环烷酸为十氢化-1-萘戊酸（图 4.9），这种环烷酸的环非芳香烃类，因此不能与沥青分子发生 π—π 重叠。因此，烷基苯基羧酸的性能可能优于环烷酸。采用动态光散射、近红外光谱和分子模拟等方法研究了沥青质在两种环烷酸、二酸甲酯和氢化二酸甲酯存在下的聚集沉降行为。环烷酸的存在，延迟了沥青质的絮凝。

图 4.9 十氢化-1-萘戊酸

其他在两亲体端基含有羧酸的沥青分散剂也获得了专利。例如，肌氨酸盐表面活性剂经室内测试验证是有效的沥青质分散剂，这类结构如图 4.10 所示。这类化合物可以通过酰胺和羧酸基团与沥青质形成氢键。

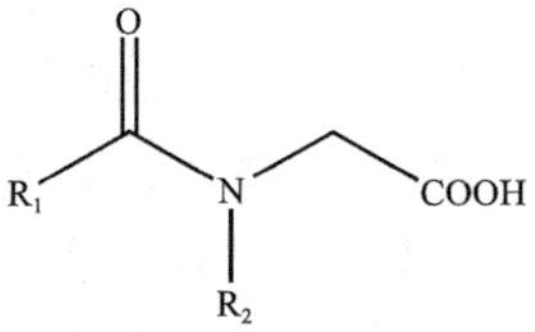

图 4.10 肌氨酸沥青质分散剂分子结构（R_1 和 R_2 为较长的烷基或烯基链）

另一类含酸性端基的两亲性沥青质分散剂是胺和不饱和有机酸的反应产物（图 4.11）。

此类沥青质分散剂中，优选的例子是油酰胺与丙烯酸以 1∶2 比例的反应产物，其结构是 Michael 加成产物，如图 4.12 所示。其与水接触时可能以两性离子的形式存在，但在如烃等非极性溶剂中则不太可能发生。

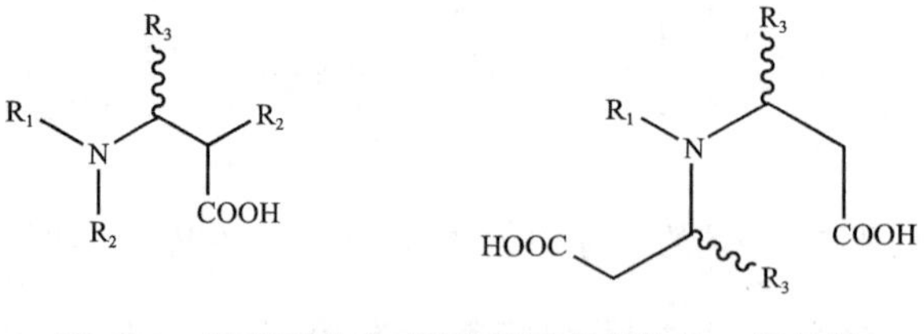

图 4.11 胺和不饱和有机酸反应产物的一般结构

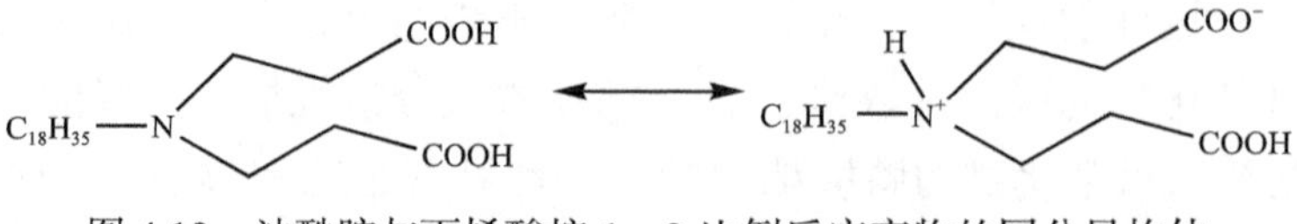

图 4.12 油酰胺与丙烯酸按 1∶2 比例反应产物的同分异构体

4.2.1.4 酰胺和酰亚胺类非聚合表面活性剂沥青质分散剂

很多带有酰胺或酰亚胺基团的非离子两亲体可以作为沥青质分散剂，其中的部分已商业化应用。已研究的最简单酰胺类只含有一个酰氨基，没有其他官能团。包括长链 N,N-二烷基酰胺和烷基吡咯烷酮（图 4.13）。烷基吡咯烷酮含有一个五元环，与沥青质中的吡咯基类似。

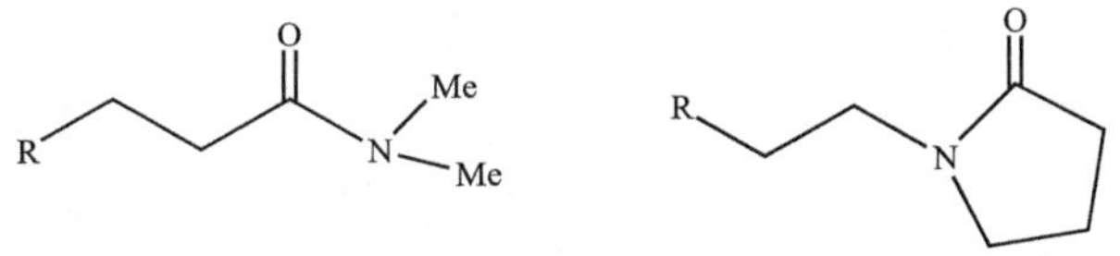

图 4.13　N,N-二烷基酰胺和烷基吡咯烷酮（R 最好大于 8 个碳）

聚异丁烯丁二酰亚胺是小型两亲体，也是沥青质分散剂。其分子含有一个酰亚胺端基，可以通过任一羰基与沥青质形成氢键，如图 4.14 所示。该专利还提出，聚异丁烯丁二酰亚胺与其他两亲物质的混合物可以重新分解沉淀的沥青质。与此相关的另一项专利提出了一种胺与聚异丁烯丁二酸酐（链上有 C_{50}—C_{70}）的反应产物。胺可以是脂肪胺、叔烷基胺或如二乙烯三胺等多胺。为了达到最好的效果，反应产物应同时含有羧基和酰氨基，及一个或多个烷基尾链。这些添加剂还可以通过防止沥青质结块来降低原油的黏度。

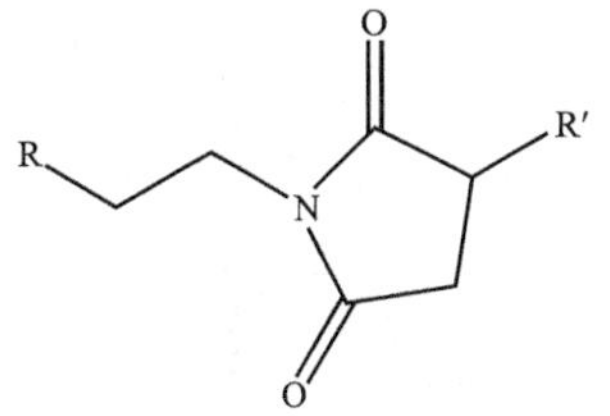

图 4.14　聚异丁烯丁二酰亚胺等烷基丁二酰亚胺的常见结构

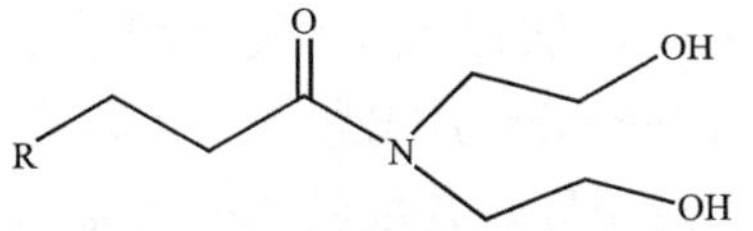

图 4.15　脂肪酸二乙醇酰胺的结构

实验表明，脂肪酸二乙醇酰胺与烷基苯基乙氧基膦酸酯的共混物是较好的沥青质分散剂。前述已经讨论过膦酸酯。脂肪酸二乙醇酰胺结构如图 4.15 所示。这类化合物有三个官能团，能与沥青质形成氢键。

更复杂酰胺产物也是沥青质分散剂，并可与烷基芳香基磺酸和乳化剂混合。其是由脂肪酸与一种多胺缩合得到，多胺分子式为 H_2N—$[(CH_2)_n$—$NH]_m$—R，其中 $n=1\sim4$，$m=1\sim6$，R=H 或烷基，如典型的多胺是二亚乙基三胺。这种三胺与 2mol 脂肪酸（如高尔酸）反应生成无环酰胺和咪唑啉环产物，如图 4.16 所示。同样，有类似于沥青质的吡咯基的环状结构。这类沥青质分散剂中的环亚胺和酰胺基团可以与沥青质中的酸性质子氢键结合，而不饱和咪唑啉环和沥青质中的芳香环之间也可能存在一些 π—π 重叠。与 N-甲基吡咯烷酮、N-乙基吡咯烷酮、二甲基甲酰胺和芳香烃等溶剂混合的类似缩合产物据称能抑制和溶解沥青质，也获得了专利授权。

由酰氨基表面活性剂等两种组分混合的沥青质分散剂也获得了专利，酰胺表面活性剂由直链 N-烷基多胺与环酐缩合而成，如马来酸酐和 N-油酰二氨基-1,3-丙烷的反应产物，或邻苯二酸酐和 N- 硬脂基甲基 -1- 二氨基 -1，3- 丙烷的反应产物（图 4.17）。两种结构都含有五元环和亚胺基团，可以通过酸碱相互作用吸附沥青质的吡啶基团。酞酰亚胺结构中还含有芳香环，可以通过 π—π 相互作用与沥青质中的芳香环进行吸附。混合物中的第二组分为乙氧基胺与 8～30 个碳原子羧酸的反应产物。例如，三乙醇胺与牛油脂肪酸以 1：2 比例进行双酯化反应得到的化合物。这些混合物能延缓燃料中沥青质的絮凝。

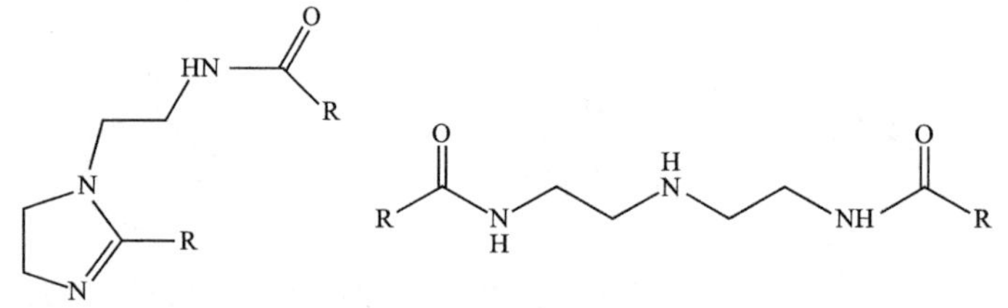

图 4.16　二亚乙基三胺与 2mol 脂肪酸缩合的可能产物（R 最好是 C_{16}—C_{18}）

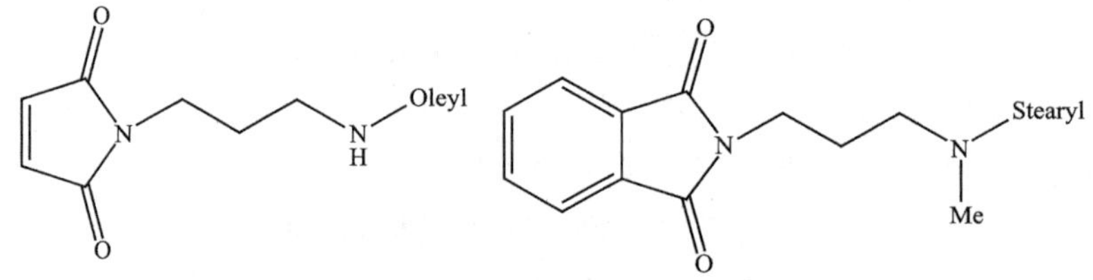

图 4.17　马来酸酐与 N- 油酰二氨基 -1，3- 丙烷反应产物及邻苯二酸酐与 N- 硬脂基甲基 -1- 二氨基 -1，3- 丙烷反应产物

4.2.1.5　烷基酚及相关沥青质分散剂

因烷基酚沥青质分散剂是一类海洋内分泌干扰物，在某些环境敏感地区不能使用（见第 1 章），但可作为下游的市售产品。几个团队研究了单体烷基酚沥青质分散剂，其弱酸性的酚端基与沥青质结构有部分相似（图 4.18）。在上文中，对单体烷基酚的部分效果与烷基芳香基磺酸进行了比较。有研究发现，十二烷基苯磺酸及其钠盐比极性较低的烷基酚（烷基链上有 2～12 个碳原子）表现得更好。烷基酚的性能随烷基尾链长的增大而提升，此类沥青质分散剂中以十二烷基酚（DDP）的性能最好。还有研究团队也对不同脂肪链长度的烷基酚进行了研究。据称，脂肪链长度较短的表面活性剂不能通过形成空间稳定层而胶溶化沥青质，只是嵌入沥青质中或者与沥青质共沉淀。反之，烷基链过长，则可能导致表面活性剂与沥青质的相互作用变差。

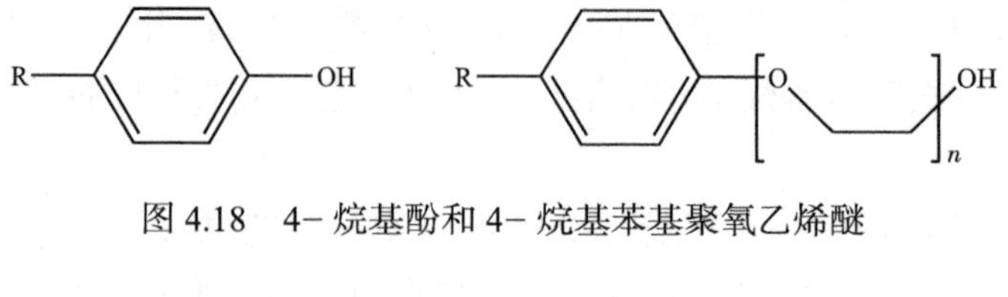

图 4.18　4- 烷基酚和 4- 烷基苯基聚氧乙烯醚

还有研究分析了壬基酚（NP）、十二烷基酚（DDP）和一种壬基酚乙氧基化物（NPE）对沥青质的稳定性影响。壬基酚（NP）、十二烷基酚（DDP）、壬基酚乙氧基化物（NPE）的沥青质抑制性能依次降低。在该研究中，通过乙氧基化使两亲体的端基极性更强，其性能比未乙氧基化的烷基酚差，但与之相反的结果也有发表。另一研究团队研究了壬基酚乙氧基化物（NPE）的沥青质胶凝能力，发现其性能优于脂肪胺和醇。显然带芳香环的表面活性剂比脂肪基表面活性剂要好。壬基酚（NP）也优于壬基酚乙氧基化物膦酸酯。用Langmuir槽和Brewster角显微镜研究了沥青质及其与非离子表面活性剂（聚氧乙烯壬基酚）混合物的界面膜性质。HLB值为14.2时，表面活性剂在防止油水界面的沥青质吸附方面最有效。

有研究报道了从腰果壳液（CNSL）提取的酚类化合物对沥青质稳定性的作用。腰果壳的提取液几乎全部由酚类化合物组成，含有15个不同饱和度的碳链，在芳香环中有间位取代。结果表明，具有C_{15}烷基链的腰果壳提取液CNSL和腰果醇的沥青稳定性能与壬基酚（NP）相当。

烷基酚的活性源于其含有一个π—相互作用的芳香环和极性氢键基苯酚。这两个特征已经被纳入新的一类单端基两亲化合物，能够抑制和溶解沥青质。这些两亲化合物是醚羧基苯基酯，其结构有苯环和极性链，极性链由一个酯基和一个或多个烷氧基链组成（图4.19）。

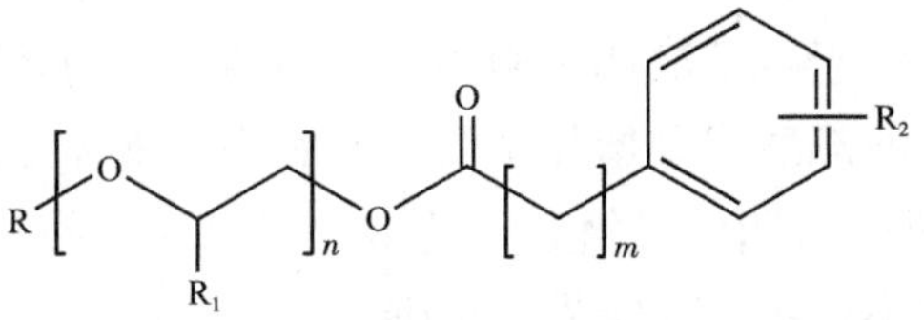

图4.19 优选的聚醚羧基苯基酯作为沥青分散剂（R=H 或 CH_3，$R_2=C_1$—C_4烷基）

4.2.1.6 离子对表面活性剂沥青质分散剂

油溶性离子对表面活性剂在20世纪90年代初首次用作沥青质分散剂。除前面讨论过的表面活性剂与沥青质之间的相互作用外，此类沥青质分散剂还可能选择性地与沥青质中的金属结合，提高其吸附性能。烷基芳香基磺酸和烷基咪唑啉的混合物就是如此。酸质子被咪唑啉夺走，形成了阴离子-阳离子对。已有多个专利申请涉及离子对表面活性剂沥青质分散剂。例如，胺和有机酸的油溶性盐反应产物已被提出，其首选化学式为$R^1R^2R^3N^+R^4COO^-$，分子也可能包含一个与羧基相距2~10个碳的极性基团。脂肪酸胺与2-羟基丁酸或水杨酸的反应产物如图4.20所示。这些盐呈油溶性，以离子对的形式存在。

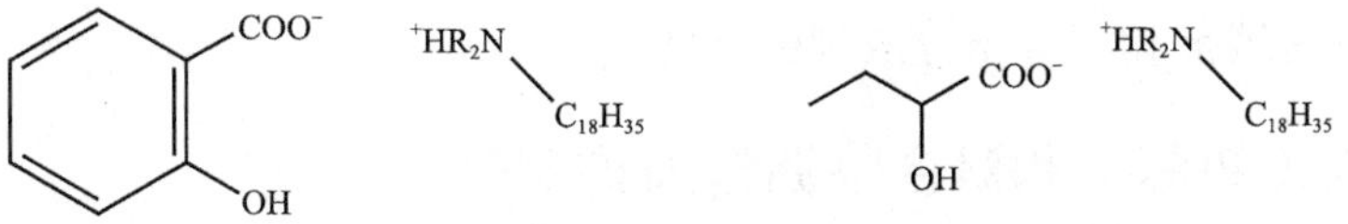

图4.20 烷基胺和有机羟基酸反应产物的离子对盐的结构

在同一专利申请人的另一项相关专利中，亚胺和有机酸的反应产物也是沥青质分散剂。其首选的结构是由长链叔烷基亚甲基亚胺和羧酸（如乙醇酸）反应形成的盐（图 4.21）。这些盐也是油溶性，也可能以离子对的形式存在。

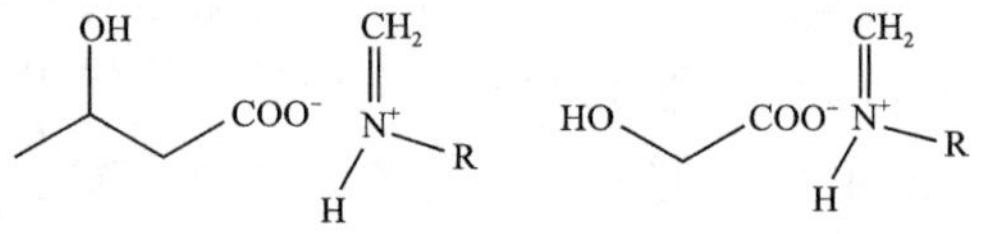

图 4.21　亚胺与有机酸反应产物离子对盐的结构

其他离子对沥青质分散剂由长链不饱和烷基咪唑啉和至少具有一个羟基或至少一个额外的羧基（如抗坏血酸或草酸）的 C_2—C_{10} 有机酸组成。烷基咪唑啉与 EDTA-叔烷基伯胺络合物和 10%~80% 的烷基双（2-羟乙基）酰胺也可以共混。

4.2.1.7　其他非聚合物沥青质分散剂

针对沥青质沉淀，已经开展了包括阳离子表面活性剂离子液体等在内的表面活性剂沥青质分散剂的一系列研究。例如，正丁基异喹啉阳离子与对烷基苯磺酸的性能相当。但阳离子表面活性剂上的长烷基链会使性能变差。最好的离子液体是高电荷密度阴离子与低电荷密度阳离子的结合物。其机理是离子液体通过破坏油层中的沥青质缔合物，有效地防止储层中析出沥青质，这源于离子液体中阳离子和阴离子电荷密度的局部非中性。研究表明，烷基咪唑离子液体中的硼酸成分增强了沥青质和离子液体之间的相互作用，极大地限制了沥青质聚集。离子液体的侧烷基链长度也是一个重要参数。离子液体-沥青质复合物形成空间稳定作用的最小碳链长度应为 8 个碳。

尽管非聚合物沥青质分散剂中包含由缩水甘油醚或环氧化物与多元醇反应而形成的醚，以及由缩水甘油醚或环氧化物与羧酸反应而形成的酯，但也已有多元醇和羧酸实验制备沥青质分散剂的报道。这类非聚合物沥青质分散剂首选的实例是十聚甘油四油酸酯和山梨醇酐单油酸酯，主要用于乳化剂配方（图 4.22）。这两种分子的极性端基都含有若干个羟基，具有比烷基酚更强的极性，而且分散沥青质的效果更好。

有研究提出了生产燃料油过程中的一系列稳定沥青质的非离子型两性化合物。例如，在沥青质沉淀试验中，发现烷基乙二醇、邻二氯苯和石脑油溶剂的混合物比熟知的十二烷基苯磺酸、膦酸酯产生的沉淀要少。另一种性能良好的混合物是溶解在乙二醇或甲醇中的四羟基对苯醌。该专利还声称，这些添加剂的混合物可以使沥青质沉淀再次胶化。多烷基或多烯基 N-羟烷基琥珀酰亚胺的恶唑啉衍生物也可作为沥青质分散剂。包括 N-烷基或 N-烯基或 N-环烷基或 N-芳香基丙酸的 1,3-恶唑啉-6-酮衍生物和多聚甲醛等的碱类化合物，兼具缓蚀剂和沥青质分散剂的性能。

4.2.2　低聚物（树脂）和聚合物沥青质抑制剂

市售的上游沥青质抑制剂大多基于聚合物表面活性剂。一些也具有沥青质分散剂的功

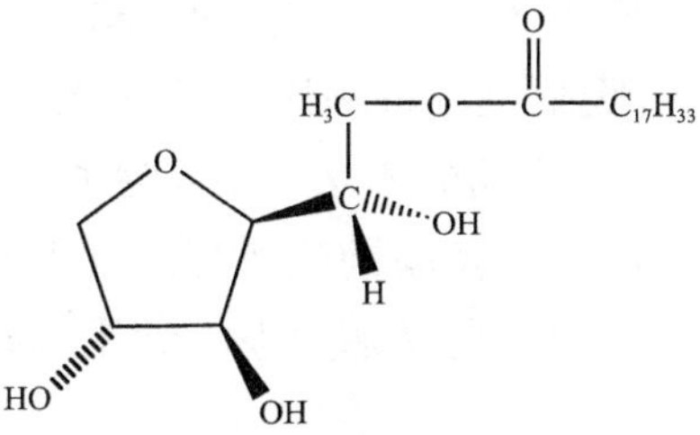

(a) 山梨醇单油酸酯

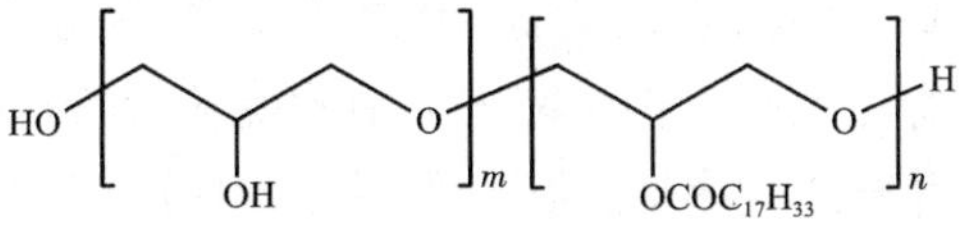

(b) 聚甘油多油酸酯

图 4.22 山梨醇单油酸酯和聚甘油多油酸酯的结构

能。聚合物表面活性剂沥青质抑制剂含有许多极性基团，每个极性基团理论上都能与沥青质单体相互作用。沥青质抑制剂和沥青质分子之间的强结合是实现良好性能的一个必要条件。本节将回顾各种类型的低聚（2～12个单体单元）和聚合（>12个单体单元）沥青质抑制剂。尽管沥青质抑制剂也有挤注处理的报道，但多数情况下，沥青质抑制剂使用方式都是连续注入。挤注处理时，沥青质抑制剂中的酸性基团能够促进沥青质抑制剂与岩石的结合，增加吸附量，以延长挤注处理的有效期。部分操作人员还将沥青质抑制剂加入破乳剂配方，因为沥青质抑制剂可覆盖沥青质团块，并将其从油水界面分离。沥青质抑制剂可以视作破乳剂的协同剂。

聚合物表面活性剂类沥青质抑制剂或沥青质分散剂包括：烷基酚醛树脂和类似的磺化树脂；具有烷基、亚烷基苯基或亚烷基吡啶基官能团的聚烯烃酯、酰胺或酰亚胺；烯基/乙烯吡咯烷酮共聚物；聚烯烃与马来酸酐或乙烯基咪唑的接枝共聚物；超支化聚酯酰胺；木质素磺酸盐；聚烷氧基化沥青质等种类。

4.2.2.1 烷基酚醛树脂低聚物

关于聚烷基酚树脂作为沥青质抑制剂的有效性，已有多篇报道。这些添加剂的性能似乎取决于制备添加剂的聚合过程。烷基酚醛树脂低聚物是在石油工业中研究最多并经常使用的聚合物沥青质抑制剂之一。例如，烷基酚与甲醛反应得到通常由2～12个烷基酚基团与亚甲基桥接而成的低聚物（图4.23）。这些低聚物的聚烷氧基酯通常用作破乳剂。

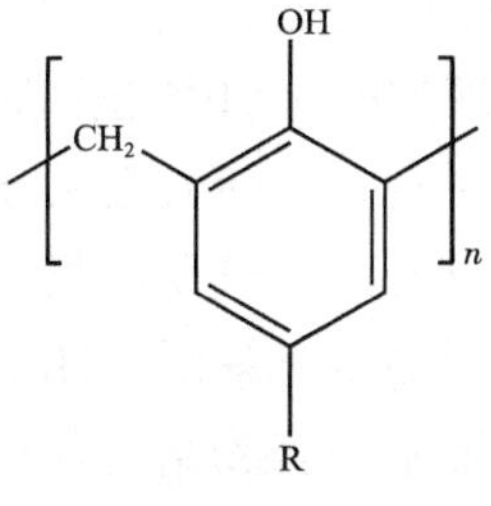

有研究明确了壬基酚（NP）、壬基酚甲醛树脂（NPR）和天然树脂（NRs）对沥青质的吸附等温

图 4.23 烷基酚醛树脂沥青质抑制剂的典型结构（其中 $R=C_3$—C_{24}，$n=2$～12）

线。发现在沥青质抑制的有效性方面，顺序是 NP＜NR＜NPR。吸附等温线符合 LS（Langmuir-S）曲线，可以用两步吸附机制来解释：第一步是表面活性剂单独吸附在沥青质表面；第二步是吸附的表面活性剂之间的相互作用成为主导，在表面开始形成两亲性聚集体。针对两亲性沥青质稳定剂活性的实验理论研究，发现两亲性物质的极化率及其偶极运动与其吸附沥青质的能力之间存在着平衡关系。如果两亲体的极性太大，也会使其不溶。

研究了腰果酚和聚腰果酚对沥青质稳定性的影响（应注意，聚腰果酚是单体以阳离子聚合法制得；而通常的烷基酚树脂是烷基酚与醛类聚合制得）。在腰果酚中，15 个碳的烷基与酚基团相邻，而壬基酚乙氧基化物（NPE）中的烷基与酚基团对位。研究发现，聚腰果酚不仅比其单体稳定沥青质的效率低，还会促进沥青质的沉淀。这是由于聚合物中存在大量苯酚基团，可能会使沥青质颗粒絮凝或极性增强，从而降低其在脂肪族溶剂中的溶解度。因此聚腰果酚与若干个沥青质分子相互作用，使其聚集，或聚腰果酚的羟基伸入溶剂中，提高被吸附沥青质的极性，降低其溶解度。但是由甲醛与腰果酚缩合制备得到的聚腰果酚，则有较好的沥青质抑制效果。图 4.24 给出了腰果酚醛树脂的可能结构。富含酚类化合物的可再生油源已用于甲醛缩聚反应。聚腰果酚和磺化聚苯乙烯在低浓度下都是沥青质絮凝剂，在高浓度下则是分散剂。

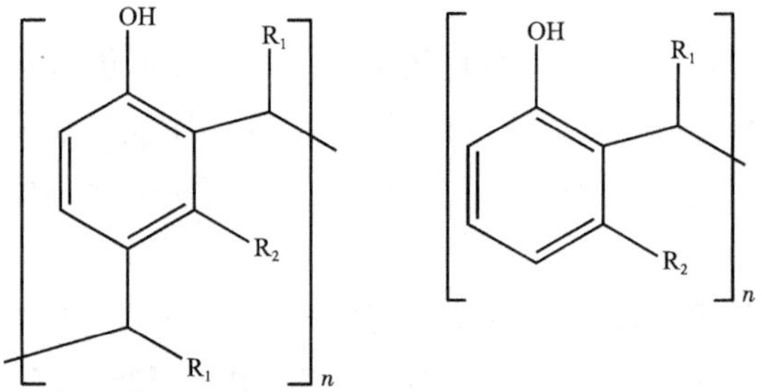

图 4.24 腰果酚醛树脂的两种可能结构（其中 R_1 变化很大，R_2 约有 15 个碳，至少有 4 个异构体的混合物）

碱性烷基酚醛树脂沥青质抑制剂的多项改进已获得专利。例如，已发现磺化烷基酚醛树脂的性能优于未磺化树脂。这些树脂的结构类似于十二烷基苯磺酸单体的串联物。因此，聚合物具有与沥青质进行酸碱和氢键相互作用的若干个基团，以及与烃类溶剂相容的若干个烷基尾链。另一组研究人员发现，经聚胺（如三乙基四胺）处理的烷基酚醛树脂在碱处理过的原油中的性能，优于十二烷基苯磺酸和聚烯烃/马来酸酐酯类共聚物。

还有一些专利是关于第二种添加剂的使用，这种添加剂可以与烷基酚醛树脂协同作用，以提高沥青质的性能。例如，研究发现加入如乙氧基三乙基四胺等烷氧基化胺可以协同提高烷基酚醛树脂的性能。同时与单独使用树脂相比，烷基酚醛树脂与亲水－亲油性乙烯基聚合物结合使用的效果更好。其他协同性专利提到特定的亲水－亲油乙烯基聚合物可以是两亲性酯共聚物，如甲基丙烯酸月桂酯/甲基丙烯酸羟乙基共聚物。后文将讨论酯共聚物。聚烷基酚醛树脂与各种化学品的反应产物组成沥青质抑制剂，可用于挤注处理。

还有一些研究深入比较了烷基酚醛树脂和单分子两亲体及其他聚合物两亲体作为沥青质分散剂的性能。例如，研究发现膦酸酯与醚羧酸（前面讨论过）协同共混物的性能优于壬基酚醛树脂。另一项研究发现，十二烷基酚醛树脂（DPR，重均分子量为200~2000）比十八烯/马来酸酐共聚物（POM，重均分子量为10000，图4.25）能更好地促进沥青质分散。两亲性聚合物的分子量被认为至关重要，过高的分子量会导致聚合物-沥青质颗粒的不必要混凝作用。十八烯/马来酸酐共聚物（POM）与沥青质的结合强度大于十二烷基酚醛树脂（DPR）和单分子两亲物，表明POM的抑制效果更好。聚合物与沥青质之间的氢键被认为是两者缔合的主要机理。研究者认为十二烷基酸酐共聚物（POM）的酸酐基比十二烷基酚醛树脂（DPR）的酚羟基，对沥青质有更好的氢键作用，但没有考虑十二烷基酚醛树脂（DPR）芳香环与沥青质之间的π-π相互作用的可能性。与这项研究相反，另一个团队发现相对于未处理的样品，十二烷基酸酐共聚物（POM）实际上增加了沥青质絮凝量。傅里叶变换红外光谱研究表明，该酸酐基与沥青质中的OH基形成氢键。十二烷基酚醛树脂（DPR）与十二烷基酸酐共聚物（POM）的对比研究结果仍让人感到困惑，但应该指出的是后一项研究中使用的是燃料油，而前一项研究使用的是原油。如果沥青质样品在化学处理之前已经聚合到一定程度，则后一项结果也可以得到合理解释。这些聚集体会有许多伸向溶剂的脂肪族烷基。这种聚合物具有许多长脂肪族烷基，不能渗透到沥青质聚合体的极性部分。但这些基团的极性酸酐基团从外部伸向烃类溶剂，可以与多个沥青质聚集物相互作用，使其絮凝成更大的颗粒。因此，面向溶剂的是极性酸酐基团而不是烷基。通过对原油的上述研究发现，C_{28} α-丙烯（辛二烯）/马来酸酐共聚物是炼油中沥青质的极佳分散剂。

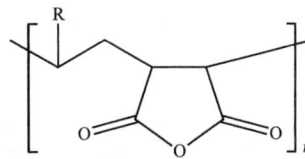

图4.25 烯烃/马来酸酐共聚物沥青抑制剂的结构（R最好是一个长烷基）

4.2.2.2 聚酯和聚酰胺/酰亚氨沥青质抑制剂

聚酯或聚酰胺/酰亚氨基的沥青质抑制剂已经被某些服务公司商业化应用。聚酯中的酯基和聚酰胺中的酰氨基通常是通过与丙烯酸和（或）顺丁烯二酸酐单体反应而得到（也可用乙烯基链烷酸酯）。这种类型的未酯化单体还可以加入自由羧基，以便在挤注应用中更好地吸附于岩石。典型的酯类共聚物有（甲基）丙烯酸酯共聚物、苯乙烯/马来酸酯共聚物和烯烃/马来酸酯共聚物，如图4.26所示。酯基可与酰氨基交换制成聚酰胺。烯烃/马来酸共聚物可以与胺反应生成琥珀酰亚胺基团（图4.27）。这些聚合物都是市售可采购到的沥青质抑制剂。

如甲基丙烯酸月桂酯/甲基丙烯酸羟乙基共聚物等以亲油性单体为主要成分和亲水性单体制备的共聚物也是一种沥青质抑制剂。使用亲水性单体可提高共聚物中侧基末端的极性，使其与沥青质粒子之间产生更强的氢键作用。在此理论基础上，有研究者提到只有亲油性单体的市售产品（与脂肪醇反应的炔/马来酸酐二酯）效果就较差。此外，他们发现未酯化的α-烯烃/马来酸酐共聚物在沥青质抑制方面的表现优于亲油性共聚物。马来酸

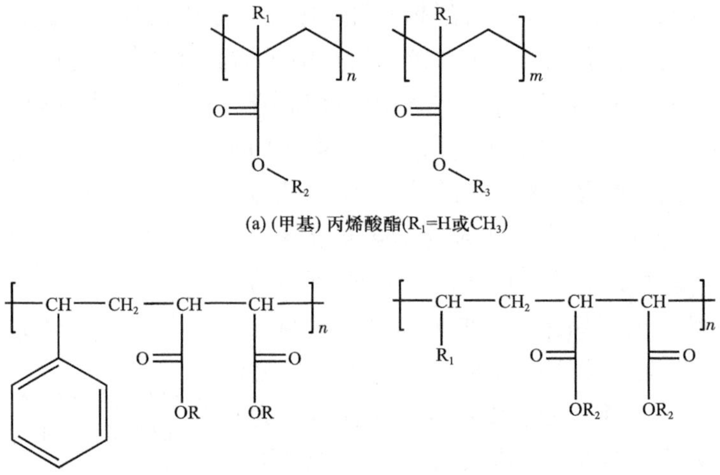

(a) (甲基) 丙烯酸酯(R_1=H或CH_3)

(b) 苯乙烯/马来酸二酯　　(c) 烯烃/马来酸二酯共聚物

图 4.26 （甲基）丙烯酸酯（R_1=H 或 CH_3）、苯乙烯 / 马来酸二酯和烯烃 / 马来酸二酯共聚物的结构

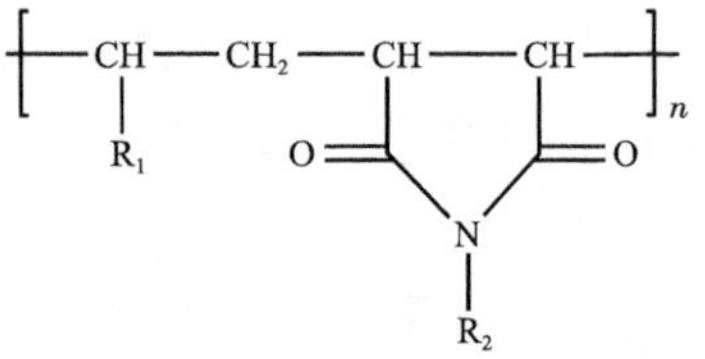

图 4.27　聚烷基丁二酰亚胺共聚物

酐的聚（乙二醇）酯和 α- 十八烯共聚物是一类用于含沥青质原油的新型梳型添加剂。这些聚合物添加剂可以降低屈服应力、分散沥青质，并减少模型原油和重油的中长链正烷烃的晶体尺寸。

有专利认为与其在聚酯和聚酰胺的侧基上使用烷基链，不如使用能与沥青质发生 π—π 相互作用的芳香环。如甲基丙烯酸甲酯和羟甲基吡啶的酯交换产物，随后被聚合（图 4.28）。其中的羟甲基吡啶可以是 2- 异构体、3- 异构体或 4- 异构体，也可以与烯烃 / 马来酸酐共聚物（而不是甲基丙烯酸酯）反应。所有这些聚合物都有邻近极性酯基的吡啶环侧基。其他如苯基或萘基等芳香环，也可与酯或酰胺间隔基合并到共聚物中。

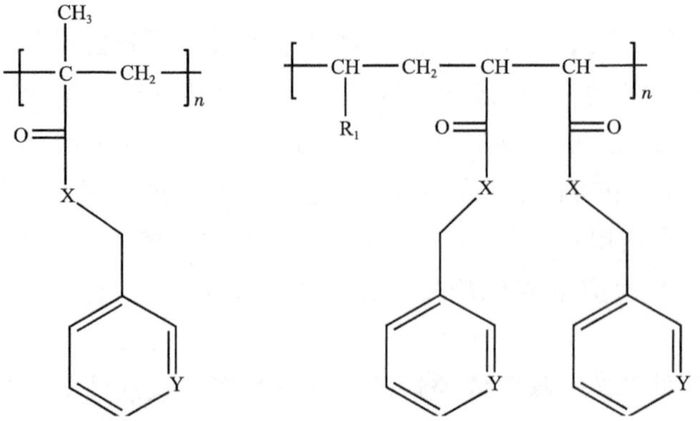

图 4.28　两类带有侧基芳香环的聚合物结构（R_1= 烷基或烯基，X=O 或 NH，Y=CH 或 N）

在同一团队的相关专利中，通过将聚羧酸与含芳香环和杂环结构的胺或醇部分衍生，制得聚酯或聚酰胺作为沥青质抑制剂。具体实例是聚甲基丙烯酸 4- 烷基苯基酯，其中烷

基 R 有 9~12 个碳（图 4.29），也可用马来酸酐酯类共聚物。

另一种方法是聚合物上接枝酰胺、亚胺或酯基，具体是先将酸酐接枝到聚乙烯的主链，然后将产物与胺或醇反应。制备并测试了以聚异丁烯为基础的聚酯酰胺，首先与马来酸酐反应，再与单乙醇胺等胺反应（图 4.30）。该共聚物皂化值低，即聚异丁烯和马来酸酐比值高。

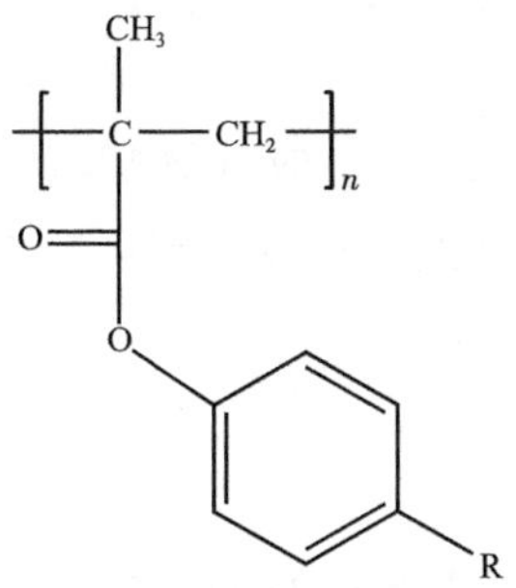

图 4.29 聚甲基丙烯酸 4-烷基苯基酯
（R 有 9~12 个碳）

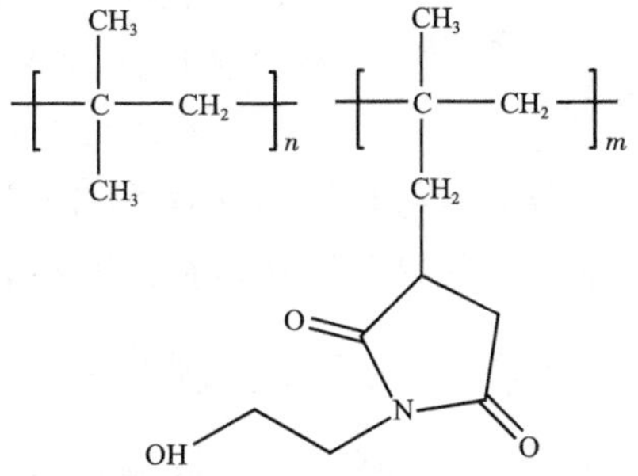

图 4.30 带有 N-羟乙基琥珀酰亚胺侧基的接枝共聚物沥青质抑制剂

还有一类聚酰胺沥青质抑制剂是 1-乙烯-4-烷基-2-吡咯烷酮聚合物和相关的 1-乙烯基吡咯烷酮/α-烯烃共聚物（图 4.31）。吡咯烷酮基有一个强的氢键羰基，并显示出与沥青质中的五环吡咯相似结构。另一类相关的沥青质抑制剂是聚乙烯烷基氨基甲酸酯。

迄今为止，所讨论的所有酯和酰胺聚合物都具有线型聚乙烯醇主干。一种新型超支化聚酯酰胺是优良的沥青质增溶剂（图 4.32）。这些聚合物具有枝状的三维结构，其中烃基指向各个方向。聚酯酰胺由环酸酐和二烷基胺按 $n:(n+1)$ 的比例缩合而成，其中 n 是整数（通过改变 n 可以改变聚合物的分子量）。这样产生了一种端部带羟基的聚合物。超支化是由带三个反应性基团的二烷醇胺形成。通过羧酸（如脂肪酸）与羟基化的超支化聚合物反应，该聚合物的尖端可以调整为弱亲水性。沥青质增溶剂的优选实例是由琥珀酸酐和二异丙醇氨以物质的量比约 5:6 组成，其中部分羟基官能团通过可可脂肪酸和聚异丁烯酰琥珀酸酐（含 ca.22 异丁烯单体）的酯化反应

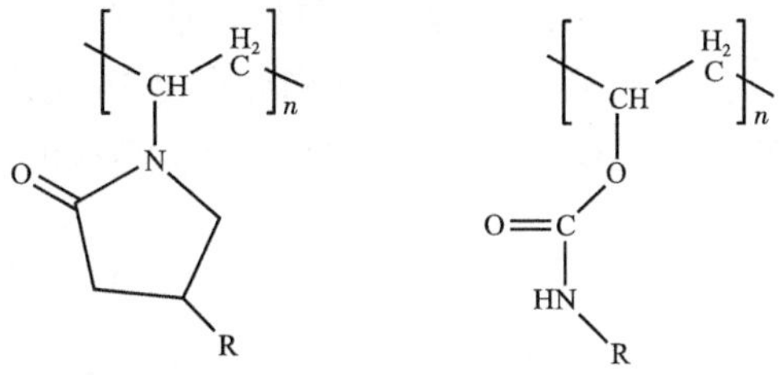

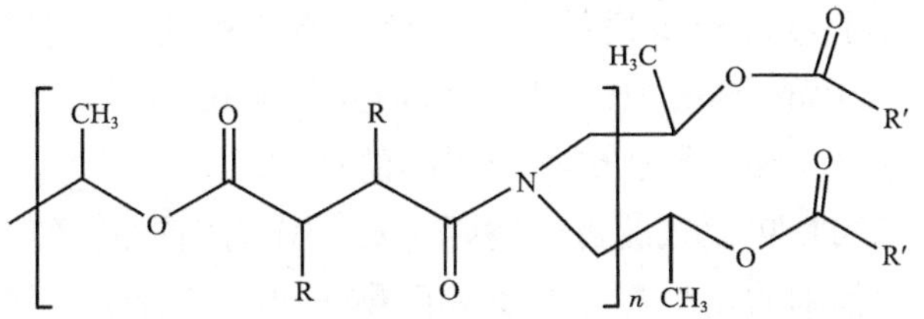

图 4.31 聚 1-乙烯-4-烷基-2-吡咯烷酮聚合物（左）和聚乙烯烷基氨基甲酸酯（右）
（R 是一个长链烷基）

图 4.32 超支化聚酯酰胺沥青质抑制剂的结构和端基（R 是 H 或烃基；R′是长烷基链）

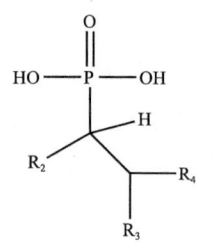

图 4.33 聚合膦酸酯沥青质分散剂

得到。第二种优选方式是基于琥珀酸酐和二异丙胺组成的结构单元，其中羟基官能团的一部分通过仅与聚异丁烯基琥珀酸酐反应而改性。芳香族聚异丁烯琥珀酰亚胺已报道是有沥青质分散能力的降黏剂，可以用于稠油和超稠油的沥青质分散。

除前面讨论的酚醛树脂类外，还有膦酸酯聚合物类沥青质分散剂（图 4.33），可以由烯烃（如聚异丁烯）自由基插入膦酸二酯的 P—H 键中来制备。

由聚合的长链多羟基酸与多胺反应生成的聚酯酰胺混合物结构复杂，也是具有分散特性的沥青质抑制剂。其中，典型的多羟基酸是聚蓖麻油酸或聚 12-羟基硬脂酸，多胺是亚烷基多胺或聚乙烯亚胺。

4.2.2.3 其他聚合沥青质抑制剂

已经有报道使用含有极性环侧基的接枝共聚物作为沥青质抑制剂。接枝共聚物是通过具有氮和（或）氧原子的乙烯基单体与聚烯烃接枝来制备。乙烯基单体的实例有 N-乙烯基咪唑和 4-乙烯基吡啶，也可以使用 N-乙烯基内酰胺（图 4.34）。

作为挤注处理的沥青质抑制剂，与芳香基磺酸盐单体相关的一类是木质素磺酸盐聚合物。木质素磺酸盐的结构如图 4.35 所示。极性端基包含酚基和磺酸基，与烷基芳基磺酸有部分相似之处，但现在与酚基连接的是其他官能团且具有更加复杂的聚合物结构。

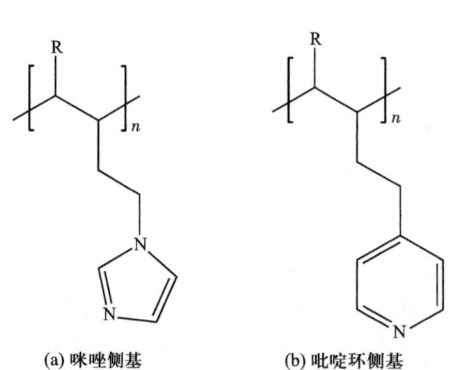

(a) 咪唑侧基　　(b) 吡啶环侧基

图 4.34 带咪唑侧基或吡啶环侧基的接枝聚合物沥青抑制剂（R 最好是烷基链）

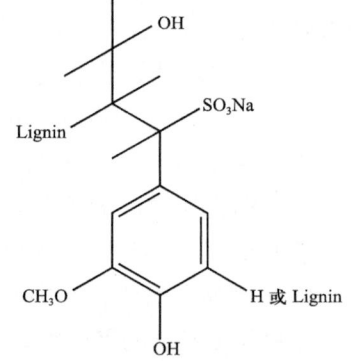

图 4.35 带有芳香烃类和极性基团的木质素磺酸盐的局部结构

下面简要介绍一种制备半合成沥青质抑制剂的新方法。可以通过对沥青质本身的膦酸烷氧基化衍生，产生抑制相同沥青质的产物。将沥青质中的酸性基团与 PCl_3 反应，然后再与聚丙二醇反应，产生混合物，其中两种结构如图 4.36 所示。这些化合物的自结合能力可能不如原来的沥青质，并且具有更类似树脂的性质。

另有专利使用具有四吡咯模式的可生物降解分子作为沥青质稳定剂。这些分子是以植物叶片中提取叶绿素为基础的衍生物。对锡和硅酞菁进行了分子工程设计，作为可示踪的沥青质稳定剂。

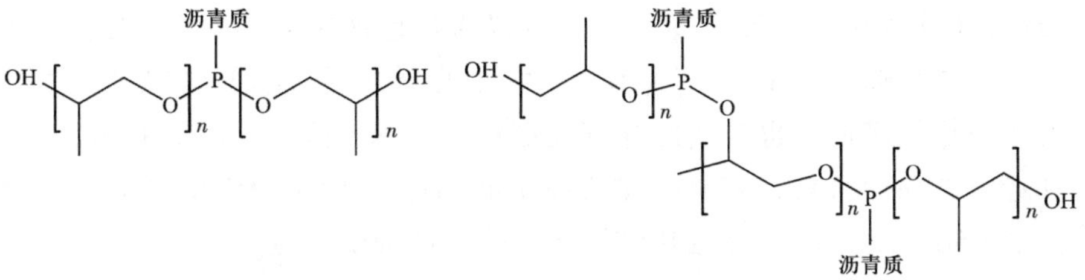

图 4.36　基于沥青质膦酸烷氧基化的沥青质抑制剂

4.2.3　沥青质分散剂和沥青质抑制剂小结

一般而言，单体表面活性剂起分散剂的作用，而聚合表面活性剂是沥青质絮凝的真正抑制剂，也有分散剂作用。沥青质分散剂、抑制剂对沥青质的作用机理可以总结为沥青质与不饱和烃或芳香烃类基团之间的 π—π 相互作用、酸碱相互作用、氢键、偶极－偶极相互作用、金属离子络合。

几乎所有的沥青质分散剂、抑制剂具有的关键特性是具有一个或多个与沥青质单体或聚集体相互作用的极性官能团，再接枝一条或多条烷基链，吸附在沥青质表面形成低极性的烷基尾链立体稳定层，并且将其增溶于原油。烷基尾链与主要成分是脂肪烃的原油相容。沥青质分散剂和沥青质抑制剂中的环状结构也很常见，特别是芳香环。在表面活性剂中，适当长度的支链烷基比直链烷基更有利于沥青质的分散。分子建模可能有助于加深对这些问题的理解。与沥青质抑制剂和沥青质分散剂一起使用的较好的溶剂是芳香族或极性溶剂，如甲苯、二甲苯、三甲基苯、1-甲基萘、四氢萘（四萘）、喹啉、异喹啉、二甲基甲酰胺、N-甲基吡咯烷酮、烷基二醇及其混合物，都有助于沥青质形成胶体溶液。

4.3　沥青质溶解剂

沥青质沉积物可出现在从井下到工艺设备等部位，在设备中的严重沉积可能需要机械清理。如果储层中存在严重的沥青质沉积问题，则可能需要进行重新射孔或重复压裂。在中国曾试验采用酶来有效降低沥青质和胶质对储层的伤害。唯一有效除去沥青质沉积物的化学方法是沥青质溶解剂或溶剂。为了清除管线中的沥青质沉积，溶解剂需要间歇加注或者循环处理受影响的区域达数个小时。增产液通过油管注入井内（也可从井口环空加注），以弥补可能与沥青有关的产能下降问题。有关文献提到的所有 12 项工作都是在低含水油井中完成，但是这些程序也适用于高含水油井的应用。

对于油管的处理，产品未掺水使用或用原油稀释，并通过油套环空或井口循环。在井下温度下，有效的溶解剂能在数小时内溶解与自身重量相当的沥青质。酸化增产作业后因为沥青质通过水层与带电矿物表面发生化学黏结，形成的沥青污泥通常用溶剂很难清除。在使用前，必须检查沥青质溶剂与体系中使用的橡胶弹性体和塑料材料的相容性。

多数沥青质溶解剂以芳香烃类溶剂为基础，有时添加增强剂。有资料研究了沥青质在不同烃类液体中的溶解度。高芳香烃含量的脱沥青油也被用作化学溶剂的低成本替代品。二甲苯（1,2-二甲苯、1,3-二甲苯和1,4-二甲苯的混合物），闪点低（28℃），可能是最常见的芳香烃类溶剂。也有使用挥发性更强的甲苯（闪点5℃），但效果稍差。从健康、安全和经济的角度来看，有研究建议使用低剂量（按油的体积计算）的7.5%甲苯/柴油的混合物（50∶50），而不是使用纯甲苯来防止沥青质沉淀。

现在多数取代芳香烃类溶剂被列为海洋污染物。芳香烃类溶剂与沥青质中的芳香组分通过π—π轨道重叠相互作用，取代沥青质—沥青质π—π相互作用，从而使其溶解。研究表明，沥青质溶解剂中石蜡（无环或环状非芳香烃）的存在不利于沥青质溶解性能。同样的研究表明，单环和双环芳香烃类溶剂的性能优于三环或多环芳香烃类化学品。另一项研究表明，双环分子四氢萘（1,2,3,4-四氢萘）和1-甲基萘在沥青质溶解量和溶解速率方面都优于 n-丙苯、甲苯和二甲苯等单环溶剂（图4.37）。较高的温度和搅拌速度有助于提高沥青质溶解速率。考虑到成本较高的因素，多数商品沥青质溶解剂都含有单环芳香烃类主剂和少量的双环芳香烃。意大利已经使用了基于原油馏分、煤焦油、萘油和甲基萘的高闪点溶解剂。与芳香烃溶剂相比，萜烯类溶剂（含 d-柠檬烯）具有更优良的人员健康、安全性和环保特性，也用于海洋领域，但其溶解沥青质的能力有限（图4.38）。

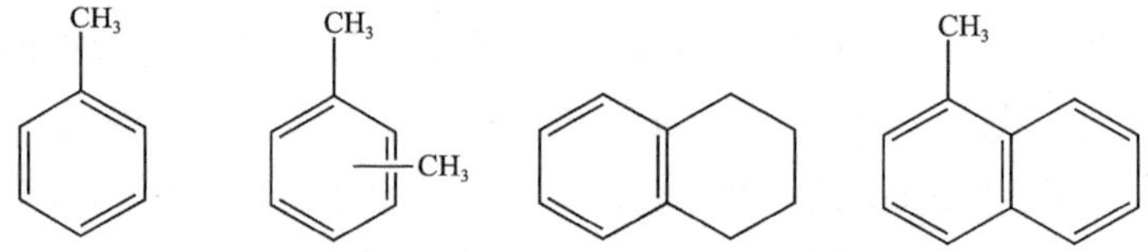

图4.37 甲苯、二甲苯、四氢萘和1-甲基萘

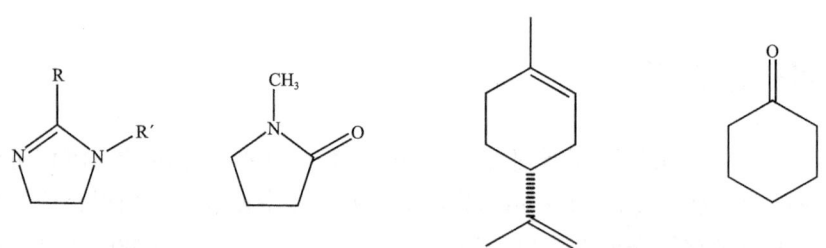

图4.38 咪唑啉、N-甲基吡咯烷酮、柠檬烯（一种萜烯）和环己酮

萜烯类溶剂对溶解蜡也很有用。由于沥青质含有杂原子，能使结构具有一定的极性，具有杂原子和极性基团的芳香烃类溶剂也可以作为沥青质的良好溶解剂。二硫化碳的闪点很低，是很好的沥青质溶解剂。吡啶是良好的沥青质溶解剂，但有毒，与许多橡胶弹性体不相容。或者，芳香烃类溶剂可与极性助溶剂结合使用。通过将芳香烃类溶剂和含有极性官能团的添加剂结合，表明助溶剂整体上的极性与油田的沥青质类型相匹配。这种混合剂增强了二甲苯作为碱性溶剂的溶解能力，并增加了沥青质从储层岩石矿物表面的脱附。有专利声称，优选添加3%～10%喹啉和异喹啉（或 C_1—C_4 烷基取代）的芳香烃溶剂，极性比普通芳香烃更强，可以提高沥青质的溶解速率（图4.39）；还有专利声称，在芳香烃类

溶剂中添加苯并三唑可提高性能；某些共烃羧酸的烷基或烯基酯（如苯甲酸异丙酯）也是很好的沥青质溶解剂，而且毒性较低、环保性也较优良。解除地层无机和有机伤害的延迟乳化酸增产液配方中，采用以上溶剂作为基础油相。含杂原子的溶剂不必须是芳香烃类溶剂才能溶解沥青质。例如，含有1%～15% N-取代咪唑啉的溶剂，最好是N-甲基吡咯烷酮（或N-乙基吡咯烷酮），也是很好的溶解剂（图4.38）。含有煤油或芳香烃类溶剂与至少一种C_4—C_{30}烯烃或其氧化产物的组合物，可作为改进型沥青质溶解剂。也筛选了其他如乙二醇醚、烷胺和酯等"更绿色"的溶剂，可能在改善环境影响方面会发现未来有更多的应用。具体实例是二基酯的混合物，其中包括二烷基戊二酸甲酯和二烷基己二酸酯或二烷基琥珀酸乙酯的至少一种、至少一种萜烯和至少一种表面活性剂。

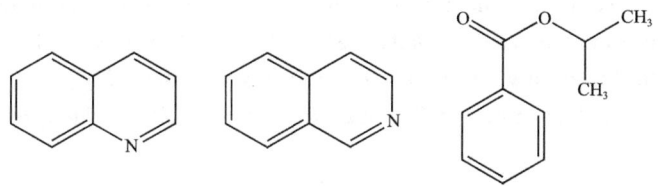

图4.39　喹啉、异喹啉和异丙酯

沥青质溶解剂中有时添加增强剂，以提高其性能，但增强剂的选择似乎取决于沥青质是吸附于岩层，或是在生产流程的其他环节。有研究认为，与仅使用甲苯相比，甲苯中含有2%（质量分数）极性基团（如聚合物）且没有提供质子基团（如烷基苯磺酸）的化学剂时，能有效地提升对沥青质固体的溶解能力。但在岩心流动实验中，聚合物、烷基苯磺酸均能增加吸附于岩石表面沥青质的溶解，其中磺酸的溶解效果最好。研究认为，沥青质从岩石中脱附的机理是添加剂与岩石活性位点的竞争性相互作用。有油田服务公司发现，向沥青质溶解剂中加入一系列的沥青质分散剂，并不能提高沥青质的溶解量和溶解速率。

还开发了高闪点的乳液型沥青质溶解剂，其中芳香烃类溶剂不使用BETX（苯、乙苯、甲苯或二甲苯）。这些乳液基溶解剂有时比二甲苯等芳香烃类溶剂的性能更好（同一文献还有一份有效清单，列出了影响持续去除沥青质的关键因素）。所述水包油乳液可包括：（1）水；（2）一种有机溶剂混合物，包括非极性有机溶剂，如萜烯混合物或已预先蒸馏出轻芳烃溶剂的原油，以及极性有机溶剂，如N-甲基吡咯烷酮或环己酮；（3）一种表面活性剂，有助于在有机溶剂混合物和水中形成乳液，并分解沉积的沥青质（图4.38）。令人意外的是，与两种溶剂单独加入混合物相比，与表面活性剂结合的混合溶剂溶解沥青质的效果更好。针对全球各地的沥青质，优化和测试了该乳液型溶解剂。在二甲苯等芳香烃溶剂中乳液化的酸，可以通过井下酸化和溶解沥青质作业，来提高油井的产能。与单用二甲苯相比，乳液体系对沥青质的溶解度更高。通常其中的水相是淡水或含铁离子稳定剂的15%盐酸。

微乳液也可以消除沥青质沉积物造成的地层伤害。一种提高原油采收率的新方法是加入如五羰基铁[$Fe(CO)_5$]的螯合剂，从沥青质颗粒中分离出较小的沥青质分子，从而减小沥青质粒径。

参 考 文 献

[1] K. J. Leontaritis, *Oil & Gas Journal* 1 (1998): 122.

[2] E. Y. Sheu and O. C. Mullins, eds., *Asphaltenes: Fundamentals and Applications*, New York: Plenum Press, 1995.

[3] J. G. Speight, *The Chemistry and Technology of Petroleum*, 3rd ed., New York: Marcel Decker, 1999, 412-467.

[4] E. Y. Sheu and O. C. Mullins, eds., *Asphaltenes*, New York: Springer, 1996.

[5] O. C. Mullins and E. Y. Sheu, eds., *Structures and Dynamics of Asphaltenes*, New York: Springer, 1999.

[6] K. J. Leontaritis and G. A. Mansoori, "Asphaltene Deposition: A Survey of Field Experiences and Research Approaches," *Journal of Petroleum Science and Engineering* 1 (1988): 229.

[7] K. Barker and M. E. Newberry, "Inhibition and Removal of Low-pH-Fluid-Induced Asphaltic Sludge Fouling of Formations in Oil and Gas Wells," SPE 102738 (paper presented at the SPE International Symposium on Oilfield Chemistry, Houston, TX, 28 February-2 March 2007).

[8] L. C. C. Marques, J. B. Monteiro, and G. González, *Journal of Dispersion Science and Technology* 28 (2007): 391.

[9] F. Trejo, G. Centeno, and J. Ancheyta, "Precipitation, Fractionation and Characterization of Asphaltenes from Heavy and Light Crude Oils," *Fuel* 83 (2004): 2169.

[10] O. C. Mullins, "Molecular Structure and Aggregation of Asphaltenes and Petroleomics," SPE 95801 (paper presented at the SPE Annual Technical Conference and Exhibition, Dallas, TX, 9-12 October 2005).

[11] J. P. Maclean and P. Kilpatrick, "Comparison of Precipitation and Extrography in the Fractionation of Crude Oil Residua," *Energy & Fuels* 11 (1997): 570.

[12] S. Asomaning, "Test Methods for Determining Asphaltene Stability in Crude Oils," *Petroleum Science and Technology* 21 (2003): 581.

[13] N. G. Graham and M. L. Gorbaty, "Sulfur K-Edge X-ray Absorption Spectroscopy of Petroleum Asphaltenes and Model Compounds," *Journal of the American Chemical Society* 111 (1989): 3182.

[14] S. Mitra-Kirtley, O. C. Mullins, J. van Elp, S. J. George, J. Chen, and S. P. Cramer, "Determination of the Nitrogen Chemical Structures in Petroleum Asphaltenes Using XANES Spectroscopy," *Journal of the American Chemical Society* 115 (1993): 252.

[15] J. Castillo, A. Fernandez, M. A. Ranaudo, and S. Acevedo, "New Techniques and Methods for the Study of Aggregation, Adsorption, and Solubility of Asphaltenes. Impact of These Properties on Colloidal Structure and Flocculation," *Petroleum Science and Technology* 19 (2001): 75.

[16] O. C. Mullins, E. Y. Sheu, A. Hammami, and A. G. Marshall, *Asphaltenes, Heavy Oils, and Petroleomics*, New York: Springer, 2006.

[17] G. W. Zajac, N. K. Sethi, and J. T. Joseph, "Molecular Imaging of Petroleum Asphaltenes by Scanning Tunneling Microscopy," *Scanning Microscopy* 8 (1994): 463.

[18] A. Sharma, H. Groenzin, A. Tomita, and O. C. Mullins, "Probing Order in Asphaltenes and Aromatic Ring Systems by HRTEM," *Energy & Fuels* 16 (2002): 490.

[19] (a) H. Groenzin and O. C. Mullins, "Molecular Size and Structure of Asphaltenes from Various Sources," *Energy & Fuels* 14 (2000): 677. (b) A. E. Pomerantz, M. R. Hammond, A. L. Morrow, O. C. Mullins, and R. N. Zare, *Energy Fuels*, 23 (2009): 1162.

[20] O. P. Strausz, T. W. Mojelsky, F. Faraji, E. M. Lown, and P. Peng, "Additional Structural Details on Athabasca Asphaltene and Their Ramifications," *Energy & Fuels* 13 (2) (1999): 207.

[21] J. Murgich, "Molecular Simulation and the Aggregation of the Heavy Fractions in Crude Oils," *Molecular Simulation* 29(2003): 451.

[22] O. P. Strausz, T. W. Mojelsky, and E. M. Lown, "The Molecular Structure of Asphaltene: An Unfolding Story," *Fuel* 71(1992): 1355.

[23] O. P. Strausz, P. Peng, and J. Murgich, "About the Colloidal Nature of Asphaltenes and the MW of Covalent Monomeric Units," *Energy & Fuels* 16(2002): 809.

[24] K. L. Gawrys, PhD thesis, Raleigh, N.C.: North Carolina State University, 2005.

[25] K. L. Gawrys and P. K. Kilpatrick, "Asphaltenic Aggregates are Polydisperse Oblate Cylinders," *Journal of Colloid Interface Science* 288(2005): 325.

[26] V. Calemma, R. Rausa, P. D'Antona, and L. Montanari, "Characterization of Asphaltenes Molecular Structure," *Energy & Fuels* 12(1998): 422.

[27] S. A. A. Castro, J. G. Negrin, A. Fernandez, G. Escobar, V. Piscitelli, F. Delolme, and G. Dessalces, "Relations Between Asphaltene Structures and Their Physical and Chemical Properties: The Rosary-Type Structure," *Energy & Fuels* 21 (4)(2007): 2165.

[28] O. P. Strausz, I. Safarik, E. M. Lown, and A. Morales-Izquierdo, "A Critique of Asphaltene Fluorescence Decay and Depolarization-Based Claims about Molecular Weight and Molecular Architecture," *Energy & Fuels* 22(2008): 1156–1166.

[29] A. A. Herod, K. D. Bartle, and R. Kandiyoti, "Characterization of Heavy Hydrocarbons by Chromatographic and Mass Spectrometric Methods: An Overview," *Energy & Fuels* 21(2007): 2176.

[30] O. C. Mullins, B. Martínez-Haya, and A. G. Marshall, "Contrasting Perspective on Asphaltene Molecular Weight: This Comment vs. the Overview of A. A. Herod, K. D. Bartle, and R. Kandiyoti," *Energy & Fuels* 22(2008): 1765.

[31] A. E. Pomerantz, M. R. Hammond, A. L. Morrow, O. C. Mullins, and R. N. Zare, "Two-Step Laser Mass Spectrometry of Asphaltenes," *Journal of the American Chemical Society* 130(2008): 7216.

[32] O. C. Mullins, S. S. Betancourt, M. E. Cribbs, F. X. Dubost, J. L. Creek, A. Ballard, and L. Venkataramanan, "The Colloidal Structure of Crude Oil and the Structure of Oil Reservoirs," *Energy & Fuels* 21 (5)(2007): 2785.

[33] M. S. Diallo, T. Cagin, J. L. Faulon, and W. A. Goddard III, "Asphaltenes and Asphalts," in *Developments in Petroleum Science 40B*, eds. T. F. Yen and G. V. Chilingarian, Amsterdam, The Netherlands: Elsevier, Chap. 5, 103, 2000.

[34] J. X. Wang, J. S. Buckley, N. E. Burke, and J. L. Creek, "A Practical Method for Anticipating Asphaltene Problems," SPE 87638, *SPE Production & Facilities* 19 (3)(2004): 152–160.

[35] S. Asomaning and A. Yen, "Prediction and Solution of Asphaltene Related Problems in the Field," *Chemistry in the Oil Industry VII*, Royal Society of Chemistry, 2002, 277.

[36] H. W. Yarranton, "Asphaltene Self-Association," *Journal of Dispersion Science and Technology* 26 (2005): 5.

[37] R. E. Guerra, K. Ladavac, A. B. Andrews, O. C. Mullins, and P. N. Sen, "Diffusivity of Coal and Petroleum Asphaltene Monomers by Fluorescence Correlation Spectroscopy," *Fuel* 86(2007): 2016.

[38] K. Rastegari, W. Y. Svrcek, and H. W. Yarranton, "Kinetics of Asphaltene Flocculation," *Industrial and Engineering Chemistry Research* 43(2004): 6861.

[39] J. Wang, J. S. Buckley, and J. L. Creek, "Asphaltene Deposition on Metallic Surfaces," *Journal of Dispersion Science and Technology* 25(2004): 287.

[40] E. Buenrostro-Gonzalez, H. Groenzin, C. Lira-Galeana, and O. C. Mullins, "The Overriding Chemical

Principles that Define Asphaltenes," *Energy & Fuels* 15 (4)(2001): 972.

[41] J. Murgich, J. Rodriguez, and Y. Aray, "Molecular Recognition and Molecular Mechanics of Micelles of Some Model Asphaltenes and Resins," *Energy & Fuels* 10 (1996): 68.

[42] T. Takanohashi, S. Sato, and R. Tanaka, "Molecular Dynamics Simulation of Structural Relaxation of Asphaltene Aggregates," *Petroleum Science and Technology* 21 (2003): 491.

[43] T. Takanohashi, S. Sato, and R. Tanaka, "Structural Relaxation Behaviors of Three Different Asphaltenes Using MD Calculations," *Petroleum Science and Technology* 22 (2004): 901.

[44] L. Carbognani and E. Rogel, "Solid Petroleum Asphaltenes Seem Surrounded by Alkyl Layers," *Petroleum Science and Technology* 21 (3-4)(2003): 537.

[45] K. L. Gawrys, M. Spiecker, and P. K. Kilpatrick, "The Role of Asphaltene Solubility and Chemistry on Asphaltene Aggregation," *Petroleum Science and Technology* 21 (2003): 461.

[46] H.-J. Oschmann, "New Methods for the Selection of Asphaltene Inhibitors in the Field," *Chemistry in the Oil Industry VII*, Royal Society of Chemistry, Manchester, 2001, 254.

[47] L. C. C. Marques, G. Gonzalez, and J. B. Monteiro, "A Chemical Approach to Prevent Asphaltenes Flocculation in Light Crude Oils: State-of-the-Art," SPE 91019 (paper presented at the SPE Annual Technical Conference and Exhibition, Houston, TX, 26-29 September 2004).

[48] D. F. Smith, G. C. Klein, A. T. Yen, M. P. Squicciarini, R. P. Rodgers, and A. G. Marshall, "Crude Oil Polar Chemical Composition Derived from FT—ICR Mass Spectrometry Accounts for Asphaltene Inhibitor Specificity," *Energy & Fuels* 22 (2008): 3112.

[49] J. Marugán, "Characterization of the Onset of Asphaltenes by Focus-Beamed Laser Reflectance: A Tool for Chemical Additives Screening" (paper presented at the 9th International Conference on Petroleum Phase Behavior and Fouling, Victoria, BC, 15-19 June 2008).

[50] M. Barcenas, P. Orea, E. Buenrostro-González, L. S. Zamudio-Rivera, and Y. Duda, "Study of Medium Effect on Asphaltene Agglomeration Inhibitor Efficiency," *Energy & Fuels* 22 (2008): 1917.

[51] S. Asomaning, "Methods for Selecting Asphaltene Inhibitors and New Insights into Inhibitor Mechanisms," Abstr. No. GEOC 142, 225th ACS National Meeting, New Orleans, LA, 23-27 March 2003.

[52] K. Karan, A. Hammami, M. Flannery, and B. A. Stankiewicz, "Evaluation of Asphaltene Instability and Chemical Control during Production of Live Oils," *Petroleum Science and Engineering* 21 (2003): 629.

[53] H. Pan and A. Firoozabadi, "Thermodynamic Micellization Model for Asphaltene Precipitation Inhibition," *AIChE Journal* 46 (2000): 416.

[54] L. Z. Pillon, "Effect of Dispersants and Flocculants on the Colloidal Stability of Asphaltene Constituents," *Petroleum Science and Engineering* 19 (2001): 863.

[55] L. F. Bandeira Moreira, E. F. Lucas, and G. González, "Stabilization of Asphaltenes by Phenolic Compounds Extracted from Cashew-Nut Shell Liquid," *Journal of Applied Polymer Science*, 73 (1)(1999): 29.

[56] J. Dunlop, "Novel High Performance Dispersants for Oil Industry Applications" (paper presented at the 4th International Conference on Petroleum Phase Behaviour and Fouling, Trondheim, Norway, 23-26 June 2003).

[57] S. J. Allenson and M. A. Walsh, "A Novel Way to Treat Asphaltene Deposition Problems Found in Oil Production," SPE 37286 (paper presented at the SPE International Symposium on Oilfield Chemistry, Houston, TX, 18-21 February 1997).

[58] L. M. Cenegy, "Survey of Successful World-Wide Asphaltene Inhibitor Treatments in Oil Production

Fields," SPE 71542 (paper presented at the SPE Annual Technical Conference and Exhibition, New Orleans, LA, 30 September-3 October 2001).

[59] O. Niemeyer, "New Squeeze Applications for Asphaltene Blocked Wells" (paper presented at the 7th International Conference on Petroleum Phase Behaviour and Fouling, Asheville, NC, 25-29 June 2006).

[60] K. Allan, "Asphaltene Inhibitor Squeezing—What Can We Learn from Scale Inhibitor Squeezing" (paper presented at the 9th International Conference on Petroleum Phase Behavior and Fouling, Victoria, BC, 15-19 June 2008).

[61] R. B. De Boer, K. Leerlooyer, M. R. P. Eigner, and A. R. D. van Bergen, "Screening of Crude Oils for Asphalt Precipitation: Theory, Practice and the Selection of Inhibitors," *SPE Production & Facilities* 10 (1995): 55-61.

[62] J. K. Borchardt, "Chemicals Used in Oil-Field Operations," in *Oil-Field Chemistry*, eds. J. K. Borchardt and T. F. Yen, *American Chemical Society Symposium Series 396*, Washington, DC: ACS, 1989.

[63] S. Takhar, "A Fast and Effective Chemical Screening Technique for Identifying Asphaltene Inhibitors for Field Deployment," Proceedings of the Second International Conference on Fluid and Thermal Energy Conversion, eds. A. Mansoori and A. Suwono (1997), S83-S90.

[64] H. J. Oschmann, "New Methods for the Selection of Asphaltene Inhibitors in the Field," Special Publication, *Royal Society of Chemistry* 280 (2002): 254-263.

[65] M. N. Bouts, R. J. Wiersma, H. M. Muijs, and A. J. Samuel, "An Evaluation of New Asphaltene Inhibitors: Laboratory Study and Field Testing," *Journal of Petroleum Technology* 782 (1995): 782.

[66] M. B. Manek, "Asphaltene Dispersants as Demulsification Aids," SPE 28972 (paper presented at the SPE International Symposium on Oilfield Chemistry, San Antonio, TX, 14-17 February 1995).

[67] A. K. M. Jamaluddin, J. Nighswander, N. B. Joshi, D. Calder, and B. Ross, "Asphaltene Characterization: A Key to Deepwater Developments" SPE 77936 (paper presented at the SPE Asia Pacific Oil and Gas Conference and Exhibition, Melbourne, Australia, 8-10 October 2002).

[68] A. Yen, R. Yin, and S. Asomaning, "Evaluating Asphaltene Inhibitors: Laboratory Tests and Field Studies," SPE 65376 (paper presented at the SPE International Symposium on Oilfield Chemistry, Houston, TX, 13-16 February 1995).

[69] H. Anfindsen, P. Fotland, and A. M. Mathisen (paper presented at the 9th International Oilfield Chemical Symposium, Geilo, Norway, 22-25 March 1995).

[70] M. A. Aquino-Olivos, E. Buenrostro-Gonzalez, S. I. Andersen, and C. Lira-Galeana, "Investigations of Inhibition of Asphaltene Precipitation at High Pressure Using Bottomhole Samples," *Energy & Fuels* 15(1) (2001): 236.

[71] T. Maqbool, I. A. Hussein, and H. S. Fogler (paper presented at the 9th International Conference on Petroleum Phase Behaviour and Fouling, Victoria, BC, Canada, 15-19 June 2008).

[72] A. Hammami, C. H. Phelps, T. Monger-McClure, and T. M. Little, "Asphaltene Precipitation from Live Oils: An Experimental Investigation of Onset Conditions and Reversibility," *Energy & Fuels* 14 (2000): 14.

[73] S. Asomaning, "Assessing the Performance of Asphaltene Inhibitors Using High Pressure Methods: The Deepwater Gulf of Mexico Experience," RSC/EOSCA Proceedings of Chemicals in the Oil Industry, Manchester, UK, 3-5 November 2003, 35.

[74] S. Asomaning and C. Gallagher, "High Pressure Asphaltene Deposition Technique for Evaluating the Deposition Tendency of Live Oil and Evaluating Inhibitor Performance, Preprints" (paper presented at the Second International Conference on Petroleum and Gas Phase Behaviour and Fouling, Copenhagen,

Denmark, 26-31 August 2000).

[75] H. Alboudwarej, W. Y. Svrcek, A. Kantzas, and H. W. Yarranton, "A Pipe Loop Apparatus to Investigate Asphaltene Deposition," *Petroleum Science and Engineering* 22 (2005): 799.

[76] J. Dunlop, "Novel High Performance Additives for Asphaltene Control in Oil Production Operations," RSC/EOSCA Chemistry in the Oil Industry VIII, Manchester, UK, 3-5 November 2003.

[77] J. Dunlop, "Low Environmental Asphaltene Inhibitors" (paper presented at the Tekna 19th International Oilfield Chemistry Symposium, Geilo, Norway, 9-12 March 2008).

[78] R. J. Gochin, U.S. Patent Application 20050082231, 2005.

[79] R. J. Gochin and A. Smith, U.S. Patent 6270653, 2005.

[80] D. Espinat, J. C. Ravey, V. Guille, J. Lambard, T. Zemb, and J. P. Cotton, "Colloidal Macrostructure of Crude Oil Studied by Neutron and X-ray Small Angle Scattering Techniques," *Journal De Physique IV* 3 (C8)(1993): 181.

[81] A. Del Bianco and F. Stroppa, European Patent EP0737798, 1996.

[82] M. E. Newberry and K. M. Barker, U.S. Patent 4414035 1983.

[83] N. Feldman, U.S. Patent 4441890, 1984.

[84] W. H. Stover and S. A. Hunter, U.S. Patent 4182613, 1980.

[85] L. C. Rocha Junior, M. S. Ferreira, and A. C. S. Ramos, "Inhibition of Asphaltene Precipitation in Brazilian Crude Oils Using New Oil Soluble Amphiphiles," *Journal of Petroleum Science* 51 (2006): 26.

[86] C. L. Chang and H. S. Fogler, "Stabilization of Asphaltenes in Aliphatic Solvents Using Alkylbenzene Derived Amphiphiles. 1. Effect of the Chemical Structure of Amphiphiles on Asphaltenes Stabilization," *Langmuir* 10 (1994): 1749-1757.

[87] C. L. Chang and H. S. Fogler, "Asphaltene Stabilization in Alkyl Solvents Using Oil-Soluble Amphiphiles," SPE 25185 (paper presented at the SPE International Symposium on Oilfield Chemistry, New Orleans, LA, 2-5 March 1993).

[88] H. H. Ibrahim and R. O. Idem, "Interrelationships between Asphaltene Precipitation Inhibitor Effectiveness, Asphaltenes Characteristics, and Precipitation Behavior during n-Heptane (Light Paraffin Hydrocarbon)-Induced Asphaltene Precipitation," *Energy & Fuels* 18 (4)(2004): 1038.

[89] L. Goual and A. Firoozabadi, "Effect of Resins and DBSA on Asphaltene Precipitation from Petroleum Fluids," *AIChE Journal* 50 (2004): 470.

[90] C. L. Chang and H. S. Fogler, "Stabilization of Asphaltenes in Aliphatic Solvents Using Alkylbenzene-Derived Amphiphiles. 2. Study of the Asphaltene-Amphiphile Interactions and Structures Using Fourier Transform Infrared Spectroscopy and Small-Angle X-ray Scattering Techniques," *Langmuir* 10 (1994): 1758-1766.

[91] P. Permsukarome, C. L. Chang, and H. S. Fogler, "Kinetic Study of Asphaltene Dissolution in Amphiphile/Alkane Solutions," *Industrial and Engineering Chemistry Research* 36 (1997): 3960.

[92] A. C. S. Ramos, C. C. Delgado, R. S. Mohamed, V. R. Almeida, and W. Loh, "Reversibility and Inhibition of Asphaltene Precipitation in Brazilian Crude Oils," SPE 38967 (paper presented in the Latin American and Caribbean Petroleum Engineering Conference, 30 August-3 September 1997).

[93] T. A. Al-Sahhaf, A. F. Mohammed, and A. S. Elkilani, "Retardation of Asphaltene Precipitation by Addition of Toluene, Resins, Deasphalted Oil and Surfactants," *Fluid Phase Equilibria* 194 (2002): 1045.

[94] A. I. Victorov and A. Firoozabadi, "Thermodynamics of Asphaltene Deposition Using a Micellization Model," *AIChE Journal* 42 (1996): 1753.

[95] H. Pan and A. Firoozabadi, "Thermodynamic Micellization Model for Asphaltene Precipitation: Micellar Growth and Precipitation," *SPE Production and Facilities* 13（2）(1998): 118.

[96] H. Pan and A. Firoozabadi, "A Thermodynamic Micellization Model for Asphaltene Precipitation from Reservoir Crudes at High Pressures and Temperatures," *SPE Production and Facilities* 15(2000): 58.

[97] O. Leon, E. Rogel, A. Urbina, A. Andujar, and A. Lucas, "Study of the Adsorption of Alkyl Benzene-Derived Amphiphiles on Asphaltene Particles," *Langmuir* 15（22）(1999): 7653.

[98] D. Miller, A. Vollmer, and M. Feustel, U.S. Patent 5925233, 1999.

[99] I. Wiehe, R. Varadaraj, T. Jermansen, R. J. Kennedy, and C. H. Brons, International Patent Application WO00/32546, 2000.

[100] R. Varadaraj and C. H. Brons, U.S. Patent Application 20040072361, 2004.

[101] I. Wiehe and T. G. Jermansen, "Design of Synthetic Dispersants for Asphaltene Constituents," *Petroleum Science and Engineering* 21(2003): 527.

[102] J. A. Ostlund, M. Nyden, H. S. Fogler, and K. Holmberg, "Functional Groups in Fractionated Asphaltenes and the Adsorption of Amphiphilic Molecules," *Colloids and Surfaces A* 234(2004): 95.

[103] D. Miller, A. Vollmer, M. Feustel, and P. Klug, U.S. Patent 6063146, 2000.

[104] N. Ikenaga, Y. Watanabe, and S. Hayashi, Japanese Patent JP63023991, 1988.

[105] C. A. Stout, Canadian Patent CA1142114, 1983.

[106] R. F. Miller, U.S. Patent 4425223, 1984.

[107] S. V. Tapavicza, W. Zoellner, C. P. Herold, J. Groffe, and J. Rouet, U.S. Patent 6344431, 2002.

[108] D. Miller, A. Vollmer, M. Feustel, and P. Klug, U.S. Patent 6204420, 2001.

[109] G. Woodward, C. R. Jones, and K. P. Davis, International Patent Application WO/2004/002994.

[110] I. H. Auflem, T. E. Havre, and J. Sjoblom, "Near Infrared Study on the Dispersive Effects of Amphiphiles and Naphthenic Acids on Asphaltenes in Model Heptane-Toluene Mixtures," *Colloid Polymer Science* 280(2002): 695.

[111] D. Miller, A. Vollmer, M. Feustel, and P. Klug, International Patent Application, WO98/16595.

[112] M. Ravindranath, European Patent Application EP1359206, 2003.

[113] J. M. Romocki, International Patent Application, WO94/18430, 1994.

[114] M. Ferrara, International Patent Application WO95/20637.

[115] M. Ravindranath and R. M. Banavali, U.S. Patent Application US2004238404.

[116] R. Banavali, "Reducing Viscosity of Asphaltenic Crudes via Chemical Additives," *Chemistry in the Oil Industry VIII*, Royal Society of Chemistry(2003): 1.

[117] A. Lesimple, C. P. Herold, D. Groffe, and W. Breuer, International Patent Application WO01/27438, 2001.

[118] H. L. Becker and B. W. Wolf, U.S. Patent 5504063, 1996.

[119] C. Bernasconi, A. Faure, and B. Thibonnet, U.S. Patent 4622047, 1986.

[120] Y. F. Hu and T. M. Guo, "Effect of the Structures of Ionic Liquids and Alkylbenzene-Derived Amphiphiles on the Inhibition of Asphaltene Precipitation from CO_2-Injected Reservoir Oils," *Langmuir* 21（18）(2005): 8168.

[121] O. León, E. Contreras, and E. Rogel, "Amphiphile Adsorption on Asphaltene Particles: Adsorption Isotherms and Asphaltene Stabilization," *Colloids and Surfaces A* 189(2001): 123.

[122] G. D. Sutton, U.S. Patent 3914132, 1975.

[123] G. Gonzalez and A. Middea, "Peptization of Asphaltene by Various Oil Soluble Amphiphiles," *Colloids and Surfaces* 52(1991): 207.

[124] T. Cox, N. Grainger, and E. G. Scovell, Canadian Patent CA2404316(TW546370B), 2003.

[125] J. Groffe, J. Rouet, and D. Chauvie, French Patent FR2679151, 1993.

[126] M. Ravindranath, U.S. Patent Application US2004232042.

[127] M. Ravindranath, U.S. Patent Application US2004232044.

[128] B. D. Chheda, U.S. Patent Application 20070124990.

[129] R. M. Banavali, B. D. Cheda, and G. M. Manari, U.S. Patent Application 20060079434.

[130] P. J. Breen, International Patent Application, WO01/55281, 2001.

[131] V. J. Mena Cervates, L. S. Zamudio-Rivera, M. Lozada y Casso, H. Beltrán Conde, E. Buenrostro-González, S. López Ramirez, Y. Douda, A. Morales Pacheco, R. Hernández Altamirano, and M. Barcenas Castañeda, Mexican Patent Request, File No. MX/E/2007/084388. Also see International Patent Application WO/2009/078694.

[132] S. S. Schantz and W. K. Stephenson, "Asphaltene Deposition: Development and Application of Polymeric Asphaltene Dispersants," SPE 22783 (paper presented at the SPE Annual Technical Conference and Exhibition, Dallas, TX, 6–9 October 1991).

[133] L. Barberis Canonico, A. Del. Bianco, G. Piro, F. Stroppa, C. Carniani, and E. I. Mazzolini, "A Comprehensive Approach for the Evaluation of Chemicals for Asphaltene Deposit Removal," *Recent Advances in Oilfield Chemistry*, Chemicals In the Oil Industry V, The Royal Society of Chemistry, Ambleside, Cumbria, UK, 1994, 220.

[134] A. L. Soldan, L. C. F. Barbosa, R. L. A. Santos, J. C. C. B. R. Moreira, S. C. Menezes, M. A. G. Teixeira, C. R. Souza, R. B. Haag, L. C. C. Marques, and A. N. Sanmartin, "1st SPE Brazil Sect. Colloid Chemistry in Oil Production," Proceedings Asphaltenes and Wax Deposition International Symposium, Rio de Janeiro, Brazil, 26–29 November 1995, 51–55.

[135] O. León, E. Contreras, E. Rogel, G. Dambakli, J. Espidel, and S. Acevedo, "The Influence of the Adsorption of Amphiphiles and Resins in Controlling Asphaltene Flocculation," *Energy & Fuels* 15 (2001): 1028.

[136] E. Rogel, E. Contreras, and O. León, "An Experimental Theoretical Approach to the Activity of Amphiphiles as Asphaltene Stabilizers," *Petroleum Science and Engineering* 20(2002): 725.

[137] D. Leinweber, M. Feustel, E. Wasmund, and H. Grundner, International Patent Application WO02/64706, 2002.

[138] A. Behler, W. Breuer, and M. Hof, International Patent Application WO03/054348, 2003.

[139] A. Behler, U.S. Patent Application 20050091915.

[140] M. B. Manek and N. K. Sawney, U.S. Patent 5494607, 1996.

[141] D. Miller, M. Feustel, A. Vollmer, R. Vybiral, and D. Hoffmann, U.S. Patent 6180683, 2001.

[142] W. K. Stephenson, B. D. Mercer, and D. G. Comer, U.S. Patent 5143594, 1992.

[143] W. K. Stephenson, B. D. Mercer, and D. G. Comer, U.S. Patent 5100531, 1992.

[144] W. Stephenson and M. Kaplan, U.S. Patent 5021498, 1991.

[145] W. Stephenson, J. Walker, B. Krupay, and S. Wolsey-Iverson, Canadian Patent CA2075749, 1993.

[146] W. Stephenson and M. Kaplan, U.S. Patent 5073248, 1991.

[147] C. L. Chang and H. S. Fogler, "Peptization and Coagulation of Asphaltenes in Apolar Media Using Oil-Soluble Polymers," *Petroleum Science and Engineering* 14(1996): 75.

[148] D. G. Comer and W. K. Stephenson, U.S. Patent 5214224 1993.

[149] S. Handa, P. Hodgson, International Patent Application WO98/58580, 1998.

[150] S. Handa, P. K. G. Hodgson, and W. J. Ferguson, U.K. Patent Application GB2337522A. 1999.

[151] W. Breuer, P. Birnbrich, D. Groffe, S. Von Tapavicza, C.-P. Herold, and M. Hof, International Patent Application WO02/18454, 2002.

[152] P. M. W. Cornelisse, International Patent Application WO02/102928, 2002.

[153] M. Wilkes and M. Davies, International Patent Application, WO/2006/047745.

[154] F. J. Boden, R. P. Sauer, I. L. Goldblatt, and M. E. McHenry, U.S. Patent 6686321.

[155] F. J. Boden, R. P. Sauer, I. L. Goldblatt, and M. E. McHenry, U.S. Patent 5663126.

[156] F. J. Boden, S. P. Sauer, I. L. Goldblatt, and M. E. McHenry, U.S. Patent 5873389.

[157] D. M. Bilden and V. E. Jones, U.S. Patent 6051535, 2000.

[158] R. L. Sung, T. F. DeRosa, D. A. Storm, and B. J. Kaufman, U.S. Patent 5207891, 1993.

[159] J. Rouet, D. Groffe, and M. Salaun, International Patent Application WO/2008/084178.

[160] Y. Wang, A. Kantzas, B. Li, Z. Li, Q. Wang, and M. Zhao, "New Agent for Formation-Damage Mitigation in Heavy-Oil Reservoir: Mechanism and Application," SPE 112355 (paper presented at the SPE International Symposium and Exhibition on Formation Damage Control, Lafayette, LA, 13–15 February 2008).

[161] S. T. Dubey and M. H. Waxman, "Asphaltene Adsorption and Desorption from Mineral Surfaces," SPE 18462, *SPE Reservoir Engineering* 6 (3)(1991): 389.

[162] G. Piro, L. B. Canonico, G. Galbariggi, L. Bertero, and C. Carniani, "Asphaltene Adsorption Onto Formation Rock: An Approach to Asphaltene Formation Damage Prevention," SPE 30109, *SPE Production & Facilities* 11 (3)(1996): 156–160.

[163] A. Del Bianco and F. Stroppa, U.S. Patent 5382728, 1995.

[164] A. K. M. Jamaluddin and T. W. Nazarko, U.S. Patent 5425422, 1995.

[165] S. R. King and C. R. Cotney, "Development and Application of Unique Natural Solvents for Treating Paraffin and Asphaltene Related Problems," SPE 35265 (paper presented at the SPE Mid-Continent Gas Symposium, Amarillo, TX, 28–30 April 1996).

[166] M. G. Trbovich and G. E. King, "Asphaltene Deposit Removal: Long-Lasting Treatment with a Co-Solvent," SPE 21038 (paper presented at the SPE International Symposium on Oilfield Chemistry, Anaheim, CA, 20–22 February 1991).

[167] M. Galoppini, "Asphaltene Deposition Monitoring and Removal Treatments: An Experience in Ultra Deep Wells," SPE 27622 (paper presented at the European Production Operations Conference and Exhibition, Aberdeen, UK, 15–17 March 1994).

[168] C. W. Benson, R. A. S. Simcox, and I. C. Huldal, "Tailoring Aromatic Hydrocarbons for Asphaltene Removal," *Chemicals in the Oil Industry IV*, Royal Society of Chemistry, 1991, 215.

[169] J. Curtis, "Environmentally Favorable Terpene Solvents Find Diverse Applications in Stimulation, Sand Control and Cementing Operations," SPE 84124 (paper presented at the SPE Annual Technical Conference and Exhibition, Denver, CO, 5–8 October 2003).

[170] C. J. Bushman, International Patent Application WO/2008/024488.

[171] W. Kleinitz, "Asphaltene Precipitates in Oil Production Wells" (paper presented at the 8th Oil Field Chemical Symposium, NIF, Geilo, Norway, 1997).

[172] L. Minssieux, "Removal of Asphalt Deposits by Cosolvent Squeeze: Mechanisms and Screening," SPE 69672, *SPE Journal* 6 (1)(2001): 39.

[173] A. Del Bianco and F. Stroppa, U.S. Patent 5690176, 1997.

[174] M. B. Lawson and K. J. Snyder, U.S. Patent 4033784, 1977.

[175] E. G. Scovell, N. Grainger, and T. Cox, International Patent Application WO/2001/074966.

[176] E. G. Scovell, N. Grainger, and T. Cox, International Patent Application WO/2001/088333.

[177] H. L. Becker and B. W. Wolf, U.S. Patent 5504063, 1996.

[178] M. L. Trimble, M. A. Fleming, B. L. Andrew, G. A. Tomusiak, P. M. Digiacinto, and L. M. Heymans, International Patent Application, WO2008/010923.

[179] S. Lightford, E. Pitoni, F. Armesi, and L. Mauri, "Development and Field Use of a Novel Solvent-Water Emulsion for the Removal of Asphaltene Deposits in Fractured Carbonate Formations," SPE 101022 (paper presented at the SPE Annual Technical Conference and Exhibition, San Antonio, TX, 24-27 September 2006).

[180] S. Lightford and F. Armesi, International Patent Application WO/2007/129348.

[181] K. A. Frost, R. D. Daussin, and M. S. van Domelen, "New, Highly Effective Asphaltene Removal System with Favorable HSE Characteristics," SPE 112420 (paper presented at the 2008 SPE International Symposium and Exhibition on Formation Damage Control, Lafayette, LA, 13-15 February 2008).

[182] W. A. Fatah and H. A. Nasr-El-Din, "Acid Emulsified in Xylene: A Cost-Effective Treatment to Remove Asphaltene Deposition and Enhance Well Productivity," SPE 117251 (paper presented at the Eastern Regional/AAPG Eastern Section Joint Meeting, Pittsburgh, PA, 11-15 October 2008).

[183] L. Quintero, T. A. Jones, D. E. Clark, A. D. Gabrysch, A. Forgiarini, and J.-L. Salager, U.S. Patent Application 20090008091.

[184] W. Frenier, M. Ziauddin, and R. Venkatesan, *Organic Deposits in Oil and Gas Production*, Society of Petroleum Engineers, 2011.

[185] N. Evdokimov, *Petroleum Science and Technology* 28 (13)(2010): 1351.

[186] J. Czarnecki, P. Tchoukov, and T. Dabros, *Energy & Fuels* 26 (9)(2012): 5782.

[187] K. P. Gierszal, J. G. Davis, M. D. Hands, D. S. Wilcox, L. V. Slipchenko, and D. Ben-Amotz, *Journal of Physical Chemistry Letters* 2 (2011): 2930.

[188] K. Nikooyeh, S. R. Bagheri, and J. M. Shaw, *Energy & Fuels* 26 (2012): 1756.

[189] E. Rogel, C. Ovalles, and M. Moir, *Energy & Fuels* 26 (5)(2012): 2655.

[190] O. C. Mullins, *Energy & Fuels* 24 (4)(2010): 2179.

[191] O. C. Mullins, *Energy & Fuels* 23 (2009): 2845.

[192] O. C. Mullins, H. Sabbah, J. Eyssautier, A. E. Pomerantz, L. Barré, A. B. Andrews, Y. Ruiz-Morales, F. Mostowfi, R. McFarlane, L. Goual, R. Lepkowicz, T. Cooper, J. Orbulescu, R. M. Leblanc, J. Edwards, and R. N. Zare, *Energy & Fuels* 26 (7)(2012): 3986.

[193] H. Sabbah, A. L. Morrow, A. E. Pomerantz, and R. N. Zare, *Energy & Fuels* 35 (2011): 1597.

[194] M. Fossen, H. Kallevik, K. D. Knudsen, and J. Sjoblom, *Energy & Fuels* 25 (8)(2011): 3552.

[195] M.-J. T. Mui Ching, A. E. Pomerantz, A. B. Andrews, P. Dryden, R. Schroeder, O. C. Mullins, and C. Harrison, *Energy & Fuels* 24 (9)(2010): 5028.

[196] S. Salmon-Vega, R. Herrera-Urbina, C. Lira-Galeana, and M. A. Valdez, *Petroleum Science and Technology* 30 (2012): 986-992.

[197] I. A. Wiehe, *Energy & Fuels* 26 (7)(2012): 4004.

[198] P. Juyal, V. Ho, A. Yen, and S. J. Allenson, *Energy & Fuels* 26 (5)(2012): 2631.

[199] C. A. Wright, G. B. Brons, and S. A. Feiller, European Patent EP2367910.

[200] C. Jian, T. Tang, and S. Bhattacharjee, *Energy & Fuels* 27 (2013): 2057-2067.

[201] W. W. Frenier and M. Ziauddin, "A Multifaceted Approach for Controlling Complex Deposits in Oil and Gas Production," SPE 132707 (paper presented at the SPE Annual Technical Conference and Exhibition, Florence, Italy, 19-22 September 2010).

[202] M. Amani and I. Najafi, *Advances in Petroleum Exploration and Development* 2（2）（2011）: 24.

[203] A. S. Shedid and S. R. Attallah, "Influences of Ultrasonic Radiation on Asphaltene Behavior with and without Solvent Effects," SPE 86473（paper presented at the SPE International Symposium and Exhibition on Formation Damage Control, Lafayette, LA, 2004）.

[204] B. Champion, F. Van der Bas, and G. Nitters, "The Application of High-Power Sound Waves for Wellbore Cleaning," SPE 82197（paper presented at the SPE European Formation Damage Conference, The Hague, Netherlands, 2004）.

[205] M. Ashtari, S. N. Ashrafizadeh, and M. Bayat, *Journal of Petroleum Science and Engineering* 82-83（2012）: 44.

[206] D. V. Satya Gupta, S. Szymczak, and J. M. Brown, "Solid Production Chemicals Added with the Frac for Scale, Paraffin and Asphaltene Inhibition," SPE 119393（paper presented at the SPE Hydraulic Fracturing Technology Conference, The Woodlands, TX, 19-21 January 2009）.

[207] M. A. McGowen, M. Van Domelen, K. A. Frost, and P. A. Curtis, International Patent Application WO/2010/029318.

[208] C. Ovalles, E. Rogel, and J. Segerstrom, "Improvement of Flow Properties of Heavy Oils Using Asphaltene Modifiers," SPE 146775（paper presented at the SPE Annual Technical Conference and Exhibition, Denver, CO, 30 October-2 November 2011）.

[209] V. González Dávila, International Patent Application WO/2012/002790.

[210] K. Kraiwattanawong, H. S. Fogler, S. G. Gharfeh, P. Singh, W. H. Thomason, and S. Chavadej, *Energy & Fuels* 23（3）（2009）: 1575.

[211] M. Barcenas, P. Orea, E. Buenrostro-González, L. S. Zamudio-Rivera, and Y. Duda, *Energy & Fuels* 22（2008）: 1917.

[212] F. M. Vargas, J. L. Creek, and W. G. Chapman, *Energy & Fuels* 24（4）（2010）: 2294.

[213] E. Rogel, *Energy & Fuels* 25（2011）: 472.

[214] K. Takabayashi, H. Maeda, Y. Miyagawa, M. Ikarashi, H. Okabe, S. Takahashi, H. R. Al-Shehhi, and H. M. Al-Hammadi, "Do Asphaltene Deposition Troubles Happen in Low Asphaltene Content of Crude Oil？," SPE 161489（paper presented at the Abu Dhabi International Petroleum Conference and Exhibition, Abu Dhabi, UAE, 11-14 November 2012）.

[215] J. S. Buckley, *Energy & Fuels* 26（7）（2012）: 4086.

[216] P. Juyal, A. T. Yen, R. P. Rodgers, S. Allenson, J. Wang, and J. Creek, *Energy & Fuels* 24（4）（2010）: 2320.

[217] M. A. Buriro and M. T. Shuker, "Asphaltene Prediction and Prevention: A Strategy to Control Asphaltene Precipitation," SPE 163129（paper presented at the SPE/PAPG Annual Technical Conference, Islamabad, Pakistan, 3-5 December 2012）.

[218] L. A. Hough, J.-C. Castaing, G. Woodward, R. T. Pabalan, and F. De Campo, International Patent Application WO/2011/106088.

[219] X. Gong, Y. Gu, F. Shu, and Y. She, "Experimental Determination of Asphaltene Precipitation from Different Live Heavy Oils and Technical Evaluation of Potential Inhibitors for the Primary Depletion Process," SPE 157835（paper presented at the SPE Heavy Oil Conference Canada, Calgary, Alberta, Canada, 12-14 June 2012）.

[220] N. Passade-Boupat, H. Zhou, and M. Rondon-Gonzalez, "Asphaltene Precipitation from Crude Oils: How to Predict It and to Anticipate Treatment？" Paper 164184（paper presented at the 18th Middle East Oil & Gas Show and Conference（MEOS）, Bahrain International Exhibition Centre, Manama, Bahrain,

10—13 March 2013).

[221] R. R. Alapati and N. Joshi, "New Test Method for Field Evaluation of Asphaltene Deposition," SPE 24168 (2013 Offshore Technology Conference, Houston, TX, 6—9 May 2013).

[222] E. F. Ghloum, M. Al-Qahtani, and A. Al-Rashid, *Journal of Petroleum Science and Engineering* 70 (2010): 99.

[223] G. B. Dickakian, U.S. Patent 4931164, 1990.

[224] D. T. Heaps, J. P. Phillips, and P. K. Madasu, "Effect of Naphthenic Acids on Particle Aggregation and Sedimentation Behavior of Asphaltene Suspensions," *Preprint Papers—American Chemical Society, Division of Petroleum Chemistry* 55 (1)(2010): 162.

[225] D. T. Heaps, P. K. Madasu, D. H. Magers, and J. P. Buchanan, *Energy & Fuels* 26 (2012): 1862.

[226] Y. R. Fan, S. Simon, and J. Sjoblom, *Langmuir* 26 (2010): 10497.

[227] M. Boukherissa, F. Mutelet, A. Modarressi, A. Dicko, D. Dafri, and M. Rogalski, *Energy & Fuels* 23(5) (2009): 2557.

[228] R. I. Hernandez Altamirano, V. Y. Mena Cervantes, L. S. Zamudio Rivera, H. I. Beltran Conde, and E. Buenrostro Gonzalez, U.S. Patent Application 20110269650.

[229] F. B. da Silva, M. M. Moreno, M. J. O. Guimaraes, and P. R. Seidl, *Abstracts of Papers of the American Chemical Society* 241 (2011): 41-Petr.

[230] A. F. Lima, C. R. E. Mansur, E. F. Lucas, and G. Gonzalez, *Energy & Fuels* 24 (4)(2010): 2369.

[231] S. Xing, J. Xu, J. Sun, and H. Qian, Paper presented at 241st ACS National Meeting, Anaheim, CA, 27—31 March 2011.

[232] R. F. Miller, International Patent Application WO/2012/039900.

[233] T. E. Chávez-Miyauchi, L. S. Zamudio-Rivera, and V. Barba-López, *Energy & Fuels* 27 (2103): 1994—2001.

[234] C. Cohrs, S. Dilsky, D. Leinweber, and M. Feustel, U.S. Patent Application 20110098507.

[235] V. Y. Mena-Cervantes, R. Hernandez-Altamirano, E. Buenrostro-Gonzalez, H. I. Beltran, and L. S. Zamudio-Rivera, *Energy & Fuels* 25 (1)(2011): 224.

[236] T. Peruzzi, T. Coulon, J. Fauria, D. Frey, H. Marechal, and R. Sloan, "Umbilical Deployed Stimulations for Asphaltene Related Damage in Offshore Oil Fields," SPE 151813 (paper presented at the SPE International Symposium and Exhibition on Formation Damage Control, Lafayette, LA, 15—17 February 2012).

[237] A. Miadonye and L. Evans, *Petroleum Science and Technology* 28 (14)(2010): 1407.

[238] A. M. Rashed, E. F. Ghloum, M. F. Al-Matrook, Gh. R. Oskui, P. Mali, M. Telang, and A. Al-Jasmi, "Continuous Solvent Flush Approach for Asphaltene Precipitation in a Kuwaiti Reservoir: Phase I—An Experimental Solvent Screening," SPE 163316 (paper presented at the 2012 SPE Kuwait International Petroleum Conference and Exhibition, Kuwait City, Kuwait, 10—12 December 2012).

[239] O. M. Falana, F. G. Zamora, E. C. Marshall, and S. R. Kakadjian, U.S. Patent Application 20120071367.

[240] D. Fluck, A. Sehgal, S. Trivedi, R. T. Pabalan, and C. Aymes, International Patent Application WO/2012/078193.

[241] R. J. Dyer, International Patent Application WO/2012/128819.

[242] S. C. Lightford, F. Armesi, and F. Reynaldi, U.S. Patent Application 20100130384 and 20100130389.

[243] D. Appicciutoli, R. Maier, D. Pasquale Strippoli, A. Tiani, and L. Mauri, "Novel Emulsified Acid Boosts Production in a Major Carbonate Oil Field with Asphaltene Problems," SPE 135076 (paper presented at the SPE Annual Technical Conference and Exhibition, Florence, Italy, 19—22 September

2010).
[244] L. Salgaonkar and A. Danait, "Environmentally Acceptable Emulsion System: An Effective Approach for Removal of Asphaltene Deposits," SPE 160877 (paper presented at the SPE Saudi Arabia Section Technical Symposium and Exhibition, Al-Khobar, Saudi Arabia, 8-11 April 2012).
[245] C. A. Franco, N. N. Nassar, M. A. Ruiz, P. Pereira-Almao, and F. B. Cortés, *Energy & Fuels* 27 (6) (2013): 2899-2907.
[246] R. Martínez-Palou, M. de Lourdes Mosqueira, B. Zapata-Rendón, E. Mar-Juárez, C. Bernal-Huicochea, J. de la Cruz Clavel-López, and J. Aburto, *Journal of Petroleum Science and Engineering* 75 (2011): 274.
[247] J. L. Daridon, M. Cassiède, D. Nasri, J. Pauly, and H. Carrier, *Energy & Fuels* 27 (8) (2013): 4639-4647.
[248] O. A. Mazyar, H. M. Shammal, and G. Agrawal, International Patent Application 20130220883.
[249] G. C. Leonard, G. T. Rivers, S. Asomaning, and P. Breen, US Patent Application 20130186629.

5 酸化增产

5.1 概述

全球油藏约 50% 是碳酸盐矿物（石灰岩/白垩/白云岩）储层，另约 50% 是砂岩储层（石英、长石等），也可能含有少量碳酸盐矿物。酸化是通过向采油井和注水井地层注入酸液，溶解岩石基质中天然存在的各种酸溶性矿物及钻完井、修井、采油过程中堵塞储层的物质，从而提高储层渗透率和油气井产能的增产措施。造成储层伤害的因素很多，其中只有部分可以用酸处理，如蜡和沥青质等有机沉积物无法用酸处理。低 pH 值酸也能去除碳酸盐和硫化物垢沉积，因此建议本章与第 3 章结垢防治的防垢剂相关内容一起阅读。为了使用化学方法去除硫酸盐垢，也使用如聚氨基羧酸盐等高 pH 值的螯合物。

酸化增产技术历史长久，可追溯到 19 世纪，酸化改造的基本原理、现代技术的介绍，可以参见有关书籍及相对简短的综述。

酸化增产在很多情况下会对储层造成暂时性或永久性伤害，包括将油井变成 100% 出水井，因此酸化增产需要充分了解井史情况，确定最佳的措施方案。这就需要考虑地层矿物的复杂、非均质性，以及其对油田常规酸化配方反应的不可预测性。本章将简要总结在用技术，着重介绍各种酸化处理所涉及的化学剂。为了对酸化有更全面的了解，本章末列出了参考文献。

酸化增产有压裂酸化和基质酸化两种基本方法。

5.2 碳酸盐岩地层压裂酸化

可以使用支撑剂或酸液实施压裂改造。这两种技术的目标都是形成从井筒渗透到储层深部的长孔道通道。在压裂过程中，以高于储层破裂压力，将酸液部分或全部泵入储层内。压裂酸化主要应用于渗透率比砂岩储层低的碳酸盐岩储层。碳酸盐岩储层（白垩、石灰岩和白云岩）的压裂酸化可以消除地层伤害，或使未受伤害的地层增产。一旦超压形成裂缝，就需要酸液刻蚀裂缝，造成通道中的高低点。这就在裂缝中形成了油气可运移的通道。所使用的酸与下面讨论的碳酸盐岩基质酸化相同。

压裂酸化过程中存在的问题是：注入的酸倾向于与易反应的岩石或首次接触的岩石发生反应。这样大部分酸会在井筒附近消耗掉，无法刻蚀深部的裂缝面。此外，酸液沿着阻力最小的路径流动，如岩石中的天然裂缝或者渗透性更强或酸溶性更强的岩石区域。这一过程经常发生在近井筒段，通常在裂缝面形成远离裂缝的长分支通道。这些高导流能力的微通道被称为"蚓孔",非常有害,因为后期注入的压裂液往往会沿溶蚀孔滤失,而无法使裂缝延伸到预期水平。为堵塞溶蚀孔,人们开发了一种"降滤失"技术。在压裂处理后,这种暂时性封堵随着溶蚀孔打开而解除,流体可以恢复流动,有利于原油产量的提高。同样的方法也可以用于酸压裂中的降滤失和基质酸化中的"转向"。可以使用黏性指进压裂或黏性酸压裂来产生酸蚀裂缝。使用黏性指进压裂时,首先使用高黏前置液形成裂缝,然后注入较低黏度的酸液,前进中刻蚀出不均匀的形状。稠化酸压裂使用如胶凝酸、乳化酸或泡沫酸体系,以及化学缓速酸。这些化学方法将在后文讨论。

5.3 基质酸化

在基质酸化过程中,酸化处理是在不高于岩石破裂压力之下,将酸液注入储层。基质酸化对砂岩储层和碳酸盐岩储层的增产都很有效。可参考2003年发表的相关综述。

碳酸盐岩基质酸化的目标是让酸溶解近井区域形成溶蚀孔,尽可能深入地层。这种方法对于未被伤害的储层,产量最多可增加一倍;对于已受伤害的储层,可以获得更高的产量。值得注意的是,碳酸盐岩基质酸化对于处理碳酸盐胶结砂岩及酸溶性物质[如碳酸钙($CaCO_3$)、漏失物质、碳酸盐岩或硫化物垢]造成的伤害也很有用。

5.4 酸化中的酸

5.4.1 用于碳酸盐岩储层的酸

在碳酸盐岩压裂或基质酸化中最常用的酸是盐酸(HCl)、醋酸(CH_3COOH)和甲酸(HCOOH)等有机酸,有时特别用于高温储层。油田使用的盐酸常见浓度是15%(质量分数),也可能使用高达28%(质量分数)的浓度[工业盐酸通常是37%(质量分数)溶液]。低浓度酸可用于前置液(去除水垢和锈)或顶替洗井。针对高压、高温井的增产,可采用氯化钙或溴盐配制的盐酸混合物。

碳酸钙岩石(石灰岩或白垩)溶解在酸中,释放出二氧化碳,形成氯化钙溶液。与盐酸的反应如下:

$$CaCO_3 + 2HCl \longrightarrow CaCl_2 + CO_2 + H_2O$$

在酸作用下白云岩会释放镁、钙阳离子。像盐酸这类的强酸主要形成无分支的溶蚀孔,而较弱的有机酸和延缓酸形成分支的溶蚀孔。

在一些高温应用中,盐酸由于穿透性差或表面反应快,不能产生预期的增产效果。引入甲酸和醋酸等有机酸,可提供一种较慢反应速率、深部改造的酸体系。这些"延缓"酸的缺点是醋酸盐或甲酸盐的溶解度有限及高温腐蚀。可以使用顺丁烯二酸或乳酸,因其盐类(如顺丁烯二酸钙)更易溶于水。有机酸的腐蚀问题比盐酸小。烷烃磺酸(如甲烷磺酸）是低腐蚀性的碳酸酸化剂。在 5.8.7 中讨论了在地层中水解生成酸的酯类和其他化学物质。

高 pH 值的基质酸化螯合剂有 EDTA 盐或羟氨基甲酸盐。通过调整流体的流速和 pH 值，根据井况定制反应较慢的螯合溶液，以最少的溶剂量形成最大的溶蚀孔。此外，使用高 pH 值溶剂可显著降低腐蚀问题。有报道称，螯合物可用于 177℃ 以上的碳酸盐岩酸化处理。EDTA 等螯合物比盐酸和有机酸的成本高很多。也研究了低腐蚀速率的长链羧酸，在高温下具有良好的溶解能力、高生物降解性和安全性。

5.4.2 用于砂岩储层的酸

砂岩基质酸化的主要目的是消除井筒和近井区域的酸溶物导致的储层伤害，从而为油气流动提供更好的流动通道。如果储层天然裂缝不发育，使用基质酸化处理未伤害的砂岩井通常不会产生好的增产效果。由碳酸盐岩垢导致的伤害，可以用与碳酸盐岩基质酸化相同的酸来去除。砂岩储层主要由石英和铝硅酸盐（如长石）组成，颗粒（细粒）迁移到近井区域的孔隙中，可能导致产量下降。这些微粒不会溶解在如盐酸等强酸中，但会溶解在氢氟酸（HF）中。

虽然 HF 具有很强的腐蚀性，但因其在水中的电离率低，视为弱酸。HF 毒性强。HF 或者更常见如氟化铵（NH_4HF_2）能释放 HF 的化学物质，与盐酸或有机酸结合用于砂岩基质酸化。HF 还会溶解钻井作业后留下的膨润土。氢氟酸和盐酸的混合物通常被称为"土酸"。用铵盐进行预冲洗，通常是去除如 Na^+、K^+ 和 Ca^{2+} 等不相容离子，这些离子可能导致不溶性的氟硅酸盐（如 Na_2SiF_6）沉淀。不同作业服务公司在砂岩和碳酸盐岩基质酸化处理中使用的酸浓度各有不同，其指导意见见参考文献。出于对砂岩储层近井段的胶结破坏考虑，通常使用的氢氟酸最高浓度为 3%（质量分数）。HCl/HF 值通常为（4~9）:1。

在砂岩酸化过程中，尤其在长时间关井的情况下，必须特别注意反应产物的二次沉淀及其可能造成储层的二次伤害。化学反应复杂，但基本是氢氟酸首先与铝硅酸盐反应生成氟硅酸盐，氟硅酸盐再与黏土矿物反应生成不溶性氟硅酸钠或氟硅酸钾。稀盐酸或氯化铵可以将潜在的沉积溶液从关键的近井区域推向储层深处。控制这种沉淀问题的另一种方法是使用缓慢生成 HF 的延迟酸配方，如黏土矿物酸（氟硼酸，HBF_4）和可以在高温下水解成酸的自生酸（酯），也可以使用能更加深入储层的缓冲酸，如 pH 值为 1.9~4 的缓冲 HF 酸溶液，含有有机酸及其盐和膦酸盐，以缓解形成硅质沉淀物。高 pH 值缓冲体系已成功应用于单级砂岩酸化处理，无须预冲洗和反复冲洗。可以使用如聚羧酸盐、膦酸盐或有机硅烷等二氧化硅沉淀抑制剂。通过添加碱金属络合剂（如冠醚），可以改善 HF 或能释放 HF 增产配方的性能。

在硅酸盐反应后，黏土矿物中的铝随后与 HF 发生反应。如果没有后冲洗稀释或提高 pH 值，氟铝酸盐将溶于残酸中。如多氨基羧酸或缓冲有机酸等螯合剂可以添加到酸液体系中，以防止产生这种沉淀，与多级土酸处理相比，这种体系可用于单级处理，尤其适用于高温井。此外，当砂岩储层中碳酸钙含量过高时，不溶性氟化钙会在残酸中沉淀，此时仅盐酸处理就可以达到效果。如果残酸的 pH 值提高到 2 以上，不溶性铁（Ⅲ）盐也会造成问题。在砂岩酸化中应用盐酸预冲洗，在某些情况下可能足以消除大部分地层碳酸钙垢伤害。这样就能避免 HF 酸化产物对地层造成的伤害。还可以添加膦酸盐防垢剂以避免碳酸盐垢的再沉淀。

岩屑砂岩储层对盐酸流体非常敏感，因为这种酸会导致微粒运移和地层伤害。已经发现了替代 HCl/HF 钻井液的螯合液，可以增产并消除这种伤害。在砾石充填井中，一种含 EDTA 二铵和少量二氟化铵的新型基质酸成功地避免了微粒运移和地层伤害。氢氟酸和甲酸对砂岩储层的酸化作用已有研究，9%（质量分数）甲酸可以破坏砂岩岩心，产生铁和铝基沉淀。Zeta 电位测定表明，甲酸可以引发黏土矿物微粒絮凝，加入 5%（质量分数）的氯化铵有助于屏蔽黏土矿物表面的负电荷。

5.5 酸化对储层的潜在伤害

对于砂岩和碳酸盐岩储层，如果酸化方法不正确，还会导致其他类型储层伤害，如：
（1）由于 HF 用量过多或浓度过大，导致近井眼抗压强度下降。
（2）由于酸和地层流体之间的不相容性，形成乳状液或沥青沉淀物。
（3）水锁和润湿性改变造成的伤害（这可以用含有表面活性剂的互溶剂处理及与水或烃类溶剂混合来恢复）。
（4）酸化后的微粒运移（这在砂岩酸化中相当常见，在油井处理后缓慢复产可以将这种伤害最小化）。

经验表明，与气井相比，开发砂岩储层的油井对基质酸化的响应方式不同。在特定的酸体积时，油井增产措施提高的渗透率达到峰值，然后随注入酸体积的增加而下降。而气井注入酸的体积与渗透率的提高大致成正比，通常比油井的效果更好。因此，砂岩储层的油井增产时，为减少残酸产物与油反应产生的乳化液或沉淀物，在酸化作业前注天然气取代原油，可改善待处理的储层。

5.6 酸化添加剂

为控制处理效果，酸液配方中含有缓蚀剂、铁离子稳定剂和亲水表面活性剂，还可以使用许多其他类别的添加剂。首先讨论上述三类添加剂，其次是其他类别的添加剂。

5.6.1 酸化用缓蚀剂

5.6.1.1 概述

使用强酸性增产液体系，会导致严重的金属腐蚀及氢和氯化物腐蚀应力开裂。有关腐蚀控制的第 8 章讨论了生产中主要类型的缓蚀剂。碳钢或含铬钢（耐腐蚀合金）在酸化过程中的腐蚀速率高于正常生产条件下的腐蚀速率。酸化过程中用于保护碳钢和含铬钢（如双相钢）的缓蚀剂与日常生产加注的缓蚀剂差异大，通常添加浓度也更高。例如，生产管道中使用的咪唑啉和膦酸酯表面活性剂通常不用于酸化作业，尽管据称可以使用胺和咪唑啉的不饱和酸（如丙烯酸）衍生物。酸化缓蚀剂必须能够防止腐蚀性酸与钢的反应，并在低pH值和高储层温度、高浓度酸溶液中保持稳定。由于酸液中缓蚀剂的浓度相当高，还必须考虑成本与效益，缓蚀剂加量通常随井温的增加而增加。在环境严格要求的地区，缓蚀剂的选用也受到限制。更环保的酸化缓蚀剂研究已经取得了一些进展，特别是在高温井中，但有时会以牺牲性能为代价。此类缓蚀剂的发展历程和试验方法可参见文献。

砷化合物曾用作酸液缓蚀剂，但因其高毒性，已经被淘汰。过去许多市售酸液缓蚀剂是以曼尼希缩合产物为基础，由酮、甲醛和胺反应生成曼尼希碱。这种反应很难完全进行，因此市售缓蚀剂产品中仍会存在一些甲醛和致癌物。可以选用替代品如季铵盐表面活性剂氨基缓蚀剂（下文有述及），但多数季铵盐表面活性剂仍有一定毒性。有研究以亚胺（希夫碱），以及更为环保的如肉桂醛和相关衍生物等炔醇和 $\alpha-$、$\beta-$ 不饱和醛等，作为酸液缓蚀剂。后续将更详细地讨论这些问题。烟头被认为是一种廉价的酸液缓蚀剂来源。菊苣植物提取物也被作为绿色缓蚀剂。

溶剂和湿润性表面活性剂也用于酸液缓蚀剂配方。一般来说，大多数用于酸化的有机缓蚀剂具有电负性原子、高度共轭的双键或三键和芳香环，以及高度的平面性。这些性质使缓蚀剂通过电子与钢的"电子雾"混合，能够紧紧吸附在管材的表面。

控制铁氧化腐蚀的一个简单方法是在HCl（碳酸盐）或HCl/HF（砂岩）配方中使用一定比例的热稳定的还原性弱酸。最简单且成本最低的是甲酸（HCOOH），还可以加入碘离子等还原剂，原位形成强还原剂——氢碘酸。碘也被认为是一种缓蚀增效剂。含有还原性酸（如甲酸）的有机酸体系对于抑制高温环境下的腐蚀问题尤为重要。

5.6.1.2 氮基缓蚀剂

为达到理想的效果，大多数成膜缓蚀剂含一个杂原子（如氮、氧或硫），该杂原子具有可与金属表面相互作用的非成键电子对。例外的是单体季铵盐表面活性剂，其耐强酸，但是性能一般，特别是单独用于高温酸化作业时。大多数用于矿物的酸化缓蚀剂（HCl或HCl/HF混合物）含有季铵盐化合物或胺与不饱和氧化物的混合物。一些季铵盐两性表面活性剂（两个季铵盐端基和两个疏水尾链）表现出良好的缓蚀性能。芳香季铵盐表面活性剂（如吡啶或喹啉）在酸化作业中缓蚀效果比烷基季铵盐表面活性剂更好（图5.1）。

季萘甲基喹啉氯与氯化锑作为缓蚀剂也有报道。2-氨基吡啶和其他氨基吡啶与所有的胺一样，会在酸性溶液中发生质子化，是低碳钢的良好缓蚀剂。具有萘基或四唑基等多种取代基的吡啶盐也是性能优良的酸化缓蚀剂。部分季铵盐表面活性剂（如苄基喹啉

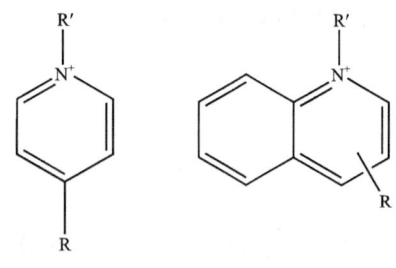

图 5.1　吡啶和喹啉离子

氯）可以通过添加长链羧酸等协同剂来提高性能。硫代硫酸盐（$Na_2S_2O_3$）或硫代乙醇酸（$HSCH_2COOH$）等硫化合物可作为有机酸组分的增效剂。季铵盐表面活性剂还可以通过添加增效剂（如钼酸盐离子）或还原剂提高性能，如常见的碘离子、亚铜、锑、铋盐和甲酸也有应用。酸溶性的乙酸铜与碘化钾配合使用，生成碘化亚铜（CuI）作为酸化缓蚀剂。铋盐的毒性比锑盐小。增效剂通常用于高温环境中。一种基于季铵盐和增效剂的缓蚀剂配方产品可用于 121～135℃条件下的碳酸盐岩储层增产。

次磷酸盐可用于弱酸的缓蚀剂配方。另一种提高性能的方法是使用聚合或低聚胺/季铵盐。含有若干个苄基喹啉基团的低聚化芳香胺或季胺衍生物，以及聚甲苯胺和季铵化聚乙烯亚胺作为酸化缓蚀剂已有研究。一种只含有部分乙氧基胺的胺类表面活性剂已推荐用于生产操作或酸化作业，如脂肪烷基胺、烷基醚胺或烷基酰胺丙胺与 1mol 环氧乙烷反应的产物 $R'C(O)NHCH_2CH_2CH_2NHCH_2CH_2OH$。一类新的单/二氨基缓蚀剂也有研究。聚烷氧基二胺可作为含硫化氢井的酸化缓蚀剂。如北海盆地的部分环境敏感地区，不允许使用在酸性溶液中季铵化的胺和烷基季铵盐，此时需要用其他酸化缓蚀剂加以替代。P,P'-双（三苯基膦酰）甲基二溴苯酮是低碳钢在 1.0mol/L 盐酸溶液中的高效缓蚀剂。

5.6.1.3 不饱和键的含氧缓蚀剂

与生产操作中的缓蚀剂不同，多数酸化缓蚀剂的端基中含有氧和不饱和键。不饱和氧化物在金属表面聚合。通常添加季铵盐或胺类化合物，除了自身可提供表面缓蚀作用，还有助于聚合物牢固吸附于表面。一些市售的酸化缓蚀剂用如 1-辛炔-3-醇、1-己炔-3-醇、2-甲基-3-丁醇和 1-丙炔-3-醇（丙炔醇，图 5.2）等乙炔醇配制。如 3,8-二羟基-1,9-癸二炔等二炔二醇也有研究，其在高温下与酸反应，在钢表面形成成膜性的低聚物。可以通过添加上述强化剂或元素碘、季铵盐表面活性剂或胺（如六亚甲基毒鼠胺）来进一步改善性能，酸化缓蚀剂也可以清除酸与金属硫化物反应释放的硫化氢。但一些小分子的乙炔醇有毒性，会给生产返出的热残酸处理中带来一些问题。乙炔醇的一个新结构是由丙炔醇和碘反应制得 2,3-二碘-2-丙烯-1-醇（用于过量的丙炔醇）。这种化合物以稳定的形式向介质提供碘，而且不会随时间的推移而降解，也可用二丙基硫化物、双（1-甲基-2-丙基）硫化物和双（2-乙基-2-丙基）硫化物等炔属硫化物代替乙炔醇。也有一些人强烈要求将丙炔醇从缓蚀剂配方中去除。

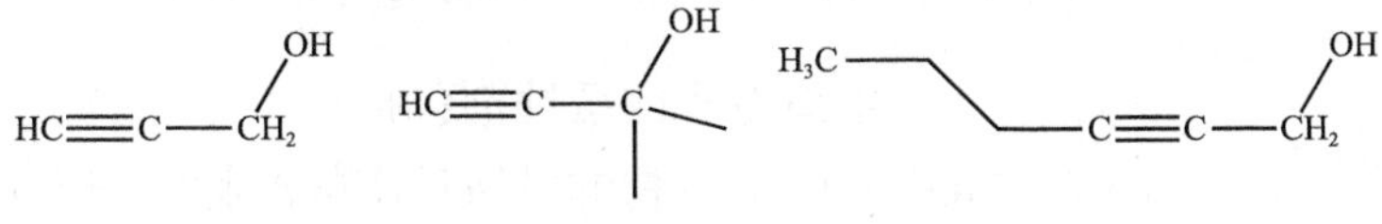

图 5.2 丙炔醇、2-甲基-3-丁醇和 1-己炔-3-醇

肉桂醛（另一种含氧分子）和相关衍生物作为酸化缓蚀剂的有效添加剂已使用多年（图 5.3），通常比季铵盐产品和曼尼希缩合化合物更环保，其毒性也低于多数乙炔醇。其结构的主要特征是含有一个醛基和一个共轭双键。在酸液中，肉桂醛通过各种有机金属

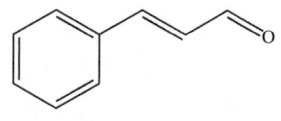

图 5.3 肉桂醛

和金属—氧的相互作用在金属表面形成一层膜。肉桂醛本身的缓蚀性能有限，但其可以与其他缓蚀剂和增效剂（如表面活性剂、其他醛类、乙炔化合物、烷基或芳基膦、锑或碘盐）混合以提高性能。肉桂醛与乙醛酸或乙二醛等脂肪醛混合也有研究。例如，2-氯-2,2-二苯乙酸和2-溴异丁等酸卤代羧酸可作为缓蚀剂的增效剂。

相关的肉桂基化合物的增强缓蚀性能研究也在开展。例如，肉桂醛与硫乙醇等有机硫化物（形成硫缩醛）和选择性季铵盐表面活性剂结合时性能表现良好。尿素与肉桂醛和相关衍生物协同工作，可以改善酸化缓蚀效果。伯胺或仲胺和烷醇胺与肉桂醛等 $\alpha-$, $\beta-$ 不饱和醛或酮的反应产物可以提升缓蚀剂性能。由肉桂醛或取代肉桂醛与 C_3—C_9 酮的反应产物，如苯乙酮、硫脲、甲醛和盐酸组成的组合物，被认为是一种环境友好型缓蚀剂。肉桂醇的生物可降解酯型季铵盐表面活性剂可以用作缓蚀剂，如［双（2-羟乙基）椰碱］肉桂醇季酯还具有抗沉淀和破乳的功能。肉桂基硫三唑在浓度最高达 3mol/L 的无机酸中性能良好（图 5.4）。

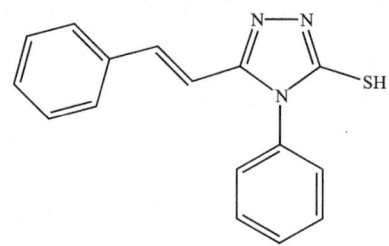

图 5.4 3-肉桂基-4-苯基-1-5-巯基-1,2,4-苯三唑（肉桂基可以被各种取代苯基取代）

肉桂醛等具有平面分子结构的共轭性缓蚀剂，能与钢表面形成良好的相互作用。例如，肉桂醛与苯二胺的希夫碱衍生物——2,4-二肉桂烯亚胺亚苯，在酸性介质中对钢具有良好的缓蚀性能（图 5.5）。另一类具有共轭性的缓蚀剂是 $\alpha-$ 烯基苯酮（图 5.6）。这几种缓蚀剂在 15% 的盐酸中形成 2-苯甲酰烯丙醇中间体。这种分子聚合形成聚苯基乙烯基酮，可在钢表面成膜。$\alpha-$ 烯基苯酮与表面活性剂共混后，性能得到改善。

图 5.5 2,4-二肉桂烯亚胺亚苯

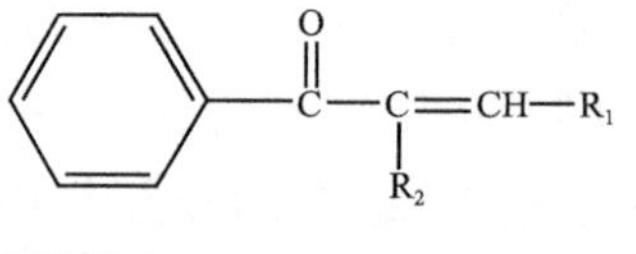

图 5.6 $\alpha-$ 烯基苯酮（R_1=H，R_2=H、CH_2OH 或 CH_2OCH_3）

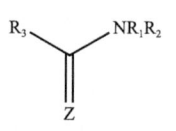

图 5.7 基于醛类和酰胺类反应产物的缓蚀剂

已申请专利的缓蚀剂包括至少一种醛和至少一种非甲酰胺或甲酰胺衍生物的酰胺反应的产物（图 5.7）。例如，肉桂醛或苯甲醛与苯甲酰胺的反应产物，如肉桂醛或 2-羟萘醛等醛与硫醇（如二巯基-L,3,4-噻二唑）或胺官能化环结构（如三嗪）的反应产物（图 5.8）。

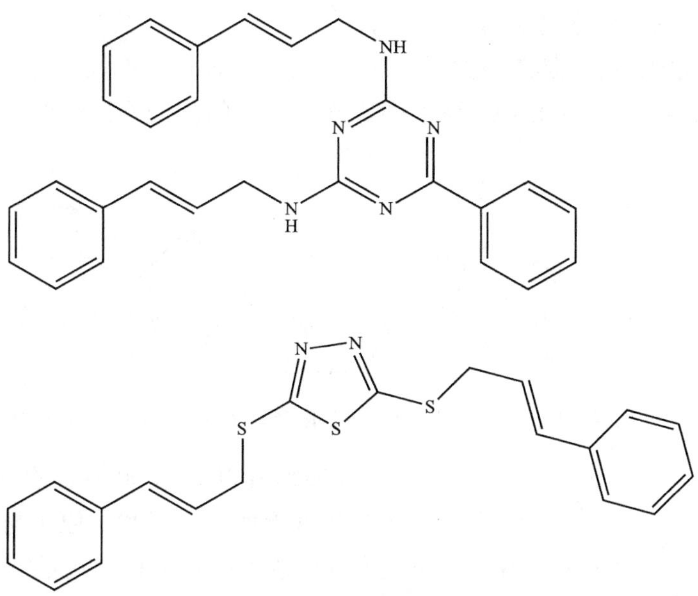

图 5.8 肉桂醛与硫醇或胺功能化环结构的反应产物

如苯乙酮或 3- 羟基 -1- 苯基 -1- 丙酮等酮类作为含氧添加剂，已被研究用于酸化缓蚀剂（图 5.9）。例如，在 120℃以下时，苯乙酮与吡啶或喹啉季铵盐和锑或铋离子的组合物是很好的酸化缓蚀剂。例如，醛（如甲醛）在存在含氮化合物（如烷基胺）和羰基化合物（如苯乙酮）时反应生成的缩合产物，以及酮、醛、脂肪酸与锑、铋化合物作为增效剂的醛醇碱催化缩合产物。松香胺组分（如脱氢枞胺）、酮组分［如苯乙酮、羟基苯乙酮和（或）二苯乙酮］、一种或多种羧酸组分（如甲酸、乙醇酸、柠檬酸）和多聚甲醛是毒性较低、性能良好的酸化缓蚀剂。萜烯和羟丙酸是缓蚀增效剂。

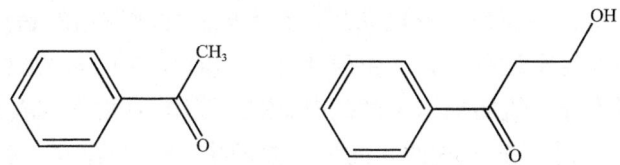

图 5.9 苯乙酮和 3- 羟基 -1- 苯基 -1- 丙酮

5.6.1.4 含硫缓蚀剂

除了氮和氧，有机酸化缓蚀剂中常见杂原子是硫。含硫缓蚀剂通常与乙炔醇、肉桂醛和季氮化合物等其他类缓蚀剂一起使用，对减少点蚀特别有效。

常见的含硫缓蚀剂是如 1，3- 二丁基硫脲和 1，3- 二烷基硫脲，但可以用更易溶于水的硫脲分子。在肉桂醛的配方中，已经提出了一系列其他硫化物。如果化合物含有巯基（—SH），可以与肉桂醛或巴豆醛等醛反应，形成一种新的化合物，也可以用作缓蚀剂。硫脲（或伯胺）与甲醛和芳香族酮（如苯乙酮）在有机酸（如乙酸）和无机酸存在下的反应产物也是一种有效的酸化缓蚀剂。

含硫含氮缓蚀剂，特别是杂环类缓蚀剂已有相当多研究。例如，恶二唑（X=O）和噻二唑（X=S）化合物在 1mol/L HCl 溶液中表现出良好的阴极缓蚀性能，未见更高浓度的报道（图 5.10）。

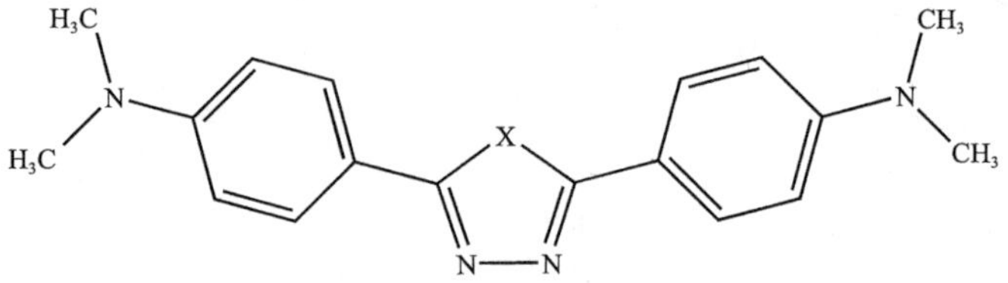

图 5.10　恶二唑（X＝O）和噻二唑（X＝S）化合物在酸性介质中具有良好的缓蚀性能

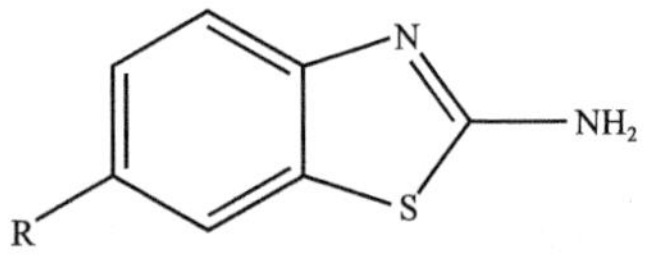

图 5.11　氨基苯丙噻唑

氨基苯丙噻唑在酸性介质中也有很好的缓蚀效果（图 5.11）。除了环上的氮原子和硫原子与钢表面的相互作用外，胺在低 pH 值下的正电荷也可能对性能有贡献。由于氯分子中的偶极矩更大，R＝Cl 的抑制效果比 R＝H 较好。有趣的是，肉桂醛衍生物（可能是希夫碱）的表现甚至比氨基苯并噻唑更好。加入碘化钾可以增强缓蚀性能。

其他含氮和硫的杂环化合物是酸性溶液中用于钢的良好缓蚀剂。一种含有三氮一硫环结构、烯烃（C＝C）和甲烷（C＝N）的共轭结构的产品在 15% 盐酸沸水中明显优于丙炔醇。其他具有良好缓蚀性能的硫氮化合物，包括硫三唑、硫二脲化合物、硫半酰肼，以及亚砜亚胺[R_1R_2S＝（NH）O]和硫取代（异）硫脲。

5.6.2　铁离子稳定剂

在油井和近井区域，特别是含硫化氢井，有形成硫化亚铁垢的趋势。这类井用酸处理能溶解硫化亚铁，但在这个过程中会产生硫化氢，不但有毒，还会加剧腐蚀。此外，随着处理液中酸消耗和液体 pH 值增加，溶解的铁倾向于以氢氧化铁或硫化亚铁的形式沉淀。铁化合物的沉淀会破坏地层的渗透性。此外，溶液中 Fe^{3+} 的释放，加剧了沥青沉淀物造成的伤害。

Fe^{3+} 引起沉淀物主要原因是其作为 HCl 的相迁移催化剂。Fe^{3+} 与油层中沥青质中的极性基团反应形成沉淀物。用 Fe^{3+} 稳定剂和破乳剂控制沉淀物，可以缓解乳状液堵塞问题。主要有三种 Fe^{3+} 稳定剂：抗淤渣表面活性剂、将 Fe^{3+} 还原为 Fe^{2+} 的还原剂、络合剂（也称螯合剂）。

含烷基芳香基磺酸或其盐、乙炔醇和烷基二苯基氧化物磺酸的混合物有助于防止在盐酸酸化过程中形成乳化液和沥青沉淀物。阴离子和阳离子抗淤渣表面活性剂都有研究。这两种类型都已在该领域得到应用，但阴离子表面活性剂对酸诱导沥青质沉淀的总体控制效果最好，还需要使用铁还原剂和非离子偶合阳离子缓蚀剂。阴离子表面活性剂比阳离子表面活性剂更容易造成原油乳化，因此需要另一种表面活性剂来防止这种情况发生。阳离子表面活性剂有利于湿润碳酸盐岩，而阴离子表面活性剂则会导致油湿润。

可以使用的还原剂包括：

（1）处于还原氧化态的金属离子，如 Sn^{2+} 或 Cu^+。

（2）碘化物盐或碘。

（3）还原酸，如甲酸、次磷酸或次磷酸前体，如由 Sb^{5+} 或 Cu^{2+} 催化的金属膦酸盐。

（4）异抗坏血酸或抗坏血酸（图 5.12）。

（5）带有催化剂如 Cu^{2+} 和 I^- 的还原性硫代酸（如硫代乙酸，$HSCH_2COOH$）或能与硫化物反应的酮。

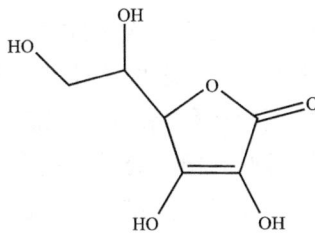

图 5.12　异抗坏血酸

碘化物和无机酸等还原剂有助于防止腐蚀。其他容易达到氧化态的金属离子（如钒或铼离子）也可以与还原剂结合使用。I^- 等还原剂首先将金属离子还原到较低的氧化态，然后将 Fe^{3+} 还原为 Fe^{2+}，使其保持可溶性。实际上，金属离子是电子转移剂和催化剂前驱体。

络合剂通常用于铁的控制。与第 3 章 3.9 节中讨论的防垢剂产品相同，柠檬酸、EDTA 或氮三乙酸等络合剂也是螯合剂，前两种螯合剂是工业用于控制铁最常用的螯合剂。120℃以上时，EDTA 通过成为还原剂，增强铁离子稳定效果。此外，还可以加入酮、醛或缩醛等硫化物改性剂与溶解的硫化物结合。也可以使用隔离铁的二硫代氨基甲酸盐组合物。另一种减少金属硫化物从酸溶液中沉淀的方法是加入肟（如乙醛肟），肟优先与溶液中的硫离子反应，从而防止硫离子与溶液中的金属离子反应。

5.6.3　亲水润湿剂

在酸化增产处理过程中，需要使用亲水润湿剂以去除岩石或垢表面的油层，使含水的酸与岩石有良好的接触。亲水润湿剂的主要作用是洗井和保持地层水湿润，并增强油气的流动。亲水润湿剂是简单的单体表面活性剂，具有高 HLB 值，如烷基乙氧基化物和烷基苯乙氧基化物等非离子表面活性剂。还可以使用互溶剂，与水和油混溶并有助于润湿地层。最常见的溶剂是乙二醇单丁醚（也称丁基二醇）、二丙二醇甲基醚。互溶剂也可以处理水基溶液处理带来的水锁伤害。酸性溶液中成膜缓蚀剂可以提供水润湿性。

5.6.4　酸化处理中的其他可选化学剂

根据地层流体和伤害类型的不同，还可以选择其他化学品。

（1）黏土矿物稳定剂：这有助于防止由砂岩酸处理引起的黏土矿物运移或膨胀。使用质量分数约 5% 的铵盐预冲洗。改进的黏土矿物稳定剂是多胺和多季铵（如聚二甲基二烯丙基氯化铵），最好添加到预冲洗中，可以在初始离子交换期间保护黏土矿物。

（2）微粒固化剂：砂岩地层中的大多数微粒不是黏土矿物，而是石英、长石和其他矿物。黏土矿物稳定剂不能控制这些微粒，有机硅剂（如 3-氨基丙基三乙氧基硅烷）用于固定这些微粒，与水原位反应形成聚硅氧烷，与硅质微粒结合。

（3）沉淀抑制剂：上文讨论的铁离子稳定剂如果可以防止形成未络合的铁离子，则可以防止沥青质的沉淀；沥青质溶剂、抑制剂和分散剂也可以用于控制沉淀，特别是在乳化酸中（见5.8.8或第4章），如二甲苯作溶剂，烷基芳香基磺酸作分散剂，其他可用于酸处理的抗沉淀表面活性剂，包括酯基季铵盐表面活性剂、乙氧基烷基酚、烷氧基取代烷基苯酚磺酸盐和酯，或磺化脂肪酸盐，油田酸化或低pH值防垢剂挤注处理之前，最好先进行沥青质的稳定处理；生物可降解的酯基季铵盐表面活性剂，也具有缓蚀性能；防止沉淀物的一项发明是结合了HCl、一种季铵缓蚀剂（首选芳香族）、阳离子胺氧化物表面活性剂和一种不与缓蚀剂反应的阴离子表面活性剂的共轭离子对，首选的共轭离子对是二甲基烷基胺氧化物和十二烷基硫酸钠。

（4）破乳剂：在使用主体酸体系溶液之前，先注入破乳剂。如前所述，在酸性溶液中使用铁离子稳定剂也可以解决一些乳化液问题。$N-$烷基化聚羟乙胺等破乳剂专门设计用于酸化作业（也可见第11章）。

（5）醇类：小分子醇类可以通过降低表面张力，帮助残酸返排，从而有利于气井改造。在深水低温气井中，可能需要甲醇或乙二醇来防止天然气水合物堵塞。

（6）硫酸钙防垢剂：如果配液用的地层水硫酸盐浓度较高或使用海水，则必须使用硫酸钙防垢剂（见第3章）或者螯合剂。

（7）碳酸钙防垢剂。

（8）H_2S清除剂：如果酸化处理接触到硫化物垢，会释放出有毒的H_2S。可以使用醛或其他H_2S清除剂（参见第15章）。

（9）起泡剂：可与氮气共用，有助于残酸返排，使气井恢复生产。

（10）减阻剂：通常是高分子量聚丙烯酰胺等水溶性聚合物。减阻剂降低流体的摩擦阻力，对处理液量大的井或深井非常有用。

（11）表面活性剂：在气井中，需要降低表面张力，可以降低地层中束缚水相的毛细管力，从而避免水锁。

5.7 井筒轴向布酸酸化

在酸化处理前，除确定每个特定油藏最有效的组合酸液和其体积外，通常还需要进行设计和优化，确保酸能够覆盖整个层段。在裸眼水平井中，由于处理层的长度和储层性质的潜在变化，在基质处理中成功均匀布酸的难度更大。一项成功的转向技术对于将酸液体系置放在伤害的位置至关重要。

为改善酸液与待处理层段的接触，可以使用机械或化学分流技术。有文献介绍了一种可回收式桥塞。机械布酸可以通过封隔器系统、球型封隔器或连续油管作业完成。化学布酸需使用转向剂。除酸化作业外，转向剂在防垢挤注作业等许多其他井作业中也有使用。转向剂通常用于预冲洗，暂时封堵渗透率较高的储层，使酸液与其他渗透性较差或伤害较严重的储层发生反应。复产后，堵塞射孔段的转向剂会被排除。目前也发展了自转向黏弹性流体，下文将对此进行讨论。

酸化用转向剂至少有四类：可在产出水或油中降解、溶解或熔化的固体颗粒，聚合物凝胶，泡沫，黏弹性表面活性剂（VESs）。

碳酸盐岩储层在酸中的溶解度高，并形成大通道，故此类储层酸化中，通常不使用固体颗粒转向剂。上述其他三种类型均已成功应用于碳酸盐岩储层酸化。在极端条件下，黏弹性表面活性剂与颗粒转向（由可降解纤维材料提供）相结合的协同转向作用，提供优越的酸液铺置效果。

除上述方法以外，相对渗透率调节剂（RPMs）也可用于提高酸化处理的转向能力。相关的疏水改性水溶性聚丙烯酰胺体系已用于生产井和注入井的基质酸化增产处理。此外，该体系还可以降低地层水产量或调整注水井吸水剖面（见第2章）。

在某个碳酸盐岩油田，与使用泡沫或聚合物凝胶转向剂相比，基质酸化处理中使用疏水改性 RPM 获得了更好的生产效果。使用乳化酸可达到更高的渗透率，其配方的高黏度也可以起到转向作用。另一种改善 HF 酸井筒轴向布酸的方法是最大压差和注入速率（MADPIR）技术。通过保持最大注入速率，同时始终提高注入速率，以保持最大允许的基质注入压力（低于压裂压力），从而消除对转向剂的需求。

5.7.1 固体颗粒转向剂

在 HCl 和非 HF 酸化处理中，最常见的转向剂是岩盐（NaCl）。岩盐可以加入主要的酸冲洗。苯甲酸薄片、油溶性树脂和蜡球是很好的颗粒转向剂，可用于砂岩酸化处理。树脂和蜡颗粒因在较低井温下不熔化，不能用于如气井或注水井等低温井。其他适宜的固体添加剂包括聚酯、聚碳酸酯、缩醛、三聚氰胺聚合物、聚氯乙烯和聚醋酸乙烯酯，当其与增黏流体结合时，可以在近井区域形成更深的屏蔽暂堵。

在过去的 20 年里，固体转向技术已经被其他转向技术所取代。但一种新型的转向剂——聚酯固体颗粒在高温下可通过水解降解，在酸化增产作业中表现出优越的性能。此外，还报道了成功使用可降解（水解）聚合物纤维技术的现场试验，该技术可在酸压裂过程中实现有效的酸转向。通过纤维丝团聚桥堵机制，前置酸的黏度显著增加，有效地堵塞被压裂层。

5.7.2 聚合物凝胶转向剂

聚合物凝胶比常规酸化溶液黏高，这意味着其首先会进入低渗透层段。有两种使用黏性溶液（如聚合物凝胶）的方法。首先，可以注入聚合物凝胶黏性前置液（或预冲洗液）。这种技术依赖预冲洗液的黏度来影响所进入井段的注入压力。当预冲洗液进入地层时，前置液的黏度将限制其他液体进入该区域。随着该井段注入压力增加，其他井段会起裂，开始接纳液体。低黏度酸液将穿透低渗透受伤害储层。其次，可以将酸性溶液本身胶凝化。酸化处理采用胶凝酸转向剂，有两个优点：（1）反应速率显著减慢，使得酸液穿透得更深（见井筒径向轴向布酸）；（2）由于流动阻力的增加，高黏度降低了滤失。

这两个优点结合起来，增加了后续处理酸转移到其他位置的趋势。通过交替泵注胶凝酸和非胶凝酸前置液，可以酸化整个层段。这在碳酸盐岩基质酸化中特别有用。破胶剂可用来降低酸化处理后稠化酸的黏度。聚合物胶凝酸也用作压裂酸化的降滤失剂。

典型的化学转向剂是凝胶化的羟乙基纤维素（HEC）小球。如果温度超过95℃，凝胶HEC的基础黏度和有效期大大降低，这种方法将受到严重限制。其他已经使用的典型聚合物是丙烯酰胺聚合物和其他天然或多糖聚合物（如瓜尔胶、黄胞胶、硬葡聚糖或琥珀聚糖）。生物可降解的硬葡聚糖或双葡聚糖在部分交联时，比大多数天然聚合物具有更好的热稳定性，也比生物可降解性差的合成聚合物（如聚丙烯酰胺）更容易分散。天然聚合物HEC和聚丙烯酰胺在高井温的酸中不稳定，耐盐性也较差。因此，针对上述环境的应用，提出了包括乙烯内酰胺、N,N-二甲基丙烯酰胺、丙烯酰胺丙磺酸和季铵盐（如丙烯酰胺乙基三甲基氯化铵和甲基丙烯酸二甲胺乙酯）聚合物和共聚物等其他合成聚合物，这些材料通常成本更高或生物降解性较差。据称通过基因工程或细菌选择制备的澄清黄胞胶可用于温度低于150℃的胶凝酸。

含有一种聚合物和另一种化学物质的主体酸溶液泵入储层后，随井温升高发生原位交联聚合，进一步增加溶液的黏度。在压裂酸化中，四价和三价金属盐［锆（Ⅳ）、钛（Ⅳ）］通常用于交联多糖（如瓜尔胶）。金属盐［如铝盐、铁（Ⅲ）盐］、酚类化合物和（或）小单醛和多胺已用作与阴离子聚合物（如部分水解丙烯酰胺聚合物）的交联剂，其中一些在成分上与堵水处理中使用的化学物质相似（见第2章）。一种新开发的铝交联剂不需要任何还原剂或破胶助剂，在酸作用下可以实现交联凝胶的完全破胶。

在井筒温度升高时，一些聚合物溶液会变稠。包括聚氧乙烯，如乙二胺与环氧丙烷反应后，再与环氧乙烷反应，产物的分子量高达30000。也有人提出了水包油的乳化剂（见第9章）。

聚合物胶凝酸的自转向也可以通过改变pH值来完成，这是由于酸被消耗后pH值升高。而pH值升高可以激活高价金属离子试剂，如锆（Ⅳ）交联聚合物。随着黏度的增加，导致了更高的流动阻力，并将未反应完的酸转移到储层其他区域。pH值进一步升高，用还原剂使金属交联剂失去活性，并将分子链断裂到最初的聚合物链。自胶凝转向酸处理称为"降滤失酸"，在碳酸盐岩储层酸化作业中形成深穿透溶蚀孔，已经非常流行。交联黄胞胶效果不好，但据称在自转向酸中，交联黄胞胶的氧化形式更稳定和有效。

使用聚合物增稠剂处理地层时，有时会遇到一个问题，即在作业后，聚合物增稠剂能否被很容易地移除。部分增稠或高黏度的聚合物溶液在作业完成后，很难从地层孔隙或裂缝中去除。有时堵塞残留物会滞留在地层的孔隙或裂缝中，可能会抑制储层流体的产出，需要昂贵的清洗作业。避免这种情况的一个方法是延迟稠化。例如，已经提出使用含有亚硫酸盐离子或硫代乙酰胺等还原剂的六价铬盐。这些成分会在原位形成Cr^{3+}，与聚合物交联。另一种依靠醛来交联聚合物的方法是使用如缩醛等醛的前体，其在井温升高时就地分解形成醛。在水力压裂中很常见的是通过注入酶和过硫酸钠等氧化剂来分解多糖液体。如果多糖聚合物与金属离子（如钛或锆）交联，注入水溶性树脂包裹的破胶剂可以延迟破胶和降低黏度。例如，金属盐断裂破胶依靠的是优先与金属离子络合的离子材料，如氟化物离子（氟石，冰晶石）、硫酸盐、磷酸盐、膦酸盐和羧酸盐（如 EDTA）。

在压裂酸化作业中，一种降滤失的方法是，首先在酸性流体中加入一种或多种化学物质，在大量的酸消耗（残酸）和 pH 值升高后，这些化学物质会形成滤饼，阻止流体的流动。另一种或多种化学物质会随着更多的酸消耗和 pH 值的进一步升高而破坏滤饼。最初的强酸性体系包括可溶的铁离子源和聚合物凝胶，其黏度低，在 pH 值约为 2 或更高时，铁离子发生交联反应，pH 值较低时则不反应。典型的聚合物是阴离子聚丙烯酰胺。但聚合物不是由 Fe^{2+} 交联。因此，该体系包括一种还原剂，在 pH 值高于 3～3.5 时，将 Fe^{3+} 还原为 Fe^{2+}。例如，当酸在溶蚀孔中消耗时，pH 值增加到 2 或更高，聚合物交联和形成非常黏稠的凝胶，抑制鲜酸进一步流入溶蚀孔。随着酸的消耗（处理后）和 pH 值继续上升，还原剂将 Fe^{3+} 转化为 Fe^{2+}，恢复到类似于水的流动状态。过去一直使用肼盐和羟胺盐作为还原剂，但其毒性和致癌性较大，有人提出用危害较小的碳酰肼、半碳酰肼、酮肟和醛肟作为替代物。

5.7.3 泡沫转向剂

泡沫酸早在 20 世纪 70 年代就被提出用于压裂酸化，但也用于转向和深穿透基质酸化。泡沫通过将酸导向井筒附近受伤害或低渗透的储层，可以帮助酸化井的增产。这种泡沫是由一种水溶性表面活性剂和一种气体（如氮气、一氧化碳、二氧化碳或天然气）制成。可选择使用如 HEC、羧甲基纤维素聚合物、聚丙烯酰胺、丙烯酸酯和多糖、VES 等增黏聚合物。

形成泡沫的表面活性剂包括阴离子表面活性剂、阳离子表面活性剂、两性和非离子表面活性剂，其性能依次递增。阴离子表面活性剂受到原油的不利影响，在盐酸等强酸存在下起泡性能严重降低；在酸性环境下，阳离子表面活性剂的起泡性能中等，但存在原油时，会产生不稳定泡沫；两性表面活性剂也是如此；非离子型表面活性剂起初泡沫很好，但在酸和原油存在的情况下，泡沫的寿命短，无法进行酸化作业。可以复配非离子表面活性剂主剂（如乙氧基化醇或聚糖苷）和阳离子表面活性剂（如氟化季铵氯盐）来改善泡沫性能。在表面活性剂溶液中加入聚合物（如多糖或部分水解聚丙烯酰胺）增强泡沫，降低流动性。另一类发泡的表面活性剂是乙氧基化脂肪胺被酸原位季铵化。已经有资料公布泡沫酸使用指南。针对高温环境，采用多种表面活性剂提升泡沫酸增产的研究正在进行。

泡沫必须有的特性如下：在酸和储层烃类化合物共存时必须稳定；必须能产生比气体流动性小（如至少 100 倍）的硬泡沫，以阻止酸的流动；在注入酸的过程中，泡沫必须保持其堵塞流体的能力（即刚度），之后泡沫自然分解，允许流体再次通过更高的渗透层。

泡沫流体可以通过其黏度、动态流动条件下的破裂和再形成泡沫的特性来堵塞地层。此外，相对于孔隙大小，泡沫中的气泡尺寸越大，泡沫流体就会更有效地堵塞地层。当存在分层（渗透率不同的层）时，在整个处理过程中，通过在高渗透性区域产生并保持稳定的泡沫来实现转向。当有很长的层段需要处理时，可以通过酸处理部分层段来实现转向，再通过泡沫来阻止后注入的酸进入该层段，随后注入更多的酸。这些交替的步骤可以重复。最终使处理液完全进入处理区域，通过酸有效地清除不同程度的储层伤害。根据所使用的表面活性剂类型和浓度及泡沫质量，泡沫会产生不同程度的屈服应力。泡沫流体也被认为可以携带固体颗粒，增强流体的稳定性和流变性性能，并且是酸化改造的最佳转向流体之一。泡沫液与非泡沫液相比（即使其含有聚合物），其所含液体较少，本质上更清洁，有助于残酸返排及净化，因为其膨胀有助于克服阻力（如静压头），以实现回流；在枯竭产层中，这种"增能"尤其重要。

下文提出了泡沫酸和黏弹性酸体系的组合。凝胶发泡 VES 系统的流动阻力比单独使用泡沫或黏弹性凝胶系统的预期要大。甜菜碱表面活性剂或烷基氨基胺氧化物表面活性剂可制备黏弹性泡沫。

5.7.4 黏弹性表面活性剂

另一种用于辅助地层改造（压裂和基质酸化）的黏性流体转向剂是 VES 或表面活性剂混合物。这些表面活性剂与用于形成泡沫的表面活性剂不同。黏弹性，也称为抗弹性，表明材料既具有黏性又具有弹性特性。VES 流体的流变性，特别是增加的溶液黏度，归因于流体中组分形成的三维结构。当表面活性剂浓度显著超过临界水平时，表面活性剂分子聚集并形成胶束或囊泡等结构，这些结构可以相互作用形成具有黏弹性行为的网状结构。事实上，可以先将化学物质包裹在 VES 形成的胶束中，达到较远的位置时可控地释放出来。虽然通常比聚合物胶凝酸更昂贵，但现代 VES 酸增产的好处是，在处理后通常很少或没有残留（对地层造成伤害）。作为胶凝剂的表面活性剂需要相当高的浓度来产生必要的黏度，比制备聚合凝胶所需的浓度要高。考虑到储层典型温度和 pH 值条件，对 VES 市售配方的流变性能进行了测试。

黏弹性表面活性剂溶液通常是在浓缩的表面活性剂溶液中加入某些试剂，由长链两性或季铵盐［如十六烷基三甲基溴化铵（CTAB）］组成。根据其离子性，许多常见的试剂或共表面活性剂可以添加到表面活性剂溶液中以产生额外的黏弹性和稳定性。可以使用氯化铵、氯化钾、氯化镁、水杨酸钠、异氰酸钠等盐类和氯仿等非离子有机分子。某些阳离子／阴离子表面活性剂与非水溶剂共混也形成黏弹性溶液。水杨酸或邻苯二甲酸与阳离子或两性表面活性剂一起使用。例如，研究了水杨酸钠对双子表面活性剂亚丁基 $-\alpha, \omega-$ 双（十六烷基二甲基溴化铵）溶液性质的影响。许多表面活性剂溶液的电解质含量也是其黏弹性行为的重要控制因素。

油田生产中，能形成黏弹性溶液的表面活性剂主要包括以下几种：

（1）阳离子表面活性剂，有芥酸甲基双（2-羟乙基）氯化铵或 4-芥酸酰胺丙基 -1，1，1-三甲基氯化铵（图 5.13），芥酸甲基双（2-羟乙基）氯化铵与水杨酸钠（0.06%）或芥酸胺辅助表面活性剂（较宽 pH 值范围内保持稳定），双子和非双子双季铵和其他聚阳离子表面活性剂。

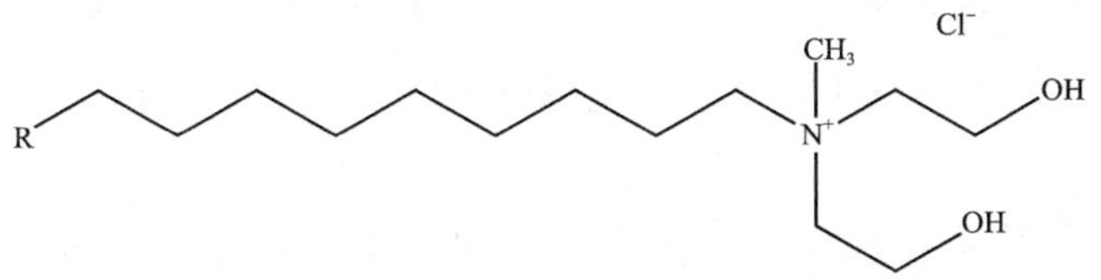

图 5.13　长链烷基甲基双（2-羟乙基）氯化铵表面活性剂

（2）两性离子/两性表面活性剂，有甜菜碱表面活性剂，如油酰胺丙基甜菜碱或芥子酰胺丙基甜菜碱（图 5.14）。

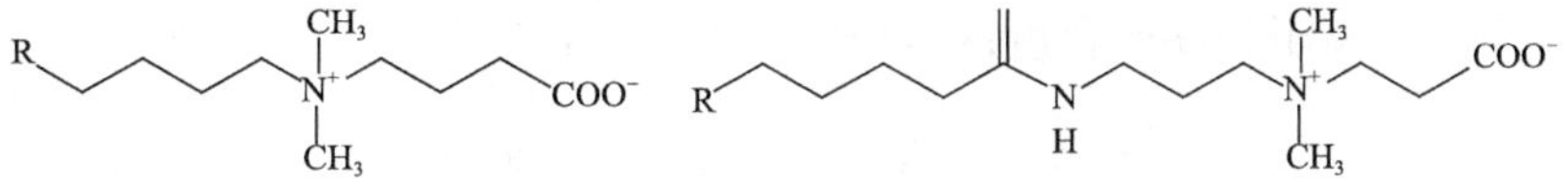

图 5.14　两性离子/两性表面活性剂

（3）阴离子表面活性剂，有烷基牛磺酸盐阴离子表面活性剂、甲酯磺酸盐、磺基琥珀酸酯。

（4）胺氧化物和氨基胺氧化物，有二甲氨基丙基牛油酰胺氧化物（图 5.15）。

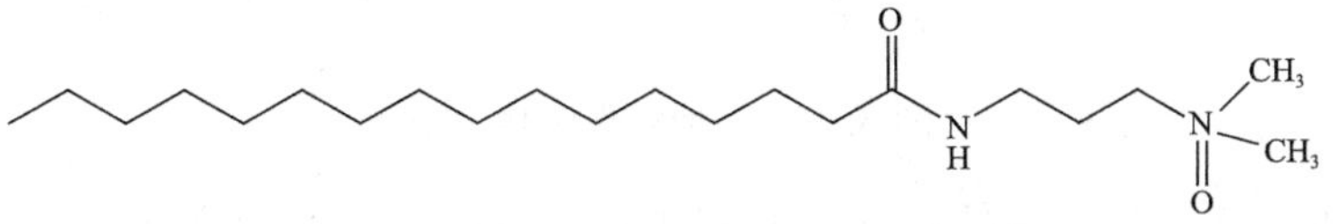

图 5.15　二甲氨基丙基牛油酰胺氧化物

（5）乙氧基化脂肪胺。

（6）阳离子表面活性剂和阴离子表面活性剂，如油酸钠的离子对。

含有表面活性剂胶束流体转向酸有两类：自转向表面活性剂流体和使用表面活性剂胶体作为转向颗粒。

自转向表面活性剂流体由新鲜酸和表面活性剂组成。在最初的混合和泵送时，鲜酸体系的黏度与水差不多。鲜酸体系与地层中的碳酸盐接触反应后损失酸度，pH 值上升，再加上 Ca^{2+} 和（或）Mg^{2+} 等二价离子的作用，体系黏度骤然上升（胶凝）。随着接触位置黏度上升，流体系统就能有效地分流后续的酸液。后续的鲜酸转向进入地层中其他层段进行增产处理。酸注入过程中转向是一个连续过程。VES 处理技术已经在世界各地开展，仅 2009—2012 年，巴西海上就开展了 40 多口井的自转向 VES 处理。

酸化用的表面活性剂胶束转向颗粒由表面活性剂和盐水组成，泵出时是黏性（或胶凝）的，在不同的阶段转向颗粒随着酸液注入。在泵送处理时，通常在转向颗粒和鲜酸间加入隔离液。与自转向表面活性剂酸系统一样，表面活性剂转向颗粒对 pH 值敏感，在低 pH 值下颗粒黏度很小。芥酸甲基双（2-羟乙基）氯化铵是一种市售的用于表面活性剂胶束转向颗粒的 VES，无须很大的反离子就可形成黏弹性胶体。也可以在很低的浓度使用而不稠化流体，但在配方中浓缩以便使流体系统胶凝。加入柠檬酸或醋酸，可以防止因铁离子带来的黏弹性表面活性剂相分离。阳离子和两性/两性离子VESs，特别是那些包含甜菜碱部分的VESs［如R（Me）$_2$N$^+$CH$_2$CH$_2$CH$_2$COO$^-$］可用于最高160℃环境，因此特别适用于中高温井。

但与上述阳离子VESs一样，其不适用于高盐环境。包括十二烷基苯磺酸钠在内的增强表面活性剂，有助于提升甜菜碱VES耐盐性、增加凝胶强度和降低剪切敏感性。其他适用于甜菜碱VES的增强表面活性剂是一些螯合剂（如羟乙基乙二胺三乙酸三钠）。据称两性离子表面活性剂的特定共混物可以提高黏度。双阳离子和多阳离子表面活性剂也可以是VESs，典型产品如图5.16所示。其与高含盐介质有良好的相容性，并在高温下具有良好的热稳定性。高温、高压和高矿化度的VESs应用已经有报道。

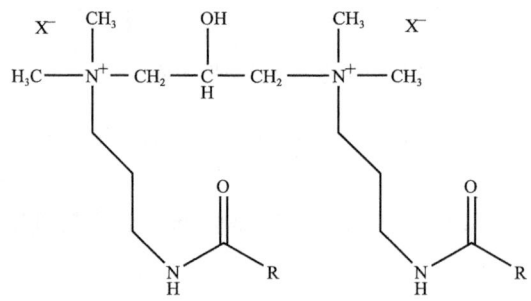

图5.16　一种双季铵盐VES

有文献报道了至少包含一种一定浓度阳离子糖醚的黏弹性化合物，阳离子糖醚具有一个或多个阳离子片段和一个或多个糖片段。例如，通过蔗糖与3-二甲基丙胺芥酸酯和环氧氯丙烷反应制成的产品。

许多VESs形成的蠕虫状胶束对烃类很敏感。利用这种敏感性，流体可以选择性地封堵含水层，而含油层则不受影响。据报道，阴离子表面活性剂具有这种效果。但只要储层含油气，一些VES流体（如甜菜碱）无法区分不同渗透率的储层。此外，与依赖于滤饼沉积的聚合物钻井液不同，VES转向剂通过胶束结构大小来控制流体在地层的滤失。胶束型黏弹性甜菜碱表面活性剂液体由于蠕虫状胶束的体积小，在地层滤失率较高。

甜菜碱表面活性剂已被证明适用于地层温度高达160℃的自转向酸体系（VDAs或SDVAs）。两性离子胺氧化物表面活性剂可在约125℃下使用。由于pH值的降低和水中二价金属阳离子浓度的增加，黏度增加，导致球状胶束转变为蠕虫状胶束。黏稠的滤饼会迫使后面的酸进入其他注入能力较低的层段，这些层段也得以有效改造。酸化处理后，由于VES系统不含固体或聚合物，当接触到产出的烃类、低矿化度的产出水、产出的前置液或后置液时，滤饼就会破坏，不会留下对储层造成伤害的固体残留物。VDA两性离子转向剂也用于碳酸盐岩气井，其在28%的盐酸中稳定，并且相容于酸与碳酸盐反应后的产物。有文献报道了一种可用于高温碳酸盐岩地层增产的有效VES体系，但没有介绍表面活性剂结构相关的细节。有文献回顾了VES转向剂的基质酸化经验教训，说明表面活性剂的负载量不能太高，否则可能会对地层造成伤害。但使用互溶剂可以解除地层伤害。此外，高浓度的缓蚀剂或铁盐也会影响 VES 的性能。

为了避免对地层造成伤害，可以通过几种机制实现水性黏弹性处理液凝胶的破胶，包括与地层中的烃类接触、pH 值变化、稀释/矿化度变化、与烷氧基化醇溶剂接触及与反应剂接触等。特别是对于干气层，依靠烃类地层或水的稀释分解凝胶的黏度，将延长处理时间、延迟生产。有 VES 作业经验的技术人员都认识到后期经常需要进行 VES 补救清理。因此，为了更可靠、更彻底地清洗流体，开发出了化学破胶剂，可以破坏一些用作转向剂的 VES 流体的胶束结构。例如，专门为钻井和水力压裂作业后的清理而设计的带有与 VESs 端基相反电荷的化合物，可以破坏 VES 流体的胶束结构，从而起到破胶作用。其他破胶剂包括高温下降解表面活性剂的化学物质，或高温下降解并原位生成胶束破胶剂的化合物。内部破胶剂包括细菌、过渡金属离子源、皂化脂肪酸、矿物油、氢化聚烯烃油、饱和脂肪酸、不饱和脂肪酸（如多烯酸和单烯酸）及其组合。据称有一种含有至少具有两种氧化态的金属离子和氧化还原试剂的 VES 流体。如可以封装 Fe^{2+} 等金属离子，酸化后在地层中释放并被氧化成 Fe^{3+}，降低凝胶的黏度。自身可破胶的铵盐也可用于自转向酸的主段塞液。但 Fe^{3+} 会与几种 VESs 形成强凝胶，导致地层的二次伤害。加入合适的螯合剂可以最大限度地减少 Fe^{3+} 对 VES 酸的影响。已经开发了用于碳酸盐岩的自转向表面活性剂/有机酸的内部破胶剂。

VESs 作为降滤失剂或转向剂的基本用途已经有了许多变化或优化。例如，泡沫和 VES 酸化都可以通过在一种溶液中结合，得到两种性质。另一种变化是使用某些阴离子表面活性剂 VES 流体体系与少量低级醇（如甲醇）混合。注入表面活性剂的浓度不足以改变地层中的流体流动。这种液体只有在流过地层时，通过从溶液中失去低级醇，才能展现出流体转向的能力。

在 HF−酸化组合物中使用黏弹性酰氨基胺氧化物表面活性剂，可以避免砂岩储层近井区域的塌陷问题。这种表面活性剂的实例如图 5.13 所示。这种自转向表面活性剂溶液随着酸的消耗而增黏，然后随着酸的进一步消耗而降低黏度。

一种改进的 VES 转向颗粒溶液是非水溶性溶剂，其由黏弹性两性表面活性剂（如卵磷脂）及能够与季铵聚电解质反应的两性表面活性剂组成，这样增加了体系的黏度。该聚合物是 HEC 的聚季铵衍生物。与许多其他胶束 VES 流体相比，这种混合物的优势在于其在溶液中形成聚集的囊泡比胶束大得多。因此，可以更好地控制流体滤失。引入聚电解质可以减少表面活性剂的负载量，并使溶液在约 170℃ 环境下比较稳定。该流体在低 pH 值或者该流体内部含有过硫酸盐等破胶剂时破胶，与单独使用表面活性剂相比，其他聚合物/表面活性剂共混物具有更好的黏弹性性能。

有一些热敏性的表面活性剂/助表面活性剂体系，黏度随温度而增加。已报道最简单的体系是含有 5−甲基水杨酸的 CTAB。这些体系还未开发出适用于油田的化学品。

表面活性剂胶凝酸化增产配方已经用于高含水稠油井增产。这种处理可以暂时堵塞水层，并通过化学转向有效地改造油层。此外，该酸体系具有一种独特的固有特性，可以限制酸进入高含水层段，使酸有效地进入富含油带。

5.8 井筒径向布酸酸化

在酸化过程中,所有地层,特别是高温地层都会遇到一个问题,即酸在进入地层之前,就会被井眼附近的反应物质迅速消耗。提出了一种假设体系,即强缓速的体系在泵入储层时不会发生任何反应,并将其作为基质处理的最终方案。

在基质酸化作业中,"良好的穿透距离"通常认为是几米,而在压裂酸化作业中,该距离为 100~200m。其他可以采用缓速酸来提高穿透距离的方法有亲油表面活性剂,用于碳酸盐岩酸化的弱有机酸,对砂岩酸化作用弱的氟化剂、缓冲有机酸、胶凝酸或稠化酸、泡沫酸,对温度敏感的生酸化学品和酶,乳化酸。

5.8.1 亲油表面活性剂

亲油表面活性剂覆盖孔隙表面,减缓酸的溶蚀速率。在 5.6 节中讨论的一些缓蚀剂和表面活性防沉淀添加剂也具有这种功能。这些体系简单,甚至适用于高温井。乳化酸也是亲油体系(见 5.8.8)。

5.8.2 弱有机酸

乙酸、甲酸、柠檬酸、乙二烯胺四乙酸、乙醇酸、酒石酸或氨基磺酸等,比盐酸弱,价格较贵,在碳酸盐岩储层酸化时,特别是在盐酸消耗较快的高温井中,可减缓酸化过程,其腐蚀性也比无机酸弱。也可以使用有机酸和无机酸的混合物。在高温井中进行的一项研究表明,因为只有部分井段得到处理,有机酸没有表现出足够的延缓作用。因此,还必须另外使用其他延缓方法。例如,除本章讨论的方法外,还可以使用包裹的固体有机酸颗粒,但酸的离子可能有时与地层中溶解的阳离子不相容,如在碳酸盐岩酸化过程中的 Ca^{2+} 与柠檬酸离子。HF、柠檬酸和一种膦酸盐已被提出用于砂岩的酸化。有文献综述了最常用的有机酸(甲酸和乙酸)的优势。最近在该领域应用的一种弱酸是谷氨酸二乙酸(GLDA)。GLDA 具有良好的生态毒理学特征和非常低的毒性水平。甲基甘氨酸 $-N,N-$二乙酸被认为是一种溶垢剂。关于溶垢剂可参考第 3 章。

单级膦酸/氢氟酸体系已经在全球范围内用于基质酸化增产,目前正在评估其在水力压裂中的应用。例如,将氟化氢铵加入磷酸,形成含膦酸铵和氟化氢的体系。通过 OECD 301 淡水试验,一种易于生物降解但结构未知的螯合剂已报道可用于碳酸盐岩地层改造。在砂岩井中使用的钻井液中,HEDTA(羟乙基二胺三乙酸)用于消除碳酸钙加重剂对地层造成的伤害。

5.8.3 对砂岩酸化作用弱的氟化剂

对砂岩进行深度酸化增产改造,将通过稳定井眼附近的微粒和沉淀酸化产物,避免产量的快速下降。有人提出使用酸性较弱的砂岩酸化氟化剂(如黏土酸、HBF_4)与螯合剂

（螯合铝）混合。硼酸（H_3BO_3）与 HF 反应生成 HBF_4。HBF_4 与砂岩的反应速率比 HF 慢，使酸能够深入地层，避免近井伤害。另一种系统使用氯化铝与 HF 反应生成的氟化铝，其酸化率低于游离的 HF。氟化氢铵与五膦酸还被认为是一种减速砂岩酸，其腐蚀性比普通土酸低。通过减缓动力学过程，人们可以将这些缓速酸用于高温环境。有文献综述了砂岩缓速酸体系的研究进展。

5.8.4 缓冲有机酸

砂岩酸化时，建议使用 pH 值为 1.9~4 的含有机酸的缓冲酸溶液，与有机酸盐和膦酸盐混合。主处理酸具有相似的 pH 值，并作为缓冲酸可以渗透更深地层。高 pH 值缓冲体系已成功应用于单级砂岩酸化处理，无须前置液和反复冲洗。对于碳酸盐岩储层，当 pH 值在优选的 4~9.5 范围内时，氨基多羧酸阴离子比 HCl 反应慢。

5.8.5 胶凝酸或稠化酸

本章前面描述的聚合物凝胶或 VESs 除了用作转向剂或自转向剂，还可以使酸更深入地层，其化学性能在这里不再赘述。这类药剂使酸性溶液增稠，增加其黏度，作为缓速酸溶液提高酸的渗透性。在碳酸盐岩压裂酸化作业中，更高黏稠度的酸液会产生更宽、更长的裂缝。反应过程中，表面活性剂或聚合物的稳定性必须足以让酸液与储层中接触足够长的时间，使组合物中的酸与地层中的酸溶性组分充分反应，在地层中建立新的油流通道或扩大现有的通道，实现增产的目的。胶凝酸的扩散速率约为非胶凝酸的 1/10。

在一项实验室研究中，聚丙烯酰胺与盐酸、甲酸或醋酸一起使用没有显示出任何缓速，而向酸中添加黄胞胶则显示出酸化缓速。

以一种特殊的聚季铵盐聚合物为基础，研制了用于碳酸盐岩压裂酸化的新型温控凝胶缓速酸。该体系在注入和渗流过程中具有黏性，当达到储层最高温度时失去黏性。

5.8.6 泡沫酸

泡沫酸除用作转向剂外，还可以特别用作压裂酸化过程中缓速酸。许多类型的表面活性剂可以用来产生泡沫，最优的是非离子表面活性剂（如乙氧基化醇）。氟化季铵盐可以作为聚合物的添加剂稳定泡沫。

5.8.7 对温度敏感的生酸化学品和酶

一种延缓碳酸盐岩酸化的方法是使用醋酸酯的水溶液，或选择可促进酯裂解并释放醋酸的酶，进行储层酸化。如水解酶、脂肪酶或酯酶等酶，可将酯类（如乙酸甲酯、乙酸乙酯、甲酸甲酯或 1，2-乙二醇二乙酸酯）生成乙酸。与 HCl 处理相比，通过酶生成的酸高效地输送到水平井的裸眼段，能够提供更好的清洗效果。但由于酶的价格高、储存困难（特别是在热带和沙漠地区）、地层温度要求及生成的酸只有理论值的一小部分等原因，使用这种处理液可能还存在一些问题。在不需要酶的情况下，也可以使用酯基体系进行井下水解。例如，1，2，3-丙三醇二乙酸酯和甲酸铵或甲酸钾的混合物。

在储层温度下，甲酸酯在水介质中水解生成比乙酸更强的甲酸，是优良的生酸产品。例如，甲酸酯（乙二醇单甲酸酯或乙二醇二甲酸酯）可用于延缓碳酸盐岩酸化中甲酸的形成，提高地层渗透率。这种方法还可以预防注入过程中的腐蚀问题。

另一种用于高温碳酸盐岩储层酸化压裂的生酸体系是固体聚乳酸和（或）聚乙醇酸。这些物质会在高温水中分解生成乳酸和乙醇酸。固体酸前体可与固体酸反应物质混合加速水解和（或）包裹以减慢其水解。还可以加入促进水解的水溶性液体化合物。这种方法确保酸可以到达远离井筒的裂缝面。

生酸的化学固体酸已经在中东进行了现场试验，酸和热的产生导致了非均匀刻蚀和导流能力提高。

5.8.8 乳化酸

乳化酸同胶凝酸的历史一样长，作为缓速酸增加渗透，用于压裂或基质酸化。乳化酸通常是油包酸的形式，由低 HLB 值（3～6）的油溶性表面活性剂配制而成乳化液。乳化酸通常在储层设计温度下破乳，释放酸。与非缓速酸和胶凝酸体系相比，乳化酸应在油藏井底温度为 135～190℃时表现更好。由于乳液的黏度比水溶液酸高，因此其具有转向性能。

各种表面活性剂已用于制作乳化剂，分为阳离子、阴离子和非离子类型。一种早期的、商业上成功的乳化酸是基于饱和烃磺酸和烷基芳香基磺酸或其水溶性盐。另一种体系使用长链胺，只有在与盐酸混合时会变成季铵化的乳化表面活性剂。相关文献还提出脂肪胺和油酸二乙醇酰胺等共混物的乳化表面活性剂。除了起到缓速酸的作用，乳化酸的黏度还使其具有转向性能。油包酸型乳化酸还具有在注入过程中避免腐蚀问题的优点。将酸化缓蚀剂由内酸相改为外油相的方法已提出。缓蚀剂处于外相会增强乳状液的缓速作用，从而使其在储层中渗透更深。此外，新方法能更好地保护油管和压裂设备。

如果有机溶剂相是由如二甲苯等沥青质溶解剂组成，乳化酸还可以用来溶解蜡和（或）沥青质沉积物。混合型二甲苯/酸配方比单独注入二甲苯和酸更有优势。

某些芳香族羧酸烷基或烯基酯，最好是苯甲酸异丙酯，可以在乳液中使用，其毒性和环境特性更好。

可以配制酸消耗后就会破乳的乳化酸。这样就可以避免高黏度乳液在地层中流动引起过高注入压力。有报道称，针对碳酸盐岩酸化，乳化酸延迟效果是盐酸的 14～19 倍。乳化酸也适用于高温井。15% 的乳化盐酸对低渗透白云岩储层具有良好的渗透性。

与常规粗乳酸相比，微乳液酸具有更好的延迟效果。微乳液是热力学稳定的、各向同性溶液。一个制备油包酸型微乳液的表面活性剂实例：2mol 2-乙基己基环氧化物与 1mol 乙醇胺反应，然后对产物进行季铵化。中东地区的酸化压裂已采用非离子表面活性剂制成的微乳液，可降低表面张力并改变接触角。该技术降低了含水饱和度，提高了采收率和气体相对渗透率，在创纪录的时间内使油井恢复生产，并正在推广到其他油井作业。

某些外酸相乳液也用作缓速酸，如盐酸、甲酸/乙酸、烃类和有机二膦酸胺表面活性剂的混合物，与碳酸盐岩岩心具有较长的反应时间。

参 考 文 献

[1] B. B. Williams, J. L. Gidley, and R. S. Schecter, "Acidizing Fundamentals," *Monograph Series*, vol. 6, Dallas: Society of Petroleum Engineers, 1979.

[2] L. Kalfayan, *Production Enhancement with Acid Stimulation*, 2nd ed., Tulsa, OK: PennWell Corporation, 2008.

[3] P. Rae and G. di Lullo, "Matrix Acid Stimulation—A Review of the State-of-the-Art," SPE 82260 (paper presented at the SPE European Formation Damage Conference, The Hague, Netherlands, 13-14 May 2003).

[4] M. Buijse, P. de Boer, M. Klos, and G. Burgos, "Organic Acids in Carbonate Acidizing," SPE 82211 (paper presented at the SPE European Formation Damage Conference, The Hague, Netherlands, 13-14 May 2003).

[5] (*a*) W. W. Frenier, C. N. Fredd, and F. Chang, "Hydroxyaminocarboxylic Acids Produce Superior Formulations for Matrix Stimulation of Carbonates at High Temperatures," SPE 71696 (paper presented at the SPE Annual Technical Conference and Exhibition, New Orleans, LA, 30 September-3 October 2001). (*b*) T. Huang, P. M. McElfresh, and A. Gabrysch, "Acid Removal of Scale and Fines at High Temperatures," SPE 74678 (paper presented at the SPE Oilfield Scale Symposium, Aberdeen, UK, 30-31 January 2002).

[6] W. W. Frenier, C. N. Fredd, and F. Chang, "Hydroxyaminocarboxylic Acids Produce Superior Formulations for Matrix Stimulation of Carbonates," SPE 68924 (paper presented at the SPE European Formation Damage Conference, The Hague, Netherlands, 21-22 May 2001).

[7] A. Husen A. Ali, W. W. Frenier, Z. Xiao, and M. Ziauddin, "Chelating Agent-Based Fluids for Optimal Stimulation of High-Temperature Wells," SPE 77366 (paper presented at the SPE Annual Technical Conference and Exhibition, San Antonio, TX, 29 September-2 October 2002).

[8] C. N. Fredd and H. S. Fogler, "The Influence of Transport and Reaction on Wormhole Formation in Porous Media," *Journal of American Institute of Chemical Engineers* 44 (1998): 1933.

[9] C. N. Fredd and H. S. Fogler, *Journal of American Institute of Chemical Engineers* 44 (1998): 1949.

[10] T. Huang, P. M. McElfresh, and A. D. Gabrysch, "Carbonate Matrix Acidizing Fluids at High Temperatures: Acetic acid, Chelating Agents, or Long-Chained Carboxylic Acids?," SPE 82268 (paper presented at the SPE European Formation Damage Conference held in The Hague, 13-14 May 2003).

[11] H. O. McLeod, "Matrix Acidizing," *Journal of Petroleum Technology* 36 (12) (1984): 2055.

[12] G. R. Coulter and A. R. Jennings, "A Contemporary Approach to Matrix Acidizing," SPE 38594 (paper presented at the SPE Annual Technical Conference and Exhibition, San Antonio, TX, 5-8 October 1997).

[13] L. J. Kalfayan and D. R. Watkins, "A New Method for Stabilizing Fines and Controlling Dissolution during Sandstone Acidizing," SPE 20076 (paper presented at the SPE California Regional Meeting, Ventura, CA, 4-6 April 1990).

[14] M. N. Al-Dahlan, H. A. Nasr-El-Din, and A. A. Al-Qahtani, "Evaluation of Retarded HF Acid Systems," SPE 65032 (paper presented at the SPE International Symposium on Oilfield Chemistry, Houston, TX, 13-16 February 2001).

[15] P. J. Rae, G. Di. L. Arias, A. B. Ahmad, and L. J. Kalfayan, U.S. Patent Application 20050016731.

[16] P. Rae and G. Di Lullo, "Single Step Matrix Acidizing with HF—Eliminating Preflushes Simplifies the Process, Improves the Results," SPE 107296 (paper presented at the European Formation Damage Conference, Scheveningen, The Netherlands, 30 May–1 June 2007).

[17] W. Frenier, M. Ziauddin, S. Davies, and F. Chang, U.S. Patent 7192908, 2007.

[18] F. E. Tuedor, Z. Xiao, M. J. Fuller, D. Fu, G. Salamat, S. N. Davies, and B. Lecerf, SPE 98314 (paper presented at the SPE International Symposium and Exhibition on Formation Damage, Lafayette, LA, 15–17 February 2006).

[19] H. A. Nasr-El-Din, S. Kelkar, and M. M. Samuel, "Investigation of a New Single Stage Sandstone Acidizing Fluid for High Temperature Formations," SPE 107636 (paper presented at the European Formation Damage Conference, Scheveningen, The Netherlands, 30 May–1 June 2007).

[20] A. N. Martin, "Stimulating Sandstone Formations with Non-HF Treatment Systems," SPE 90774 (paper presented at the SPE Annual Technical Conference and Exhibition, Houston, TX, 25–29 September 2004).

[21] M. A. Aggour, M. Al-Muhareb, S. A. Abu-Khamsin, and A. A. Al-Majed, "Improving Sandstone Matrix Simulation of Oil Wells by Gas Preconditioning," *Petroleum Science and Technology* 20 (3) (2002): 425.

[22] D. A. Williams, J. R. Looney, D. S. Sullivan, B. I. Bourland, J. A. Haslegrave, P. J. Clewlow, N. Carruthers, and T. M. O'Brien, U.S. Patent 5322630, 1994.

[23] J. Hall and S. Almond, "OSPAR Regulators Drive Design: How Regulations Have Impacted a Service Company's Product Development Program," Proceedings of the Chemistry in the Oil Industry IX, Manchester, UK: Royal Society of Chemistry, 2005, 61.

[24] (a) W. W. Frenier, "Review of Green Chemistry Corrosion Inhibitors for Aqueous Systems" (paper presented at the 9th European Symposium on Corrosion Inhibitors, University of Ferrara, Ferrara, Italy, September 2000). (b) D. G. Hill and H. Romijn, "Reduction of Risk to the Marine Environment from Oilfield Chemicals: Environmentally Improved Acid Corrosion Inhibition for Well Stimulation," Paper 00342 (paper presented at the NACE CORROSION Conference, 2000). (c) A. Rostami, H. Nasr-El-Din, "Review and Evaluation of Corrosion Inhibitors Used in Well Stimulation," SPE 121726 (paper presented at the SPE International Symposium on Oilfield Chemistry, The Woodlands, TX, 20–22 April 2009).

[25] W. Frenier and M. Ziauddin, Formation, Removal, and Inhibition of Scale in the Oilfield Environment, eds. N. Wolf and R. L. Hartman, Society of Petroleum Engineers, 2008.

[26] (a) A. J. Saukaitis and G. S. Gardner, U.S. Patent 2758970, 1956. (b) R. C. Mansfield, J. G. Morrison, and C. J. Schmidle, U.S. Patent 2874119, 1959.

[27] M. N. Desai, M. B. Desai, C. B. Shah, and S. M. Desai, "Schiff Bases as Corrosion Inhibitors for Mild Steel in Hydrochloric Acid Solutions," *Corrosion Science* 26 (1986): 827.

[28] D. G. Hill, K. Dismuke, W. Shepherd, I. Witt, H. Romijn, W. Frenier, and M. Parris, "Development Practices and Achievements for Reducing the Risk of Oilfield Chemicals," SPE 80593 (paper presented at the SPE/EPA/DOE Exploration and Production Environmental Conference, San Antonio, TX, 10–12 March 2003).

[29] M. Vorderbruggen and H. Kaarigstad, "Meeting the Environmental Challenge: A New Acid Corrosion Inhibitor for the Norwegian Sector of the North Sea," SPE 102908 (paper presented at the SPE Annual Technical Conference and Exhibition, San Antonio, TX, 24–27 September 2006).

[30] W. W. Frenier and A. Iob, Paper 150 (paper presented at the NACE CORROSION Conference, 1988).

[31] D. A. Williams, L. A. McDougall, and J. R. Looney, U.S. Patent 5543388, 1996.

[32] E. P. da Motta, M. H. V. Quiroga, A. F. L. Aragao, and A. Pereira, "Acidizing Gas Wells in the Merluza Field Using an Acetic/Formic Acid Mixture and Foam Pigs," SPE 39424 (paper presented at the SPE International Symposium on Formation Damage Control, Lafayette, LA, 18–19 February 1998).

[33] M. S. Van Domelen and A. R. Jennings Jr., "Alternate Acid Blends for HPHT Applications," SPE 30419 (paper presented at the SPE Offshore Europe Conference, Aberdeen, UK, 5–8 September 1995).

[34] V. Sharma, M. Borse, S. Jauhan, K. B. Pai, and S. Devi, *Surfarctants and Detergents* 42 (2005): 163.

[35] (a) G. Schmitt and K. Bedbur, *Werkstoffer und Korrosion* 1985, 38, 575. (b) W. W. Frenier, U.S. Patent 5096618, 1992.

[36] (a) O. O. Adeyemi and S. O. Oluwafemi, "2-Aminopyridine as an Effective Inhibitor for the Corrosion of Mild Steel in Acidic Solutions," *Bulletin of Electrochemistry* 22 (2006): 317. (b) S. N. Hettiarachchi, C. Subhash, and D. D. Macdonald, International Patent Application WO/1990/001478.

[37] M. M. Brezinski, U.S. Patent 5763368, 1998.

[38] M. M. Brezinski, U.S. Patent 5976416, 1999.

[39] M. L. Walker, U.S. Patent 5441929, 1995.

[40] M. M. Brezinski and B. Desai, U.S. Patent 5697443, 1997.

[41] D. A. Williams, P. K. Holifield, J. R. Looney, and L. A. McDougall, U.S. Patent 5200096, 1993.

[42] J. M. Cassidy, J. L. Lane, and C. E. Kiser, International Patent Application WO/2007/141524.

[43] M. M. Brezinski, U.S. Patent 5756004, 1998.

[44] (a) P. Manivel and G. Venkatachari, "The Inhibitive Effect of Poly (p-Toluidine) on Corrosion of Iron in 1M HCl Solutions," *Journal of Applied Polymer Science* 104 (2007): 2595. (b) B. Gao, X. Zhang, and Y. Sheng, "Studies on Preparing and Corrosion Inhibition Behavior of Quaternized Polyethyleneimine for Low Carbon Steel in Sulfuric Acid," *Materials Chemistry and Physics* 108 (2008): 375.

[45] K. Overkempe, W. J. E. Parr, and J. C. Speelman, World Patent Application WO/2003/054251.

[46] A. A. AlTaq, S. A. Ali, and H. A. Nasr-El-Din, "Inhibition Performance of a New Series of Mono/Diamine-Based Corrosion Inhibitors for HCl Acid Solutions," SPE 114087 (paper presented at the SPE 9th International Conference on Oilfield Scale, Aberdeen, UK, 28–29 May 2008).

[47] R. L. Hoppe, R. L. Martin, M. K. Pakulski, and T. D. Schaffer, U.S. Patent Application 20070261853.

[48] J. M. Cassidy, J. L. Lane, K. A. Frost, and C. E. Kiser, International Patent Application WO/2007/034155.

[49] D. S. Sullivan III, C. E. Strubelt, and K. W. Becker, U.S. Patent 4039336, 1977.

[50] G. P. Funkhouser, J. M. Cassidy, J. L. Lane, K. Frost, T. R. Gardner, and K. L. King, U.S. Patent 6192987, 2001.

[51] A. Cizek and A. Hackerott, International Patent Application WO/01/79590.

[52] W. W. Frenier, Paper 96154 (paper presented at the 51st NACE International Corrosion Forum, Denver, CO, March 1996).

[53] W. Frenier and F. Growcock, U.S. Patent 4734259, 1988.

[54] (a) M. A. Vorderbruggen and D. A. Williams, U.S. Patent 6117364, 2000. (b) J. M. Cassidy, C. E. Kiser, J. L. Lane, and K. A. Frost, U.S. Patent Application 20070071887, 2007.

[55] J. M. Cassidy, C. E. Kiser, and J. L. Lane, U.S. Patent Application 20080139414.

[56] W. W. Frenier and D. G. Hill, U.S. Patent 6399547, 2002.

[57] T. D. Welton and J. M. Cassidy, World Patent Application, WO/2005/075707.

[58] A. Punet Plensa and L. Lozano Salvatella, World Patent Application WO/2006/136262.

[59] J. M. Cassidy and K. A. Frost, U.S. Patent Application 20050123437.

[60] (a) G. A. Scherubel, R. Reid, A. L. Fauke, and K. Schwartz, U.S. Patent 5854180, 1998. (b) A. Cizek, International Patent Application WO/2002/103081.

[61] J. M. Cassidy, J. L. Lane, and C. E. Kiser, U.S. Patent 7163056, 2007.

[62] (a) M. A. Quraishi and R. Sardar, "Aromatic Triazoles as Corrosion Inhibitors for Mild Steel in Acidic Environments," NACE CORROSION Conference, September 2002, 748. (b) M. A. Quraishi, Paper 04421 (paper presented at the NACE CORROSION Conference, 2004).

[63] K. C. Pilai and R. Narayan, *Corrosion Science*, 23 (1983): 151.

[64] W. W. Frenier, D. G. Hill, F. B. Growcock, and V. R. Lopp, "α-Alkenylphenones, A New Class of Corrosion Inhibitors, Provide Improved Inhibition in Strong HCl" (paper presented at the RSC 3rd Chemicals in the Oil Industry Symposium, 19−29 April 1988).

[65] R. J. Jasinski and W. W. Frenier, U.S. Patent 5120371, 1992.

[66] M. D. Coffey, M. Y. Kelly, and W. C. Kennedy Jr., U.S. Patent 4493775, 1985.

[67] M. L. Walker, U.S. Patent 5591381, 1997.

[68] D. R. McCormick and J. P. Bershas, U.S. Patent Application 20070018135, 2007.

[69] (a) A. Penna, G. F. Di Lullo Arias, and P. J. Rae, U.S. Patent Application 20060264335, 2006. (b) T. D. Welton and J. M. Cassidy, U.S. Patent Application 20070010404, 2007.

[70] (a) J. D. Anderson, E. S. Hayman Jr., and E. A. Rodzewich, U.S. Patent 3992313, 1976. (b) J. D. Nichols, R. Derby, G. T. Von dem Bussche, and D. A. Hannum, U.S. Patent 4557838, 1985.

[71] F. Bentiss, M. Traisnel, H. Vezin, H. F. Hildebrand, and M. Lagrenee, "2,5-Bis(4-Dimethylaminophenyl)-1,3,4-Oxadiazole and 2,5-Bis(4-Dimethylaminophenyl)-1,3,4-Thiadiazole as Corrosion Inhibitors for Mild Steel in Acidic Media," *Corrosion Science* 46 (2004): 2781.

[72] M. Ajmal, M. A. W. Khan, S. Ahmad, and M. A. Quraishi, Paper 217 (paper presented at the NACE CORROSION, 1996).

[73] M. A. Quaraishi and D. Jamal, "CAHMT: A New Eco-Friendly Acidizing Corrosion Inhibitor," *Corrosion* 56 (10) (2000): 983.

[74] H.-N. Lin, R. D. Martin, J. M. Brown, and G. F. Brock, U.S. Patent 6132619, 2000.

[75] M. Girgis-Ghaly and J. R. Delorey, U.S. Patent 6308778, 2001.

[76] J. P. Feraud, H. Perthuis, and P. Dejeux, U.S. Patent 6306799, 2001.

[77] M. M. Brezinski, T. R. Gardner, K. L. King, and J. L. Lane Jr., U.S. Patent 6225261, 2001.

[78] M. M. Brezinski, U.S. Patent 6415865, 2002.

[79] C. Smith, D. Oswald, D. Skibinski, and N. Sylvestre, U.S. Patent Application 20060281636, 2006.

[80] M. M. Brezinski, U.S. Patent 6653260, 2002.

[81] C. W. Crowe, U.S. Patent 4633949, 1987.

[82] W. R. Dill and M. L. Walker, U.S. Patent 4888121, 1989.

[83] C. D. Williamson, U.S. Patent 5126059, 1992.

[84] I. C. Jacobs and N. E. S. Thompson, U.S. Patent 5112505, 1992.

[85] M. M. Brezinski and R. D. Gdanski, U.S. Patent 5264141, 1993.

[86] C. Smith, D. Oswald, and M. D. Daffin, U.S. Patent Application 20060289164, 2006.

[87] R. E. Himes and E. F. Vinson, U.S. Patent 4842073, 1989.

[88] D. R. Watkins, L. J. Kalfayan, and G. S. Hewgill, U.S. Patent 5039434, 1991.

[89] F. O. Stanley, S. A. Ali, and J. L. Boles, "Laboratory and Field Evaluation of Organosilane as a Formation

Fines Stabilizer," SPE Paper 29530 (paper presented at the SPE Production Operations Symposium, Oklahoma City, OK, April 1995).

[90] B. S. Douglass and G. E. King, "A Comparison of Solvent/Acid Workovers in Embar Completions-Little Buffalo Basin Field," SPE 15167 (paper presented at the SPE Rocky Mountain Regional Meeting, Billings, MT, 19–21 May 1986).

[91] (a) R. J. Dyer, U.S. Patent 5622921, 1997. (b) C. Smith and D. Skibinski, U.S. Patent Application 20070062698, 2007.

[92] (a) A. R. Mokadam, C. E. Strubelt, D. A. Williams, and K. M. Webber, U.S. Patent 5543387, 1996. (b) W. G. F. Ford, U.S. Patent 4823874, 1989. (c) A. R. Mokadam, U.S. Patent 5797456, 1998.

[93] K. M. Barker and M. E. Newberry, "Inhibition and Removal of Low-pH Fluid-Induced Asphaltic Sludge Fouling of Formations in Oil and Gas Wells," SPE 102738 (paper presented at the SPE International Symposium on Oilfield Chemistry, Houston, TX, 28 February–2 March 2007).

[94] J. M. Cassidy, C. E. Kiser, and J. L. Lane, U.S. Patent Application 20060201676, 2006.

[95] J. M. Cassidy, C. E. Kiser, and J. L. Lane, U.S. Patent Application 20060040831, 2006.

[96] D. S. Treybig, D. Williams, and K. T. Chang, International Patent Application WO/2003/053536.

[97] I. R. Collins, S. P. Goodwin, J. C. Morgan, and N. J. Stewart, U.S. Patent 6225263, 2001.

[98] W. W. Frenier and D. G. Hill, U.S. Patent 6068056, 2000.

[99] A. Ahrenst, B. Lungwitz, C. N. Fredd, C. Abad, N. Gurmen, Y. Chen, J. Lassek, P. Howard, W. T. Huey, Z. Azmi, D. Hodgson III, and O. Bustos, U.S. Patent Application 20080064614.

[100] H. A. Nasr-El-Din, A. M. Al-Othman, K. C. Taylor, and A. H. Al-Ghamdi, "Surface Tension of Acid Stimulating Fluids at High Temperature," *Journal of Petroleum Science and Engineering* 43 (2004): 57.

[101] D. Dalrymple, L. Eoff, B. R. Reddy, and J. Venditto, U.S. Patent 7182136, 2007.

[102] (a) B. Garcia, E. Soriano, W. Chacon, and L. Eoff, "Novel Acid-Diversion Technique Increases Production in the Cantarell Field, Offshore Mexico," SPE 112413, (paper presented at the SPE International Symposium and Exhibition on Formation Damage Control, Lafayette, LA, 13–15 February 2008). (b) B. Garcia, E. Soriano, W. Chacon, and L. Eoff, "Novel Acid Diversion Technique Boosts Production," *World Oil* 229 (2001): 121.

[103] G. Paccaloni, "A New, Effective Matrix Stimulation Diversion Technique," *SPE Production and Facilities* 10 (3)(1995): 151.

[104] C. Abad, J. C. Lee, P. F. Sullivan, E. Nelson, Y. Chen, B. Baser, and L. Lin, U.S. Patent Application 20070032386.

[105] G. Glasbergen, B. Todd, M. Van Domelen, and M. Glover, "Design and Field Testing of a Truly Novel Diverting Agent," SPE 102606 (SPE Annual Technical Conference and Exhibition, San Antonio, TX, 24–27 September 2006).

[106] J. R. Solares, J. J. Duenas, M. Al-Harbi, A. Al-Sagr, V. Ramanathan, and R. Hellman, "Field Trial of a New Non-Damaging Degradable Fiber-Diverting Agent Achieved Full Zonal Coverage during Acid Fracturing in a Deep Gas Producer in Saudi Arabia," SPE 115525 (paper presented at the SPE Annual Technical Conference and Exhibition, Denver, CO, 21–24 September 2008).

[107] N. A. Menzies, E. J. Mackay, and K. S. Sorbie, "Modelling of Gel Diverter Placement in Horizontal Wells," SPE 56742 (paper presented at the SPE Annual Technical Conference and Exhibition, Houston, TX, 3–6 October 1999).

[108] J. F. Tate, U.S. Patent 3749169, 1973.

[109] (a) T. D. Welton, R. W. Pauls, and I. D. Robb, U.S. Patent Application 20060247135. (b) T. D. Welton,

R. W. Pauls, L. Song, J. E. Bryant, S. R. Beach, and I. D. Robb, International Patent Application WO/2008/096164.

[110] (a) B. L. Swanson and L. E. Roper; U.S. Patent 4205724, 1980. (b) L. D. Burns and G. A. Stahl, U.S. Patent 4690219, 1987.

[111] L. R. Norman, M. W. Conway, and J. M. Wilson, "Temperature Stable Acid Gelling Polymers: Laboratories Evaluation and Field Results," SPE 10260 (paper presented at the SPE 56th Annual Fall Technical Conference, San Antonio, TX, 5–7 October 1981).

[112] D. J. Poelker, J. McMahon, and D. Harkey, U.S. Patent 6855672, 2005.

[113] R. W. Pauls and T. D. Welton, International Patent Application WO/2009/022107.

[114] B. L. Swanson, U.S. Patent, 4103742, 1978.

[115] W. Abdel Fatah, H. A. Nasr-El-Din, T. Moawad, and A. Elgibaly, "Effects of Crosslinker Type and Additives on the Performance of In-Situ Gelled Acids," SPE 112448 (paper presented at the SPE International Symposium and Exhibition on Formation Damage Control, Lafayette, LA, 13–15 February 2008).

[116] C. G. Inks, U.S. Patent 4163727, 1979.

[117] E. Clark Jr. and B. L. Swanson, U.S. Patent 4997582, 1991.

[118] (a) S. Mukherjee and G. Gudney, SPE 25395, *Journal of Petroleum Technology* 45 (1993): 102. (b) A. Saxon, B. Chariag, and M. Rahman, SPE 37734 (paper presented at the SPE Middle East Oil Show, Bahrain, 15–18 March 1997).

[119] T. D. Welton, International Patent Application WO/2008/102138.

[120] J. D. Lynn and H. A. Nasr-El-Din, "A Core Based Comparison of the Reaction Characteristics of Emulsified and In-Situ Gelled Acids in Low Permeability, High Temperature, Gas Bearing Carbonates," SPE 65386 (paper presented at the SPE International Symposium on Oilfield Chemistry, Houston, TX, 13–16 February 2001).

[121] B. L. Swanson, U.S. Patent 4055502, 1977.

[122] R. L. Clampitt and J. E. Hessert, U.S. Patent 4068719, 1978.

[123] C. B. Josephson, U.S. Patent 4476033, 1984.

[124] J. L. Boles, A. S. Metcalf, and J. Dawson, U.S. Patent 5497830, 1996.

[125] D. G. Hill, U.S. Patent Application 20050065041, 2005.

[126] R. Gdanski, "Experience and Research Show Best Designs for Foam-Diverting Acidizing," *Oil & Gas Journal* (9) (1993): 85.

[127] H. A. Nasr-El-Din, *Surfactant Use in Acid Stimulation*, Surfactants: *Fundamentals and Applications in the Petroleum Industry*, ed. L. L. Schramm, Cambridge, UK: Cambridge University Press, 2000, 329.

[128] H. A. Volz, U.S. Patent 4044833, 1977.

[129] P.-A. Francini, K. Chan, M. Brady, and C. Fredd, U.S. Patent Application, 20050020454, 2005.

[130] S. Thach, U.S. Patent 5529122, 1996.

[131] D. R. Watkins, U.S. Patent 4737296, 1988.

[132] L. R. Norman and T. R. Gardner, U.S. Patent 4324669, 1982.

[133] K. Thompson and R. Gdanski, "Laboratory Study Provides Guidelines for Diverting Acid with Foam," *SPE Production and Facilities* 8 (4) (1993): 285.

[134] S. Siddiqui, S. Talabani, J. Yang, S. T. Saleh, and M. R. Islam, "An Experimental Investigation of the Diversion Characteristics of Foam in Berea Sandstone Cores of Contrasting Permeabilities," *Chemical Engineering Science* 37 (2003): 51.

[135] S. I. Kam, W. W. Frenier, S. N. Davies, and W. R. Rossen, "Experimental Study of High Temperature Foam for Acid Diversion," *Journal of Petroleum Science and Engineering* 58(2007): 138.

[136] K. E. Cawiezel and J. C. Dawson, U.S. Patent Application 20050067165, 2005.

[137] C. Zeiler, D. Alleman, and Q. Qu, "Use of Viscoelastic Surfactant-Based Diverting Agents for Acid Stimulation: Case Histories in GOM," SPE 90062 (paper presented at the SPE Annual Technical Conference, Houston, TX, 26-29 September 2004).

[138] B. Lungwitz, C. Fredd, M. Brady, M. Miller, S. Ali, and K. Hughes, "Diversion and Cleanup Studies of Viscoelastic Surfactant-Based Self-Diverting Acid," SPE 86504, *SPE Production & Operations* 22 (1)(2007): 121.

[139] J. Yang, "Viscoelastic Wormlike Micelles and Their Applications," *Current Opinion in Colloid Interface Science* 7(2002): 276.

[140] D. P. Acharya and H. Kunieda, "Wormlike Micelles in Mixed Surfactant Solutions," *Advances in Colloid and Interface Science* 123-126(2006): 401.

[141] H. A. Nasr-El-Din, T. D. Welton, L. Sierra, and M. S. van Domelen, "Optimization of Surfactant-Based Fluids for Acid Diversion," SPE 107687 (paper presented at the European Formation Damage Conference, Scheveningen, The Netherlands, 30 May-1 June 2007).

[142](a) D. S. Treybig, G. N. Taylor, and D. K. Moss, U.S. Patent Application 20060025321, 2006. (b) R. Franklin, M. Hoey, and R. Pramachandran, "The Use of Surfactants to Generate Viscoelastic Fluids," Proceedings of the Chemistry in the Oil Industry VII, Manchester, UK: Royal Society of Chemistry, 2002.

[143] D. Fu, M. Panga, S. Kefi, and M. Garcia-Lopez de Victoria, U.S. Patent 7237608, 2007.

[144](a) M. M. Samuel, K. I. Dismuke, R. J. Card, J. E. Brown, and K. England, U.S. Patent 6306800, 2001. (b) P. W. Knox, International Patent Application WO/2007/056393.

[145] M. S. Dahayanake, J. Yang, J. H. Y. Niu, P.-D. Derian, R. Li, and D. Dino, U.S. Patent, 6258859, 2001.

[146] D. Fu and F. Chang, U.S. Patent 6929070, 2005.

[147] R. S. Hartshorne, T. L. Hughes, T. G. J. Jones, and G. J. Tustin, U.S. Patent Application 20050124525, 2005.

[148] T. D. Welton, S. J. Lewis, and G. P. Funkhouser, U.S. Patent 7159659, 2007.

[149] P. M. McElfresh and C. F. Williams, U.S. Patent 7216709, 2007.

[150] R. E. Dobson Sr., D. K. Moss, and R. S. Premachandran, U.S. Patent 7060661, 2006.

[151] T. Huang and J. B. Crews, "Do Viscoelastic-Surfactant Diverting Fluids for Acid Treatments Need Internal Breakers?," SPE 112484 (paper presented at the SPE International Symposium and Exhibition on Formation Damage Control, Lafayette, LA, 13-15 February 2008).

[152] P. D. Berger and C. H. Berger, U.S. Patent Application 20060084579, 2006.

[153] D. Fu, Diankui, Y. Chen, Z. Xiao, M. Samuel, and S. Daniel, U.S. Patent 7148185, 2006.

[154] G. Di Lullo, A. Ahmad, P. Rae, L. Anaya, and R. Ariel Meli, "Toward Zero Damage: New Fluid Points the Way," SPE 69453 (paper presented at the SPE Latin American and Caribbean Petroleum Engineering Conference, Buenos Aires, Argentina, 25-28 March 2001).

[155](a) F. F. Chang, Q. Qu, and M. J. Miller, U.S. Patent 6399546, 2002. (b) F. F. Chang, Q. Qu, and W. Frenier, "A Novel Self-Diverting Acid Developed for Matrix Simulation of Carbonate Reservoirs," SPE 65033 (paper presented at the SPE International Symposium on Oilfield Chemistry, Houston, TX, 13-16 February 2001).

[156] D. Taylor, P. Santhana Kumar, D. Fu, M. Jemmali, H. Helou, F. Chang, S. Davies, and M. Al-Mutawa, "Viscoelastic Surfactant Based Self-Diverting Acid for Enhanced Simulation in Carbonate Reservoirs," SPE 82263 (paper presented at the SPE European Formation Damage Conference, The Hague, Netherlands, 13–14 May 2003).

[157] H. A. Nasr-El-Din, J. B. Chesson, K. E. Cawiezel, and C. S. Devine, SPE 102468 (paper presented at the SPE Annual Technical Conference and Exhibition, San Antonio, TX, 24–27 September 2006).

[158] B. Lungwitz, C. Fredd, M. Brady, and T. Bui, "Application of Viscoelastic Surfactant Based Self-Diverting Acid in Gas Wells," Proceedings of the Chemistry in the Oil Industry IX, Manchester, UK: Royal Society of Chemistry, 31 October–2 November 2005, 214.

[159] J. B. Crews, "Internal Phase Breaker Technology for Viscoelastic Surfactant Gelled Fluids," SPE 93449 (paper presented at the SPE International Symposium on Oilfield Chemistry, The Woodlands, TX, 2–4 February 2005).

[160] E. B. Nelson, B. Lungwitz, K. Dismuke, M. Samuel, G. Salamat, T. Hughes, J. Lee, P. Fletcher, D. Fu, R. Hutchins, M. Parris, and G. J. Tustin, U.S. Patent Application 20020004464, 2002.

[161] J. B. Crews and T. Huang, "Internal Breakers for Viscoelastic Surfactant Fracturing Fluids," SPE 106216 (paper presented at the SPE International Symposium on Oilfield Chemistry, Houston, TX, 28 February–2 March 2007).

[162] J. B. Crews and T. Huang, U.S. Patent Application 20070151726.

[163] J. B. Crews and T. Huang, U.S. Patent Application 20070299142.

[164] T. D. Welton, R. D. Gdanski, and R. W. Pauls, U.S. Patent Application 20070060482.

[165] D. Fu and M. Garcia-Lopez De Victoria, U.S. Patent 7341107, 2008.

[166] A. R. Al-Nakhli, H. A. Nasr-El-Din, and A. A. Al-Baiyat, "Interactions of Iron and Viscoelastic Surfactants: A New Formation-Damage Mechanism," SPE 112465 (paper presented at the SPE International Symposium and Exhibition on Formation Damage Control, Lafayette, LA, 13–15 February 2008).

[167] P.-A. Francini, K. Chan, M. Brady, and C. Fredd, U.S. Patent 7148184, 2006.

[168] M. Garcia-Lopez De Victoria, Y. Christanti, G. Salamat, and Z. Xiao, U.S. Patent Application 20060131017, 2006.

[169] K. E. Cawiezel and C. S. Devine, U.S. Patent Application 20050137095, 2005.

[170] K. E. Cawiezel and C. S. Devine, "Nonpolymer Surfactant Enhances High-Strength Hydrofluoric Acid Treatments," SPE 95242 (paper presented at the SPE Annual Technical Conference and Exhibition, Dallas, TX, 9–12 October 2005).

[171] Q. Qu and D. Alleman, U.S. Patent 7115546, 2006.

[172] D. Alleman, Q. Qu, and R. Keck, "The Development and Successful Field Use of Viscoelastic Surfactant-Based Diverting Agents for Acid Stimulation," SPE 80222 (paper presented at the SPE International Symposium on Oilfield Chemistry, Houston, TX, 5–7 February 2003).

[173] T. S. Davies, A. M. Ketner, and S. R. Raghavan, "Self-Assembly of Surfactant Vesicles that Transform into Viscoleastic Wormlike Micelles upon Heating," *Journal of the American Chemical Society* 128 (2006): 6669.

[174] M. A. Samir, I. Elnashar, M. Samuel, and M. Jemmali, "Smart Chemical Systems for the Stimulation of High-Water-Cut Heavy Oil Wells," SPE 116746 (paper presented at the 2008 SPE Annual Technical Conference and Exhibition, Denver, CO, 21–24 September 2008).

[175] B. B. Williams, J. L. Gidley, and R. R. Schechter, "Acidizing Fundamentals," SPE Monograph No. 6,

New York: SPE, 1979.
[176] J. A. Knox and W. R. Dill, U.S. Patent 3343602, 1967.
[177] C. C. Bombardieri, U.S. Patent 3434545, 1969.
[178] J. L. Boles and M. Usie, U.S. Patent 6443230, 2002.
[179] W. Frenier and F. F. Chang, U.S. Patent 6806236, 2004.
[180] G. Di Lullo and P. Rae, "A New Acid for True Stimulation of Sand Stone Reservoirs," SPE 37015 (paper presented at the SPE International 6th Asia Pacific Oil and Gas Conference, Adelaide, Australia, 28–31 October 1996).
[181] W. Frenier, U.S. Patent Application 20020170715, 2002.
[182] M. M. Amro, "Extended Matrix Acidizing Using Polymer-Acid Solutions," SPE 106360 (paper presented at the SPE Technical Symposium of Saudi Arabia Section, Dhahran, Saudi Arabia, 21–23 May 2006).
[183] F. Zhou, Y. Liu, C. Xiong, J. Peng, X. Yang, X. Liu, Y. Lian, C. Qian, J. Yang, and F. Chen, SPE 104446 (paper presented at the SPE International Oil & Gas Conference and Exhibition, Beijing, China, 5–7 December 2006).
[184] D. L. Holcombe, "Foamed Acid as a Means for Providing Extended Retardation," SPE 6376 (paper presented at the SPE Permian Basin Oil and Gas Recovery Conference, Midland, TX, 10–11 March 1977).
[185] R. E. Harris and I. D. McKay, "New Applications for Enzymes in Oil and Gas Production," SPE 50621 (paper presented at the SPE European Petroleum Conference, The Hague, Netherlands, 20–22 October 1998).
[186] V. Moses and R. E. Harris, U.S. Patent 5678632, 1997.
[187] P. Leschi, G. Demarthon, E. Davidson, and D. Clinch, "Delayed-Release Acid System for Cleanup of Al Khalij Horizontal Openhole Drains," SPE 98164 (paper presented at the SPE International Symposium and Exhibition on Formation Damage Control, Lafayette, LA, 15–17 February 2006).
[188] R. E. Harris, I. D. McKay, J. M. Mbala, and R. P. Schaaf, "Stimulation of a Producing Horizontal Well Using Enzymes that Generate Acid In-Situ—Case History," SPE 68911 (paper presented at the SPE European Formation Damage Conference, The Hague, Netherlands, 21–22 May 2001).
[189] B. L. Todd and E. Davudson, U.S. Patent Application 20040163814.
[190] J. W. Still, K. Dismuke, and W. Frenier, U.S. Patent Application 20040152601.
[191] H. A. Nasr-El-Din, A. Al-Zahrani, J. Still, T. Lesko, and S. Kelkar, "Laboratory Evaluation of an Innovative System for Fracture Stimulation of High-Temperature Carbonate Reservoirs," SPE 106054 (paper presented at the SPE International Symposium on Oilfield Chemistry, Houston, TX, 28 February–2 March 2007).
[192] M. A. Buijse and M. S. Van Domelen, "Novel Application of Emulsified Acids to Matrix Stimulation of Heterogeneous Formations," SPE 39583 (paper presented at the SPE International Symposium on Formation Damage Control, Lafayette, LA, 18–19 February 1998).
[193] H. A. Nasr-El-Din and M. M. Samuel, "Development and Field Application of a New, Highly Stable Emulsified Acid," SPE 115926 (paper presented at the Annual Technical Conference and Exhibition, Denver, CO, 2008, 22–24 September 2008).
[194] C. W. Crowe, U.S. Patent 3779916, 1973.
[195] C. W. Crowe, U.S. Patent 3962102, 1976.
[196] G. A. Scherubel, U.S. Patent 4140640, 1979.

[197] E. G. Scovell, N. Grainger, and T. Cox, International Patent Application WO/2001/088333.

[198] D. K. Sarma, P. Agarwal, E. Rao, and P. Kumar, "Development of a Deep-Penetrating Emulsified Acid and Its Application in a Carbonate Reservoir," SPE 105502 (paper presented at the 15th SPE Middle East Oil and Gas Conference, Bahrain, 11–14 March 2007).

[199] R. C. Navarrete, B. A. Holms, S. B. McDonnell, and D. E. Linton, "Emulsified Acid Enhances Well Production in High-Temperature Carbonate Formations," SPE 50612 (paper presented at the SPE European Petroleum Conference, The Hague, Netherlands, 20–22 October 1998).

[200] A. T. Jones, C. Rodenburg, D. G. Hill, A. H. Akbar Ali, and P. de Boer, "An Engineered Approach to Matrix Acidizing HTHP Sour Carbonate Reservoirs," SPE 68915 (paper presented at the SPE European Formation Damage Conference, The Hague, Netherlands, 21–22 May 2001).

[201] P. Kasza, M. Dziadkiewicz, and M. Czupski, "From Laboratory Research to Successful Practice: A Case Study of Carbonate Formation Emulsified Acid Treatments," SPE 98261 (paper presented at the SPE International Symposium and Exhibition on Formation Damage Control, Lafayette, LA, 15–17 February 2006).

[202] E. M. Andreasson, F. Egeli, K. A. Holmberg, B. Nystrom, K. G. Stridh, and E. M. Sterberg, U.S. Patent 4650000, 1987.

[203] W. R. Dill, U.S. Patent 4322306, 1982.

[204] M. Al-Khaldi, H. A. Nasr-El-Din, S. Metha, and A. Aamri, "Reactions of Citric Acid with Calcite," *Chemical Engineering Science* 62 (2007): 5880.

[205] R. E. Harris and I. D. Mckay, International Patent Application WO/2004/007905.

[206] R. Taylor, G. C. Fyten, and F. McNeil, "Acidizing-Lessons from the Past and New Opportunities," SPE 162238 (paper presented at the SPE Canadian Unconventional Resources Conference, Calgary, Alberta, Canada, 30 October–1 November 2012).

[207] A. Z. I. Pereira, M. G. F. da Silva, L. C. A. da Paixao, T. J. L. de Oliveira, and P. D. Fernandes, "Used Approaches for Carbonates Acidizing Offshore Brazil," SPE 151797 (paper presented at the SPE International Symposium and Exhibition on Formation Damage Control, Lafayette, LA, 15–17 February 2012).

[208] R. Ya. Kharisov, A. E. Folomeev, A. R. Sharifullin, G. T. Bulgakova, and A. G. Telin, *Energy & Fuels*, 26 (5)(2012): 2621–2630.

[209] X. Cheng, Y. Li, Y. Ding, M. Che, F. Zhang, and J. Peng, "Study and Application of High Density Acid in HPHT Deep Well," SPE 142033 (paper presented at the SPE European Formation Damage Conference, Noordwijk, Netherlands, 7–10 June 2011).

[210] X. W. Qiu, F. F. Chang, and G. Tustin, U.S. Patent Application 20090209439.

[211] C. O. E. Jimenez Bueno, G. R. Ramirez, M. A. Quevedo Z., J. T. Resendiz Torres, and F. Tellez Cisneros, "Pushing the Limits: HT Carbonate Acidizing," SPE 151740 (SPE International Symposium and Exhibition on Formation Damage Control, Lafayette, LA, 15–17 February 2012).

[212] M. J. Fuller, I. Couillet, and R. Hartman, U.S. Patent Application 20090233819.

[213] M. A. Mahmoud, H. A. Nasr-El-Din, and C. A. DeWolf, "Removing Formation Damage and Stimulation of Deep Illitic-Sandstone Reservoirs Using Green Fluids," SPE 147395 (paper presented at the Annual Technical Conference and Exhibition, Denver, CO, 30 October–2 November 2011).

[214] F. Armirola, M. Machacon, C. Pinto, Cepcolsa, A. Milne, M. Lastre, and E. Miquilena, "Combining Matrix Stimulation and Gravel Packing Using a Non-acid Based Fluid," SPE 143788, (paper presented at the SPE European Formation Damage Conference, Noordwijk, Netherlands, 7–10 June 2011).

[215] F. Yang, H. A. Nasr-El-Din, and B. Al-Harbi, "Acidizing Sandstone Reservoirs Using HF and Formic Acids," SPE 150899 (paper presented at the SPE International Symposium and Exhibition on Formation Damage Control, Lafayette, LA, 15-17 February 2012).

[216] B. G. Al-Harbi, M. N. Al-Dahlan, and M. H. Al-Khaldi, "Aluminum and Iron Precipitation During Sandstone Acidizing Using Organic-HF Acids," SPE 151781 (paper presented at the SPE International Symposium and Exhibition on Formation Damage Control, Lafayette, LA, 15-17 February 2012).

[217] C. Sitz, W. Frenier, and C. Vallejo, "Acid Corrosion Inhibitors with Improved Environmental Profiles," SPE 155966 (paper presented at the SPE International Conference and Exhibition on Oilfield Corrosion, Aberdeen, UK, 28-29 May 2012).

[218] J. Zhao, N. Zhang, C. Qu, X. Wu, J. Zhang, and X. Zhang, *Industrial & Engineering Chemistry Research* 49 (2010): 3986.

[219] Y. K. Choudhary, A. Sabhapondit, and A. Kumar, "Application of Chicory as Corrosion Inhibitor for Acidic Environment," SPE 155725 (paper presented at the SPE International Conference and Exhibition on Oilfield Corrosion, Aberdeen, UK, 28-29 May 2012).

[220] M. E. Palomar, C. O. Olivares-Xometl, N. V. Likhaniva, and J.-B. Perez-Navarrete, *Journal of Surfactants and Detergents* 14 (2011): 211-220.

[221] M. A. Malwitz and D. K. Woloch, International Patent Application WO/2013/070550.

[222] B. Evans, K. Seth, A. D. Gabrysch, P. A. Kelly, and D. N. Horner Jr., European Patent EP2496789.

[223] A. Nahle, M. Al-Khayat, I. Abu-Abdoun, and I. Abdel-Rahman, *Anti-corrosion Methods and Materials* 60 (1)(2013): 20-27.

[224] M. I. Walker, International Patent Application WO/2011/032032.

[225] J. M. Cassidy, C. E. Kiser, and M. J. Wilson, U.S. Patent Application 20090156432.

[226] J. Cassidy and C. E. Kiser, International Patent Application WO/2012/072986.

[227] J. M. Cassidy, C. E. Kiser, and J. L. Lane, International Patent Application WO/2010/119235.

[228] A. Jenkins, U.S. Patent Application 20110028360.

[229] A. M. Gomaa and H. A. Nasr-El-Din, "Effect of Residual Oil Saturation on the Propagation of Regular, Gelled, and *In-Situ* Gelled Acids Inside Carbonate Formations," SPE 143643 (paper presented at the SPE European Formation Damage Conference, Noordwijk, Netherlands, 7-10 June 2011).

[230] M. S. Knopp, U.S. Patent Application 20090181868.

[231] B. J. O'Neil, D. M. Maley, and C. A. Lalchan, "Prevention of Acid Induced Asphaltene Precipitation: A Comparison of Anionic vs. Cationic Surfactants," SPE 1640087 (paper presented at the 2013 SPE International Symposium on Oilfield Chemistry, The Woodlands, TX, 8-10 April 2013).

[232] S. L. Berry, J. L. Boles, and K. L. Smith, "Cost-Effective Acid Stimulation of Carbonates Using Seawater-Based Systems Without Calcium Sulfate Precipitation," SPE 149997 (paper presented at the SPE International Symposium and Exhibition on Formation Damage Control, Lafayette, LA, 15-17 February 2012).

[233] J. He, I. M. Mohamed, and H. A. Nasr-El-Din, "Mixing Hydrochloric Acid and Seawater for Matrix Acidizing: Is It a Good Practice?," SPE 143855 (paper presented at the SPE European Formation Damage Conference, Noordwijk, Netherlands, 7-10 June 2011).

[234] S. Al-Ghamdi, A. Al-Najim, A. Bouyabes, F. S. Al-Hadyani, I. Nugraha, and S. Hamid, "A Novel Stimulation Approach for Scale Control in Marrat Carbonate Reservoir—Case Studies from Joint Operations, PZ Kuwait," SPE 143649 (paper presented at the SPE European Formation Damage Conference, Noordwijk, Netherlands, 7-10 June 2011).

[235] P. B. Entchev, C. Shuchart, and J. Burdette, "Methods of Isolation and Diversion for High Rate Carbonate Acidizing," SPE 156920 (paper presented at the SPE International Production and Operations Conference & Exhibition, Doha, Qatar, 14–16 May 2012).

[236] F. Martin, M. Quevedo, F. Tellez, A. Garcia, T. Resendiz, O. Jimenez Bueno, and G. Ramirez, "Fiber-Assisted Self-Diverting Acid Brings a New Perspective to Hot Deep Carbonate Reservoir Stimulation in Mexico," SPE 138910 (paper presented at the SPE Latin American and Caribbean Petroleum Engineering Conference, Lima, Peru, 1–3 December 2010).

[237] W. Nunez-Garcia, J. A. Leal-Jauregui, A. R. Malik, J. A. Solares, Y. A. Al-Abdulmohsen, M. Al-Mumen, and G. A. Izquierdo, "Achieving Successful Diversion in Acid Stimulation Treatments: Case Study of Excellent Results Achieved Using Associative Polymer Treatment (APT) in Highly Heterogeneous Carbonate Reservoirs in Saudi Arabia," SPE 125955 (paper presented at the SPE International Symposium and Exhibition on Formation Damage Control, Lafayette, LA, 10–12 February 2010).

[238] K. Kritsanaphak and J. Vasquez, "Novel Acid-Diversion Technique for Matrix Acidizing in Production and Injection Wells: Case Histories of Alger," SPE 15223 (paper presented at the International Petroleum Technology Conference, Bangkok, Thailand, 7–9 February 2012).

[239] P. Patil, A. Sarda, S. George, Y. K. Choudhary, and R. Kalgaonkar, "Non-Iron-Based Composition for In-Situ Crosslinked Gelled Acid System," SPE 156190 (paper presented at the SPE International Production and Operations Conference & Exhibition, Doha, Qatar, 14–16 May 2012).

[240] J. B. Crews and T. Huang, U.S. Patent Application 20090192053.

[241] F. Lechuga, C. Mansur, M. Bezerra, L. Barbosa, and E. Lucas, *Journal of Petroleum Science Research* 1 (2012): 4–9.

[242] Z. Ye, L. Han, H. Chen, L. Shi, and P. Luo, *Journal of Surfactants and Detergents* 13 (2010): 287.

[243] R. S. Hartshorne, T. L. Hughes, T. G. J. Jones, G. J. Tustin, and J. Zhou, U.S. Patent Application 20090291864.

[244] R. van Zanten and D. J. Harrison, International Patent Application WO/2010/116117.

[245] A. T. Jardim Neto, C. M. Silva, R. S. Torres, F. G. Prata, L. M. Souza, A. Z. Pereira, A. Calderon, and E. F. Sandes, "Self-Diverting Acid for Effective Carbonate Stimulation Offshore Brazil: A Successful History," SPE 165089 (paper presented at the 10th SPE International Conference and Exhibition on European Formation Damage, Noordwijk, Netherlands, 5–7 June 2013).

[246] A. A. Al-Taq, A. R. Nakhli, H. H. Haji, and J. A. Saleem, U.S. Patent Application 20120181022.

[247] M. Madyanova, R. Hezmela, P. Artola, C. R. Guimaraes, Schlumberger, and B. Iriyanto, "Effective Matrix Stimulation of High-Temperature Carbonate Formations in South Sumatra through the Combination of Emulsified and Viscoelastic Self-Diverting Acids," SPE 151070 (paper presented at the SPE International Symposium and Exhibition on Formation Damage Control, Lafayette, LA, 15–17 February 2012).

[248] G. Degre and M. Morva, International Patent Application WO/2010/105879.

[249] P. W. Knox, International Patent Application WO/2007/056393.

[250] L. Li, J. F. Gadberry, J. Zhou, and S. Holt, "A New HTHP Viscoelastic Surfactant System for Completion in High Density Brines," Chemistry in the Oil Industry XIII—New Frontiers, Manchester, UK, November 2013.

[251] S. Holt, J. Zhou, F. Gadberry, H. Nasr-El-Din, and G. Wang, "IBP1034-12, A Novel Viscoelastic Surfactant Suitable For Use in High Temperature Carbonate Reservoirs for Diverted Acidizing Stimulation

Treatments," Rio Oil & Gas Expo and Conference, 2012.

[252] P. W. Knox and N. F. Perreault, International Patent Application WO/2007/059266.

[253] A. M. Gomaa, J. Cutler, Q. Qu, and K. E. Cawiezel, "Acid Placement: An Effective VES System to Stimulate High-Temperature Carbonate Formations," SPE 157316 (paper presented at the SPE International Production and Operations Conference & Exhibition, Doha, Qatar, 14-16 May 2012).

[254] Y. Shu, G. Wang, H. A. Nasr-El-Din, and J. Zhou, "Impact of Fe (III) on the Performance of VES-Based Acids," SPE 165149 (paper presented at the 10th SPE International Conference and Exhibition on European Formation Damage, Noordwijk, Netherlands, 5-7 June 2013).

[255] W. Braun, C. A. de Wolf, and H. A. Nasr-El-Din, "Improved Health, Safety and Environmental Profile of a New Field Proven Stimulation Fluid (Russian)," SPE 157467 (paper presented at the SPE Russian Oil and Gas Exploration and Production Technical Conference and Exhibition, Moscow, Russia, 16-18 October 2012).

[256] M. A. Mahmoud, H. A. Nasr-El-Din, and C. A. De Wolf, "Novel Environmentally Friendly Fluids to Remove Carbonate Minerals from Deep Sandstone Formations," SPE 143301 (paper presented at the SPE European Formation Damage Conference, Noordwijk, Netherlands, 7-10 June 2011).

[257] H. Nasr-El-Din, C. A. De Wolf, M. A. Nasr-El-Din Mahmoud, A. Bouwman, M. Jacobus, and N. T. George, International Patent Application WO/2012/080296.

[258] C. A. De Wolf, H. Nasr-El-Din, and M. A. Nasr-El-Din Mahmoud, International Patent Application WO/2012/080298.

[259] C. A. De Wolf, J. N. Lepage, H. Nasr-El-Din, M. A. Nasr-El-Din Mahmoud, E. R. A. Bang, and N. T. George, International Patent Application WO/2012/080299.

[260] C. A. De Wolf, H. Nasr-El-Din, M. A. Nasr-El-Din Mahmoud, J. N. Lepage, A. Bouwman, and G. Wang, International Patent Application WO/2012/080463.

[261] M. A. Mahmoud, H. A. Nasr-El-Din, and C. A. De Wolf, "Stimulation of Sandstone and Carbonate Reservoirs Using an Environmentally Friendly Chelating Agent," RSC Chemicals in the Oil Industry, Manchester, UK, November 2011.

[262] R. Melo, J. Curtis, J. Gomez, A. Melo, F. Garcia, and H. Pedrosa, "Phosphonic/Hydrofluoric Acid: A Promising New Weapon in the Tortuosity Remediation Arsenal for Fracturing Treatments," SPE 152624 (paper presented at the SPE Hydraulic Fracturing Technology Conference, The Woodlands, TX, 6-8 February 2012).

[263] E. A. Reyes, A. Smith, and A. Beuterbaugh, "Carbonate Stimulation with Biodegradable Chelating Agent Having Broad Unique Spectrum (pH, Temperature, Concentration) Activity," SPE 164380 (paper presented at the 18th Middle East Oil & Gas Show and Conference (MEOS), Bahrain International Exhibition Centre, Manama, Bahrain, 10-13 March 2013).

[264] E. A. Reyes, A. Smith, and A. Beuterbaugh, "Properties and Applications of an Alternative Aminopolycarboxylic Acid for Acidizing of Sandstones and Carbonates," SPE 165142 (paper presented at the 10th SPE International Conference and Exhibition on European Formation Damage, Noordwijk, Netherlands, 5-7 June 2013).

[265] B. S. Bageri and M. A. Mahmoud, "A New Diversion Technique to Remove the Formation Damage From Maximum Reservoir Contact and Extended Reach Wells in Sandstone," SPE 165163 (paper presented at the 10th SPE International Conference and Exhibition on European Formation Damage, Noordwijk, Netherlands, 5-7 June 2013).

[266] F. O. Garzon, H. M. Al-Marri, J. R. Solares, and C. A. Franco Giraldo, SPE, Saudi Aramco, and V.

Ramanathan, "Long Term Evaluation of an Innovative Acid System for Fracture Stimulation of Carbonate Reservoirs in Saudi Arabia," IPTC 12668 (International Petroleum Technology Conference, Kuala Lumpur, Malaysia, 3-5 December 2008).

[267] D. W. Boswood and K. A. Kreh, "Fully Miscible Micellar Acidizing Solvents vs. Xylene, The Better Paraffin Solution," SPE 140128 (paper presented at the SPE Production and Operations Symposium, Oklahoma City, OK, 27-29 March 2011).

[268] M. A. Sayed, H. A. Nasr-El-Din, J. Zhou, S. Holt, and H. Al-Malki, "A New Emulsified Acid To Stimulate Deep Wells in Carbonate Reservoirs," SPE 151061 (paper presented at the SPE International Symposium and Exhibition on Formation Damage Control, Lafayette, LA, 15-17 February 2012).

[269] M. Paterniti, "Microemulsion Surfactant Increases Production in the Codell Formation of the DJ Basin," SPE 116237 (paper presented at the SPE Rocky Mountain Petroleum Technology Conference, Denver, CO, 14-16 April 2009).

[270] J. A. Leal, J. Duarte, F. Al-Ghurairi, G. A. Izquierdo, and J. Soriano, "Post Stimulation Fluid Recovery, Every Drop Counts: Case Histories from Saudi Arabia," SPE 164004 (paper presented at the 2013 SPE Middle East Unconventional Gas Conference & Exhibition, 28-30 January 2013).

[271] A. A. Al-Zahrani, "Innovative Method to Mix Corrosion Inhibitor in Emulsified Acids," IPTC 16946 (6th International Petroleum Technology Conference, Beijing, China, 26-28 March 2013).

[272] W. Bertkau and N. Steidl, U.S. Patent Application 20120222863.

[273] S. Holt, J. Zhou, and F. Gadberry, A Novel Viscoelastic Surfactant (VES) Suitable for Use in High-temperature Carbonate Diverted Acidizing Treatments for Reservoir Stimulation, *Chemistry in the Oil Industry XIII: Oilfield Chemistry—New Frontiers*, Manchester Conference Centre, UK, 4-6 November 2013.

[274] E. A. Reyes and L. A. Smithe, International Patent Application WO2013154711.

6 油井防砂

6.1 概述

全球很多油气井都存在出砂（或微粒）问题。砂粒在油井和生产管线的流动会造成设备磨蚀，而且出砂可能会影响生产中的油水分离。机械防砂工艺包含的技术有很多，包括应用筛管、砾石充填、压裂填砂及完井阶段的射孔技术优化等多种技术。对于胶结较差大量出砂的油层，机械法防砂效果有限，因此可以选择化学防砂技术，这种方式可避免成本高昂的措施作业，对海上油井防砂尤其有利。

6.2 化学防砂

6.2.1 树脂固结砂技术

树脂或环氧树脂可以使松散的砂粒硬化，应用此类材料的化学防砂方法已开展多年，典型的树脂包括基于双酚 A-环氧氯丙烷树脂、聚环氧树脂、聚酯树脂、酚醛树脂、脲醛树脂、呋喃树脂、聚氨酯树脂、丙烯酸树脂和缩水甘油醚等体系。对于浅层天然气储层，化学固砂液和相应的现场试验也已见报道。

如果树脂主剂包括双酚 A-环氧氯丙烷聚合物，则优选 4,4-亚甲基苯胺作为固化剂；如果树脂由聚氨酯组成，则固化剂优选二异氰酸酯。呋喃树脂是最常见的一种体系，其关键原料糠醇在酸性催化剂条件下可以自聚合（图 6.1），因此反应不需要固化剂。这类体系的研发和设计旨在使地层保持生产所需的足够渗透率。有研究认为自转向树脂基固砂液比传统树脂处理剂的处理间隔更长。但研究发现大多数树脂基化学品都不是环境友好型，因此在具体应用时，也应重点考察和优选该类体系的环保特性指标。

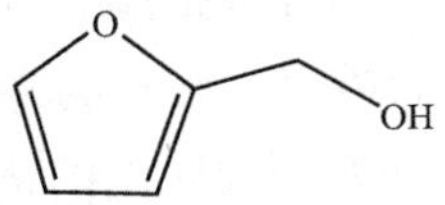

图 6.1 糠醇

水基和硅基改性聚酰胺等非水基增黏化学剂可以黏结砂粒而阻碍其移动。广义的增黏剂范围宽泛，包括如树脂、凝胶、硅酸盐、乳液、聚合物和单体等水溶性和油溶性产品。

泡沫和非泡沫水基体系都已成功应用于固砂领域。在北海地区，研究人员用水溶性活化剂将可固化环氧树脂分散在盐水中，用以处理固结性差的地层。为避免措施影响产量，建议在渗透率高于 500mD 的试验井上开展该项试验。可固化的固结组分被制成水包油乳液，以便活性材料溶于盐水溶液并进行输送。为更好置放处理液，有时也应用泡沫转向剂。

一些聚合物配方在室内填砂管实验评价中显示出良好的效果。这些材料是带有分散剂的水溶性氨基醛树脂和无机交联剂。此外，现场应用时也会涉及苯乙烯/丙烯酸共聚物等商业固砂产品。

一些与堵水作业相似的交联聚合物凝胶体系也建议用于固砂处理。据研究报道，相较于树脂体系，如聚丙烯酰氨基的聚合物凝胶体系用于地层固砂作业时，其失败的概率会更低。

除了上述高分子化学体系，用于固砂的无机化学剂体系也有研发，如基于不溶性二氧化硅源和氢氧化钙源（如氯化钙和氢氧化钠的水溶液）的体系。水溶液体系成分反应产生硅酸钙水合物凝胶，在地层孔隙中具有胶结性。此外，还有研究提出了一种以酶为基础的碳酸钙固结砂工艺。反应需要氯化钙、尿素和脲酶等，反应机制为脲酶催化尿素分解为氨和二氧化碳，从而提高体系的 pH 值。当存在可溶性钙离子时，反应形成的不溶性碳酸钙会沉积于石英砂和岩心表面，并将其结合在一起。

除上述提到的酶促方法，还有一种非酶促的矿物固结地层的方法。该固结材料是一种由环境友好和组分廉价的碱性处理液产生得到的碳酸盐。该体系通过新颖的脱羧反应，在一系列条件下以可控的反应速率从水溶液中沉积碳酸钙，以实现和改善砂粒之间的胶结。这些反应大多要采用甘氨酸盐等氨基酸盐。

为了控制出砂以提高油井产量，使用硝酸钙而非氯化钙的改进型准自然固结砂技术已在北海成功应用。

6.2.2 有机硅烷固砂

大约从 2005 年开始，开发出一种基于有机硅烷的化学固砂新方法，并在现场使用。与其他处理措施相比，这种方法只是将储层残余强度略微提高。由于有机硅烷类产品是油溶性的，加入后不会改变储层的相对渗透率，所以可有效降低因含水饱和度变化而增大表皮系数的风险。这种体系在低压储层的油田特别有优势。该方法只需通过简单的井口作业即可完成措施，并且具有自转向的特点。室内测试研究表明，与前文讨论的水溶性胶凝聚合物和 $CaCl_2$/尿素/酶体系等措施方法相比，有机硅烷技术在固砂和适度控制渗透率降低等方面表现出的整体性能更佳。

目前已优选出的油溶性有机硅烷包括 3-氨基丙基三乙氧基硅烷、双[3-（三乙氧基硅）丙基]胺或二者的混合物（图 6.2）。使用时通常将其混配入柴油，随后从井口注入井内。研究者认为氨基官能团修饰的作用主要有两个：（1）氨基存在时似乎可以使有机硅烷更好地吸附于砂粒表面；（2）氨基可能有助于形成具有黏弹性的凝胶状结构。此外还建议将有机硅烷化合物与水反应使之水解，这样生成的化合物可以与地层中的硅质表面（如硅砂的表面）反应，包裹砂粒，并通过形成束缚砂粒移动的硅酸盐桥，将砂粒黏合在适当的位置，从而起到更好的固砂效果。双[3-（三乙氧基硅）丙基]胺等双功能有机硅烷的优势在于其能将两个颗粒结合在一起。这种有机硅烷具有低的生物富集潜力和较高的生物降解性，因而合乎环境要求。

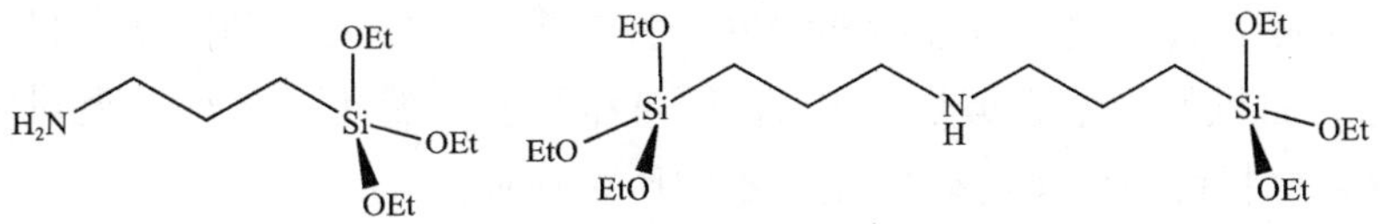

图 6.2　3-氨基丙基三乙氧基硅烷和双 [3-(三乙氧基硅) 丙基] 胺

有几种类型的油井已经采用活性组分体积分数为 5%～7% 的有机硅烷体系进行了固结砂处理。在减少产砂方面的结果有好有坏，其中以海底油井的措施效果最佳。一些油井处理后出现了渗透率下降、油井生产指数（PI）降低 10%～15% 的不利结果，但考虑到这些井的生产受到出砂量的限制，PI 降低的结果仍可接受。在提高油井（尤其是水平井）的最大无砂率（MSFR）方面，找到正确的处理位置（出砂位置）是实现预期性能的关键所在。另有研究认为，当有机硅烷的使用浓度比固砂所需的浓度高时，同时可起到堵水作用。此外，也有研究提出利用可固化树脂和有机硅烷偶联剂来固结砂粒。

除小分子的有机硅烷外，硅烷化聚合物也证实可用于砂粒固结。该类聚合物属于共聚物，如图 6.3 所示，其中心部分 B 为羰基聚合物，如聚（甲基）丙烯酸酯、聚氧乙烯或聚氨酯；A 是烷氧基连接；R 基最好是小分子烷基。

$$(RO)_{3-n}R_nSi \text{——} A \text{——} B \text{——} A \text{——} Si(OR)_{3-n}R_n$$

图 6.3　用于固砂的硅烷化聚合物

6.2.3　其他化学固砂方法

使用带正电荷的水溶性聚合物也是一种固砂方法，其对地层施加的增量力较小或只导致相对较弱的残余强度。典型的包括聚氨基酸 [如聚天冬氨酸、由天冬氨酸和脯氨酸和（或）组氨酸得到的共聚物]，以及聚二烯丙基铵盐（如聚二甲基二烯丙基氯化铵和其混合物）（图 6.4）。研究人员认为，凭借聚合物链长度和多重正电荷，这类聚合物体系可以与地层诸多不同粒子发生静电相互作用，从而将其固定或结合在一起。在该过程中，聚合物链有望跨越地层中砂粒之间的间隙，形成不阻碍流体流动的"筛状"或"网状"结构，因此，按照该方法处理前后对比，地层的渗透性基本不发生改变。

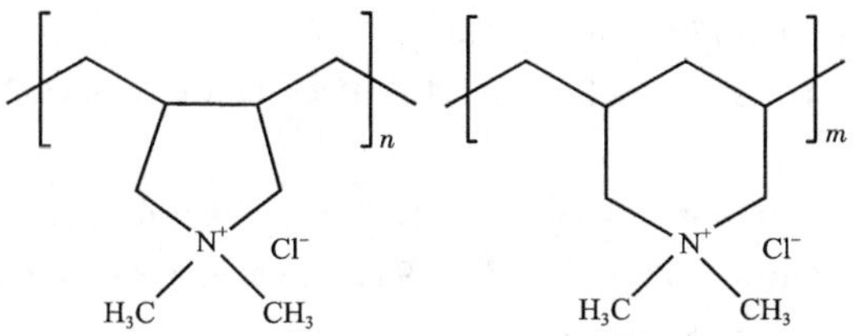

图 6.4　聚二烯丙基二甲基二烯丙基氯化铵（五环的吡咯烷单体是主要成分，六环的哌啶单体是次要成分）

据报道，纳米颗粒也可以用于固砂作业，由于纳米颗粒具有较高的表面力，可以黏附在压裂充填作业的支撑剂表面，从而减少细小砂粒的产出；同时纳米颗粒还可以吸附迁移的地层细砂，保持油井产能。

基于此，研究人员研发了一种能够缓释一种或多种油井处理剂的复合材料，该材料是一种具有高比表面积的纳米级煅烧多孔基材（吸附剂），因此可以有效吸附并缓释油井处理剂。这类复合材料可适用于防砂作业及水力压裂作业。

参 考 文 献

[1] P. D. Nguyen, J. A. Barton, and O. M. Isenberg, U.S. Patent 7013976, 2006.

[2] B. W. Surles, P. D. Fader, R. H. Friedman, and C. W. Pardo, U.S. Patent 5199492, 1993.

[3] M. Parlar, S. A. Ali, R. Hoss, D. J. Wagner, L. King, C. Zeiler, and R. Thomas, "New Chemistry and Improved Placement Practices Enhance Resin Consolidation: Case Histories from the Gulf of Mexico," SPE 39435（paper presented at the SPE Formation Damage Control Conference, Lafayette, LA, 18-19 February 1998）.

[4] J. A. Ayoub, J. P. Crawshaw, and P. W. Way, U.S. Patent 6632778, 2003.

[5] P. D. Nguyen, L. Sierra, E. D. Dalrymple, and L. S. Eoff, International Patent Application WO/2007/010190.

[6] S. G. James, E. B. Nelson, and F. J. Guinot, U.S. Patent 6450260, 2002.

[7] E. B. Nelson, S. Danican, and G. Salamat, U.S. Patent 7111683, 2006.

[8] R. E. Harris and I. D. McKay, "New Applications for Enzymes in Oil and Gas Production," SPE 50621（paper presented at the European Petroleum Conference, the Hague, Netherlands, 20-22 October 1998）.

[9] H. K. Kotlar and F. Haavind, International Patent Application WO/2005/124100.

[10] H. K. Kotlar, F. Haavind, M. Springer, S. S. Bekkelund, and O. Torsaeter, "A New Concept of Chemical Sand Consolidation: From Idea and Laboratory Qualification to Field Application," SPE 95723（paper presented at the SPE Annual Technical Conference and Exhibition, Dallas, TX, 9-12 October 2005）.

[11] H. K. Kotlar, F. Haavind, M. Springer, S. S. Bekkelund, A. Moen, and O. Torsaeter, "Encouraging Results with a New Environmentally Acceptable, Oil-Soluble Chemical for Sand Consolidation: From Laboratory Experiments to Field Application," SPE 98333（paper presented at the International Symposium and Exhibition on Formation Damage Control, Lafayette, LA, 15-17 February 2006）.

[12] F. Haavind, S. S. Bekkelund, A. Moen, H. K. Kotlar, J. S. Andrews, and T. Haaland, "Experience with Chemical Sand Consolidation as a Remedial Sand-Control Option on the Heidrun Field," SPE 112397（paper presented at the SPE International Symposium and Exhibition on Formation Damage Control, Lafayette, LA, 13-15 February 2008）.

[13] A. Jordan and B. Comeaux, "Keeping Fines in Their Place to Maximise Inflow Performance," *World Oil* 228（2007）: 115-122.

[14] H. K. Kotlar, A. Moen, T. Haaland, and T. Wood, "Field Experience with Chemical Sand Consolidation as a Remedial Sand-Control Option," OTC 19417（paper presented at the Offshore Technology Conference, Houston, TX, 5-8 May 2008）.

[15] H. K. Kotlar, International Patent Application WO/2005/124099.

[16] H. K. Kotlar and P. Chen, International Patent Application WO/2005/124097.

[17] V. Chaloupka, Q. Tran, R. Descapria, and M. Haekal, "Sand Consolidation in the Mahakam Delta: From Trials to Field Application," SPE 151488, Offshore Brazil, Rio de Janeiro, Brazil, 14-17 June 2011.

[18] M. R. Talaghat, F. Esmaeilzadeh, and D. Mowla, *Journal of Petroleum Science and Engineering* 67（1-2）（2009）: 34-40.

[19] M. J. Fuller, R. A. Gomez, J. Gil, C. Guimaraes, A. F. Abdurachman, V. Chaloupka, and R. Descapria, "Development of New Sand Consolidation Fluid and Field Application in Shallow Gas Reservoirs," SPE 145409（paper presented at the SPE Asia Pacific Oil and Gas Conference and Exhibition, Jakarta, Indonesia, 20-22 September 2011）.

[20] P. D. Nguyen, R. G. Dusterhoft, and Ronald, International Patent Application WO/2011/117578.

[21] P. D. Nguyen and R. D. Rickman, "Foaming Aqueous-Based Curable Treatment Fluids Enhances Placement and Consolidation Performance," SPE 151002（paper presented at the SPE International Symposium and Exhibition on Formation Damage Control, Lafayette, LA, 15-17 February 2012）.

[22] J. Villesca, G. Hurst, P. Bern, P. Nguyen, R. Rickman and R. Dusterhoft, "Development and Field Applications of an Aqueous-Based Consolidation System for Remediation of Solids Production," OTC 20970（paper presented at the Offshore Technology Conference, Houston, TX, 3-6 May 2010）.

[23] Y. Christanti, G. Ferrara, T. Ritz, B. Busby, J. Jeanpert, C. Abad, and B. Gadiyar, "A New Technique to Control Fines Migration in Poorly Consolidated Sandstones: Laboratory Development and Case Histories," SPE 143947（paper presented at the SPE European Formation Damage Conference, Noordwijk, Netherlands, 7-10 June 2011）.

[24] R. Bhasker, A. F. Foo-Karna, and I. Foo, "Successful Application of Aqueous-Based Formation Consolidation Treatment Introduced to the North Sea," SPE 163880（paper presented at the SPE/ICoTA Coiled Tubing & Well Intervention Conference & Exhibition, The Woodlands, TX, 26-27 March 2013）.

[25] J. Villesca, S. Loboguerrero, J. Gracia, A. Hansford, P. D. Nguyen, R. D. Rickman, and R. G. Dusterhoft, "Development and Field Applications of an Aqueous-Based Consolidation System for Proppant Remedial Treatments," SPE 128025（paper presented at the SPE International Symposium and Exhibition on Formation Damage Control, Lafayette, LA, 10-12 February 2010）.

[26] M. Lahalih and E. F. Ghloum, "Polymer Compositions for Sand Consolidation in Oil Wells," SPE 136024（paper presented at the SPE Production and Operations Conference and Exhibition, Tunis, Tunisia, 8-10 June 2010）.

[27] N. Fleming, E. Berge, M. Ridene, T. Østvold, L. O. Jøsang, and H. C. Rohde, "Controlled Use of Downhole Calcium Carbonate Scaling for Sand Control: Laboratory and Field Results on Gullfaks," SPE 144047, *SPE Production & Operations* 27（2）（2012）: 223.

[28] J. D. Weaver, P. D. Nguyen, and P. A. Curtis, International Patent Application WO/2010/128271.

[29] M. S. Aston, D. Aytkhozhina, and I. Gray, International Patent Application WO/2009/071876.

[30] T. Huang, B. A. Evans, J. B. Crews, and C. K. Belcher, "Field Case Study on Formation Fines Control with Nanoparticles in Offshore Wells," SPE 135088（paper presented at the SPE ATCE, Florence, Tuscany, Italy, 20-22 September 2010）.

[31] D. V. S. Gupta, International Patent Application WO/2012/148819.

[32] D. Holdsworth, Poster at *Chemistry in the Oil Industry XIII: Oilfield Chemistry—New Frontiers'*, Manchester Conference Centre, UK, 4-6 November 2013.

[33] D. Holdsworth, International Patent Application WO/2013/064823.

7 环烷酸盐和其他羧酸盐垢的控制

7.1 概述

在原油生产的上游领域中,有机羧酸盐和(或)环烷酸盐结垢问题不如无机垢或有机质沉积(蜡和沥青质)普遍,但随着西非等环烷酸盐沉积物严重地区的原油产量增加,这个问题才得到特别关注。原油中的油溶性脂肪族羧酸和环烷酸与产出水中的金属阳离子接触后,形成易沉积的盐类,导致油水乳状液难破乳、油水界面不平齐,最终导致油水分离困难,排放的水质变差。在下游炼油厂中,原油中的环烷酸还会与脱盐流程的高浓度钙离子反应,给后续原油加工带来问题。

羧酸盐(俗称皂类)通常是线型长链的羧酸钠盐。在分离器中,羧酸钠皂积聚在油水界面,形成厚乳化层并导致乳化问题。总可溶性固体含量低的产出水更容易受到羧酸盐皂类的乳化问题影响。随着长链羧酸链长、盐水 pH 值的增大,羧酸盐的沉积量增加。长链羧酸的钙盐比钠盐更易溶于油,因此问题要小得多。

环烷酸是一种羧酸,其烷基与一个或多个饱和环己烷或环戊烷连接。在环烷酸的化学异质性混合物中,其毒性变化很大。环烷酸皂通常是钙等二价阴离子的盐。与 Ca^{2+} 相比,Mg^{2+} 的水化程度更高,其与环烷酸的结合力较弱。环烷酸盐皂比羧基皂更容易形成沉积垢,但也会导致乳化问题。即使是总酸值、Ca^{2+} 浓度都相当低的油田,如果大量存在导致沉积的有机酸,也会出现环烷酸盐沉积。同一油田可能同时发现羧酸盐和环烷酸盐的生产问题,但更常见的是两个问题中只有一个占主导,如马来西亚油田面临的主要是羧酸盐问题,而西非油田面临的主要是环烷酸盐问题。

传统上认为,"破坏性"环烷酸的分子量为 200~500。对来自广泛领域的环烷酸盐沉积分析表明,环烷酸的主要成分是 C_{80} 四元酸,含有 4~8 个环,有时被称为 ARN 酸(图 7.1)。

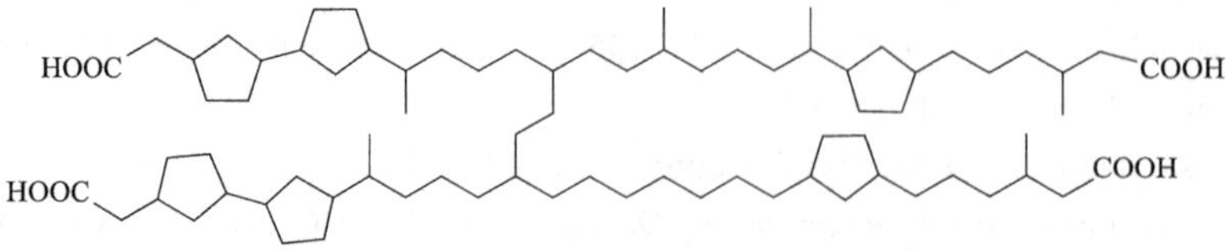

图 7.1 一种环烷酸(ARN)

在与产出水接触时,四元酸会与 Ca^{2+} 形成聚合环烷酸钙黏性固体,与空气接触后变硬。这种环烷酸盐固体会堵塞管道和加工设备,导致生产流量减少,甚至会导致经常性的非计划停产。而许多含有环烷酸的原油并不会出现环烷酸盐固体堵塞问题。环烷酸离子的表面活性会加剧乳状液问题，相比而言，羧酸钠皂带来的乳化问题更为严重。

环烷酸钙沉积的问题通常发生在海上平台，在海底管线中很少发现。这是因为平台的压力下降导致二氧化碳释放和 pH 值上升，这反过来又导致更多的环烷酸成为环烷酸阴离子，并与 Ca^{2+} 结合。有研究提出，Ca^{2+} 在环烷酸盐沉积中的主导地位取决于离子的可用性和选择性。还有研究探讨了一个区块的碱性水突破，并与另一区块的地层水和原油混合后，导致的环烷酸钙垢生成的可能性。

7.2 使用酸控制环烷酸盐沉积

环烷酸是一种弱酸，与环烷酸阴离子存在电离平衡。pH 值越高，酸的解离度越大，在水—油界面就越容易形成皂类。因此，避免环烷酸盐沉积的传统方法是通过添加 pK_a 值低于环烷酸的酸来降低 pH 值。现场经验表明，将 pH 值降低到 6.0 左右可以防止环烷酸盐沉积，进一步降低 pH 值没有额外益处，且会加重腐蚀问题。可用的典型酸包括：

（1）无机矿物酸，如磷酸。
（2）小分子有机酸，如乙酸或乙醇酸。
（3）表面活性剂酸，如十二烷基苯磺酸（DDBSA）。

加入酸使电离平衡从环烷酸阴离子转移到环烷酸，环烷酸的表面活性较低、水溶性较差，不会与金属离子结合成盐。防止环烷酸盐沉积使用最广泛的酸是乙酸（CH_3COOH），其次是磷酸等较强的酸。乙酸易挥发，可造成管线的顶部腐蚀。北海某油田对已使用的乙醇酸（$HOCH_2COOH$）调查发现，分离器中产生的乳化层比乙酸少。盐酸（HCl）等矿物酸也被用来暂时消除环烷酸盐结垢问题。还可以加入如单丁二醇醚（$C_4H_9OCH_2CH_2OH$）或异丙醇［$CH_3CH（OH）CH_3$］等互溶剂，来增加溶解度。也曾使用烃类溶剂。可以用磷酸（H_3PO_4）控制形成羧酸钠。十二烷基苯磺酸常与乙酸组合使用，以控制环烷酸皂。

酸也可以用于去除环烷酸盐沉积物。溶解剂的配方中还可以加入与乙酸互溶的溶剂组分。溶剂可以是烃类，首选是更环保的乳酸乙酯/甲基酯溶剂混合物。

7.3 低剂量的环烷酸盐抑制剂

防止环烷酸或羧酸皂问题所需的酸量是基于总水相量及其 pH 值，而非环烷酸或羧酸的浓度。因此，通过注入酸来避免这些问题的成本很高，还必须仔细控制加量，以避免过度腐蚀问题。最近防止环烷酸盐沉积的方法是添加低加量环烷酸盐抑制剂（LDNI）。已有一些专利详细说明环烷酸盐抑制剂的化学结构，并有现场应用。

低加量环烷酸盐抑制剂可以用界面流变学（和乳液倾向）、界面张力（吊坠法）或有机酸耗竭法进行测试。其中，有机酸耗竭法可以通过加入碱性水，消耗油中的环烷酸，测试剩余酸量。还可以在毛细管路动态评价装置中直接测试环烷酸钙沉积。如果一种环烷酸盐抑制剂通过了前三项测试，就可以进行动态评价。

一种未知的非磷水溶性脱钙剂据称可以从油中去除 Ca^{2+}，以防止环烷酸盐沉积。但这种脱钙剂与 Ca^{2+} 的应用比例是 2∶1，更可能是螯合剂而非低加量环烷酸盐抑制剂。

早期获得专利的低加量环烷酸盐抑制剂是醇类或烷基乙氧基化物与五氧化二磷（P_2O_5）以 2∶1 的比例反应，得到单膦酸酯和二膦酸酯混合物（图7.2）。多元醇的膦酸三酯与正酯、甘油三酯、三乙醇胺烷氧基化物和硝酸三乙酸酯也可以联合使用（图7.3）。此类低加量环烷酸盐抑制剂表现出表面活性特性，使抑制剂在油—水界面上排列并集中，从而防止油相中的有机酸与水中的阳离子或阳离子复合物之间相互作用，即低加量环烷酸盐抑制剂比环烷酸更具有界面活性。在现场应用中，加入了一种破乳剂，以防止环烷酸或低加量环烷酸盐抑制剂引起的乳化问题。环烷酸抑制剂的物理定位和几何形状会阻止环烷酸盐晶体的生长。低加量环烷酸盐抑制剂还能避免形成水包油和油包水的乳状液。药剂使用浓度最高可以达到约 100mg/L。

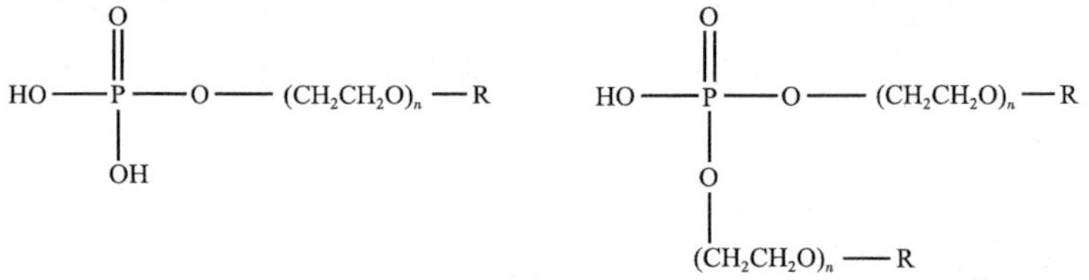

图 7.2 含单膦酸酯和二膦酸酯的低加量环烷酸盐抑制剂（$n=0\sim9$，R= 烷基）

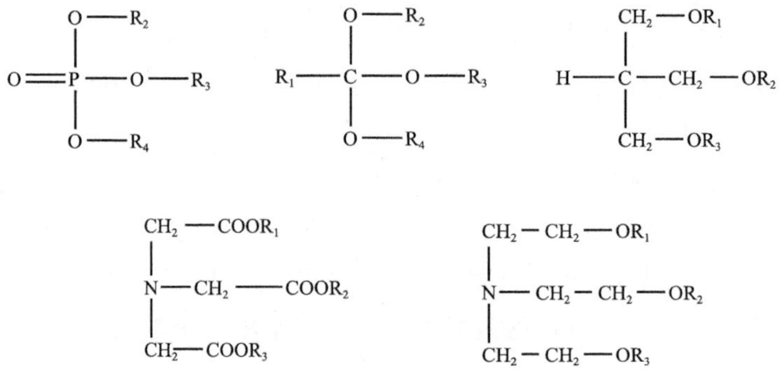

图 7.3 三酯类和烷氧基化物的低加量环烷酸盐抑制剂

膦酸盐末端封端的水溶性聚合物据称也是一种皂化控制剂。这类有效的聚合物将清除原油中的钙并迁移到水相中。因此，皂化控制剂的功效与 Ca^{2+} 浓度成正比。这些聚合物是含有磺酸盐和（或）羧酸盐基团的聚乙烯聚合物，同时也可以防止无机垢的沉积。

具有表面活性的丙烯酰胺（甲基）丙磺酸（AMPS）与丙烯酸、丙烯酸酯、乙烯基芳香烃或乙烯基乙氧基化物等共聚单体的共聚物可抑制环烷烃结垢。例如，低分子量的 2-乙基己基丙烯酸、AMPS、苯乙烯以 1∶1∶1 的比例形成的三元共聚物，其性能比膦酸酯或十二烷基苯磺酸更好，可分布在油—水界面，抑制油中的环烷酸等有机酸与水中的阳离子或阳离子复合物之间相互作用。当 pH 值和压力条件适合于有机酸电离时，就会形成有机酸羧酸盐，而该聚合物还能抑制有机酸羧酸盐的聚集。

季铵或季鏻化合物也是低加量环烷酸盐抑制剂。例如，椰油基甲基—双（2-羟乙基）氯化铵、椰油基甲基［聚氧乙烯（15）］氯化铵和四（羟甲基）硫酸鏻。前两种化合物是表面活性剂，通过取代油水界面上的环烷酸离子而发挥作用。后一种化合物也是一种杀菌剂和硫化物垢溶解剂。同一专利还要求具有至少两个羧酸或丙烯酸功能分子的线型化合物，如聚丙烯酸或聚马来酸盐，也可作为防垢剂使用。事实上，可以使用任何能够与环烷酸相互作用的表面活性剂，以防止随后与金属离子相互作用产生固体或乳液。

取代的伯胺也是低加量环烷酸盐抑制剂。例如，基于聚亚烷基多胺或聚亚烷基多胺-氢化琥珀酰亚胺残留物，如四乙烯戊胺与聚异丁烯琥珀酰亚胺的反应产物。

其他被称为低加量环烷酸盐抑制剂的胺包括松香胺和季铵盐衍生物（图7.4）。如基础松香胺，特别是其乙氧基化物。这些胺最好与十二烷基苯磺酸等其他产品和溶剂混合以提高性能。松香胺具有类似的多环烷基化学结构，但与环烷酸的极性相反。

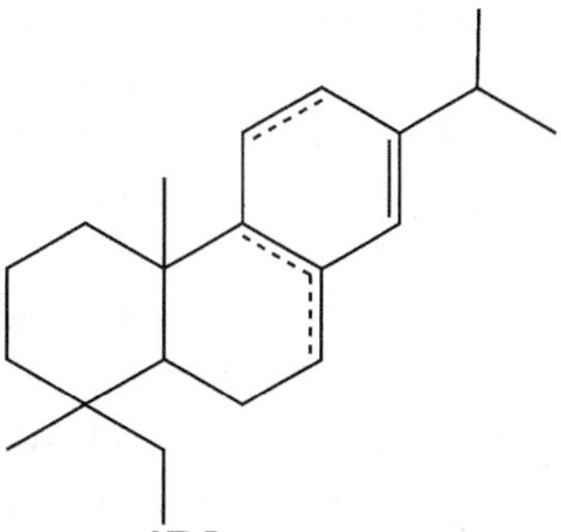

更简单的烷氧基化脂肪胺也可作为低加量环烷酸盐抑制剂使用，可选择与酸和醇混合使用。

多年来，原油下游加工中使用碱性化合物（如碱性盐或胺）以除去原油中的环烷酸，避免腐蚀和其他问题，但下游的内容不是本节的主题。

图7.4　基于松香胺及其铵盐衍生物的低加量环烷酸盐抑制剂

参 考 文 献

［1］(a) R. A. Rodriguez and S. J. Ubbels, "Understanding Naphthenate Salt Issues in Oil Production," *World Oil* 228 (8) (2007): 143–145. (b) C. Hurtevent and B. Brocart, *Journal of Dispersion Science and Technology* 29 (2008): 1496.

［2］S. J. Dyer, G. M. Graham, and C. Arnott, "Naphthenate Scale Formation—Examination of Molecular Controls in Idealised Systems," SPE 80395 (paper presented at the International Symposium on Oilfield Scale, Aberdeen, UK, 29–30 January, 2003).

［3］A.-M. Dahl Hanneseth, M. Fossen, A. Silset, and J. Sjoblom, "Naphthenic Acid/Naphthenate Stabilized Emulsions and the Influence of Crude Oil Components" (paper presented at the 8th International Conference on Petroleum Phase Behavior and Fouling, Pau, France, 10–14 June 2007).

［4］K. S. Sorbie, A. Shepherd, C. Smith, M. Turner, and R. A. Westacott, "Naphthenate Formation in Oil Production, General Theories and Field Observations," Proceedings of the Chemistry in the Oil Industry IX Symposium, Manchester, UK: Royal Society of Chemistry, 2005, 289.

［5］T. Baugh, K. V. Grande, H. Mediaas, J. E. Vindstad, and N. O. Wolf, "The Discovery of High Molecular Weight Naphthenic Acids (ARN Acid) Responsible for Calcium Naphthenate Deposits," *Chemistry in the Oil Industry IX*, Manchester, UK: Royal Society of Chemistry, 2005, 275.

［6］K. B. Melvin, C. Cummine, J. Youles, H. Williams, G. M. Graham, and S. Dyer, "Optimising Calcium Naphthenate Control in the Blake Field," SPE 114123 (paper presented at the SPE 9th International

Conference on Oilfield Scale, Aberdeen, UK, 28−29 May 2008).

[7] B. E. Smith, G. Fowler, J. Krane, B. Lutnaes, and S. J. Rowland, "Separation and Identification of High Molecular Weight Tetra Acids Responsible for Calcium Naphthenate Deposition" (paper presented at the 8th International Conference on Petroleum Phase Behavior and Fouling, Pau, France, 10−14 June 2007).

[8] B. F. Lutnaes, O. Brandal, J. Sjoblom, and J. Krane, *Organic & Biomolecular Chemistry* 4 (2006): 616.

[9] J. E. Vindstad, K. V. Grande, K. R. Hoevik, H. Kummernes, and H. Mediaas, "Applying Laboratory Techniques and Test Equipment for Efficient Management of Calcium Naphthenate Deposition at Oil Fields in Different Life Stages" (paper presented at the 8th International Conference on Petroleum Phase Behavior and Fouling, Pau, France, 10−14 June 2007).

[10] B. Brocart, M. Bourrel, C. Hurtevent, J.−L. Volle, and B. Escoffier, "ARN−Type Naphthenic Acids in Crudes: Analytical Detection and Physical Properties," *Journal of Dispersion Science and Technology* 28 (2007): 331.

[11] R. A. Rodriguez and S. J. Ubbels, "Understanding Naphthenate Salt Issues in Oil Production," *World Oil* 228 (2007): 143−145.

[12] C. Hurtevent and S. Ubbels, "Preventing Naphthenate Stabilised Emulsions and Naphthenate Deposits on Fields Producing Acidic Crude Oils," SPE 100430 (paper presented at the SPE International Oilfield Scale Symposium, Aberdeen, UK, 31 May−1 June 2006).

[13] M. S. Turner and P. C. Smith, "Controls on Soap Scale Formation, including Naphthenate Soaps—Drivers and Mitigation," SPE 94339 (paper presented at the SPE International Symposium on Oilfield Scale, Aberdeen, UK, 11−12 May 2005).

[14] A. Goldszal, C. Hurtevent, and G. Rousseau, "Scale and Naphthenate Inhibition in Deep−Offshore Fields," SPE 74661 (International Symposium on Oilfield Scale, Aberdeen, UK, 30−31 January 2002).

[15] S. J. Ubbels, "Preventing Naphthenate Stabilized Emulsions and Naphthenate Deposits during Crude Oil Processing," Proceedings from the 5th International Conference on Petroleum Phase Behaviour and Fouling, 13−17 June 2004.

[16] J. E. Vindstad, A. S. Bye, K. V. Grande, B. M. Hustad, E. Hustvedt, and B. Nergård, "Fighting Naphthenate Deposition at the Heidrun Field," SPE 80375 (paper presented at the International Symposium on Oilfield Scale, Aberdeen, UK, 29−30 January 2003).

[17] J. S. Ubbels, P. J. Venter, and V. M. Nace, World Patent Application WO/2006/025912.

[18] C. R. Jones, International Patent Application WO/2005/085392.

[19] C. Gallagher, J. D. Debord, S. Asomaning, J. Towner, and P. Hart, World Patent Application WO/2007/065107.

[20] M. M. Mapolelo, L. A. Stanford, R. P. Rodgers, A. T. Yen, J. D. Debord, S. Asomaning, and A. G. Marshall, *Energy & Fuels* 23 (2009): 349.

[21] M. Hellsten and I. Uneback, International Patent Application WO/2008/155333.

[22] A. G. Shepherd, S. Poteau, S. Dubey, G. J. Zabaras, M. Van Dijk, and M. Grutters, "Flow Assurance in Oil Systems: Lessons Learned on the Role and Impact of Naphthenic Acids," SPE 157295 (paper presented at the SPE International Production and Operations Conference & Exhibition, Doha, Qatar, 14−16 May 2012).

[23] V. Alvarado, X. Wang, and M. Moradi, *Energy & Fuels* 25 (2011): 4606−4613.

[24] D. Arla, L. Flesisnki, P. Bouriat, and C. Dicharry, *Energy & Fuels* 25 (2011): 1118–1126.

[25] J. Junior, L. J. Borges, C. Carmelino, P. Hango, J. D. Milliken, and S. Asomaning, "Calcium Naphthenate Mitigation at Sonangol's Gimboa Field," SPE 164069 (paper presented at the SPE International Symposium on Oilfield Chemistry, The Woodlands, TX, 8–10 April 2013).

[26] K. E. Tollefsen, K. Petersen, and S. J. Rowland, *Environmental Science & Technology* 4 (9) (2012): 5143–5150.

[27] L. L. Ge, M. Vernon, S. Simon, Y. Maham, J. Sjoblom, and Z. H. Xu, *Colloids and Surfaces A: Physicochemical and Engineering Aspects* 396 (2012): 238–245.

[28] S. J. Rowland, A. G. Scarlett, D. Jones, C. E. West, and R. A. Frank, *Environmental Science & Technology* 45 (7) (2011): 3154–3159.

[29] J. V. Headley, K. M. Peru, and M. P. Barrow, *Mass Spectrometry Reviews* 28 (2009): 121–134.

[30] J. V. Headley, K. M. Peru, M. P. Barrow, and P. J. Derrick, *Analytical Chemistry* 79 (2007): 6222–6229.

[31] M. P. Barrow, J. V. Headley, K. M. Peru, and P. J. Derrick, *Energy & Fuels* 23 (5) (2009): 2592–2599.

[32] O. Sundman, S. Simon, E. L. Nordgård, and J. Sjöblom, *Energy & Fuels* 2 (11) (2010): 6054.

[33] S. L. Berry, H. L. Becker, J. L. Boles, F. De Benedictis, and D. Galvan, U.S. Patent Application 20100029514.

[34] C. U. Igwebueze, L. Oduola, O. Smith, P. Vijn, and A. G. Shepherd, "Calcium Naphthenate Solid Deposit Identification and Control in Offshore Nigerian Fields," 164055 (paper presented at the SPE International Symposium on Oilfield Chemistry, The Woodlands, TX, 8–10 April 2013).

[35] S. Caird, private communication.

[36] D. Han, Y. Wang, D. Li, and Z. Cao, *Petroleum Science and Technology* 28 (2010): 826.

[37] J. P. Vijn and P. In't Veld, International Patent Application WO/2012/154378.

[38] D. Smith, C. Smith, and D. Watson, U.S. Patent Application 20090301936.

[39] C. Khandekar, S. Gopal, and J. Smith, International Patent Application WO/2011/063459.

[40] C. Khandekar, R. Wilson, and J. Smith, International Patent Application WO/2010/017575.

[41] K. Sandengen, K. Solbakken, and L. O. Jøsang, *Energy & Fuels* 27 (6) (2013): 3595–3601.

8 生产过程中的腐蚀控制

8.1 概述

井下油套管和设备、海底或地面管道、压力容器和储罐等的内外腐蚀是当前石油和天然气工业生产过程中面临的主要问题之一。这些由电化学反应导致的均匀或局部腐蚀不仅会造成金属损耗,而且会使材料脆化或开裂,最终可能导致设备服役失效。钢材中的铁腐蚀要求有水和能被还原的水溶性物质,铁同时被氧化。油气田产出流体中的氧气、酸性气体(如CO_2、H_2S)及天然有机酸等都会促进金属腐蚀。单质硫也会加剧腐蚀,这在全球多个油气田中有发现。腐蚀过程的本质是在金属表面发生局部阳极和阴极反应的电化学氧化还原过程,腐蚀基本过程如图8.1所示。氢原子进入金属是造成其脆化的一个原因。金属的脆化和开裂往往不可预测,并增加了产生灾难性后果的风险。

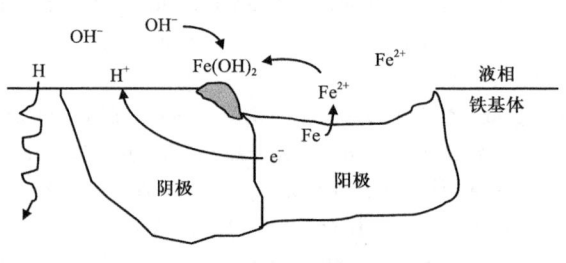

图 8.1 腐蚀原电池反应

在酸性溶液中,阴极反应如下:

$$2H^+ + 2e^- \longrightarrow 2H$$

$$H + H \longrightarrow H_2$$

氢原子形成氢气的反应会受到硫化物的毒化作用,这样产出流体(酸性液体)中的硫化物会加快氢原子渗入金属基体的速度,进而可能导致硫化物应力开裂。这种类型的腐蚀通常发生在服役寿命的后期,很难通过加注缓蚀剂等方式来防控。

溶液中去除H^+后会剩余OH^-。在中性或碱性溶液中,阴极反应如下:

$$O_2 + 2H_2O + 4e^- \longrightarrow 4OH^-$$

在阳极,铁被氧化:

$$Fe \longrightarrow Fe^{2+} + 2e^-$$

产生的OH^-与Fe^{2+}反应,形成不溶性的氢氧化亚铁。

$$Fe^{2+} + 2OH^- \longrightarrow Fe(OH)_2$$

当体系中有氧时，$Fe(OH)_2$ 会被氧化为 $Fe(OH)_3$。油井产出液中通常不含氧，因此很少发生氧腐蚀。而在压力较低的老油田，氧气会更容易侵入井环空、储罐的蒸汽空间及随泵注引入，导致产出液中出现微量的氧。有室内研究表明，CO_2/O_2 系统的腐蚀控制比单独 CO_2 系统更困难，但是依然可以通过缓蚀剂来抑制。由于海水中的溶解氧量约为 9mg/L，注海水井的氧腐蚀更为普遍。解决油气生产过程中氧腐蚀的最好办法是除去水中的氧或采用除氧剂。除了氧腐蚀，CO_2 腐蚀（甜气腐蚀）也是石油和天然气生产中常见的腐蚀。H_2S 腐蚀（酸气腐蚀）比 CO_2 腐蚀更严重。CO_2 腐蚀中关键性的被还原物质是存在于水相中的 H^+、HCO_3^- 和 H_2CO_3。腐蚀发生后，会导致溶解性的铁质及 $FeCO_3$ 等垢产物沉积在钢表面。这种保护性的膜层降低了表观的均匀腐蚀速率，但垢层下仍会发生局部腐蚀（所谓的垢下腐蚀）。对于 H_2S 腐蚀，H_2S、H^+ 和 HS^- 等相关被还原物质与铁反应后，形成如 FeS 垢产物。与 $FeCO_3$ 相似，FeS 能在金属表面成膜，从而抑制腐蚀的进一步发展。CO_2 和 H_2S 的腐蚀速率通常会随着产出液的温度、压力和矿化度（电解质浓度）的增高而增大。

据报道，CO_2 和 H_2S 这两种酸性气体被捕获在水合物沉积物中，可导致比游离气态更高的腐蚀速率。此外，水合物分解时，会生成更高浓度的酸性气体，造成更严重的腐蚀。

油田中发生的腐蚀类型包括均匀腐蚀、局部腐蚀、点蚀和缝隙腐蚀、电偶腐蚀、冲刷腐蚀、微生物腐蚀（MIC）和腐蚀开裂等。

均匀腐蚀指油管或集输管线作为整体的金属均匀损失，属于最容易控制的一类腐蚀。局部腐蚀更为常见，并且通常发生在特定的位置。电偶腐蚀指当异种金属直接电连接，并同时与电解质溶液接触时，一种金属会优先地发生腐蚀的现象。点蚀和缝隙腐蚀比较类似，属于局部腐蚀的极端类型，都会导致金属产生小孔，点蚀在高 Cl^- 浓度下特别普遍。冲刷腐蚀也被称为流动诱导局部腐蚀，是在机械冲刷和电化学腐蚀综合作用下，材料的一类复杂降解机制，主要源于高流速下的高剪切应力。微生物腐蚀是一类非常普遍的腐蚀，源自产出流体中厌氧微生物新陈代谢引发的化学过程。在厌氧条件下，需要特别注意硫酸盐还原菌（SRB）对硫酸根的还原。大多数微生物腐蚀的腐蚀形式是在细菌群（也称生物膜）下形成坑，经常在矿物和生物沉积物中发展。生物膜的保护性环境会增强腐蚀性，加速腐蚀。表面垢的形成是决定局部腐蚀速率的关键因素之一。

8.2 腐蚀控制的方法

腐蚀管理通常需要将监（检）测、检查和系统建模及各种控制策略相结合，通过设定关键性能指标（如年度的泄漏数量），审查检测数据并改进提升，以降低腐蚀失效数量。腐蚀控制有多种方法，如使用耐蚀合金（CRA）、除水（通过清管器或脱水）、阴极保护、涂料、缓蚀剂、除氧剂和 H_2S 清除剂、防止 MIC 的杀菌剂（见第 14 章）、稳定 pH 值、降低阻力等。

尽管含铬的不锈钢等耐蚀合金价格昂贵，但当井下使用普通碳钢的腐蚀速率过高、难以接受时，可采用耐蚀合金。北海和墨西哥湾的油气井管柱材质通常是13Cr或22Cr，高温井则倾向于选用双相不锈钢。在其他的生产区域，考虑到应用耐蚀合金相对高昂的一次性投入与后期间歇加注缓蚀剂保护的运行成本，则可能会选择碳钢管柱。产出流体中的H_2S也会影响管柱选材。在陆上油田，经常从井口定期和不定期加注缓蚀剂来保护井下碳钢管柱。据称，乳液缓蚀剂的性能更好。

通过清管器的清管作业，可以最大限度地降低水在地面管道低洼部位聚集所造成的局部腐蚀。石油工业中，有时还会使用涂料和塑料等涂层防腐技术，特别是在一些腐蚀敏感场合。这一方面是出于成本的考虑，另一方面涂层可能会因细沙/微粒和其他颗粒的磨蚀，使钢材重新裸露而失去保护。

阴极保护有外加电流和铝、锌牺牲阳极（称为电偶）两种方式，常用于保护管线或水下结构（如平台支撑腿）的外表面。牺牲阳极会因为构筑物内部流体的侵蚀和化学反应而迅速消耗，较少用于内腐蚀控制，但可以用于除油罐的内防腐。使用涂层控制内腐蚀时，也要考虑长期磨蚀情况下的涂层损伤。

还可以通过使用除氧剂和H_2S清除剂来降低氧腐蚀和H_2S腐蚀（分别见第15章和第16章）。但无法采用此方法消除CO_2腐蚀，因为产出流体中CO_2的含量通常比O_2和H_2S高得多，相应的消除成本很高。加注杀菌剂可以降低因细菌产生的腐蚀性化学物质和次生H_2S浓度（见第14章）。杀菌剂可以在分离器上游和下游使用，其最主要用途是在注水系统中降低微生物腐蚀。

湿气输送管道的腐蚀控制方法之一是采用pH值稳定剂。金属氢氧化物、金属碳酸盐和碳酸氢盐、胺等pH值稳定剂可以提高液相产出流体的pH值，促进金属表面形成$FeCO_3$等固体沉积物的致密保护层。pH值稳定技术可以与乙二醇等水合物抑制剂组合使用，因为pH值稳定剂会随着乙二醇的循环再生而保留，这样就不需要持续补充pH值稳定剂。该方法已在挪威和荷兰的海上油田使用。pH值稳定技术主要用于输送介质中不含H_2S的湿气管道，但目前在波斯湾地区高含H_2S和CO_2的油气输送管道也有应用。有文章介绍了为抑制湿气输送管道的严重腐蚀，采用乙二醇组合缓蚀剂的研发进展。

使用减阻剂，可以降低流动诱导局部腐蚀的严重程度，同时部分成膜缓蚀剂也具有减阻特性（见第17章）。水溶性高分子减阻聚合物和表面活性剂型成膜缓蚀剂的组合，可以有效抑制与油包水乳液接触的钢铁腐蚀。

8.3 缓蚀剂类型

缓蚀剂可分为钝化（阳极）型、阴极型、气相或挥发性、成膜型四种类型。

为保护原油、凝析油和天然气的采集输管道（基本为无氧环境），最常采用的缓蚀剂类型是成膜型缓蚀剂，有时还使用增效剂。本节先简要讨论其他类型的缓蚀剂，再详细讨论成膜型缓蚀剂。适用于酸化作业的缓蚀剂在第 5 章中单独讨论。

钝化型缓蚀剂不适用于油气生产系统，这类缓蚀剂在如公用系统（淡水或冷凝水系统）等低矿化度环境中应用效果最好。钝化型缓蚀剂会在金属表面形成非反应性的薄膜层，阻止腐蚀性物质接触金属，抑制后续的腐蚀。当加注浓度不足时，所有的钝化型缓蚀剂都会加速腐蚀。部分钝化型缓蚀剂使用时，需要介质中存在氧气。这类缓蚀剂包括磷酸盐（PO_4^{3-}）和多聚磷酸盐、钨酸盐（WO_4^{2-}）、硅酸盐（SiO_3^{2-}）。

不需要氧气存在的缓蚀剂包括：铬酸盐（CrO_4^{2-}），亚硝酸盐（NO_2^-），钼酸盐（MoO_4^{2-}），偏钒酸盐、正钒酸盐和焦钒酸盐（如 $NaVO_3$、Na_3VO_4 和 $Na_4V_2O_7$）。

部分钝化型缓蚀剂是金属含氧酸盐。铬酸盐具有致癌性，目前已不建议使用；磷酸盐或多聚磷酸盐是无毒的，但由于磷酸钙等的溶解度有限，在许多情况下难以保持足够的有效浓度；钼酸盐在厌氧条件下也能起作用，但在有氧条件下效果更好。

亚硝酸盐是一种有效的钝化型或阳极型缓蚀剂。阳极型缓蚀剂的使用浓度如果过低，会与局部阳极形成不均匀钝化层，加剧点蚀。亚硝酸盐常用于减少热水管束腐蚀。在 H_2S 腐蚀环境下，钒酸盐与 2，4- 二氨基 -6- 巯基嘧啶硫酸盐混合物是一种效果良好的碳钢缓蚀剂。

阴极型缓蚀剂不适用于油气生产过程，但可用于钻井液。例如，氧化锌中的 Zn^{2+} 可以通过抑制水还原为氢气来延缓腐蚀。

缓蚀剂对电偶腐蚀的影响仍然是相对新的研究课题。针对电偶腐蚀的控制，有文章测试了市售缓蚀剂和不同配比下的腐蚀速率。

气相缓蚀剂是在常压下具有足够的蒸气压，通过气体扩散和物理吸附在金属表面起保护作用的有机化合物，如二环己胺亚硝酸、二环己胺碳酸盐、二乙胺磷酸盐、三甲胺等小分子挥发性胺类和苯并三唑。

潮湿环境下，气相缓蚀剂分子会电离为不同的阴、阳离子（如环己胺亚硝酸盐会形成环己胺阳离子和亚硝酸根阴离子），吸附到金属的阳极和阴极。阳离子吸附于金属，分子的疏水部分形成阻隔污染物（如氧气、水、氯化物和其他腐蚀促进剂）的保护膜。这样腐蚀原电池无法形成，腐蚀就会停止，阴离子也起到缓蚀剂作用。

起中和作用的小分子挥发性胺类气相缓蚀剂偶尔也用于湿气管道或气体冷却器，但由于需要相对较高的浓度，考虑经济性方面无法实现连续加注。小分子胺的使用目的不是中和气流中的所有酸性气体，而是中和冷凝水中溶解的酸性气体。小分子胺也用于密闭环路环境，轻质胺和咪唑啉的混合物（8.4 节有讨论）有很好的抑制气相输气管道腐蚀作用。胺类混合物的实例有甲氧基 -3- 丙胺或 N，N- 二异丙基乙胺、辛胺。氨基羧酸气相缓蚀剂与成膜型缓蚀剂可以一起用于油气管道。含硫的气相缓蚀剂也有报道，如 2- 巯基乙烷 -1- 醇［也称二巯基乙醇（$HSCH_2CH_2OH$）］或其甲基衍生物，如 2- 巯基丙烷 -1- 醇［$HSCH(CH_3)CH_2OH$］。

8.4 成膜型缓蚀剂

成膜型缓蚀剂对于防止氯化物腐蚀、CO_2 腐蚀和 H_2S 腐蚀特别有效,可以通过井下或井口装置连续或间歇加注。有文献讨论了间歇加注的适用性条件。成膜型缓蚀剂中的活性组分加量通常为 10~100mg/L。最近报道了几种成膜型缓蚀剂加量优化的新方法,如现场检测胶束浓度的荧光实验新方法以辅助优化缓蚀剂加量,还有采用基于 DNA 诊断试剂盒的方法来快速判断 MIC。

包覆成膜型缓蚀剂的缓释胶囊已用于井下,生产井和水源井的缓蚀剂挤注处理也有报道。有专利介绍了控制油井均匀腐蚀和氢脆的乳液缓蚀剂配方,还有硫代磷酸酯和焦磷酸酯的乳液混合物等缓蚀剂介绍。

8.4.1 成膜型缓蚀剂的工作原理

成膜型缓蚀剂的有效性部分取决于其对金属表面(或菱铁矿、碳酸亚铁等铁质垢表面)的吸附强度,形成保护层,从物理上防止水和 Cl^- 等渗透到金属表面。有研究表明,N80 钢表面是否存在碳酸亚铁产物膜对成膜型缓蚀剂(巯基乙酸、二乙烯三胺和环烷酸咪唑)的缓蚀效率有明显影响。成膜型缓蚀剂可以是小分子或聚合物。许多成膜型缓蚀剂是分子结构上带有一个极性端基和一个疏水尾基的有机两亲性表面活性剂。极性端基被设计与表面的铁原子相互作用,而疏水尾基吸引液体烃类形成油膜,进一步防止腐蚀性水相渗透到金属表面(图 8.2)。如果疏水基很长,在有或没有助表面活性剂或溶剂的情况下,均可以形成表面活性剂成膜型缓蚀剂的保护性双层。有研究表明,在某些多相流条件下,部分管壁部位因表面活性剂成膜型缓蚀剂无法全部覆盖,可能出现局部腐蚀。例如,管道顶部因缓蚀剂无法到达,就经常因含 CO_2、H_2S、有机酸的冷凝水而腐蚀。

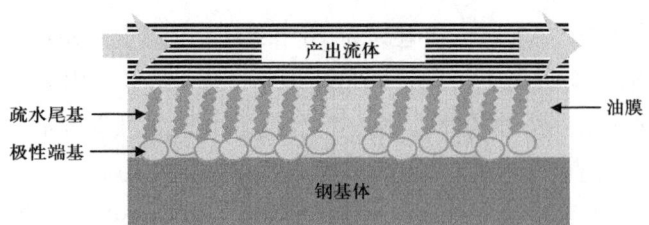

图 8.2 表面活性剂成膜型缓蚀剂的作用机理

还有很多小分子和聚合物类的缓蚀剂吸附在金属表面,但没有大的疏水尾基,因此不会将液态烃吸附在金属表面。此类"膜"可能是由单独的缓蚀剂或以铁复合物的形式构成,但仍能提供良好的缓蚀作用,这些缓蚀剂也被称为成膜型缓蚀剂,本节后面将讨论一些实例。计算机建模和定量构效关系(QSAR)分析已用于辅助设计改进或研究更环保的成膜型缓蚀剂。

缓蚀剂在含 H_2S 系统中已成功应用了 40 多年。虽然效果显著,但人们对其抑制机理

及其与 H_2S 腐蚀产物层的相互作用了解甚少。缓蚀剂的性能不仅取决于其在钢材表面的吸附，还取决于其结合到产物层中提供保护的能力，以及改变成垢晶体结构。

有些成膜型缓蚀剂在防止固体沉积物（如 $FeCO_3$ 或焊缝腐蚀）的腐蚀方面比其他缓蚀剂好，还有些成膜型缓蚀剂抑制冲刷腐蚀的性能更好。有文献报道称，一种新型的多功能缓蚀剂配方对清除管道内表面的油性沉积物极为有效，但是并没有给出这种成膜型缓蚀剂的具体结构。

8.4.2 缓蚀剂测试

石油工业中，有一系列的缓蚀剂测试方法，如鼓泡或瓶试法、旋转圆柱电极（RCE）法、旋转圆盘电极（RDE）法、喷射冲击试验法、高剪切高压釜、旋转笼试验法、循环流动试验法、转轮测试法、静态测试法。

腐蚀试验的目的是测量金属随时间变化的失重，观察表面变化或测量/解析腐蚀电流的变化。在上述许多试验中，与管道成分相同的金属试样被放置在仪器中。腐蚀速率通常使用线性极化电阻或电化学阻抗谱法进行监测，以每年损失的金属毫米数来表示。点蚀可以用光学显微镜或扫描电镜来观察。实际应用中是将成膜型缓蚀剂暴露在湍流中，如果不能很好地吸附，就会从管壁剥离。因此，一些场合是以湍流试验来评价缓蚀剂性能。一般认为管壁剪切应力会去除如碳酸亚铁或缓蚀剂等保护性表面膜层。对于低成本、低剪切力条件下的初步化学筛选，通常采用鼓泡试验。RCE 和 RDE 试验是相当简单的低成本方法，可以在中等剪切应力或湍流状态下测试缓蚀剂，而循环流动试验更贴近实际工况条件，但是成本也更高。喷射冲击试验或高剪切高压釜主要适用于高剪切条件的测试，此时产生的如段塞流等流态会引起局部腐蚀。当然也可以采用流动环路进行评价。流动环路试验方法评价冲刷腐蚀也有报道。而对点蚀及焊接处腐蚀的评价手段也有报道。对于多相流，还需要考虑液态烃相的效应，因为成膜缓蚀剂会在液态烃相和水相之间分配。也可能会出现油相优先润湿金属表面的情况。还可能需要评估固相物质通过吸附作用，从溶液中移除缓蚀剂的影响。关于成膜型缓蚀剂测试方法的更多信息，可参阅两篇有用的简短评论，文中有多篇参考文献，比较了评估成膜型缓蚀剂的实验室测试方法。此外，还有一些文章介绍了可以快速和经济的方式测试和选择 CO_2/H_2S 腐蚀缓蚀剂的测试程序。最后，有文献综述了成膜型缓蚀剂的分析方法。

有专利报道了一种测试缓蚀剂有效性的方法，将缓蚀剂置于至少一种流体中，然后，将试样悬挂在含有缓蚀剂的流体及至少另外一种流体中，两种流体在同一个容器内。已报告了一种新的 MIC 测试方法。该方法使用含甲烷的微生物来测试金属合金的抗 MIC 能力。

现实中往往需要对成膜型缓蚀剂与现场使用的其他化学剂进行兼容性测试。例如，许多成膜型缓蚀剂会对防垢剂和动力学水合物抑制剂的性能产生不利影响（见第 3 章和第 9 章）。对成膜型缓蚀剂还需进行其他兼容性测试，如热稳定性、发泡、溶剂闪蒸及与弹性密封件等材料的兼容性。例如，在天然气生产系统中，要求成膜型缓蚀剂具有较低的发泡特性。此外，成膜型缓蚀剂不应加剧乳化，使油水分离困难。成膜型缓蚀剂在使用过程中，含水率变化会极大地改变乳液的稳定性。

8.4.3 开发更环保的成膜型缓蚀剂

油气生产过程中，倾向使用油溶性成膜型缓蚀剂。油溶性成膜型缓蚀剂热稳定性好，比水溶性成膜型缓蚀剂在井下连续或间歇注入的效果都更好，其缺点是有一定毒性，且生物积累潜力较高，已开发出更环保的（通常是生物降解性更好）油溶性成膜型缓蚀剂。由于大部分油田已处于高含水开发后期，产出水量多于烃类含量，现已发展为研发更好的水分散型或水溶性缓蚀剂。以油气井中的主要流体为载体，可以提高缓蚀剂的利用率。此外，油溶性成膜型缓蚀剂中的有毒芳香烃载体溶剂，可以被水或醇类等更环保的水溶性溶剂取代。成膜型缓蚀剂的表面活性剂溶解度可以通过改变端基的亲水性或疏水尾基的长度来实现。但如果烷烃链太短（通常少于12个碳），则在金属表面不会形成防腐效果良好的烃类油膜。相反，在高剪切力下，疏水端过长（大于20个碳）的成膜型缓蚀剂也可能从管壁冲走。通过使用助表面活性剂也可以改变成膜型缓蚀剂的水溶性或分散性，但有时可能降低成膜性和膜的持久性，并有可能加剧乳化问题。

传统表面活性剂类成膜型缓蚀剂，特别是含氮表面活性剂，大多有很高的毒性。因此，要求开发毒性较低的成膜型缓蚀剂或容易生物降解为弱毒性小分子的产品。降低毒性的一种方法是将表面活性剂类成膜型缓蚀剂的疏水链长度减少到8个或9个碳原子以下，这样得到的产品水溶性更强并且生物积累潜力较低。但这种链长的减少将大大降低缓蚀剂的成膜特性，使其几乎失去缓蚀作用。通过与具有8～10个碳原子的乙氧基化醇等润湿剂结合，可以提高成膜型缓蚀剂的功效。该方法已用于乙氧基咪唑啉。对于酸性介质中可能被质子化的季铵盐成膜型缓蚀剂或咪唑类，减少其毒性的方法是通过添加阴离子基团（如羧酸盐基）来中和头部的正电荷，使分子成为两性离子。据报道，一种二癸基二甲基季铵碳酸盐/碳酸氢盐成膜型缓蚀剂比传统的季铵成膜型缓蚀剂更环保，还有开发易生物降解的酯类季铵聚烷氧基化烷醇胺和低聚季铵酯成膜型缓蚀剂（见8.4.4.3）。还开发了单体或聚合成膜型缓蚀剂（如可生物降解的聚氨基酸或各种硫氮化合物），这种缓蚀剂几乎没有疏水尾基，其毒性比多数表面活性剂类成膜型缓蚀剂低。这些本质上亲水的聚合物本身可以充分覆盖金属表面以防止腐蚀，而非吸引烃类保留在金属表面形成疏水膜。例如，聚氨基酸用作可生物降解的复合缓蚀防垢剂。下面将讨论成膜型缓蚀剂的类别及其环境特性。

8.4.4 成膜型缓蚀剂的类别

成膜型缓蚀剂的分子结构大多含有一个或多个杂原子（氮、磷、硫和氧），这些杂原子通过孤对电子与金属表面的铁原子结合。常见的表面活性剂类成膜型缓蚀剂有膦酸酯类、各种含氮化合物、经常与其他杂原子（如氮）结合的含硫化合物。

可生物降解和低毒性的聚氨基酸已用于环境敏感地区。这种含氮化合物有（聚）羧酸的胺盐、季铵盐和甜菜碱（两性离子）、胺类化合物和咪唑类化合物、多羟基和乙氧基化的胺/氨基甲烷、酰胺及其他杂环类。

有疏水尾基的脂肪烷基二胺和聚胺等胺类也用作成膜型缓蚀剂。除了成膜，胺类还有助于中和水相中的腐蚀性碳酸和硫化氢。在早期专利中，恶唑类、吡咯啉类和松香胺类也作为成膜型缓蚀剂，但当前未普遍使用。最近对松香酰胺咪唑啉进行了研究。有些成膜型缓蚀剂可以协同起效，如咪唑类和膦酸酯类经常一起使用。还有包括两种或多种缓蚀剂复配的成膜型缓蚀剂，既可以保护铁基金属表面的阴极，又可以保护阳极，产生更好的保护效果。价廉和环境友好的烟草提取物、糖蜜、树叶和植物提取物等天然缓蚀剂已广泛研究，但要使其具有商业竞争力，还需要做更多的工作。通过添加小分子无机化合物（如钝化型缓蚀剂）及其他一些增效剂（如硫代硫酸酯和硫代膦酸酯）来增强成膜型缓蚀剂的缓蚀性能。在某些应用中，SRB 存在时，硫代硫酸钠会引起铁基合金的点蚀、缝隙腐蚀和应力腐蚀开裂，其还可以作为氧化剂，被 SRB 还原，进一步加剧局部腐蚀。

有许多文章和会议论文都有关于新型成膜型缓蚀剂结果的讨论，但很少有化学方面的相关描述。因此，以下关于成膜型缓蚀剂的参考文献大多取自专利文献。

8.4.4.1 膦酸酯类

包括单酯和双酯的膦酸酯，都是很好的成膜型缓蚀剂，经常与其他类别的成膜型缓蚀剂复配使用。膦酸酯类缓蚀剂是由醇或烷基酚或其烷氧基化衍生物与磷化剂（如五氧化二磷或正磷酸）反应而成。形成的单酯和双酯膦酸酯（图 8.3）混合物具备不同的亲水性，会在液态烃与水相中分布。环状聚磷酸也可用于制备膦酸酯。已证实含有疏水性壬基酚基的膦酸酯比线型或支链脂肪族膦酸酯效果更好，且壬基酚膦酸二酯比相应的单酯更有效。膦酸酯会形成相当难溶的 Fe(II) 和 Ca(II) 盐，沉积在管壁上，阻碍腐蚀继续。多氧烷基硫醇的膦酸酯（如与不同数量的环氧乙烷反应的辛基或十二烷基硫醇）作为成膜型缓蚀剂对抑制点蚀特别有效，尤其适用于深井气相环境含铁金属。聚（氧-1,2-乙二基）十三烷基羟基膦酸酯与双十烷基二甲基氯化铵等季铵盐类成膜型缓蚀剂和硫代羰基化合物组成的成膜型缓蚀剂效果较好。膦酸酯与各种胺（如酰化多胺、吗啉和乙氧基化脂肪胺）的胺盐反应产物作为均匀腐蚀和开裂型腐蚀的缓蚀剂，效果较好。

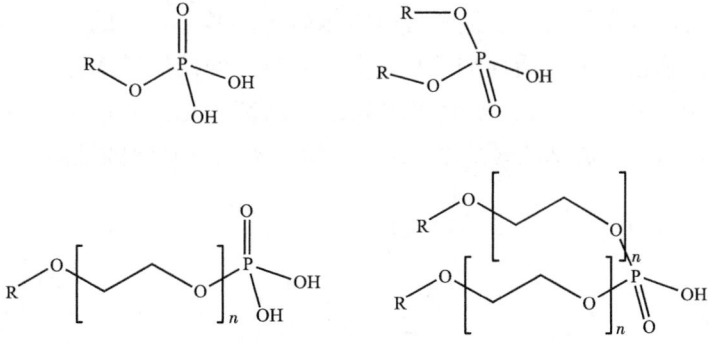

图 8.3 典型膦酸酯类成膜型缓蚀剂的结构

8.4.4.2 （聚）羧酸胺盐

长链羧酸的胺盐经常用于成膜型缓蚀剂配方中。以三烷基胺、烷基吡啶、烷基喹啉或咪唑啉为代表。其与巯基羧酸铵盐混合，能进一步提升缓蚀性能。油溶性二/三聚酸基成膜型缓蚀剂是通过功能化的 C_{18} 脂肪酸（含有一个或两个双键，如油酸和亚油酸）的缩聚产生，得到不同数量的 C_{36}（二聚）和 C_{54}（三聚）脂肪酸。这些二聚体或三聚体脂肪酸用适当的胺中和可得到缓蚀剂。一项相关的专利称，用适当的胺中和马来酸酐或富马酸与妥尔油脂肪酸的反应产物，可以产生油溶性缓蚀剂。二聚体/三聚体混合物通常由脂肪酸咪唑和某些油共同配制。有报道称，用氨基乙基乙醇胺、咪唑啉或酰胺所中和的 C_{22} 三羧酸形成一种水溶性的成膜型缓蚀剂。有专利提到一种或多种脂肪酸（或二聚体/三聚体酸）、烷醇胺、烷基胺和有机磺酸复配的成膜型缓蚀剂。成膜性优良、持久的成膜型缓蚀剂可通过以下方法生成：首先由多元酸或马来酸化的脂肪酸与多元醇反应，形成羧酸偏酯，然后与咪唑啉或脂肪二胺反应，使其成盐。产物中添加助表面活性剂，可调整其水溶性或油溶性。环境友好的胺盐成膜型缓蚀剂，可以通过脂肪酸酸酐和 C_{21} 二元酸的一组酸与烷基胺或咪唑啉反应而制成。该产品的配方是将缓蚀剂溶解在脂肪酸油或酯中，加入由磺酸盐和长链乙氧基化醇组成的水分散剂，并用异丙醇等醇调节黏度。

8.4.4.3 季铵盐和亚氨基盐及两性离子

季铵盐表面活性剂作为成膜型缓蚀剂很少单独使用，一般与其他成膜型缓蚀剂混合使用。其有毒性，也用作杀菌剂，防止形成生物膜及垢下腐蚀。阴离子表面活性剂脂肪酸羧酸盐随着烷基羧酸盐中的烷基链增加，抑菌毒性降低。典型的表面活性剂包括苯扎氯铵，如脂肪烷基苄基二甲基氯化铵、烷基吡啶季铵、乙氧基化季铵和具有两个长烷基链的表面活性剂，如双十二烷基二甲基氯化铵（图 8.4）。二癸基二甲基季铵盐据称是首选的季铵盐表面活性剂成膜型缓蚀剂，可单独或与其他成分（如膦酸酯和硫代碳酰化合物）混合用于井下。烷基二甲基萘酮盐也是商品化的成膜型缓蚀剂，双季铵表面活性剂及基于乙二胺的胺类氧化物等也可作为缓蚀剂使用。季铵盐表面活性剂的生物降解源于长烷基链和季铵盐氮原子之间弱连接，最常见的弱连接是可以被水解的酯基。较小的、表面活性较低的季铵化合物降解缓慢。季铵盐表面活性剂抗聚集剂（AAs）的低加量水合物抑制剂也具有成膜性能。因此，连续加注 AAs 时，无须使用其他缓蚀剂（见第 9 章）。一些季铵型 AAs 可通过添加含硫增效剂（如巯基乙酸或 2-巯基乙醇）来增强其缓蚀性能。

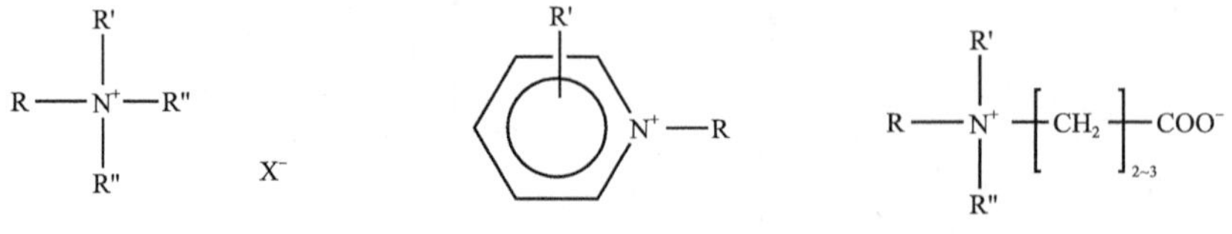

图 8.4 季铵盐、烷基吡啶季铵盐和两性离子甜菜碱成膜型缓蚀剂

有两性离子的甜菜碱类也可以作为成膜型缓蚀剂，且毒性较季铵盐表面活性剂低（咪唑啉两性离子将在后面讨论）。两性的甜菜碱还含有一个季铵中心，但作为羧基的反离子，以共价键链接在表面活性剂上。长链烷基丙二胺成膜型缓蚀剂通过迈克尔加成，与至少 1mol 丙烯酸反应，表现出低于原始二胺的海洋生物毒性。在水溶液中其可能也是两性的。这些产品的毒性一般随着丙烯酸取代的增加而降低。

两性水溶性亚胺化合物也是成膜型缓蚀剂（图 8.5）。常见的是亚胺与丙烯酸反应的产物，季亚胺化合物也有报道。二胺（如多烷氧基化二胺及其季铵化衍生物）作为成膜型缓蚀剂，尤其耐硫化氢腐蚀。在某些情况下，某些双季铵盐表面活性剂比单季铵盐表面活性剂的性能更好。这些双季铵盐表面活性剂可以通过环氧卤化物与 2mol 的叔胺（如二甲基十二烷基胺）反应制成（其中 1mol 叔胺可以用酸质子化）。油溶性聚亚烷基多胺的季铵盐或烷氧基化衍生物显示出良好的耐 CO_2 腐蚀作用。通过二醇和二胺或通过烷醇胺可获得聚亚烷基多胺的新催化路线（图 8.6）。这些聚合物可用作各种缓蚀剂的中间体。

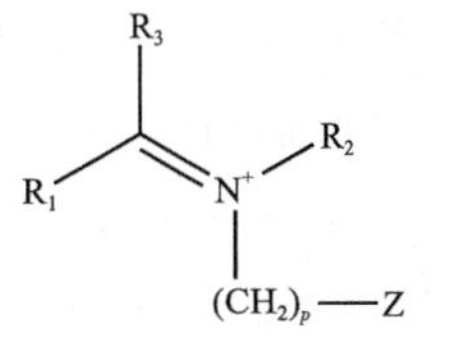

图 8.5　两性离子型水溶性亚胺化合物（Z 是羧酸盐或其他各种阴离子基团，R_1、R_2 或 R_3 中至少有一个是疏水性基团）

聚环氧氯丙烷与叔胺反应制成的可生物降解聚季铵盐杀菌剂也可作为成膜型缓蚀剂使用。聚环氧氯丙烷是由环氧氯丙烷在单体多元醇（如甘油）存在下通过聚合反应制得。典型的胺是十二烷基二甲胺或咪唑啉与 4mol 环氧乙烷或烷基吡啶缩合成胺。聚环氧氯丙烷的噻嗪类季铵盐也是一种成膜型缓蚀剂。

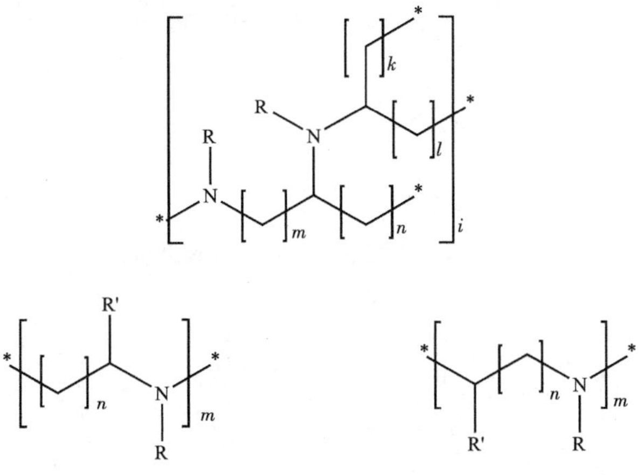

图 8.6　聚亚烷基多胺缓蚀剂的中间体

氮原子上连接有四个烷基或芳香基的季铵盐表面活性剂的生物降解性很差。若疏水端没有被降解，则其急性毒性将存在。通过在疏水尾基引入如酯基或酰氨基的弱连接，可以制成生物降解性更好的季铵盐成膜型缓蚀剂。这类产物已用作新型表面活性剂型的织物护理剂，在成膜型缓蚀剂中的应用也越来越普遍。双 N- 烷氧基化和羰基化的铵盐是性能良好的成膜型缓蚀剂，具有良好的生物降解性。在一个优选实例中，长链烷基胺或二胺经过双烷氧基化，与氯乙酸进行酯化反应，最后季铵化得到相应缓蚀剂（图 8.7）。对聚烷氧基化三醇胺进行酯化及季铵化，能形成具有良好生物降解性、有效的成膜型缓蚀剂（图 8.8）。与有更多单酯成分的表面活性剂相比，含有较高比例的二酯、三酯且无单酯的表面活性剂缓蚀性能较差。少量硫代硫酸钠的加入，会显著提升缓蚀性能。吡啶类和喹啉类化合物的单酯和双酯衍生物也是环境友好的成膜型缓蚀剂。甲硝唑衍生化制成的季铵盐表面活性剂型杀菌剂兼具良好的缓蚀性能（图 8.9）。低聚的聚季铵酯缓蚀剂也兼具杀菌功能，显示出良好的生物降解性，并且可以制成油溶性或水溶性药剂。由双羧酸（或酸酐）、烷醇胺和脂肪酸制成的化合物的分子结构更优，如图 8.10 所示。二烯丙基 −1，12− 二氨基十二烷基 − 环状聚合物的季铵化产物是生物降解性相对较差的成膜型缓蚀剂，如由 N，$N-$ 二烯丙基 $-N-$ 丙炔基 $-$（12$-N'-$ 甲酰氨基）$-1-$ 十二烷基胺制成的环状聚合物。

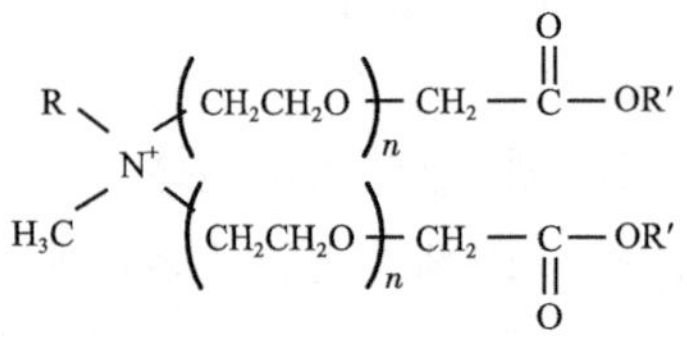

图 8.7　双 $N-$ 烷氧基化和羧基化的铵盐 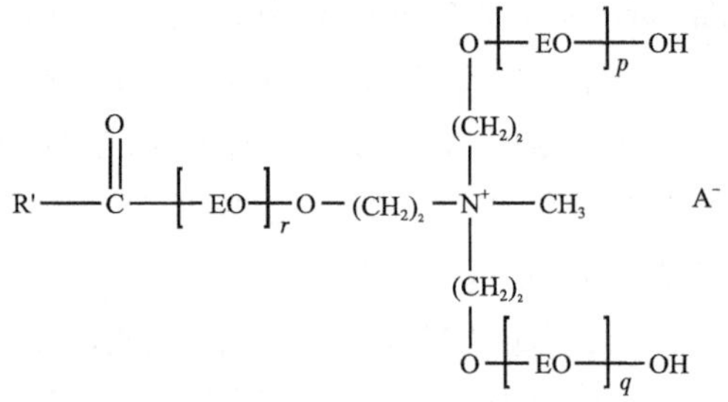　　　　图 8.8　聚氧乙烯三烷醇胺季铵盐

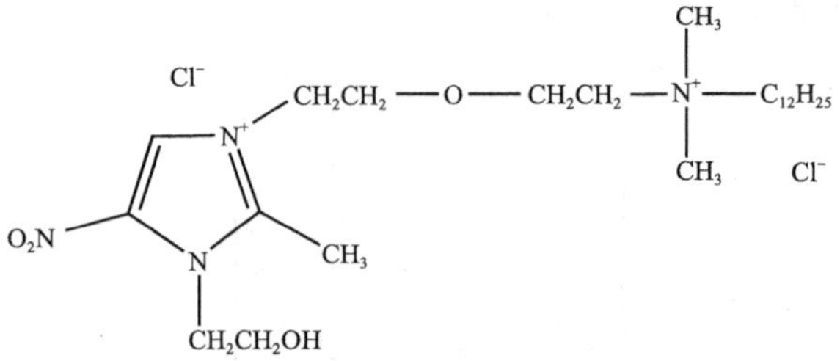

图 8.9　基于杀菌剂甲硝唑的成膜型缓蚀剂

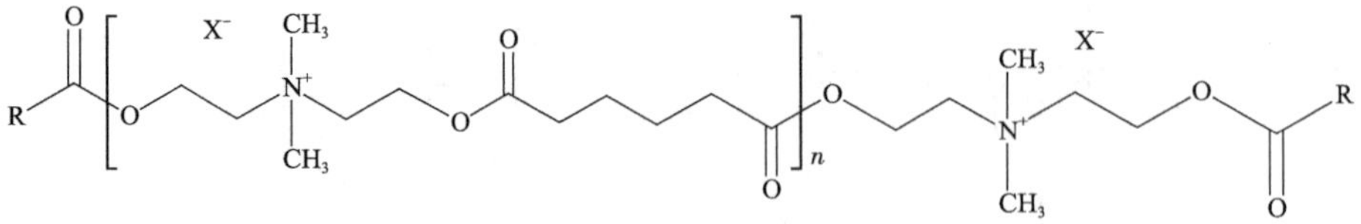

图 8.10　低聚酯类季铵盐表面活性剂

8.4.4.4　酰氨基胺类和咪唑类化合物

咪唑类可能是石油和天然气行业中最常见的，也是研究最多的一类成膜型缓蚀剂。某些咪唑啉基成膜型缓蚀剂甚至在高压高温下性能良好。虽然基础咪唑类化合物在高温高压条件下只具有较差到中等的性能，但改性后的成膜型缓蚀剂也可满足高温高压条件。咪唑

啉类化合物是由含有 1, 2- 二氨基乙烷官能团的多胺（如二乙烯三胺）与羧酸缩合而成：首先形成酰胺，保留产物是 2- 烷基咪唑啉（图 8.11）。如果 1, 2- 二氨基乙烷上的一个氮原子发生取代，就会形成 N- 取代的 2- 烷基咪唑啉产物。用 2mol 的羧酸，就会形成酰氨基咪唑啉，这也是知名的成膜型缓蚀剂（图 8.12）。制备咪唑啉时常采用脂肪胺，也可以使用醚类羧酸。非环状酰氨基胺是反应的副产品，通常仍存在于咪唑啉基成膜型缓蚀剂中。有研究称制备了一种水溶性的脂肪酸/胺类缓蚀剂。如果多胺分子足够大，也可以形成双咪唑类，一种由分散剂、咪唑啉或双咪唑啉、酰胺、烷基吡啶和重芳香烃溶剂组成的缓蚀剂配方已获得专利（图 8.13）。咪唑啉产品的质子化或烷基化可形成亲水更好的咪唑啉盐缓蚀剂（图 8.14）。加注后咪唑啉也可能与产出水中的酸反应而发生质子化。二聚体或三聚体脂肪酸也可用于制备咪唑啉。使用过量多胺制备出的咪唑啉产品，性能据说比单体的咪唑啉更好。

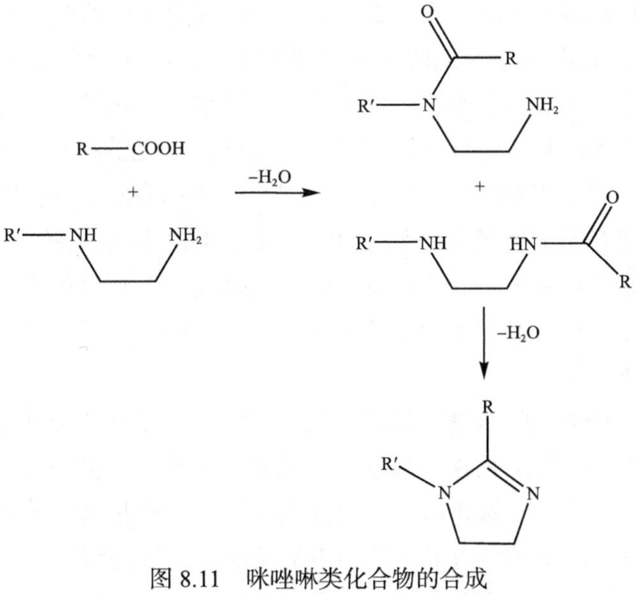

图 8.11　咪唑啉类化合物的合成

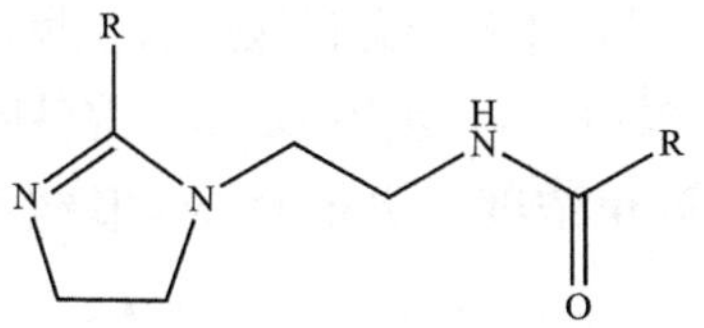

图 8.12　氨乙基咪唑啉

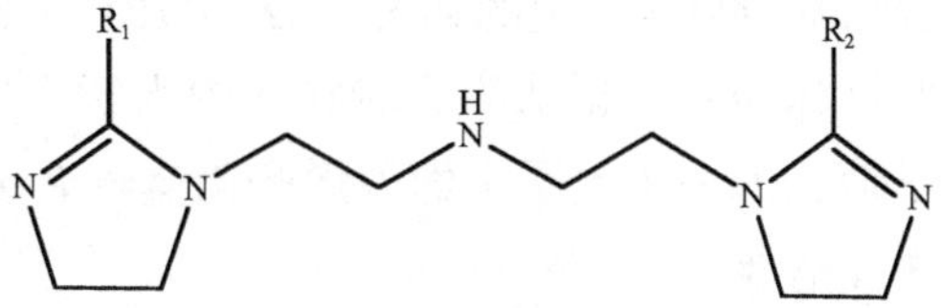

图 8.13　双咪唑啉结构

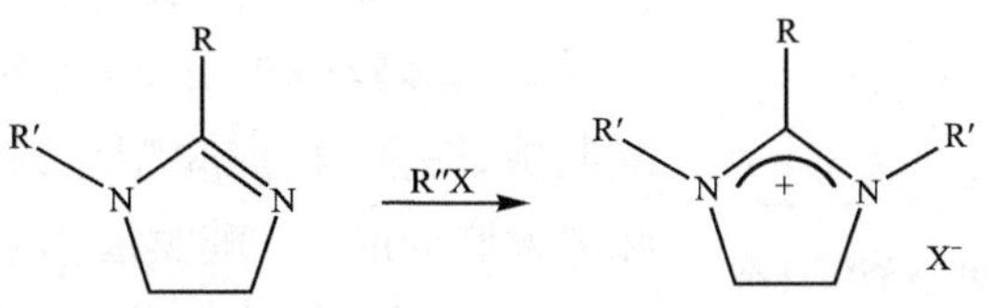

图 8.14　咪唑啉的质子化（R″=H）或烷基化（R= 烷基）

四氢嘧啶、咪唑啉的六环类似物及其羟甲基衍生物也是有效的成膜型缓蚀剂。通过二胺与二聚酸反应制成的双酰胺也是一种成膜型缓蚀剂，如2mol $N-$ 油基 -1，$3-$ 丙二胺和1mol 二聚酸的反应产物。

利用缓蚀剂测试、表面二次谐波分析和分子建模技术对油酸咪唑啉进行了研究。结果表明，该分子主要通过五元氮环进行键合，而氮环平铺在金属表面；长烃链在缓蚀机制中起着重要作用；改变侧基链的化学性质并不会显著影响该分子的性能。相反，密度泛函理论和蒙特卡洛模拟表明，咪唑啉类化合物倾向于在金属表面垂直吸附，而其质子化（或烷基化）产物在金属表面的平行位置吸附。理论和电化学试验都表明，$N-$ 取代的 $2-$ 烷基咪唑啉类化合物似乎比未取代的 $2-$ 甲基咪唑啉类化合物性能更好。已经开发了一种用于预测咪唑啉性能的 QSAR 方法。另一项使用 RCE 的研究发现，咪唑啉的最小有效浓度和其烃链长度之间存在线性关系。疏水链长度小于 12 个碳时，没有观察到缓蚀效果。咪唑啉分子的疏水基团对其缓蚀作用有显著影响，这与其吸附胶束的双层内聚能有关。

一项研究表明，常规咪唑啉成膜型缓蚀剂的疏水端链长为 C_{18} 时可以提供最佳的性能，但环境性能最差。乙氧基化咪唑啉是咪唑啉成膜型缓蚀剂的一种常见亚类，其分子的环或侧链胺中的氮被不同数量的环氧乙烷所乙氧基化，以提供具有较低生物积累潜力和低毒性的更好水溶性产品。咪唑啉最好由脂肪酸与 2，$2-$ 氨基乙胺乙醇或二乙基四胺反应制成，已经通过乙氧基咪唑啉的制备验证了该方法。随着脂肪酸链中碳原子数量的减少，通过与润湿剂（最好是具有 $8\sim10$ 个碳原子的乙氧基化醇）结合使用，缓蚀剂的功效会增加。咪唑啉碳酸氢盐为具有低环境影响的可转换表面活性剂，碳酸氢盐阴离子的稳定性取决于存在的 CO_2 浓度。

水溶性烷氧基咪唑啉成膜型缓蚀剂可与低聚膦酸酯成膜型缓蚀剂（乙氧基化多元醇的膦酸酯）发挥协同作用，特别是在低加量时。烷氧基咪唑啉（可能由于膦酸酯的质子化作用而成为烷氧基咪唑啉离子）保护电化学电池的阴极，而低聚膦酸酯则保护阳极，这种双重机制确保即使在其中一个成分可能失效的极端情况下，混合缓蚀剂仍能继续提供保护。

使咪唑啉成膜型缓蚀剂具有更易溶于水且更低毒性的另一种方法是使咪唑啉中间体的侧烷氨基与化学计量的如丙烯酸（$CH_2=CH_2COOH$）（图 8.15）等有机羧酸发生反应。产物是两性咪唑啉，据认为可能在水中水解产生酰胺。据报道，$1-$ 羧甲基 $-3-$ 烷基咪唑两性离子盐比少一个碳原子的丙烯酸衍生物有更低的临界胶束浓度，更长的长烷基链可以提高其表面活性。

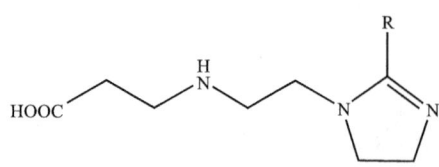

图 8.15 氨基咪唑啉与 1mol 丙烯酸反应举例

季铵化酰氨基咪唑啉类、季铵化亚氨基咪唑啉类和季铵化取代的二乙氨基咪唑啉类（如季铵化二丙烯氨基咪唑啉类），均由胺取代的咪唑啉类和丙烯酸制成，是一种成膜型缓蚀剂。有文章报道了一种咪唑啉膦酸酯/二酯类成膜型缓蚀剂，其在水中以两性离子的形式存在。胺、酰氨基胺和与羧基（如丙烯酸）反应得到的咪唑啉衍生物，与具有2～6个碳原子的巯基羧酸[HS(CH$_2$)$_n$COOH]复配，增效、低毒。例如，阿拉斯加北坡油田的措施作业用的就是巯基乙酸（TGA，HSCH$_2$COOH）复配咪唑啉类化合物。也有用巯基酸（也称巯基或硫代羧酸）和聚（乙烯氨基）咪唑啉盐替代的。氨基咪唑啉混合物也可以与巯基羧酸（如TGA）反应（而不仅仅是在室温下简单混合）来生产有效的成膜型缓蚀剂。有研究表明，使咪唑类或胺类和酰氨基胺类的所有活性氮原子（与氢原子连接的氮原子）与丙烯酸和氯乙酸反应，用碱将pH值调整到8～9，就可以制成低毒的成膜型缓蚀剂。咪唑啉或酰胺中的侧胺与SO$_2$反应，可得到含有—NSO$_2$基的更易溶于水的产品。

许多基于咪唑啉的成膜型缓蚀剂在环状结构中的一个氮原子上有侧基（图8.11中的R'）。因为制备咪唑啉时，常采用侧基含有氮原子的多胺（如二亚乙基三胺或三亚乙基四胺）。这些侧基中的氮原子（或其他具有非键合孤对电子的杂原子）与金属表面相互作用，增强了吸附力。随后发现，上述侧基并非缓蚀性能绝佳的必要条件。有报道称，两性离子丙烯酸咪唑啉在咪唑啉环的3号位含有未取代烷基，没有杂原子或可用的未成键电子，依然展现出良好的缓蚀效果（图8.16）。如（N-丙基-2-十七烯基）咪唑啉丙烯酸盐，以咪唑鎓的形式存在。

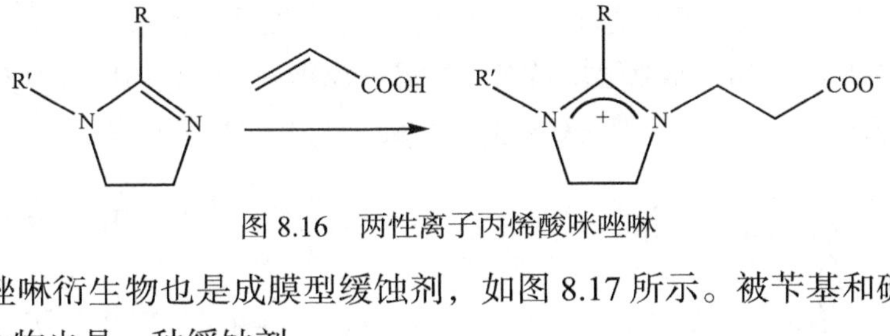

图 8.16 两性离子丙烯酸咪唑啉

含硫的咪唑啉衍生物也是成膜型缓蚀剂，如图8.17所示。被苄基和硫脲基团取代的油基咪唑类化合物也是一种缓蚀剂。

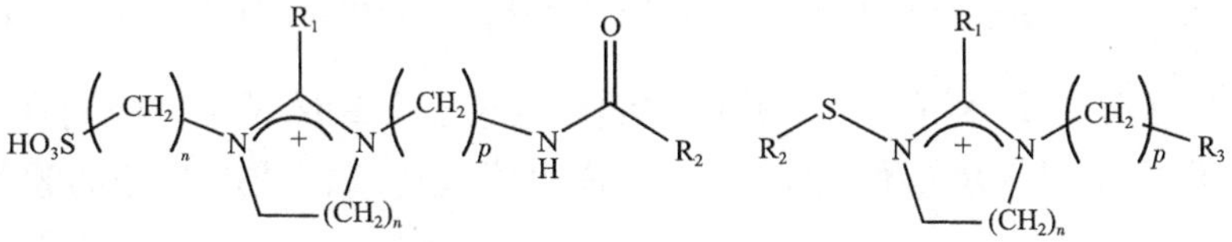

图 8.17 含硫的咪唑啉衍生物

咪唑啉类缓蚀剂通常显示出较高的急性毒性。然而发现C$_5$—C$_{21}$烷基羟乙基咪唑啉化合物的开环衍生物的急性毒性比传统的咪唑啉缓蚀剂小。特别有用的一组化合物通常被称为两性乙酸盐、烷基酰氨基氨基甘氨酸盐或两性羧基甘氨酸盐。另外两组特别有意义的相关化合物为二醋酸盐和两性磺酸盐，其与季铵盐的混合物据称是特别有效的缓蚀剂。

8.4.4.5 酰胺类

长链胺的酰胺衍生物可作为环境可接受的成膜型缓蚀剂，但其配制困难，并对油水分离过程有不利影响。如前所述，酰氨基胺是作为聚乙烯胺与羧酸反应的副产品存在，咪唑啉是主要产品。聚亚甲基多胺二丙酰胺被认为是低海洋毒性的 CO_2 缓蚀剂，是由丙烯酰胺与二乙烯三胺、三乙烯四胺或双（丙胺）乙二胺等多胺缩合而成，不含疏水基团，如图 8.18 所示，其与如 TGA 等巯基酸的混合，可发挥协同作用提高缓蚀性能。

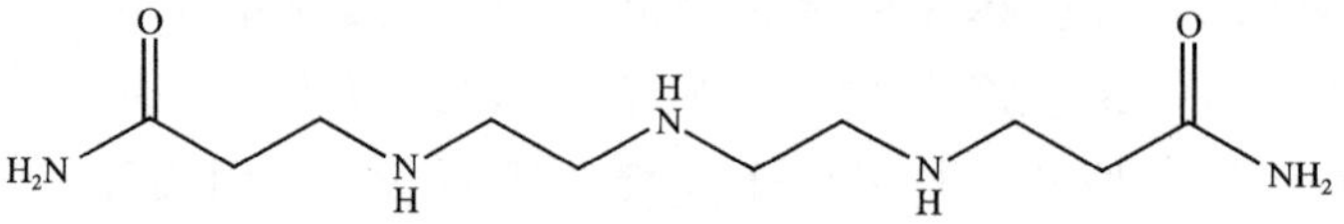

图 8.18 聚亚甲基多胺二丙酸酰胺成膜型缓蚀剂

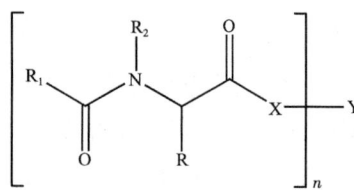

图 8.19 氨基酸的酰化衍生物通过连接 X 基团，附加到适当的 Y 主干

具有较高生物降解性的酰胺成膜型缓蚀剂是在骨架上附加了氨基酸的酰基化衍生物（图 8.19）。例如，$N-$癸酰$-L-$天冬氨酸可以（通过酸酐）与多元醇、多胺和羟胺反应，形成有效的成膜型缓蚀剂。

8.4.4.6 多羟基和乙氧基胺/酰胺

通常作为成膜型缓蚀剂配方组分的乙氧基化脂肪胺，可以通过脂肪胺或二胺用环氧乙烷进行乙氧基化制得（图 8.20）。在分子链的疏水性端基和乙氧基化氮原子之间，接入酰氨基等可生物降解基团，满足环保要求。例如，成膜型缓蚀剂牛脂—$CONH(CH_2)_3NHCH_2CH_2OH$ 的性能，就优于 3mol 环氧乙烷与牛脂—$NH(CH_2)_3NH_2$ 的乙氧基化反应产物。

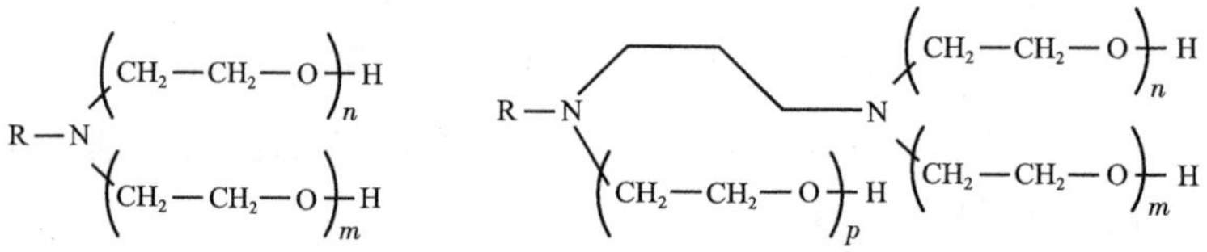

图 8.20 乙氧基化的脂肪胺和二胺成膜型缓蚀剂 [R 可以是烷基或 $R'C(=O)$]

乙氧基化使胺分子更易溶于水，并提供额外的位点（氧原子）以吸附在金属表面。另一种以氧原子官能团引入水溶性的方法是在胺中接入羟基，如烷基胺的脱氧氨基衍生物（图 8.21）。

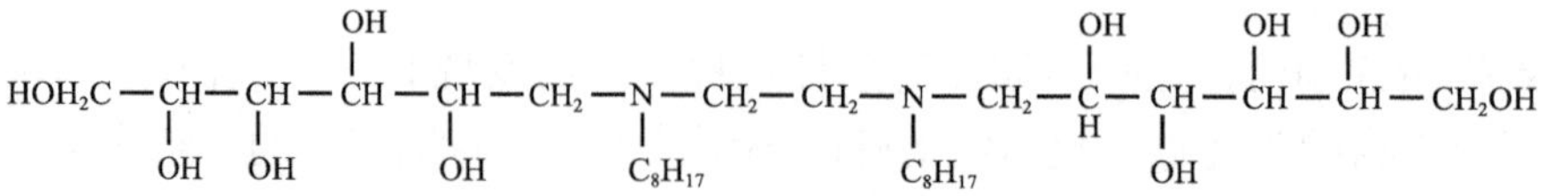

图 8.21 $N,N'-$二辛基$-N,N'-$双（1-脱氧葡萄糖基）乙二胺

8.4.4.7 其他氮杂环化合物

甲基取代的含氮芳香杂环化合物（如 2,5-二甲基吡嗪、2,3,5,6-四甲基吡嗪或 2,4,6-三甲基吡啶），与醛（如 1-十二烷醛）或酮（如 5,7-二甲基-3,5,9-癸三烯-2-

酮）反应，已用于制备油气井缓蚀剂。同样的含氮芳香杂环化合物与二羧酸—酐（如2-十二烯-1-基琥珀酸酐）反应产物也有相同的用途。

井下成膜型缓蚀剂可以通过主要是杂环化合物（由不饱和醛和多胺，如丙烯醛和乙二胺制成）的混合含氮物与羧酸、有机卤化物或环氧化物反应而制成。

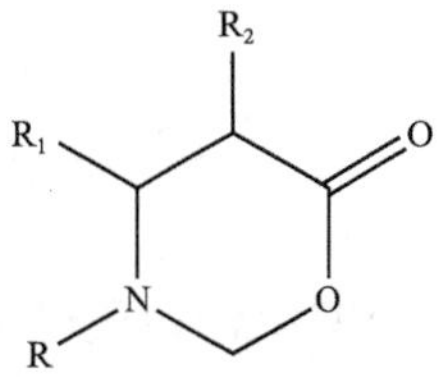

碱性化合物，包括 N-烷基或 N-烯基或 N-环烷基或 N-芳香基丙酸的1,3-恶嗪类衍生物和多聚甲醛，可作为成膜型缓蚀剂（图8.22），也具有沥青质抑制剂的功能。典型的化合物是由脂肪胺、丙烯酸和多聚甲醛合成制得。

图8.22 基于1,3-氧化氮-6-1取代衍生物的缓蚀剂

还有报道称，从六氢三嗪的脱硫剂废液中分离出了高效的缓蚀剂。

8.4.4.8 含硫化合物

前文已经提到硫代硫酸根离子和巯基羧酸是含氮成膜型缓蚀剂的增效剂。事实上，一些含硫化合物在防止开裂腐蚀方面的效果特别好。含有水溶性巯基羧酸的成膜型缓蚀剂（如毒性相对较低的TGA）已成功地用于高剪切应力环境，但如果单独使用，这些物质在二氧化碳饱和的介质环境中部分有效（图8.23）。巯基化合物的缓蚀机理已证明是巯基（—SH）被氧化成二硫化物（—S—S—），在金属表面与铁离子形成复合物。因此，尽管二硫化物3,3′-二硫代二丙酸（DTDPA）是一种氧化剂，而TGA是一种还原剂，但DTDPA应对均匀腐蚀的性能与TGA同样好（图8.23）。事实上，DTDPA预计比TGA控制局部腐蚀的效果更好。此外，巯基醇（MA）对均匀腐蚀和局部/点腐蚀的抑制作用都比TGA或DTDPA好。MA主要是一种阴极缓蚀剂，而TGA是阳极缓蚀剂。硫代羰基化合物（如二甲基二硫代氨基甲酸钾）也是成膜型缓蚀剂的增效剂（图8.23）。

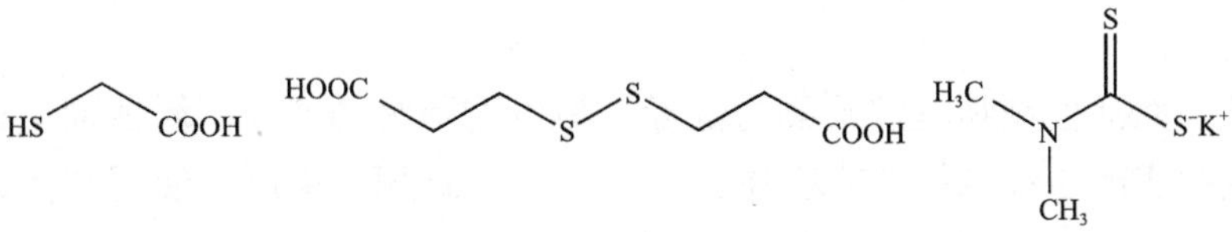

图8.23 巯基乙酸、3,3′-二硫代二丙酸和二甲基二硫代氨基甲酸钾

很少有专利声称将没有其他杂原子的硫化合物作为成膜型缓蚀剂。但有三个硫原子的三硫酮和适当的分散剂被认为是含 CO_2 井的良好缓蚀剂（图8.19）。三硫酮（如4-新戊基-5-叔丁基-1,2-二硫烯-3-硫酮的季铵盐）尤其适用于应力开裂失效的环境。在少量含氧系统的成膜型缓蚀剂配方中，三硫酮与硫代磷酸盐、季铵盐、聚磷酸酯和环状脒（如咪唑啉的脂肪羧酸盐）混合使用。硫代磷酸盐通过烷氧基化的脂肪醇与 P_2S_5 反应制得。其他声称仅含硫的成膜型缓蚀剂包括具有一个或多个—SH基团的硫醇基产物和选择性的含硫基团如2-巯基乙基硫醚（图8.24）。有文章介绍了有较高抗剪切作用的含硫缓蚀剂，还有助于提高水质。

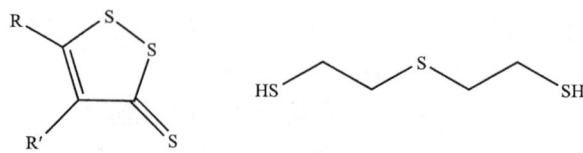

图 8.24　三硫酮和 2- 巯基乙基硫醚的结构

硫磷化合物还可以抑制炼化高温环境下的有机酸与环烷酸的腐蚀。环烷酸腐蚀缓蚀剂包括磷酸和某些用氮取代的无硫和无磷的芳香族化合物，其中在 5- 位或 3- 位含有官能团。

硫氮基缓蚀剂常用于黑色金属。其中大多没有疏水端，难以在金属表面吸附液态烃类形成油膜。例如，天然氨基酸中的半胱氨酸和胱氨酸及其脱羧反应得到的半胱胺和胱胺，是聚氨基酸成膜型缓蚀剂的良好增效剂。硫氮化合物（如苯并噻唑）更常用于铜等金属的缓蚀。烷氧基化巯基苯并噻唑的醚羧酸可抑制黑色金属的腐蚀（图 8.25）。此外用含有羰基（如单醛）、胺（如烷基单胺）和硫氰酸盐（如硫氰酸铵）化合而形成的硫氮化合物也是一种成膜性缓蚀剂。

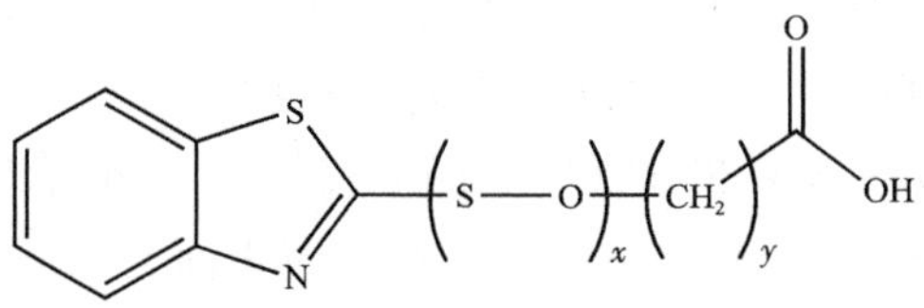

图 8.25　基于烷氧基化巯基苯并噻唑的醚羧酸，可以使用酸或酸的胺盐

硫代羧酸因气味难闻，研究人员一直在寻找替代品。2，5- 二氢噻唑类是用于气井的挥发性缓蚀剂（图 8.26），但其水溶性有限。噻唑烷类化合物是水溶性更强的成膜型缓蚀剂（图 8.21），其合成方法是将二氢噻唑［如 2，5- 二氢 -5，5- 二甲基 -（1- 甲基乙基）噻唑］与甲酸和醛的混合物反应。优选的噻唑烷产品没有大分子疏水基团，因此不是典型的表面活性剂。

小分子非表面活性剂咪唑烷硫酮为低毒、易生物降解的成膜型缓蚀剂。如由硫脲和二乙烯三胺反应制成的 1-（2- 氨基乙基）-2- 咪唑烷硫酮，也可以使用其他聚亚烷基多胺（图 8.27）。

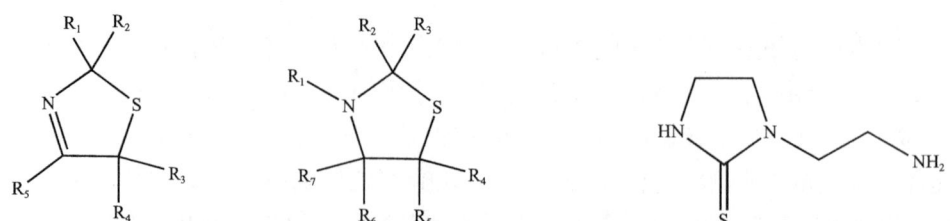

图 8.26　2，5- 二氢噻唑类和噻唑烷类的结构图　　图 8.27　1-（2- 氨基乙基）-2- 咪唑烷硫酮

据称，存在十二烷基硫醇（$C_{12}H_{25}SH$）等硫醇链转移剂时，通过聚合乙烯单体（如烯酸及其酯类，如 N- 羟乙基丙烯酸酯）制备的低分子量多官能聚合物对井下 CO_2 腐蚀的缓

蚀效果特别好。

酰氨基蛋氨酸衍生物和胺的盐类据称是生物降解性更强和毒性更低的成膜型缓蚀剂（图 8.28）。酰氨基蛋氨酸的实例是椰油酰基或辛酰基蛋氨酸。其中适合的胺包括吗啉、三乙醇胺和二丁胺。蛋氨酸表面活性剂与两性表面活性剂、阴离子表面活性剂、阳离子表面活性剂和非离子表面活性剂的混合体也是一种成膜型缓蚀剂。

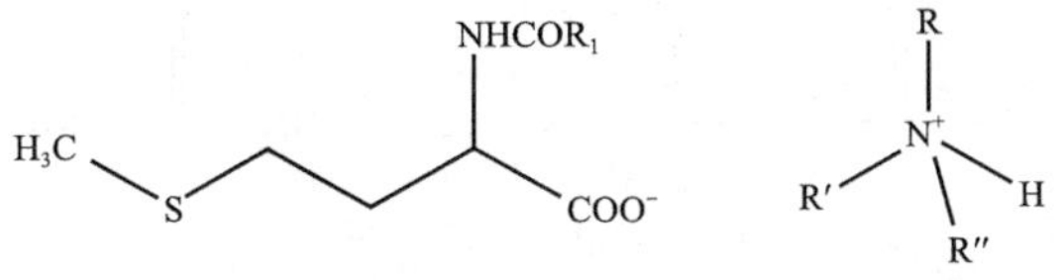

图 8.28　酰氨基蛋氨酸衍生物和胺的盐类

8.4.4.9　聚氨基酸和其他聚合物水溶性缓蚀剂

磷酸盐和有机膦酸盐｛如 2-羟基-2-磷酰基乙酸［$(HO)_2POCH(OH)COOH$］｝长期以来作为防垢剂，也经常用于水处理系统的腐蚀保护（见第 3 章）。但其在石油和天然气生产的苛刻环境中，通常缓蚀效果有限。烷基膦酸与四氢乙基磷盐的混合物也是一种成膜型缓蚀剂。2-膦酸-1,2,4-三羧酸丁烷（一种已知的碳酸盐防垢剂）与 β-环糊精的包合物据称对钢铁腐蚀有良好的缓蚀效率。

某些取代的羧基甲氧基琥珀酸化合物、氨基羟基琥珀酸或酒石酸的低聚物和聚合物（也被称为聚环氧琥珀酸）等有机羧酸和聚羧酸也能提供一些保护。这些有机聚羧酸中的许多种类也起到防垢剂作用。

随后的研究发现带有羧酸侧基的聚氨基酸作为防垢剂的效果良好，还能起到防 CO_2 腐蚀的作用，尽管不能和最好的成膜型缓蚀剂媲美。最知名且便宜的聚氨基酸是聚天冬氨酸盐（图 8.29），也可以使用谷氨酸。聚天冬氨酸在低氯、低 pH 值条件下单独使用时，表现出极好的防垢、缓蚀能力，但该条件在石油生产中比较少见。还可以通过添加氨基硫醇或氨基二硫化合物，来提高聚天冬氨酸的性能。天然氨基酸—半胱氨酸和胱氨酸及其脱羧类似物半胱胺和胱胺与聚天冬氨酸的组合也是特别有效的缓蚀剂，并且环保性好、低毒、生物降解性良好。另有缓蚀剂专利声称使用一种由基本上水溶的酸性氨基酸聚合物和至少一种水溶性钼盐（钼酸盐）或锌（Ⅱ）盐。含天冬氨酸的聚合物也可以与烷基聚葡萄糖苷协同组合为一类新的绿色缓蚀剂。基于聚天冬氨酸，已开发出添加增效剂的缓蚀和阻垢复合剂，并在北海油田应用。对 30 种实验室合成的多肽的研究得出的结论是，没有一种多肽比聚天冬氨酸类型的商品聚氨基酸更好。另外，有序聚合物比无序聚合物的缓蚀作用更好，而且在聚合物分子量小于 1000 时，其性能会大幅下降。此外，还指出有更好的市售低毒性非肽类的成膜型缓蚀剂。

聚天冬氨酸是通过水解不溶于水的聚琥珀酰亚胺而制成。环保型聚琥珀酰亚胺的酰胺衍生物，以及带有 C_{12}—C_{18} 尾基的乳酸酰胺的酰胺衍生物，已被证明具有良好的酸性气体缓蚀剂功能。聚琥珀酰亚胺也可以与羟基胺（如乙醇胺）或羟胺反应，得到分别带有羟乙基侧基或异羟肟基侧基的聚合物［R—C（═O）NHOH］。尽管 2-羟乙基天冬酰胺或 2-羟乙基谷酰胺的均聚物不可生物降解，但其中一些聚合物是可生物降解的防垢剂和缓蚀剂。其他含异羟肟酸基侧基的水溶性聚合物的生物降解性稍差，也可作为缓蚀剂使用，如聚丙烯酰胺与羟胺的反应产物。

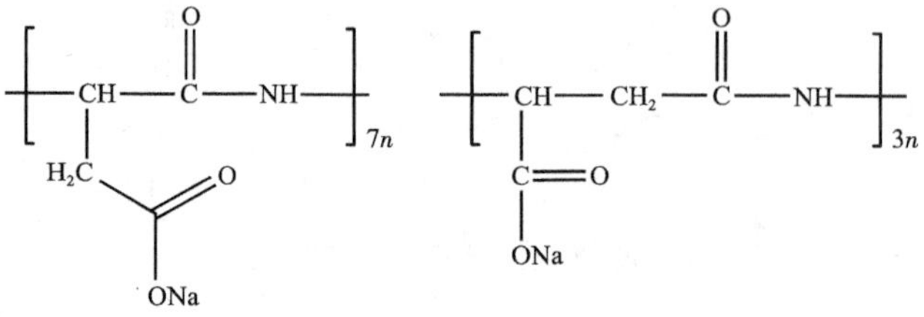

图 8.29　聚天冬氨酸钠

有已授权的专利提出以聚合物作为成膜型缓蚀剂和其他几种生产化学剂应用。这些乙烯基聚合物在为不同应用而设计的侧链中可以包含多种功能基团，如聚羧酸酯或氨基烷基膦酸酯、烷醇氨基或烷基氨基乙氧基、脂肪季铵基团、杂环，如咪唑类或由三烷基六氢三嗪分子组成的基团。

工业废物也可作为低成本缓蚀剂的来源，如烟头提取物对酸性介质中的 N80 钢有一定的缓蚀作用。

参 考 文 献

［1］(a) A. W. Peabody, *Control of Pipeline Corrosion*, 2nd ed., ed. R. L. Bianchetti. Houston: NACE International, 2001.（b）B. Craig, *Oilfield Metallurgy and Corrosion*, 3rd ed. Houston: MetCorr （NACE），2004.

［2］H. H. Uhlig and R. W. Revie, *Corrosion and Corrosion Control*, 3rd ed. New York: Wiley-Interscience, 1985.

［3］(a) G. V. Chilingar, R. Mourhatch, and G. Al-Qahtani, *The Fundamentals of Corrosion and Scaling: A Handbook for Petroleum and Environmental Engineers*, Houston: Gulf Publishing Company, 2008.（b）M. Davies and P. J. B. Scott, *Oilfield Water Technology*, Houston: National Association of Corrosion Engineers（NACE），2006.

［4］R. L. Martin, "Corrosion Consequences of Oxygen Entry into Sweet Oilfield Fluids," SPE 71470（paper presented at the SPE Annual Technical Conference and Exhibition, New Orleans, LA, 30 September-3 October 2001）.

［5］E. Dayalan, F. D. de Moraes, J. R. Shadley, S. A. Shirazi, and E. F. Rybicki, "CO_2 Corrosion Prediction in Pipeflow under $FeCO_3$ Scale-Forming Conditions," Paper 51（paper presented at the NACE CORROSION Conference, 1998）.

［6］R. Nyborg, "Controlling Internal Corrosion in Oil and Gas Pipelines. Business Briefing: Exploration and Production," *Oil and Gas Review* 2（2005）: 71.

［7］J. A Billman, "Antibiofoulants: A Practical Methodology for Control of Corrosion Caused by Sulfate-Reducing Bacteria," *Materials Performance* 36（1997）: 43.

[8] W. Sun and S. Nesic, "A Mechanistic Model of H_2S Corrosion of Mild Steel," Paper 07655 (paper presented at the NACE CORROSION Conference, 2007).

[9] J.-L. Crolet, "Microbial Corrosion in the Oil Industry: A Corrosionist's View," in *Petroleum Microbiology*, eds. B. Ollivier and M. Magot. Washington, DC: ASM Press, 2005, 143.

[10] D. Pope, *Microbiologically Influenced Corrosion in Pipelines*, Houston, TX: Gulf Publishing Co., 2000.

[11] S. W. Borenstein, *Microbiologically Influenced Corrosion Handbook*. New York: Woodhead, 1994.

[12] (*a*) R. Javaherdashti, *Microbiologically Influenced Corrosion: An Engineering Insight*, London: Springer, 2008. (*b*) B. J. Little and J. S. Lee, *Microbiologically Influenced Corrosion*, Hoboken, NJ: Wiley-Interscience, 2007.

[13] W. Sun and S. Nesic, "Basics Revisited: Kinetics of Iron Carbonate Scale Precipitation in CO_2 Corrosion," Paper 06365 (paper presented at the NACE CORROSION Conference, 2006).

[14] H. G. Byars, *Corrosion Control in Petroleum Production*, TCP 5, 2nd ed. Houston, TX: Forbes Custom Publication, 1999.

[15] P. R. Rincon, J. P. McKee, C. E. Tarazon, and L. A. Guevara, "Biocide Stimulation in Oilwells for Downhole Corrosion Control and Increasing Production," SPE 87562 (paper presented at the SPE International Symposium on Oilfield Corrosion, Aberdeen, UK, 28 May 2004).

[16] A. Dugstad and M. Seiersten, "pH-Stabilization, a Reliable Method for Corrosion Control of Wet Gas Pipelines," SPE 87560 (paper presented at the SPE International Symposium on Oilfield Corrosion, Aberdeen, UK, 28 May 2004).

[17] S. Ramachandran, S. Mancuso, K. A. Bartrip, and P. Hammonds, "Inhibition of Acid Gas Corrosion in Pipelines Using Glycol for Hydrate Inhibition," *Materials Performance* 45 (2006): 44-47.

[18] G. Schmitt, "Drag Reduction by Corrosion Inhibitors—A Neglected Option for Mitigation of Flow Induced Localized Corrosion," *Materials and Corrosion* 52 (5) (2001): 329.

[19] G. Schmitt, M. Bakalli, and M. Hörstemeier, "Contribution of Drag Reduction to the Performance of Corrosion Inhibitors in One- and Two-Phase Flow," Paper 615 (paper presented at the NACE CORROSION Conference, 2007).

[20] J. D. Johnson, S.-L. Fu, M. J. Bluth, and R. A. Marble, U.S. Patent 5939362, 1999.

[21] J. Palmer, W. Hedges, and J. Dawson, eds., *Working Party Report on the Use of Corrosion Inhibitors in Oil and Gas Production*, EFC39 (European Federation of Corrosion), Maney, 2004.

[22] (*a*) V. S. Sastri, *Corrosion Inhibitors: Principles and Applications*. Chichester, UK: Wiley, 1998. (*b*) I. L. Rosenfeld, *Corrosion Inhibitors*. Moscow: M. Khimia Publisher, 1977.

[23] A. M. S. El Din and L. Wang, "Mechanism of Corrosion Inhibition by Sodium Molybdate," *Desalination* 107 (1996): 29.

[24] M. Saremi, C. Dehghanian, and M. Mohammadi Sabet, "The Effect of Molybdate Concentration and Hydrodynamic Effect on Mild Steel Corrosion Inhibition in Simulated Water Cooling," *Corrosion Science* 48 (2006): 1404.

[25] E. Sletfjerding, A. Gladsø, S. Elsborg, and H. Oskarsson, "Boosting the Heating Capacity of Oil-Production Bundles Using Drag-Reducing Surfactants," SPE 80238 (paper presented at the International Symposium on Oilfield Chemistry, Houston, TX, 5-7 February 2003).

[26] T. A. Ramanarayanan and H. L. Vedage, U.S. Patent 5279651, 1994.

[27] B. Boyle, "A Look at Developments in Vapor Phase Corrosion Inhibitors," *Metal Finishing* 102 (5) (2004): 37.

[28] R. L. Martin, Paper 337 (paper presented at the NACE CORROSION Conference, 1997).

[29] M. Kharshan and A. Furman, "Incorporating Vapor Corrosion Inhibitors (VCIs) in Oil and Gas Pipeline Additive Formulations," Paper 236 (paper presented at the NACE CORROSION Conference, 1998).

[30] S. E. Campbell, U.S. Patent 7135440, 2006.

[31] S. J. Weghorn, C. W. Reese, and B. Oliver, "Field Evaluation of an Encapsulated Time Release Corrosion Inhibitor," Paper 07321 (paper presented at the NACE CORROSION Conference, 2007).

[32] S. Kokal, K. Raju, and A. Biedermann, "Cost Effective Design of Corrosion Inhibitor Squeeze Treatments for Water Supply Wells," SPE 53143 (paper presented at the Middle East Oil Show and Conference, Bahrain, 20–23 February 1999).

[33] H. A. Nasr-El-Din, H. R. Rosser, and M. S. Al-Jawfi, "Formation Damage Resulting from Biocide/Corrosion Inhibitor Squeeze Treatments," SPE 58803 (paper presented at the SPE International Symposium on Formation Damage Control, Lafayette, LA, 23–24 February 2000).

[34] E. C. French, W. F. Fahey, and J. G. Harte, U.S. Patent 5027901, 1991.

[35] R. L. Martin, J. P. Mullen, P. E. Brown, and T. G. Braga, U.S. Patent 5753596, 1998.

[36] (a) S. Ramachandran and K. Bartrip, "Molecular Modeling of Binary Corrosion Inhibitors," Paper 3624 (paper presented at the NACE CORROSION Conference, 2003). (b) A. Swift, A. J. Paul, and J. C. Vickerman, "Investigation of the Surface Activity of Corrosion Inhibitors by XPS and ToF SIMS," *Surface Interface Analysis* 20 (1) (1993): 27.

[37] S. S. Shah, T. G. Braga, B. A. O. Alink, and J. Mathew, U.S. Patent 5456767, 1995.

[38] H. Wang, H. Wang, H. Shi, C. Kang, and P. W. Jepson, "Why Corrosion Inhibitors Do Not Perform Well in Some Multiphase Conditions?: A Mechanistic Study," Paper 2276 (paper presented at the NACE CORROSION Conference, 2002).

[39] W. P. Singh, J. Ahmed, G. H. Lin, Y. Kang, and J. O'M Bockris, "About a Chemical Computational Approach to the Design of Green Inhibitors," Paper 33 (paper presented at the NACE CORROSION Conference, 1995).

[40] W. P. Singh, G. Lin, J. O'M. Bockris, and Y. Kang, "Designing Green Corrosion Inhibitors Using Chemical Computational Methods," Paper 208 (paper presented at the NACE CORROSION Conference, 1998).

[41] W. H. Durnie, "Modeling the Functional Behavior of Corrosion Inhibitors," Paper 4401 (paper presented at the NACE CORROSION Conference, 2004).

[42] W. H. Durnie, M. A. Gough, and J. A. M. de Reus, Paper 5290 (paper presented at the NACE CORROSION Conference, 2005).

[43] J. A. M. de Reus, E. L. J. A. Hendriksen, M. E. Wilms, Y. N. Al-Habsi, W. H. Durnie, and M. A. Gough, "Test Methodologies and Field Verification of Corrosion Inhibitors to Address under Deposit Corrosion in Oil and Gas Production Systems," Paper 5288 (paper presented at the NACE CORROSION Conference, 2005).

[44] A. E. Jenkins, W. Y. Mok, C. G. Gamble, and G. E. Dicken, "Development of Green Corrosion Inhibitors for Preventing under Deposit and Weld Corrosion," SPE 87558 (paper presented at the SPE International Symposium on Oilfield Corrosion, Aberdeen, UK, 28 May 2004).

[45] S. Ramachandran, Y. S. Ahn, K. A. Bartrip, V. Jovancicevic, and J. Bassett, "Further Advances in the Development of Erosion Corrosion Inhibitors," Paper 5292 (paper presented at the NACE CORROSION Conference, 2005).

[46] D. I. Horsup, T. S. Dunstan, and J. H. Clint, "Breakthrough Corrosion Inhibitor Technology for Heavily

Fouled Systems," *Corrosion* 65（8）（2009）: 527.

［47］M. Stern and A. L. Geary, "Electrochemical Polarization. A Theoretical Analysis of Shape of Polarization Curves," *Journal of Electrochemical Society* 104（1957）: 56.

［48］T. Hong, Y. H. Sun, and W. P. Jepson, "Study on Corrosion Inhibitor in Large Pipelines under Multiphase Flow Using EIS," *Corrosion Science* 44（2002）: 101.

［49］A. E. Jenkins, W. Y. Mok, C. G. Gamble, and S. R. Keenan, "Development of Green Corrosion Inhibitors for High Shear Applications," Paper 4370（paper presented at the NACE CORROSION Conference, 2004）.

［50］D. Abayarantha, A. Naraghi, and N. Grahmann, "Inhibitor Evaluations Using Various Corrosion Measurement Techniques in Laboratory Flow Loops," Paper 21（paper presented at the NACE CORROSION Conference, 2000）.

［51］M. Tandon, K. P. Roberts, J. R. Shadley, S. Ramachandran, E. F. Rybicki, and V. Jovancicevic, "Flow Loop Studies of Inhibition of Erosion-Corrosion in CO_2 Environments with Sand," Paper 6597（paper presented at the NACE CORROSION Conference, 2006）.

［52］S. Papavinasam, R. W. Revie, M. Attard, A. Demoz, and K. Michaelian, "Comparison of Laboratory Test Methodologies to Evaluate Corrosion Inhibitors for Oil and Gas Pipelines," *Corrosion* 59（10）（2003）: 897.

［53］S. Papavinasam, R. W. Revie, and M. Bartos, "Testing Methods and Standards for Oil Field Corrosion Inhibitors," Paper 4424（paper presented at the NACE CORROSION Conference, 2004）.

［54］S. Papavinasam, R. W. Revie, T. Panneerselvam, and M. Bartos, "Standards for Laboratory Evaluation of Oil Field Corrosion Inhibitors," *Materials Performance* 46（2007）: 46.

［55］(a) S. D. Kapusta, "Corrosion Inhibitor Testing and Selection for Exploration and Production: A User's Perspective," *Materials Performance* 38（1999）: 56.（b）S. Stewart, V. Jovancicevic, C. M. Menedez, and J. Maloney, "New Corrosion Inhibitor Evaluation Approach for Highly Sour Conditions," Paper 09360, NACE, Corrosion 2009 Conference and Exposition, Atlanta, GA, 22–26 March 2009.

［56］A. J. Son, "Developments in the Laboratory Evaluation of Corrosion Inhibitors: A Review," Paper 7618（paper presented at the NACE COROSION, 2007）.

［57］C. W. Bowman, W. Y. Mok, S. R. Keenan, C. G. Gamble, and S. Jarrett, "Environmental Constraints of Oil Soluble Corrosion Inhibitors: Challenges and Opportunities," Proceedings of Chemistry in the Oil Industry IX, Royal Society of Chemistry, 2005, 95.

［58］V. Jovancicevic, S. Ramachandran, and P. Prince, "Inhibition of CO_2 Corrosion of Mild Steel by Imidazolines and their Precursors," Paper 18（paper presented at the NACE CORROSION Conference, 1998）.

［59］T. G. Braga, R. L. Martin, J. A. McMahon, B. A. O. Alink, and B. T. Outlaw, U.S. Patent 6338819, 2002.

［60］A. H. Schroeder, T. A. Ching, S. Suzuki, and K. Katsumoto, International Patent Application WO/1988/005039.

［61］S. Taj, A. Sidekkha, S. Papavinasam, and E. W. Revie, "Some Natural Products as Green Corrosion Inhibitors," Paper 7630（paper presented at the NACE CORROSION Conference, 2007）.

［62］R. L. Martin, "Unusual Oilfield Corrosion Inhibitors," SPE 80219（paper presented at the International Symposium on Oilfield Chemistry, Houston, TX, 5–7 February 2003）.

［63］N. J. Phillips, J. P. Renwick, J. W. Palmer, and A. J. Swift, "The Synergistic Effect of Sodium

Thiosulfate on Corrosion Inhibition"（paper presented at the 7th International Symposium on Oil Field Chemicals, Geilo, Norway, 1996）.

[64] R. L. Martin, U.S. Patent 3959177, 1976.

[65] V. Jovancicevic, Y. S. Ahn, J. Dougherty, and B. Alink, "CO_2 Corrosion Inhibition by Sulfur-Containing Organic Compounds," Paper 7（paper presented at the NACE CORROSION Conference, 2000）.

[66]（a）A. Naraghi, U.S. Patent 5611991, 1997.（b）A. Naraghi and N. Grahmann, U.S. Patent 5611992, 1997.

[67] B. A. Alink, B. Outlaw, V. Jovancicevic, S. Ramachandran, and S. Campbell, "Mechanism of CO_2 Corrosion Inhibition by Phosphate Esters," Paper 99037, CORROSION 99, San Antonio, TX, 25-30 April 1999.

[68] B. Alin, B. Outlaw, V. Jovancicevic, S. Ramachandran, and S. Campbell, "Mechanism of CO_2 Corrosion Inhibition by Phosphate Esters," Paper 37（paper presented at the NACE CORROSION Conference, 1999）.

[69] H. Yu, J. H. Wu, H. R. Wang, J. T. Wang, and G. S. Huang, "Corrosion Inhibition of Mild Steel by Polyhydric Alcohol Phosphate Ester（PAPE）in Natural Sea Water," *Corrosion Engineering Science and Technology* 41（2006）: 259.

[70] T. J. Bellos, U.S. Patent 4311662, 1982.

[71] B. T. Outlaw, B. A. O. Alink, J. A. Kelley, and C. S. Claywell, U.S. Patent 4511480, 1985.

[72] R. L. Martin, G. F. Brock, and J. B. Dobbs, U.S. Patent 6866797, 2005.

[73] R. L. Martin, U.S. Patent 4722805, 1988.

[74] R. L. Martin, European Patent EP567212, 1993.

[75] A. Naraghi and P. Prince, International Patent Application WO/1997/008264.

[76] A. B. Gainer, U.S. Patent 4197091, 1980.

[77] D. E. Knox and E. R. Fischer, U.S. Patent 4927669, 1990.

[78] E. R. Fischer and P. G. Boyd, U.S. Patent 5759485, 1998.

[79] J. A. Alford, P. G. Boyd, and E. R. Fischer, U.S. Patent 5174913, 1992.

[80] E. R. Fischer, J. A. Alford, and P. G. Boyd, U.S. Patent 5292480, 1994.

[81] B. A. Miksic, A. Furman, and M. Kharshan, U.S. Patent 6800594, 2004.

[82] R. J. Goddard and M. E. Ford, U.S. Patent Application 0180794, 2006.

[83] D. Leinweber and M. Feustel, International Patent Application WO/2006/040013.

[84] A. Chalmers, I. G. Winning, D. McNaughtan, and S. McNeil, "Laboratory Development of a Corrosion Inhibitor for a North Sea Main Oil Line Offering Enhanced Environmental Properties and Weld Corrosion Protection," Paper 6487（paper presented at the NACE CORROSION Conference, 2006）.

[85] P. J. Clewlow, J. A. Haslegrave, N. Carruthers, D. S. Sullivan II, and B. Bourland, U.S. Patent 5427999, 1995.

[86] G. R. Meyer, U.S. Patent 6171521, 2001.

[87] K. M. Henry and K. D. Hicks, International Patent Application WO/2006/019585.

[88] K. M. Henry, R. Meyer, K. D. Hicks, and D. I. Horsup, "The Design and Synthesis of Improved Corrosion Inhibitors," Paper 5282（paper presented at the NACE CORROSION Conference, 2005）.

[89] A. W. Ho, International Patent Application WO/1993/007307.

[90] A. Naraghi and N. Obeyesekere, International Patent Application WO/2006/034101.

[91] P. M. Quinlan, U.S. Patent 4371497, 1983.

[92] (a) U. Dahlmann and M. Feustel, International Patent Application WO/2003/008668. (b) L. Tiwari, International Patent Application WO/2008/157234.

[93] J. Y. Huang, L. S. Zheng, C. Y. Fu, U. E. Qu, and J. G. Liu, "The Inhibition Effects of a New Heterocyclic Bisquaternary Ammonium Salt in Simulated Oilfield Water," *Anti-corrosion Methods and Materials* 51 (2004): 272.

[94] S. Ramachandran, Y. S. Ahm, M. Greaves, V. Jovancicevic, and J. Bassett, "Development of High Temperature, High Pressure Corrosion Inhibitors," Paper 6377 (paper presented at the NACE CORROSION Conference, 2006).

[95] H. Chen, T. Hong, and W. P. Jepson, "High Temperature Corrosion Inhibition Performance of Imidazoline and Amide," Paper 35 (paper presented at the NACE CORROSION Conference, 2000).

[96] N. Obeyesekre, A. Naraghi, L. Chen, S. Zhou, and S. Wang, "Novel Corrosion Inhibitors for High Temperature Applications," Paper 5636 (paper presented at the NACE CORROSION Conference, 2005).

[97] M. Feustel and P. Klug, U.S. Patent 6372918, 2002.

[98] S. Kanwar and P. Eaton, International Patent Application WO/1997/007176.

[99] L. Xiao, W. Qiao, H. Guo, and J. Qu, "Synthesis of an Imidazoline Phosphate Surfactant and Its Application on Corrosion Inhibition," *Tenside Surfactants Detergents* 5 (2008): 244.

[100] (a) B. A. O. Alink, U.S. Patent 4212843, 1980. (b) B. A. O. Alink and B. T. Outlaw, U.S. Patent 4343930, 1982.

[101] J. Levy, U.S. Patent 4344861, 1982.

[102] A. Edwards, C. Osborne, D. Klenerman, M. Joseph, O. Ostovar, and M. Doyle, "Mechanistic Studies of the Corrosion Inhibitor Oleic Imidazoline," *Corrosion Science* 36 (2) (1994): 315.

[103] D. Turcio-Ortega, T. Pandiyan, J. Cruz, and E. Garcia-Ochoa, "Interaction of Imidazoline Compounds with Fe_n ($n=1-4$ Atoms) as a Model for Corrosion Inhibition: DFT and Electrochemical Studies," *Journal of Physical Chemistry C* 111 (2007): 9853.

[104] Y. Duda, R. Govea-Rueda, M. Galicia, H. I. Beltrán, and L. S. Zamudio-Rivera, "Corrosion Inhibitors: Design, Performance, and Simulations," *Journal of Physical Chemistry B* 109 (47) (2005): 22674.

[105] W. H. Durnie and M. A. Gough, "Characterization, Isolation and Performance Characteristics of Imidazolines: Part II. Development of Structure-Activity Relationships," Paper 03336 (paper presented at the NACE CORROSION Conference, 2003).

[106] V. Jovancicevic, S. Ramachandran, and P. Prince, "Inhibition of Carbon Dioxide Corrosion of Mild Steel by Imidazolines and their Precursors," *Corrosion* 55 (5) (1999): 450.

[107] R. L. Martin, B. A. Alink, J. A. McMahon, and R. Weare, "Further Advances in the Development of Environmentally Acceptable Corrosion Inhibitors," Paper 98 (paper presented at the NACE CORROSION Conference, 1999).

[108] R. L. Martin, J. A. McMahon, and B. A. O. Alink, U.S. Patent 5393464, 1995.

[109] W. M. McGregor, "Novel Synergistic Water Soluble Corrosion Inhibitors," SPE 87570 (paper presented at the SPE International Symposium on Oilfield Corrosion, Aberdeen, UK, 28 May 2004).

[110] N. E. Byrne and J. D. Johnson, U.S. Patent 5322640, 1994.

[111] P. J. Clewlow, J. A. Haselgrave, N. Carruthers, and T. M. O'Brien, European Patent Application EP526251, 1994.

[112] G. R. Meyer, U.S. Patent 6303079, 2001.

[113] G. R. Meyer, U.S. Patent 6448411, 2002.

[114] G. R. Meyer, U.S. Patent 6599445, 2003.

[115] J. D. Watson and J. G. Garcia Jr., U.K. Patent Application GB2319530, 1997.

[116] B. A. M. O. Alink and B. T. Outlaw, U.S. Patent 6419857, 2002.

[117] T. E. Pou and S. Fouquay, International Patent Application WO/1998/041673.

[118] G. R. Meyer, U.S. Patent 6696572, 2004.

[119] A. Naraghi, H. Montgomerie, and N. U. Obeyesekere, European Patent EP1043423, 2000.

[120] A. Naraghi, U.S. Patent 6063334, 2000.

[121] G. R. Meyer, International Patent Application WO/2004/092447.

[122] T. Gu, Y. Hu, Y. Tang, Z. Liu, L. Huang, Z. Yang, Y. Huang, J. Wang, H. Chang, H. Yu, G. Li, and J. Cao, International Patent Application WO/2007/112620.

[123] D. Darling and R. Rakshpal, "Green Chemistry Applied to Scale Inhibitors," *Materials Performance* 37 (1998): 42.

[124] T. E. Pou and S. Fouquay, U.S. Patent 6365100, 2002.

[125] T. E. Pou and S. Fouquay, International Patent Application WO/1999/039025.

[126] R. M. Thompson, International Patent Application WO/1999/059958.

[127] K. Overkempe, E. Parr, W. John, and J. C. Speelman, International Patent Application WO/2003/054251.

[128] D. S. Treybig, U.S. Patent 4676834, 1987.

[129] (a) D. S. Treybig and J. L. Potter, U.S. Patent 4725373, 1988. (b) R. G. Martinez, D. S. Treybig and T. W. Glass, U.S. Patent 4762627, 1988.

[130] R. H. Hausler, B. A. Alink, M. E. Johns, and D. W. Stegmann, European Patent Application EP0275651, 1988.

[131] R. L. Martin and E. W. Purdy, U.S. Patent 4339349, 1982.

[132] M. D. Greaves, C. M. Menendez, and Q. Meng, International Patent Application WO/2008/091429.

[133] M. J. Zetlmeisl, U.S. Patent 5863415, 1999.

[134] G. Sartori, D. C. Dalrymple, S. C. Blum, L. M. Monette, M. S. Yeganeh, and A. Vogel, U.S. Patent 6706669, 2004.

[135] M. S. Yeganeh, S. M. Dougal, G. Sartori, D. C. Dalrymple, C. Zhang, S. C. Blum, and L. M. Monette, U.S. Patent 6593278, 2003.

[136] J. C. Fan, L.-D. G. Fan, and J. Mazo, International Patent Application WO/2000/075399.

[137] U. Dahlmann, M. Feustel, and R. Kupfer, International Patent Application WO/2002/092583.

[138] P. R. Petersen, L. G. Coker, and D. S. Sullivan III, U.S. Patent 4938925, 1990.

[139] B. A. M. O. Alink, R. L. Martin, J. A. Dougherty, and B. T. Outlaw, U.S. Patent 5197545, 1993.

[140] P. Prince, International Patent Application WO/1998/051902.

[141] Y. Wu and R. A. Gray, U.S. Patent 5135999, 1992.

[142] D. Leinweber and M. Feustel, International Patent Application WO/2007/087960.

[143] V. Jovancicevic and D. Hartwick, "Recent Developments in Environmentally-Safe Corrosion Inhibitors," Paper 226 (paper presented at the NACE CORROSION Conference, 1996).

[144] L. W. Jones, U.S. Patent 4554090, 1985.

[145] G. Woodward, G. P. Otter, K. P. Davis, and R. E. Talbot, Patent 6814885, 2004.

[146] K. P. Davis, G. P. Otter, and G. Woodward, International Patent Application WO/2001/057050.

[147] C. G. Carter and V. Jovancicevic, U.S. Patent 5135681, 1992.

[148] C. G. Carter, V. Jovancicevic, J. A. Hartman, and R. P. Kreh, U.S. Patent 5183590, 1992.

[149] C. G. Carter, L.-D. G. Fan, J. C. Fan, R. P. Kreh, and V. Jovancicevic, U.S. Patent 5344590, 1994.

[150] A. J. McMahon and D. Harrop, "Green Corrosion Inhibitors: An oil Company Perspective," Paper 32 (paper presented at the NACE CORROSION Conference, Orlando, 1995).

[151] D. I. Bain, G. Fan, J. Fan, and R. J. Ross, "Scale and Corrosion Inhibition by Thermal Polyaspartates," Paper 120 (paper presented at the NACE CORROSION Conference, 1999).

[152] W. J. Benton and L. P. Koskan, U.S. Patent 5607623, 1997.

[153] J. C. Fan and L.-D. G. Fan, U.S. Patent 6277302, 2001.

[154] H. A. Craddock, S. Caird, H. Wilkinson, and M. Guzmann, "A New Class of 'Green' Corrosion Inhibitors: Development and Application," SPE 104241 (paper presented at the SPE International Oilfield Corrosion Symposium, Aberdeen, UK, 30 May 2006).

[155] M. Guzmann, U. Ossmer, and H. Craddock, International Patent Application WO/2007/063069.

[156] N. Obeyesekre, A. Naraghi, and J. S. McMurray, "Synthesis and Evaluation of Biopolymers as Low Toxicity Corrosion Inhibitors for North Sea Oil Fields," Paper 1049 (paper presented at the NACE CORROSION Conference, 2001).

[157] G. Schmitt and A. O. Saleh, "Evaluation of Environmentally Friendly Corrosion Inhibitors for Sour Service," Paper 335 (paper presented at the NACE CORROSION Conference, 2000).

[158] J. Tang, R. T. Cunningham, and B. Yang, U.S. Patent 5750070, 1998.

[159] J. Tang, S.-L. Fu, and D. H. Emmons, U.S. Patent 6022401, 2000.

[160] D. W. Fong and B. S. Khambatta, U.S. Patent 5308498, 1994.

[161] C. Patton, "Corrosion," *SPE Reprint Series 46*. Society of Petroleum Engineers, 1997.

[162] X. Tang, J. F. Liu, and J. Moore, "Corrosion Inhibitors in the Presence of Elemental Sulfur," Paper 09363, CORROSION 2009, Atlanta, GA, 22-26 March 2009.

[163] A. Morshed, "Corrosion Management for Seawater Injection Systems," *Materials Performance* August (2009): 68.

[164] D. Jingen, Y. Wei, L. Xiaorong, and D. Xiaoqin, *Petroleum Science and Technology* 29 (2011): 1387-1396.

[165] A. Davoodi, M. Pakshir, M. Babaiee, and G. R. Ebrahimi, *Corrosion Science* 53 (1) (2011): 399-408.

[166] E. O. Obanijesu, Curtin University of Technology, "Hydrate Formation and its Influence on Natural Gas Pipeline Internal Corrosion Rate," 128544, Oil and Gas India Conference and Exhibition, Mumbai, India, 20-22 January 2010.

[167] R. Heidersbach, *Metallurgy and Corrosion Control in Oil and Gas Production*. Singapore: Wiley, 2011.

[168] L. A. Nolasco, A. MacDonald, and H. Wang, "Galvanic Corrosion Inhibition: Validation of Simple Laboratory Testing," Paper 09274, CORROSION 2009, Atlanta, GA, 22-26 March 2009.

[169] T. E. Pou and S. Boito, International Patent Application WO/2013/038100.

[170] S. Ramachandran, I. Ahmed, V. Jovancicevic, and M. Al-Waranbi, International Patent Application WO/2013/062952.

[171] J. J. Wylde, M. Reid, A. Kirkpatrick, N. Obeyesekere, and D. Glasgow, "When to Batch and When Not to Batch: An Overview of Integrity Management and Batch Corrosion Inhibitor Testing Methods and Application Strategies," SPE 164080 (paper presented at the SPE International Symposium on Oilfield Chemistry, The Woodlands, TX, 8-10 April 2013).

[172] C. Mackenzie, C. Rowley-Williams, A.-M. Fuller, C. L. H. Wilson, D. Blumer, and M. H. Achour,

"Introducing a New Approach to Aid Optimisation of Corrosion Inhibitor Dosing," *RSC Chemicals in the Oil Industry*, Manchester, November 2011.

[173] C. D. Mackenzie, V. Magdalenic, E. Perfect, M. Achour, D. J. Blumer, M. W. Joosten, and M. Rowe, "Development of a New Corrosion Management Tool—Inhibitor Micelle Presence as an Indicator of Optimum Dose," SPE 130285 (paper presented at the SPE International Conference on Oilfield Corrosion, Aberdeen, UK, 24-25 May 2010).

[174] T. L. Skovhus, J. Larsen, M. Jensen, K. Sorensen, and K. Rasmussen, "Rapid Determination of MIC in Oil Production Facilities with a DNA-based Diagnostic Kit," SPE 130744 (paper presented at the SPE International Conference on Oilfield Corrosion, Aberdeen, UK, 24-25 May 2010).

[175] J. E. Wong and N. Park, Paper 09569, CORROSION 2009, Atlanta, GA, 22-26 March 2009.

[176] D. Liu, Y. B. Qiu, Y. Tomoe, K. Bando, and X. P. Guo, *Materials and Corrosion-Werkstoffe und Korrosion* 62 (12) (2011): 1153-1158.

[177] M. Foss, E. Gulbrandsen, and J. Sjoblom, "Effect of Corrosion Inhibitors and Oil on Carbon Dioxide Corrosion and Wetting of Carbon Steel with Ferrous Carbonate Deposits," *Corrosion* 65 (1) (2009): 3.

[178] M. Pähler, J. J. Santana, W. Schuhmann, and R. M. Souto, *Chemistry: A European Journal* 17 (2011): 905.

[179] R. Martin, "Control of Top-of-Line Corrosion in a Sour Gas Gathering Pipeline with Corrosion Inhibitors," Paper 09288, CORROSION 2009, Atlanta, GA, 22-26 March 2009.

[180] E. E. Oguzie, Y. Li, S. G. Wang, and F. H. Wang, *RSC Advances* 1 (5) (2011): 866-873.

[181] L. Morello, N. Park, and G. M. Abriam, "Understanding Inhibition of Sour Systems with Water Soluble Corrosion Inhibitors," Paper 09362, CORROSION 2009, Atlanta, GA, 22-26 March 2009.

[182] J. Moloney, S. Stewart, C. M. Menendez, and V. Jovancicevic, "New Corrosion Inhibitor Evaluation Approach for Highly Sour Service Conditions," Paper 09360, CORROSION 2009, Atlanta, GA, 22-26 March 2009.

[183] B. Miksic, M. Kharshan, and A. Furman, "Effectiveness of Corrosion Inhibitors for the Petroleum Industry Under Various Flow Conditions," Paper 09573, CORROSION 2009, Atlanta, GA, 22-26 March 2009.

[184] S. Nesic, *Energy & Fuels* 26 (7) (2012): 4098-4111.

[185] S. A. Al-Jutaily, U.S. Patent Application 20110283783.

[186] J. Jensen, S. Juhler, T. Lundgaard, and K. Brensen, "New Test For Material Resistance against Microbiologically Influenced Corrosion," SPE 157388 (paper presented at the SPE International Production and Operations Conference & Exhibition, Doha, Qatar, 14-16 May 2012).

[187] J. Jackson, R. Harrington, and D. Manko, "Emulsion Tendency Studies—Understanding Method, Inhibitors and Water Cut," Paper 2012-1223, CORROSION 2012, Salt Lake City, UT, 11-15 March 2012.

[188] P. C. Okafor, C. B. Liu, Y. J. Zhu, and Y. G. Zheng, *Industrial & Engineering Chemistry Research* 50 (2011): 7273.

[189] Q. Ye and S. Yan, *Journal of Surfactants and Detergents* 13 (2010): 349.

[190] G. R. Meyer and K. A. Monk, U.S. Patent Application 20120149608.

[191] H. Yan, Q. Li, T. Geng, and Y. Jiang, *Journal of Surfactants and Detergents* 15 (2012): 593.

[192] E. J. Acosta, P. A. Webber, and K. A. Monk, U.S. Patent Application 20100219379.

[193] E. J. Acosta and J. C. Clark, U.S. Patent Application 20100084612.

[194] M. M. A. El-Sukkary, I. Aiad, A. Deeb, M. Y. El-Awady, H. M. Ahmed, and S. M. Shaban,

Petroleum Science and Technology 28（2010）: 1158.

[195] R. L. Hoppe, R. L. Martin, M. K. Pakulski, and T. D. Schaffer, U.S. Patent 7481276, 2009.

[196] M. Mahdavian, A. R. Tehrani-Bagha, and K. Holmberg, *Journal of Surfactants and Detergents* 14（2011）: 605.

[197] J. Strautmann, T. Schaub, S. Hüffer, S. Maas, and C. Wood, International Patent Application WO/2013/076025（see also WO/2013/076053 and WO/2013/076023）.

[198] A. Jenkins, N. Grainger, M. Blezard, and M. Pepin, International Patent Application WO/2010/128313.

[199] L. Tiwari, G. R. Meyer, and D. Horsup, International Patent Application WO/2009/076258.

[200] P.-E. Hellberg and N. Gorochovceva, International Patent Application WO/2011/000895.

[201] P.-E. Hellberg and N. Gorochovceva, International Patent Application WO/2012/028542.

[202] P. Hellberg, "Environmentally Acceptable Polymeric Corrosion Inhibitors," SPE 140780（paper presented at the SPE International Symposium on Oilfield Chemistry, The Woodlands, TX, 11–13 April 2011）.

[203] P.-E. Hellberg, "Polymeric Corrosion Inhibitors—A New Class of Versatile Oilfield Formulation Bases," *Chemistry in the Oil Industry XIII—New Frontiers*, Manchester, UK, November 2013.

[204] S. A. Ali, M. T. Saeed, and A. M. Z. El-Sharif, *Polymer Engineering & Science* 52（12）（2012）: 2588–2596.

[205] S. A. Ali, S. M. J. Zaidi, A. M. Z. El-Sharif, and A. A. Al-Taq, *Polymer Bulletin* 69（4）（2012）: 491–507.

[206] V. Jovancicevic, J. Long, and S. Ramachandran, "Development of a New Water Soluble High Temperature Corrosion Inhibitor," Paper 09237, CORROSION 2009, Atlanta, GA, 22–26 March 2009.

[207] J. Yang and V. Jovancicevic, U.S. Patent Application 20090181867.

[208] W. Qiao, Z. Zheng, and Q. Shi, *Petroleum Science* 9（1）（2012）: 75–81.

[209] B. Ni, J. Hu, X. Liu, H. Chen, and Y. Fang, *Journal of Surfactants and Detergents* 15（6）（2012）: 729–734.

[210] I. A. Aiad, A. A. Hafiz, M. Y. El-Awady, and A. O. Habib, *Journal of Surfactants and Detergents* 13（2010）: 247.

[211] P.-E. Hellberg, "Structure-Property Relationships for Novel Low-Alkoxylated Corrosion Inhibitors," *RSC Chemistry in the Oil Industry XI*, Manchester, UK, 2–4 November 2009.

[212] R. Hernandez Altamirano, V. Y. Mena Cervantes, L. S. Zamudio Rivera, H. I. Beltran Conde, and E. Buenrostro Gonzalez, U.S. Patent Application 20110269650.

[213] G. N. Taylor, "The Isolation and Formulation of Highly Effective Corrosion Inhibitors from the Waste Product of Hexahydrotriazine Based Hydrogen Sulphide Scavengers," *Chemistry in the Oil Industry XIII—New Frontiers*, Manchester, UK, November 2013.

[214] G. Taylor, "An Example of Chemical Recycling in the Oil and Gas Industry—A By-Product from Hydrogen Sulphide Scavenging is Identified, Isolated and Formulated into a Highly Effective Corrosion Inhibitor," SPE 140439（paper presented at the SPE International Symposium on Oilfield Chemistry, The Woodlands, TX, 11–13 April 2011）.

[215] S. Ramachandran, V. Jovancicevic, G. Williams, K. Smith, and C. McAfee, "Development of a New High Shear Corrosion Inhibitor with Beneficial Water Quality Attributes," Paper 10375, CORROSION 2010, San Antonio, TX, 14–18 March 2010.

[216] D. Leinweber and M. Feustel, U.S. Patent Applications 20090057615 to 20090057618.

[217] B. B. Karthik, P. Selvakumar, and C. Thangavelu, *Asian Journal of Chemistry* 24(8)(2012): 3303-3308.

[218] C. Jones and J. Hardy, U.S. Patent Application 20090170817.

[219] C. J. Zoua, Q. W. Tanga, P. W. Zhaoa, X. Wud, and H. Yee, *Journal of Petroleum Science and Engineering* 103(2013): 29-35.

[220] H. L. Becker, U.S. Patent 8114819, 2012.

[221] J. Zhao, N. Zhang, C. Qu, X. Wu, J. Zhang, and X. Zhang, "Impact of Molybdate and Nitrite Anions on the Corrosion of Mild Steel," *Industrial & Engineering Chemistry Research* 49(8)(2010): 3986-3991.

[222] R. A. Cottis, A. Al-Refaie, and R. Lindsay, "Impact of Molybdate and Nitrite Anions on the Corrosion of Mild Steel," Paper 09282, CORROSION 2009, Atlanta, GA, 22-26 March 2009.

[223] Y. Duan, F. Yu, D. Zhao, X. Cui, and Z. Cui., *Petroleum Science and Technology* 31(19)(2013): 1959-1966.

[224] F. C. Yu, Y. F. Duan, S. Z. Peng, C. X. Li, and H. M. Wang, "Performance is good but not compared to other more well-known CIs." Chinese patent application CN 101280222A, 2008.

[225] A. F. Miles, N. Bretherton, K. Richterova, and A. Naragi, U.S. Patent Application US20130228095.

9 天然气水合物控制

9.1 概述

天然气水合物是由水和小分子烃类在高压和低温下形成的冰状固体。水合物形成的温度随着压力的增加而增加，可高达 25~30℃。天然气水合物的典型压力—温度条件如图 9.1 所示。天然气水合物最常见于海底或寒冷气候下的湿气或多相（油—水—气）管道，也可能在钻井、完井和修井作业过程及在天然气处理设施、注水或注气管线中形成。当压力—温度条件合适时，甚至还可以在注入水溶性化学剂的气举管线中形成天然气水合物。

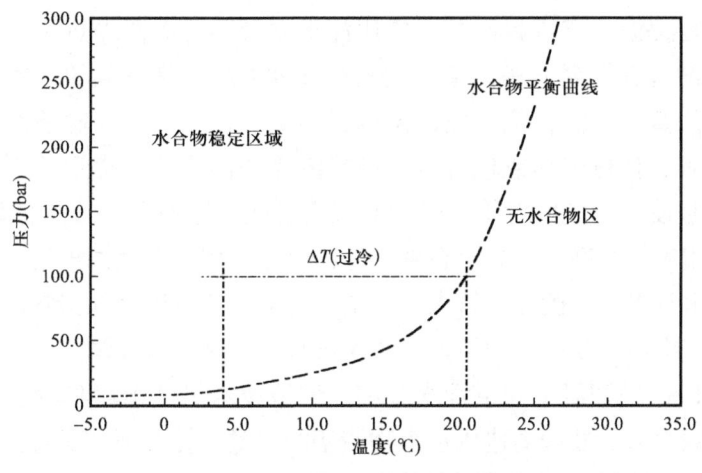

图 9.1 典型天然气水合物的压力—温度图

除由自由水形成的水合物外，在没有液态水相时，形成水合物（或冰）段塞还有以下两种情况：（1）当超过液态水饱和度曲线时，水的凝聚作用（如管道停输并冷却后，管道顶部的冷凝水）；（2）液态烃相中的溶解水。

许多输送多相流的管道在设计上不用担心形成水合物，但如果发生停产，液体未经处理就冷却到水合物形成的范围内，就会出现问题。另外，当海底油井关井时，若不采取必要的预防措施，也会在井口下方形成水合物段塞。深海油田开发中，集输管线的天然气水合物堵塞防治是需要处理的主要生产问题之一。

在天然气水合物中，水分子形成由氢键连接的开放式笼形结构。笼形结构被烃类等小分子占据，通过范德华相互作用力来稳定笼合物的结构。天然气水合物可以形成有不同压力—温度平衡曲线的三种结构：SⅠ型（结构Ⅰ）、SⅡ型（结构Ⅱ）及SH型。当天然气富含甲烷、几乎不含 C_3—C_4 烃类成分时，可形成 SⅠ型结构；SⅡ型结构是现场最常见的水合物结构，只要天然气中除甲烷外还含有一些丙烷或丁烷，结构就很稳定；SH 型结构在石油工业中很少遇见，由含甲烷的小笼形结构和含甲基环戊烷或苯等大分子烃类的大笼形结构结合，组成稳定结构。

防止天然气水合物的形成和沉积的方法包括以下几点：

（1）保持在水合物稳定区之外的低压区；

（2）在系统压力下，通过被动保温或主动加热，保持温度高于水合物平衡温度；

（3）分离除水（脱水）；

（4）用另一种气体调节气相组分；

（5）在不使用化学品的情况下，将水转换为可输送的水合物颗粒；

（6）化学处理。

连续保持低压的做法很少见，尽管在停产期间可以进行减压，但多数情况下，生产效率低且很不经济。深海油气田环境因为水的静压力始终保持流体处于水合物形成区域，无法有效减压。有若干种方法可以通过提高温度，避免管道内形成天然气水合物，最简单的是对管道进行保温。有资料回顾了应用深水立管和封隔器隔热保温液体，这也是目前许多深海项目的首选保温方法。管道埋地在某种程度上是有益的，也可以在管道外敷设保温材料或采用真空保温，但真空保温价格昂贵，且在长期停输条件下也会失效。另一种方法是用电阻、伴热电缆或包含可循环热流体的伴热管缠绕管道壁，给管道壁提供热量。还有方法是使用交流电对管道进行直接电加热，这种方法已在北海数个油田应用，但该方法连续使用时成本高昂，只有在停产停输时才应用。北海油田还使用了管束管道系统的方法，将热水从平台一侧注入管束的环空，使多相流体温度保持在水合物形成区域之外，可以根据生产中停产或投产的需要，注入热水以加热集输管线。

在北海某油田已经开展了井下油水分离，但该技术仍处于起步阶段。油水分离技术上的主要优势是减少需要处理的造成水合物堵塞的水量。另一种"反直觉"的方法是注入 N_2 或 CO_2 等气体，以提高水合物形成的压力阈值。

一种可能代表深海水合物控制重大突破的技术被称为"冷流"（"CONWHYP"或"Hydraflow"），是在不使用化学品的情况下将产出水转化为可输送的水合物颗粒，但不是所有油田都适用。"冷流"现在特指在使用或不使用化学品的情况下，通过长输管线输送水合物浆液。

非化学技术的基础是使用含有水合物颗粒的再循环流。这些颗粒被送入快速冷却（冲击冷却）的含水产出流体，其中的水被混合器分散或乳化在液态烃中。水合物颗粒晶种促进水滴中水合物由内而外生长，并以可控方式快速生长。干燥的水合物颗粒不沉积、不结块，并将消除集输系统其他部分的自由水。该技术似乎在高过冷度下效果最好。据称也有一种不需要回收水合物的冷流方法。冷流的可能问题是自由水（在到达混合器之前）是否在非计划停产期间结成团块。在湿气或高含水气田，因为水合物颗粒无法流动，该技术不适用。此外，作为天然气水合物的自由水被清除后，溶液中的离子浓度更高，需要解决可能增加的无机垢沉积问题，有运营商在单向管道中使用冷流方法的现场试验表明，只有很有限的作业区域没有堵塞。这些结果表明，当前的水合物冷流技术需要解决水合物在管壁的生长和沉积问题。

对于非常高的含水率，有人建议无论是否有液态烃存在，额外注入水或盐水，以使气体完全转化为水合物，并使水合物浆液在过剩的水中运移。此时可能需要配套特定类型的水合物抗凝结剂（AAs）。也有人提出用同样的冲击冷却原理来避免蜡的沉积（见第10章）。

AAs的商业化应用、非堵塞油（两者将在后面讨论）及对冷流方法的持续研究等都使业界更多地思考"天然气水合物管理"，而非只是预防，即如何能与天然气水合物共存并避免其堵塞管线。水合物颗粒流动或堵塞风险评估模型、水合物安全边界监测及早期水合物检测系统都作为水合物管理工艺包的组成部分。还有专利提出采用光滑、憎水的管道内涂层方法（包括原油中添加特定表面活性剂制成的涂层），以降低水合物黏附和冻堵风险。有人提出将烃类产出流体经过换热器冷却并形成固体，以确保海底管道生产的系统方案。固体沉积物可通过管道的闭环清管器收发系统定期清除。

9.2 水合物堵塞的化学预防

水合物堵塞的化学预防可以通过热力学水合物抑制剂（THIs）、动力学水合物抑制剂（KHIs）、抗凝结剂（AAs）三种类型的化学剂来实现。

后两类被统称为低加量水合物抑制剂（LDHIs），与THIs相比的加量要求低得多［通常小于1%（质量分数）的活性成分］。有石油公司使用阈值水合物抑制剂（缩写为THIs）来表述KHIs。在本书中，THIs特指热力学水合物抑制剂。

尽管在环境可接受性和性能方面仍需改进，目前LDHIs是一项成熟技术。与使用THIs或其他水合物控制策略相比，LDHIs比THIs更昂贵，但可以节省大量的投资和运营成本。一些油田也在使用KHIs/THIs混合物。有专利提出了一种从THIs过渡到KHIs的改进方法。

9.2.1 热力学水合物抑制剂（THIs）

THIs有时被称为水合物防冻剂，是目前最常采用的类型。其作用是改变流体系统的整体热力学特性，从而将天然气水合物形成的平衡条件移到较低温度或较高压力。因此，THIs既可以防止水合物的形成，也可以"融化"已沉积水合物。THIs的添加浓度非常高，有时可以达到两倍的水体积量。

最常用的THIs是醇类、二醇类和盐类。甲醇（CH_3OH）和乙二醇（MEG，$HOCH_2CH_2OH$）广泛用于防止生产、修井和工艺操作中的水合物，并用于熔化水合物段塞。二甘醇（DEG）和三甘醇（TEG）有时也用于水合物防治，但作用不大。三甘醇主要用于吸附天然气管线或处理设施中的水分，起到干燥气体的作用。在南美洲，乙醇因为成本低，而且可从糖发酵中大量获得，通常被用作THIs，北海油田现在也有使用。但乙醇加量超过约5.6%（摩尔分数）后，会与甲烷形成二元水合物，并使其抑制水合物能力显著降低，无法达到融冰抑制的预期。甘油和其他来自生物柴油废液处理的极性化合物也可作为THIs。

虽然甲醇和二醇类等THIs的成本相对低，但因加量大，其回收及再利用往往是经济上划算的。目前广泛使用乙二醇再生设施，但甲醇再生相对少。因为需要处理大量的水，甲醇和二醇类很少在油田中连续使用，但二醇类在凝析油和气田中可以特别地连续使用（MEG最常见）。正确掌握THIs的加量很重要，因为加量不能完全防止水合物形成时，会增大堵塞的风险。

除醇类和二醇类以外，防止水合物形成的常见化学品还有氯化钠、氯化钙、甲酸钾和乙酸钠等盐类。这些盐类通常用于抑制钻井液的水合物形成，有时还与二醇类结合使用。最近还有使用硝酸盐和磷酸盐的盐类（如磷酸氢二钾）。卤化物盐类不太适合注入生产管线，因为其需要高浓度加注，会与产出水不配伍并增大腐蚀风险。还有人建议使用低腐蚀性的甲酸钾盐。据资料称，乙酸钠的腐蚀性更低。

其他被作为THIs研究的化学品包括二甲基甲酰胺和N-甲基吡咯烷酮、乙醇胺，以及除乙醇外与水形成共沸物的其他醇类（如异丙醇）等的水溶性溶剂。这些化学品都比甲醇和乙二醇更昂贵，热力学抑制效果也较差，目前没有在本领域使用。

THIs的性能通常表示为在一定压力和抑制剂浓度下水合物平衡曲线的温度变化（抑制）。例如，20%（质量分数）的甲醇水溶液可使水合物平衡温度降低约10℃，而需要32%（质量分数）的MEG才能达到同样的温度抑制。计算避免水合物形成所需的THIs量的粗略指南，最先由Hammerschmidt制定，其公式为

$$\Delta T = \frac{Ks}{M(100-s)} \tag{9.1}$$

式中 ΔT——水合物抑制值，℉；

K——常数，取决于抑制剂的性质和水合物的形成热，此处为Hammerschmidt的华氏常数，2335；

s——水相中THIs的质量分数，%；

M——抑制剂的分子量。

抑制剂的分子量对其性能至关重要，通常抑制剂分子量越低，性能就越高。这就是甲醇比乙二醇效果更好，而乙二醇又比二甘醇更好的原因。同理，尽管THIs密度会影响性能，但甲醇融化水合物段塞的能力应该比乙二醇更强。上述Hammerschmidt公式没有考虑THIs在气相或液相之间的分布及系统的压力，在抑制剂浓度较高的情况下，结果准确性较差。因为甲醇在气态和液态烃相中都有明显的分布，比二醇类的分布要大得多，用该公式反而会得出特别差的效果。乙二醇几乎只分散到水相中，在较高气油比而含水较低的多相

集输系统中，二醇类通常比甲醇效果更好。据报道，有工具可以预测甲醇在油相中的损失。

还有方法通过在线测量电导率和声速确定 THIs 的浓度。一种在线检测器已经安装在海上油田。另一种方法是使用包括传感元件的插入式探针，可以更有效地给气井或油井提供水合物抑制剂的剂量。插入式探针可以首先检测井内多相流中的水，测量井内水中的抑制剂含量，确定准确的含水率，并测量矿化度等其他参数。MEG 的再生也比较简单。在实践中，通常会以比要求高一些的剂量注入 THIs，以绝对确保不会形成任何水合物。已经声称有一种方法可以从 THIs（如 MEG）中去除造成问题的有机羧酸盐。

有研究者开发了更精确的公式，以计算热力学抑制剂提供的过冷度。目前最简单的方法是使用 PVT 模型，该模型考虑到抑制剂在各相间的分布。一些公司、研究机构和大学提供了附加水合物计算模块的 PVT 模型商业软件。但这些模型有时也会给出不充分的结果，特别是在高压、高抑制剂浓度或含有高浓度盐水溶液的情况下，但也在努力减少这些问题。预测水合物的数值模型和经验关联应精确考虑电解质和有机抑制剂之间的相互作用，以准确预测水合物的抑制效果。表 9.1 使用两个商业软件生成的数据，给出了部分 THIs 性能的粗略指南。该表说明了水相中，THIs 不同浓度下的过冷度值，即没有考虑气体或液体烃相中的损失。其中的 HCOOK（甲酸钾）相关数据源自文献。

如果操作正确，确定形成水合物的平衡点［无论是否添加 THIs（但不包括 LDHIs）］的最佳方法是实验测试。具体是通过解离预先制备的水合物，并确定其形成温度（因为水合物的形成是随机过程，温度值存在波动，不能通过冷却来测得水合物形成的起始温度）。这样一旦形成部分水合物，就对系统进行加热。加热至接近预期的平衡温度时，升温必须非常缓慢（最好小于 0.2℃/h），并且所有的相必须充分混合。当最后一个水合物颗粒解离，压力曲线返回到冷却时形成水合物前的曲线点，即认为是水合物形成的平衡点。逐步加热的方法相比于持续加热，结果更可靠，测量时间也更短。

表 9.1 水相中各种热力学抑制剂的计算过冷度值

质量分数（%）	计算过冷度（℃）						
	MeOH	EtOH	MEG	DEG	TEG	NaCl	HCOOK
5	2.0	1.4	1.05	0.63	0.46	1.96	
10	4.2	3.0	2.25	1.4	1.05	4.3	2.5
20	9.3	6.6	5.2	3.3	2.7	10.7	7.1
30	15.3	10.7	9.0	5.9	5.0	15.0	12.9
35	18.6	13.0	11.35	7.5	6.5	—	
40	22.2	15.4	14.0	9.3	8.2		

使用 THIs 时，要考虑很多操作问题。前面已强调了 THIs 的相间分布问题（THIs 在气体或液态烃相中的损失）及由此确定的加量比例问题。其他潜在的问题简述如下：

（1）抑制剂用量大，需要高额运费、大型储罐和注入管线等。

（2）甲醇的毒性和易燃性。

（3）醇类或二醇类再生设施的建设和维护成本。再生设施中常存在结垢、结盐和腐蚀等问题。

（4）烃类（气体或液体）对下游处理厂的污染问题。甲醇和MEG处理系统都有可能发生污染，如果THIs浓度超过规定，会影响烃类价值。

（5）THIs增加了形成垢和环烷酸盐的可能性。氯化钠可在高矿化度下沉积，增加碳酸盐垢或硫酸盐垢的可能性，超过初始状态所需的防垢剂最低有效浓度。甲醇和乙醇比二醇类的副作用更大。

（6）部分甲醇会分散到油相中，并起到降低析蜡点的蜡质沉淀剂作用。

（7）发生意外停产时，很难将抑制剂泵送到所需位置。泵和注入管线的容积可能不足以处理预期的大量水。在深水钻井或海底多相流集输期间，可能出现此问题。

（8）在非常寒冷的环境（天然气处理）和细长的低温注入管道中，特别是更高分子量的二醇类抑制剂的黏度变大。

（9）抑制剂在天然气处理系统中的冻结。

（10）甲醇和二醇类的生物/化学需氧量较高，受到排放许可的限制。

对于长距离的海底多相流集输，由于热力学抑制剂的加量大及储存、注入和再生设施的成本高，促使人们寻找其他方法来防治天然气水合物，这带动了20世纪90年代LDHIs的发展。有综述广泛总结了至2005年前的相关研究，后续还有关于KHIs的综述。

提高水合物抑制剂（THIs或LDHIs）加注泵效的润滑剂研究也有所报道，有C_1—C_{36}脂肪酸或脂肪酸的衍生物（如胺盐、酯、酰胺等）。有实例表明，几百毫克每升的油酸或油酸的$N,N-$二甲基环己胺盐就能产生很好的效果。

9.2.2　动力学水合物抑制剂（KHIs）

9.2.2.1　KHI及其机理介绍

所有已知的商业KHIs配方的主要成分都是水溶性聚合物，通常添加其他较小分子量的有机物作为增效剂。一般来说，这类聚合物在液态烃相的分散较少，但这些相却经常影响KHIs性能。有若干研究团队已经尝试研发有效的如烷基化柠檬酰胺等非聚合物类KHIs。其他非聚合物类产品还在开发中。一些亲水的AAs也具有KHIs特性。

动力学水合物抑制剂可延缓天然气水合物的成核及晶体的生长，其时间取决于系统的过冷度和相应的压力。只要有THIs充分降低过冷度，在76MPa的高压下，KHIs也能发挥作用。这样可以保障在天然气水合物形成和沉积之前，海底多相流集输管道系统可将产出流体输送到处理设施。因此，任何长时间停产对确定KHIs的现场适用性都至关重要。通常多数商用KHIs的现场应用仅限于生产管线中最大过冷度为9~10℃的SⅡ型天然气（SⅠ型天然气的过冷度略低），因为延迟水合物形成的时间通常以数天计。更高的过冷度（驱动力）会缩短该延迟时间。因此，KHIs不适用于多数过冷度和压力都很高的深海油田。

如果产出流体的停留时间较短，在过冷度高于 9～10℃ 的情况下，就可能使用 KHIs。已经研究了层流态下，管道顶部因凝结水，存在后续形成水合物的风险，而水相中的 KHIs 不能抑制管道顶部形成水合物。

自 1995 年以来，KHIs 已得到了商业化应用，其添加浓度很低，在水相中浓度低于 1%（质量分数），通常为 0.1%～0.3%（质量分数）（季节性的温度变化会改变过冷度，对应加量也随之变化）。与之相比的是甲醇或二醇类等 THIs 的加量为 20%～60%。在某个油田通过选择 KHIs 技术而非甲醇的注入和再生，节省了 4000 万美元的资本支出。但 KHIs 仍然是非常昂贵的油田化学剂，只是在低产水油田应用时比较经济。有文章报道了优化 KHIs 注入率的方法。

许多水溶性聚合物已证实可用作 KHIs。KHIs 聚合物有两个关键的结构特征。首先，聚合物需要有与水分子或天然气水合物颗粒表面氢键相连接的功能基团，通常是酰胺基团。其次，有与每个酰胺基团相邻或直接相连的憎水基团。例如，聚乙烯吡咯烷酮（PVP）（图 9.2）就是首个被发现的 KHIs。KHIs 的性能有时被指代为在特定压力下，多相流体可以在 48h 内（保持时间）流动不形成水合物的过冷度。在没有任何增效剂的情况下，7～8MPa 时的 PVP 过冷度只有 3～4℃。PVP 和其他改进型 KHIs 的工作机制仍存在争议。人们提出了不同条件下可能的两种主要机制。第一种机制是，KHIs 聚合物通过疏水相互作用，扰动水的结构，使水合物颗粒不能增长到核开始自发生长的临界尺寸。来自含有抗冻蛋白（AFPs）的冰成核实验的证据表明，在蛋白质周围的溶液中有长程效应。分子建模研究表明，这种情况发生在包括 PVP 在内的部分 KHIs。但有中子衍射研究表明，在水合物形成之前和期间，PVP 并不影响丙烷—水系统中的水结构。另一个模拟试验表明未添加 KHIs 时，水合物的临界成核尺寸极大。有拉曼光谱研究表明，聚乙烯己内酰胺（PVCap）在水合物形成的早期阶段，阻止了大空腔的封装速度。采用 KHIs 增效剂的试验验证也有报道。

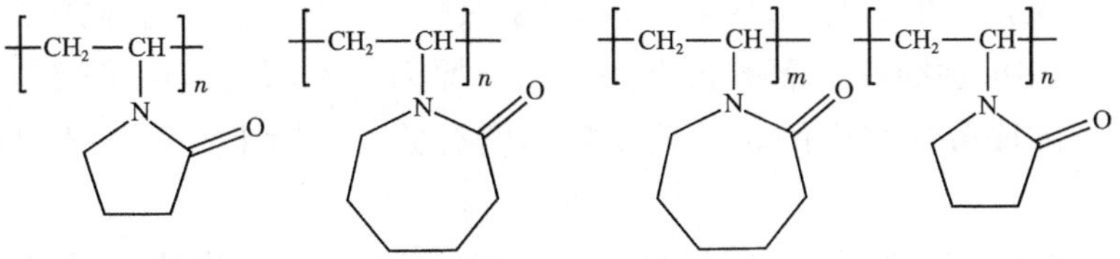

图 9.2　聚 -N- 乙烯基内酰胺聚合物 / 聚乙烯吡咯烷酮、聚乙烯己内酰胺及聚乙烯吡咯烷酮 / 聚乙烯己内酰胺共聚物

第二种机制表明，KHIs 聚合物吸附在不断增长的水合物颗粒表面，限制其增长，并可能使水合物空腔变形。作为成核和晶体生长的抑制剂，其作用可以发生在水合物颗粒达到临界成核尺寸之前或之后。KHIs 聚合物类可能以不同的方式吸附在水合物表面。通常人们认为聚合物上的疏水基团类似于小分子烃类，作用于水合物表面的空腔。酰胺基团通过氢键将聚合物锚定在水合物表面。四氢呋喃（THF）水合物抑制研究和中子小角散射及分子模拟研究，明确了部分 KHIs 聚合物的第二种机制。后续讨论的超支化聚酯酰

胺、聚乙烯恶唑啉和聚天冬酰胺等聚合物都归类为性能较差的四氢呋喃水合物晶体生长抑制剂，这表明还会有作用于水合物的其他机制。对于某些聚合物，两种机制可能同时起效。

在高于特定浓度后，PVCap 的环戊烷水合物会有多层吸附，而 PVP 不会有类似情况。这种多层吸附使 PVCap 比 PVP 更有效地减少了水合物从体相向表面的扩散，而水合物的生长更倾向于在表面。

有分子建模研究表明，甲烷水合物最初的成核相与常见的块状晶体结构都不相符，而是包含所有晶体的结构单元。后续的模拟研究表明，$5^{12}6^3$ 水晶体笼不是 SⅠ型或 SⅡ型晶体所固有的，在其与溶液的界面上频繁出现，并在笼合物的交叉成核机制中发挥核心作用。预测在 SⅡ型水合物成核过程中会形成同样的笼。在其他模型研究中发现了一系列的笼结构。另一项关于亚稳态条件水合甲烷溶解度的分子动力学研究发现，降低温度比增加压力能更有效地促进水合物成核。核磁共振（NMR）和高压粉末 X 射线衍射（PXRD）的实验室研究表明，在有或没有 KHIs 的情况下，固相中可能同时存在 SⅠ型和 SⅡ型相。所有这些观察表明，与仅仅防止形成纯 SⅡ型水合物相比，SⅡ型的动力学水合物抑制机制可能更复杂：SⅠ型也是如此。

KHIs 的另一个机制是有疏水侧链的聚合物可结合水相中形成水合物的溶解性气体。另一个可能机制是，性能最好的 KHIs 强烈地吸附在原有形成水合物的异核点表面（如管壁或 SiO_2 等产出颗粒）。

许多 KHIs 聚合物在水中有浊点和沉积点。虽然低浊点可能有助于提高 KHIs 性能，但聚合物会从溶液中析出，并在沉积点以上条件，失去抑制水合物的活性。沉积点还可能导致有热产出液的井口注入点处凝固。例如，PVCap 在淡水中的浊点为 35～40℃，预计高于此温度时会出现沉淀。因此，大多数运营商坚持要求 KHIs 与注入点的热产出液兼容。

在含高浓度硫化氢的天然气系统中，KHIs 的性能会大大降低。高浓度的二氧化碳、低 pH 值都会对 KHIs 性能有显著的负作用。以上原因尚不清楚，但可能是与小分子烃类相比，与这些气体成分在水中的相对较高溶解度有关。实际上这些气体也是笼合物的形成剂。

多数 KHIs 室内研究都是针对天然气混合物，将 SⅡ型作为热力学最稳定的水合物（大多数建模研究却都是基于 SⅠ型的甲烷水合物，因为混合气体的 SⅡ型水合物成核建模存在困难）。

多数 KHIs 的现场应用是基于乙烯基内酰胺聚合物和共聚物、超支化聚酯酰胺、异丙基甲基丙烯酰胺（IPMA）聚合物和共聚物三类聚合物中的一类。

这些类别的聚合物可以混合后协同使用。KHIs 也可以与 THIs 联合使用，以增加对过冷度的保护。在某些情况下，由于产水量增加，THIs 可能已经达到了其注入能力，而加入少量 KHIs 就可以提供所需的保护效果。另一种现场情况下，KHIs/THIs 混合物在稳定运行期间作为传统的 KHIs 使用，而在长期停产和关停冷井复产期间作为 THIs 使用。另一类在低过冷度环境应用的商用 KHIs 是可生物降解的聚酯焦谷氨酸聚合物。上述 KHIs 都将在后续章节中讨论。

9.2.2.2 乙烯基内酰胺 KHIs 聚合物

聚乙烯吡咯烷酮属于乙烯基内酰胺类聚合物，但是其性能较差，其商业应用已被在现场 9～10℃条件可应用的 VCap 聚合物所替代。六环聚乙烯哌啶酮（PVPip）和八环聚乙烯环辛酮（PVACO）尽管合成成本较高，但将其与五环 PVP、七环 PVCap 的 KHIs 性能进行比较的结果表明，随着环尺寸增大，其性能提升。这种趋势可能是由于内酰胺环的尺寸增大使疏水性增加，导致聚合物在水中的浊点降低。

另一种提高聚乙烯基内酰胺的疏水性及 KHIs 性能的方法，是在内酰胺环上添加烷基或使用更多的疏水性单体。对 3- 烷基 PVP 的研究显示，其性能比 PVP 更优，甚至在优化后，优于 PVCap。部分丁基化的 PVP（一种市售的聚合物）和乙烯吡咯烷酮（VP）与小分子烷基丙烯酸酯的共聚物是比 PVP 更好的 KHIs，但不如最好的 VCap 聚合物。尽管 VP/VCap 等共聚物浊点更高，VCap 聚合物中最简单的均聚物 PVCap 也有商业上的使用（图 9.2）。N- 甲基 -N- 乙烯基乙酰胺（VIMA）/VCap 共聚物已证明比 PVCap 性能更好，并曾被商业化应用，但由于 VIMA 单体的成本过高，现在已不再使用（图 9.3）。实际上，VCap/ 二甲氨基乙基甲基丙烯酸酯（DMAEMA）共聚物也是高性能的 KHIs 聚合物（图 9.3）。以 DMAEMA 为基础与甲基丙烯酰胺或 N, N- 二甲基丙烯酰胺的共聚物，没有任何乙烯基内酰胺成分，也是一种 KHIs 共聚物。VCap/ 乙烯基吡啶共聚物被认为是具有缓蚀剂（I）性能的 KHIs。包括 VCap 的乙烯酰胺与酸性单体或其盐的共聚物，也是具有缓蚀剂功能的 KHIs。酸性单体包括乙烯基膦酸钠和丙烯酰胺丙烷磺酸（AMPS）单体的盐。VCap 或 VP 与短聚乙氧基化（甲基）丙烯酸酯的共聚物也是一种 KHIs。

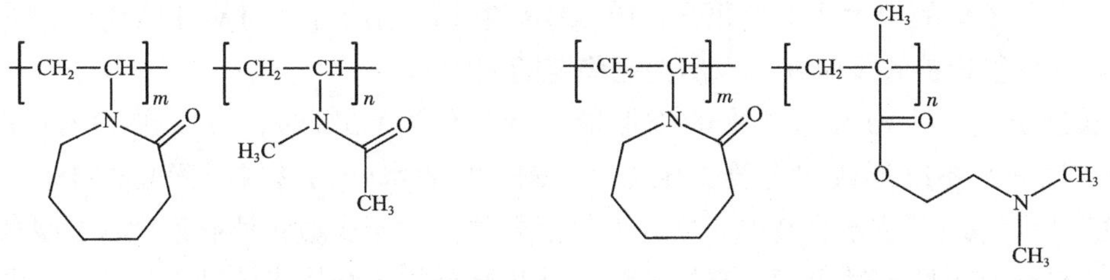

图 9.3　N- 甲基 -N- 乙烯基乙酰胺 / 乙烯基己内酰胺共聚物，乙烯基己内酰胺 / 二甲基氨基乙基甲基丙烯酸酯共聚物

乙烯基内酰胺聚合物的生物降解性通常较差，即使在低分子量时也是如此。已经开发出聚乙二醇和 VCap 单体的接枝共聚物，也可选择与其他乙烯基单体（如醋酸乙烯酯）接枝，在 OECD 306 超过 60 天的海水测试中，其生物降解率超过 20%。使用具有蛋白质或肽骨架的聚合物，接枝了正乙烯基内聚胺和其他乙烯基单体，已经可作为 KHIs 应用。实例包括明胶（从沸腾骨头中提取的胶原蛋白）与马来酸酐和氢氧化钠的反应。该产品与 VCap（或与 VP 或 VIMA 的混合物）及其他添加剂和偶氮引发剂反应，得到接

枝聚合物。这类接枝聚合物已经商业化，在 28 天内海水生物降解率为 53%，60 天内为 98%。聚乳酸链也可以接枝到 N- 乙烯基己内酰胺或其他乙烯基酰胺共聚物中的 HEMA 或乙烯基醇的羟基链上，从而得到比母体 VCap 共聚物更具生物降解性的聚合物（图 9.4）。VCap 共聚物有从乙烯醇基团或羟基烷基（甲基）丙烯酸酯等获得的羟基，是性能良好的 S I 型水合物 KHIs。

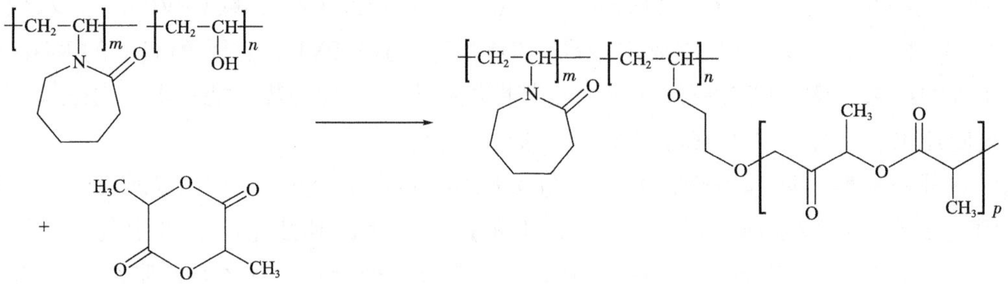

图 9.4　聚乳酸接枝到 N- 乙烯基己内酰胺 / 聚乙烯醇共聚物的聚合物

性能最优的 KHIs 聚合物（或低聚物）的理想重均分子量为 1500～3000。当分子量低于 1000 时，防止水合物生成的性能会急剧下降，而当分子量增加到 3000～4000 以上时，性能会缓慢下降但不会消失。最佳分子量范围与聚合物的比表面积（给予周围水最大的扰动）及聚合物在溶液中的高流动性有关。低分子量聚合物具有保持注入配方的低黏度优点，市售 PVCap（目前性能最高的是乙烯基己内酰胺聚合物）样品的分子量为 1000～4000。有说法和证据表明，分子量的双峰分布可以提高性能。可以通过一次聚合或将两种分子量分布不等的聚合物混合实现双峰分布。但两种具有相同双峰分布和经验公式的聚合物不一定会有相同的 KHIs 性能。与只使用低分子量的聚合物相比，商业上使用低分子量和高分子量 VCap 聚合物的混合物以提高性能。使用不同的引发剂所产生的聚合物链末端基占分子量的很大比例，也会影响 KHIs 性能。制造方法对最佳性能至关重要。一些供应商发现很难扩大规模并生产出性能稳定的 KHIs。

超低分子量 PVCap 在水相中的添加量质量分数为 0.5% 时，可以在 13℃过冷度下，延迟天然气—水系统形成水合物，压力为 7MPa 时的水合物形成时间超过 48h。此时的"天然气"指形成 S II 型水合物的正常天然气混合物。对于少数预测会产生 S I 型水合物的油田，PVCap 有更低的过冷度。有报道称，某些聚合物（结构未公开）对 S I 型水合物具有更好的性能。在流动循环系统及各种流体条件下，测试了这种超低分子量 PVCap 聚合物。根据不同的流体类型（仅含天然气，或伴随有凝析油或原油），在持续 2～3 天的恒压测试中，过冷度在 11～19℃之间变化。结果的变化强调了在生产条件下，用恰当的产出液体测试 KHIs 性能的重要性。基于 VCap 的 KHIs 除了可用于多相流生产系统，据称也可用于压裂液。

VCap 聚合物的聚合工艺将影响其性能和水溶液浊点。有研究表明，在考虑到 2- 丁氧基乙醇作为增效剂的影响后，由 VCap 在丁二醇（2- 丁氧基乙醇）中聚合而成的

PVCap 比用异丙醇制成的 PVCap 有更好的 KHIs 性能。高浊点聚合物在蒸馏水中的浊点为 35~40℃(取决于分子量和聚合步骤)，由于盐化作用，其不溶于高盐水。VCap 与图 9.2 和图 9.3 中的更亲水单体的高浊点共聚物，已在市场上出售。目前，市场上最常见的高浊点共聚物是 VP/VCap 的 1∶1 低分子量共聚物，其性能略差于相同分子量的 PVCap。但 KHIs 在温度高于浊点但低于沉积点的情况下，仍有良好的性能。更耐高温和高总溶解固体（TDS）的 VCap 共聚物已被报道。含有 VCap、AMPS 钠和 VP 的共聚物已在市场有销售。含有 VCap 和可选的 VP 的乙烯基甲酰胺共聚物也是高浊点 KHIs。

为提高 VCap 聚合物的性能，已经提出了多种增效剂。除了上面讨论的 2-丁氧基乙醇，2-异丁氧基乙醇也被认为是一种更好的增效剂。小的阴离子有机化合物也可作为 VCap 聚合物的增效剂，如溴化四丁基铵（TBAB）等小分子的季铵盐多年来已作为 VCap 增效剂在现场应用。TBAB 也是 VP/乙烯基吡啶共聚物的增效剂。据称，TBAB 有生物降解性更好的酯连接替代物。如三丁基丙基磺酸铵等小的季铵盐两性离子分子也具有增效剂作用。所有这些季铵型分子的关键特征是一个季氮（或磷）原子带有三个或多个丁基或戊基，其结构与季铵型抗凝结表面活性剂有关（见 9.2.3）。如三丁胺氧化物等胺类氧化物，也是 VCap 聚合物的良好增效剂，但尚未有商业应用。双胺氧化物和多胺氧化物也是 VCap 聚合物 KHIs 的良好增效剂。多胺氧化物单独使用时也是有效的 KHIs。聚烷基胍盐也是 VCap 聚合物的良好增效剂（图 9.5）。

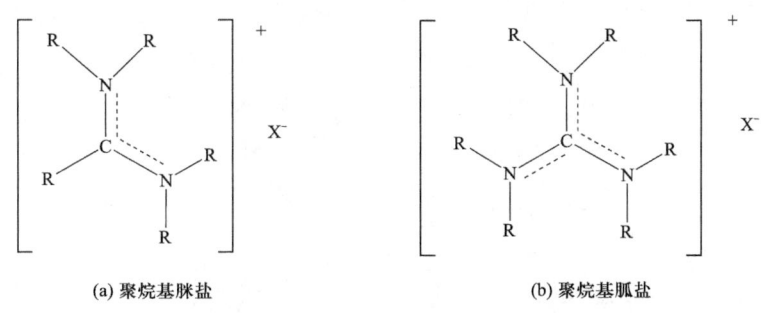

图 9.5 聚烷基脒盐和聚烷基胍盐

某些基于 VCap 聚合物的取代胺、亚烷基二胺或多胺及其衍生物已获得专利，可作为 KHIs 的增效剂。如 PVCap 等小分子聚醚胺已在现场用作 VCap 聚合物的增效剂。聚醚胺也用作气井的抗凝结剂（见 9.2.3）。此外，各种类型的阳离子或非离子表面活性剂或一种糖，也可作为 VCap 聚合物的增效剂。具体的表面活性剂有 N-烷基吡咯烷酮、聚乙二醇或聚环氧乙烷—环氧丙烷，糖类具体有山梨醇、甘露醇、果糖或蔗糖。

聚环氧乙烷也是 PVCap 的增效剂。有研究者称 KHIs 组合物包括聚乙烯基聚合物 [如 VCap 或烷基（甲基）丙烯酰胺聚合物]、溶剂和至少一种无卤素的杀菌剂。

9.2.2.3 超支化聚酯酰胺 KHIs

除了乙烯基己内酰胺聚合物，另一类已投入商业应用的主要 KHIs 聚合物是超支化聚酯酰胺。其与 VCap 聚合物类似，也适用于现场应用的 10℃ 过冷度。其中某些聚合物据称比 PVCap 的生物降解性更好，OECD 306 测试的 28 天生物降解率甚至超过了 20%。还开发了高浊点的聚合物，在含盐高达 12% 的系统中很有用，其性能下降被盐对水合物的抑制作用所抵消。聚酯酰胺是由环状酸酐与二异丙醇胺按 $n:(n+1)$ 的比例缩合而成，其中 n 是一个整数（通过改变 n，可以改变聚合物的分子量）。这就得到了一种在端基有羟基的聚合物。超支化由二乙醇胺引发（其有三个反应性基团）。通过在反应混合物中加入如亚胺等第三个分子，可以改变聚合物的端基，使其更亲水。环状酸酐的首选是顺式 -1，2-环己烷二羧酸酐（或六氢苯酐），而亚胺可以是 3，3-亚氨基双（$N, N-$ 二甲基丙胺）或 1-甲基哌嗪（图 9.6）。将聚酯酰胺与其他 KHIs 聚合物区分的特征是其超支化。这些其他 KHIs 聚合物都是线性蛇形分子，其中大多数都只有一个聚乙烯骨架。用 KHIs 与水合物表面的结合来比喻，聚酯酰胺就像一只手攥着一个水合物球，而其他 KHIs 就像一根长手指连接在球上。聚合物的分子量也不需要很高。例如，分子量为 1500~2000 的聚合物可提供良好的 KHIs 性能。这与聚乙烯 KHIs（如 PVCap）的分子量大致相同。在图 9.4 所示的聚酯酰胺情况下，能够与开放的水合物笼相互作用[和（或）扰乱水结构]的疏水基团是环己基环，尽管二异丙醇胺的甲基可能也有贡献。酰胺基团及酯基团（较弱）都参与了与水分子的氢键。对于 SⅠ型水合物，超支化聚酯酰胺比 VCap 聚合物的效果更好。

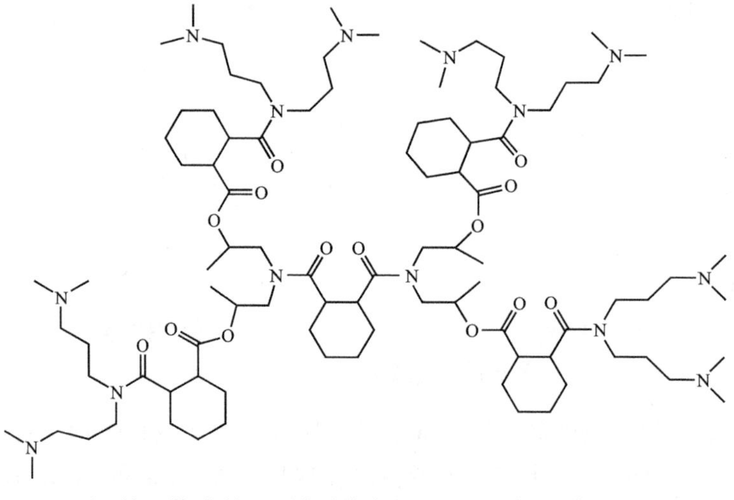

图 9.6　由顺式 -1，2-环己烷二羧酸酐、二异丙醇胺和 3，3-亚氨基（$N, N-$ 二甲基丙胺）形成的超支化聚酯酰胺结构

聚酯酰胺在应用中，有时与 VCap 聚合物协同混合使用，也有各种其他的增效剂提高超支化聚酯酰胺的效果。例如，聚乙烯亚胺与甲醛和己内酰胺的反应产物，这就产生了具有己内酰胺环侧基的聚合物，正如 PVCap 的结构。另一种 KHIs 增效剂是由 $N-$ 甲基丁胺与甲醛和聚丙烯酰胺反应而成。其他提到的增效剂是具有己内酰胺或烷基酰胺端基的非聚合表面活性剂。也可以用 KHIs 混合物配制少量的季铵抗凝结剂，以提高性能。由于结构不同，VCap 聚合物可能是超双链聚酯酰胺的增效剂，但文献中没有实例。

9.2.2.4　KHIs 与其他生产化学品的相容性

研究表明，某些表面活性剂成膜型缓蚀剂会对商用 KHIs 的性能产生不利影响。目前已经开发出了兼容的缓蚀剂。反之，KHIs 也会对缓蚀剂的性能产生负面影响。这些影响可能是由于 KHIs 在管壁上的优先吸附或缓蚀剂—KHIs 表面活性剂—聚合物的相互作用所致。此外，许多类别的表面活性剂（非离子型、阳离子型和两性型）可以加速水合物的形成，特别是如果其有良好的发泡或乳化特性，使气体（或油中的溶解性气体）能与水相更好地接触。与 KHIs 兼容的缓蚀剂可避免出现类似情况。防垢剂通常不会损害中性 KHIs 的性能，在某些情况下还可以提高其性能。KHIs 与防蜡剂可以复配在同一个注剂体系配方中。

9.2.2.5　焦谷氨酸 KHIs 聚合物

已经开发出含焦谷氨酸侧基的聚酯或聚乙烯亚胺，作为高凝固点、可生物降解的 KHIs。其中一种聚合物已经在欧洲的一个低过冷度小型油田中应用。具体示例如图 9.7 所示。尽管与聚合物骨架的连接位置不尽相同，但其侧基与 PVP 中的相似。因此，预计这类聚合物的过冷度不是很高，但据称比 PVP 效果略好。这类含脂肪酸酯基等的聚合物，还表现出一些水合物抗凝结剂的特性。聚酯是由二元醇或多元醇与取代的二羧酸缩合，然后与焦谷氨酸反应制成。

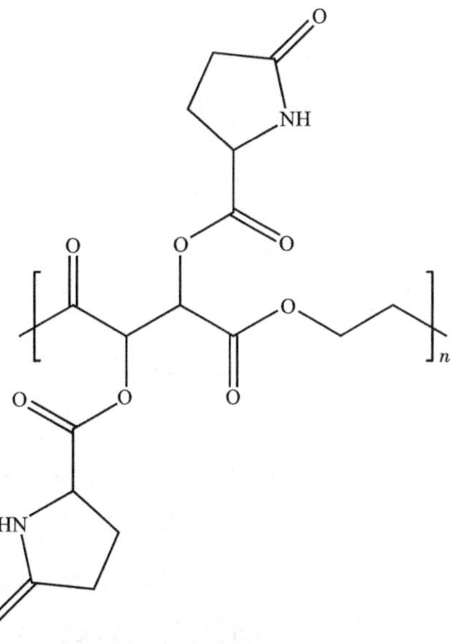

9.2.2.6　聚（二）烷基（甲基）丙烯酰胺 KHIs

由于成本和可获得性问题，还有许多其他类别的 KHIs 聚合物尚未实现商业化。其中研究最多的是聚烷基（甲基）丙烯酰胺和聚二烷基（甲基）丙烯酰胺，其中一种已经商业化（图 9.8）。特别是甲基丙烯酰胺单体的聚合物比丙烯酰胺单体具有更好的性能。因此，可在聚乙烯骨架上接入额外的甲基，降低聚合物的柔韧性，使烷基酰胺侧基在水中的移动减少（熵降低），改善其动力学抑制作用。这一想法也证明对其他 KHIs 聚合物类别有效。已发表的一篇论文表明，聚 IPMA 链在水溶液中的构象变化比聚异丙基丙烯酰胺（聚 IPAM）小。

图 9.7　焦谷氨酸聚酯动力学水合物抑制剂

最好的甲基丙烯酰胺聚合物似乎是由 IPMA 制成，其次是由甲基丙烯酰吡咯烷制成。与 PVCap 的观察结果类似，由 IPMA 在 2-丁氧基乙醇中聚合而成的聚 IPMA 性能优于通过异丙醇中制备的聚 IPMA。事实上，在未优化的分子量下，IPMA 与 VIMA 的 1∶1 共聚物性能优于 IPMA 均聚物（图 9.9）。之前讨论过含有 VIMA 的共聚物效应，PVCap 与 1∶1 的 VIMA/VCap 共聚物相比，后者是更好的 KHIs。但分子量非常低时，聚 IPMA（或寡聚 IPMA）与 VIMA 共聚物的效果相同。目前寡聚 IPMA 已销售并用于现场。化学品服务公司未公布的实验室测试表明，其性能优于市售的 PVCap 技术，在淡水中的浊点约为 35℃。由更强亲水性单体制备的 IPMA 共聚物具有更高的浊点和耐盐性，也已用作商业 KHIs。制备 IPMA 单体的新工艺也已有报道。

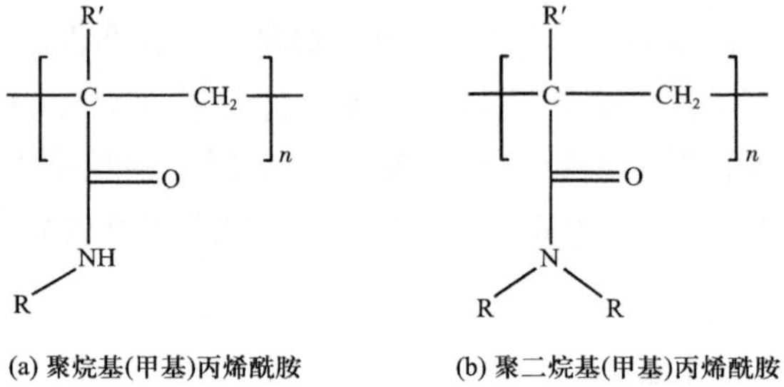

(a) 聚烷基(甲基)丙烯酰胺　　(b) 聚二烷基(甲基)丙烯酰胺

图 9.8　聚烷基（甲基）丙烯酰胺和聚二烷基（甲基）丙烯酰胺（R= 烷基或 RNR= 环酰亚胺，R'=H 或 CH_3）

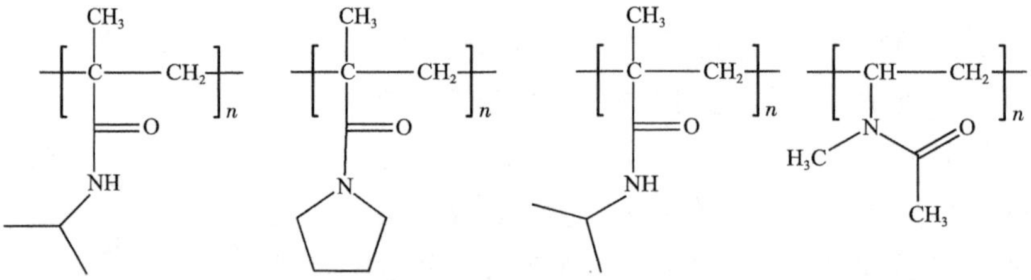

图 9.9　聚异丙基丙烯酰胺、聚甲基丙烯酰吡咯烷和 N− 乙烯基 −N− 甲基乙酰胺 / 异丙基甲基丙烯酰胺的 1∶1 共聚物

骨架上有甲基的聚 IPMAM 与没有甲基的聚 IPAM 相比，是更好的 KHIs，至少有两个原因。首先，甲基的空间体积打开主链，使聚 IPAM 的比表面积更大。第二，已知由于 IPMAM 单体上的甲基，IPMAM 的聚合比聚 IPAM 有更多的间规聚合物结构。还有研究表明，与等规或随机的无规聚合物相比，为了获得最佳的 KHIs 性能，更倾向于采用间规性。

针对正丙烯酰吡咯烷与 VP 的同聚物和共聚物的 KHIs 试验也有报道。含有丙烯酰吗啉和乙烯基咪唑及可选的高达 98% 的 VCap 或 VP 的共聚物，是特别适用于高矿化度水或高温注入环境的 KHIs。

9.2.2.7　其他类型的 KHIs

图 9.10 至图 9.13 显示了其他类别非商业化的性能良好的 KHIs 聚合物，包括马来酸共聚物的烷基酰胺衍生物、聚马来酰亚胺、聚亚烷基恶唑啉（或聚酰基乙基亚胺）、聚乙烯恶唑啉、聚烯丙基酰胺、聚天冬酰胺、淀粉衍生物、乙烯烷酸酯/VIMA 共聚物、聚乙烯烷酰胺、聚胺氧化物和改性丙烯酰胺丙基磺酸（AMPS）聚合物。应注意到聚乙烯恶唑啉不包含酰胺基团，而其他 KHIs 聚合物则含有酰胺基团。另外，乙烯基链烷酸酯 /VIMA 共聚物中的酰胺基团作为烷酸酯功能基团的侧基，与疏水基团不在同一个单体中。

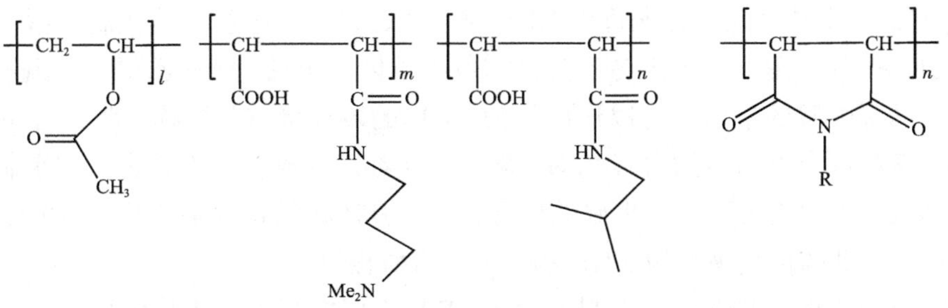

图 9.10 马来酸共聚物和聚马来酰亚胺的烷基酰胺衍生物（通常与亲水单体共聚，以提高水溶性）

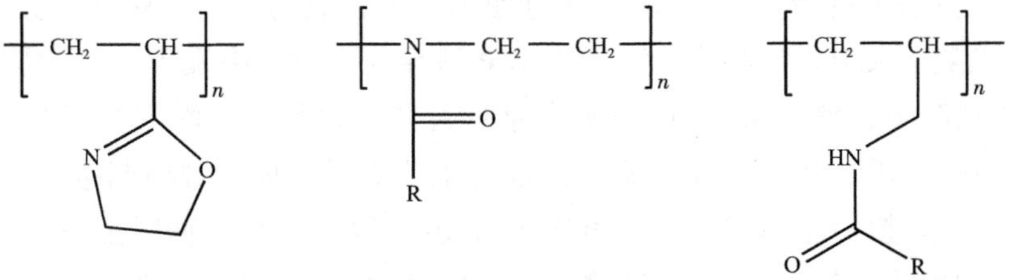

(a) 聚乙烯基恶唑啉　　(b) 聚亚烷基恶唑啉
　　　　　　　　　　　（或聚酰基乙基亚胺）　　(c) 聚烯丙基酰胺

图 9.11　聚乙烯基恶唑啉、聚烷基恶唑啉（或聚酰基乙基亚胺）和聚烯丙基酰胺

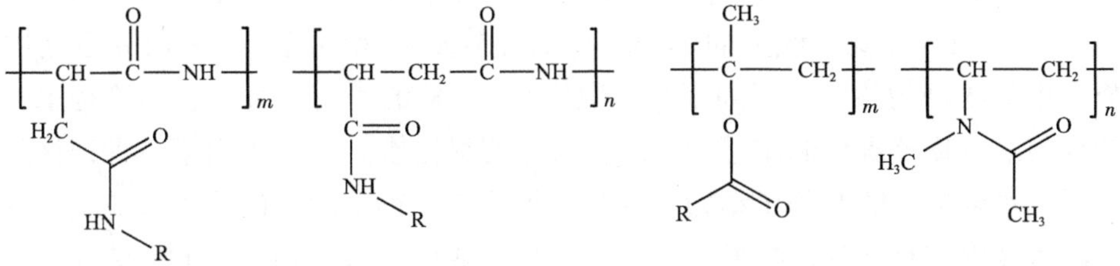

图 9.12　聚天冬酰胺和乙烯烷酸酯 /N- 甲基 -N- 乙烯基乙酰胺共聚物

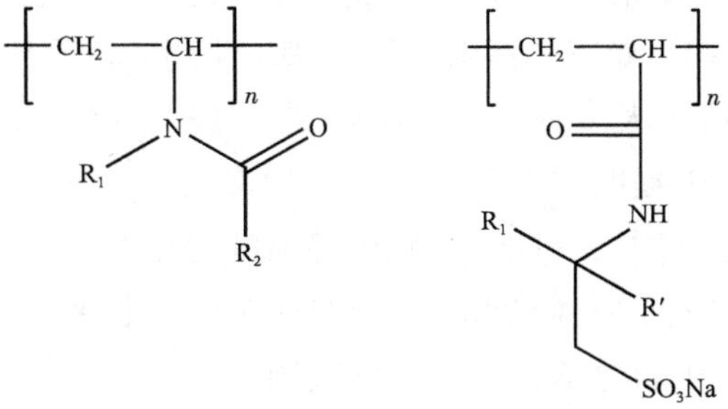

图 9.13　聚乙烯烷酰胺和改性 AMPS 聚合物（$R=C_1—C_6$，$R'=H$ 或 CH_3）

当疏水的 R 基为 5 个碳时，改性 AMPS 聚合物的性能最好。这表明此类聚合物的主要机制是干扰水结构，抑制水合物成核，而不是附着在水合物颗粒表面（因为 R 基不是这种相互作用的理想尺寸）。

聚天冬酰胺也显示出良好的生物降解性，但在 SⅡ型天然气水合物上的表现比商用

1∶1 VP/VCap 共聚物稍差。在典型的 KHIs 浓度下，聚天冬酰胺也能控制结垢。在水介质中自由基引发剂的作用下，聚天冬氨酸钠与各种单体［包括与 N- 乙烯基内酰胺和 N- 烷基（甲基）丙烯酰胺的单体］可以形成作为 KHIs 的多种接枝共聚物。将乙烯不饱和单体（如 VCap）接枝到自然衍生的含羟基的链转移剂（如麦芽糊精）上，也可用来制备 KHIs。将聚乙烯链接枝到生物可降解性更好的聚合物上，其产物的缺点是这些侧链的生物可降解性不高。将壳聚糖的衍生物作为 KHIs 的研究也有所报道。

自 20 世纪 90 年代初以来，人们就知道 AFPs 不仅抑制水合物的生长，还能抑制其形成。AFPs、AFPs 的活性片段及其模拟物也作为 KHIs 进行了研究。

有研究提供了进一步的证据，表明通过使用人工合成的混合天然气，会同时产生 SⅠ型与 SⅡ型水合物。KHIs 甚至可以在没有 AFP 活性的蛋白质中发现。最初从鱼类、昆虫、植物、真菌或细菌中分离出来的 AFPs 或其活性片段，可能是在酶反应过程中产生。因为从以上来源中分离 AFPs 往往异常昂贵，酶反应工艺则可实现大规模制造。

除了 AFPs，天然多元醇或糖类也可作为 KHIs。例如，从一种甲虫身上获得的多种木甘露聚糖。有趣的是，某些 AFPs 能从熔化的水合物中消除记忆效应。记忆效应指熔化的水合物形成的水比没有经过这种处理的水，冷却时有更容易形成水合物的能力。通常水合物的熔化温度不应超过水合物平衡温度的若干摄氏度，这样才能保留记忆效应：温度进一步升高，记忆效应就会消失。关于这种记忆效应的假设是，水合物熔化后形成的水，在分子水平上保留了如部分笼状或多面体团块等的部分水合物结构。但有研究者利用中子散射，分析了甲烷水合物从熔化至高于平衡温度小于 1K 的甲烷水合物，结论是水合物形成前的水结构与水合物分解后的水结构之间，没有明显区别。有人提出第二种记忆效应的理论，通过甲烷水合物分解的分子动力学计算模拟研究，得出的结论是该效应更多的是与水合物中的高浓度甲烷和其延迟扩散有关，而不是与亚稳态水合物前体的持续存在有关。一项未发表的针对四氢呋喃水合物和甲烷水合物的核磁共振光谱研究，也得出了类似结论。

一种最初设计用于化妆品的商业化三元共聚物已证实有良好的 KHIs 性能，但尚未有现场应用报告（图 9.14）。如三乙醇胺的丙氧基化衍生物等聚烷氧基化胺，也证实具有 KHIs 活性，尽管其活性低于大多数市售 KHIs（图 9.15），但可作为其他 KHIs 的良好增效剂。有报告称，从腰果壳中提取的聚醚二醇（一种聚乙氧基化的苯酚—甲醛树脂）可作为甲烷水合物的 KHIs。还研究了 2- 四氢呋喃丙烯酸酯均聚物，或其与 VCap 的共聚物。

聚季铵盐已证明对四氢呋喃水合物具有 KHIs 作用。非聚合的咪唑基离子液体适用于相当极端条件下的热力学和动力学的水合物抑制剂。

9.2.2.8　KHIs 的性能测试

在实验室或现场研究中，重要的是尽可能地模拟现场流体。所有流体如水、气体和油相的组成等都会影响 KHIs 的性能。前面已经讨论了盐水、H_2S 和 CO_2 及原油组分等的问题。即使有相同的压力—温度平衡曲线，两种不同成分烃类气体也会得出不同的 KHIs 测试结果。

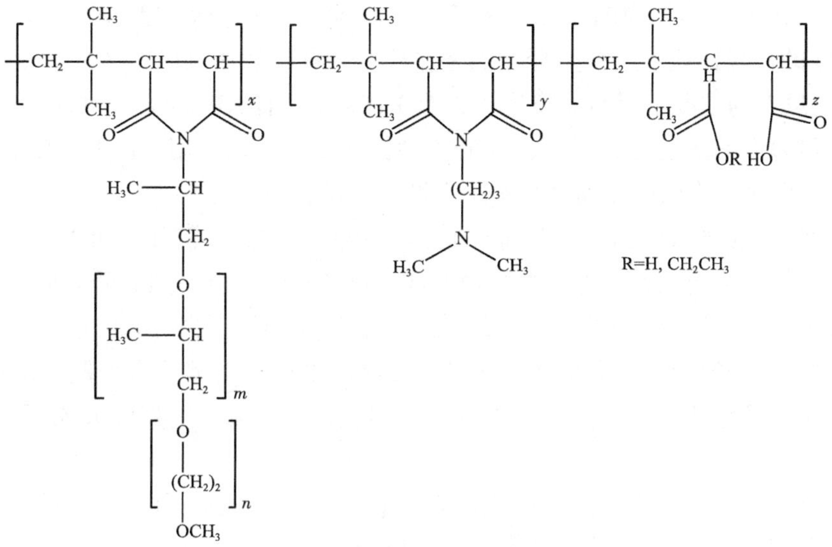

图 9.14　马来酸基三元高聚物 KHIs

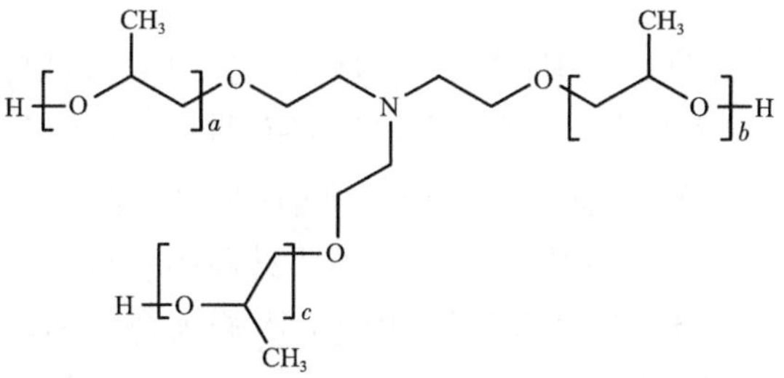

图 9.15　聚烷氧基胺的优选结构（其中 $a+b+c=14.9$）

测试 KHIs 性能有若干种方法。如果操作者关注 KHIs 现场使用的达标能力，通常要测试在现场极端过冷度条件下的最小保持时间（保持时间指系统从水合物稳定域到开始形成水合物的时间）。因为这一阶段的管线并非一直处于最大过冷度和压力，该方法将给出消除水合物所需的保守加量。如前所述，测试应在系统压力下进行，因为有研究表明，即使在相同的过冷度下，保持时间也会随着压力的变化而变化。还必须包括无流动的停产测试，以便在计划或未计划的停产条件下进行现场验证，以及特别有缓蚀剂（已知许多成膜型缓蚀剂会大幅降低 KHIs 的性能）等其他生产化学剂时，进行测试。这些测试可以在高压釜、摇瓶、老化罐或具有实际现场流体和气体组分的循环管路中进行。

有资料介绍了两种类型的台式管路循环。高压釜是多年来的标准筛选设备，至今仍在使用。使用多组摇瓶设备是当前筛选 KHIs 的首选方法。鉴于天然气水合物形成的随机性，人们筛选 KHIs 时，可能从一台设备中获得多个结果。

大型环路或管路中的测试，通常是现场试验或实施 KHIs 前的最后和最佳方法。这是因为小型、"干净"的实验室设备的测试结果受水合物形成随机性的影响最大。因此，需要多次重复测试以判断其性能。从理论上讲，应该在实验室重复无数次的 KHIs 测试，并将最差的结果作为现场相同温压和流体条件下的预期性能。而实践中，运营商通常不会冒险在 KHIs 的极限性能下运行生产，而是在比实验室或现场测试中达标过冷度低 1~2℃（或更高的剂量）的情况下运行。事实上，有 KHIs 比没有 KHIs 时，形成的水合物段塞更难分离。在与海上油田已建成气井相连的流动保障设计中，利用对已有 KHIs 水合物抑制作用的了解，可快速应用而无须进行实验室性能测试。

如果只对 KHIs 的性能排序感兴趣，也可采用其他方法。例如，可以在没有搅拌/流动的情况下，将系统冷却到热力学水合物形成区域的预设温度，然后开始搅拌。然后可以测量每种 KHIs 的水合物形成的诱导时间。该方法如果操作不当，可能会得出非常不同的结果。重要的是，流体要与设备的所有部分接触，以避免局部不含抑制剂的冷凝水形成水合物。在循环管路测试中，流态（如层流、湍流、段塞流等）将极大地影响 KHIs 保持时间。

另外，可以在搅拌/流动的情况下持续冷却至水合物区域，并测量开始形成水合物的时间：起始温度越低，KHIs 越好。也可以进行梯度冷却，如保持温度不变若干小时，然后再冷却到更高的过冷度，如此反复。因为在每个温度梯度下，压力都保持不变，这样更容易检测水合物开始形成。

由于水合物形成的随机性，还开发出一种新方法，以降低诱导时间（或保持时间）数据的分散性。这种方法需要形成水合物，然后在短时间内将水合物熔化到略高于平衡温度，以便在水相中保留水合物"前体"结构（所谓的记忆效应），然后冷却到预设的水合物形成区域温度。因为水合物的形成主要是由异质结晶引发，这种方法的保持时间更具有可重复性。而不同的 KHIs 聚合物会在不同程度上影响水合物的解离速度和温度，所以在所有的冷却实验过程中，没有恒定的前体量，可能会对 KHIs 的排序产生负面影响。在水合物解离后，注入 KHIs 可以克服该问题。还有研究者发现，如果混合气体中含有 3% 以上的二氧化碳，该方法的结果就无法重复。这是由于首次测试后，水相中的二氧化碳过饱和，会在第二次测试中更快地引发生成水合物。

最近开发了一种水合物晶体生长抑制（CGI）的新方法。这种方法表明，KHIs 会诱导产生一些高度可重复的、定义明确的水合物 CGI 区域（是过冷度的函数），范围从"完全"抑制，到严重及适度降低增长率，之后随着过冷度的增加，出现最终的快速/灾难性增长。这些区域的划分为评价 KHIs 在最苛刻模拟条件下的相对性能提供了更可靠和快速的手段。该方法假设 KHIs 的抑制机制主要是 CGI，而非成核抑制。到目前为止，这种方法的相关文献似乎表明，测试采用的 KHIs 只有基于 VCap 的聚合物（这种聚合物已知是一种性能良好的 CGI）。未发表的更宽泛范围的 KHIs 研究表明，在没有预先形成水合物的情况下，持续冷却或梯度冷却试验的结果基本一致。在过冷度试验环境下以 CGI 方法完全没有晶体生长，可以理解为使用 KHIs 的安全下限，即更高过冷度下可能出现晶体生长，会导致管道的缓慢堵塞。差示扫描量热法（DSC）也已用于筛选 KHIs。使用 DSC 测试 KHIs 的另一种方法是制备稳定的油包水乳液。有乳液时，首次检测到水合物形成的延迟时间大于没有乳液的情况，而且晶体更加分散。还有人提议将吊环张力测量法作为筛选

KHIs 的一种新方法。

在常压下对 THF 水合物（SⅡ型）的诱导时间和 CGI 实验，可以提供部分 KHIss（和调节晶体生长的抗凝结剂）的性能信息，但对其他 KHIss 可能会产生误导，因为 THF 是强水溶性，抑制机制可能与实际气体系统不同。例如，在真正的天然气水合物体系中，低分子量 PVP 的效果最好，而高分子量 PVP 却在抑制 THF 水合物晶体生长方面的效果最好。此外，有些聚合物不溶于 THF—水的混合物，但可单独溶于水。

9.2.2.9　KHIs 的回收或处置

有人提出从采出液中回收和再利用 KHIs 聚合物（无论是否添加 THIs）的方法。但还未见现场应用。只要 KHIs 不沉淀，并且在 160～200℃ 的高温下保持热稳定性，就可以在再生设施中随 MEG 循环使用。另有专利声称有方法从采出水相中去除 KHIs。还可以使用膜过滤器和高级氧化工艺，以减少采出水中的 KHIs 量。

有研究者称通过添加含疏水尾基和含羟基亲水端基的液态化学剂来回收 KHIs。该化学剂应只能与水相部分混溶，并吸附水相中的大部分 KHIs。含 KHIs 的液相被分离后，可除去溶剂，回收 KHIs 并再利用。有研究介绍了一种含 KHIs 产出水回注中防止沉淀的方法。该方法包括加入一种极性指数大于 3 的不溶于水的溶剂，并在采出水回注之前除去和破坏 KHIs。

9.2.3　抗凝结剂

抗凝结剂（AAs）属于 LDHIs 的一种类型，其中最好的抗凝结剂可以在比 KHIs 更高的过冷度下防止水合物堵塞。因此，抗凝结剂适用于更严重的或深海环境，或通常因计划外的停产，在管线中长时间停留（数周）的情况。部分抗凝结剂也具有 KHIs 效果。抗凝结剂有生产或输油管线用抗凝结剂、气井用抗凝结剂。

所有的 AAs 都允许形成水合物，但会防止水合物结团块及随后的积聚成大块。管线用 AAs 可以使水合物形成可运移的、分散在液态烃相中的非黏性水合物颗粒浆液。气井用 AAs 将水合物颗粒分散在过量的水中。这两类产品都已商业化应用。有人说明了在使用 AAs 后对天然气水合物浆液再汽化的工艺流程。

因为需要有液态烃相，管线用 AAs 不能用于气田（而该问题有可能的解决方法，本节将予以讨论）。通常管线用 AAs 应用时，含水率应低于约 50%；否则，剩余液体中超过 50%（体积分数）的水合物颗粒浆液会过于黏稠，而难以运输。已报道了对应的解决方案，即向流体混合物中添加足量的水，以降低气水比，从而获得可泵送的水合物浆液。在高含水率和 AA 浓度足够的情况下，与油相乳液相比，水相乳液可以输送更多的水合物。

虽然已知 KHIs 的性能取决于超过特定值的绝对压力，但 AAs 的相应研究很少。有循环管路研究表明，市售 AAs 在中压（20MPa）条件下测试失败，但在高压（52MPa）下表现良好。这种现象的原因尚不清楚。

多数 AAs 的现场应用都是在低温水合物形成条件下的刚投产的油气井。在海上油田连续注入 AAs 正变得越来越普遍。

9.2.3.1 管线用 AAs 乳液

实现管线 AAs 的效果有两种机理。此处讨论的一种机理是注入表面活性剂，形成特殊的油包水乳液。这种乳液限制了水滴中水合物的形成，而且水合物不会积聚。最终产品是分散于烃类相中的水合物颗粒浆液，基于水相的加量为 0.8%~1.0%（质量分数）。包括二乙醇酰胺、二辛基磺基琥珀酸盐、山梨糖醇、乙氧基化多元醇、乙氧基化脂肪酸和乙氧基化胺在内的多种表面活性剂都可用于制备 AAs 乳液。用矿物油 70T、Span 80、二-2-乙基硫代琥珀酸钠（AOT）和水开发了稳定的油包水乳液，其中水的体积分数为 10%~70%。模型乳液的高压釜实验表明，当系统中存在表面活性剂混合物时，会形成松散的水合物浆液。

AAs 乳液的最佳例子是基于聚烯烃琥珀酸酐的聚亚烷基二醇衍生物或被取代或未被取代的羧酸羟基碳酰胺，以及含有 3~6 个碳原子的羧酸单乙醇酰胺或二乙醇酰胺等的聚合物表面活性剂。由至少一个酯及聚合［二聚体和（或）三聚体］羧酸类型的非离子助表面活性剂等组成的组合物，也是性能良好的 AAs 乳液。这种 AAs 乳液的添加剂还可以回收，其产品的先导试验和现场试验都取得了成功，但目前还没有进行全面的现场推广。此类 AAs 的研究团队也已解散。AAs 乳液技术有若干缺点，首先，在进入水合物形成条件之前，水相必须被彻底乳化；否则，水合物就可能结块和沉积。这在现场很难得到保证。其次，在层流或停输期间，管道上壁的冷凝水会形成水合物。当流动是非层流时（较高的液体量和流速），凝结水问题及浆液输送困难可能会被克服。

9.2.3.2 亲水合物的管线用 AAs

另一种管线用 AAs 的工作机理发现于 20 世纪 90 年代初，表面活性剂分子有一个亲水（作用于水合物晶体表面）的端基和一个疏水（亲油）的尾基。当若干个这样的表面活性剂分子附着在水合物晶体表面时，水合物的生长被破坏，晶体变得疏水，然后很容易分散于液态烃相。亲水合物的 AAs 不需要形成乳液就可以正常运作。相反，不良的乳化能力有助于提高外排水的水质。因此，多数商用亲水合物 AAs 不是乳液。

AAs 的加量通常表示为 AAs 占水相的质量分数。但需要注意的是，亲水合物 AAs 的加量取决于其在水油界面的活性。对于主要分配于油相的 AAs，基于水相的 AAs 用量随含水率变化而变化。有专利提出了基于不同离子强度下，水与有机溶剂之间的分散比例来回收 AAs 的方法。

许多非亲水合物的表面活性剂（特别是阴离子型），实际上会加速水合物的形成，并已研究用于天然气储存和运输的天然气水合物。有几类表面活性剂已被证明具有亲水合物 AAs 特性，如己内酰胺和烷基酰胺表面活性剂，但只有一类得以商业化应用。这些是季铵盐和膦酸盐表面活性剂，其头部基团在大多数情况下含有两个或多个正丁基、正戊基或异戊基（图 9.16）。这些是在 20 世纪 90 年代早期发现的，现在有许多不同来源的季铵盐 AAs 实现商业化应用。有研究表明，异己基比其他烷基能提供更好的水合物 CGIs，但在 LDHIs 中使用异己基成本太高。AAs 的性能取决于影响亲水亲油平衡值和界面吸附性能的疏水尾端长度。季铵端基也可以制备为膦酸盐基，但这样的 AAs 价格昂贵。只研究过少数的膦酸 AAs，其中如十四烷基或十六烷基三丁基膦酸盐等是商业化杀菌剂（见第 14 章）。

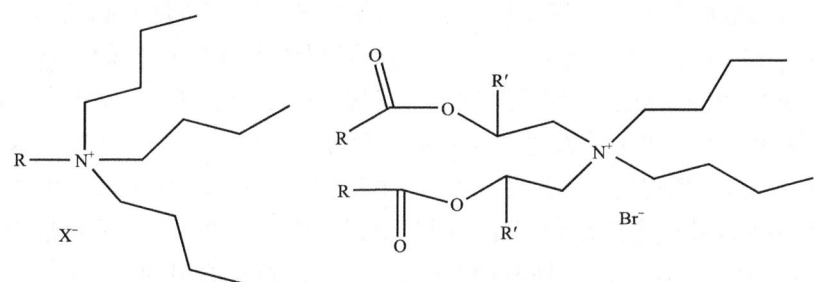

图 9.16　丁基化单尾和双尾季铵盐 AAs 的一般结构（R= 长烷基链，可选一个间隔基，R'=H 或 CH_3，X= 反离子）

由于易于制造和成本低的原因，通常商业管线用 AAs 中使用丁基，而不是戊基。与 KHIs 一样，季铵盐的性能也取决于烃类流体的组成。随着原油中环烷酸含量增加，AAs 的性能下降。假设所研究的 AAs 是阳离子表面活性剂，可能原因是环烷酸阴离子与 AAs 形成离子对，增加了 AAs 的油溶性，使其远离了需要发挥作用的油水界面。原油中的高含量酚也会对 AAs 的性能产生不利影响。季铵盐型管线用 AAs 显示出相当好的缓蚀性，且因其加注量高达每升几千毫克，有时不需要额外的缓蚀剂。但季铵盐中所含的氯离子会加速腐蚀。低腐蚀且无点蚀风险的 AAs 已有供货。AAs 配方中的残留水会导致注入管道的腐蚀，特别是当存在氯离子时，有研究者称可以通过添加有机酸酐等干燥剂的方法来防止这种情况。

关于季铵盐的机制，有证据表明，丁基或戊基等铵盐渗入开放的水合物空腔并嵌入晶体表面，防止表面活性剂脱附。这可能是其效果好于其他表面活性剂的原因，因为后者只能与水合物表面作用，无法嵌入晶体。季铵盐表面活性剂 AAs 在高矿化度下的表现更佳。作为亲水合物型管线 AAs 的聚合物表面活性剂（如连接长的疏水尾端的 PVCap 或超支化聚酯酰胺）似乎不能很好地发挥作用。

市场上所有类别的季铵盐表面活性剂 AAs 都有一个疏水尾端。一种双尾的季铵盐表面活性剂 AAs 一直在商业上使用，但现在不再生产（图 9.16）。双尾 AAs 有酯类连接，具有相当好的生物降解性，主要在油相中分布，而单尾 AAs 则主要在水相中分布。最好的单尾 AAs 可以在非常高的过冷度下工作（如果加量足够高，可能没有限制），而双尾 AAs 基于水相用量仅为 0.25%~0.3%（质量分数）（活性表面活性剂）时，似乎可以在过冷度 14~15℃内工作。单尾 AAs 需要更高的加量，在淡水中的表现也不尽如人意；在相同的过冷度下，通常随着矿化度增加，性能会更好。尽管已经开发出能改善水质的 AAs，但总体上此类 AAs 会对排放的水质产生不利影响。

尽管添加盐类通常会改善季铵盐表面活性剂 AAs 的性能，但有证据表明二价阳离子会使 AAs 的性能变差。此外，AAs 在停产或冷却条件下的性能似乎在很大程度上取决于系统的冷却速度。单尾季铵盐的现场应用自 2000 年已有报道，在高含量硫化氢气体系统也可以有效作用。

第一个版本的双尾季铵盐没有酯类间隔基，生物降解性差。第二个版本的双尾季铵盐由 N- 丁基二乙醇胺制成，在酯基和季铵盐的氮原子之间有乙烯间隔基团。还发现在间隔基团中加入一个小的烷基分支（甲基或乙基）可以改善 AAs 的性能。具体来说，通过预制水合物浆液的关闭或重启循环管路试验，发现性能得到了提高。较长的间隔基团（如丙烯），会使性能变差。双尾季铵盐的成品中还含有一定量的未季铵化胺，这有助于提高 AAs 的性能和破乳，相关应用已有报道。由于酯基的降解，这种 AAs 的保质期只有 1 年左右。这种双尾季铵盐的生产已经停止，因为根据欧盟新的 REACH 环境法规，批准这种化学品需要增加成本。运营商计划改用 KHIs，因为该凝析油田的最大过冷度已经下降，处于该类 LDHIs 的性能范围内。

最初的单尾季铵盐表面活性剂 AAs 生物降解性很差，而且毒性相当大。但通过添加阴离子、非离子或两性表面活性剂或聚合物（阴离子表面活性剂似乎是最理想的），其毒性会有所降低。例如，十二烷基硫酸钠和烷基醚硫酸铵。这同时也改善了 AAs 的性能，或允许使用较低的活性剂量。另有公司声称，使用单尾季铵盐与阴离子表面活性剂反离子作为离子对两亲混合物。这使其更易溶于油。例如，苄基二甲基椰油胺/妥尔油脂肪酸离子对，已有现场应用。通过添加阴离子表面活性剂改善季铵盐的性能，可能是由于阴离子表面活性剂作为离子对跟随季铵盐表面活性剂进入水合物晶体表面，产生额外的覆盖，并增强晶体的疏水性。其他的离子对也包括在一项专利中，还提出将不含卤化物的无机酸、有机酸和有机胺的反应产物作为 AAs，如马来酸酐/二甲基棕榈酰胺和水杨酸/椰油二胺这样的混合物。通过使用二元反离子，可以设想形成三元季铵盐 AAs（图 9.17）。这些季铵盐据称从乳液中易于分离、热稳定性好（没有早期季铵盐中的霍夫曼消除现象）、低腐蚀（与氯盐相比）和低毒。将两个季铵中心拉到水合物表面可能是有利的，因为与单季铵盐相比，双季铵盐已证明是更优良的水合物生长抑制剂。

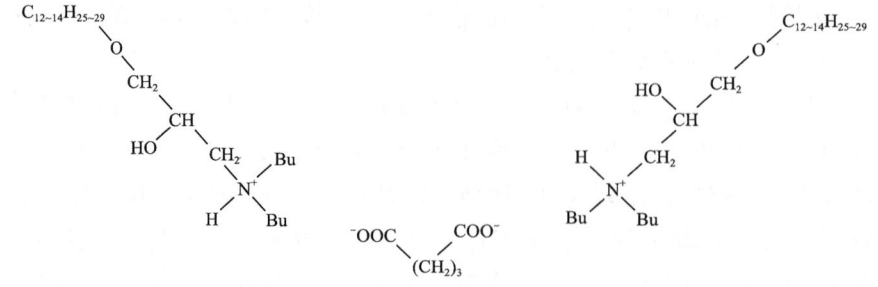

图 9.17 离子三元季铵盐 AAs

一家服务公司报告了对高含水和过冷度达 16.7℃ 的情况很有用的油溶性季铵盐。同一公司还有一项关于 AAs 的国际专利申请，声称有多种季铵盐和膦酸盐表面活性剂，其中一些具有可降解的酯或酰氨基连接，不过也提到了一些甜菜碱和氧化胺表面活性剂。还有一项更具体的专利声称某些氨基季铵盐表面活性剂可作为 AAs。例如，N，N-正二丁基—椰油丙氨基甲氧基甜菜碱。这种商业化的 AAs 据称比其他市售的 AAs 毒性更小，比原来的季铵盐 AAs 用量更少，能提供更好的油或水生产质量，而且不会破坏沥青质的稳定性。实验室测试表明，该产品甚至能在含水率高达 80%～90% 下防止水合物堵塞。然而，人们普遍认为，AAs 的应用仅限于 50% 以下的含水率；否则，浆液就会过于黏稠而无法运输。如果矿化度很高，理论上 AAs 可以在更高的含水率下工作。在这种情况下，并非所有的水都能转化为水合物。这是因为未转化的水将有更高的盐浓度，并可能受到足够的热力学抑制，不会进一步形成水合物。这种高矿化度水与液态烃可共同作为水合物颗粒的运输介质。另外，如果将 THIs 与 AAs 同时注入，THIs 可以成为水合物晶体的润滑剂。

据称将单尾季铵盐与胺盐和可选的溶剂混合，可以改善原有的性能。胺盐最好含有 1～3 个碳原子的烷基或羟烷基，或者可以使用铵盐。其他获得专利授权的季铵化表面活性剂 AAs，有二烷氧基季铵盐、季铵化 N，N'-二丁基烷基醚羧酸酯和季铵化烷基胺烷基二酯等。通过迈克尔加成，将长尾烷基丙烯酸酯添加到哌嗪或二烷基氨基丙胺中，然后进行季铵化处理，还可以得到另一类季铵盐 AAs（图 9.18）。

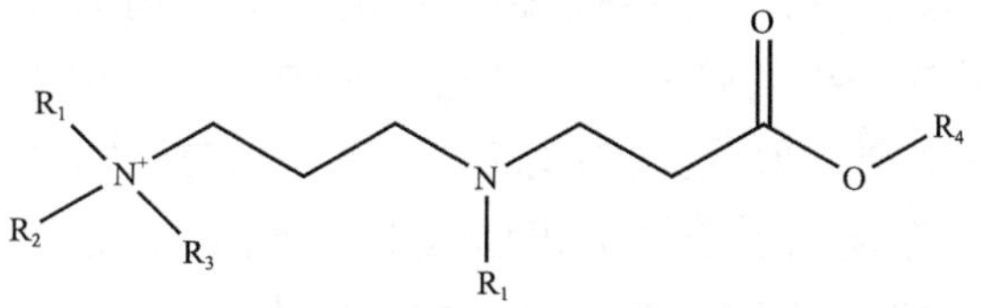

图 9.18 丙烯酸烷基迈克尔加成法制得的季铵盐 AAs

另有专利描述了从卤醇中制造羟基季铵盐表面活性剂 AAs 的方法（图 9.19）。其中，受阻胺（如三丁胺）与烷基卤化物的反应很困难，经常会得到产率很低的单尾季铵盐 AAs。卤代烃（如环氧氯丙烷）与长链醇反应，1，2-羟基卤化物反应产物将更容易与三丁胺反应，因为反应是通过质子化的环氧化物进行，可得到高产的季铵盐 AAs。

基于恶唑啉化合物的季铵盐表面活性剂 AAs 也已获得专利。特别有效的化合物如图 9.20 所示。恶唑啉化合物由卤代烃或环氧化物与二级胺和醛或酮反应形成。恶唑啉化合物不需要预先形成的恶唑烷与烷基化剂的反应，而是直接形成。

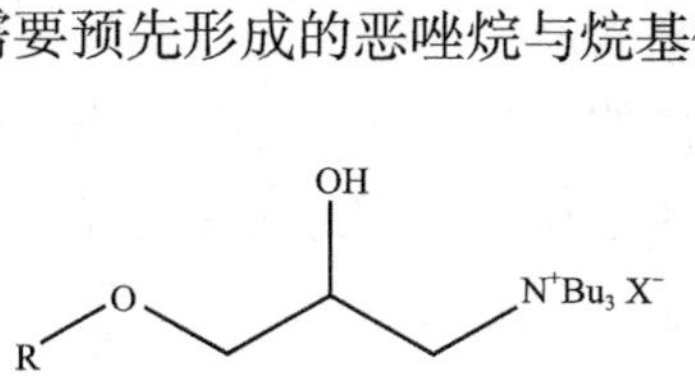

图 9.19 由卤代素制成的羟基季铵盐 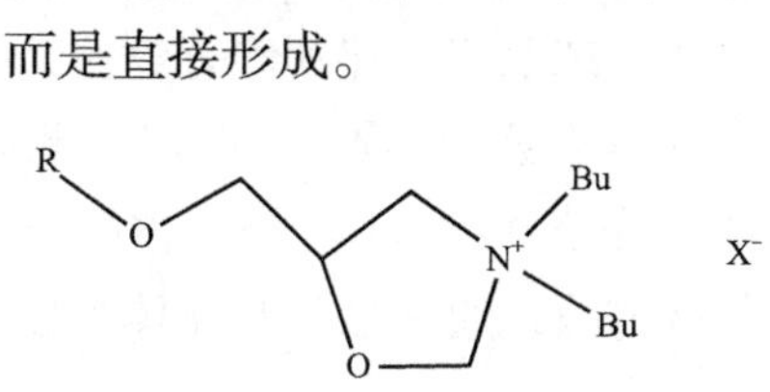　　图 9.20 恶唑啉类 AA 化合物示例（其中 R 优选为 C_{12}—C_{14}）

由于季铵型表面活性剂 AAs 通常有毒性，而且只能部分生物降解，因此试图寻找更环保的非季铵型 AAs 一直是研究主题，尽管过冷性能似乎会受到影响。这方面的例子有己内酰胺和烷基酰胺表面活性剂，其端基与 VCap 和烷基丙烯酰氨基 KHIs 中的侧基相同。对两性离子季铵盐 AAs 也进行了研究。另外，某些烷氧基化或酰基化非季铵的含氮化合物也有报道。如鼠李糖脂等可生物降解的生物表面活性剂，已证实在模型油和四氢呋喃水合物中表现出 AAs 行为。通过添加醇类共表面活性剂，性能得到优化。这些类别的 AAs 目前尚无市售产品。

椰油酰胺丙基二甲胺与包括甘油在内的少量其他化学品，可以作为 S I 型甲烷水合物的 AAs，甚至在含水率为 60%～100% 时也能发挥作用。但高含量的二氯化物会限制形成天然气水合物的最高含水率为 25%。非常高的盐浓度（如 20%～25% 的氯化钠溶液）可以分别分散 S I 型或 S II 型天然气水合物，而不需要使用 AAs。

山梨糖醇表面活性剂和与其他表面活性剂的混合剂已经成功地作为低含水率的 AAs。聚丙氧酸和其他一些破乳剂在搅拌条件下可作为 AAs，但在停产或再启动情况下无效。皂苷是由一个或多个亲水糖苷基团与亲脂性三萜衍生物结合组成。除季铵盐表面活性剂外，这些 AAs 目前还未上市。

9.2.3.3　管线用 AAs 的性能测试

可以在摇瓶、高压釜、转轮和循环管路中，测试抗凝剂的性能。摇瓶是各种测试条件下筛选 AAs 候选产品的一种很好方法，并且经常被服务公司用作确定产品资格的第一步。一个金属球在含有水合物形成流体的瓶子里来回摇晃，瓶的末端有一个观察窗（如蓝宝石），瓶底端的传感器检测水合物沉积物阻止球接触底端的时间。在试验过程中，视觉观察最重要。也有采用带观察窗和扭矩测量的搅拌式高压釜。一个水平转轮中流体被外部控制的磁铁推开。循环管道中的泵会破坏水合物颗粒，有助于 AAs 的效果。如果只对研究水合物在管环壁上的积聚感兴趣，可以通过在每次循环后，加热流体以熔化可运移的水合物，并在泵之后的管段中重新冷却。转轮法则完全避免使用泵。转轮的扭矩变化表明水合物沉积，与转轮的视窗观察同步。通常需要在不同含水率时，测试三种情况下的 AAs 性能：

（1）流动条件下，冷却到水合物区域；
（2）停产后，用预先形成的水合物浆液重新启动；
（3）在停产期间冷却到水合物区，一段时间后重新启动。

冷却速度会影响 AAs 的性能。此外，由于气体膨胀和管线中的焦耳—汤姆逊效应，冲击冷却会降至 0℃ 以下。类似情况下，过冷度会快速增加。已经为这种情况下的 LDHIs 设计了测试方法，通常是先对流体加压，然后将测试瓶迅速浸入冷水。

9.2.3.4　天然表面活性剂和非堵塞油

在实验室的不同含水率测试中，一些油已经证实不会形成水合物段塞。可能是其含有天然表面活性剂，可以防止水合物聚集和沉积，但不确定是 AAs 乳液机制，或是亲水合物机制在起作用。另外，尽管怀疑环烷酸、胶质和沥青质是 AAs 行为的原因，但没有研究团队能够确定这些非堵塞油品中活性成分的确切结构。这些成分即使含量很高，也不一定能阻止在多相流系统中的油品形成水合物堵塞。此外，原油的非堵塞特性似乎是其整体特性，而不是源于某些"神奇"组分。环戊烷水合物的研究结果表明，模型两亲聚合物和市售环烷酸混合物可能在水合物颗粒上具有表面活性，大大降低了颗粒间的积聚趋势。一些不堵塞的油可以在低含水率（<15%）的淡水中工作。通常含水率极限值随着矿化度的增加而增大。另有研究表明，除了油相或水相中的化学成分，高剪切力可以通过防止水合物颗粒聚集，而在水合物区域实现无堵塞。

理论上，应该可以用不堵塞的油进行多相集输，尽管目前还没有采用这种水合物堵塞预防策略的报道。有运营商的水合物预防策略已从非堵塞油的知识中受益。此外，一家北海运营商发现即使海底多相流管道在停产或启动后，当无意中停加 THIs 或加量不足时，也不会出现水合物堵塞，故不在该管道上使用 THIs。

9.2.3.5 天然气井用 AAs

如聚醚二胺（特别是氧丙烯型）等聚醚多胺可以在过量的水中分散水合物（图 9.21）。这类聚合物的分子量很低（通常低于 500）。在这种测试中，季铵化衍生物的性能比 PVCap 或几种亲水合物的季铵盐 AAs 更好。

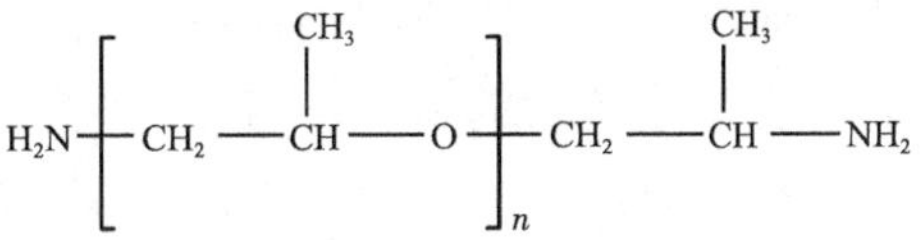

图 9.21 典型的聚醚二胺 AAs 的结构

近年来，聚醚胺已被用于 100 多口气井（多数在陆上）防止井下水合物堵塞。发现一定比例的甲醇和聚醚胺具有较强的协同效应。在完井液和压裂液中使用聚醚胺也已经获得了专利。如本章前面提到的，聚醚胺是基于 VCap 的 KHIs 的良好增效剂。一种聚醚胺类 AAs 的管道现场应用也已有报道。这种情况下，先注入 AAs，如果"检测"到管线内水合物堆积，就注入 THIs，直到管线畅通为止，然后再继续注入 AAs。THIs 注入的速率也会在停产时增加，以提供热力学保护。聚醚胺及其衍生物也是很好的缓蚀剂，特别是针对 H_2S 的腐蚀（见第 8 章）。聚醚/KHIs/THIs 新组合显示出改进的协同作用。与聚醚胺化学性质有关，只要流体有良好的搅拌，某些聚氧丙烯聚胺已证明在油/盐/气系统中显示出 AAs 特性。

9.3 水合物堵塞物的清除

在正常的生产作业中，水合物段塞可以在极端寒冷气温下或停产期间或海底油井形成，或在管道顶部形成。除去水合物堵塞有减压、加长型连续油管或牵引器、加热型电缆、注热油和化学处理等多种方式。

用氮气或其他一些不形成水合物，或只在比轻质油气水合物更高的压力下形成水合物的气体进行清洗，可以有效地清除轻烃气体—水的水合物。也有报道采用特制的推进型清管器强力清除水合物沉积物。

最近在多相流建模能力方面取得的进展，使操作人员能够对井筒或管道内任何一点存在水合物堵塞时的井喷影响进行详细评估。工业界也一直在开发新的工具，以在短时间内实现管道或井筒的减压。建议对段塞两边同时减压，以避免水合物段塞突然松动，成为危险弹射体。在北海一个海底温度低于0℃的深水油田，试验发现水合物段塞在0℃以下的减压过程中总会转化为冰塞。目前，还没有任何技术可以保证清除位于管道入口下游20km的冰塞。清除冰或水合物段塞的加长型连续油管正在开发，这对长距离深水回接特别有用。对油管—套管环空的注热油已证实可以熔化采气树的水合物塞。

有两种化学处理方法可以除去水合物堵塞：使用热力学抑制剂（THIs），使用发热化学品。第一种方法在现场通常使用，但第二种方法只在很少情况下使用。

9.3.1　使用 THIs

正如前面关于 THIs 的章节所提到的，这些化学品不仅有助于抑制天然气水合物的形成，而且可以熔化已经形成的水合物段（或冰）塞。甲醇和 MEG 是熔化水合物段塞的首选。而其在补救作业中，并不总是都有效。这是因为水合物段塞的性质是决定其熔化效率的关键。堵塞段的特性、THIs 的密度和黏度及 THIs 和塞子之间的接触面积是决定熔化效率的最重要因素。因此，THIs 与堵塞段的良好搅动有利于新鲜的 THIs 总能接触到堵塞段。对于非常多孔的段塞，低黏度甲醇似乎有利于熔化过程。否则，比油更高密度的 MEG 可能对其他类型的段塞更有效。MEG 不仅能渗透进堵塞段，还可以渗过水合物堵塞段表面的油。甲醇/MEG 和 $CaCl_2$/甲醇的抑制剂混合物并没有提高熔化效率。甲酸钾可能是一种潜在的低腐蚀性水合物解堵剂，其水溶液质量分数可以达到75%，密度很大。

9.3.2　发热化学品

发热化学品已用于除去水合物和蜡质沉积物。第一种方法最初是为除蜡而开发，使用铵盐和亚硝酸钠的混合物（制成亚硝酸铵）、酸催化剂（如乙酸）或前催化剂，在原位产生酸。如果使用酸酐等前催化剂，则在内置的时间延迟后产生热量。这种化学系统被称为"SGN 工艺"。酸催化了亚硝酸铵的分解，产生氯化钠、氮、水和热量。该方法已被用于除去海底采油树内的水合物堵塞。第二种方法依靠简单的原位酸碱反应（如盐酸加氢氧化钠）来产生热量。由于反应动力学迅速，酸和碱必须在如井上部的水合物塞顶部等预设的位置混合。如果设计和放置不当，使用发热化学品会带来一定风险。在最坏的情况下，对放热反应失控可能会导致管道材料受到意外的应力，并可能出现故障。

9.2.3.5 中讨论的聚醚胺，可以与酸原位混合产生放热反应，产生的聚醚胺化合物可作为 LDHIs，提供进一步保护。

参 考 文 献

[1] E. D. Sloan Jr. and C. A. Koh, *Clathrate Hydrates of Natural Gases*, 3rd ed. Boca Raton, FL: CRC Press, 2008.

[2] Y. F. Makogon, *Hydrates of Hydrocarbons*, Tulsa, OK: PennWell Publishing Company, 1997.

[3] (*a*) J. Carroll, *Natural Gas Hydrates—A Guide for Engineers*, Boston: Gulf Professional Publishing, Elsevier Science, 2003. (*b*) E. D. Sloan, "A Changing Hydrate Paradigm—From Apprehension to Avoidance to Risk Management," *Fluid Phase Equilibria* 228–229 (2005): 67.

[4] S. Mokhatab, R. J. Wilkens, and K. J. Leontaritis, "A Review of Strategies for Solving Gas-Hydrate Problems in Subsea Pipelines," *Energy Sources, Part A: Recovery, Utilization and Environmental Effects* 29 (2007): 39.

[5] O. Urdahl, A. H. Boernes, K. J. Kinnari, and R. Holme, "Operational Experience by Applying Direct Electrical Heating for Hydrate Prevention," SPE 85015, *SPE Production & Facilities* 19 (3) (2004): 161.

[6] R. G. Harris and J. Clapham, *Offshore* 59 (2) (1999).

[7] *Offshore* 66 (9) (2006).

[8] K. J. Kinnari, C. Labes-Carrier, K. Lunde, and L. Aaberge, International Patent Application WO/2006/027609.

[9] R. Larsen, A. Lund, V. Andersson, and K. W. Hjarbo, "Conversion of Water to Hydrate Particles," SPE 71550 (paper presented at the SPE Annual Technical Conference and Exhibition, New Orleans, LA, 30 September–3 October 2001).

[10] A. Lund, D. Lysne, R. Larsen, and K. W. Hjarbo, International Patent Application WO/2000/025062.

[11] L. D. Talley, D. J. Turner, and D. K. Priedeman, International Patent Application WO/2007/095399.

[12] L. Talley, "Hydrate Inhibition via Cold Flow: No Chemicals or Insulation," Proceedings of the 6th International Conference on Gas Hydrates Vancouver, British Columbia, Canada, 6–10 July 2008.

[13] D. Merino-Garcia and S. Correra, "Cold Flow: A Review of a Technology to Avoid Wax Deposition," *Petroleum Science Technology* 26 (2008): 446.

[14] H. Haghighi, R. Azarinezhad, A. Chapoy, R. Anderson, and B. Tohidi, "Hydraflow: Avoiding Gas Hydrate Problems," SPE 107335 (paper presented at the SPE Europec/EAGE Annual Conference and Exhibition, London, 11–14 June 2007).

[15] F. J. P. C. M. G. Verhelst, A. Twerda, J. P. M. Smeulers, M. C. Peters, M. Adrianus, S. P. C. Belfroid, and W. Schiferli, International Patent Application WO/2008/023979.

[16] R. Azarinezhad, A. Chapoy, F. Ahmadloo, R. Anderson, and B. Tohidi, "Hydraflow: A Novel Approach in Addressing Flow Assurance Problems," Proceedings of the 6th International Conference on Gas Hydrates Vancouver, British Columbia, Canada, 6–10 July 2008.

[17] R. Anderson, A. Chapoy, J. Tanchawanich, H. Haghighi, J. Lachwa-Langa, and B. Tohidi, "Binary Ethanol-Methane Clathrate Hydrate Formation in the System $CH_4-C_2H_5OH-H_2O$: Experimental Data and Thermodynamic Modelling," Proceedings of the 6th International Conference on Gas Hydrates Vancouver, British Columbia, Canada, 6–10 July 2008.

[18] U. C. Klomp and A. P. Mehta, "An Industry Perspective on the State-of-the Art of Hydrates Management," Proceedings of the 5th International Conference on Gas Hydrates, Trondheim, Norway, 12–16 June 2006.

[19] M. H. Yousif, "Effect of Underinhibition with Methanol and Ethylene Glycol on the Hydrate-Control Process," SPE 50972, *SPE Production & Facilities* 13 (3) (1998): 184.

[20] P. Hemmingsen and X. Li, "Hydrate Plugging Potential in Underinhibited Systems," Proceedings of the 6th International Conference on Gas Hydrates Vancouver, British Columbia, Canada, 6–10 July 2008.

[21] F. H. Fadnes, T. Jakobsen, M. Bylov, A. Holst, and J. D. Downs, "Studies on the Prevention of Gas

Hydrates Formation in Pipelines Using Potassium Formate as a Thermodynamic Inhibitor," SPE 50688 (paper presented at the SPE European Petroleum Conference, The Hague, Netherlands, 20-22 October 1998).

[22] E. G. Hammerschmidt, "Formation of Gas Hydrates in Natural Gas Transmission Lines," *Industrial & Engineering Chemistry* 26 (1934): 851.

[23] K. K. Østergaard, R. Masoudi, B. Tohidi, A. Danesh, and A. C. Todd, *Journal of Petroleum Science and Engineering* 48 (2005): 70.

[24] M. D. Jager, A. L. Ballard, and E. D. Sloan, "Comparison between Experimental Data and Aqueous-Phase Fugacity Model for Hydrate Prediction," *Fluid Phase Equilibria* 232 (2005): 25.

[25] A. L. Ballard and E. D. Sloan, "The Next Generation of Hydrate Prediction: Part III. Gibbs Energy Minimization Formalism," *Fluid Phase Equilibria* 218 (2004): 15.

[26] R. Masoudi, B. Tohidi, A. Danesh, A. C. Todd, and J. Yang, "Measurement and Prediction of Salt Solubility in the Presence of Hydrate Organic Inhibitors," SPE 87468, *SPE Production & Operations* 21 (2) (2006): 182.

[27] M. B. Tomson, A. T. Kan, and G. Fu, "Inhibition of Barite Scale in the Presence of Hydrate Inhibitors," SPE 87437, *SPE Journal* 10 (3) (2005): 256.

[28] M. A. Kelland, "History of the Development of Low Dosage Hydrate Inhibitors," *Energy & Fuels* 20 (2006): 825.

[29] M. Arjmandi, B. Tohidi, A. Danesh, and A. C. Todd, "Is Subcooling the Right Driving Force for Testing Low-Dosage Hydrate Inhibitors？," *Chemical Engineering Science* 60 (2005): 1313-1321.

[30] J.-P. Peytavy, P. Glénat, and P. Bourg, "Kinetic Hydrate Inhibitors—Sensitivity towards Pressure and Corrosion Inhibitors," IPTC 11233 (paper presented at the International Petroleum Technology Conference, Dubai, UAE, 4-6 December 2007).

[31] M. A. Kelland, J.-E. Iversen, K. Moenig, and K. Lekvam, "Feasibility Study for the Use of Kinetic Hydrate Inhibitors in Deep-Water Drilling Fluids," *Energy & Fuels* 22 (2008): 2405.

[32] N. J. Phillips and M. Grainger, "Development and Application of Kinetic Hydrate Inhibitors in the North Sea," SPE 40030 (paper presented at the Proceedings of the Annual Gas Technology Symposium, Calgary, Alberta, Canada, 15-18 March 1998).

[33] M. Varma-Nair, C. A. Costello, K. S. Colle, and H. E. King, "Thermal Analysis of Polymer-Water Interactions and Their Relation to Gas Hydrate Inhibition," *Journal of Applied Polymer Science* 103 (4) (2007): 2642.

[34] E. D. Sloan, U.S. Patent 5420370, 1995.

[35] J. Long, J. Lederhos, A. Sum, R. L. Christiansen, and E. D. Sloan, Proceedings of the 73rd Annual GPA Convention, New Orleans, LA, 7-9 March 1994.

[36] R. Hawtin and P. M. Rodger, "Polydiversity in Oligomeric Low Dosage Gas Hydrate Inhibitors," *Journal of Materials Chemistry* 16 (2006): 1934.

[37] N. Aldiwan, Y. Lui, A. Soper, H. Thompson, J. Creek, R. Westacott, E. D. Sloan, and C. Koh, "Neutron Diffraction and EPSR Simulations of the Hydrate Structure around Propane Molecules before and during Gas Hydrate Formation," Proceedings of the 6th International Conference on Gas Hydrates Vancouver, British Columbia, Canada, 6-10 July 2008.

[38] (a) T. Y. Makogon, R. Larsen, C. A. Knight, and E. D. Sloan Jr., "Melt Growth of Tetrahydrofuran Clathrate Hydrae and Its Inhibition," *Journal of Crystal Growth* 179 (1997): 258. (b) H. E. King, J. L. Hutter, M. Y. Lin, and T. Sun, *Journal of Chemical Physics* 112 (2000): 2523.

[39] (a) B. J. Anderson, J. W. Tester, G. P. Borghi, and B. L. Trout, "Properties of Inhibitors of Methane Hydrate Formation via Moecular Dynamics," *Journal of the American Chemical Society* 127 (2005): 17852. (b) A. Cruz-Torres, A. Romero-Martinez, and A. Galano, "Computational Study on the Antifreeze Glycoproteins as Inhibitors of Clathrate-Hydrate Formation," *Chemical Physics and Physical Chemistry* 9 (11) (2008): 1630.

[40] T. J. Carver, M. G. B. Drew, and P. M. Rodger, "Inhibition of Crystal Growth in Methane Hydrate," *Journal of the Chemical Society, Faraday Transactions* 91 (1995): 3449.

[41] (a) D. A. Gomez Gualdron and P. B. Balbuena, "Classical Molecular Dynamics of Clathrate-Methane-Water-Kinetic Inhibitor Composite Systems," *Journal of Physical Chemistry* 111 (2007): 15554. (b) R. W. Hawtin, D. Quigley, and P. M. Rodger, "Gas Hydrate Nucleation and Cage Formation at a Water/Methane Interface," *Physical Chemistry Chemical Physics* 10 (2008): 4853.

[42] H. Zeng, H. Lu, E. Huva, V. K. Walker, and J. A. Ripmeester, "Differences in Nucleator Adsorption May Explain Distinct Inhibition Activities of Two Gas Hydrate Kinetic Inhibitors," *Chemical Engineering Science* 63 (2008): 4026.

[43] U. C. Klomp and A. P. Mehta, "Validation of Kinetic Inhibitors for Sour Gas Fields," IPTC 11374 (paper presented at the International Petroleum Technology Conference, Dubai, UAE, 4–6 December 2007).

[44] B. Huang, Y. Wang, S. Zhang, and Y. Ao, "Kinetic Model of Fixed Bed Reactor with Immobilized Microorganisms for Removing Low Concentration SO," *Journal of Natural Gas Chemistry* 16 (2007): 81.

[45] S. Szymczak, K. Sanders, M. Pakulski, and T. Higgins, "Chemical Compromise: A Thermodynamic and Low-Dose Hydrate-Inhibitor Solution for Hydrate Control in the Gulf of Mexico," SPE 96418, *SPE Project Facilities & Construction* 1 (4) (2006): 1.

[46] L. W. Clark and J. Anderson, "Low Dosage Hydrate Inhibitors (LDHI): Further Advances and Developments in Flow Assurance Technology and Applications Concerning Oil and Gas Production Systems," IPTC 11538 (paper presented at the International Petroleum Technology Conference, Dubai, UAE, 4–6 December 2007).

[47] E. D. Sloan, U.S. Patent 5432292, 1995.

[48] E. D. Sloan, R. L. Christiansen, J. Lederhos, V. Panchalingam, Y. Du, A. K. W. Sum, and J. Ping, U.S. Patent 5639925, 1997.

[49] C. B. Argo, R. A. Blaine, C. G. Osborne, and I. C. Priestly, "Commercial Deployment of Low Dosage Hydrate Inhibitors in a Southern North Sea 69 Kilometer Wet-Gas Subsea Pipeline," SPE 37255 (paper presented at the SPE International Symposium on Oilfield Chemistry, Houston, TX, February 1997).

[50] L. D. Talley and G. F. Mitchell, "Application of Kinetic Hydrate Inhibitor in Black-Oil Flowlines," SPE 56770 (paper presented at the SPE Annual Technical Conference and Exhibition, September 1999).

[51] S. B. Fu, L. M. Cenegy, and C. Neff, "A Summary of Successful Field Applications of A Kinetic Hydrate Inhibitor," SPE 65022 (paper presented at the SPE International Symposium on Oilfield Chemistry, Houston, TX, 13–16 February 2001).

[52] (a) K. Bakeev, R. Myers, J.-C. Chuang, T. Winkler, and A. Krauss, U.S. Patent 6242518, 2001. (b) M. Angel, S. Stein, and K. Neubecker, International Patent Application WO/2001/066602.

[53] T. Namba, Y. Fujii, T. Saeki, and H. Kobayashi, International Patent Application WO96/37684, 1996.

[54] M. Angel, K. Neubecker, and S. Stein, International Patent Application WO/2004/042190, 2004.

[55] L. D. Talley and M. Edwards, "First Low Dosage Hydrate Inhibitor Is Field Proven in Deepwater," *Pipeline and Gas Journal* 226 (1999): 44.

[56] V. Thieu, K. Bakeev, and J. S. Shih, U.S. Patent 6359047, 2002.

[57] D. J. Freeman, D. J. Irvine, J. Kitching, and C. S. Rogers, International Patent Application WO/2006/051265.

[58] K. Bakeev, J.-C. Chuang, T. Winkler, M. A. Drzewinski, and D. E. Graham, U.S. Patent 6281274, 2001.

[59] U. Dahlmann, M. Feustel, and C. Kayser, U.S. Patent 7297823, 2007.

[60] (a) A. Maximilian, K. Neubecker, and A. Sanner, U.S. Patent 6867262, 2005. (b) R. Widmaier, L. Wegmann, A. Mauri, K. Mathauer, W. Jahnel, L. H. Taboada, K. Neubecker, and A. Khvorost, U.S. Patent Application 20080255326.

[61] K. Colle, L. D. Talley, and J. M. Longo, World Patent Application WO/2005/005567, 2005.

[62] B. Fu, S. Neff, A. Mathur, and K. Bakeev, "Application of Low-Dosage Hydrate Inhibitors in Deepwater Operations," SPE 78823, *SPE Production and Facilities* 17 (2002): 133.

[63] M. Jurek, M. Alexandre, and S. Bell, "New Approaches to Low Dose Gas Hydrate Treatment," *Chemicals in the Oil Industry*, Royal Society of Chemistry, Manchester, 5-7 November 2007.

[64] A. Rasch, A. Mikalsen, T. Austvik, L. H. Gjertsen, and X. Li, "Evaluation of a Kinetic Inhibitor with Focus on the Pressure and Fluid Dependency," Proceedings of the 4th International Conference on Gas Hydrates, Yokohama, Japan, 19-23 May 2002.

[65] M. Pakulski and J. C. Dawson, U.S. Patent 7067459, 2006.

[66] B. Fu, "The Development of Advanced Kinetic Hydrate Inhibitors," *Proceedings of Chemistry in the Oil Industry VII*, Royal Society of Chemistry, Manchester, UK, 13-14 November 2002, 264.

[67] J. M. Cohen, P. F. Wolf, and W. D. Young, U.S. Patent Application 5723524, 1998.

[68] P. Klug and F. Holtrup, European Patent Application EP0933415A2, 1999.

[69] S. Duncum, A. R. Edwards, and C. G. Osborne, International Patent Application WO96/04462, 1996.

[70] D. Leinweber and M. Feustel, International Patent Application WO/2006/092208, 2006.

[71] M. Arjmandi, S. Ren, and B. Tohidi, "Anti-agglomeration and Synergism Effect of Quaternary Ammonium Zwitterions," Proceedings of the 5th International Conference on Gas Hydrates, Trondheim, Norway, 12-16 June 2005.

[72] P. Klug, M. Feustel, and V. Frenz, International Patent Application WO98/03615, 1998.

[73] I. K. Meier, R. J. Goddard, and M. E. Ford, U.S. Patent Application 7452848, 2008.

[74] D. Hurd and M. Pakulski, "Uncovering a Dual Nature of Polyether Amines Hydrate Inhibitors," Proceedings of the 5th International Conference on Gas Hydrates, Trondheim, Norway, 12-16 June 2006.

[75] K. Bakeev, R. Myers, and D. E. Graham, U.S. Patent 6180699, 2001.

[76] M. J. Jurek, International Patent Application WO/2007/143489.

[77] J. D. Lee and P. Englezos, "Enhancement of the Performance of Gas Hydrate Kinetic Inhibitors with Polyethylene Oxide," *Chemical Engineering Science* 60 (2005): 5323.

[78] U. C. Klomp, WO Patent Application 01/77270, 2001.

[79] G. T. Rivers and D. L. Crosby, International Patent Application WO/2004/022909.

[80] G. T. Rivers and D. L. Crosby, WO Patent Application WO/2004/022910.

[81] J. Anderson and L. W. Clark, "Development of Effective Combined Kinetic Hydrate Inhibitor/Corrosion Inhibitor (KHI/CI) Products," Proceedings of the 5th International Conference on Gas Hydrates, Trondheim, Norway, 12-16 June 2006.

[82] B. Fu, "Development of Non-interfering Corrosion Inhibitors for Sour Gas Pipelines with Co-injection of Kinetic Hydrate Inhibitors," Paper 07666 (paper presented at the NACE CORROSION Conference, Nashville, TN, 11-15 March 2007).

[83] A. W. R. MacDonald, SPE; M. Petrie, J. J. Wylde, SPE; A. J. Chalmers, and M. Arjmandi, "Clariant Oil Services, Field Application of Combined Kinetic Hydrate and Corrosion Inhibitors in the Southern North Sea: Case Studies," SPE 99388 (paper presented at the SPE Gas Technology Symposium, Calgary, Alberta, Canada, 15-17 May 2006).

[84] J. Moloney, W. Mok, and C. Gamble, "Corrosion and Hydrate Control in Wet Sour Gas Transmission Systems," SPE 115074 (paper presented at the SPE Asia Pacific Oil and Gas Conference and Exhibition, Perth, Australia, 20-22 October 2008).

[85] M. Pakulski, "Accelerating Effect of Surfactants on Gas Hydrates Formation," SPE 106166 (paper presented at the International Symposium on Oilfield Chemistry, Houston, TX, 28 February-2 March 2007).

[86] R. Masoudi and B. Tohidi, "Experimental Investigation on the Effect of Commercial Oilfield Scale Inhibitors on the Performance of Low Dosage Hydrate Inhibitors (LDHI)," Proceedings of the 5th International Conference on Gas Hydrates, Trondheim, Norway, 12-16 June 2006.

[87] T. A. Swanson, M. Petrie, and T. R. Sifferman, "The Successful Use of Both Kinetic Hydrate and Paraffin Inhibitors Together in a Deepwater Pipeline with a High Water Cut in the Gulf of Mexico," SPE 93158, Proceedings of the SPE International Symposium on Oilfield Chemistry, Houston, TX, 2-4 February 2005.

[88] D. Leinweber and M. Feustel, International Patent Application WO/2006/084613.

[89] D. Leinweber and M. Feustel, International Patent Application WO/2007/054226.

[90] (a) M. Arjmandi, D. Leinweber, and K. Allan, "Development of A New Class of Green Kinetic Hydrate Inhibitors" (paper presented at the 19th International Oilfield Chemical Symposium, Geilo, Norway, March 2008). (b) D. Leinweber and M. Feustel, U.S. Patent Application 20080214865. (c) D. Leinweber, A. R. Roesch, and M. Feustel, U.S. Patent Application 20090054268.

[91] K. S. Colle, C. A. Costello, L. D. Talley, J. M. Longo, R. H. Oelfke, and E. Berluche, International Patent Application WO96/08672, 1996.

[92] T. Namba, Y. Fujii, T. Saeki, and H. Kobayashi, International Patent Application WO96/38492, 1996.

[93] M. A. Kelland, T. M. Svartaas, and J. Ovsthus, "A New Class of Kinetic Hydrate Inhibitor," Proceedings of the 3rd Natural Gas Hydrate Conference, Salt Lake City, UT, July 1999.

[94] M. A. Kelland, P. M. Rodger, and T. Namba, International Patent Application WO98/53007, 1998.

[95] K. S. Colle, C. A. Costello, L. D. Talley, R. H. Oelfke, and E. Berluche, International Patent Application WO96/41786, 1996.

[96] Y. Tang, Y. Ding, and G. Zhang, "Role of Methyl in Phase Transition of Poly (N-Isopropylmethcrylamide)," *Journal of Physical Chemistry B* 112 (29) (2008): 8447.

[97] V. Thieu, K. Bakeev, and J. S. Shih, U.S. Patent 6451891, 2002.

[98] L. D. Talley and R. H. Oelfke, International Patent Application WO97/07320, 1997.

[99] M. A. Kelland and P. Klug, International Patent Application WO98/23843, 1998.

[100] K. S. Colle, C. A. Costello, and L. D. Talley, Canadian Patent Application 96/2178371, 1996.

[101] K. S. Colle, L. D. Talley, R. H. Oelfke, and E. Berluche, International Patent Application WO96/08673, 1996.

[102] K. S. Colle, C. A. Costello, L. D. Talley, R. H. Oelfke, and E. Berluche, International Patent

Application WO96/41834, 1996.

[103] (a) M. A. Kelland, L. Del Villano, and R. Kommedal, "A Class of Kinetic Hydrate Inhibitor with Good Biodegradability," *Energy & Fuels* 22 (2008): 3143. (b) M. A. Kelland, "Additives for Inhibiting Gas Hydrate Formation," International Patent Application WO/2008/023989.

[104] K. Kannan and A. D. Punase, "Low-Dosage, High Efficiency, and Environment-Friendly Inhibitors: A New Horizons in Gas Hydrates Mitigation in Production Systems," SPE 120905 (paper presented at the SPE International Symposium on Oilfield Chemistry, The Woodlands, TX, 20–22 April 2009).

[105] K. S. Colle, L. D. Talley, R. H. Oelfke, and E. Berluche, International Patent Application WO96/41784, 1996.

[106] D. G. Peiffer, C. A. Costello, L. D. Talley, and P. J. Wright, International Patent Application WO99/64718, 1999.

[107] (a) V. Walker, J. A. Ripmeester, and H. Zeng, International Patent Application WO03/087532, 2003. (b) H. Zeng, A. Brown, B. Wathen, J. A. Ripmeester, and V. K. Walker, "Antifreeze Proteins: Adsorption to Ice, Silica and Gas Hydrates," Proceedings of the 5th International Conference on Gas Hydrates, Trondheim, Norway, 13–16 June 2005.

[108] V. Walker, H. Zeng, R. Gordienko, M. Kuiper, E. Huva, Z. Wu, D. Miao, and J. Ripmeester, "The Mysteries of the Memory Effect and Its Elimination with Antifreeze Proteins," Proceedings of the 6th International Conference on Gas Hydrates Vancouver, British Columbia, Canada, 6–10 July 2008.

[109] M. A. Kelland, unpublished results.

[110] (a) C. R. Burgazli, World Patent Application WO/2004/111161, 2004. (b) C. R. Burgazli, R. C. Navarrete, and S. L. Mead, "New Dual Purpose Chemistry for Gas Hydrate And Corrosion Inhibition," Paper 2003-070 (paper presented at the Petroleum Society's Canadian International Petroleum Conference, Calgary, Alberta, Canada, 10–12 June 2003) (Also *Journal of Canadian Petroleum Technology* 44 (2005): 47).

[111] C. Xiao and H. Adidharma, *Chemical Engineering Science*, 64 (2009): 1522.

[112] U. Klomp, "The World of LDHI: From Conception to Development to Implementation," Proceedings of the 6th International Conference on Gas Hydrates Vancouver, British Columbia, Canada, 6–10 July 2008.

[113] M. Arjmandi, S.-R. Ren, J. Yang, and B. Tohidi, "Anti-agglomeration and Synergism Effect of Quaternary Ammonium Zwitterions," Proceedings of the 4th International Conference on Natural Gas Hydrates, Yokohama, Japan, 19–23 May 2002.

[114] M. A. Kelland, T. M. Svartaas, and L. A. Dybvik, "Control of Hydrate Formation by Surfactants and Polymers," SPE 28506, Proceedings of the SPE 69th Annual Technical Conference and Exhibition, New Orleans, LA, October 1994.

[115] (a) C. Duchateau, J.-L. Peytavy, P. Glenat, T.-E. Pou, M. Hidalgo, and C. Dicharry, "Laboratory Evaluation of Kinetic Hydrate Inhibitors: A New Procedure for Improving the Reproducibility of Measurements," Proceedings of the 6th International Conference on Gas Hydrates Vancouver, British Columbia, Canada, 6–10 July 2008. (b) C. Duchateau, J.-L. Peytavy, P. Glenat, T.-E. Pou, M. Hidalgo, and C. Dicharry, *Energy & Fuels* 23 (2) (2009): 962–966.

[116] L. D. Talley, G. F. Mitchell, and R. H. Oelfke, *Annals of the New York Academy of Sciences* 912 (2000): 314.

[117] M. J. Anselme, M. J. Reijnhout, and U. C. Klomp, International Patent Application WO93/25798, 1993.

[118] L. D. Talley and K. Colle, International Patent Application WO/2006/110192, 2006.

[119] L. D. Talley, International Patent Application WO/2007/111789.

[120] A. Sugier and J. P. Durand, U.S. Patent 5244878, 1993.

[121] J. P. Durand, A. S. Delion, P. Gateau, and M. Velly, European Patent Application 740048, 1995.

[122] M. Velly, M. Hillion, A. Sinquin, and J. P. Durand, European Patent Application EP905350.

[123] A. Sinquin, C. Dalmazzone, A. Audibert, and V. Pauchard, U.S. Patent Application US2006015102.

[124] A. Rojey, M. Thomas, A.-S. Delion, and J.-P. Durand, U.S. Patent 5816280, 1998.

[125] T. Palermo, A. Sinquin, H. Dhulesia, and J. M. Fourest, Proceedings of Multiphase, BHR Group, 1997, 1333.

[126] T. Palermo, C. B. Argo, S. P. Goodwin, and A. Henderson, "Flow Loop Tests on a Novel Hydrate Inhibitor to Be Deployed in the North Sea ETAP Field," Proceedings of the 3rd International Conference on Natural Gas Hydrates, *Annals of the New York Academy of Sciences*, 2000.

[127] U. C. Klomp, V. C. Kruka, and R. Reijnhart, "Low Dosage Hydrate Inhibitors and How They Work," Proceedings of Symposium *Controlling Hydrates, Waxes and Asphaltenes*, IBC Conference, Aberdeen, UK, October 1997.

[128] U. C. Klomp, V. C. Kruka, and R. Reijnhart, WO Patent Application 95/17579, 1995.

[129] R. Azarinezhad, A. Chapoy, R. Anderson, and B. Tohidi, "HYDRAFLOW: A Multiphase Cold Flow Technology for Offshore Flow Assurance Challenges," OTC 19485 (paper presented at the Offshore Technology Conference, Houston, TX, 5–8 May 2008).

[130] U. C. Klomp and R. Reijnhart, International Patent Application WO96/34177, 1996.

[131] L. H. Gjertsen and F. H. Fadnes, "Measurements and Predictions of Hydrate Equilibrium Conditions, Gas Hydrates: Challenges for the Future," *Annals of the New York Academy of Sciences* 912 (2000): 722–734.

[132] B. Tohidi, R. W. Burgass, A. Danesh, K. K. Ostergaard, and A. C. Todd, "Improving the Accuracy of Gas Hydrate Dissociation Point Measurements," *Annals of the New York Academy of Sciences* 912 (2000): 924.

[133] L. M. Frostman, C. G. Gallagher, S. Ramachandran, and K. Weispfennig, "Ensuring Systems Compatibility for Deepwater Chemicals," SPE 65006 (paper presented at the SPE International Symposium on Oilfield Chemistry, Houston, TX, 13–16 February 2001).

[134] L. M. Frostman, "Anti-agglomerant Hydrate Inhibitors for Prevention of Hydrate Plugs in Deepwater Systems," SPE 63122 (paper presented at the SPE Annual Technical Conference and Exhibition, Dallas, 1–4 October 2000).

[135] L. M. Frostman and J. L. Przybylinski, "Successful Applications of Anti-agglomerant Hydrate Inhibitors," SPE 65007 (paper presented at the SPE International Symposium on Oilfield Chemistry, Houston, TX, 13–16 February 2001).

[136] A. P. Mehta, P. B. Herbert, E. R. Cadena, and J. P. Weatherman, "Successful Applications of Antiagglomerant Hydrate Inhibitors," OTC 14057, Proceedings of the Offshore Technology Conference, Houston, TX, 6–9 May 2002.

[137] V. Thieu and L. M. Frostman, "Use of Low-Dosage Hydrate Inhibitors in Sour Systems," SPE 93450 (paper presented at the SPE International Symposium on Oilfield Chemistry, Houston, TX, 2–4 February 2005).

[138] A. Buijs, G. van Gurp, T. Nauta, R. Smakman, and A. M. Wit-Van Grootheest, U.S. Patent 6379294, 2002.

[139] U. C. Klomp, International Patent Application WO99/13197, 1999.

[140] U. C. Klomp, M. Le Clerq, and S. Van Kins, "The First Use of Hydrate Anti-agglomerant for a Fresh Water Producing Gas/Condensate Field," Proceedings of the 2nd Petromin Deepwater Conference, Shangri-La, Kuala Lumpur, Malaysia, 18-20 May 2004.

[141] G. T. Rivers, L. M. Frostman, J. L. Pryzbyliski, and J.-A. McMahon, U.S. Patent 6620330, 2003.

[142] D. L. Crosby, G. T. Rivers, and L. M. Frostman, International Patent Application WO/2005/116399.

[143] B. Fu, C. Houston, and T. Spratt, "New Generation LDHI with an Improved Environmental Profile," Proceedings of the 5th International Conference on Gas Hydrates, Trondheim, Norway, 13-16 June 2005.

[144] P. A. Spratt, International Patent Application WO/2006/052455.

[145] L. Cowie, W. Shero, N. Singleton, N. Byrne, and L. Kauffman, *Deepwater Technology*. Gulf Publishing Co., 2003.

[146] (a) R. Alapati and A. Davies, "Oil-Soluble LDHIs Represents New Breed of Hydrate Inhibitor," *Journal of Petroleum Technology* (2007): 28. (b) R. Alapati, J. Lee, and D. Beard, "Flow Assurance Chemistry Found Effective at High Watercuts," *World Oil* (2008): 39.

[147] R. Alapati, J. Lee, and D. Beard, "Two Field Studies Demonstrate that New AA LDHI Chemistry Is Effective at High Water Cuts without Impacting Oil/Water Quality," OTC 19505 (paper presented at the Offshore Technology Conference, Houston, TX, 5-8 May 2008).

[148] (a) V. Panchalingam, M. G. Rudel, and S. H. Bodnar, International Patent Application WO/2005/042675. (b) V. Panchalingam, M. G. Rudel, and S. H. Bodnar, U.S. Patent 7, 381, 689, 2008.

[149] J. L. Przybylinski and G. T. Rivers, U.S. Patent 6596911B2, 2003.

[150] U. Dahlmann and M. Feustel, U.S. Patent 7183240, 2007.

[151] U. Dahlmann and M. Feustel, U.S. Patent 7214814, 2007.

[152] U. Dahlmann and M. Feustel, U.S. Patent 7323609, 2008.

[153] G. T. Rivers, International Patent Application WO/2008/008697.

[154] G. T. Rivers, J. Tian, and J. A. Hackerott, International Patent Application WO/2008/063794.

[155] M. A. Kelland, T. M. Svartaas, J. Ovsthus, T. Tomita, and K. Mizuta, "Studies on Some Alkylamide Surfactant Gas Hydrate Anti-agglomerants," *Chemical Engineering Science* 61 (2006): 4290.

[156] Z. Huo, E. Freer, M. Lamar, B. Sannigrahi, D. M. Knauss, and E. D. Sloan, "Hydrate Plug Prevention by Anti-agglomeration," *Chemical Engineering Science* 56 (2001): 4979.

[157] M. A. Kelland, T. M. Svartaas, J. Ovsthus, T. Tomita, and J. Chosa, "Studies on Some Zwitterionic Surfactant Gas Hydrate Anti-agglomerants," *Chemical Engineering Science* 61 (2006): 4048.

[158] M. Hellsten and H. Oskarsson, International Patent Application WO/2007/107502.

[159] (a) A. Firoozabadi and J. D. York, "Comparing Effectiveness of Rhamnolipid Biosurfactant with a Quaternary Ammonium Salt Surfactant for Hydrate Anti-agglomeration," *Journal of Physical Chemistry B* 112 (2008): 845. (b) J. D. York and A. Firoozabadi, "Alcohol Cosurfactants in Hydrate Anti-agglomeration," *Journal of Physical Chemistry B* 112 (2008): 10455.

[160] A. Firoozabadi and J. D. York, "Use of Biosurfactants in Hydrate Anti-agglomeration," SPE 116214 (paper presented at the SPE Annual Technical Conference and Exhibition, Denver, CO, 22-24 September 2008).

[161] O. Urdahl, A. Lund, P. Mork, and T. Nilsen, "Inhibition of Gas Hydrate Formation by Means of Chemical Additives: I. Development of an Experimental Set-Up for Characterization of Gas Hydrate

Inhibitor Efficiency with Respect to Flow Properties and Deposition," *Chemical Engineering Science* 50 (5) (1995): 863.

[162] D. Lippmann, D. Kessel, and I. Rahimian, "Gas Hydrate Nucleation and Growth Kinetics in Multiphase Transport Systems," Proceedings of the 5th International Offshore and Polar Engineering Conference, The Hague, Netherlands, 11–16 June 1995.

[163] A. Sinquin, X. Bredzinsky, and V. Beunat, "Kinetic of Hydrates Formation: Influence of Crude Oils," SPE 71543, Proceedings of the SPE Annual Technical Conference and Exhibition, New Orleans, LA, 30 September–3 October 2001.

[164] P. V. Hemmingsen, X. Li, J.-L. Peytavy, and J. Sjoblom, "Hydrate Plugging Potential of Original and Modified Crude Oils," *Journal of Dispersion Science and Technology* 28 (2007): 371.

[165] K. Erstad, S. Hoeiland, T. Barth, and P. Fotland, "Isolation and Molecular Identification of Hydrate Surface Active Components in Petroleum Acid Fractions," Proceedings of the 6th International Conference on Gas Hydrates (ICGH 2008), Vancouver, British Columbia, Canada, 6–10 July 2008.

[166] T. Palermo, A. Mussumeci, and E. Leporcher, "Could Hydrate Plugging Be Avoided Because of Surfactant Properties of the Crude and Appropriate Flow Conditions," OTC0 16681 (paper presented at the Offshore Technology Conference Houston, TX, 2004).

[167] (a) R. M. T. Camargo, M. A. L. Goncalves, J. R. T. Montesami, C. A. B. R. Cardoso, and K. Minami, "A Perspective View of Flow Assurance in Deepwater Fields in Brazil," OTC 16687 (paper presented at the Offshore Technology Conference, Houston, TX, 2004). (b) K. Kinnari and P. Fotland, Statoil, personal communication.

[168] M. Pakulski, U.S. Patent 6331508, 2001.

[169] M. Pakulski, U.S. Patent 5741758, 1998.

[170] M. Pakulski, "Twelve Years of Laboratory and Field Experience for Polyether Polyamine Gas Hydrate Inhibitors," Proceedings of the 6th International Conference on Gas Hydrates, Vancouver, British Columbia, Canada, 6–10 July 2008.

[171] M. Pakulski, U.S. Patent European Patent Application 6025302, 2000.

[172] D. Lovell and M. Pakulski, "Hydrate Inhibition in Gas Wells Treated with Two Low Dosage Hydrate Inhibitors," SPE 75668 (paper presented at the SPE Gas Technology Symposium, Alberta, Canada, 2002).

[173] D. Budd, D. Hurd, M. Pakulski, and T. D. Schaffer, "Enhanced Hydrate Inhibition in Alberta Gas Field," SPE 90422 (paper presented at the SPE Annual Technical Conference and Exhibition, Houston, TX, 26–29 September 2004).

[174] M. Pakulski and J. C. Dawson, U.S. Patent 6756345, 2004.

[175] S. R. Davies, J. A. Boxall, C. Koh, E. D. Sloan, P. V. Hemmingsen, K. J. Kinnari, and Z.-G. Xu, "Predicting Hydrate Plug Formation in a Subsea Tieback," SPE 115763 (paper presented at the SPE Annual Technical Conference and Exhibition, Denver, CO, 21–24 September 2008).

[176] A. F. Harun, T. E. Krawietz, and M. Erdogmus, "When Flow Assurance Fails: Melting Hydrate Plugs in Dry-Tree Wells," *World Oil* 228 (2007): 51.

[177] A. F. Harun, T. E. Krawietz, and M. Erdogmus, "Hydrate Remediation in Deepwater Gulf of Mexico Dry-Tree Wells: Lessons Learned," *SPE Production & Operations* 22 (4) (2007): 472.

[178] T. Austvik, X. Li, and L. H. Gjertsen, "Hydrate Plug Properties—Formation and Removal of Plugs," Proceedings of Gas Hydrates: Challenges for the Future, *Annals of the New York Academy of Sciences* 912 (2000): 294.

[179] X. Li, L. H. Gjertsen, and T. Austvik, "Thermodynamic Inhibitors for Hydrate Plug Melting," Proceedings of Gas Hydrates: Challenges for the Future, *Annals of the New York Academy of Sciences* 912 (2000): 822.

[180] X. Li, L. H. Gjertsen, and T. Austvik, "Melting Hydrate Plugs by Thermodynamic Inhibitors—Plug Properties and Melting Efficiencies," Proceedings of the 4th International Conference on Gas Hydrates, Yokohama, Japan, 19–23 May 2002.

[181] C. N. Khalil, European Patent Application EP0909873.

[182] C. N. Khalil, N. D. O. Rocha, and L. C. F. Leite, U.S. Patent 6035933, 2000.

[183] L. C. C. Marques, C. A. Pedroso, and L. F. Neumann, "A New Technique to Solve Gas Hydrate Problems in Subsea Christmas Trees," SPE 77572, *SPE Production & Facilities* 19 (4) (2004): 253.

[184] J. Chatterji and J. E. Griffith, U.S. Patent Application 5713416, 1998.

[185] R. Von Flatern, "Pulling the Plug from Deep Pipe," *Offshore Engineer* September (2006): 71.

[186] A. C. Gulbrandsen and T. M. Svartaas, "Influence of Formation Temperature and Inhibitor Concentration on the Dissociation Temperature for Hydrates Formed with Poly Vinyl Caprolactam," Proceedings of the 6th International Conference on Gas Hydrates, Vancouver, British Columbia, Canada, 6–10 July 2008.

[187] S. Gao, "Investigation of Interactions between Gas Hydrates and Several Other Flow Assurance Elements," *Energy & Fuels* 22 (5) (2008): 3150.

[188] P. Buchanan, A. K. Soper, H. Thompson, R. E. Westacott, J. L. Creek, G. Hobson, and C. A. Koh, "Search for Memory Effects in Methane Hydrate: Structure of Water Before Hydrate Formation and after Hydrate Decomposition," *Journal of Chemical Physics* 123 (2005): 164507.

[189] P. M. Rodger, "Methane Hydrate: Melting and Memory," *Annals of the New York Academy of Sciences* 912 (2000): 474.

[190] L. Ding, C. Geng, Y. Zhao, X. He, and H. Wen, "Molecular Dynamics Simulation for Surface Melting and Self-Preservation Effect of Methane Hydrate," *Science in China Series B: Chemistry* 51 (7) (2008): 651.

[191] W. G. Chapman, S. Gao, M. Yarrison, K. Song, and W. House, "Equilibrium and Dynamics of Gas Hydrates," Paper 7, Proceedings of the 10th International Conference on PPEPPD, Snowbird, UT, Engineering Conferences International, NY, 16–21 May 2004.

[192] M. A. Kelland, T. M. Svartaas, and L. Dybvik Andersen, *Journal of Petroleum Science and Engineering* 64 (2009): 1.

[193] J. W. Lachance, C. Koh, and E. D. Sloan, *Chemical Engineering Science* 64 (2009): 180.

[194] V. F. Rodrigues, L. G. Loures, F. D. Siqueira, and H. M. Frota, "Hydrate Blockage in Subsea Water Injection Wells—Causes and Preventive Procedures," SPE 120273 (paper presented at the 8th European Formation Damage Conference, Scheveningen, Netherlands, 27–29 May 2009).

[195] A. Nengkoda, H. Reerink, A. Hase, I. Prasetyo, and S. Purwono, "Hydrate Problems in Gas Lift Production: Experiences and Integrated Inhibition," SPE 126323 (paper presented at the Kuwait International Petroleum Conference and Exhibition, Kuwait City, Kuwait, 14–16 December 2009).

[196] J. W. Nicholas, C. A. Koh, E. D. Sloan, L. Nuebling, H. He, and B. Horn, *AIChE Journal* 55 (2009): 1882.

[197] E. D. Sloan Jr., *SPE Monograph Series*, vol. 21. Hydrate Engineering, Society of Petroleum Engineers, 2000.

[198] C. Koh, A. K. Sum, and E. D. Sloan, *Natural Gas Hydrates in Flow Assurance*. Boca Raton, FL:

CRC Press, 2010.

[199] R. F. Stoisits, D. C. Lucas, L. D. Talley, D. P. Shatto, and J. Cai, International Patent Application WO/2009/042319.

[200] X. Wang, Q. Qu, P. Joavora, and R. Pearcy, "New Trend in Oilfield Flow-Assurance Management: A Review of Thermal Insulating Fluids," SPE 103829, *SPE Production & Facilities* 24 (1) (2009): 35-42.

[201] R. R. Roth, "Direct Electrical Heating of Flowlines—A Guide to Uses and Benefits," OTC 22631, OTC Brasil, Rio de Janeiro, Brazil, 4-6 October 2011.

[202] P. B. Baugh and B. F. Baugh, U.S. Patent Application 20100051279.

[203] T. Grimseth, I. Wold, J. D. Friedemann, and C. Borchgrevink, U.S. Patent Application 20100044053.

[204] C. A. Broussard, T. A. Fowler, D. P. Shatto, D. J. Turner, L. D. Talley, J. W. Lachance, and D. Greaves, International Patent Application WO/2011/062720.

[205] C. A. Broussard, T. A. Fowler, D. P. Shatto, D. J. Turner, L. D. Talley, J. W. Lachance, and D. Greaves, International Patent Application WO/2011/062793.

[206] A. Lund, R. Larsen, J. H. Kaspersen, E. O. Straume, M. Fossen, and K. W. Hjarbo, U.S. Patent Application 20110220352.

[207] (*a*) S. Nuland and M. Foss, U.S. Patent Application 20100236634. (*b*) F. J. P. C. M. G. Verhelst, A. Twerda, J. P. M. Smeulers, M. C. A. M. Peters, S. P. C. Belfroid, and W. Schiferli, U.S. Patent Application 20100180952.

[208] A. Lund, B. Wittgens, and P. Skjetne, U.S. Patent Application 20100145115.

[209] J. W. Lachance and D. J. Turner, International Patent Application WO/2011/109118.

[210] J. W. Lachance, L. D. Talley, D. P. Shatto, D. J. Turner, and M. W. Eaton, *Energy & Fuels* 26 (7) (2012): 4059-4066.

[211] J. L. Creek and D. A. Estanga, "New Method for Managing Hydrates in Deepwater Tiebacks," OTC 22017 (paper presented at the Offshore Technology Conference, Houston, TX, 2-5 May 2011).

[212] L. E. Zerpa, Z. M. Aman, S. Joshi, I. Rao, E. D. Sloan, C. A. Koh, and A. K. Sum, "Predicting Hydrate Blockages in Oil, Gas and Water-Dominated Systems," OTC 23490 (paper presented at the Offshore Technology Conference, Houston, TX, 30 April-3 May 2012).

[213] A. Sum, L. E. Zerpa, E. D. Sloan, and C. Koh, "Hydrate Risk Assessment and Restart Procedure Optimization of an Offshore Well Using a Transient Hydrate Prediction Model," OTC 22406, OTC Brasil, 2011 Preliminary Technical Program, Rio de Janeiro, Brazil, 4-6 October 2011.

[214] S. R. Davies, J. A. Boxall, L. E. Dieker, A. K. Sum, C. A. Koh, E. D. Sloan, J. L. Creek, and Z.-G. Xu, *Journal of Petroleum Science and Engineering* 72 (2010): 302-309.

[215] A. K. Sum, C. A. Koh, and E. D. Sloan, *Energy & Fuels* 26 (7) (2012): 4046-4052.

[216] B. Tohidi, R. Anderson, A. Chapoy, J. Yang, and R. W. Burgass, *Energy & Fuels* 26 (7) (2012): 4053-4058.

[217] G. Bhatnagar, D. L. Crosby, G. J. Hatton, and Z. Huo, International Patent Application WO/2012/058144.

[218] J. G. Broze, J. O. Esparza, A. P. Mehta, and G. J. Hatton, International Patent Application WO/2006/068929.

[219] T. Huff, S. Cook, Baker Hughes; R. Trebing, M. Glover, T. Garza, and V. Thieu, "Easy-to-Implement and Cost-Effective Hydrate Prevention for Onshore Unconventional Gas Well Flowback after Hydraulic Fracturing: Kinetic Hydrate Inhibitor (KHI)/Methanol Mixture," SPE 166364 (paper presented at

the SPE Annual Technical Conference and Exhibition, New Orleans, LA, 30 September-2 October 2013).

[220] J. Tian and T. Z. Garza, International Patent Application WO/2010/011804.

[221] S. S. Awbrey and H. C. Riney, U.S. Patent Application 20090149683.

[222] B. Kaasa and P. H. Billington, International Patent Application WO/2010/084323.

[223] X. Li, P. V. Hemmingsen, and K. Kinnari, "Use of Under-Inhibition in Hydrate Control Strategies," Proceedings of the 7th International Conference on Gas Hydrates (ICGH 2011), Edinburgh, Scotland, UK, 17-21 July 2011.

[224] G. E. Casey, A. Hoover, and A. J. Barden, U.S. Patent Application 20120103431.

[225] S. Brustad, K.-P. Løken, and J. G. Waalmann, "Hydrate Prevention Using MEG Instead of MeOH: Impact of Experience from Major Norwegian Developments on Technology Selection for Injection and Recovery of MEG," OTC 17355 (paper presented at the Offshore Technology Conference, Houston, TX, 2-5 May 2005).

[226] A. Bahadori and H. B. Vuthaluru, *Energy & Fuels* 24 (2010): 2999-3002.

[227] A. Bahadori, *Petroleum Science and Technology* 27 (2009): 943-951.

[228] J. Yang and B. Tohidi, *Energy & Fuels* 27 (2013): 736-742.

[229] A.-M. Fuller, F. Mackay, C. Rowley Williams, and E. Perfect, "Development of an Analytical Method Suitable for On-Site Measurement of Thermodynamic Hydrate Inhibitors in Oil Field Produced Fluids," Tekna 23rd International Oil Field Chemistry Symposium, Geilo, Norway, 18-21 March 2012.

[230] C. MacPherson, P. Glenat, S. Mazloum, and I. Young, "Successful Deployment of a Novel Hydrate Inhibition Monitoring Systems in a North Sea Gas Field," Tekna 23rd International Oil Field Chemistry Symposium, Geilo, Norway, 18-21 March 2012.

[231] M. K. Hsieh, Y. T. Yeh, Y. P. Chen, P. C. Chen, S. T. Lin, and L. J. Chen, *Industrial & Engineering Chemistry Research* 51 (5) (2012): 2456-2469.

[232] R. Masoudi and B. Tohidi, *Journal of Petroleum Science and Engineering* 74 (3-4) (2010): 132-137.

[233] M. A. Kelland, "A Review of Kinetic Hydrate Inhibitors—Tailor-Made Water-Soluble Polymers for Oil and Gas Industry Applications," *Advances in Materials Science Research*, Chapter 5, vol. 8, ed. M. C. Wytherst. New York: Nova Science Publishers Inc., 2011.

[234] A. Perrin, O. M. Musa, and J. W. Steed, *Chemical Society Reviews* 42 (5) (2013): 1996-2015.

[235] P. J. Biggerstaff, M. N. Lehmann, A. M. Dhuet, and J. J. Weers, International Patent Applications WO/2013/074360.

[236] R. Rodríguez Gonzáles and J. Djuve, International Patent Application WO/2010/101477.

[237] M. A. Kelland and J. E. Iversen, *Energy & Fuels* 24 (5) (2010): 3003.

[238] M. A. Kelland, K. Mønig, J. E. Iversen, and K. Lekvam, *Energy & Fuels* 22 (2008): 2405.

[239] M. Nazeri, B. Tohidi, and A. Chapoy, "An Evaluation of Risk of Hydrate Formation at the Top of a Pipeline," SPE 160404 (paper presented at the SPE Asia Pacific Oil and Gas Conference and Exhibition, Perth, Australia, 22-24 October 2012.

[240] J. L. Creek, *Energy & Fuels* 26 (7) (2012): 4112-4116.

[241] O. Lavallie, A. Al Ansari, S. O'Neil, O. Chazelas, P. Glénat, and B. Tohidi, IPTC 13765, International Petroleum Technology Conference, Doha, Qatar, 7-9 December 2009.

[242] M. Varma-Nair, C. A. Costello, K. S. Colle, and H. E. King, *Journal of Applied Polymer Science* 103

(2007): 2642-2653.

[243] S. Ebbinghaus, K. Meister, B. Born, A. L. de Vries, M. Gruebele, and M. Havenith, *Journal of the American Chemical Society* 132 (35) (2010): 12210-12211.

[244] B. C. Knott, V. Molinero, M. F. Doherty, and B. Peters, *Journal of the American Chemical Society* 134 (48) (2012): 19544-19547.

[245] S. Y. Hong, J. I. Lim, J. H. Kim, and J. D. Lee, *Energy & Fuels* 26 (11) (2012): 7045-7050.

[246] P. C. Chua and M. A. Kelland, *Energy & Fuels* 26 (2012): 1160.

[247] M. A. Kelland and L. Del Villano, *Chemical Engineering Science* 64 (2009): 3197.

[248] J. S. Zhang, C. Lo, A. Couzis, P. Somasundaran, J. Wu, and J. W. Lee, *Journal of Physical Chemistry C* 113 (40) (2009): 17418-17420.

[249] A. H. Nguyen, L. C. Jacobson, and V. Molinero, *Journal of Physical Chemistry C* 116 (37) (2012): 19828-19838.

[250] M. R. Walsh, C. A. Koh, E. D. Sloan, A. K. Sum, and D. T. Wu, *Science* 326 (5956) (2009): 1095-1098.

[251] A. K. Sum and D. T. Wu, "Advancing the Science of Clathrate Hydrates with Molecular Simulations: Past, Present, and Future," Proceedings of the 7th International Conference on Gas Hydrates (ICGH 2011), Edinburgh, Scotland, UK, 17-21 July 2011.

[252] M. R. Walsh, J. D. Rainey, P. G. Lafond, D. H. Park, G. T. Beckham, M. D. Jones, K. H. Lee, C. A. Koh, E. D. Sloan, D. T. Wu, and A. K. Sum, *Physical Chemistry Chemical Physics* 13 (2011): 19951-19959.

[253] H. Ohno, I. Moudrakovski, R. Gordienko, J. Ripmeester, and V. K. Walker, *Journal of Physical Chemistry A* 116 (5) (2012): 1337-1343.

[254] F. A. Agizah, T. A. Ali, M. Baydoon, S. Allenson, and A. Scott, "Pioneer Challenge Reduction of MEG Consumption Using KHI for Hydrate Control in a Deepwater Environment Offshore Egypt," OTC 20338 (paper presented at the Offshore Technology Conference, Houston, TX, 3-6 May 2010).

[255] P. A. Webber, N. Morales, P. Conrad, K. McNamee, R. Jones, G. de Vries, and J. Garming, "Development of a Dual Functional Kinetic Hydrate Inhibitor for a Novel North Sea Wet Gas Application," SPE 164107 (paper presented at the SPE International Symposium on Oilfield Chemistry, The Woodlands, TX, 8-10 April 2013).

[256] P. C. Chua, R. O'Reilly, N. S. Leong, and M. A. Kelland, *Energy & Fuels* 25 (2011): 4595.

[257] P. C. Chua and M. A. Kelland, *Energy & Fuels* 26 (2012): 4481-4485.

[258] F. T. Reyes and M. A. Kelland, *Energy & Fuels* 27 (2013): 1314-1320.

[259] F. T. Reyes and M. A. Kelland, *Energy & Fuels*, submitted for publication.

[260] M. O. Musa and C. C. Lei, U.S. Patent Application 20130123147.

[261] D. Leinweber, A. Rösch, and C. Schaeffer, International Patent Application WO/2010/149253.

[262] R. Reichenbach-Klinke and K. Neubecker, International Patent Application WO/2009/083377.

[263] S. Frenzel, A. Assmann, and R. Reichenbach-Klinke, "'Green' Polymers for the North Sea—Biodegradability as Key towards Environmentally Friendly Chemistry for the Oilfield Industry," *RSC Chemistry In The Oil Industry XI*, Manchester, UK, 2-4 November 2009.

[264] O. M. Musa and L. Cuiyue, International Patent Application WO/2010/114761.

[265] O. M. Musa, L. Cuiyue, J. Zheng, and M. M. Alexandre, "Advances in Kinetic Gas Hydrate Inhibitors," *RSC Chemistry in the Oil Industry XI*, Manchester, UK, 2-4 November 2009.

[266] D. K. Hood, U.S. Patent Application 20100041846.

[267] O. M. Musa and L. Cuiyue, International Patent Application WO/2010/117660.

[268] O. M. Musa and C. Lei, U.S. Patent Application 20120077717.

[269] H. Guan, "Kinetic Gas Hydrate Inhibitors and Synergy of the Inhibiting Molecules," Proceedings of the 7th International Conference on Gas Hydrates (ICGH 2011), Edinburgh, Scotland, UK, 17–21 July 2011.

[270] H. Guan, "The Inhibition of Gas Hydrates and Synergy of the Inhibiting Molecules," SPE 131314 (paper presented at the International Oil and Gas Conference and Exhibition in China, Beijing, China, 8–10 June 2010).

[271] (a) M. Musa, J.-C. Chuang, Y. Zhang, and J. Zheng, International Patent Application WO/2012/054569. (b) J. Zheng, O. M. Musa, C. Lei, Y. Zhang, M. Alexandre, and S. Edris, "Innovative KHI Polymers for Gas Hydrate Control," OTC 21275 (paper presented at the Offshore Technology Conference, Houston, TX, 2–5 May 2011).

[272] J. Zheng, O. M. Musa, M. Alexandre, and S. Edris, "Development of Innovative Polymers as Kinetic Hydrate Inhibitors," Proceedings of the 7th International Conference on Gas Hydrates (ICGH 2011), Edinburgh, Scotland, UK, 17–21 July 2011.

[273] M. O. Musa, C. Lei, and K. S. Narayanan, U.S. Patent Application 20110277844.

[274] P. C. Chua and M. A. Kelland, *Chemical Engineering Science* 66 (2011): 2050.

[275] J. Hu, S. J. Li, Y. H. Wang, X. M. Lang, Q. P. Li, and S. S. Fan, *Journal of Natural Gas Chemistry* 21 (2012): 126–131.

[276] M. A. Kelland, International Patent Application WO/2013/053770.

[277] M. A. Kelland, International Patent Application WO/2013/053766.

[278] M. A. Kelland, N. Moi, and M. Howarth, *Energy & Fuels* 27 (2013): 711–716.

[279] L.-T. Chen, C.-Y. Sun, B.-Z. Peng, and G.-J. Chen, "The Synergism of PEG to Kinetic Hydrate Inhibitor," Proceedings of the Twentieth International Offshore and Polar Engineering Conference, Beijing, China, 20–25 June 2010.

[280] N. Daraboina, C. Malmos, and N. von Solms, *Fuel* 108 (2013): 749–757.

[281] S. Bauer, F. Fischer, and E. Bohrer, International Patent Application WO/2013/060679.

[282] J. Moloney, C. G. Gamble, and W. Y. Mok, "Compatible Corrosion and Kinetic Hydrate Inhibitors for Wet Sour Gas Transmission Lines," Paper 09350, CORROSION 2009, Atlanta, GA, 22–26 March 2009.

[283] J. A. Moore, L. Ver Vers, and P. Conrad, "Understanding Kinetic Hydrate Inhibitor and Corrosion Inhibitor Interactions," SPE 19869 (paper presented at the Offshore Technology Conference, Houston, TX, 4–7 May 2009).

[284] J. F. Garming, J. Anthony, G. J. de Vries, N. L. Morales, P. A. Webber, and R. A. Trompert, "A Mature Southern North Sea Asset Considers Conversion to Wet Gas Operation which Requires the Development of Compatible and Novel Chemistries for Flow-Assurance and Asset Integrity," OTC 24092 (paper presented at the Offshore Technology Conference, Houston, TX, 6–9 May 2013).

[285] J. Tian, C. Bailey, J. F. Fontenot, and M. Nicholson, "Low Dosage Hydrate Inhibitors (LDHI): Advances and Developments in Flow Assurance Technology for Offshore Oil and Gas Productions," OTC 21442 (paper presented at the Offshore Technology Conference, Houston, TX, 2–5 May 2011).

[286] R. Alapati, E. A. Sanford, E. Kilhne, and E. Vita, "Proper Selection of LDHI for Gas-Condensate Systems in the Presence of Corrosion Inhibitors," OTC 20896 (paper presented at the Offshore Technology Conference, Houston, TX, 3–6 May 2010).

[287] P. G. Conrad, E. J. Acosta, K. P. McNamee, B. M. Bennett, O. E. S. Lindeman, and J. R. Carlise, International Patent Application WO/2010/045523.

[288] J. R. Carlise, O. E. S. Lindeman, P. E. Reed, P. G. Conrad, and L. M. Ver Vers, International Patent Application WO/2010/045520.

[289] J. Knebel, W. Karnbrock, and V. Kerscher, U.S. Patent Application 20110218312.

[290] J. D. Morris, International Patent Application WO/2010/021956.

[291] L. Del Villano, M. A. Kelland, G. M. Miyake, and E. Y.-X. Chen, *Energy & Fuels* 24 (4) (2010): 2554.

[292] P. C. Chua, M. A. Kelland, T. Hirano, and H. Yamamoto, *Energy & Fuels* 26 (2012): 4961-4967.

[293] P. C. Chua, M. A. Kelland, K. Ishitake, K. Satoh, M. Kamigaito, and Y. Okamoto, *Energy & Fuels* 26 (2012): 3577-3585.

[294] A. P. Semenov, P. A. Gushchin, E. V. Ivanov, V. A. Vinokurov, and D. A. Sapozhnikov, *Chemistry and Technology of Fuels and Oils* 46 (6) (2011): 417-423.

[295] H. J. Spencer, R. Virdee, M. P. Squicciarini, G. T. Rivers, and M. N. Lehmann, International Patent Application WO/2013/059058.

[296] L. Del Villano, R. Kommedal, R. Hoogenboom, M. W. M. Fijten, and M. A. Kelland, *Energy & Fuels* 23 (2009): 3665.

[297] M. Kelland and F. M. Reyes, *Energy & Fuels*, in press.

[298] H. Ajiro, Y. Takemoto, M. Akashi, P. C. Chua, and M. A. Kelland, *Energy & Fuels* 24 (2010): 6400.

[299] P. C. Chua, M. Sæbø, A. Lunde, and M. A. Kelland, *Energy & Fuels* 25 (2011): 5165.

[300] D. R. Patil, G. L. Fan, J. C. Fan, and J. Yu, U.S. Patent Application 20090326165.

[301] S. Holt and J. S. Thomaides, International Patent Application WO/2012/089654.

[302] S. Holt and I. Unebäck, "Kinetic Hydrate Inhibitors with Improved Biodegradation," Tekna 23rd International Oil Field Chemistry Symposium, Geilo, Norway, 18-21 March 2012.

[303] Y. J. Xu, M. L. Yang, and X. X. Yang, *Journal of Natural Gas Chemistry* 19 (4) (2010): 431-435.

[304] N. Daraboina, J. Ripmeester, V. K. Walker, and P. Englezos, *Energy & Fuels* 25 (2011): 4392.

[305] N. Daraboina, P. Linga, J. Ripmeester, V. K. Walker, and P. Englezos, *Energy & Fuels* 25 (2011): 4384.

[306] L. Jensen, K. Thomsen, and N. von Solms, *Energy & Fuels* 25 (2011): 17.

[307] L. Jensen, H. Ramløv, K. Thomsen, and N. von Solms, *Industrial & Engineering Chemistry Research* 49 (4) (2010): 1486-1492.

[308] H. Ohno, R. Susilo, R. Gordienko, J. Ripmeester, and V. K. Walker, *Chemistry: A European Journal* 16 (34) (2010): 10409-10417.

[309] N. Daraboina, J. Ripmeester, V. K. Walker, and P. Englezos, *Energy & Fuels* 25 (2011): 4398.

[310] D. Myran, A. Middleton, J. Choi, R. Gordienko, H. Ohno, J. A. Ripmeester, and V. K. Walker, "Genetically-Engineered Mutant Antifreeze Proteins Provide Insight into Hydrate Inhibition," Proceedings of the 7th International Conference on Gas Hydrates (ICGH 2011), Edinburgh, Scotland, UK, July 18-21, 2011.

[311] J. Ivall and P. Servio, "Next-Generation Anti-freeze Hydrate Inhibitors Based on the Xylomannan Compound," Proceedings of the 7th International Conference on Gas Hydrates (ICGH 2011), Edinburgh, Scotland, UK, 17-21 July 2011.

[312] J. C. Ferreira, A. Teixeira, and P. M. Esteves, *Journal of the Brazilian Chemical Society* 23 (1) (2012):

11-U326.

[313] X. Lou, A. Ding, N. Maeda, S. Wang, K. Kozielski, and P. G. Hartley, *Energy & Fuels* 26 (2012): 1037.

[314] A. K. Norland and M. A. Kelland, *Chemical Engineering Science* 69 (2012): 483.

[315] H. K. Abay and T. M. Svartaas, *Energy & Fuels* 25 (2011): 42-51.

[316] J. Tian, U.S. Patent Application 20090175774.

[317] A. Lone and M. A. Kelland, *Energy & Fuels* 27 (2013): 2536-2547.

[318] K. McNamee and P. Conrad, Proceedings of the 7th International Conference on Gas Hydrates (ICGH 2011), Edinburgh, Scotland, UK, 17-21 July 2011.

[319] C. Koh, A. K. Sum, E. D. Sloan, M. Eaton, J. Lachance, and L. Talley, Natural Gas Hydrates in Flow Assurance, Chapter 6. Boca Raton, FL: CRC Press, 2010.

[320] C. Duchateau, P. Glénat, T.-E. Pou, M. Hidalgo, and C. Dicharry, *Energy & Fuels* 24 (1) (2010): 616-623.

[321] P. Glénat, SPE; P. Bourg, and M.-L. Bousqué, "Selection of Commercial Kinetic Hydrate Inhibitors Using A New Crystal Growth Inhibition Approach Highlighting Major Differences Between Them," SPE 164258 (paper presented at the 18th Middle East Oil & Gas Show and Conference (MEOS), Bahrain International Exhibition Centre, Manama, Bahrain, 10-13 March 2013).

[322] P. Glénat, R. Anderson, H. Mozaffar, and B. Tohidi, "Application of a New Crystal Growth Inhibition Based KHI Evaluation Method to Commercial Formulation Assessment," Proceedings of the 7th International Conference on Gas Hydrates (ICGH 2011), Edinburgh, Scotland, UK, 17-21 July 2011.

[323] S. Cadger, "Investigation of Induction Time Method on the Crystal Growth Inhibition Regions of Kinetic Hydrate Inhibitors (KHIs) Using a High Pressure Stirred Autoclave," Oilfield Chemistry Symposium, Geilo, Norway, 17-20 March 2013.

[324] K. McNamee, "Evaluation of Hydrate Nucleation Trends and Kinetic Hydrate Inhibitor Performance by High-Pressure Differential Scanning Calorimetry," Proceedings of the 7th International Conference on Gas Hydrates (ICGH 2011), Edinburgh, Scotland, UK, 17-21 July 2011.

[325] C. Duchateau, T.-E. Pou, M. Hidalgo, P. Glénat, and C. Dicharry, *Chemical Engineering Science* 71 (2012): 220-225.

[326] B. Kaasa and P. V. Hemingsen, International Patent Application WO/2013/041143.

[327] G. A. Schrader, International Patent Application WO/2012/041785.

[328] H. J. Spencer, M. N. Lehmann, T. C. Ionescu, T. Z. Garza, and S. J. Jackson, International Patent Application WO/2012/135116.

[329] A. Hussain, S. Gharfeh, and S. Adham, "Study of Kinetic Hydrate Inhibitor Removal Efficiency by Physical and Chemical Processes," SPE 157146 (paper presented at the SPE International Production and Operations Conference & Exhibition, Doha, Qatar, 14-16 May 2012).

[330] J. Tian and C. R. Bailey, International Patent Application WO/2011/123341.

[331] C. B. Argo, R. N. Harper, D. C. King, M. B. Power, and P. Willcox, U.S. Patent Application 20090230025.

[332] H. Moradpour, A. Chapoy, and B. Tohidi, "Transportability of Hydrate Particles at High Water Cut Systems and Optimisation of Anti-agglomerant Concentration," Proceedings of the 7th International Conference on Gas Hydrates (ICGH 2011), Edinburgh, Scotland, UK, 17-21 July 2011.

[333] R. Larsen, E. Straume, A. Lund, V. Andersson, G. Shoup, and C. B. Argo, "On the Effects of

Extreme Pressure on LDHI Performance and Hydrate Behaviour in Realistic Pipe Flow Situations," The 20th International Oil Field Chemistry Symposium, Oslo, Norway, 22-25 March.

[334] Z. Patel, M. Dibello, K. Fontentot, A. Guillory, and R. M. Hesketh-Prichard, "Continuous Application of Anti-agglomerant LDHI for Gas-Condensate Subsea Tieback Wells in Deepwater Gulf of Mexico," OTC 21836 (paper presented at the OTC Brasil, Rio de Janeiro, Brazil, 4-6 October 2011).

[335] P. Webber, P. Conrad, D. Lewis, and R. Jagneaux, "Continuous Anti-agglomerant LDHI Application for Deepwater Subsea Tieback: A Study on Water Quality and Low Water Cut Scenarios," OTC 23397 (paper presented at the Offshore Technology Conference, Houston, TX, 30 April-3 May 2012).

[336] P. A. Webber, "Fundamental Understanding on the Effects of Anti-agglomerants towards Overboard Water Quality," OTC 20841 (paper presented at the Offshore Technology Conference, Houston, TX, 3-6 May 2010).

[337] G. C. Blytas and V. R. Kruka, International Patent Application WO01/38695.

[338] P. C. Chua and M. A. Kelland, *Energy & Fuels* 27 (2013): 1285-1292.

[339] F. Maccioni and C. Passucci, "Torque Moment as Indicator of Low Dosage Hydrates Inhibitors: Effects on Multiphase Systems. Experimental Study on Quaternary Ammonium and Phosphonium Compounds," Proceedings of the 7th International Conference on Gas Hydrates (ICGH 2011), Edinburgh, Scotland, UK, 17-21 July 2011.

[340] H. A. Suarez, R. J. Franco, W. E. Bond, and R. S. Pakalapati, "Chemical Induced Pitting Corrosion of Super Duplex Stainless Steel Umbilical Tubes," CORROSION 2011, Houston, TX, 13-17 March 2011.

[341] Y. H. Tsang, T. Garza, C. A. Burger, D. Neptune, C. M. Menendez, P. Stead, and M. Lehmann, "Localized Corrosion Testings of Stainless Steel in Low-Dose Hydrate Inhibitor by Cyclic Potentiodynamic Polarization," OTC 22539 (paper presented at the OTC Brasil, Rio de Janeiro, Brazil, 4-6 October 2011).

[342] T. Garza, Y. H. Tsang, V. Thieu, P. Stead, and M. Lehmann, "Development of an Improved, Nonpitting Anti-agglomerant Low-Dose Hydrate Inhibitor (AA LDHI) with Minimal Corrosion Potential," Proceedings of the 7th International Conference on Gas Hydrates (ICGH 2011), Edinburgh, Scotland, UK, 17-21 July 2011.

[343] D. Durham, J. Russum, N. Davis, and C. Conkle, U.S. Patent Application 20120103422.

[344] J. Tian and C. Walker, "Non-emulsifying, New Anti-agglomerant Developments," Offshore Technology Conference, Houston, TX, 30 April-3 May 2012.

[345] J. D. York and A. Firoozabadi, *Energy & Fuels* 23 (6) (2009): 2937-2946.

[346] D. Durham, C. Conkle, and J. Russum, International Patent Application WO/2013/048365.

[347] J. Tian and C. R. Bailey, U.S. Patent Application 20120190893.

[348] S. Gao, *Energy & Fuels* 23 (4) (2009): 2118-2121.

[349] R. Alapati, E. Sanford, E. Kiihne, and E. Vita, "SS FA—Why Use Kinetic Hydrate Inhibitor," OTC 22896 (paper presented at the Offshore Technology Conference, Houston, TX, 3-6 May 2010).

[350] (a) P. A. Webber, International Patent Application WO/2012/082815. (b) P. A. Webber, P. G. Conrad, and A. K. Flatt, International Patent Application WO/2013/089802.

[351] E. J. Acosta and P. A. Webber, International Patent Application WO/2010/101853.

[352] E. J. Acosta, U.S. Patent Application 20100087339 and 201000087338.

[353] A. Firoozabadi, D. York, and L. Xiaokai, International Patent Application WO/2010/111226.

[354] M. Sun and A. Firoozabadi, *Journal of Colloid and Interface Science* 402 (2013): 312-319.

[355] K. MacNamee, private communication.

[356] S. Fan, F. Long, J. Du, Y. Wang, X. Lang, and D. Bi, Chinese Patent CN101666427, 2010.

[357] S. Fan, L. Shuanshi, X. Lang, Y. Wang, J. Du, D. Bi, and L. Yang, Chinese Patent CN101608112, 2009.

[358] C. Guangjin, L. Wenzhi, L. Qingping, S. Changyu, M. Liang, C. Jun, P. Baozi, Y. Yuntao, and M. Hao, Chinese Patent Application CN201110096579.2. Publication No. CN102746361A.

[359] C. Guangjin, L. Wenzhi, L. Qingping, S. Changyu, M. Liang, C. Jun, P. Baozi, Y. Yuntao, and M. Hao, Chinese Patent Application CN201110227519.X. Publication No. CN102925126A.

[360] J. Tian, International Patent Application WO/2012/047821.

[361] M. Johansen and B. Knapstad, "Joule-Thompson Cooling and Hydrate Risk During Cold Restart of a Subsea Production Well," Oilfield Chemical Symposium, Geilo, Norway, 14-17 March 2010.

[362] E. A. Sanford, T. S. Golczynski, B. Hampton, R. R. Alapati, and J. Lee, "Successful LDHI Qualification for Produced Fluids in a Flexible Riser at 10°F," OTC 20189 (paper presented at the Offshore Technology Conference, Houston, TX, 4-7 May 2009).

[363] K. Erstad, I. V. Hvidsten, K. M. Askvik, and T. Barth, *Energy & Fuels* 23 (8) (2009): 4068-4076.

[364] Z. M. Aman, L. E. Dieker, G. Aspenes, A. K. Sum, E. D. Sloan, and C. A. Koh, *Energy & Fuels* 24 (2010): 5441-5445.

[365] J. Sjoblom, B. Ovrevoll, G. Jentoft, C. Lesaint, T. Palermo, A. Sinquin, P. Gateau, L. Barre, S. Subramanian, J. Boxall, S. Davies, L. Dieker, D. Greaves, J. Lachance, P. Rensing, K. Miller, E. D. Sloan, and C. A. Koh, *Journal of Dispersion Science and Technology* 31 (8) (2010): 1100-1119.

[366] M. Pakulski, "Development of Superior Hybrid Gas Hydrate Inhibitors," OTC 21747 (paper presented at the Offshore Technology Conference, Houston, TX, 2-5 May 2011).

[367] A. L. Ballard, N. D. McMullen, and G. J. Shoup, U.S. Patent Application WO/2009/055525.

[368] J. L. Panter, A. L. Ballard, A. K. Sum, E. D. Sloan, and C. A. Koh, *Energy & Fuels* 25 (6) (2011): 2572-2578.

[369] K. Kinnari, C. Labes-Carrier, J. B. Crawford, L. Kirspel, and B. Torrance, European Patent Application EA010044.

[370] F. Dong, X. Zang, D. Li, S. Fan, and D. Liang, *Energy & Fuels* 23 (2009): 1563-1567.

[371] M. K. Pakulski and Q. Qi, U.S. Patent Application 20090325823.

[372] C.-G. Xie and P. S. Hammond, U.S. Patent Application 20130009048.

[373] P. Kondapi and R. Moe, "Today's Top 30 Flow Assurance Technologies: Where Do They Stand?," OTC 24250 (paper presented at the Offshore Technology Conference, Houston, TX, 6-9 May 2013).

[374] K. Moen, International Patent Application WO/2013/093789.

[375] M. Cha, K. Shin, J. Kim, D. Chang, Y. Seo, H. Lee, and S.-P. Kang, *Chemical Engineering Science* 99 (2013): 184.

[376] M. A. Kelland, "Production Chemicals and their Future," *Chemistry in the Oil Industry XIII: Oilfield Chemistry—New Frontiers*, Manchester Conference Centre, UK, 5-6 November 2013.

[377] E. G. Dirdal, A. Grinrød and M. A. Kelland, to be presented at ICGH8, International Gas Hydrate Conference, Beijing China, 28 July-1 August 2014.

[378] A. Hase and M. A. Kelland, unpublished results.

[379] E. Luna-Ortiz, K. Szklarczyk, M. Healey, E. Sørhaug, "Fast-Track Flow Assurance Design for

Kinetic Hydrate Inhibitors Without Laboratory Testing: A Case Study, 2013-D4," 16th International Conference on Multiphase Production Technology, Cannes, France, 12-14 June 2013.

[380] G. J. Guo and P. M. Rodger, *Journal of Physical Chemistry B*, 117 (21) (2013), 6498-6504.
[381] O. M. Falana, M. Morrow, and F. G. Zamora, US Patent Application 20130178399.
[382] R. Anderson, International Patent Application, WO2013121217.
[383] D. Al Mutawa, "Produced Water Management for Sustainable Reinjection—Bench Scale Tests to Remove and Destroy KHI, IPTC 17321," International Petroleum Technology Conference, Doha, Qatar, 19-22 January 2014.
[384] J. G. Delgado-Linares, A. A. A. Majid, E. D. Sloan, C. A. Koh and A. K. Sum, *Energy Fuels* 27 (8) (2013): 4564-4573.
[385] C. M. Menendez, A. Bhattacharya, S. Ramachandran, and V. Jovancicevic, "New Sour Gas Corrosion Inhibitor Compatible with Kinetic Hydrate Inhibitor," IPTC 17440 (paper presented at the International Petroleum Technology Conference, Doha, Qatar, 19-22 January 2014).
[386] N. Feasey, N. Grainger, I. T. Helgeland and S. Mainali, "New Anti-Agglomerant Chemistry for Gas Hydrate Control," *Chemistry in the Oil Industry XIII: Oilfield Chemistry—New Frontiers*, Manchester Conference Centre, UK, 4-6 November 2013.
[387] A. Deshmukh, International Patent Application WO/2013/16807.

10 蜡（石蜡）控制

10.1 概述

油田井下和地面系统都会存在蜡沉积问题，原油中蜡冷却后会阻碍油品流动。英语中特指油田蜡的术语是"paraffin"或"paraffin wax"。蜡是由天然存在于原油及某些凝析油中的直链或支链的长链烷烃（$>C_{18}$）化合物组成的固体，也可能存在一些环状烷烃和芳香烃。已有研究确定直链烷烃（正构烷烃蜡）是造成管道中蜡沉积的主要原因。原油中的蜡属于长链烷烃，防控难度比凝析油中的蜡更大。链长为16~25个碳原子的长链烷烃，是软质糊状蜡，链长为25~50个或更多碳原子的长链烷烃是硬质结晶蜡。随着蜡质分子链长的升高，蜡熔点增加。一般来说，石蜡熔点越高，其沉积物防治就越困难。

在高温和高压的油藏环境，蜡都溶于原油。随着原油温度的下降，蜡开始以针状和板状的微观形貌从原油中逐步析出。此外，随着生产过程中压力下降，低分子烃类（轻质）组分逸散到气相中，降低了原油中蜡的溶解度。析蜡温度（WAT）或浊点是原油中的蜡晶体开始析出的温度，是一个非常重要的度量值。析蜡温度取决于压力、油品成分（特别是轻质油的浓度）和泡点，某些油品的析蜡温度可能高达50℃。通常发生蜡的析出或沉积的温度高于形成天然气水合物的温度。原油输至外销储罐时，常含有石蜡固体，这是因为原油的轻质组分损失及较低的温度环境，使其丧失了很多保持溶解态蜡质的能力。蜡沉积造成的典型问题包括：

（1）降低输量或堵塞管道，阻碍流动（如果油井温度较低，可能出现在井下或在寒冷气候下的平台顶部或海底集输过程）；

（2）流体黏度增加，导致泵送压力增加；

（3）蜡胶凝强度导致的恢复生产问题；

（4）操作效率降低，生产中断或停产造成的工艺问题；

（5）清除成本高，技术难度大，特别是深水管道的结蜡；

（6）沉积物干扰阀门和仪器的操作，造成安全隐患；

（7）与积蜡有关的排放处置问题。

图10.1为蜡沉积的典型相图，其中位置A代表储层压力，油品处于未饱和状态；随着液体产出，压力下降，轻质组分与溶解蜡的比例增大，增加了蜡的溶解度，析蜡温度较低；在泡点B，轻质组分与重质组分的体积比达到最大，此时的析蜡温度最低；进一步降低压力至C，会导致溶解气释放和轻质组分进入气相，蜡的溶解度降低，析蜡温度增加。与0.1MPa的储罐原油相比，这种影响可使析蜡温度的改变高达15℃。

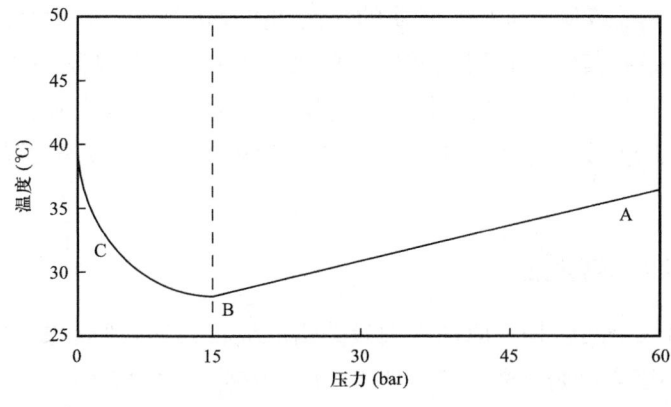

图 10.1　蜡沉积的典型相图

蜡质沉积有三种影响：

（1）沉积在管壁导致管道流动逐渐受限。蜡沉积使管径减少和管壁粗糙度增加，二者叠加导致管道输量减少，完全堵塞的情况很少见。

（2）流体黏度增加，造成管道内压力损失。在最坏的情况下，流体胶结可导致完全停产。

（3）蜡质凝胶的形成。如果管道停输，流体冷却至倾点以下，会形成凝胶，剪切速率为 0 时产生凝胶。如果胶凝屈服应力高于管道最大压力，管道就无法重启。

10.1.1　蜡的沉积

蜡的沉积机制主要有分子扩散机制和剪切分散机制两种观点。

（1）分子扩散机制：如果管壁温度低于析蜡温度，会在管壁上形成沉积蜡。即使宏观流体温度高于析蜡温度，也会发生这种情况。

（2）剪切分散机制：靠近管壁的沉积蜡会移动到管壁的低速区域并沉淀。

机制 1 的蜡沉积可以发生在析蜡温度附近，机制 2 只出现在低于析蜡温度段。蜡沉积的速度主要取决于机制 1 及流体动力学（蜡从管壁剥离的速度也很重要）。尽管文献中针对固体或液体蜡向冷壁面的运移提出了多达 8 种不同的机制，但唯一能解释蜡向冷壁面显著通量变化的机制是分子扩散，即液体蜡受浓度梯度的驱动而向冷壁移动。但可靠的通用蜡沉积模型仍有待商榷。例如，一项基于多相流体的实验室研究认为，对于某些流体有不同于传统扩散理论的蜡沉积机理。另一项环路试验研究表明，蜡沉积可能源于流变因素造成的流动模式变化。已有确定蜡沉积速度和沉积量的商业化软件模型。所有的蜡沉积模型都是基于经验性的关联，需要对原油中 C_{50} 以下的脂肪族和芳香族成分进行全面的分析。

油井中的原油在温度足够低时，从井下部开始蜡沉积。可以通过使用化学剂、热油或水、钢丝刮蜡器或清管工具间歇性地关井以除蜡。除非管道保温良好或加热，主要的蜡沉积问题可能发生在海底输油管和立管位置，特别是在壁面温度很低的冷水或深水中。蜡沉积和析蜡温度都受原油或凝析油中沥青质的含量和类型的影响。一般来说，沥青质比例较高的原油，可以观察到其蜡沉积明显减少。有观点认为析蜡温度是沥青质表面积的函数，并得到了实验支持。但据报道，具有相似的饱和烃、芳香烃、胶质和沥青质化学特征的两种轻质原油，其析蜡温度相差多达15℃。压力（溶解气体的量）对于新鲜原油析蜡温度的影响非常明显。

10.1.2 黏度增加和蜡质胶凝

与蜡有关的第二个问题是，原油中大量的蜡沉淀，会导致原油黏度增加，甚至胶凝，高含蜡原油更容易出现这种问题。在冷却时，原油中析出片状蜡晶，相互作用形成三维网络，液态油被包裹在其中，导致原油黏度增加，流动性下降，管输压降增大。已经观察到油水分层流动过程中，胶凝化是蜡质沉积的主要机制。

如果集油管道停输，流体可能进一步冷却，在无剪切时会形成胶凝。如果胶凝的屈服应力高于管道的最大压力，管道就无法重新启动。对24种原油的实验结果表明，质量分数约2%的沉淀蜡就足以导致原始蜡质原油胶凝。实验测量的倾点是对原油开始胶凝温度的粗略测量，通常低于析蜡温度10~30℃，而且在连续生产期间不会出现，只在停产期间才会出现。为确定潜在的蜡质问题，需要先测量析蜡温度，但倾点（更容易测量）可以快速反映现场操作期间系统的变化。相比于常压下形成的凝胶，恒定上游压力（为尽量减少形成气体空穴）下形成的凝胶更难清除。已报道了含蜡原油再启动的新方法。沥青质的存在会降低原油的倾点，而环烷酸则会提高倾点，流动状态和水等其他因素会影响黏度和胶凝点。

10.2 蜡质控制策略

为避免管道中的蜡质问题，有时向含蜡原油（尤其是重质原油）中掺混稀释剂。稀释剂可以是天然气凝析油、液化天然气，或具有较低析蜡温度或倾点的轻质原油。掺混后，含蜡原油的蜡含量被稀释，从而使其析蜡温度或倾点降低到较低温度。

控制井下和管道中蜡质胶凝及沉积的其他方法包括：保温，机械清除（集输管道的清管器或井下的有缆刮蜡器），加热井下、管道，溶蜡剂、防蜡剂、降凝剂（PPD）和分散剂，磁防蜡，冲击冷却（冷流），超声波防蜡，微生物防蜡。

本章只详细讨论化学除蜡和化学防蜡技术。

有可能在产蜡油井附近有不产蜡油井，这些井的产出液混合不一定会产生蜡的问题。这样先启动不产蜡井的生产，可以缓解采出液混合处的结蜡问题，但在更远的下游管线可能因温度降低而析出蜡。

通常为防止管道内的蜡沉积物堆积，需要定期清管。可以使用蜡沉积模拟软件来确定清管频率和管道沿程压降。清管通常与连续加注防蜡剂同步，可以降低蜡沉积速度并软化沉积的蜡。井下的机械清刮蜡工艺相对简单，但作业关井时间较长，此外，经常会有大量的蜡残留在井内，除蜡效率很低。

通过管道保温技术，可以保持管输流体温度高于蜡沉积温度。很多海底管道采用了真空保温。管道加热虽然成本高昂，但却是海底多相流集输的一个可行选择，特别是如果加热能同时缓解更严重的天然气水合物问题。无论是在管束或管中管系统，加热方式可以用电感应式加热或热水。持续电加热的成本较高，通常仅在关井状态下使用。

已有研究探讨防止原油集输中蜡沉积的冲击冷却或低温流体方法，特别是在深水温度很低、蜡沉积严重的环境。其原理是尽可能快速地冷却含蜡原油，使蜡尽可能地沉积在流体中，而不是在管壁。最初形成的蜡晶种微粒成为蜡晶进一步生长的位点，最终所有的蜡成为可运移的微粒分散体，不再或很少沉积在管壁。这种非化学方法仍处于研究阶段，同一方法也被探索用于防止天然气水合物堵塞（见第9章）。还有研究采用电场或电磁场诱导方法，使蜡不沉积到管壁，而沉积在垢颗粒上。

永磁体和电磁体（或流体磁性调节法，MFC）主要用于控制井下蜡沉积。该技术在亚洲使用较多，在西方使用较少。据称1995年前，中国已经应用了14440套磁性防蜡设备，成功率高。多路磁场效果更好。脉冲电场或磁场也显示可以降低原油的黏度。而这种技术在实验室和现场抑垢的有效性有差异。操作者在超出产品性能的现场条件下使用该技术，可能会导致结垢加剧。一些实验室研究表明，磁铁在减少蜡沉积和降低原油黏度方面具有积极作用，其在凝析油系统的效果优于原油系统。尽管磁防蜡主要作用于纯烃类体系，水和含盐量也会影响防蜡效果。另有研究表明，不同原油中，顺磁过渡金属离子的浓度不同，对磁场响应不同，对该技术的效果也有影响。由石油公司联合体赞助的深水研发计划DeepStar项目，研究发现MFC确实降低了析蜡温度。MFC系统尚未得到优化时，降幅较小（约为1.1℃）。

有关在井下放置加热元件以熔化和除蜡的方法也有报道。一种价格较低、使用电感加热元件诱导加热和避免变阻加热的方法已申报专利。采用极性水润湿玻璃内衬油管或环氧酚醛涂层化合物，可以防止井下管壁蜡附着。

微生物处理也可以去除油井中的蜡，4个月的现场试验表明，油井热洗处理减少了16次，清防蜡剂加注减少了44次，实现了增产和可观的经济利润。微生物清防蜡大规模使用的成本较高，但据称在中国油井的相应措施有效。有机酸和生物表面活性剂等代谢副产物也有分散蜡质和其他有机沉积物的作用，并增加其在采出液中的溶解度。还有研究筛选了针对C_{20}—C_{30}烷烃的特定酶，将酶用于海底原油集输防蜡。初步研究表明酶可以将原油的倾点降低若干摄氏度，更引人关注的是在海底温度下的降黏作用。

室内实验证明超声波处理技术可以去除蜡沉积，并提出将其作为井下处理的新方法。建议将这种方法用于化学处理法可能成本过高的长段油层。超声空化和磁场耦合处理的实验室研究表明这类技术可显著降低高倾点原油的黏度，现场应用延长了油井的结蜡周期。

10.3 化学除蜡

使用化学剂去除沉积蜡的方法有注热油、溶蜡剂和热化学段塞液。

10.3.1 注热油及相关技术

原油生产以来就开始实施注热油方法，以熔化和溶解蜡质沉积物。在热油法中，产出的原油被加热到远高于蜡熔点温度，从油井环形空间循环注入，通过油管返回到热油加热系统。目的是热油熔化或溶解蜡，使其以液体形式从井中产出。作业时，原油需要加入破乳剂并需通过加热器，以除去原油中的固体和水，其成本较高。因此，通常海底管道不采用此方法。有报道称结合毛细管注入法除去深海油井的石蜡堵塞。

在注热油作业过程中，常向原油中加入石油磺酸盐类的蜡分散剂，以将熔化的蜡分散于热油相。有文章讨论了热油带来的储层伤害问题。注热油技术有一定危险性，特别是在产出低闪点原油的油井作业时。为避免过高温度热油带来的安全风险，采用了水、烷基芳烷基聚氧乙烯膦酸酯表面活性剂、由醇（选自脂肪醇、乙二醇、聚乙二醇等）和乙二醇酯组成的互溶剂及芳香烃（如甲苯或二甲苯）的混合物配方。所述混合物被加热到高于待清除蜡熔点 15～20℃，就可起到清蜡效果。热蒸汽或热水也可用于溶解油井蜡，但可能会导致井筒腐蚀和产出液乳化问题。

10.3.2 溶蜡剂

实践中常用溶蜡剂来清除井下和管道中的蜡沉积物。曾使用如苯、氯化溶剂和其组合等有毒溶剂及其他烃类化合物，在美国陆上油田仍偶尔使用低闪点的二硫化碳清蜡剂。目前，最常用的溶剂包括热水（含分散剂）、取代的芳香烃（如甲苯和二甲苯，或高芳香烃含量的馏出物）及与天然气凝析油的混合物。有专利提出一种溶剂组合物，其中包括纯芳香烃（如甲苯或低闪点二甲苯），以及脂肪族或脂环族烃（如石脑油）。蜡溶剂组分中可能包含表面活性剂。还有室内研究表明，如二甲苯等热溶剂（而不是注热油）对井下除蜡有最大的潜在优势，因为热油可能只除去轻质蜡，残留的蜡更难溶解或分散，会给下次注热油作业带来更多的问题，而二甲苯等芳香族溶剂还可以去除沥青质沉积物。使用井内产出气吹扫的二甲苯段塞成功地清除了某海底管线中的沉积蜡，测试表明问题源于乙二醇和产出的冷凝物的相互作用，乙二醇的过度处理和低产水量导致石蜡沉淀和最终堵塞管道。

近些年，芳香烃和多数烷基取代的芳香烃被列为海洋污染物。有专利提出基于柠檬烯（一种萜烯）的更环保且无害型清蜡剂，还有柠檬烯与烷基乙二醇醚和各种其他极性化学品。柠檬烯生物降解性良好，对鱼类有一定的毒性，闪点为 45～50℃。在北海的海底出油管道上，采用良好环境评级的萜类提取物混合物规模地进行了溶蜡作业。但有研究认为，在海底温度（4～6℃）下，溶剂能完全溶解由冷凝物组成的沉积蜡，必须辅以加热。

一些具有可生物降解、生态友好、不易燃和无挥发性的含氧有机溶剂，已经作为替代性溶蜡剂上市。这些溶剂包括 2- 甲基戊二酸二甲酯、乳酸乙酯和 2- 乙基己基乳酸酯、N, N- 二甲基辛酰胺和 N, N- 二甲基癸酰胺。目前尚未有报告引用相关溶蜡剂配方。

某深海管道的蜡堵问题，先前处理措施失效，并使问题更复杂化。采用化学分散剂—溶剂组合，结合正确的机械处理，成功清除了堵塞物。

烃类化合物溶剂已用于井下清蜡或挤注防蜡剂作业。但在溶剂的转向方法方面所做的工作很少。化学转向方法多数用于酸化处理，而烃类溶剂作业则主要依靠机械封隔手段。使用延迟增黏剂可以优化流体布放，但不能使用带有金属交联剂的膦酸酯，因为混合物会瞬间胶结。只要避免使用互溶剂和十二烷基苯磺酸，带有脂肪酸衍生物交联剂的辛酸铝就会逐渐稠化流体。可以采用煤油为溶剂，分散的氧化镁粉末作为降黏剂。

10.3.3　热化学段塞液

可以通过化学反应原位生热，熔化蜡质沉积物。这种熔蜡技术如果设计和实施不当，会带来一定的风险。在最坏的情况下，放热反应失控可能导致管材超过预期的应力和损坏。

酸和碱反应等一系列热化学反应可产生热量。例如，将水与乙酸酐混合产生的乙酸，与氢氧化钠溶液分开计量并注入井下管柱或管道。但这种酸碱中和反应无法控制或延迟，也会造成腐蚀，在实践中可能很难应用。腐蚀性较低的中和反应首选是腐蚀性较低的磺酸（如十二烷基苯磺酸）与胺（如异丙胺）的反应。

更有前景的热化学反应是原位生成的亚硝酸铵在酸催化作用下分解，这是井下和海底清蜡的热化学段塞液基础。例如，据报道该体系已用于美国陆上和海上油井的增产。这些油井之前用溶蜡剂和酸体系的增产尝试并不成功。亚硝酸铵的放热分解反应采用了亚硝酸钠和硝酸铵或氯化铵在水溶液中的混合物，反应产物是氮气、水和硝酸钠或氯化铵（图10.2）。井下应用时，控制反应速率以在预定井深产生预定热量。以盐酸作为催化剂时，

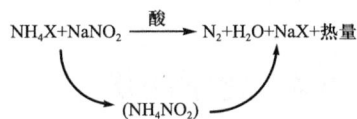

图 10.2　热化学除蜡的分解过程
（X=Cl 或 NO_3）

形成的盐类一经混合就会发生反应，将溶液 pH 值缓冲在 5.0～8.0 可控制反应速率，在较低的 pH 值时，反应速率会加快。控制方式是使反应逐渐开始，随着管柱内溶液以恒定速度移动而缓慢进行。作业时，射孔段上部的反应速率加快并产生大量热，温度达到最高值，热量散失到环境中，废液温度随之降低。因为钙质地层可能与酸催化剂发生反应，这种工艺仅限于非钙质地层。其成本比注热油溶剂更高，作为低产油井的增产手段可能不经济。

类似的热化学反应体系，在巴西深海油田的井下或深水管道清蜡方面也有成功应用，其基础是高价和低价氮化合物的氧化还原分解，释放出氮气。该体系被称为"生氮气体系"（葡萄牙语简作 SGN）。使用表面活性剂将高浓度氯化铵（或硫酸铵）和亚硝酸钠制成油包水的乳状液，在酸催化剂作用下，发生氧化还原反应，原位生成的亚硝酸铵

在低 pH 值下不稳定，释放出氮气、水和氯化钠。早期的 SGN 方法使用乙酸等有机酸作为催化剂，但分解可控性不好。更好的是使用酸酐，与乳剂中的水缓慢反应，原位生成酸。溶解在适当溶剂中的酸酐可以制成乳液以便于泵送。据称，酸酐包括聚（己二酸）酸酐等聚酸酐，或者是溶于氯仿等极性有机溶剂的聚（己二酸－共癸二酸）酸酐等共聚酸酐。乳液泵入管道后，在原位生成酸。SGN 反应产生的热足以熔化蜡，乳液的有机溶剂也溶解蜡，释放出的氮气为液体系统提供了搅拌作用，加速了溶解过程。有研究使用带涂层胶囊的缓释型酸催化剂。井下应用时，乳液从油套环空注入含氮基的 SGN 氧化还原系统、非极性有机溶剂、乳化液激活剂（过硫酸钠）和黏弹性聚合物的井内。随着乳液温度的升高，激活剂分解，pH 值下降到 4，然后开始发生氧化还原反应，产生热和高质量的稳定泡沫。泡沫对多孔介质的渗透足以确保充分除蜡和提高产量。

还有相关研究试验了组合溶剂/酸的发热体系，使用强酸（如硫酸）与含有碳碳双键的萜烯类溶剂反应。另一种热化学配方在推荐加量下，还可以起到破乳剂和缓蚀剂的作用。

固体材料控释热化学系统使用与催化剂混合能够水化的聚合物材料。在重新水化时，溶液中产生释放 H^+ 的催化剂，对生成亚硝酸铵的化学剂分解起到催化作用。过氧化氢被过渡金属离子（如 Fe^{3+}）分解时也会产生大量热（芬顿反应）。过氧化物被分解为氢氧根离子和羟基自由基。羟基自由基是主要的氧化物，可氧化和分解有机分子，目前还没有现场应用的报道。另一种也未见现场应用的热化学除蜡方法是以过饱和溶液（如乙酸钠）的快速放热性结晶来产热熔蜡。还有一种井下生热的体系是将咪唑啉酮衍生物（如 2-咪唑啉酮）与硫酸等酸进行原位反应，产生热量、震击和二氧化碳，据称可起到温和的压裂增产作用。

10.4　化学防蜡

10.4.1　测试方法

下述讨论的实验方法主要是关于防蜡剂和倾点抑制剂的需求确认和性能。有文章评价了原油总蜡含量的不同测定方法。蜡的碳数分布通常通过高温气相色谱法（HTGC）确定，针对其技术局限性，相应改进方法已有报道。

为确定原油或凝析油的蜡沉积潜力，首先要测量析蜡温度或浊点。有资料讨论了相关的多种方法，黏度测量和过滤堵塞法只有在非常理想的情况下才能使用，差示扫描量热法也是如此，还可以用高压微量热法。此外，常规的美国材料与试验协会（ASTM）浊点测定方法不适用于深色原油，也没有考虑蜡的潜在过冷度影响。在墨西哥管道的石蜡沉积潜力研究中，应用以上方法均未取得成功。

确定析蜡温度的可靠方法包括测定温热流动原油的冷却表面开始结蜡的温度。其他好的技术包括交叉偏振显微镜，激光和准直器固体检测系统，傅里叶变换红外光谱，近红外散射，高压、冷台显微镜技术，声波测试装置，核磁共振。

还设计了一个带有视频显微镜的流动池，以观察蜡结块和沉积。在未加防蜡剂的原油中可以观察到紧密的蜡质聚集，而在加防蜡剂的原油中观察到了松散黏结的团块。应注意的是，与传统的方法相比，使用该技术观察到的浊点更低。

蜡的沉积速度测试通常采用冷指法、奇尔顿—科尔本类比法和单一模型调谐数法。有文章介绍了多路冷指装置，也有专利介绍了带同轴剪切仓的蜡沉积设备。有文章介绍了一种可改变剪切条件和温度的小体积蜡样沉积测试系统，能够测量和观察沉积物的高度、成分、微观形貌和宏观结构。在蜡沉积试验台中，增大市售防蜡剂加量的实验，对当前的蜡沉积模型提出了挑战。其结果显示，防蜡剂加入后并未改变结蜡预测的输入参数（析蜡温度、蜡含量、溶解度曲线）。此外，预测未考虑防蜡剂对蜡质孔隙度参数的显著影响，而只是将其作为一项调整参数。

测量蜡沉积率的更复杂仪器是流动环路或毛细管装置，后者可测量毛细管内壁蜡沉积导致的微孔毛细管压力变化。毛细管或环路可以置于夹套中或冷浴中。还有使用大直径的试验环路。有文章研究了流动管路中的蜡沉积硬化过程和形貌变化。流动环路沉积实验证明，与冷盘实验的量化结果有较好的一致性。还有文章对比了现场蜡沉积与冷指法沉积的差异。

蜡质原油的倾点测量传统上使用简单的 ASTM D-97（或 IP 15）倾点测试法，不过也应该测量黏度。该方法的重现性差，但可以对倾点抑制剂的效率起到排序作用。实验室测试所需的倾点抑制剂加量总是高于现场所需量，因此不具有代表性。ASTM 倾点测量为在逐渐下降的温度范围内，定期倾斜装有内套温度计和加剂或未加剂油样的试管，并观察移动情况。这种技术所产生的剪切力极低，可以反映黏度的微变化。因此，测量黏度可弥补这种方法的模糊性。倾点最好用流变力学机械分光仪来测量。蜡质凝胶的强度可以通过使用黏度计的屈服应力测试来确定。也已开展了倾点抑制剂作用形成的蜡晶形貌和微观结构的研究。

需要研究防蜡剂或倾点抑制剂与溶剂混配的黏度，特别是低温或深水海底环境通过脐带缆和毛细管注入药剂时。高压黏度测定法有利于产品质量评价。高压和低温下，防蜡剂配方的聚合物溶液相行为不稳定，可能导致部分脐带缆失效。这种不稳定性取决于防蜡剂的化学成分及其浓度，多数有良好溶剂的防蜡剂在低温环境的溶解度也不高。通过弱至中等溶蜡剂和另一种强溶蜡剂的组合，可以使防蜡剂在低温下有很高的溶解度。弱至中等溶剂包括苯、甲苯、二甲苯、乙基苯、丙基苯、三甲基苯及其混合物，强溶剂包括环戊烷、环己烷、二硫化碳、癸烷及其混合物。

10.4.2 防蜡剂和倾点抑制剂

防控蜡沉积及胶凝过程所采用的化学剂会分别影响析蜡温度和蜡的倾点，还会改变蜡的晶体结构。影响析蜡温度的化学剂通常被称为防蜡剂或蜡晶改进剂，影响倾点的化学剂被称为倾点抑制剂或流动改进剂。这两类化学品都以某种方式作用于蜡结晶过程，其化学特征和机理有很多重叠，因此多数防蜡剂也具有倾点抑制剂的功能。蜡分散剂的作用方式则不同。防蜡剂和倾点抑制剂必须在流体温度降至析蜡温度前，加注到管道中。

防蜡剂或倾点抑制剂的加注浓度取决于结蜡问题的严重程度，多数应用条件的加注浓度是100～2000mg/L。某油田的原油沥青质含量增加后，倾点抑制剂的量只得加倍，这可能是源于倾点抑制剂与沥青质的相互作用。

很多情况下，使用防蜡剂并不能完全防止蜡在整条管道中的沉积，但可以减少清管或注热油等除蜡作业的频次。当流体冷却到倾点以下时，常用倾点抑制剂以防止管道停输期间流体冷却至倾点以下时的蜡胶凝。某条亚洲管道使用倾点抑制剂的年花费约1200万美元，后续调查发现，原油的冷却程度没有预估的那么显著，无须倾点抑制剂也能安全集输。有文章回顾了井下气举系统中防蜡剂的应用。

为实现干扰蜡的成核和结晶过程，化学剂有多种分子结构设计方法。简而言之，部分分子结构与蜡相互作用或共晶，以改变或干扰蜡的结晶，另一部分则可以覆盖新蜡分子本应附着的位点，阻止蜡的生长，两部分作用结合可防止管壁上形成结构化的蜡晶格。正如一项冷指沉积的研究表明，至少有两种机制：一类化学剂大幅减少冷指表面的蜡沉积，留下透明的溶液；第二类化学剂作用于溶液中的蜡，得到不透明的溶液。一些研究表明，防蜡剂有效时生成的沉积物往往较弱，并更容易被流场中的剪切力清除。在微观尺度上，防蜡剂对蜡晶形貌有巨大影响：形成高度支化的微晶网，而非纯蜡的板状生长。

防蜡剂和倾点抑制剂的主要类别有乙烯聚合物和共聚物、梳状聚合物、带有长烷基的各种支链聚合物。

因为高浓度的防蜡剂或倾点抑制剂聚合物的凝固温度低，可以将这些产品配制到芳香族溶剂中提高其性能，并克服其在寒冷气候中难以加注的问题。可以使用水包油乳化法，或在同一管线内布放多种生产化学剂来防止胶凝。也可混合使用弱到中等的溶剂（如甲苯或二甲苯）和强溶剂（如环戊烷或环己烷），以避免寒冷气候的防蜡剂胶凝。有文章研究了乙烯-醋酸乙烯酯（EVA）共聚物在各种溶剂中的相行为。与其他溶剂相比，在强溶剂（环己烷）中的流动改进剂对原油流动性的改善效果更好。据称，至少一种不溶于水的有机溶剂、水、至少一种多环羧酸的烷醇胺盐、可选的至少一种水溶性有机溶剂和倾点抑制剂或流动改进剂组成的分散体，其黏度低、易使用且性能更优，另见10.4.8。

标准的防蜡剂与更低加量的有机硅氧烷（有机硅）复配使用，能更有效地降低原油黏度、倾点和蜡沉积量。

梳状聚合物是最有效的一类防蜡剂，也可以与乙烯共聚物（如EVA共聚物）和表面活性剂协同起效。梳状聚合物，顾名思义就是类似于梳子，聚乙烯的分子结构主链上带有许多长链侧基（图10.3）。实际上，更好的描述是"卷发器"，因为在无规聚合物中，主链上的侧链指向许多不同方向。防蜡剂或倾点抑制剂在典型加量50～200mg/L时，可以将蜡沉积温度降到析蜡温度以下10～15℃，通常的倾点下降幅度大于析蜡温度下降，倾点抑制剂在非常高的加量下，观察到倾点下降高达30℃。梳状聚合物除了能够通过范德华力相互作用与蜡形成共晶体外，还能在蜡晶上形成空间位阻，干扰后续蜡分子的正常排列，从而终止生长。这样可使蜡晶无法聚集，有时也无法黏附于管壁，倾点通常也会降低。梳状聚合物防蜡剂已使用二十八烷上的聚丙烯酸十八烷基酯进行建模。该模型预测，在低过饱和度下，聚丙烯酸十八烷基酯将阻止蜡的生长，此时蜡的生长主要发生在阶梯状缺陷上，而在高过饱和度下，蜡的生长减慢，此时的岛状成核很重要。梳状聚合物中侧链的最佳长度取决于蜡中烷烃的长度。一般高分子量的蜡最好采用长侧链的梳状聚合物防蜡剂，蜡的烷基长度与侧链长度匹配较好。但对于很长碳链的蜡（C_{30+}），无法通过高效低成本的合成方法将同样长度的烷基链引入梳状聚合物。实验室研究表明，一些最好的市售梳状聚合物产品及 EVA 都无法应对这类长链烷烃蜡。梳状聚合物中长侧链的比例也很关键，有研究表明，聚丙烯酸酯中 C_{18} 侧链占比 60% 时，对某种蜡的性能最优。其中 C_{18} 酯是由聚丙烯酸酯甲酯在碱催化条件下转酯化而成，其余侧链为甲基。利用核磁共振光谱法开展不同原油的蜡沉积研究表明，较黏软蜡的碳链具有支链，反之则主要是直链石蜡。因此，聚合物防蜡剂的烷基侧链支化有益于提高防蜡效果。

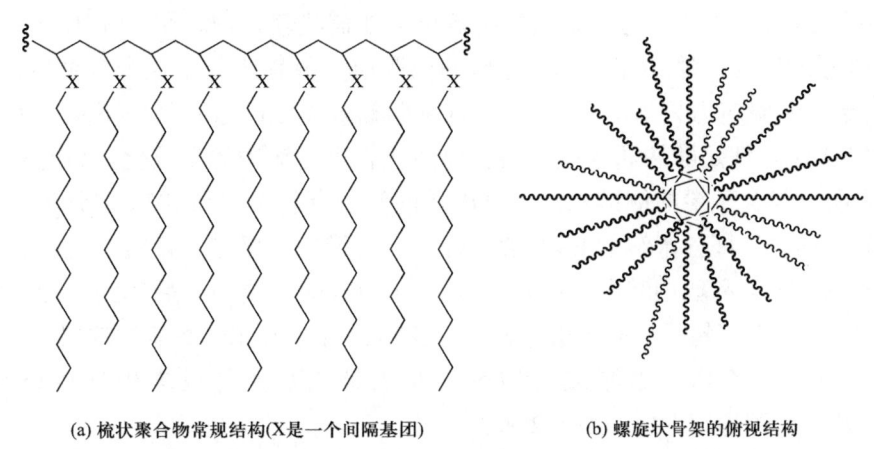

(a) 梳状聚合物常规结构(X是一个间隔基团)　　(b) 螺旋状骨架的俯视结构

图 10.3　梳状聚合物常规结构和螺旋状骨架的俯视结构

10.4.3　乙烯聚合物和共聚物

无定形的高分子量聚乙烯一直用作蜡晶改进剂或倾点抑制剂，而更好的倾点抑制剂是由乙烯与较大的单体共聚而成。较大的单体也起到类似于支链的作用，扰乱蜡结晶，如乙烯/小分子烯烃共聚物、EVA 共聚物、乙烯-丙烯腈共聚物。

研究了聚（乙烯丁烯）和聚（乙烯-b-丙烯）作为蜡沉积抑制剂。使用 1000mg/L 的聚（乙烯丁烯）可以将胶凝原油的屈服应力降低 90%，实际上增加了初始沉积率。

目前，乙烯共聚物中最常见和最知名的是 EVA。EVA 共聚物最好是无规共聚物（图 10.4）且分子量较低。决定 EVA 共聚物有效性的关键参数是醋酸乙烯酯在共聚物中的百分比。纯聚乙烯预计会与结构相似的蜡形成共晶，对结晶过程的影响很小，而增加醋酸乙烯酯的含量会降低结晶度，并由于极性较高而有助于溶解。有些醋酸乙烯酯甚至可以部分水解。醋酸乙烯酯基团中的侧链会破坏蜡的结晶过程，降低析蜡温度或倾点。但醋酸乙烯酯含量过高会降低与蜡的共结晶，并对性能产生负面影响。通常 EVA 共聚物的醋酸乙烯酯含量为 25%～30%，性能最佳。也有研究将醋酸乙烯酯/苯乙烯/丙烯酸丁酯三元共聚物作为倾点抑制流动性改进剂。有趣的是，已证明 EVA 既是一种蜡晶成核剂，又是生长抑制剂，通过分析浊点、晶体的化学成分和对晶习的观察，确定了这两种机制。原油中的沥青质会影响 EVA 的性能。

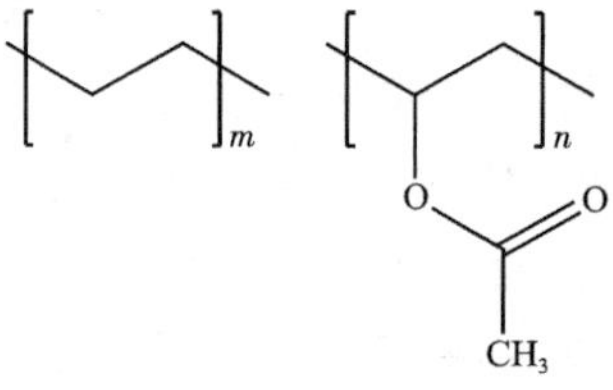

(a) 乙烯-醋酸乙烯酯 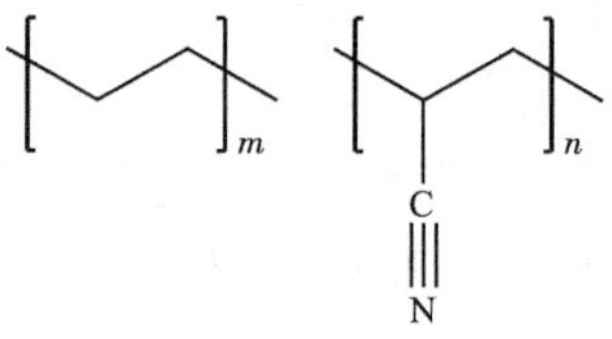　　　　　(b) 乙烯-丙烯腈共聚物

图 10.4　乙烯－醋酸乙烯酯和乙烯－丙烯腈共聚物

乙烯－醋酸乙烯酯共聚物对倾点抑制效果不如梳状聚合物。在一项研究中，EVA 共聚物的加量为 200mg/L 时，C_{36} 蜡和 C_{32} 蜡的屈服应力分别降低了 3 个数量级、1 个数量级，而 C_{28} 蜡的屈服应力只为原来的 1/3。这种防蜡效率随碳数降低而下降，表明在管输过程中 EVA 无法降低足够的屈服应力值，以防止管道运输中的凝胶化。另有实验和分子模拟研究表明，与含丁烯的侧链相比，含丙烯的侧链有利于提高 EVA 型分子与蜡平面上烷烃之间的亲和力。部分水解的 EVA 含有羟基，防蜡性能更好。长链二烷基胺和环状酸酐（如邻苯二甲酸酐和琥珀酸酐）的加合物，与 EVA 聚合物混合后作为冷却中间馏分燃料的蜡分散剂。只有当蜡晶尺寸已被 EVA 共聚物减小后，酰胺/胺盐的效果才会显现。

另一种可以与乙烯共聚制成防蜡剂或倾点抑制剂的低成本极性单体是丙烯腈（图 10.4），优选的丙烯腈占比 10%～20%。

10.4.4　梳状聚合物

通常认为梳状聚合物是最有效的一类防蜡剂，通常由（甲基）丙烯酸、马来酸酐两类单体之一或两者合成制成，但在高析蜡温度的严重结蜡情况下，也可能无法解决长期沉积问题。因为此时大部分蜡的烷烃比梳状聚合物中的烷基链要长得多。这样还是需要定期清管或其他除蜡技术。

10.4.4.1　（甲基）丙烯酸酯类聚合物

有许多关于使用丙烯酸酯或甲基丙烯酸酯聚合物作为防蜡剂或倾点抑制剂的报道（图 10.5）。酯基由长链醇制成，长度应至少为 16 个碳。与丙烯酸酯聚合物相比，甲基丙烯酸酯聚合物的侧链间隔不同，有更好的降倾点和防沉积效果。因为大多数蜡中的烷烃要长得多，烷基侧链长大于 18 个碳的聚合物在多数情况下防蜡效果表现最好。但醇的链长大于 18 个碳（硬脂醇）后，成本很高，导致防蜡剂的成本更高。一项室内研究认为，对

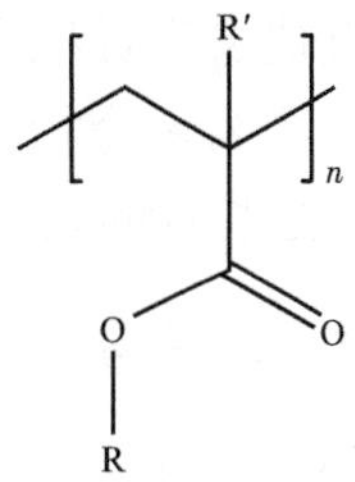

于链长主要是 20~29 个碳的蜡，烷基酯链的最佳链长是 20~24 个碳，最佳分子量是 30000~40000。此外，为限制晶体生长，建议倾点抑制剂的熔点应与蜡熔点匹配。另一项研究表明，聚甲基丙烯酸酯防蜡剂（烷基侧链最多为 18 个碳原子）以 100mg/L 的加量进行测试，可以降低低分子量石蜡（C_{24}）溶液的析蜡温度，但对高分子量石蜡（C_{36}）溶液基本不起作用。如前所述，并非所有的（甲基）丙烯酸酯基团都需要是长链，才能获得最佳性能。在聚丙烯酸酯中，大约 60% 的 C_{18} 侧链（其余的侧链为甲基酯）能使析蜡温度得到最大幅度的下降。

图 10.5 丙烯酸酯（R'=H）和甲基丙烯酸酯（R'=CH_3）聚合物（R 最好是长线型烷基链）

蜡烷基链的长度和比例因原油而异，对防蜡剂性能有影响，油服公司需要有一系列防蜡的梳状聚合物。为提供更一致的性能，有专利提出了烷基链长度呈 U 形分布的梳状聚合物。

一类改良型流动改进剂是丙烯酸硬脂酸酯与低比例的丙烯酸羟乙酯共聚，然后用硬脂酸氯将羟基酯化。这种方式可以避免使用如山嵛醇等昂贵的 C_{20} 以上醇类，实现将分子中引入较长的侧链。也可以如 EVA 的制备方式，将（甲基）丙烯酸烷基酯链接枝到聚乙烯主链。

据称，通过使用共聚物对聚烷基甲基丙烯酸酯进行了系列优化。例如，C_{16+} 醇的（甲基）丙烯酸酯与低比例的亲水性（甲基）丙烯酸、乙烯基吡啶或 N-乙烯基吡咯烷酮的共聚物作为改良型的倾点抑制剂和流动改进剂。丙烯酸硬脂基酯/烯丙基聚乙二醇共聚物可用作流动改进剂，最好与聚异丁烯和烷基酚甲醛树脂等增效剂复配使用。聚合物中极性基的分散作用避免了管壁的蜡沉积。将特定烷基链长度的（甲基）丙烯酸酯混合物，以及乙烯基共聚单体（如 2- 或 4- 乙烯基吡啶、苯乙烯、醋酸乙烯或苯甲酸乙烯）等组成的三元共聚物也可作为防蜡剂。双子表面活性剂可以作为苯乙烯/甲基丙烯酸硬脂基酯共聚物的增效剂。

两种聚烷基（甲基）丙烯酸酯的协同共混物可作为倾点抑制剂。其中一种的结晶起始温度在 15℃以上，另一种的不大于 15℃，且两种聚合物间的起始结晶温度差值至少为 5~10℃。后续对聚烷基（甲基）丙烯酸酯的研究产生一项发明，即链长不同的两种此类聚合物相比于单一链长聚合物，防蜡性能更好。这种聚酯最好为碱催化的聚甲基丙烯酸甲酯与烷醇的酯交换化反应且分子量宜为 20000~30000。市售产品有 7% 的 C_{18}、58% 的 C_{20}、30% 的 C_{22} 和 6% 的 C_{24} 烷醇混合物。在聚酯侧链中使用不同的碳链长度，可能与蜡晶中不同长度的烷烃有关。这些聚合物还可以与长链聚乙烯亚胺衍生物和油溶性成膜型表面活性剂协同使用。在聚烷基（甲基）丙烯酸酯中含氮官能团支化可提高其倾点抑制剂的性能。支化可通过添加极少量的二乙烯基单体（如二乙烯基苯或丁烯 -1，4- 二丙烯酸）实现，这样的交联程度最小。

市售的聚合物防蜡剂中添加二甲苯可以提高产品性能，如聚（丙烯酸二十二酯）与二甲苯或其他去除沥青质的芳香族溶剂混合时，可以降低倾点。三氯乙烯—二甲苯（TEX）的二元体系不含聚合物防蜡剂，但在一系列含蜡原油的冷点试验中，取得了显著的降低倾点和改善输送特性的效果。对比市售化学防蜡产品，TEX 添加剂提升了抑制蜡沉积的效果。

10.4.4.2　马来酸共聚物

制备梳状聚合物的反应物中，除（甲基）丙烯酸单体外，另一种最重要且成本同样低廉的单体是马来酸酐。马来酸酐自身不易聚合，但在有乙烯基共聚单体存在时很容易聚合。为了得到带长烷基的侧链，可以用长链醇、烷基硫醇和烷基胺将酸酐衍生为单酯、二酯、硫酯、单酰胺或酰亚氨基。在此类的多种聚合物中，有一种是始于马来酸酐和长链 α-烯烃（如1-十八烯）的共聚物。这样骨架上每隔4个碳原子就接枝一条长链，再衍生酸酐基团，以增加长侧链的密度。最终马来酸酐的共聚物就有 ABABAB 等规律的交替单体结构，使其具有比无规共聚物更好的防蜡性能。因此，（甲基）丙烯酸酯与 C_{16+} 醇、马来酸酐的共聚物是比聚烷基（甲基）丙烯酸酯更好的防蜡剂（图 10.6）。通过添加 EVA 共聚物作为增效剂和表面活性剂基的蜡质稳定剂，可以进一步提高防蜡性能。

图 10.6　马来酸酐-（甲基）丙烯酸酯共聚物的单酯（R= 长烷基链，R′=H 或 CH_3）

有关马来酸酐共聚物的研究发现当可能有至多18个碳原子的侧链时，针对 C_{24+} 烷烃蜡质的防蜡性能很差。因此除非防蜡剂与蜡质有类似长度的烷基链，否则不能起到应有效果。有研究主要针对 α-烯烃/马来酸酐聚合物的衍生物，揭示了新发现。使用支链醇制备的马来酸酯比使用普通醇的防蜡效果更差。在降低浊点方面，马来酸聚合物的单酯显然比双酯表现得更好，而这两类酯的倾点降低性能相似。此外，侧链为20个或更少碳原子的单酯比侧链较长酯的倾点抑制性能更好，推断其原因是酯类形成了蜡晶生长的屏障，因此不需要能渗透深入蜡晶的长侧链。高分子量的聚合物作为倾点抑制剂表现最好，高分子量的十八烯/马来酸酐共聚物衍生为短链醇的单酯后，将起到优良的防蜡和倾点抑制综合作用。丙烯酸、二十二烷基酯、马来酸酐和苯乙烯的共聚物相较于某些市售降黏剂，降黏效果更好。

为增加长侧链的概率，可以用长烷基胺对马来酸酐-α-烯烃共聚物进行衍生。例如，马来酸酐-α-烯烃共聚物与 C_{18} 烷基胺反应，制成马来酰亚胺的倾点抑制剂（图10.7）。十八烷基胺改性的聚（马来酸酐十八烯）也是一种很好的蜡沉积抑制剂，也可以改变流动改进剂中的马来酸基和马来酰亚氨基的比例。此外，亚胺化程度对流动改进剂的性能有很大影响，酰亚胺化程度越高，流动改进剂效果越好。这种影响归因于马来酰氨基和马来酰亚氨基之间的极性差异。有文章研究了系列聚（正烷

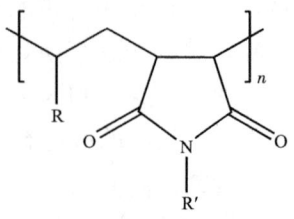

图 10.7　烷基马来酰亚胺/α-烯烃共聚物（R 和 R′最好是长烷基链）

基壬二酸 -N- 十六烷基马来酰亚胺）作为倾点抑制剂和流动改进剂的效果。

另一类倾点抑制剂是甲基丙烯酸酯与 C_{16+} 醇和含氮烯烃（包括烷基马来酰亚胺）的共聚物。

还有通过烷基乙烯基醚，将长烷基链引入马来酸共聚物的方法。十八烷基乙烯基醚/马来酸酐共聚物及其衍生物也可作为倾点抑制剂。脂肪族乙二醇醚溶剂是这类梳状聚合物及乙烯基吡咯烷酮/二十碳烯共聚物的增效剂。据报道，用二十二烷基胺改性的聚（马来酸酐/乙基乙烯基醚）也有良好的防蜡作用。

聚烯烃聚合物可以接枝马来酸酐等不饱和单体。如果使用马来酰亚胺衍生马来酸酐，是在侧链中引入长烷基的另一种方式（图 10.8）。马来酸衍生物也可以接枝到 EVA 共聚物上。

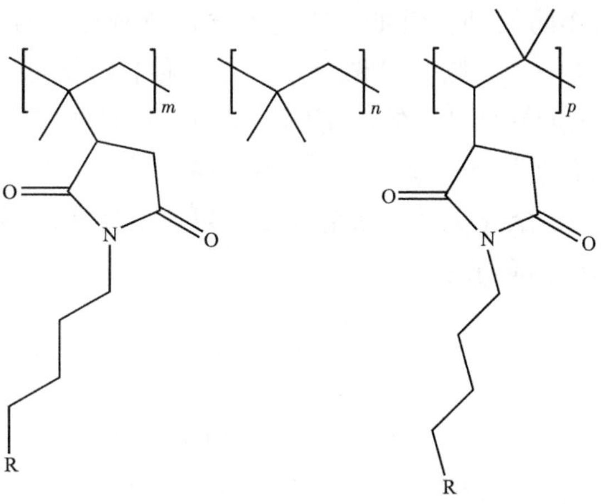

图 10.8　接枝的聚异丁烯/烷基马来酰亚胺聚合物（R 最好为 C_{12+}）

除了马来酸，其他不饱和羧酸单体也可用于制备防蜡剂。例如，长链的富马酸烷基酯－醋酸乙烯酯共聚物作为印度高蜡质原油的流动改进剂（富马酸是马来酸的反式形式）。新型的一氧化碳－富马酸二烷基酯共聚物可作为蜡晶改性剂和流动改进剂。富马酸烷基酯共聚物可以与烷醇胺（如三乙醇胺）与长链酰化剂（如 C_8—C_{20} 烷基琥珀酸酐）的反应产物协同起效。

10.4.5　杂项聚合物

烷基酚醛树脂作为一种倾点抑制剂和流动改进剂（图 10.9），其效果虽不如前面提到的梳状聚合物，但能起到增效剂作用。制备此类树脂时，苯酚首先与长链 α- 烯烃反应，然后与醛缩合。这类带有较短烷基的树脂还可作为沥青分散剂，并在多烷氧基化后作为破乳剂使用。

据报道，长链膦酸酯表面活性剂与铝酸钠反应得到高分子量化合物，可以降低含蜡原油的析蜡温度。其中酯类混合物的烷基链长度最大只有 12 个碳。尚未有资料报道有更长烷基链的酯，但此类酯应该有更好的效果。由环状酰胺和长链烷基组成的聚合物可以作为含蜡流体的流动改进剂，比较典型的是 N-乙烯基吡咯烷酮-α-烯烃共聚物（图 10.10），对含长碳链（尤其是 C_{25+}）蜡的原油的流动改进效果最好。

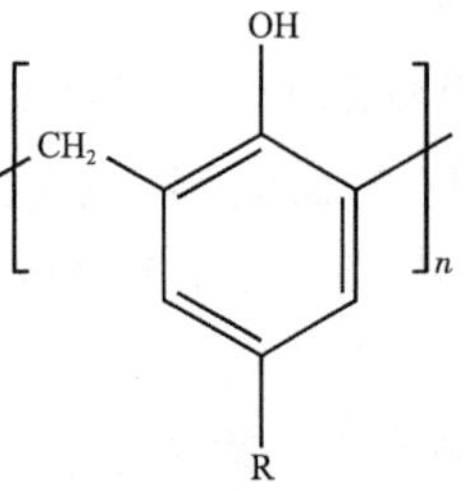

图 10.9　烷基酚醛树脂

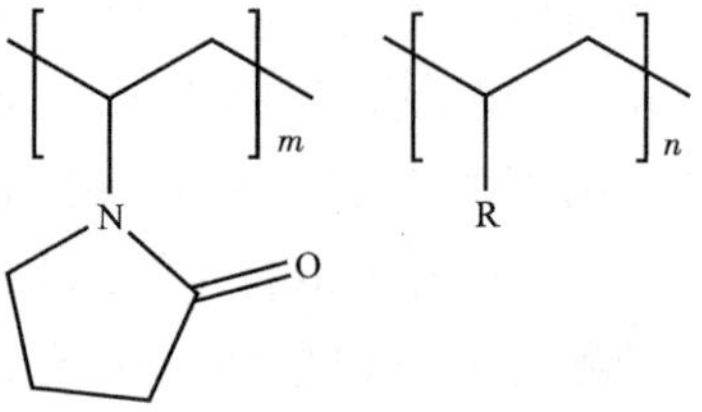

图 10.10　N-乙烯基吡咯烷酮-α-烯烃共聚物
（R 是一个较长的烷基链）

支链聚合物也可用作防蜡剂。例如，支化的聚乙烯亚胺（分子量为 1800）与 1,2-环氧十八烷反应得到具有多 C_{18} 侧基的聚合物，是一种较好的蜡沉积抑制剂（图 10.11）。枝状的超支化聚酯酰胺（最好具有长的烷基侧链）也是一种防蜡剂。众多实例中的一个是琥珀酸和二异丙醇胺的缩合产物（分子量为 3000～4000），其中羟基端基几乎都与一种脂肪酸（如硬脂酸或山嵛酸）酯化。在聚烷醇胺中也可以发现非乙烯基支链的骨架，如六三乙醇胺油酸酯是良好的倾点抑制剂。

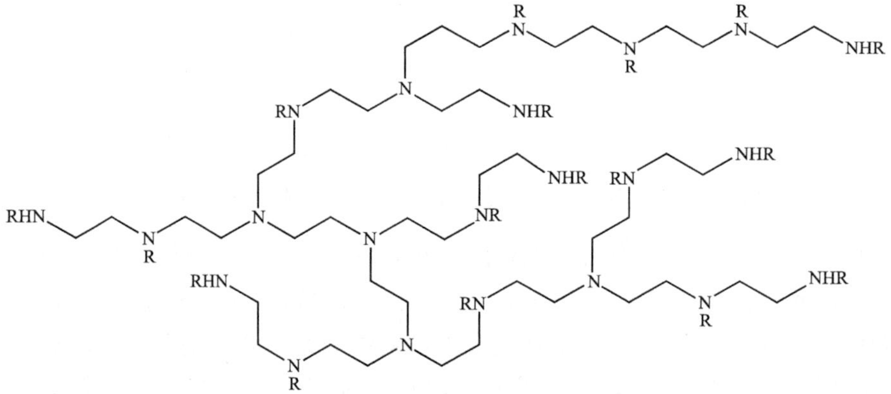

图 10.11　衍生化的聚乙烯亚胺防蜡剂 [R=R′CH(OH)CH$_2$，其中 R′是有 10～22 个碳的烷基；三级胺：二级胺：一级胺约为 1:2:1]

另一类据称可作为防蜡剂的超支链聚合物是由 2,2-二甲基苯酚丙酸合成的多元醇的脂肪酸酯衍生物（图 10.12）。一种线型多元醇（聚乙烯醇，可选交联）据称可降低含蜡烃类采出液的表观黏度。这种聚合物可能不是真正的防蜡剂，使用时聚合物的水溶液注入产出液，只是在油水混相系统中发挥降黏作用。

基于乙酰乙酸酯和十二胺缩合产物的线型和交联甜菜碱型两性聚电解质，随后与丙烯酸反应并聚合，是蜡晶生长的延缓剂（图 10.13）。这些聚甜菜碱防蜡剂已作为挤注处理的备选药剂进行了评价。与常规防蜡剂相比，羧酸基团无疑有助于聚甜菜碱防蜡剂在岩石表面的吸附。

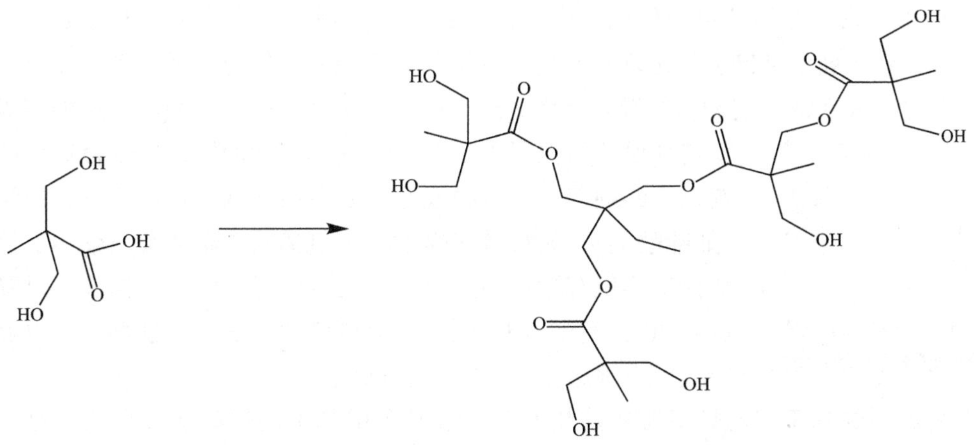

图 10.12 由二甲基酚丙酸聚合而成的多元醇

内酯/环氧烷聚合物也是一种防蜡剂（图 10.14）。这类聚合物是由含羟基或胺（包括烷醇胺）与至少有一个内酯单体和至少一个环氧烷单体的加成反应，制成的无规或嵌段聚合物。

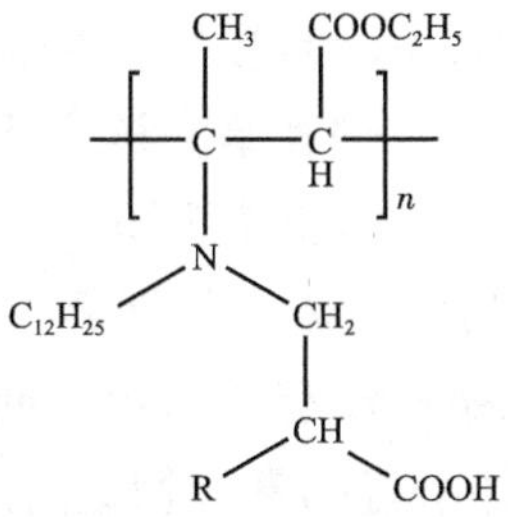

图 10.13　甜菜碱型两性聚电解质蜡晶生长抑制剂 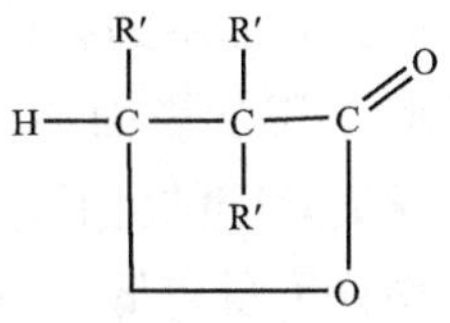　　图 10.14　制备内酯/环氧烷聚合物的内酯

10.4.6　蜡质分散剂

蜡质分散剂是一种可吸附在管道表面、减少蜡附着力的表面活性剂。可以将蜡吸附壁面的表面润湿性改变为水润湿性，或形成一个弱作用层，蜡晶在表面生长后，易被湍流剥离。有些蜡质分散剂可以通过吸附于管道表面和使其水润湿来发挥作用，也可能吸附在生长中的蜡晶表面，从而减少蜡晶黏聚趋势。好的分散剂配方还可以穿透沉积蜡，吸附于单个晶粒上并使其能够移动到周围油中。总体效果还是减少管壁上的积聚。蜡质分散剂典型加量为 50～300mg/L，可采用以防蜡为目的连续投加或高剂量的间歇加注，以达到弥补效果。分散剂经常与聚合物防蜡剂混合使用，以提高其性能。在现场单独使用蜡质分散剂的成功率有限。但有报道在新墨西哥州的某油田，EVA 和其他蜡晶改进剂无效，而蜡分散剂的评价和现场试验取得了成功。分散剂有水溶性表面活性剂和油溶性表面活性剂，典型低成本表面活性剂有烷基磺酸盐、烷基芳香基磺酸盐、脂肪胺乙氧基化和其他烷氧基化产品。有文章介绍了环保蜡质分散剂应用于北海。

还有几篇论文介绍了单独使用表面活性类蜡质分散剂以防止蜡沉积或胶凝。一篇论文表明，一些表面活性剂能够破坏蜡晶的三维网络来降低原油倾点。另一篇论文介绍了使用成膜型离子表面活性剂的协同混合物作为蜡分散剂和抗粘连剂。表面活性剂能够吸附到管道和设备的裸露表面上，使其具有憎油性，但未报道表面活性剂结构的细节。而成膜性好的表面活性剂也是性能良好的缓蚀剂（如咪唑类），如长链咪唑啉类（包括其二聚体和三聚体）等表面活性剂不仅有除抗粘连效果，还可以降低倾点，据称可以降低倾点达30℃（图10.15）。使用如咪唑啉类表面活性剂，可以克服寒冷气候中倾点抑制剂的胶凝问题。

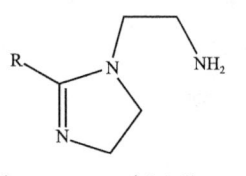

图10.15 2-氨乙基-2-烷基咪唑啉表面活性剂

上文提到的基于聚乙烯亚胺的防蜡剂与聚烷基（甲基）丙烯酸酯聚合物和可选的油溶性成膜型表面活性剂或缓蚀剂结合使用，可以发挥协同作用。油溶性成膜型表面活性剂的例子有N-妥尔基-1，3-丙二胺或其油溶性乙氧基化物，咪唑啉有如N-2-氨基乙基-2-烷基咪唑啉和膦酸酯。有研究证实了油酸咪唑啉缓蚀剂改善了两种防蜡剂的性能。分子建模项目表明，如油酸咪唑啉的缓蚀剂膜会产生有序的长链烷烃分子层，其结构类似于在烷烃蜡晶中发现的结构。虽然这样的烷烃层会提高缓蚀率，但有人认为该层可能成为蜡沉积的成核点。如果蜡确实在表面活性剂层上成核，蜡就不会直接黏附在表面，而表面活性剂和表面之间的结合力很弱，随着时间的推移，表面活性剂和蜡晶核会从表面冲刷剥离。也有资料称，原油中的高分子量馏分会吸附在金属表面，防止蜡沉积。

有水存在时，添加表面活性剂促进乳化可以降低蜡沉积趋势。如低界面张力和高剪切速率，可形成更紧密的乳状液。此外，乳状液中的沉积蜡比没有任何化学剂情况下的沉积蜡更软（平均分子量较低）。还观察到的现象是，在某些市售的聚合物基防蜡剂存在时，沉积的蜡甚至比没有化学添加剂情况下的更硬（平均分子量更高）。

有报道提出使用离子液体来防止结蜡或蜡沉积。离子液体最好是聚胺盐衍生物或聚丙二醇衍生物，仅有的例子是使用二甲苯和某种未知结构聚胺盐的混合物来溶解蜡。该试验似乎不符合宣称的防蜡作用。

10.4.7 极性原油馏分作为流动改进剂

原油中的天然成分也可以作为倾点抑制剂。源自沥青质的某些成分证明非常有效，而其性能在很大程度上取决于沥青质的来源——原油。相反的是，沉积实验发现C_{36}烷烃溶液中加入硬脂酸并不能减缓沉积，高浓度的硬脂酸实际上会增加沉积物。

一种流动改进剂的低成本制备方法是使用超临界气体（二氧化碳、乙烯、丙烯或C_1—C_3氟烷烃）从原油和馏分油中抽提极性提取物。萃取物含有沥青质、胶质和芳香烃。这些天然流动改进剂可以与EVA共聚物或梳状聚合物协同使用。除了这些极性较强的提取物，原油中高浓度的环状/支链烷烃（质量分数大于50%）可增强蜡晶改进剂的活性，其原因可能是环烷烃和支链结构的立体效应导致晶体的松散堆积。

在脱蜡催化剂作用下产出的石蜡润滑油，其重馏分可以作为蜡晶改进剂。适用作为蜡晶改进剂的石蜡润滑油的重馏分源于被催化除蜡的费托合成产品。选择性部分氧化的烯烃蜡产物可以有效地降低烃类的倾点或减少蜡质沉淀物中的平均蜡颗粒尺寸。

10.4.8 防蜡剂和倾点抑制剂的技术研发

防蜡剂和倾点抑制剂最常见的加注方法是井口注入。但如果井下有结蜡问题而且也有环空毛细管串，也可以从井下注入这些化学剂。中国常用的技术是将圆柱形固体防蜡剂放置在管状容器中随油管下入泵底部。防蜡剂作为缓释体系随着时间推移逐渐溶解。挤注防蜡工艺从理论上讲不经济，因为防蜡剂必须与原油的极性相似，其在储层岩石的吸附力很小，往往会迅速返排，有效处理期很短。也有防蜡剂挤注处理成功的报告，部分井的有效期超过 6 个月，但没有给出化学成分和吸附机理的细节。两个需要控制的重要方面是润湿性和产出液体中的最低防蜡剂浓度。

由于大多数防蜡剂和倾点抑制剂的聚合物性质，其溶液可能非常黏稠，难以处理，特别是在深海、海底或寒冷气候下应用时。因此，使用的组合物通常是溶于溶剂（通常是芳香烃）的低浓度混合物溶液，这些低活性溶剂型产品需要高加注量。提高活性成分的浓度并保持低黏度的一种方法是使用乳状液。有专利描述了使用水性外相分散体，其中包括通过连续水相分散的蜡质分散剂和蜡晶改进剂组合物，分散剂包括非离子表面活性剂（如乙氧基化脂肪醇），并且其在分散体中的量足以保障乳状液至少具有亚稳定性。另见10.4.2。

还有使用防蜡剂分散体的方法。在现场已成功试验适用于海底注入的新一代环境友好型分散体产品，并在越来越多的油田中持续使用。

有报道称有方法解决了防蜡剂的活性负载，以及其在低温和长距离输送中的可泵性问题。

参 考 文 献

[1] S. Misra, S. Baruah, and K. Singh, "Paraffin Problems in Crude Oil Production and Transportation: A Review," *SPE Production and Facilities* 10（1995）: 50.

[2] E. D. Burger, T. K. Perkins, and J. H. Striegler, "Studies of Wax Deposition in the Trans Alaska Pipeline," *SPE Journal of Petroleum Technology* 33（1981）: 1075.

[3] J. R. Becker, *Crude Oil: Waxes Emulsions and Asphaltenes*, Tulsa, OK: PennWell Publishing, 1997, Chapter 13.

[4] W. D. McCain Jr., *The Properties of Petroleum Fluids*, Tulsa, OK: PennWell Publishing Company, 1990.

[5] N. F. Carnahan, "Paraffin Deposition in Petroleum Production," *Journal of Petroleum Technology* 41（1989）: 1024.

[6] (*a*) D. Merino-Garcia and S. Correra, "Kinetics of Waxy Gel Formation from Batch Experiments" (paper presented at the 7th International Conference on Petroleum Phase Behavior and Fouling, 25–29 June 2006). (*b*) A. Bruno, C. Sarica, H. Chen, and M. Volk, "Paraffin Deposition during the Flow of Water-in-Oil and Oil-in-Water Dispersions in Pipes," SPE 114747 (paper presented at the SPE Annual Technical Conference and Exhibition, Denver, CO, 21–24 September 2008). (*c*) A. Benallal, P.

Maurel, J. F. Agassant, M. Darbouret, G. Avril, and E. Peuriere, "Wax Deposition in Pipelines: Flow-Loop Experiments and Investigations on a Novel Approach," SPE 115293 (paper presented at the SPE Annual Technical Conference and Exhibition, Denver, CO, 21–24 September 2008).

[7] P. Kriz and S. I. Andersen, "Effect of Asphaltenes on Crude Oil Wax Crystallization," Energy & Fuels 19 (2005): 948.

[8] (a) J. J. Magda, H. El-Gendy, K. Oh, M. D. Deo, A. Montesi, and R. Venkatesan, Energy & Fuels 23 (2009): 1311. (b) H. Li, J. Zhang, and D. Yan, "Correlations between the Pour Point/Gel Point and the Amount of Precipitated Wax for Waxy Crudes," Petroleum Science and Technology 23 (2005): 1313.

[9] G. E. Oliveira, C. R. E. Mansur, E. F. Lucas, G. González, and W. F. de Souza, "The Effect of Asphaltenes, Naphthenic Acids, and Polymeric Inhibitors on the Pour Point of Paraffin Solutions," Journal of Dispersion Science and Technology 28 (2007): 349.

[10] Q. Wang, C. Sarica, and T. Chen, "An Experimental Study on Mechanics of Wax Removal in Pipeline," SPE 71544 (paper presented at the SPE Annual Technical Conference and Exhibition, New Orleans, LA, 30 September–3 October 2001).

[11] D. R. Galloway, U.S. Patent 5168929, 1992.

[12] S. Feeney, "Project Case Histories and Future Applications of Vacuum Insulated Tubing" (paper presented at the AIChE Spring National Meeting, Houston, TX, 14–18 March 1999).

[13] R. C. Sarmento, G. A. S. Ribbe, and L. F. A. Azevedo, "Wax Blockage Removal by Inductive Heating of Subsea Pipelines," Heat Transfer Engineering 25 (2004): 2.

[14] K. A. Esakuhl, G. Fung, G. Harrison, and R. Perego, "Active Heating for Flow Assurance Control in Deepwater Flowlines," OTC 15188 (paper presented at the Offshore Technology Conference, Houston, TX, 5–8 May 2003).

[15] L. D. Brown, J. Clapham, C. Belmear, R. Harris, A. Loudon, S. Maxwell, and J. Stout, "Design of Britannia's Subsea Heated Bundle for a 25 Year Service Life," OTC 11017 (paper presented at Offshore Technology Conference, Houston, TX, 2–5 May 1999).

[16] A. K. Mehrotra, "'Cold Flow' Deposition Experiments with Wax-Solvent Mixtures under Laminar Flow in a Flow-Loop with Heat Transfer" (paper presented at the 9th International Conference on Petroleum Phase Behavior and Fouling, Victoria, BC, 15–19 June 2008).

[17] (a) D. Merino-Garcia and S. Correra, "Cold Flow: A Review of a Technology to Avoid Wax Deposition," Petroleum Science and Technology 26 (2008): 446. (b) C. B. Argo, P. Bollavaram, T. Y. Makogon, N. Oza, M. Wolden, R. Larsen, A. Lund, and K. W. Hjarbo, International Patent Application WO/2004/05918.

[18] (a) M. Juenke and L. Rzeznik, U.S. Patent Application 20080067129. (b) D. Stefanini, International Patent Application WO/2008/017849.

[19] B. Wang and L. Dong, "Paraffin Characteristics of Waxy Crude Oils in China and the Methods of Paraffin Removal and Inhibition," SPE 29954 (paper presented at the SPE International Meeting on Petroleum Engineering, Beijing, China, 14–17 November 1995).

[20] J. W. McDonald, K. J. Humphreys, R. D. Humphreys, K. R. Kopecky, and G. W. Adams, U.S. Patent 5804067, 1998.

[21] R. Tao and X. Xu, "Reducing the Viscosity of Crude Oil by Pulsed Electric or Magnetic Field," Energy & Fuels 20 (5) (2006): 2046.

[22] N. P. Tung, N. Van Vuong, B. Q. K. Long, N. Q. Vinh, P. V. Hung, V. T. Hue, and L. D. Hoe,

"Studying the Mechanism of Magnetic Field Influence on Paraffin Crude Oil Viscosity and Wax Deposition Reductions," SPE 68749 (paper presented at the SPE Asia Pacific Oil and Gas Conference and Exhibition, Jakarta, Indonesia, 17–19 April 2001).

[23] N. Rocha, C. González, L. C. do C. Marques, and D. S. Vaitsman, "A Preliminary Study on the Magnetic Treatment of Fluids," *Petroleum Science and Technology* 18 (2000): 33.

[24] W. A. Cañas-Marin, J. D. Ortiz-Arango, U. E. Guerrero-Aconcha, and C. Lira-Galeana, *AIChE Journal* 52 (8) (2006): 2887.

[25] L. C. C. Marques, N. O. Rocha, A. L. C. Machado, G. B. M. Neves, L. C. Vieira, and C. H. Dittz, "Study of Paraffin Crystallization Process under the Influence of Magnetic Fields and Chemicals," SPE 38990 (paper presented at the SPE Latin American and Caribbean Petroleum Engineering Conference, Rio de Janeiro, Brazil, 30 August–3 September 1997).

[26] W. Furlow, "Magnetic Fluid Conditioners' Success Depends on Factors," *Offshore* 58 (9) (1998).

[27] J. W. McManus, E. Winckler, and J. Backus, U.S. Patent 4538682, 1985.

[28] (a) Z. He, B. Mei, W. Wang, J. Sheng, S. Zhu, L. Wang, and T. F. Yen, *Petroleum Science and Technology* 21 (2003): 201. (b) K. Duncan, L. Gieg, and I. Davidova, "Paraffin Control in Oil Wells Using Anaerobic Microorganisms" (paper presented at the 14th Annual International Petroleum Environmental Conference, Houston, 8 November 2007). (c) B. Soni and B. Lal, U.S. Patent Application 20090025931.

[29] (a) I. Lazar, A. Voicu, C. Nicolescu, D. Mucenica, S. Dobrota, I. G. Petrisor, M. Stefanescu, and L. Sandulescu, "The Use of Naturally Occurring Selectively Isolated Bacteria for Inhibiting Paraffin Deposition," *Journal of Petroleum Science and Engineering* 22 (1999): 161. (b) H. K. Kotlar, A. Wentzel, M. Throne-Holst, S. Zotchev, and T. Ellingensen, "Wax Control by Biocatalytic Degradation in High Paraffinic Crude Oils," SPE 106420 (paper presented at the SPE International Symposium on Oilfield Chemistry, Houston, TX, 28 February–2 March 2007).

[30] (a) P. M. Roberts, A. Venkitaraman, and M. M. Sharma, "Ultrasonic Removal of Organic Deposits and Polymer-Induced Formation Damage," *SPE Drilling & Completion* 15 (1) (2000): 19. (b) B. Wang and C. Wang, International Patent Application WO/2009/000177.

[31] K. M. Barker, "Formation Damage Related to Hot Oiling," SPE 16230, *SPE Production Engineering* 4 (6) (1989): 371.

[32] W. B. Walton, U.S. Patent 4813482, 1989.

[33] C. L. Thierheimer, U.S. Patent 4925497, 1990.

[34] T. J. Straub, S. W. Autry, and G. E. King, "An Investigation into Practical Removal of Downhole Paraffin by Thermal Methods and Chemical Solvents," SPE 18889 (paper presented at the SPE Production Operations Symposium, Oklahoma City, OK, 13–14 March 1989).

[35] J. C. Bailey and S. J. Allenson, "Paraffin Cleanout in a Single Subsea Flowline Environment: Glycol to Blame?," OTC 19566 (paper presented at the Offshore Technology Conference, Houston, TX, 5–8 May 2008).

[36] C. J. Bushman, International Patent Application WO/2008/024488.

[37] J. A. Blunk, U.S. Patent 6176243, 2001.

[38] H. A. Craddock, K. Mutch, K. Sowerby, S. McGregor, J. Cook, and C. Strachan, "A Case Study in the Removal of Deposited Wax from a Major Subsea Flowline System in the Gannet Field," SPE 105048 (paper presented at the SPE International Symposium on Oilfield Chemistry, Houston, TX, 28 February–2 March 2007).

[39] R. A. Molland, P. Fotland, P. Clark, A. Valle, and B. H. Ovreas, "Handling Wax Deposition in a 116 km Sub-Sea, Single Phase Condensate Pipeline" (paper presented at the 19th Oilfield Chemical Symposium, Geilo, Norway, 9–12 March 2008).

[40] S. Szymczak, G. L. Poole, G. Brock, and G. Casey, "Successful Pipeline Cleanout: Lessons Learned from Cleaning Paraffin Blockage from a Deepwater Pipeline," SPE 115658 (paper presented at the Annual Technical Conference and Exhibition, Denver, CO, 22–24 September 2008).

[41] J. S. Als, U.S. Patent 6984614, 2006.

[42] P. R. Hart and M. J. Brown, U.S. Patent 5484488, 1996.

[43] E. A. Richardson and R. F. Scheuerman, U.S. Patent 4178993, 1979.

[44] J. P. Ashton, L. J. Kirspel, H. T. Nguyen, and D. J. Credeur, "In-Situ Heat System Stimulates Paraffinic-Crude Producers in Gulf of Mexico," SPE 15660, *SPE Production Engineering* 4 (3) (1989): 157.

[45] N. O. Rocha and C. N. Khalil, "Controlling Wax Deposition in Offshore Production Systems—Petrobras Experience" (paper presented at the IBC Global Conferences Ltd: Focus on Hydrates, Waxes and Asphaltenes, Aberdeen, UK, 1999).

[46] C. N. Khalil, U.S. Patent 5639313, 1997.

[47] C. R. De Souza and C. N. Khalil, U.S. Patent 6003528, 1999.

[48] D. A. Nguyen, "Fused Chemical Reactions to Remediate Paraffin Plugging in Sub-Sea Pipelines," PhD thesis, University of Michigan, 2004.

[49] N. O. Rocha, C. N. Khalil, L. C. F. Leite, and R. M. Bastos, "A Thermochemical Process for Wax Damage Removal," SPE 80266 (paper presented at the SPE International Symposium on Oilfield Chemistry, Houston, TX, 5–7 February 2003).

[50] D. Sarkar, S. T. Arrington, R. J. Powell, I. D. Robb, and B. L. Todd, International Patent Application WO/2008/032067.

[51] D. G. Clarke, U.S. Patent 6348102, 2002.

[52] J. J. Habeeb and R. L. Espino, U.S. Patent 6354381, 2002.

[53] T. G. Monger-McClure, J. E. Tackett, and L. S. Merrill, "Comparisons of Cloud Point Measurement and Paraffin Prediction Methods," SPE 54519, *SPE Production & Facilities* 14 (1) (1999): 4.

[54] J. A. P. Coutinho and J.-L. Daridon, "The Limitations of the Cloud Point Measurement Techniques and the Influence of the Oil Composition on Its Detection," *Petroleum Science and Technology* 23 (2005): 1113.

[55] S. U. Amadi, A. Y. Dandekar, G. A. Chukwu, S. Khataniar, S. L. Patil, W. F. Haslebacher, and J. Chaddock, "Energy Sources, Part A, Recovery," *Utilization and Environmental Effects* 27 (2005): 831.

[56] J. M. Letoffe, P. Claudy, M. Garcin, and J. L. Volle, "Crude Oils: Characterization of Waxes Precipitated on Cooling by DSC and Thermomicroscopy," *Fuel* 74 (1995): 810.

[57] D. D. Erickson, V. G. Niesen, and T. S. Brown, "Thermodynamic Measurement and Prediction of Paraffin Precipitation in Crude Oil," SPE 26604 (paper presented at the SPE Annual Technical Conference and Exhibition, Houston, TX, 3–6 October 1993).

[58] V. R. Kruka, E. R. Cadena, and T. E. Long, "Cloud-Point Determination for Crude Oils," *Journal of Petroleum Technology* 47 (8) (1995): 681.

[59] O. O. Bello, S. O. Fasesan, C. Teodoriu, and K. M. Reinicke, "An Evaluation of the Performance of Selected Wax Inhibitors on Paraffin Deposition of Nigerian Crude Oils," *Petroleum Science and*

Technology 24 (2006): 195.

[60] C. E. Ijeomah, A. Y. Dandekar, G. A. Chukwu, S. Khataniar, S. L. Patil, and A. L. Baldwin, "Measurement of Wax Appearance Temperature under Simulated Pipeline (Dynamic) Conditions," *Energy & Fuels* 22 (2008): 2437.

[61] M. C. Garcia, L. Carbognani, A. Urbina, and M. Orea, "Correlation between Oil Composition and Paraffin Inhibitors Activity," SPE 49200 (paper presented at the SPE Annual Technical Conference and Exhibition, New Orleans, LA, 27–30 September 1998).

[62] A. Hammami and M. A. Raines, "Paraffin Deposition from Crude Oils: Comparison of Laboratory Results With Field Data," *SPE Journal* 4 (1) (1999): 9.

[63] K. Karan, J. Ratulowski, and P. German, "Measurement of Waxy Crude Properties Using Novel Laboratory Techniques," SPE 62945 (paper presented at the SPE Annual Technical Conference and Exhibition, Dallas, 1–4 October 2000).

[64] K. A. Ferworn, A. Hammami, and H. Ellis, "Control of Wax Deposition: An Experimental Investigation of Crystal Morphology and an Evaluation of Various Chemical Solvents," SPE 37240 (paper presented at the SPE International Symposium on Oilfield Chemistry, Houston, TX, 18–21 February 1997).

[65] K. S. Wang, C. H. Wu, J. L. Creek, P. J. Shuler, and Y. C. Tang, "Evaluation of Effects of Selected Wax Inhibitors on Wax Appearance and Disappearance Temperatures," *Petroleum Science and Technology* 21 (2003): 359.

[66] R. M. Roehner and F. V. Hanson, "Determination of Wax Precipitation Temperature and Amount of Precipitated Solid Wax Versus Temperature for Crude Oils Using FT-IR Spectroscopy," *Energy & Fuels* 15 (3) (2001): 756.

[67] T. S. Brown, V. G. Niesen, and D. D. Erickson, "The Effects of Light Ends and High Pressure on Paraffin Formation," SPE 28505 (paper presented at the SPE Annual Technical Conference and Exhibition, New Orleans, LA, 25–28 September 1994).

[68] J. R. Becker, "Paraffin-Crystal Modifier Studies in Field and Laboratory," SPE 70030 (paper presented at the SPE Permian Basin Oil and Gas Recovery Conference, Midland, TX, 15–16 May 2001).

[69] N. F. Magri, B. Kalpakci, and L. Nuebling, "Evaluation of Paraffin Crystal Modifiers by Dynamic Videomicroscopy," SPE 37241 (paper presented at the SPE International Symposium on Oilfield Chemistry, Houston, TX, 18–21 February 1997).

[70] K. S. Wang, C. H. Wu, J. L. Creek, P. J. Shuler, and Y. C. Tang, "Evaluation of Effects of Selected Wax Inhibitors on Paraffin Deposition," *Petroleum Science and Technology* 21 (2003): 369.

[71] K. Weispfennig, "Advancements in Paraffin Testing Methodology," SPE 64997 (paper presented at the SPE International Symposium on Oilfield Chemistry, Houston, TX, 13–16 February 2001).

[72] D. W. Jennings and K. Weispfennig, "Effects of Shear on the Performance of Paraffic Inhibitors: Coldfinger Investigation with Gulf of Mexico Crude Oils," *Energy & Fuels* 20 (6) (2006): 2457.

[73] L. Dong, H. Xie, and F. Zhang, "Chemical Control Techniques for the Paraffin and Asphaltene Deposition," SPE 65380 (paper presented at the SPE International Symposium on Oilfield Chemistry, Houston, TX, 13–16 February 2001).

[74] S. Vaage, "A Novel Approach on Wax Inhibitor Qualification" (paper presented at the 19th Oilfield Chemical Symposium, Geilo, Norway, 9–12 March 2008).

[75] S. N. Duncum, K. James, and C. G. Osborne, International Patent Application WO98/021446, 1998.

[76] N. Fong and A. K. Mehrotra, "Deposition under Turbulent Flow of Wax-Solvent Mixtures in a Bench-Scale Flow-Loop Apparatus with Heat Transfer," *Energy & Fuels* 21 (3) (2007): 1263.

[77] I. Gjermundsen and M. Duenas Diez, "Wax Deposition: A Comparison between Measurements and Predictions from a Commercial Model" (paper presented at the 7th International Conference on Petroleum Phase Behavior and Fouling, 25–29 June 2006).

[78] N. V. Bhat and A. K. Mehrotra, "Modeling the Effects of Heat Transfer and Shear on Composition and Growth of Deposit-Layer from 'Waxy' Mixtures in a Pipeline under Laminar Flow" (paper presented at the 7th International Conference on Petroleum Phase Behavior and Fouling, 25–29 June 2006).

[79] B. F. Towler and S. Rebbapragada, "Mitigation of Paraffin Wax Deposition in Cretaceous Crude Oils of Wyoming," *Journal of Petroleum Science and Engineering* 45 (2004): 11.

[80] R. Brockmann, "HPHT Flow Loop for Testing Wax and Scale Solid Formation" (paper presented at the 7th Int. Oil Field Chemicals Symposium, Geilo, 17–20 March 1996).

[81] R. Brockmann, "Experimental Study of Pressure and Gas Content Effects on Wax Appearance Temperatures" (paper presented at the IBC Conference: Controlling Hydrates, Waxes and Asphaltenes, Aberdeen, UK, 16–17 September 1996).

[82] J. L. Creek, H. J. Lund, J. P. Brill, and M. Volk, "Wax Deposition in Single Phase Flow," *Fluid Phase Equilibria* 158 (1999): 801.

[83] J. F. Tinsley, R. K. Prud'homme, and X. Guo, "Effect of Polymer Additives Upon Waxy Deposits" (paper presented at the 7th International Conference on Petroleum Phase Behavior and Fouling, 25–29 June 2006).

[84] D. W. Jennings and M. E. Newberry, "Application of Paraffin Inhibitor Treatment Programs in Offshore Developments," OTC 19154 (paper presented at the Offshore Technology Conference, Houston, TX, 5–8 May 2008).

[85] D. W. Jennings, U.S. Patent Application 20070213231.

[86] J. Prasad, V. Sharma, P. C. Philip, S. R. Nimoria, and P. K. Verma, "Flow Assurance in 30″ Subsea Pipeline without the usage of PPD," OTC 19166 (paper presented at the Offshore Technology Conference, Houston, TX, 5–8 May 2008).

[87] M. Greenaway, "Analytical Solutions for Wax Deposition," *Chemistry in the Oil Industry VIII*, Royal Society of Chemistry, Manchester, UK, 2003.

[88] J. L. Hutter, S. Hudson, C. Smith, A. Tetervak, and J. Zhang, "Banded Crystallization of Tricosane in the Presence of Kinetic Inhibitors during Directional Solidification," *Journal of Crystal Growth* 273 (2004): 292.

[89] J. S. Manka, J. S. Magyar, and R. P. Smith, "A Novel Method to Winterize Traditional Pour Point Depressants," SPE 56571 (paper presented at the SPE Annual Technical Conference and Exhibition, Houston, TX, 3–6 October 1999).

[90] M.-F. Delamotte, D. Faure, and D. Tembou N'Zudie, U.S. Patent Application 20070062101, 2007.

[91] (a) A. Capelle, U.S. Patent 4110283, 1978. (b) M. Guzmann, Y. Liu, R. Konrad, and D. Franz, International Patent Application WO/2008/125588.

[92] D. W. Jennings, U.S. Patent Application 20040058827.

[93] M. J. Wisotsky, U.S. Patent 4153423, 1979.

[94] K. S. Pedersen and H. P. Rønningsen, "Influence of Wax Inhibitors on Wax Appearance Temperature, Pour Point, and Viscosity of Waxy Crude Oils," *Energy & Fuels* 17 (2) (2003): 321.

[95] D. M. Duffy and P. M. Rodger, "Wax Inhibition with Poly (Octadecyl Acrylate)," *Physical Chemistry Chemical Physics* 4 (2002): 328.

[96] D. M. Duffy and P. M. Rodger, "Modeling the Activity of Wax Inhibitors: A Case Study of Poly

(Octadecyl Acrylate)," *Journal of Physical Chemistry B* 106（43）（2002）：11210.

[97] D. M. Duffy, C. Moon, and P. M. Rodger, "Computer-Assisted Design of Oil Additives," *Molecular Physics* 102（2004）：203.

[98] A. Hennessy, A. Neville, and K. J. Roberts, "*In Situ* SAXS/WAXS and Turbidity Studies of the Structure and Composition of Multihomologous n-Alkane Waxes Crystallized in the Absence and Presence of Flow Improving Additive Species," *Crystal Growth & Design* 4（5）（2004）：1069.

[99] S. Y. Cho and H. S. Fogler, "Efforts on Solving the Problem of Paraffin Deposit：I. Using Oil-Soluble Inhibitors," *Journal of Industrial and Engineering Chemistry* 5（1999）：123.

[100] D. M. Duffy, C. Moon, J. L. Irwin, A. F. Di Salvo, P. C. Taylor, M. Arjmandi, A. Danesh, S. R. Ren, A. Todd, B. Tohidi, M. T. Storr, L. Jussaume, J.-P. Montfort, and P. M. Rodger, "Chemistry in the Oil Industry"（paper presented at the Symposium VIII, Manchester, England, 2003）.

[101] S. Gao, Y. Jiang, and A. Rodriguez, "Characterization of Oil Field Waxes Using Nuclear Magnetic Resonance," AIChE, Philadelphia, Pennsylvania, November 2008.

[102] S. M. Bucaram, "An Improved Paraffin Inhibitor," *Journal of Petroleum Technology* 19（2）（1967）：150.

[103] J. F. Tinsley, R. K. Prud'homme, X. Guo, D. H. Adamson, S. Callahan, D. Amin, S. Shao, R. M. Kriegel, and R. Saini, "Novel Laboratory Cell for Fundamental Studies of the Effect of Polymer Additives on Wax Deposition from Model Crude Oils," *Energy & Fuels* 21（3）（2007）：1301.

[104] C. J. Dorer Jr. and K. Hayashi, U.S. Patent 4623684, 1986.

[105]（a）O. E. Lindeman and S. J. Allenson, "Theoretical Modeling of Tertiary Structure of Paraffin Inhibitors," SPE 93090（paper presented at the SPE International Symposium on Oilfield Chemistry, The Woodlands, TX, 2-4 February 2005）.（b）J. B. Taraneh, G. Rahmatollah, A. Hassan, and D. Alireza, *Fuel Processing Technology* 89（2008）：973.

[106] L. A. McDougall, A. Rossi, and M. J. Wisotsky, U.S. Patent 3693720, 1972.

[107]（a）A. L. C. Machado, E. F. Lucas, and G. Gonzalez, "Poly（Ethylene-co-Vinyl Acetate）（EVA）as Wax Inhibitor of a Brazilian Crude Oil：Oil Viscosity, Pour Point and Phase Behavior of Organic Solutions," *Journal of Petroleum Science and Engineering* 32（2001）：159.（b）E. Marie, Y. Chevalier, F. Eydoux, L. Germanaud, and P. Flores, "Control of n-Alkanes Crystallization by Ethylene-Vinyl Acetate Copolymers," *Journal of Colloid Interface Science* 290（2005）：406.

[108] A. L. C. Machado and E. F. Lucas, "Influence of Ethylene-co-Vinyl Acetate Copolymers on the Flow Properties of Wax Synthetic Systems," *Journal of Applied Polymer Science* 85（2002）：1337.

[109] H. S. Ashbaugh, X. Guo, D. Schwahn, R. K. Prud'homme, D. Richter, and L. J. Fetters, "Interaction of Paraffin Wax Gels with Ethylene/Vinyl Acetate Co-polymers," *Energy & Fuels* 19（2005）：138.

[110] C. Wu, J.-L. Zhang, W. Li, and N. Wu, "Molecular Dynamics Simulation Guiding the Improvement of EVA-Type Pour Point Depressant," *Fuel* 84（2005）：2039.

[111] K. L. Motz, R. A. Latham, and R. J. Statz, European Patent EP345008, 1989.

[112] M. Del Carmen García, "Crude Oil Wax Crystallization：The Effect of Heavy n-Paraffins and Flocculated Asphaltenes," *Energy & Fuels* 14（5）（2000）：1043.

[113] H. K. Singhal, G. C. Sahai, G. S. Pundeer, and L. Chandra, "Designing and Selecting Wax Crystal Modifier for Optimum Field Performance Based on Crude Oil Composition," SPE 22784（paper presented at the SPE Annual Technical Conference and Exhibition, Dallas, 6-9 October 1991）.

[114] J.-F. Brunelli and S. Fouquay U.S. Patent 6218490, 2001.

[115] H. Wirtz, S.-P. Von Halasz, M. Feustel, and J. Balzer, U.S. Patent 5349019, 1994.

[116] G. Meunier, R. Brouard, B. Damin, and D. Lopez, U.S. Patent 4608411, 1986.

[117] W. Ritter, C. Meyer, W. Zoellner, C.-P. Herold, and S. Tapavicza, U.S. Patent 5039432, 1991.

[118] P. Gateau, A. Barbey, and J. F. Brunelli, U.S. Patent 6750305.

[119] (a) M. Feustel, M. Krull, and H.-G. Oschmann, U.S. Patent 6821933, 2004. (b) O. E. Shmakova-Lindeman, International Patent Application WO/2005/098200.

[120] O. E. Shmakova-Lindeman, U.S. Patent Application 20050215437, 2005.

[121] M. Mueller and H. Gruenig, U.S. Patent 5281329, 1994.

[122] R. Eckert and B. Vos, Canadian Patent 1231659, 1988.

[123] D. Chanda, A. Sarmah, A. Borthakur, K. V. Rao, B. Subrahmanyam, and H. C. Das, "Combined Effect of Asphaltenes and Flow Improvers on the Rheological Behavior of Indian Waxy Crude Oil," *Fuel* 77 (1998): 1163.

[124] V. A. Adewusi, "An Improved Inhibition of Paraffin Deposition from Waxy Crudes," *Petroleum Science and Technology* 16 (1998): 953.

[125] E. Barthell, A. Capelle, M. Chmelir, and K. Dahmen, U.S. Patent 4663491, 1987.

[126] H. P. M. Tomassen, C. van de Kamp, M. J. Reynhout, and J. Lin, U.S. Patent 5721201, 1998.

[127] J. A. Day, M. J. Reynhout, and H. P. M. Tomassen, U.S. Patent 5585337, 1996.

[128] A. J. Son, R. B. Graugnard, and B. J. Chai, "The Effect of Structure on Performance of Maleic Anhydride Copolymers as Flow Improvers of Paraffinic Crude Oil," SPE 25186 (paper presented at the SPE International Symposium on Oilfield Chemistry, New Orleans, LA, 2-5 March 1993).

[129] H. T. Le, U.S. Patent 4992080, 1991.

[130] X. H. Guo, M. Herrera-Alonso, J. F. Tinsley, and R. K. Prudhomme, Preprint Papers—*American Chemical Society, Division of Petroleum Chemistry* 50 (2005): 318.

[131] B. Wahle, C.-P. Herold, W. Zoellner, L. Schieferstein, and D. Oberkobusch, U.S. Patent 5006621, 1991.

[132] A. M. Robinson, J. R. Stromberg, M. J. Jurek, and K. Bakeev, U.S. Patent Application 20020166995, 2002.

[133] J. Balzer, M. Feustel, M. Krull, and W. Reimann, U.S. Patent 5439981, 1995.

[134] A. Borthakur, D. Chanda, S. R. Dutta Choudhury, K. V. Rao, and B. Subrahmanyam, "Alkyl Fumarate-Vinyl Acetate Copolymer as Flow Improver for High Waxy Indian Crude Oils," *Energy & Fuels* 10 (3) (1996): 844.

[135] A. O. Patil, S. Zushma, E. Berluche, and M. Varma-Nair, U.S. Patent 6444784, 2002.

[136] J. S. Manka, K. L. Ziegler, and D. R. Nelson, U.S. Patent 6017370, 2000.

[137] D. J. Martella and J. J. Jaruzelski, European Patent EP0311452, 1989.

[138] D. O. Gentili, C. N. Khalil, N. O. Rocha, and E. F. Lucas, "Evaluation of Polymeric Phosphoric Ester-Based Additives as Wax Deposition Inhibitors," SPE 94821 (paper presented at the SPE Latin American and Caribbean Petroleum Engineering Conference, Rio de Janeiro, Brazil, 20-23 June 2005).

[139] S. N. Duncum, K. James, and C. G. Osborne, U.S. Patent 6140276, 2000.

[140] P. F. Van Bergen, M. A. Van Dijk, and A. J. Zeeman, International Patent Application WO/2006/056578.

[141] A. A. Hafiz and T. T. Khidr, "Hexa-Trithanolamine Oleate Esters as Pour Point Depressant for Waxy Crude Oils," *Journal of Petroleum Science and Engineering* 56 (2007): 296.

[142] G. G. McClaflin and D. L. Whitfil, "Control of Paraffin Deposition in Production Operations," SPE 12204 (paper presented at the Annual Technical Conference and Exhibition, San Francisco, October

1983).

[143] L. K. Verma and S. N. Mukhdeo, "Effects of Surfactants on Pour Point of Crude Oil" (paper presented at the 1st Natl. Conv. of Chem. Eng., Calcutta, 21–23 February 1986).

[144] D. Groffe, P. Groffe, S. Takhar, S. I. Andersen, E. H. Stenby, N. Lindeloff, and M. Lundgren, "A Wax Inhibition Solution to Problematic Fields: A Chemical Remediation Process," *Petroleum Science and Technology* 19 (2001): 205.

[145] R. L. Martin, H. L. Becker, and D. Galvan, U.S. Patent Application 20070051033, 2007.

[146] M. A. San-Miguel and P. M. Rodger, "The Effect of Corrosion Inhibitor Films on Deposition of Wax to Metal Oxide Surfaces," *Journal of Molecular Structure: THEOCHEM* 506 (2000): 263.

[147] J. J. Hanke, "An Experimental Study on the Nature of the Film Forming Characteristics of Crude Oil Fractions on Steel Surfaces and Their Influence on Paraffin Deposition," PhD thesis, Pet. Eng., University of Texas, 1967.

[148] B. F. Birdwell, "Effects of Various Additives on Crystal Habit and Other Properties of Petroleum Wax Solutions," PhD thesis, University of Texas, 1964.

[149] S. Ahn, K. S. Wang, P. J. Shuler, J. L. Creek, and Y. Tang, "Paraffin Crystal and Deposition Control by Emulsification," SPE 93357 (paper presented at the SPE International Symposium on Oilfield Chemistry, The Woodlands, TX, 2–4 February 2005).

[150] M. Feustel, H.-G. Oschmann, and U. Kentschke, U.S. Patent 6803492, 2004.

[151] M. Del Carmen García, L. Carbognani, M. Orea, and A. Urbina, *Journal of Petroleum Science and Engineering* 25 (2000): 99.

[152] A. R. Bishop, A. L. Ansell, W. B. Genetti, M. A. Daage, D. F. Ryan, E. B. Sirota, J. W. Johnson, and P. Brant, U.S. Patent Application 20060219597.

[153] E. J. Baralt and H. Yang, U.S. Patent Application 20070095723, 2007.

[154] J. B. Dobbs, "A Unique Method of Paraffin Control in Production Operations," SPE 55647 (paper presented at the SPE Rocky Mountain Regional Meeting, Gillette, WY, 15–18 May 1999).

[155] D. J. Poelker, T. J. Baker, and J. W. Germer, U.S. Patent 5858927, 1999.

[156] B. Wang, C. Yu, and D. Gao, Chinese Patent CN1487048, 2004.

[157] H. R. Rønningsen, *Energy & Fuels* 26 (7) (2012): 4124–4136.

[158] C. Sarica and E. Panacharoensawad, *Energy & Fuels* 26 (7) (2012): 3968–3978.

[159] J. F. Tinsley, J. P. Jahnke, D. H. Adamson, X. Guo, D. Amin, R. Kriegel, R. Saini, H. D. Dettman, and R. K. Prud'home, *Energy & Fuels* 23 (2009): 2056.

[160] M. C. Khalil de Oliveira, A. Teixeira, L. C. Vieira, R. M. de Carvalho, A. B. Melo de Carvalho, and B. C. do Couto, *Energy & Fuels* 26 (2012): 2688.

[161] P. Juyal, T. Cao, A. Yen, and R. Venkatesan, *Energy & Fuels* 25 (2011): 568–572.

[162] L. C. Vieira, M. B. Buchuid, and E. F. Lucas, *Energy & Fuels* 24 (4) (2010): 2213–2220.

[163] R. Hoffmann, L. Amundsen, Z. Huang, S. Zheng, and H. S. Fogler, *Energy & Fuels* 26 (2012): 3416–3423.

[164] J. J. Magda, A. Elmadhoun, P. Wall, M. Jemmett, M. D. Deo, K. L. Greenhill, and R. Venkatesan, *Energy & Fuels* 27 (2013): 1909–1913.

[165] D. A. Phillips, I. N. Forsdyke, I. R. McCracken, and P. D. Ravenscroft, *Journal of Petroleum Science and Engineering* 77 (2011): 237.

[166] R. Hoffmann and L. Amundsen, International Patent Application WO/2009/051495.

[167] M. Margarone, A. Bennardo, C. Busto, and S. Correra, *Energy & Fuels*, (2013): 27, published

online.

[168] M. R. Jemmett, M. Deo, J. Earl, and P. Mogenhan, *Energy & Fuels* 26 (5) (2012): 2641-2647.

[169] J. L. Gonçalves, A. J. F. Bombard, D. A. W. Soares, and G. B. Alcantara, *Energy & Fuels* 24 (2010): 3144.

[170] J. L. Gonçalves, A. J. F. Bombard, D. A. W. Soares, R. D. M. Carvalho, A. Nascimento, M. R. Silva, G. B. Alcantara, F. Pelegrini, E. D. Vieira, K. R. Pirota, M. I. M. S. Bueno, G. M. S. Lucas, and N. O. Rocha, *Energy & Fuels*, 25 (2011): 3537.

[171] B. Wang and C. Wang, International Patent Application WO/2009/000177.

[172] G. Leia, B. F. Towler, L. Goual, and J. F. Schabron, Preprint Papers—American Chemical Society, *Division of Petroleum Chemistry* 56 (2) (2011): 129.

[173] C. Yaoa, C. Wua, X. Jiaa, and X. Zhanga, *Petroleum Science and Technology* 29 (2011): 2077.

[174] M. R. Embrey and J. Larke, "Paraffin-Plug Remediation in Deepwater Wells via Capillary Tubing: A Cost Effective Alternative," SPE 135136 (paper presented at the SPE Annual Technical Conference and Exhibition, Florence, Italy, 19-22 September 2010).

[175] J. P. Acunto, U.S. Patent Application 20100022417.

[176] C. Wiggins, H. Quintero, and L. Ubana, "A Novel, Chemistry-Based Approach to Fluid Diversion in Hydrocarbon Based Paraffin Solvent Treatments," IPTC 16939 (paper presented at the 6th International Petroleum Technology Conference, Beijing, China, 26-28 March 2013).

[177] D. W. Jennings, S. Asomaning, and M. E. Newberry, U.S. Patent Application 20110114323.

[178] M. D. Robustillo, B. Coto, C. Martos, and J. J. Espada, *Energy & Fuels* 26 (10) (2012): 6352-6357.

[179] B. Coto, J. A. P. Coutinho, C. Martos, M. D. Robustillo, J. J. Espada, and J. L. Pena, *Energy & Fuels* 25 (2011): 1153-1160.

[180] R. Venkatesan and J. L. Creek, "Wax Deposition and Rheology: Progress and Problems from an Operator's View," OTC 20668 (paper presented at the Offshore Technology Conference, Houston, TX, 3-6 May 2010.

[181] B. Coto, C. Martos, J. J. Espada, M. D. Robustillo, D. Merino-Garcia, and J. L. Pena, *Energy & Fuels* 35 (2011): 1707.

[182] L. C. Vieira, M. B. Buchuid, and E. F. Lucas, *Energy & Fuels* 24 (4) (2010): 2208-2212.

[183] K. Paso, H. Kallevik, and J. Sjoblom, *Energy & Fuels* 23 (2009): 4988.

[184] Y. Cheng and A. D. Kharrat, International Patent Application WO/2009/051849.

[185] Y. Zhang, J. Gong, Y. Ren, and P. Wang, *Energy & Fuels* 24 (2) (2010): 1146-1155.

[186] J. F. Tinsley and R. K. Prud'homme, *Journal of Petroleum Science and Engineering* 72 (2010): 166.

[187] Q. Huang, J. Wang, and J. Zhang, *Petroleum Science* 6 (2009): 64.

[188] B. F. Towler, O. Jaripatke, and S. Mokhatab, *Petroleum Science and Technology* 29 (2011): 468.

[189] S. Seth, B. F. Towler, and S. Mokhatab, *Petroleum Science and Technology* 29 (2011): 378.

[190] J. Gong, Y. Zhang, L. Liao, J. Duan, P. Wang, and J. Zhou, *Energy & Fuels* 35 (2011): 1624.

[191] Z. Guozhong and L. Gang, *Journal of Petroleum Science and Engineering* 70 (2010): 1.

[192] R. Hoffmann and L. Amundsen, *Energy & Fuels* 24 (2) (2010): 1069-1080.

[193] Sh. Masoudi, M. Vafaie Sefti, H. Jafari, and H. Modares, *Petroleum Science and Technology* 28 (2010): 1598.

[194] R. Venkatesan, V. Sampath, and L. A. Washington, "Study of Wax Inhibition in Different Geometries," OTC 23624 (paper presented at the Offshore Technology Conference, Houston, TX, 30

April-3 May 2012).

[195] S. A. Garner, P. Juyal, C. Hart, D. Podgorski, A. M. Mckenna, C. M. Ziglio, R. P. Rodgers, S. J. Allenson, and A. G. Marshall, "Analysis and Comparison of Paraffinic Field Deposits to Cold Finger Deposits on a Brazilian Campos Basin Crude Oil," OTC 22660 (paper presented at the Offshore Technology Conference, Brazil, 2011).

[196] S. Yi and J. Zhang, *Energy & Fuels* 35 (2011): 1686.

[197] A. Hunton, private communication.

[198] J. J. Wylde and J. L. Slayer, "Considerations and Lessons Learned on Development of Gas-Lift Paraffin Inhibitors for Subsea Wells in the Gulf of Mexico," OTC 23206 (paper presented at the Offshore Technology Conference, Houston, TX, 30 April-3 May 2012).

[199] D. W. Jennings and J. Breitigam, *Energy & Fuels* 24 (4) (2010): 2337-2349.

[200] T. Jafari Behbahani, A. Dahaghin, and K. Kashefi, *Petroleum Science and Technology* 29 (2011): 933.

[201] M. Feustel, M. Krull, C. Kayser, and M. Loew, U.S. Patent Application 20130023453.

[202] L. V. Castro, E. A. Flores, and F. Vazquez, *Energy & Fuels* 25 (2011): 539.

[203] P. Van der Meij and A. Buitelaar, U.S. Patent 3598736, 1971.

[204] M. N. Maithufi, D. J. Joubert, and B. Klumperman, *Energy & Fuels* 25 (2011): 162-171.

[205] H. Qian, J. Xu, J. Sun, Y. Shu, L. Li, and X. Guo, *Preprint Papers—American Chemical Society, Division of Petroleum Chemistry* 57 (1) (2012): 115.

[206] S. Chen, J. Xu, J. Sun, Y. Shu, H. Qian, J. Huang, S. Xing, L. Li, and X. Guo, *Preprint Papers—American Chemical Society, Division of Petroleum Chemistry* 57 (1) (2012): 117.

[207] Y. Wu, G. Ni, F. Yang, C. Li, and G. Dong, *Energy & Fuels* 26 (2012): 995.

[208] J. Xu, H. Qian, S. Xing, L. Li, and X. Guo, *Energy & Fuels* 25 (2011): 573-579.

[209] T. T. Khidr, *Petroleum Science and Technology* 29 (2010): 19.

[210] H. P. Soni, Kiranbala, and D. P. Bharambe, *Energy & Fuels* 22 (6) (2008): 3930-3938.

[211] J. Guo, H. Wang, C. Chen, Y. Chen, and X. Xie, *Petroleum Science* 7 (4) (2010): 536.

[212] R. M. Kriegel, J. F. Tinsley, R. K. Saini, R. Prud'homme, and I. D. Robb, U.S. Patent Application 20090233817.

[213] K. Cao, X.-X. Wei, B.-J. Li, J.-S. Zhang, and Z. Yao, *Energy & Fuels* 27 (2) (2013): 640-645.

[214] D. J. Moreton, A. Mastrangelo, and M. Macduff, International Patent Application WO/2009/064827.

[215] R. Rodriguez Gonzalez, J. Djuve, and A. Grinrod, International Patent Application WO/2010/003892.

[216] C. Y. Khandekar, T. Nordvik, and A. Grinrod, International Patent Application WO/2013/019704.

[217] J. Forsyth and P. Fletcher, International Patent Application WO/2006/106300.

[218] D. Nguyen and V. Balsamo, *Energy & Fuels* 27 (2013): 1736.

[219] A. G. Didukh, R. B. Koizhaiganova, L. A. Bimendina, and S. E. Kudaibergenov, *Journal of Applied Polymer Science* 92 (2004): 1042-1048.

[220] N. Halim, S. Ali, M. Nadeem, P. Abdul Hamid, and I. Mohd Tan, "Synthesis of Wax Inhibitor and Assessment of Squeeze Technique Application for Malaysian Waxy Crude," SPE 142288 (paper presented at the SPE Asia and Pacific Oil and Gas Conference, Jakarta, Indonesia, 20-22 September 2011).

[221] J. Excoffon, H. Oschmann, and M. M. Huijgen, "Novel Technologies: Environmentally Acceptable Surfactants Applied for Water Based Paraffin Control," Tekna 23rd International Oil Field Chemistry Symposium, Geilo, Norway, 18-21 March 2012.

[222] H. Oschmann, personal communication.

[223] M. Senra, E. Paracharoensawad, T. Scholand, and H. S. Fogler, *Energy & Fuels* 23（12）（2009）：6040-6047.

[224] H. A. Craddock, Patent Application 1020439.4—Silicon Materials as Additives in Wax Inhibitors, Filing date 2 December 2010.

[225] H. A. Craddock and H. Blackwood, unpublished results.

[226] H. J. Oschmann, M. C. Huijgen, and H. F. Grondman, "Production Chemicals Based on Active Dispersions—Alternatives to Conventional Solvent Based Products," *RSC Chemistry in the Oil Industry XII*, Manchester, UK, November 2011.

[227] J. Dunlop, "'Easy Flow' Wax deposition Inhibitors for Subsea Developments," *RSC Chemistry in the Oil Industry XIII*, Manchester, UK, November 4-6, 2013.

[228] Z. Huo, T. M. Shea, C. A. T. Kuijvenhoven, and Y. Zhao, International Patent Application WO/2013/096217.

[229] J. L. Sonne and M. Hilfiger, International Patent Application WO/2013/090347.

[230] K. Singh, A. Saidu Mohamed, S. Sheykh Alian, M. Ismail, M. Anwar, W. Wan Mohamad, and S. Abdul Ghani, "Thermo Chemical In-Situ Heat Generation Technique to Remove Organic Solid Deposition: Effective Tool for Production Enhancement and Flow Assurance," 2013 Offshore Technology Conference, 6-9 May 2013, Houston, TX, USA.

[231] R. Hoffmann and L. Amundsen, *Journal of Petroleum Science Engineering*, 107（2013）：12-17.

[232] S. Desmukh and D. P. Bharambe, "Energy Sources, Part A, Recovery," *Utilization and Environmental Effects*, 34（12）（2013）：1121-1129.

[233] E. Marie, Y. Chevalier, S. Brunel, F. Eydoux, L. Germanaud and P. Flores, *Journal of Colloid Interface Science*, 269（2004）：117-125.

11 破乳剂

11.1 概述

乳状液是液相液滴分散在另一种液相中形成的胶体分散物。油井产出的原油基本是油包水乳状液（水滴稳定分布在连续的原油相中）形式（图11.1）。随着原油的含水率变化，也会出现游离态的产出水。从原油乳状液中分离出水和溶解盐类的处理流程被称为破乳或脱水，此后的低含水原油才能运输或在炼油厂炼制加工。在炼油厂还有被称为"脱盐"的处理流程，以从原油中去除含盐洗涤水。销售至炼油厂的商品原油限定了最高含水率和固含量（分离后沉积在底部的水量和固体），通常可接受的最高含水率为0.2%～0.5%，最高含盐量为28.6～71.4mg/L，但炼油厂对含水率和含盐量的规定值可能比这更严格。

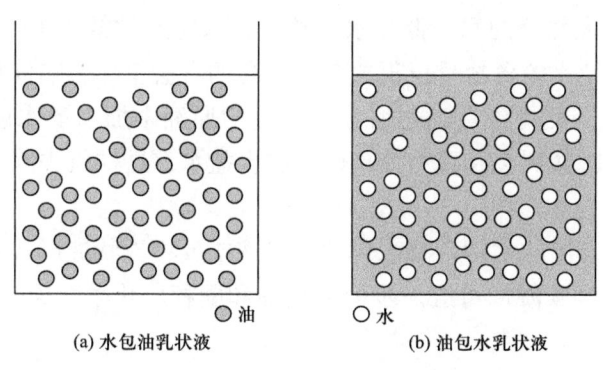

图11.1 水包油（水为连续相）和油包水乳状液示意图

从油包水乳状液中分离出的水中通常含有分散的油，以水包油乳状液形式存在（图11.1）。这时需要使用絮凝剂进行除油处理，使水中残余油含量低于规定水平后（通常为10～30mg/L，视地区情况而定），才能获批排放（见第13章）。

产出流体在油井油管和地面管道中的湍流流动会形成乳状液，特别是流经如井口等阻塞时易形成乳状液。原油中的固体颗粒、胶质（天然表面活性剂）或沥青质使油包水乳状液稳定。胶质是原油中含有S、O或N原子的一大类极性分子。由油溶性羧酸或环烷酸与水相中的阳离子相互作用而形成的皂类，也会沉积并稳定乳状液。有研究表明，与公认的观点相反，从油包水乳状液的液滴中收集的物质中，主要的极性组分并不是沥青，而是小分子极性化合物，有一些还没有芳香环。水溶液取样技术可能已经排除了沥青质。通过计算机模拟发现，芳香烃集中在油水界面区域，而其他烃类则均匀地分布于整个油相。成膜型缓蚀剂是表面活性剂，也能稳定乳状液，酸化增产或作业返排液都有可能形成很难处理的乳状液，需要破乳剂和絮凝剂处理。

粉末、垢的胶体微粒、腐蚀产物及沉积蜡等固体对乳液的稳定很有帮助，而且会形成皮克林乳液。有研究表明，衡量乳状液稳定性的最佳预测因素不是沥青质含量或任何原油参数，而是固相杂质含量。在原油处理流程中加热乳状液会溶解蜡状固体，并一定程度上破坏乳状液的稳定性。加热通常有利于破乳，但是如果存在羧酸盐或环烷酸盐稳定的乳状液，加热使其pH值上升（水相中的CO_2减少），表面活性剂成分的极性增强，使乳化更加严重。除加热方法以外，乳状液破乳的最重要方法是添加破乳剂，其使用浓度通常是基于水相的5～500mg/L。必须充分地混合或分散使破乳剂到达油水界面，并且必须有足够的时间以分离凝聚的液滴相。无法被破坏的乳状液称为老化油或废油。破乳工艺的最后阶段通常采用电破乳（静电凝聚）。应该注意某些破乳剂的过度使用会使乳状液重新稳定，而使用α-烯烃磺酸盐等简单的磺化表面活性剂则是例外。因此，破乳剂有一个最佳用量。

良好破乳的先决条件是油—水乳状液应尽可能脱气。如果原油中含有大量的气体，气泡导致的扰动会限制化学破乳剂产生清晰界面的能力。上游的专用分离器可对原油脱气。

破乳后，分离出大部分的水，低含水原油（含水率通常低于1%）就可以进一步运输到炼油厂。原油中残留水的含盐量也很关键，过高的含盐量会影响催化精炼过程，导致腐蚀和热交换器结垢。因此炼油厂通过脱盐工艺，向原油中加入淡水以稀释残留水的盐浓度，并通过破乳剂（脱盐剂）和重力分离以去除含盐水，这样就去除了大部分盐，原油就可以进入炼制环节。

从加工设施中分离出的水中含有水包油乳状液或反相乳状液形式的残余油和固体。通过机械方式或使用化学絮凝剂可进一步分离出油和固相。化学絮凝剂也被称为水澄清剂、除油剂或反相破乳剂。

11.2 破乳方法

油包水乳状液最广泛采用的破乳脱水技术是加热条件下，加注化学破乳剂结合重力分离的方式。其他使用的技术有静电破乳、水力旋转法和离心分离法。据称，也可采用热闪蒸法和超声破乳法。加热对乳状液的破乳有两方面促进作用：首先，降低原油黏度，水滴更容易沉降；其次，更高的温度使乳状液稳定性变差，如果达到了乳状液的相转换温度，将从油包水乳状液变成水包油乳状液。

还有研究将微波技术用于老化油层等难处理的乳状液。从原油或沥青中分离水和固相的过程中，往往会在油包水乳化层和水层之间形成含有大量有机物的老化油层。原油中天然和后期人工添加的表面活性剂严重影响难处理的老化油层的形成。

有资料表明，通过使用Gore-Tex膜的过滤工艺可以从油包水乳状液中分离原油，并且不改变原油特性。

11.3 油包水破乳剂

11.3.1 理论与实践

在海上油气生产中,通常在分离器前的处理设施中加注破乳剂。从上游的井口加注破乳剂也有附加优势(这种方式越来越普遍),如果井下有毛细管串,也可以从井下加注。乳状液的缓慢破乳需要经历足够长的时间,因此乳状液破乳时间是一个重要参数。只有在确定了最佳的破乳时间后,才可以调整破乳剂用量。上游产出流体的温度越高,越有利于快速破乳。此外,乳状液破乳后,多相流体的黏度降低,也会降低摩阻,提高输量。

决定油包水乳状液稳定性的重要界面特性是剪切黏度、动态界面张力和扩张弹性。单一的界面张力并不是反映乳状液稳定性的参数。已经有很多研究试图将破乳剂的效果与控制乳状液稳定性的一些物理特性相关联,但人们对于这一领域的认识仍比较有限。破乳剂选用在很大程度上仍然是一项实践性技术,通常在工艺流程中需要开展一系列破乳剂和混合剂的试错法筛选。需要特别强调的是,必须采用处理设施中的新鲜乳状液进行测试,以优化筛选最终的破乳剂配方。

为使乳状液失去稳定,可以改变其环境的黏度、密度、含水率、老化时间和乳化剂加量等参数。高黏度的原油能包纳更多、更大的水滴。提高温度、添加稀释剂或某些化学品可以降低乳状液的黏度。降低黏度会增大水滴沉降的速度及其流动性,即水滴凝聚和分离的速度。另外,加热乳状液使油的密度比水的密度下降得更快,从而使水更快地沉降。高矿化度产出水密度较高,水滴的沉降更快。重油密度更接近于水的密度,所以不太容易脱水。乳状液的稳定性随含水率的变化而变化,低含水率时水滴间距更大,破乳更难。乳状液老化时间越长越稳定,乳化液形成后应尽早破乳,如在井口加注破乳剂可能比在后端的处理设施处加注能更好或更快地分解乳状液。乳化剂可以是烃类中天然存在的沥青、胶质和环烷酸,也可以是添加到产出流体中的合成化学品。成膜型缓蚀剂通常是表面活性剂,可以稳定乳状液,因此谨慎选择缓蚀剂可有助于减少下游破乳剂的用量。

破乳的控制因素有界面流变学、质量转移或破乳剂分子到界面的速度(以抑制界面张力梯度),以及吸附于油水界面的胶体或大分子的空间位阻效应。特征弛豫时间、界面黏度和弹性的降低,与乳状液滴的半衰期和破乳剂的效果直接相关。因此,性能良好的破乳剂表现出短弛豫时间,并显著降低油水界面的扩张黏弹性。

破乳剂作用于乳状液有三个主要过程:

(1) 絮凝。水滴像一群鱼卵聚集在一起。

(2) 凝聚。乳状液中稳定水滴的乳化膜破裂,水滴增大并分离成为单独的相。大水滴的表面张力较小,因此任何增加水滴尺寸的方法都有助于分离过程。

(3) 固体润湿。稳定乳状液的固相分散到烃相中,或被水润湿,并随水排出。

工业上主要使用的是不同比例的絮凝破乳剂、凝聚破乳剂和润湿剂等溶于溶剂中的组合物。沥青质稳定剂（分散剂）和基础破乳剂的协同混合物对沥青质稳定的乳状液有良好破乳效果。有关破乳剂性能评价的观点是基于所需破乳剂的浓度应能抵消给定量沥青质的影响。

11.3.2 选择破乳剂的测试方法和参数

瓶试法是评估破乳剂产品的最佳方法，但需要大量的实验室测试且现场试验耗时，而且有时会得到不准确的结果。为了建立一种更简单、更准确、更快速的方法来评估破乳剂，提供符合客户需求的正确产品，已经有大量的策略被提出。

许多大型破乳剂供应商提供相对溶解指数（RSN）值，以便客户用于优选破乳剂性能参数。RSN值最常用于二级脱盐时破乳剂的分类，如原油乳状液中的破乳剂经过一级分离器脱水后，在二级脱盐过程仍然有一定残留量。RSN值与破乳剂产品的亲水亲油特性有关。RSN值也被称为水数，于1956年最早提出。测试时使用二氧六环与苯的比例为96:4的混合液，后来调整为使用更安全的甲苯和乙二醇二甲醚（EGDE）比例为2.6:97.4的溶剂体系。RSN值在某种程度上与亲水亲油平衡（HLB）值有关，但RSN值的测定方法、解释和使用各有不同。RSN值的测定方法如下：称量1.0g待评价产品（含有或不含破乳剂）移入含有30mL标准溶液的烧杯中（2.6%甲苯和97.4%EGDE），使用磁力搅拌器溶解产品；像滴定一样逐滴加水，当溶液出现持续的混浊时到达终点。RSN值的结果是达到浊点所需的水量（以mL计）。因此，随着产品的亲水性增加，RSN值也增加。通常不推荐使用亲水性很强的破乳剂产品，因为在分离过程中破乳剂会迁移到水相中，需要额外的水处理。

从不同类别的破乳剂中优选合适的产品时，仅考虑RSN值是不够的，还需要特别考虑破乳剂分子的化学结构。RSN值可用于选取相同化学类别的破乳剂产品。在性能评价的多数案例中，比较不同化学类别的RSN值可能不是一个好方法。另一个重要的化学特性是破乳剂在水/辛醇中的分配系数，这与乳状液的水质和处理工艺关系密切，也与RSN值和HLB值有关。

破乳剂测试时基本都采用简单的瓶试法。有经验的技术人员通过对水滴和界面性状的简便视觉评估，能为处理站筛选提供有效的破乳剂混合物。现场通常会测试一系列不同结构、类别的破乳剂，并确定最优加量。常规的瓶试法是将一定剂量的破乳剂加入盛有乳状液的瓶子或量筒中，并观察油和水分离的速度和程度。现在的技术人员常使用数码照片来记录试验和展示效果。有研究者针对瓶试法的10项性能参数（4项描述水滴，3项描述原油脱水率，3项描述油水界面），通过方差分析、多变量相关性、聚类分析和主成分分析等多种统计方法研究了不同破乳剂的效果。好的破乳剂能够快速分离乳状液中的水，提供相对平齐的油水界面和低含水率的达标净化油。

有许多更先进的分析技术可用于表征乳状液或破乳剂的效果。光学显微图像的数字化处理可以测量油包水乳状液的液滴粒径分布。界面张力测试表明，水—原油界面破乳剂的吸附动力学参数（或者吉布斯弹性模量）与相分离速率相关：吸附动力学参数越大，分离速率越快。另一种方法是测量乳状液样品中电极对之间的电流值。电流变化越快速意味着油包水乳状液的破乳剂越有效。有研究者研究发现介电常数可以作为破乳剂筛选、排序和选择的标准，证明了该方法的有效性。还有研究者研发了临界电场技术以评价破乳剂的性能，该技术在破乳剂配方研究中也发挥了重要作用。有研究在 E_{Crit} 池中使用低场核磁共振（NMR）以确定破乳剂对乳状液稳定性的影响，并采用振荡悬滴法测试破乳剂的界面响应。E_{Crit} 池测量诱导形成自由水所需的电场，而 NMR 监测分散水滴的垂直运动。还有文章提及了破乳剂性能评价的其他改进方法。

11.3.3 油包水破乳剂的类别

为破坏乳状液的稳定性，一直以来化学品供应商开发了越来越多的产品，因此很难对油包水破乳剂进行分类。许多油包水乳状液的破乳剂是非离子型聚合物，很多具有复杂的梳状或支链结构，分子量为 2000~50000。而阴离子和阳离子聚合物的使用取决于待处理液体中稳定乳状液的化合物性质，其也可作为润湿剂。油包水乳状液的常见破乳剂包括：聚烷氧基的嵌段共聚物和酯类衍生物，烷基酚醛树脂烷氧基化物，多元醇或缩水甘油醚的聚烷氧基化物，多胺聚烷氧基化物和相关的阳离子聚合物（主要用于水包油乳状液），聚氨酯（氨基甲酸酯）和聚烷氧基化物的衍生物，超支化聚合物，烯基聚合物，聚硅氧烷（也可作为破乳剂的增效剂）。

后续将讨论以上破乳剂类型的实例。还会讨论很多潜在的可生物降解破乳剂，其中只有部分属于以上的类型。破乳剂和聚乙二醇醚的混合物据称可以获得更好的破乳效果，后者可能对固体起到润湿作用。尽管固体可以稳定乳状液，但有证据表明，将纳米材料（如二氧化硅颗粒）应用于原油聚醚破乳剂，可以极大地提高破乳剂性能。破乳剂配方中还可使用其他结构更简单的表面活性剂。例如，包含阴离子表面活性剂（如烷基磺基琥珀酸酯或烷基膦酸，及其盐）、非离子表面活性剂（如环氧乙烷－环氧丙烷共聚物、聚乙二醇的乙氧基化脂肪酸、萜烯烷氧基化物或改性烷醇酰胺）和包含二元酯混合物的基础溶剂的破乳剂，这类破乳剂都有潜在的环保特性。还有研究者研究了双子表面活性剂型破乳剂。

大多数油包水破乳剂都是油溶性的，以烃类为溶剂制成溶液。溶剂通常以烃类为基础，因为典型的破乳剂将包含几种油溶性的活性成分（低 RSN 值）。破乳剂的溶剂对乳状液的破乳有重要影响，因此，应在多种溶剂中测试同一破乳剂的性能。通常采用的溶剂是芳香烃／低级醇的混合液。由于水包油乳状液很复杂，通常具有超过一种以上的稳定机制，因此许多市售破乳剂是由两类或更多的化学品协同作用的混合物。油溶性破乳剂可以通过添加水溶性表面活性剂使其分散在水溶液中，这样就可以避免使用有毒或易燃易爆的有机溶剂。相反地，更常见的水溶性破乳剂成分可以通过添加偶合溶剂（如乙醇或乙二醇）而使其具有油溶性。针对寒冷气候的配方中经常用到甲醇（"抗冻剂"），以避免异常高黏度。已经开发出基于微乳剂的复合破乳剂及相关使用方法，该方法可以避免使用对环境不友好的溶剂，另外用于井下时，还具有提升井筒清洁度的优势。

上面列出的大多数破乳剂呈中性或微碱性。而一些乳状液处理时，最好采用衍生于上述破乳剂的酸性破乳剂，通常是用磷氧化物或含氧酸对上述类别中的一种进行衍生化。因为酸性化学剂会导致腐蚀，这些破乳剂不能过度处理乳状液，另外，破乳剂的磷酸部分会与盐水中的钙反应，形成羟基磷灰石和其他固体垢，过度破乳处理时会导致乳化液再次稳定。如前所述，当矿物固相存在时，重度油包水乳状液处理会很困难。有研究表明，乳化剂在固相表面的吸附改变了其润湿性，促进了油滴在固相的黏附，从而降低了油和水之间的有效密度差，阻碍了油水分离。该研究主要专注于稠油破乳。针对稠油乳状液的新型破乳剂已有报道。

从前面的破乳剂介绍中可以看出，许多破乳剂都含有聚烷氧基链。聚烷氧基化物可以通过使用碱（如胺或醇）对环氧乙烷（EO）、环氧丙烷（PO）、环氧丁烷（BO）或四氢呋喃（THF）开环而制成（图11.2）。使用这种方法的好处是可以选择各种底物与聚烷氧基化物链进行偶联，比较容易改变分子的HLB值及分子量。聚环氧乙烷链非常亲水，聚环氧丙烷链较为憎水，而聚环氧丁烷链非常憎水。因此，在相同的醇或氨基上搭配不同的EO、PO和BO侧链分子，可制成分配系数和界面活性各不相同的系列产品。EO和PO由于成本较低，是当前最常用的聚烷氧基化物的单体。有两项相关研究表明，当聚烷氧基化物破乳剂分配系数为1.0时（油相和水相之间的分配量相等），具有最佳的破乳性能。

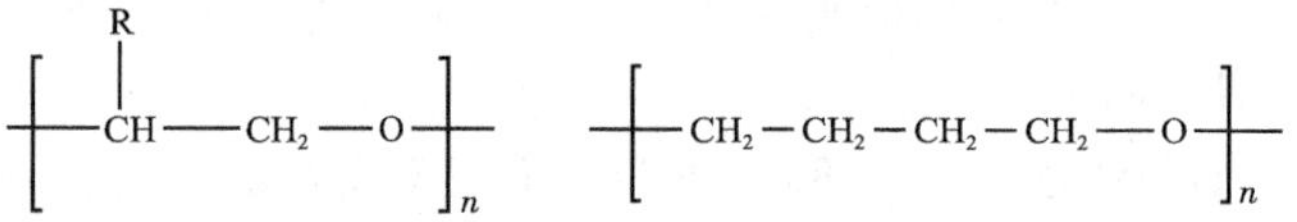

图 11.2 聚环氧基化物和聚四氢呋喃的结构 [R=H（EO），R=Me（PO），R=Et（BO）]

许多聚烷氧基化物类破乳剂可以衍生为更高的分子量和不同HLB值的化合物。例如，可以与二异氰酸酯、二羧酸、双缩水甘油醚、二羟甲基苯酚和三羟甲基苯酚等多官能团试剂交联。在分子量不变的情况下，增加支链可以改善破乳剂的性能。

下面将更详细地讨论最重要的几类破乳剂。在结尾处，针对北海盆地等地区的环境政策及环保要求，专门介绍了可生物降解破乳剂的进展，汇总了相关文章和专利。

11.3.3.1 聚烷氧基嵌段共聚物和酯类衍生物

环氧丙烷可以在碱催化下进行开环聚合，得到最大分子量约为4000的聚丙二醇。这些聚合物末端的羟基可以与环氧乙烷进行乙氧基化，形成EO/PO/EO嵌段共聚物的线型结构破乳剂。这些共聚物的破乳性能不佳。溶剂体系也会影响到性能。亲水性更好的EO/PO/EO嵌段共聚物比亲油性的PO/EO/PO嵌段共聚物性能更好。通过特殊催化工艺形成的更高分子量聚丙二醇或聚丁二醇（6000<分子量<26000）也可以被乙氧基化，从而得到性能更好的破乳剂。四氢呋喃可与环氧化物进行开环聚合，得到聚亚烷基二醇嵌段共聚物。所有这些嵌段聚合物都可以与二羧酸（如马来酸、富马酸、己二酸和氨基羧酸）或均苯四甲酸二酐反应，生成性能更好、分子量更高的聚烷氧基酯。马来酸酐—油酸加合物所形成的聚烷氧基酯也是良好的破乳剂，在一些测试中比壬基酚甲醛树脂烷氧基化物性能更好。甲基

丙烯酸酯和丙烯酸酯的烷氧基化物也是一种破乳剂。

聚烷氧基嵌段共聚物的酯可以阴离子化或阳离子化。例如，聚亚烷基二醇也可以与阴离子二元酸单体或二酯（如5-磺基异酞酸二甲酯）发生酯交换反应，从而获得阴离子官能团。EO/PO 嵌段共聚物、烷氧基脂肪胺和二羧酸的线型三元缩聚物也可以起到破乳剂作用。氮原子可以被季铵化，得到阳离子聚合物。通过加入多官能团的 EO/PO 聚合物，有可能产生高度支化的聚酯胺。

11.3.3.2 烷基酚醛树脂烷氧基化物

最常见的破乳剂类别是烷基酚醛树脂烷氧基化物（图 11.3）。由于其性能高且易于制备，已经应用了几十年。在苯酚环上选择不同长度的烷基可以改变疏水尾基。选定的烷基苯酚与醛（通常是甲醛）缩合，形成聚合物树脂，然后用不同数量的 EO 和 PO 进行烷氧基化，制成一系列的树脂烷氧基化物。烷基酚的合成主要通过异烷基酚分子结构的邻位和对位取代反应。间位上含有一个不饱和烯基的天然产物腰果酚也可以被用作原料。

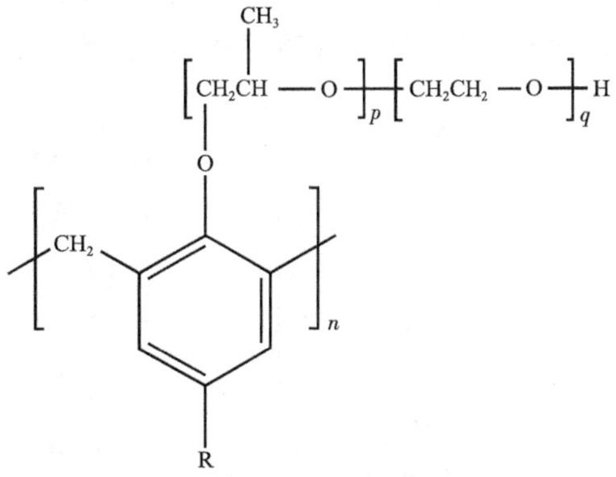

烷基酚醛树脂烷氧基化物的分子量最好在 5000~50000 之间。分子量低于 4000 的破乳性较差。研究表明树脂烷氧基化物的多分散系数（$Q=M_w/M_n$）应该至少为 1.7，在 1.7~5.0 范围内则更好，能够表现出优异的破乳性能。早期通过直接缩合反应生产的树脂产品，由于其高交联度导致分子量过高，无法获得最佳破乳剂性能。避免这种情况的合成方法已经获得专利。这种烷基酚醛树脂通常为线型，而通过在非极性溶剂（如二甲苯）中使用烷基苯酚和多聚甲醛（不是甲醛），可以高产率合成环状的四聚体。烷氧基化的 C_7 烷基取代的苯酚—甲醛树脂和至少一种聚烷基环氧化物三嵌段共聚物的混合物被认为是有协同效果的组合破乳剂。

图 11.3 烷基酚醛树脂烷氧基化物

关于烷基酚醛树脂烷氧基化物破乳剂对环境的影响，已经有很多讨论。作为乙氧基化烷基酚的降解产物，烷基酚（如壬基酚）是海洋物种的内分泌干扰物。极少量未反应的烷基酚单体原料，存在于破乳剂终产物中，或可能是排放水中的降解产物。烷基酚醛树脂烷氧基化物基本是油溶性的，对人体的毒性很低。重组酵母法测定也表明，其化学成分与海洋环境中潜在的内分泌干扰之间没有联系。但未对生物降解产物进行测试。

从 20 世纪 90 年代中期开始，已经有研究试图找到烷基酚醛树脂烷氧基化物的替代物，该替代物不以烷基酚制备，因而无雌激素活性。例如，某些不含烷基酚的芳香烃醛树脂，既有能够烷氧基化的官能团，又在芳香环上没有烷基，对水包油乳状液的破乳效果良好，并且不存在类似激素的作用（没有明确的环境数据）。芳香环前体上的典型取代基可以是—NHR、—COOR、—OR 或—CONHR（R=H 或烷基）。具体的例子有间苯二酚、对苯二酚、水杨酸乙酯、对 N,N- 二丁基氨基苯酚、对羟基苯甲酸丁酯、间苯二酚十八烷基醚和对甲氧基苯酚。

烷基酚-甲醛树脂烷氧基化物有许多已获专利授权的变体。可以用乙醛酸（HOOCCHO）代替甲醛，以获得带有羧酸基团的更亲水的烷基酚——甲醛树脂烷氧基化物。相反地，通过使用苯甲醛等醛类，可以增加疏水性。如乙二醛等二醛可用以得到更复杂的结构。在另一个例子中，树脂烷氧基化物中的羟基可以与烯基单体（如马来酸酐或丙烯酸）进行酯化，然后进行聚合，形成分子量更高、结构更复杂的破乳剂。二异氰酸酯与烷基酚醛树脂烷氧基化物的反应也可以得到一系列更高分子量的交联破乳剂。硅氧烷交联破乳剂可以通过使烷基酚醛树脂烷氧基化物或聚亚烷基二醇，与一种或多种硅基交联剂反应来制备，如四乙氧基硅烷 $[(EtO)_4Si]$。烷基酚醛树脂烷氧基化物可以与五氧化二磷、三氯氧磷或磷酸反应，产生酸性磷酸酯破乳剂。这类破乳剂据称可以使水分离速度更快速，以及降低原油中的碱性沉积物和水，比目前（专利授权时）使用的破乳剂效果更好。如 2,2-双（4-羟基苯）丙烷等双酚，可用于制备更复杂的烷基酚醛树脂烷氧基化物。氨基双酚环氧树脂也可以被烷氧基化以生成破乳剂。

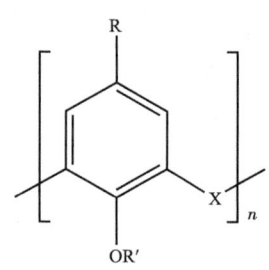

图11.4 硫代杯芳香烃（X=S，SO 或 SO_2）

乙二胺（EDA）等二胺可以作为原料用于树脂缩合过程中，制成烷氧基化的烷基酚-甲醛-二胺聚合物。有报道称，基于乙氧基化的酚甲醛希夫碱（亚氨基）聚合破乳剂能够应用于原油乳液的电脱盐和脱水。烷氧基化树脂可以与碳酸乙烯酯反应，产生一系列的破乳剂。烷氧基化的硫代杯芳香烃是烷基酚醛树脂烷氧基化的巯基等价物，也可作为破乳剂（图11.4）。自动浊度仪已用于快速筛选破乳剂。浊度仪测量水中的油含量。油的干燥度不能直接测量，但可以间接计算。

11.3.3.3 多元醇或缩水甘油醚的聚烷氧基化物

聚烷氧基化物中的支链化可以提高破乳剂的效率。与由乙二醇形成的直链 EO/PO/EO 嵌段共聚物相比，具有两个以上—OH 的多元醇的烷氧基化作用将产生支链结构。典型的多元醇有甘油、季戊四醇和三羟甲基丙烷。二缩水甘油醚的聚烷氧基化物是常见的破乳剂。典型的二缩水甘油醚是双酚的二缩水甘油醚，被称为双酚A（图11.5）。该结构的一些变体已获得专利。例如，烷氧基化的多元醇与烷基酚（如腰果酚）的缩水甘油醚的反应据称可以制得有效的破乳剂。可选地，多元醇在与芳香烃反应之前可以通过交联剂二环氧乙烷进行交联。

多元醇的聚烷氧基化物可以与烯基单体（如丙烯酸）部分交联，以增加分子量并提高破乳效率。例如，EO/PO/EO 嵌段共聚物和 2-氨基-2-羟甲基-1,3-丙二醇的烷氧基化共聚物可以与丙烯酸部分交联。在另一项发明中，环氧化脂肪酸酯可以通过醇或羧酸开环，形成多元醇，所产生的羟基与烷基环氧化物、EO 和 PO 反应。

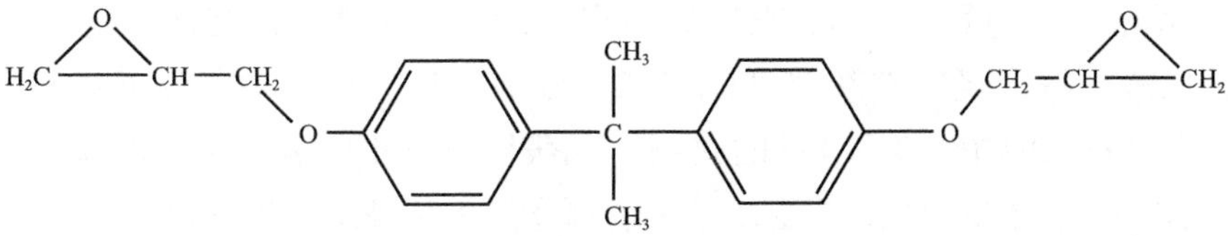

图 11.5 双酚 A（一种二缩水甘油醚）

二环氧缩水甘油或二缩水甘油化合物可用于制造许多其他破乳剂。这些化合物与胺和可选的第二种含胺基团（包括一个叔氨基）发生反应，随后的烷基化（季铵化）生成一系列季铵化脂肪族聚羟基醚胺阳离子破乳剂。

11.3.3.4 多胺聚烷氧基化物和相关的阳离子聚合物

许多市售的小分子聚亚烷基胺，用不同数量的 EO 和 PO 进行衍生，产生支化的破乳剂。多乙烯多胺有 EDA、二乙烯三胺（DETA）、三乙烯四胺（TETA）和四乙烯五胺（TEPA）（图 11.6）。市售的 TETA 和更高分子量的聚乙烯胺还含有少量的环状化合物，如氨基乙基哌啶。有关 DETA 基破乳剂的研究表明，侧链中 EO 和 PO 的含量大致相等时，破乳性能最佳。

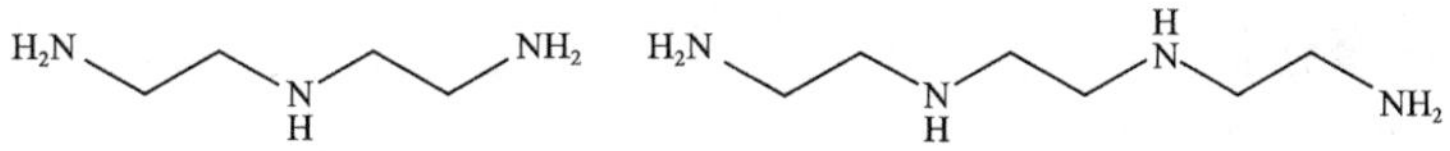

图 11.6 DETA 和 TETA（氮原子上的每个质子都可以被烷氧基化）

具有较高分子量的聚亚烷基胺至少有 50 个反复出现的乙烯亚胺或丙烯亚胺单元，可与不同数量的 EO 和 PO 反应以生产破乳剂。据称，这些多胺聚烷基化合物在与异烷基苯酚—甲醛树脂烷氧基化物等混合时，性能更佳。多胺烷氧基化合物必须具有多分散性，（Q 等于 M_w/M_n）至少为 1.7，最好为 1.7～5.0，以获得最佳的破乳剂性能。

基于环氧化脂肪酸酯与胺、二胺或多胺的开环反应的产品，在进一步烷氧基化后，即使在极低浓度下也有很好的破乳效果。典型的产品是胺（如椰油胺或 TEPA）与大豆油环氧化物反应，然后用不同数量的 EO 和 PO 对中间产物进行烷氧基化。

多胺（如 DETA）可以先与二羧酸（如己二酸）反应，再烷氧基化，以生产含有酰氨基的破乳剂。这些产品也能被硫酸二甲酯进一步进行季铵化。烷氧基化的环状二胺（如哌嗪、环己烷二胺和异佛尔酮二胺）可作为破乳剂。季铵盐阳离子聚合物破乳剂也可以由聚氧亚烷基二醇、环氧氯丙烷、含有至少两个羟基的多元醇（其中一个或多个可选择烷氧基化）和封端的聚氧化烯二胺或三胺的反应产物制成。小分子、季铵化脂肪胺乙氧基化物也是一种油包水破乳剂。丙烯酸甲酯与脂肪胺发生迈克尔加成反应，其产物再与 EDA 反应，最后和 EO/PO 反应，得到支链聚醚表面活性剂破乳剂。

11.3.3.5 聚氨酯（氨基甲酸酯）和聚烷氧基化物的衍生物

聚氨酯烷氧基化物是一类知名的破乳剂，其含有氨基甲酸酯官能团，由多异氰酸酯

（如甲苯二异氰酸酯）与聚乙二醇或带有末端羟基的聚烷氧基化物缩合而成（图 11.7）。如果使用聚乙二醇，两个羟基端基都可以与二异氰酸酯反应，产生高分子量的聚氨酯破乳剂。可以改变聚乙二醇中 EO 和 PO 的比例，以获得具有不同表面活性的系列产品。另一种改进是使用实质上含 EO 的聚氨酯，并在亲水骨架上接枝疏水基团。

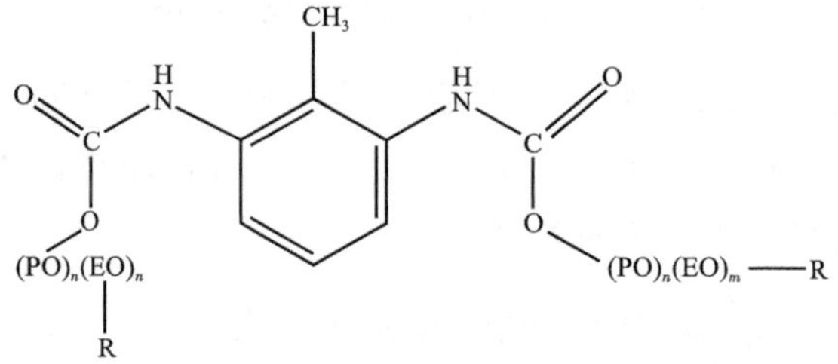

图 11.7　2，6-甲苯二异氰酸酯形成的聚氨酯聚烷氧基化物（二氨基甲酸酯）（也可使用 2，4-异构体）

烷基酚醛树脂烷氧基化物和聚氨酯在同时加入时发挥了协同作用，使水分离率明显高于单独使用时。聚氨酯在中等浓度（约 200mg/L）下通过"桥接"附近的液滴加速水的沉积，但当添加到很高的浓度时，会延缓凝聚，即使烷基酚-甲醛树脂烷氧基化物存在时也有类似情况。

11.3.3.6　超支化聚合物

各种非枝形的、高功能的、超支化的聚合物已证实有破乳剂功能，包括超支化聚碳酸酯、超支化聚酯、超支化聚醚、超支化聚氨酯、超支化聚脲聚氨酯、超支化聚脲、超支化聚酰胺、超支化聚醚胺和超支化聚酯酰胺。超支化聚合物的一个例子是通过己二酸、甘油和单硬脂酸甘油酯反应，随后使羟基与烷基异氰酸酯反应而制成。一个超支化聚碳酸酯的实例是由己二酸、烷氧基化的三羟甲基丙烷、油酰胺乙氧基化物和碳酸二乙酯制成。超支化聚酯和聚酯酰胺可使用烯基丁二酸酐与三官能团醇或氨基醇制成。特别是聚酯将表现出合理的、甚至是良好的生物降解性。

聚乙烯亚胺（PEI）是一种低成本的超支化聚合物，分子量可以在 1000～1000000 之间变化。PEI 可与 EO、PO 或 BO 通过烷氧基化形成破乳剂。这类产品与氧乙基化的异烷基苯酚-甲醛树脂有协同作用。

另一类超支化聚合物可以通过使丙烯酸酯（如甲基丙烯酸甲酯）与氨和 EDA 反应而制成。这类产品被称为聚酰胺-胺，其可以被环氧氯丙烷或 2-羟基-3-氯丙基三甲基氯化铵等进行季铵化，形成具有季铵侧基的聚酰胺-胺。这些破乳剂对于油包水型和水包油型的乳状液都可以起到相应效果。

11.3.3.7　烯基聚合物

各种烯基聚合物都可作为破乳剂，一般来说其结构中都含有亲水和疏水的部分。具体的合成方式是先通过（甲基）丙烯酸或马来酸酐、（甲基）丙烯酸羟乙酯或（甲基）烯丙基醇等烯基单体聚合，然后在碱性条件下与 EO、PO 和 BO 进行烷氧基化而得到（图 11.8）。另外，（甲基）丙烯酸或马来酸酐可以与本章前面描述的多种聚烷氧基化物中的一种反应，将新的酯类单体聚合或接枝到现有的聚合物上。另一项专利声称，聚合甲基丙烯酸酯的混合物，其中部分通过乙氧基化而具有亲水性，另一部分基于醇类而具有疏水性，可以制得良好的破乳剂。如果马来酸酐与聚烷氧基化物（如 EO/PO/EO 嵌段共聚物）反应并与丙烯酸聚合，剩余的羧酸基团可与第二种聚烷氧基化物（如烷基酚 - 甲醛树脂烷氧基化物）进行酯化。

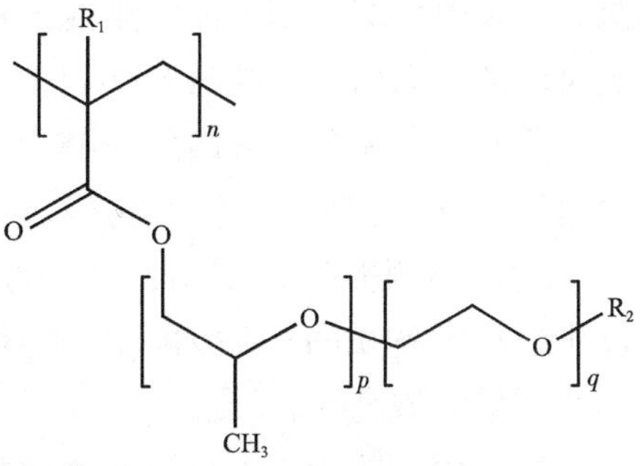

图 11.8 烷基聚烷氧基（甲基）丙烯酸酯聚合物 [R_1=H 或 CH_3，R_2=H 或烷基]

通过使用烯基羧酸酯或丙烯酸酯等单体也可以引入憎水基团。烷基聚烷氧基化物 [如 $C_6H_{13}(EO)_6OH$] 以甲基丙烯酸酯化，以及随后聚合，可得到很好的破乳剂，降水快速、零残留乳化、油水界面清晰，水相清澈适于排放或重新注入，且原油中的盐含量低。亲水单体也可使破乳剂更易溶于水，如 N- 异丙基丙烯酰胺和 2- 丙烯酰氨基 -2- 甲基 -1- 丙磺酸等 N- 烷基丙烯酰胺的共聚物作为破乳剂使用。

含有芳香基团、亲油基团、可离子化基团和亲水基团四种不同烯基单体的共聚物可作为破乳剂。典型的共聚物由苯乙烯、甲基丙烯酸烷基酯、（甲基）丙烯酸和 2- 羟乙基（甲基）丙烯酸酯制成，加入表面活性剂完善配方。这些聚合物也可以制成胶乳，即聚合物微粒分散于水中形成的稳定分散体。由甲基丙烯酸甲酯、丙烯酸丁酯、丙烯酸和甲基丙烯酸制成的一类特殊的水溶性四元共聚物（分子量为 2000～50000）可作为油包水型的破乳剂。苯乙烯可以被连接在甲基丙烯酸甲酯上，形成一种五元共聚物。

11.3.3.8 聚硅氧烷

聚氧化烯 - 聚硅氧烷嵌段共聚物是一类性能优良的主要破乳剂。聚硅氧烷 [如二甲基甲基（聚氧化乙烯）硅氧烷和二甲基硅氧烷]、乙氧基化 3- 羟基丙基封端产物，可作为破乳剂的促进剂。系列化的改性有机硅破乳剂已有相关报道。聚硅氧烷也可以起到消泡剂的作用（见第 12 章）。基于硅烷基聚醚的可生物降解聚硅氧烷（聚有机硅氧烷）也是一类破乳剂。其他可生物降解的破乳剂将在下文讨论。

11.3.3.9 改性生物降解的破乳剂

自 20 世纪 90 年代末以来，开发低毒和高生物降解的更环保破乳剂越来越受到重视，特别是适应环保要求严格的地区（如北海盆地）。一般来说，破乳剂由于其高分子量，毒性不大。但如前面指出的烷基酚醛树脂烷氧基化物破乳剂，可能含有微量的有毒单体，或者可能降解为单体或分子量更低的有毒化学品。

许多传统的破乳剂含有主要由 EO 和 PO 组成的聚烷氧基化物链。由于存在一个甲基侧基，聚丙氧基化物链的降解速度比聚乙氧基化物慢。由 BO 开环制成的聚丁氧基化物链更难降解。而由四氢呋喃开环制成的直链聚丁氧基化物（聚四亚甲基二醇）更容易生物降解。因此，主要含有聚四氢呋喃和聚乙氧基化物的破乳剂是可生物降解的破乳剂。在这些聚合物中接枝酰胺或酯，是引入生物降解性的一个好方法。例如，聚四氢呋喃、EO/PO/EO 嵌段共聚物或脂肪胺烷氧基化物及己二酸的酯化反应产物。在 OECD 306 生物降解测试中，该类物质的 28 天生物降解率为 23%～52%。其他几项专利提出了具有可生物降解的酯连接的破乳剂。

环氧化脂肪酯（如豆油环氧化物）也可用于合成潜在可生物降解的破乳剂。例如，环氧化物和含有 2～6 个—OH 的多元醇与一种胺（如椰油胺或 TETA）反应，所得的多元胺与不同数量的 EO、PO 或 BO 发生烷氧基化。

前面讨论过的超支化聚合物和树枝状聚合物如果含有酯或酰胺作为连接基团，就容易实现生物降解。例如，以多元醇为主的 2，2-羟甲基丙酸聚合得到具有很多—OH 的超支化聚酯。这些聚合物与 EO、PO 或 BO 进行烷氧基化，可以得到一类可生物降解的破乳剂。通过多元羧酸和多元醇反应还可以制成其他复杂的酯基破乳剂，如酯化柠檬酸、三乙二醇和 C_{12}—C_{14} 醇以 1∶3∶1 比例混合的反应生成物。

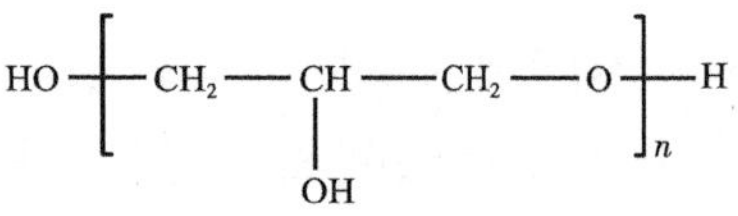

另一种向破乳剂聚合物中引入更多生物降解链的方法是使用聚甘油。市售的聚甘油有线型聚合物和超支化聚合物两种形式（图 11.9）。聚甘油与 EO 和 PO 反应生成可生物降解破乳剂，交联的聚甘油效果更好。交联反应物可以是二缩水甘油醚、二羧酸和多元羧酸、烷基琥珀酸酐、烷氧基硅烷或二异氰酸酯。在交联改性的多元醇中，加入内酯也能产生一类有效的破乳剂。

图 11.9　线型聚甘油

另一类潜在的可生物降解破乳剂是烷氧基化的烷基聚糖苷（图 11.10）。由于合成的原因，烷基聚糖苷还可能含有如残余的醇、单糖、寡糖和寡烷基聚糖等其他物质。这些聚合物可以选择性地与双官能团交联剂（如二异氰酸酯或二羧酸）反应，以增加支化程度和分子量。

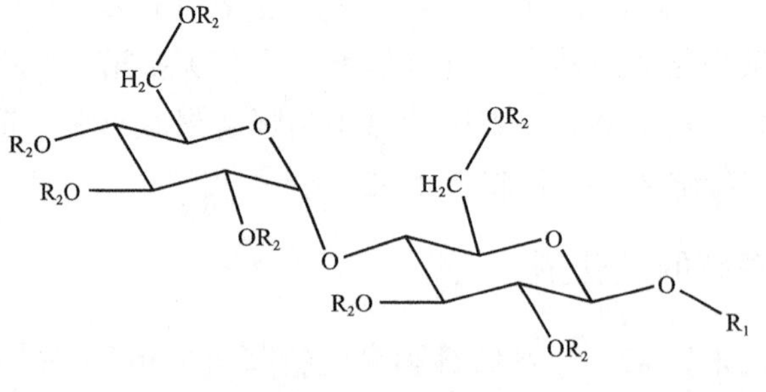

图 11.10　基于 β-麦芽糖骨架的烷基聚糖苷示例［R_1=烷基，R_2=聚亚烷基二醇。破乳剂中的六环分子占比很高］

据称，另一类改性生物降解的破乳剂是基于原酸酯（如原甲酸三乙酯）的聚合物，原酸酯结构如图 11.11 所示。典型的原酸酯是原甲酸三乙酯。这些分子与聚乙二醇、聚丙二醇中的羟基或氨基反应，或与一些氨基醇反应，形成一种高分子量、含有亲水基和疏水基的交联聚合物。内酯/亚烷氧基聚合物可作为破乳剂使用（图 11.12）。含羟基或胺的碱性化合物与至少一个内酯单体和至少一个亚烷氧基单体发生加成反应而制成的无规或嵌段聚合物可作为破乳剂。酯键或酰胺键，有助于提高其生物降解性。

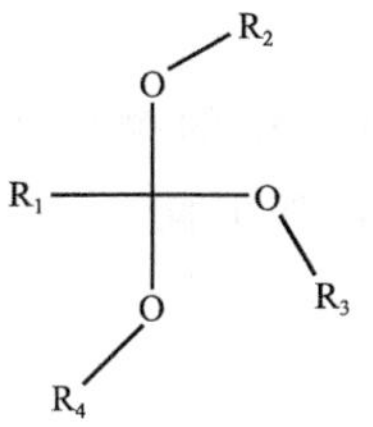

图 11.11　原酸酯的结构 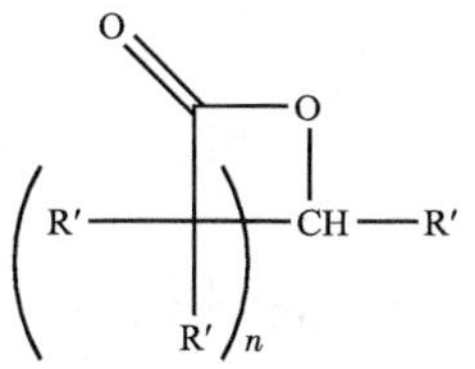　　　　图 11.12　制备烷氧基化破乳剂的内酯类单体

嵌段、星状或枝状的两亲性羟基聚酯是具有优异的环境特性的破乳剂。羟基酯键是具有至少两个羧酸基的化合物与单环氧化物反应而成，或由有多环氧基的环氧化物与至少一个羧酸化合物反应而成。例如，柠檬酸、聚乙二醇二元酸或羧甲基纤维素与 2-环氧十二烷或缩水甘油基十六烷醚的反应产物生成羟基聚酯。萜烯类烷氧基化物也可作为破乳剂，特别是作为壬基苯基烷氧化物的更环保替代品。

有文章报道了可生物降解的高性能烷氧基化聚酯枝状破乳剂（图 11.13），具体是聚甘油和二羟甲基丙酸的缩合产物的烷氧基化衍生物，分子量超过 20000。OECD 306 试验表明，其在 28 天的生物降解率为 45%～51%，而且还在增加。据报道，绿色混合破乳剂还有部分的絮凝特性。絮凝特性也意味着部分破乳剂必须溶入水相。

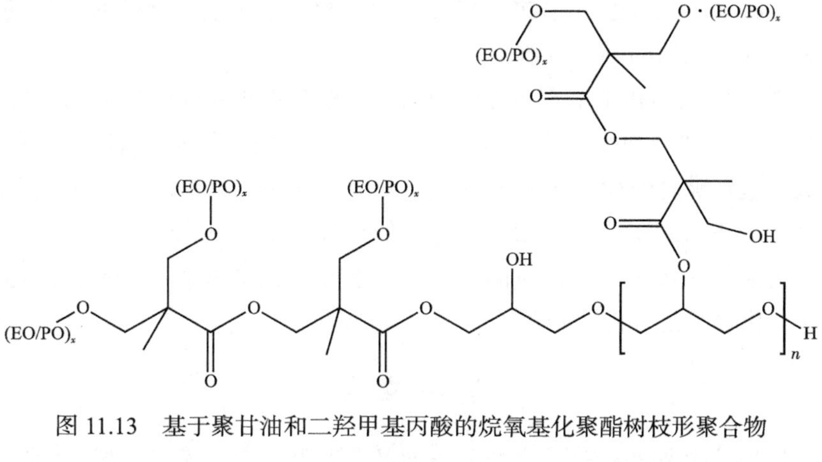

图 11.13　基于聚甘油和二羟甲基丙酸的烷氧基化聚酯树枝形聚合物

为获得良好的破乳剂性能和较低的环境影响，对表面活性剂的混合物进行了系统研究，但未提及化学细节。

一种基于烷氧基化多元醇或多元醇酯与二羧酸的反应产物的新型绿色破乳剂也有报道，具体实例是乙氧基化山梨醇的酯类（图 11.14）。

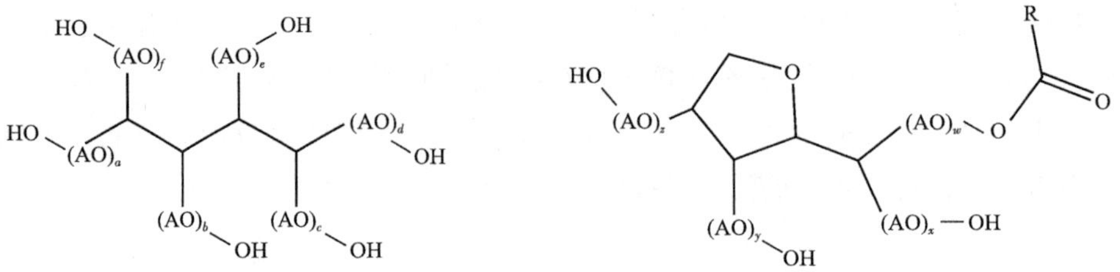

(a) 烷氧基化多元醇　　　　　　　　(b) 乙氧基化山梨醇的脂肪酯

图 11.14　烷氧基化多元醇和乙氧基化山梨醇的脂肪酯

11.3.3.10　双功能破乳剂

某些破乳剂还有辅助功能，如烷基胺和烷基芳香基磺酸所形成的盐还兼具缓蚀特性。具体实例是甲基、二椰油胺和烷基芳香基磺酸组成的盐。与之相关的是由 2mol 乙氧基化脂肪胺和 1mol 磺化油酸的三尾反应产物（图 11.15），以离子对的形式存在于非极性溶剂中。

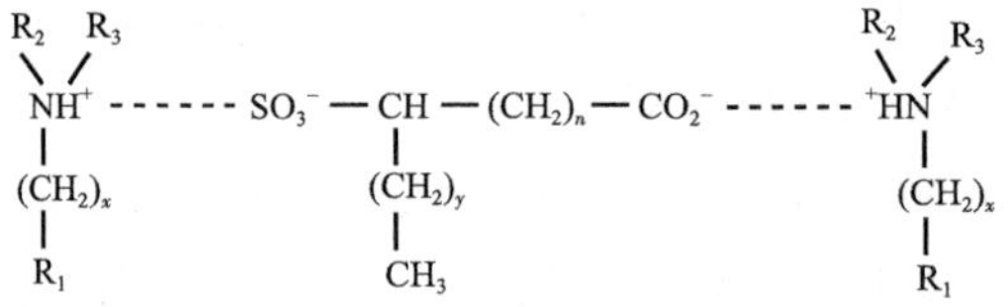

图 11.15　三尾离子对破乳剂

阳离子型的分子和聚合物破乳剂也可以具有缓蚀剂的功效，如 4，4′－双（氯甲基）二苯醚和叔十二胺聚乙二醇醚的季铵化缩合产物及相关衍生物。脂肪胺烷氧基化物也可作为破乳剂、缓蚀剂或原油的倾点抑制剂。

含有至少 2 个六元环和 1 个至少 16 个碳原子的烷基芳香族化合物的磺酸衍生物最初的设计是作为沥青质抑制剂（见第 4 章）。这些分子与辅助溶剂共同作用，稳定易致乳化的沥青质，从而起到破乳作用。稳定沥青质实际上是实现良好破乳的一个关键因素。

分子式为 $H(CH_2)_zCOO[C_2H_4O]_xC_yH_{2y+1}$ 的化合物（其中，$z=0\sim2$，$x=1\sim5$，$y=4\sim9$）具有破乳剂和消泡剂功效，是由相应的含羟基、烷基的乙氧基化物和有机酸通过简单酯化反应制成。这些化合物的环保性更好。

参 考 文 献

[1] F. Leal-Calderon, V. Schmitt, and J. Bibette, *Emulsion Science: Basic Principles*, New York: Springer, 2007.

[2] L. L. Schramm, *Emulsions, Foams and Suspensions: Fundamentals and Applications*, Weinheim: Wiley-VCH, 2005.

[3] P. J. Breen, D. T. Wasan, Y.-H. Kim, A. D. Nikolov, and C. S. Shetty, *Emulsions and Emulsion Stability*, 2nd ed., *Surfactant Science Series*, ed. J. Sjöblom, New York: Marcel Dekker, 2005, 235.

[4] S. Mukherjee and A. P. Kushnick, "Effect of Demulsifiers on Interfacial Properties Governing Crude Oil Demulsification," *Oilfield Chemistry—Enhanced Recovery and Production Stimulation*, ACS Symposium Series, eds. J. K. Borchardt and T. F. Yen, Washington, DC: American Chemical Society, 1989, 364.

[5] D. Arla, A. Sinquin, C. Hurtevent, and C. Dicharry, "Acidic Crude Oil Emulsions: Influence of pH and Water Cut on the Type and Stability of Emulsions" (paper presented at the 7th International Conference on Petroleum Phase Behavior and Fouling, 25-29 June 2006).

[6] M. Grutters, M. van Dijk, S. Dubey, R. Adamski, F. Gelin, and P. Cornelisse, "Asphaltene Induced W/O Emulsion: False or True?," *Journal of Dispersion Science and Technology* 28 (2007): 357.

[7] (a) J. Czarnecki and K. Moran, *Energy & Fuels*, 19 (2005): 2074; (b) J. Czarnecki, *Energy Fuels*, 23 (2009): 1253.

[8] (a) R. A. Rodriguez and S. J. Ubbels, "Understanding Naphthenate Salt Issues in Oil Production," *World Oil* 228 (8) (2007): 143. (b) D. Arla, A. Sinquin, T. Palermo, C. Hurtevent, A. Graciaa, and C. Dicharry, "Influence of pH and Water Content on the Type and Stability of Acidic Crude Oil Emulsions," *Energy & Fuels* 21 (3) (2007): 1337.

[9] G. A. Davies, M. Yang, and A. C. Stewart, "Interactions Between Chemical Additives and Their Effects on Emulsion Separation," SPE 36617 (paper presented at the SPE Annual Technical Conference and Exhibition, Denver, CO, 6-9 October 1996).

[10] D. K. Durham, S. A. Ali, and P. J. Stone, "Causes and Solutions to Surface Facilities Upsets Following Acid Stimulation in the Gulf of Mexico," SPE 29528, *SPE Production & Facilities* 12 (1) (1997): 16.

[11] D. Tambe, J. Paulis, and M. M. Sharma, "Factors Controlling the Stability of Colloid-Stabilized Emulsions: III. Measurements of the Rheological Properties of Colloid-Laden Interfaces," *Journal of Colloid and Interface Science* 171 (1993): 456.

[12] J. J. Oren and D. M. Mackay, "Electrolyte and pH Effect on Emulsion Stability of Water-in-Petroleum Oils," *Fuel* 56 (1977): 382.

[13] J. Sjöblom, P. V. Hemmingsen, A. Hannisdal, and A. Silset, "Stability Mechanisms of Crude Oil Emulsions—A Review" (paper presented at the 7th International Conference on Petroleum Phase Behavior and Fouling, 25-29 June 2006).

[14] M. K. Poindexter, S. Chuai, R. A. Marble, and S. C. Marsh, "The Key to Predicting Emulsion Stability: Solid Content," SPE 93008 (paper presented at the SPE International Symposium on Oilfield Chemistry, The Woodlands, TX, 2-4 February 2005).

[15] D. Graham, "Crude Oil Emulsions—Their Stability and Resolution," *Chemicals in the Oil Industry*, Manchester, UK: Royal Society of Chemistry, 1988, 155.

[16] F. S. Manning and R. E. Thompson, *Oilfield Processing, Volume Two: Crude Oil*, Oklahoma City, OK: PennWell Publishing, 1995.

[17] *Petroleum Extension Service, Treating Oilfield Emulsions*, 4th ed., Austin, TX: University of Texas at Austin, 1990.

[18] R. Varadaraj, U.S. Patent Application 20030155307.

[19] D. G. Nahmad, I. Kmiec, A. Nasir, and I. Udau, "X-O-T Technology for the Treatment of Crude Oil Emulsions SPE," SPE 115222 (paper presented at the SPE Asia Pacific Oil and Gas Conference and Exhibition, Perth, Australia, 20-22 October 2008).

[20] H. H. Kartchner, U.S. Patent 6086830, 2000.

[21] H. JianZhong, "Reducing the Drag Force of the Multiple Phases Flow in Gathering Lines by Injecting Demulsifiers at Wellhead," SPE 19576 (paper presented at the International Petroleum Exhibition and

Technical Symposium, Beijing, China, 17–24 March 1982).

[22] T. J. Jones, E. L. Neustadter, and K. P. Whittingham, "Water-in-Crude Oil Emulsion Stability and Emulsion Destabilization by Chemical Demulsifiers," *Journal of Canadian Petroleum Technology* 17 (1978): 107.

[23] S. Kokal, "Crude Oil Emulsions: A State-of-the-Art Review," SPE 77497 (paper presented at the SPE Annual Technical Conference and Exhibition, San Antonio, TX, 29 September–2 October 2002).

[24] R. J. Mikula and V. A. Munoz, "Characterization of Demulsifiers," in *Surfactants: Fundamentals and Applications in the Petroleum Industry*, ed. L. L. Schramm, Cambridge, UK: Cambridge University Press, 2000, 51.

[25] R. Grace, "Commercial Emulsion Breaking, in Emulsions," in *Fundamentals and Applications in the Oil Industry*, ed. L. L. Schramm, Washington, DC: ACS, 1992, 313.

[26] D. Tambe, J. Paulis, and M. M. Sharma, "Factors Controlling the Stability of Colloid-Stabilized Emulsions: IV. Evaluating the Effectiveness of Demulsifiers," *Journal of Colloid and Interface Science* 171 (1995): 463.

[27] Y. Wang, L. Zhang, T. Sun, S. Zhao, and J. Yu, "A Study of Interfacial Dilational Properties of Two Different Structure Demulsifiers at Oil-Water Interfaces," *Journal of Colloid and Interface Science* 270 (2004): 163.

[28] G. Leopold, "Breaking Produced-Fluid and Process Stream Emulsions," in *Emulsions, Fundamentals and Applications in the Oil Industry*, ed. L. L. Schramm, Washington, DC: American Chemical Society, 1992, 341.

[29] J. L. Stark and S. Asomaning, "Synergies Between Asphaltene Stabilizers and Demulsifying Agents Giving Improved Demulsification of Asphaltene-Stabilized Emulsions," *Energy & Fuels* 19 (2005): 1342.

[30] H. L. Greenwald, G. L. Brown, and M. N. Fineman, "Determination of Hydrophile-Lipophile Character of Surface Active Agents and Oils by Water Titration," *Analytical Chemistry* 28 (11) (1956): 1693.

[31] J. Wu, Y. Xu, T. Dabros, and H. Hamza, "Development of a Method for Measurement of Relative Solubility of Nonionic Surfactants," *Colloids and Surfaces A: Physicochemical and Engineering Aspects* 232 (2004): 229.

[32] M. K. Poindexter, S. Chuai, R. A. Marble, and S. C. Marsh, "Classifying Crude Oil Emulsions Using Chemical Demulsifiers and Statistical Analyses," SPE 84610 (paper presented at the SPE Annual Technical Conference and Exhibition, Denver, CO, 5–8 October 2003).

[33] A. Goldszal and M. Bourrel, "Demulsification of Crude Oil Emulsions: Correlation to Microemulsion Phase Behavior," *Industrial & Engineering Chemistry Research* 39 (8) (2000): 2746.

[34] W. B. Allen, J. W. Harrell, and W. W. Webster, U.S. Patent 4134799, 1979.

[35] J. A. Ajienka, N. O. Ogbe, and B. C. Ezeaniekwe, *Journal of Petroleum Science and Engineering* 9 (1993): 331.

[36] J. H. Beetge and B. O. Horne, "Chemical Demulsifier Development Based on Critical Electric Field Measurements," SPE 93325 (paper presented at the SPE International Symposium on Oilfield Chemistry, The Woodlands, TX, 2–4 February 2005). See also SPE Journal, 13 (3) (September 2008): 346 (same authors).

[37] J. H. Beetge, "Emulsion Stability Evaluation of SAGD Product with the IPR-CEF Technique," SPE 97785 (paper presented at the SPE/PS-CIM/CHOA International Thermal Operations and Heavy Oil Symposium, Calgary, Alberta, Canada, 1–3 November 2005).

[38] H. J. Oschmann, "The DEMCON method, A New Way Forward for Evaluating Demulsifier Performance" (paper presented at the 4th International Conference on Petroleum Phase Behavior and Fouling, Trondheim, Norway, 23–26 June 2003).

[39] W. Knauf, K. Oppenlander, and W. Slotman, U.S. Patent 5759409, 1998.

[40] S. Radhakrishnan, S. Ananthasubramanian, and R. A. Marble, International Patent Application WO/2000/013762.

[41] D. L. Gallup, P. C. Smith, J. F. Star, and S. Hamilton, "West Seno Deepwater Development Case History—Production Chemistry," SPE 92969 (paper presented at the SPE International Symposium on Oilfield Chemistry, The Woodlands, TX, 2–4 February 2005).

[42] C. W. Angle, T. Dabros, and H. A. Hamza, "Demulsifier Effectiveness in Treating Heavy Oil Emulsion in the Presence of Fine Sands in the Production Fluids," *Energy & Fuels* 21 (2) (2007): 912.

[43] T. Balson, "The Unique Chemistry of Polyglycols," in *Chemistry in the Oil Industry VI*, Manchester, UK: Royal Society of Chemistry, 1998, 71–79.

[44] P. D. Berger, C. Hsu, and J. P. Arendell, "Designing and Selecting Demulsifiers for Optimum Field Performance on the Basis of Production Fluid Characteristics," SPE 16285, *SPE Production Engineering* 3 (6) (1988): 522.

[45] M. A. Krawczyk, D. T. Wasan, and C. S. Shetty, "Chemical Demulsification of Petroleum Emulsions Using Oil–Soluble Demulsifiers," *Industrial & Engineering Chemistry Research* 30 (1991): 367.

[46] C. R. E. Ransur, S. P. Barboza, G. Gonzalez, and E. F. Lucas, *Journal of Colloid and Interface Science* 271 (2004): 232.

[47] J. Wu, Y. Xu, T. Dabros, and H. Hamza, *Colloids and Surfaces A: Physicochemical and Engineering Aspects* 252 (2005): 79.

[48] G. N. Taylor and R. Mgla, U.S. Patent 5407585, 1995.

[49] W. K. Langdon and R. L. Camp, U.S. Patent 4183821, 1980.

[50] M. Guzmann, P. Neumann, K.-H. Büchner, and A. Oftring, International Patent Application WO/2006/134145.

[51] A. M. Al–Sabagh, A. M. Badawi, and M. R. Noor El–Den, *Petroleum Science and Technology* 20 (2002): 887.

[52] A. M. Al–Sabagh, N. E. Maysour, N. M. Naser, and M. R. Noor El–Din, "Synthesis and Evaluation of Some Modified Polyoxyethylene–Polyoxypropylene Block Polymer as Water–in–Oil Emulsion Breakers," *Journal of Dispersion Science and Technology* 28 (2007): 537.

[53] C. W. Hahn, U.S. Patent Application 20060030491.

[54] F. Staiss, R. Bohm, and R. Kupfer, "Improved Demulsifier Chemistry: A Novel Approach in the Dehydration of Crude Oil," SPE 14841, *SPE Production Engineering* 6 (3) (1991): 334.

[55] J. Wu, Y. Xu, T. Dabros, and H. Hamza, "Effect of Demulsifier Properties on Destabilization of Water–in–Oil Emulsion," *Energy & Fuels* 17 (2003): 1554.

[56] G. Elfers, W. Sager, H.-H. Vogel, and K. Oppenlaender, U.S. Patent 5401439, 1997.

[57] M. Lancaster, D. J. Moreton, and A. F. Psaila, U.S. Patent 5272226, 1993.

[58] R. S. Buriks, A. R. Fauke, F. E. Mange, U.S. Patent 4032514, 1977.

[59] J. Beyer, A. Skadsheim, M. A. Kelland, K. Alfsnes, and S. Sanni, "Ecotoxicology of Oilfield Chemicals: The Relevance of Evaluating Low–Dose and Long–Term Impact on Fish and Invertebrates in Marine Recipients," SPE 65039 (paper presented at the SPE International Symposium on Oilfield Chemistry, Houston, TX, 13–16 February 2001).

[60] P. Jacques, I. Martin, C. Newbigging, and T. Wardell, "Alkylphenol Based Demulsifier Resins and their Continued Use in the Offshore Oil and Gas Industry," *Chemistry in the Oil Industry VII*, Royal Society of Chemistry, 2002, 56.
[61] F. Holtrup, E. Wasmund, W. Baumgartner, and M. Feustel, U.S. Patent 6646016, 2003.
[62] F. Holtrup, E. Wasmund, W. Baumgartner, and M. Feustel, U.S. Patent 6465528, 2002.
[63] D. Leinweber and E. Wasmund, U.S. Patent Application 20040102586, 2004.
[64] J. B. Byron, P. M. Lindemuth, and G. N. Taylor, U.S. Patent Application 20040266973, 2003.
[65] D. Leinweber and E. Wasmund, U.S. Patent Application 20040014824, 2004.
[66] K. Barthold, R. Baur, S. Crema, K. Oppenlaender, and J. Lasowski, U.S. Patent 5472617, 1995.
[67] H. Diaz-Arauzo, U.S. Patent 5460750, 1995.
[68] R. G. Sampson, U.S. Patent 3640894, 1972.
[69] M. B. Martin, International Patent Application WO/2008/036910.
[70] F. T. Lang, International Patent Application WO/2004/082604.
[71] C. Myers, S. R. Hatch, and D. A. Johnson, International Patent Application WO/2006/116175.
[72] K. Barthold, K. Oppenlaender, J. Lasowski, and R. Baur, U.S. Patent 4814394, 1989.
[73] G. R. Meyer, U.S. Patent 20050080221.
[74] W. K. Stephenson and J. D. DeShazo, U.S. Patent 5205964, 1993.
[75] M. Groote and S. Kwan-Ting, U.S. Patents 2792352-6, 1957.
[76] V. L. Seale, B. R. Moreland, and J. D. Shazo, U.S. Patent 3383326, 1968.
[77] P. J. Breen and J. Towner, U.S. Patent 6225357, 2001.
[78] A. A. Toenjes, M. R. Williams, and E. A. Goad, U.S. Patent 5102580, 1992.
[79] S. Podubrin, W. Breuer, C.-P. Herold, A. Heidbreder, T. Foerster, and M. Hollenbrock, International Patent Application WO/1999/007808.
[80] D. S. Treybig, D. A. Williams, and K. T. Chang, International Patent Application WO/2003/053536.
[81] Y. Xu, J. Wu, T. Dabros, H. Hamza, and J. Venter, "Optimizing the Polyethylene Oxide and Polypropylene Oxide Contents in Diethylenetriamine-Based Surfactants for Destabilization of a Water-in-Oil Emulsion," *Energy & Fuels* 19 (2005): 916.
[82] G. Liebold, K. Oppenlaender, E. Buettner, R. Fikentscher, and R. Mohr, U.S. Patent 3907701, 1975.
[83] K. Oppenlaender, R. Fikentscher, E. Buettner, W. Slotman, E. Schwartz, and R. Mohr, U.S. Patent 4537701, 1985.
[84] D. Leinweber, M. Feustel, H. Grundner, and H. Freundl, International Patent Application WO/2003/102047.
[85] K. Barthold, R. Baur, R. Fikentscher, J. Lasowski, and K. Oppenlaender, U.S. Patent 4935162, 1990.
[86] A. Lindert and M. S. Wiggins, U.S. Patent 6172123, 2001.
[87] P. R. Hart, U.S. Patent 5250174, 1993.
[88] R. M. Gipson, C. L. LaBerge, D. R. McCoy, and K. B. Young, Hoechst, British Patent 1213392, 1967.
[89] R. S. Buriks, F. E. Mange, and P. M. Quinlan, U.S. Patent 3594393.
[90] P. F. D. Reeve, U.S. Patent 6348509, 2002.
[91] A. A. Peña, G. J. Hirasaki, and C. A. Miller, "Chemically Induced Destabilization of Water-in-Crude Oil Emulsions," *Industrial and Engineering Chemistry Research* 44 (2005): 1139.
[92] B. Bruchmann, K.-H. Büchner, M. Guzmann, G. Brodt, and S. Frenzel, International Patent

Application WO2006084816.

[93] D. A. Tomalia and J. R. Dewald, U.S. Patent 4507466, 1985.

[94] G. R. Killat and J. R. Conklin, U.S. Patent 4448708, 1984.

[95] L. R. Wilson and J. R. Conklin, U.S. Patent 4457860, 1984.

[96] R. S. Buriks and J. G. Dolan, U.S. Patent 4626379, 1986.

[97] R. S. Buriks and J. G. Dolan, U.S. Patent 4877842, 1989.

[98] J. Fock and H. Rott, U.S. Patent 4678599, 1987.

[99] D. Faul, J. Roser, H. Hartmann, H.-H.Vogel, W. Slotman, and G. Konrad, U.S. Patent 5661220, 1997.

[100] C. Auschra, H. Pennewiss, U. Boehmke, and M. Neusius, U.S. Patent 6080794, 2000.

[101] G. N. Taylor, U.S. Patent 5609794, 1997.

[102] W. K. Stephenson, U.S. Patent 4968449, 1990.

[103] H. Becker, U.S. Patent 7018957, 2006.

[104] J. Behles, International Patent Application WO/2007/121165.

[105] B. Bhattacharyya, U.S. Patent 5100582, 1992.

[106] G. Koerner and D. Schaefer, U.S. Patent 5004559, 1991.

[107] C. Dalmazzone and C. Noïk, "Development of New 'Green' Demulsifiers for Oil Production," SPE 65041 (paper presented at the SPE International Symposium on Oilfield Chemistry, Houston, TX, 13-16 February 2001).

[108] K. Koczo and S. Azouani, "Organomodified Silicones as Crude Oil Demulsifiers" (paper presented at the Chemistry in the Oil Industry: Oilfield Chemistry, Royal Society of Chemistry, Manchester, UK, 5-7 November 2007).

[109] P. S. Newman, C. Hahn, and R. D. McClain, International Patent Application WO/2006/068702.

[110] D. Leinweber, M. Feustel, E. Wasmund, and H. Rausch, International Patent Application WO/2005/003260.

[111] J. Senior, L. H. Smith Sr., T. Algroy, and L. H. Smith Jr., International Patent Application WO/2004/050801.

[112] D. Leinweber, F.-X. Scherl, E. Wasmund, and H. Rausch, International Patent Application WO/2002/066136.

[113] D. Leinweber, F.-X. Scherl, E. Wasmund, and H. Rausch, International Patent Application WO/2004/108863.

[114] R. Berkhof, H. Kwekkeboom, D. Balzer, and N. Ripke, U.S. Patent 5164116, 1992.

[115] P.-E. Hellberg, "Environmentally Adapted Demulsifiers Containing Weak Links" (paper presented at the Chemistry in the Oil Industry X: Oilfield Chemistry, Royal Society of Chemistry, Manchester, UK, 5-7 November 2007).

[116] P.-E. Hellberg and I. Uneback, International Patent Application WO/2007/115980.

[117] W. Wang, International Patent Application WO/2008/103564.

[118] R. Golden, U.S. Patent 6727388, 2004.

[119] R. Varadaraj, D. W. Savage, and C. H. Brons, International Patent Application WO/2000/050541.

[120] L. Heiss and M. Hille, U.S. Patent 3974220, 1976.

[121] M. Hille, R. Kupfer, and R. Bohm, U.S. Patent 5421993, 1995.

[122] R. Varadaraj and C. H. Brons, U.S. Patent Application 20030092779, 2003.

[123] S. Kokal and J. Al-Juraid, "Reducing Emulsion Problems by Controlling Asphaltene Solubility and

Precipitation," SPE 48995 (paper presented at the SPE Annual Technical Conference and Exhibition, New Orleans, LA, 27-30 September 1998).

[124] K. W. Smith, J. Miller, and L. W. Gatlin, U.S. Patent Application 20050049148, 2005.

[125] R. Talingting-Pabalan, G. Woodward, M. Dahanayake, and H. Adam, International Patent Application WO/2009/023724.

[126] P. K. Kilpatrick, *Energy & Fuels* 26 (7) (2012): 4017-4026.

[127] A. L. Nenningsland, B. Gao, S. Simon, and J. Sjöblom, *Energy & Fuels* 25 (2011): 5746.

[128] M. Kunieda, K. Nakaoka, Y. Liang, C. R. Miranda, A. Ueda, S. Takahashi, H. Okabe, and T. Matsuoka, *Journal of the American Chemical Society* 132 (2010): 18281-18286.

[129] M. Stewart and K. Arnold, *Emulsions and Oil Treating Equipment, Emulsions and Oil Treating Equipment: Selection, Sizing and Troubleshooting*. Gulf Professional Publishing, 2008.

[130] B. M. S. Ferreira, J. B. V. S. Ramalho, and E. F. Lucas, *Energy & Fuels* 27 (2013): 615.

[131] S. N. Hudaa and A. H. Nour, *International Journal of Chemical and Environmental Engineering* 2 (1) (2011): Article 11.

[132] L. A. Kovaleva, R. Z. Minnigalimov, and R. R. Zinnatullin, *Energy & Fuels* 25 (2011): 3731.

[133] M.-J. T. Mui Ching, A. E. Pomerantz, A. Ballard Andrews, P. Dryden, R. Schroeder, O. C. Mullins, and C. Harrison, *Energy & Fuels* 24 (2010): 5028-5037.

[134] J.-L. Salager and A. M. Forgiarini, *Energy & Fuels* 26 (7) (2012): 4027-4033.

[135] J. C. Pereira, J. Delgado-Linares, C. Scorzza, M. Rondon, S. Rodriguez, and J.-L. Salager, *Energy & Fuels* 25 (2011): 1045-1050.

[136] M. Moradi, V. Alvarado, and S. Huzurbazar, *Energy & Fuels* 25 (2011): 260-268.

[137] N. van der Tuuk Opedal, I. Kralova, C. Lesaint, and J. Sjöblom, *Energy & Fuels* 25 (2011): 5718.

[138] F. H. Wang, L. B. Shen, H. Zhu, and K. F. Han, *Petroleum Science and Technology* 29 (2011): 2521.

[139] R. Talingting-Pabalan, G. Woodward, M. Dahanayake, and H. Adam, International Patent Application WO/2010/019172.

[140] Z. Huang, H. Lu, T. Zhang, R. Wang, and D. Qing, *Petroleum Science and Technology* 28 (2010): 1621-1631.

[141] F. Alvarez, E. A. Flores, L. V. Castro, J. G. Hernandez, A. Lopez, and F. Vazquez, *Energy & Fuels* 25 (2011): 562-567.

[142] V. F. Pacheco, L. Spinelli, E. F. Lucas, and C. R. E. Mansur, *Energy & Fuels* 35 (2011): 1659.

[143] C. Schaefer, C. Cohrs, and S. Dilsky, International Patent Application WO/2010/124772.

[144] C. Schaefer, C. Cohrs, and S. Dilsky, International Patent Application WO/2010/124773.

[145] A.-A. A. Azim, A.-R. M. Abdul-Raheim, R. K. Kamel, and M. E. Abdel-Raouf, *Journal of Petroleum Science and Engineering* 78 (2011): 364-370.

[146] C. R. McDaniel, K. Kuklenz, and K. Ginsel, U.S. Patent Application 20110253598.

[147] I. Riff, T. Alonso, and R. Van Voorst, International Patent Application WO/2012/068099.

[148] S. Dilsky, U.S. Patent Application 20120172270.

[149] C. Cohrs, S. Dilsky, D. Leinweber, and M. Feustel, International Patent Application WO/2011/035854.

[150] J. Wang, C.-Q. Li, N. An, and Y. Yang, *Separation Science and Technology* 47 (2012): 1583.

[151] M. Guzmann, and W. Gaschler, U.S. Patent Application 20110272327.

[152] B. Bruchmann, A. Eichhorn, and M. Guzmann, U.S. Patent Application 20100240857.

[153] J. D. Debord, U.S. Patent Application 20110257328.

[154] A. Saxena, M. Phukan, U. Senthilkumar, I. Procter, S. Gonzalez, K. Koczo, S. Azouani, and V. Kumar, U.S. Patent Application 20090192234.

[155] M. Phukan, A. Saxena, M. J. Dubey, A. Palumbo, and K. Koczo, International Patent Application WO/2012/177836.

[156] M. Hilfiger, M. P. Squicciarini, C. A. Blundell, D. M. Stepien, and P. J. Breen, International Patent Application WO/2012/030600.

[157] A. Kaiser, "Environmentally Friendly Emulsion Breakers: Vision or Reality?," SPE 164073 (paper presented at the SPE International Symposium on Oilfield Chemistry, The Woodlands, TX, 8–10 April 2013).

[158] H. Zhou, K. I. Dismuke, N. L. Lett, and G. S. Penny, "Development of More Environmentally Friendly Demulsifiers," SPE 151852 (paper presented at the SPE International Symposium and Exhibition on Formation Damage Control, Lafayette, LA, 15–17 February 2012).

[159] H. S. Bevinakatti, N. Grainger, and T. J. Wardell, International Patent Application WO/2013/041876.

[160] C. Temple-Heald, C. Davies, N. Wilson, and H. Sarginson, "The Development of New Green Demulsifiers for Use in the North Sea," *Chemistry in the Oil Industry XIII—New Frontiers*, Manchester, UK, November 2013.

[161] K. Koczo, B. Falk, A. Palumbo, and M. Phukan, "New Silicone Copolymers as Demulsifier Boosters," *RSC Chemistry in the Oil Industry XII*, Manchester, UK, November 2011.

[162] C. Temple-Heald, C. Davies, N. Wilson, and N. Readman, "The Development and Field Application of New Surfactant Chemistries for Application to Heavy Oils," SPE 164109 (paper presented at the SPE International Symposium on Oilfield Chemistry, The Woodlands, TX, 8–10 April 2013).

[163] M. Hilfiger, P. Breen, C. A. Blundell, and J. Sonne, U.S. Patent Application 20130231418.

[164] A. M. Atta, *International Journal of Electrochemical Science*, 8 (7) (2013): 9474–9498.

[165] D. T. Nguyen, International Patent Application, WO/2013/158989.

[166] C. Davies, C. Temple-Heald, N. Wilson, and H. Sarginson, "The Development of New Green Demulsifiers for Use in the North Sea," *Chemistry in the Oil Industry XIII: Oilfield Chemistry—New Frontiers*, Manchester Conference Centre, UK, 4–6 November 2013.

12 泡沫控制

12.1 概述

泡沫是由不溶性气体分散在液体或熔融固体中所形成的分散体系。许多油田工艺流程中都会发生液体起泡问题，如分离器内的气体从原油中分离的过程，或天然气处理厂的胺液脱硫脱碳、乙二醇脱水工艺过程。水系统也可能因化学剂作用或气体除氧/真空脱气而产生泡沫。两相和三相分离器中的泡沫会造成液位控制不佳导致平台关停、气体出口中的液体残留导致下游洗涤器和压缩机被淹停、液体出口中的携气导致压缩要求增加等运行方面的问题。高产量和高气油比有利于泡沫的形成。

泡沫是由天然表面活性剂（胶质、沥青质、环烷酸等）或加注的生产化学剂（如成膜型缓蚀剂）等表面活性剂稳定的气泡（薄膜）所组成。马兰戈尼效应（即液膜表面或内部因表面张力差异而产生的质量转移）使泡沫稳定。实验条件下黏度对原油是否产生泡沫起到主要作用。

12.2 消泡剂和抑泡剂

可以通过添加消泡剂或抑泡剂来控制泡沫。抑泡剂指防止或延缓泡沫形成的化学剂，而消泡剂指消除已经形成的泡沫的化学剂。然而，因为许多常见的消泡剂也是抑泡剂，这两个术语在石油工业中经常混用。因此，许多抑泡剂不仅能够防止或延缓泡沫形成，还能消除已经形成的泡沫。在原油分离器中，最好的消泡剂和抑泡剂使用浓度为 1~10mg/L，但也有使用剂量低至 0.1mg/L、高至 100mg/L 的报道。过量使用消泡剂有可能形成稳定泡沫，必须确定其最佳浓度。消泡剂和抑泡剂将表面活性剂从气泡的气—液界面置换出来，使气泡中的液体凝聚，而气体则逸出。

消泡剂或抑泡剂的最普遍特点是具有表面活性且基本不溶于水，常被配制为乳液以便使用时分散为微液滴态。在油气工业中，主要有有机硅和氟硅氧烷、聚乙二醇两类消泡剂/抑泡剂，目前最常用的是第一类。

上述两类的混合物也可在市场上买到，有时与疏水性二氧化硅微粒联合使用，以增强消泡效果。使用的油通常是矿物油或硅基油，固体颗粒是二氧化硅或疏水聚合物。这些混合物可以在溶液和水包油乳状液中使用。也可以将消泡剂涂覆在气相二氧化硅等气相金属氧化物颗粒表面。这些颗粒的功能是当消泡剂在液膜界面扩散时，能刺入泡沫，使其凝聚。如乙氧基化脂肪酯等破乳剂也可以作为消泡剂使用。

如文献介绍，实验室内测试消泡剂和抑泡剂时，常使用量筒进行计量测试。消泡剂的测试方法是，首先向装有液体的量筒中通气，制造出泡沫，再加入消泡剂，测量完全消泡时长。抑泡剂的测试方法是先将化学剂加入液体，然后通气鼓泡，记录形成一定量泡沫的延迟时间。原油消泡剂的一种筛选方法是受控减压法。

12.2.1 有机硅和氟硅氧烷

有机硅是油田最常用的抑泡剂，是一类无毒但生物降解性差的硅氧烷聚合物。使用时可以采用纯品、溶液或乳液，还可以像前文提到的那样，制成硅质颗粒混合物使用。最简单的有机硅是聚二甲基硅氧烷（PDMS），其基本结构如图12.1所示，但也可使用其他聚二取代硅氧烷。原油体系抑泡剂的聚合物最佳分子量峰值分布，建议为15000~130000。有机硅的缺点是对炼油厂的催化剂表面可能造成损害，但这可能与过量使用有关。

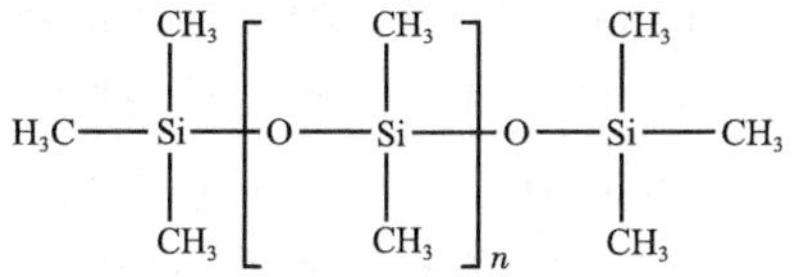

图12.1 聚二甲基硅氧烷的结构式（n=200~1500）

带有聚氧化烯基团的聚二甲基硅氧烷衍生物也被归为抑泡剂。其疏水性可以通过引入不同的氧化烯基团和改变其比例来调节，也可以使用交联有机聚硅氧烷—聚氧烷。聚硅氧烷均聚物与特定聚硅氧烷共聚物的混合物被认为是改进型的抑泡剂。聚硅氧烷共聚物可通过以下方式获得：第一步，将包含至少一个 Si—H 的有机聚硅氧烷与通式 R—（A—C_nH_{2n}）$_m$—A^1—H 的线型低聚物或聚合物反应，其中 A 和 A^1 是含有氧或氮原子的基团。聚二取代硅氧烷与聚氧化烯改性硅油的混合物据称具有更好的抑泡性能、水分散性和稳定性。另外，含氯端基的 PDMS 可以与聚乙二醇反应，形成乙二醇封端的三嵌段共聚物。另外，聚乙二醇也可以接枝在硅氧烷链的任何位置。接枝聚醚硅氧烷共聚物可以改善抑泡剂在水分散体系的低温稳定性。烷基丙烯酸酯、烯丙基聚醚可以接枝到含 Si—H 基团的硅氧烷上，而 EO/PO 嵌段共聚物可以接枝到二氯二甲基聚硅氧烷上。乙烯基或甲基丙烯酸酯基封端的有机硅聚合物也可作为消泡剂。也可以使用其他有机改性硅油。

聚二甲基硅氧烷在包括烃类等有机溶剂中具有一定的溶解性，其在原油分离器中的抑泡效果，可以通过氟化烷基取代一些甲基而得到改善。氟硅氧烷分子的全氟烷基化可降低其在烃类化合物中的溶解度，从而提高其性能。典型的氟硅化合物如图12.2所示。3,3,3-三氟丙基硅氧烷基团也被接枝到氟硅化合物抑泡剂中。氟硅氧烷比 PDMS 等普通有机硅更昂贵，但其使用浓度通常更低。氟硅化合物的加注位置很重要，制造商建议加注点应在形成高剪切作用的设备之前，或在高压分离器的原油入口端。与 PDMS 等常规有机硅相比，氟硅氧烷对液滴的粒径分布不敏感，而且能够承受高速湍流环境。

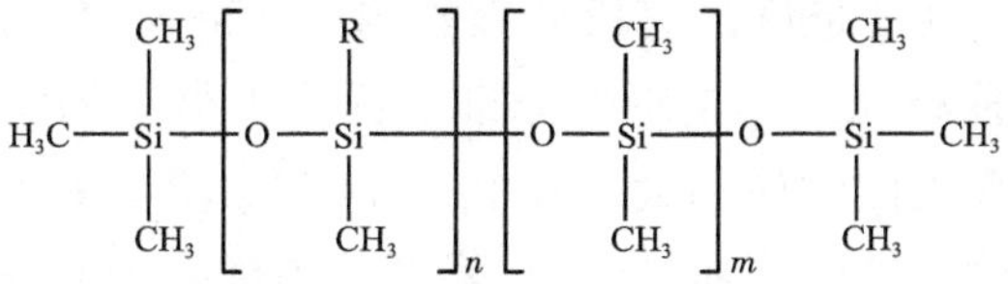

图12.2 含二甲基硅氧基的氟硅消泡剂结构式（R=C_nF_{2n+1} 或 $CF_3CH_2CH_2$）

氟硅氧烷与非氟硅氧烷（如 PDMS）的混合使用会产生协同效应，比起单独使用其中任何一种的性能更好，剂量更低。在气液分离过程中，该混合物既能减少进入气流的携液量，又能减少进入液流的携气量，分离效果更好。

12.2.2 聚乙二醇

聚乙二醇是全能型消泡剂，特别适用于去除水溶液中的泡沫，也可以应用于原油分离器或乙二醇系统。相对于聚硅氧烷，聚乙二醇的加注浓度通常高得多，如为 10~200mg/L，其优点是更符合北海等地区的环保标准要求。此外，这些聚合物会对加工原油（主要）生产沥青的炼油厂带来负面影响。一项旨在寻找有机硅消泡剂更环保型替代品的研究评价了包括磷酸盐产品、乙氧基化和丙氧基化酯、聚乙二醇酯和油酸酯、醇类、脂肪醇及乙氧基化和丙氧基化醇等的多种化学剂。为钻井和固井作业开发的更环保型消泡剂技术，在生产作业中有一定的应用前景。

聚乙二醇与其他类型的消泡剂和抑泡剂一样，在极端情况下的过量加注可能会导致泡沫稳定和分离器/乙二醇系统失效等影响。聚乙二醇等非硅类抑泡剂通常是可生物降解的，毒性低，不会对炼油厂的催化剂造成任何损害。聚乙二醇是通过将单体 EO、PO 或 BO 添加到胺或醇中反应生成。聚乙二醇消泡剂需要含有高比例的疏水性单体（PO 或 BO），才能具有良好的性能。

这类消泡剂大多以聚丙氧基化合物为基础，因为 PO 是比 BO 更便宜的疏水性单体。这种聚合物可以是线型或支化结构（图 12.3）。用作乳化剂的聚乙二醇类可以被定制为消泡剂，包括烷基酚树脂烷氧基化物和双环氧树脂。

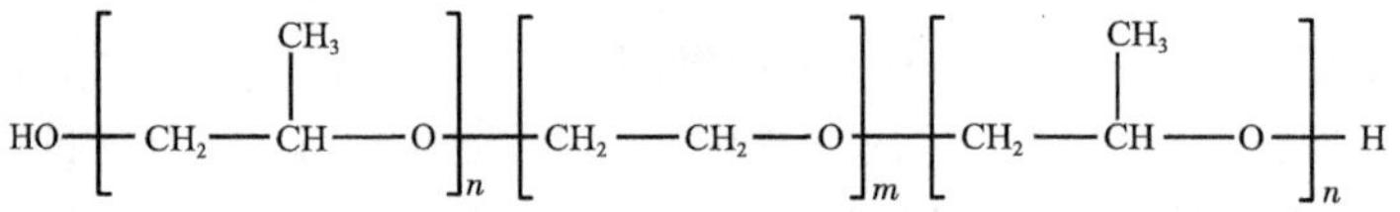

图 12.3　线型二嵌段 EO/PO 共聚物消泡剂（其中 $n>m$）

含不同数量 EO 和 PO（或 BO）的聚乙二醇在水中通常有浊点。浊点是发生相分离的特定温度，高于这个温度后，亚烷基氧化物的醚基和水分子之间的氢键会发生断裂。一些 EO/PO 共聚物或 EO 接枝的聚硅氧烷在高于浊点的水溶液中起到抑泡作用，此时处于非均相状态，被称为浊点抑泡剂。多元醇也可以通过烷氧基化来生成消泡剂，例如，甘油与不同数量的 PO 和 EO 缩合，与 25% 的二甲苯磺酸钠水溶液混合。其他消泡剂的配方是聚烷氧基化的表面活性剂与多元醇脂肪酸酯的混合物。

对聚环氧乙烷/聚环氧丙烷（PEO/PPO）嵌段共聚物和接枝 PEO/PPO 链的聚醚有机硅进行了抑泡性能测试。性能最好的有机硅聚醚的极性最强，在介质中能形成异相颗粒，破坏泡沫的稳定性。这种添加剂也是降低原油表面张力最小的添加剂，表明这种硅聚醚在介质中的作用比在液膜表面的作用大。在 PEO/PPO 嵌段共聚物中，效率最高的是对水亲和力最大的嵌段共聚物（极性更强），也是降低油类样品表面张力最少的添加剂。实验用的高黏度原油样阻碍了添加剂的分散，延长了完全消泡的所用时间。这种行为似乎更重要。

其他烷氧基化的表面活性剂和聚合物也被归类为消泡剂，如聚丙烯酸的混合烷氧基聚亚烷基二醇酯（图12.4）。这些与聚丙烯酸酯破乳剂（见第11章）在结构上有相似之处。也有资料提及基于甲基丙烯酸酯单体的聚合物消泡剂配方。

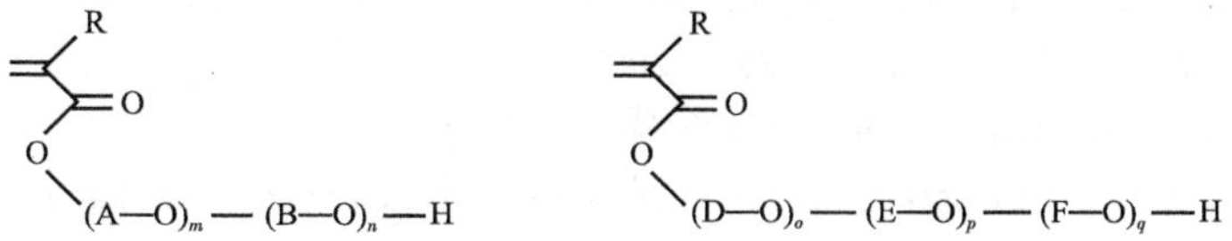

图 12.4　聚丙烯酸的混合烷氧基聚亚烷基二醇酯

参 考 文 献

[1] R. K. Prud'homme, and S. A. Khan, eds., *Foams: Theory, Measurements and Applications*, Surfactant Science Series, New York: Marcel Dekker, 57 (1996).

[2] L. L. Schramm, ed., *Foams: Fundamentals and Applications in the Petroleum Industry*, Advances in Chemistry Series 242, Washington, DC: ACS, 1994.

[3] S. Ross, "Profoams and Antifoams," *Colloids and Surfaces A: Physicochemical and Engineering Aspects* 118 (1996): 187–192.

[4] M. K. Poindexter, N. N. Zaki, P. K. Kilpatrick, S. C. Marsh, and D. H. Emmons, "Factors Contributing to Petroleum Foaming. 1. Crude Oil Systems," *Energy & Fuels* 16 (3) (2002): 700–710.

[5] I. C. Callaghan, "Anti-foams for Nonaqueous Systems in the Oil Industry," in *Defoaming: Theory and Industrial Applications*, ed. P. R. Garrett. New York: Marcel Dekker, (1993): 119–150.

[6] F. Cassani, P. Ortega, A. Davila, W. Rodriguez, S. A. Lagoven, and J. Seranno, "Evaluation of Foam Inhibitors at the Jusepin Oil/Gas Separation Plant, El Furrial Field, Eastern Venezuela," SPE 23681 (paper presented at the SPE Latin America Petroleum Engineering Conference, Caracas, Venezuela, 8–11 March 1992).

[7] R. W. Chin, H. L. Inlow, T. Keja, P. B. Hebert, J. R. Bennett, and T. C. Yin, "Chemical Defoamer Reduction with New Internals in the Mars TLP Separators," SPE 56705 (paper presented at the SPE Annual Technical Conference and Exhibition, Houston, TX, 3–6 October 1999).

[8] B. K. Jha, S. P. Christiano, and D. O. Shah, "Silicone Antifoam Performance: Correlation with Spreading and Surfactant Monolayer Packing," *Langmuir* 16 (26) (2000): 9947–9954.

[9] D. K. Durham, J. Archer, and G. Thornton, International Patent Application WO/2007/14296.

[10] K. W. Smith, J. Miller, and L. W. Gatlin, U.S. Patent Application 20050049148, 2005.

[11] P. G. Pape, "Silicones: Unique Chemicals for Petroleum Processing," *Journal of Petroleum Technology* 35 (1983): 1197–1204.

[12] M. J. Owen, "The Surface Activity of Silicones: A Short Review," *Industrial and Engineering Chemical Production Research and Development* 19 (1980): 97–103.

[13] H. Nakahara and K. Aizawa, U.S. Patent 5153258, 1992.

[14] H. Shouji and K. Aizawa, U.S. Patent 5556902, 1996.

[15] I. C. Callaghan, H.-F. Fink, C. M. Gould, G. Koerner, H.-J. Patzke, and C. Weitemeyer, U.S. Patent

4557737, 1985.
[16] I. C. Callaghan, C. M. Gould, and W. Grabowski, U.S. Patent 4711714, 1987.
[17] I. C. Callaghan, C. M. Gould, A. J. Reid, and D. H. Seaton, "Crude-Oil Foaming Problems at the Sullom Voe Terminal," SPE 12809, *Journal of Petroleum Technology* 37 (12) (1985): 2211.
[18] R. A. Elms and M. A. Servinski, U.S. Patent 6512015, 2003.
[19] K. C. Fey and C. S. Combs, U.S. Patent 5397367, 1995.
[20] W. Burger, C. Herzig, and J. Wimmer, International Patent Application WO/2006/128624.
[21] Y. Aoki and A. Itagaki, U.S. Patent 6417258, 2002.
[22] H. Kobayashi and T. Masatomi, U.S. Patent 5454979, 1995.
[23] R. Berger, H.-F. Fink, G. Koerner, J. Langner, and C. Weitemeyer, U.S. Patent 4626378, 1986.
[24] A. S. Taylor, G.B. Patent 2244279, 1991.
[25] E. R. Evans, U.S. Patent 4329528, 1982.
[26] C. T. Gallagher, P. J. Breen, B. Price, and A. F. Clemmit, U.S. Patent 5853617, 1998.
[27] J. H. Beetge, P. J. Venter, R. Cleary, J. Kuzyk, B. Shand, D. A. Davis, and D. W. Matalamaki, U.S. Patent Application 20060025324, 2006.
[28] G. P. Sheridan, U.S. Patent 5071591, 1991.
[29] J. J. Svarz, U.S. Patent 4968448, 1990.
[30] R. J. Pugh, "Foam Breaking in Aqueous Systems," in Handbook of Applied Surface and Colloid Chemistry, Chapter 8, ed. K. Holmberg. UK: Wiley, 1 (2002): 143.
[31] D. N. Denkov, *Langmuir* 20 (2004): 9463.
[32] A. Farzad, A. M. Mohammad, and V. Ali, *Journal of Applied Polymer Science* 5 (2005): 1122.
[33] G. P. Pape, *Journal of Petroleum Science and Engineering* 35 (1983): 1197.
[34] B. Kekevi, H. Berber, and H. Yıldırım, *Journal of Surfactants and Detergents* 15 (2012): 73-81.
[35] D. A. Rezendea, R. R. Bittencourta, and C. R. E. Mansur, *Journal of Petroleum Science and Engineering* 76 (2011): 172-177.
[36] J. Venzmer, S. Herrwerth, S. Maslek, C. Mund, and P. Schwab, U.S. Patent Application 20090093598.
[37] M. Zhao and J. Fang, United States Patent Application 20100015441.
[38] J. Wylde, "Successful Field Application of Novel, Nonsilicone Antifoam Chemistries for High-Foaming Heavy-Oil Storage Tanks in Northern Alberta," SPE 117176, *SPE Production & Operations* 25 (1): 25-30.
[39] C. A. Bonfillon and D. Langevin, *Langmuir* 13 (1997): 599.
[40] Z. S. Nemeth, G. Y. Racz, and K. Koczo, *Colloids and Surfaces A: Physicochemical and Engineering Aspects* 127 (1997): 151.
[41] C. Schaefer and C. Mogck, International Patent Application WO/2011/103970.
[42] C. Dalmazzone, C. Blazquez-Egea, S. Schneider, E. Emond, and V. Bergeron, "Formation and Breaking of Crude-Oil Foams," *Chemistry in the Oil Industry XIII: Oilfield Chemistry-New Frontiers*', Manchester Conference Centre, UK, 4-6 November 2013.
[43] L. Bava, A. Mahmoudkhani, R. Wilson, and L. Levy, New Generation of "Green" Defoamers for Challenging Drilling and Cementing Applications, SPE 164504 (paper presented at the SPE Production and Operations Symposium, Oklahoma, USA, 23-26 March 2013).
[44] J. Martin, R. Wilson, S. Rosencrance, and D. Previs, U.S. Patent Application 20120264863.

13 絮 凝 剂

13.1 概述

在油田集输系统油水分离的破乳阶段，分离出的水含有残余油和固体悬浮颗粒。油以分散在水中或水包油的乳状液形式存在（反相乳液）。残余油浓度通常过高，不允许直接排放到环境中，而且残油还具有经济价值。例如，2007年1月1日生效的东北大西洋和北海地区采出水的油和油脂的新排放标准为30mg/L（以前为40mg/L）。在其他地方的排放指标可能设定为很难达到要求的5~10mg/L。此外，采出水回注时，这些固体可能会堵塞注水井近井地带的孔喉或地面过滤器，提高背压，造成能源浪费和设备损坏，甚至可能导致停产。因此，需要对水进行处理，以去除油和分散的固体物质。处理方法中的化学方法是添加絮凝剂，絮凝剂也被称为"水澄清剂""脱氧剂""水包油破乳剂""反相破乳剂"或"聚电解质"。絮凝剂在井筒清理作业中也很有用。

在石油和天然气生产现场，絮凝剂与如水力旋流器、离心机、浮选或过滤设备等重力沉降设备一起配合使用，形成絮状物，将油滴和颗粒吸附于其上。絮状物随后被分离并返回到原油生产中。大多数絮状物呈黏性，会附着在设备内部表面。水处理系统在相对较短的时间后，内部絮状物堆积会造成损害，并可能需要关停和清洗。最好是絮凝剂形成"可接受"的絮凝物，不会因黏附、堵塞和界面堆积造成系统中的操作问题。絮凝剂有时用于水力旋流器的上游，此时絮凝剂会增大絮凝物尺寸。如果分散的油/固体粒径低于约10μm，水力旋流器就不能正常运转。有文献报道了在北海和墨西哥湾深水区的产出水处理系统对比絮凝剂技术的效果。

近年来，在注入破乳剂后，向分离器的上游注入絮凝剂，可以进一步改善分离性能。原因可能是絮凝剂的聚合性质，需要时间来完全"解卷"并达到其最佳构象以发挥其功效。

还开发了去除采出水中的油和其他污染物等的其他技术。有一种工艺使用组合式脱气和浮选池来分离含有大量石油和气体的进水。在水箱中产生旋流，使油滴和气滴等较轻成分向内侧的同心圆柱凝聚，上升到液体表面，并通过出口被排出。而重质组分则沉降，沉淀到底部的重颗粒作为污泥被清除。在海上使用一种大孔聚合物萃取技术。含烃污水通过装有多孔聚合物珠子的柱子，珠子内含有特制的萃取液。固定化的萃取液可从工艺流程的水中去除烃类成分。最近还有研究发现被称为"白色石墨烯"的多孔氮化硼纳米片可以有效地去除水中的原油和其他有机污染物。另一种新型无机吸附剂（未提供结构信息）被设计为具有去除总油和油脂的高亲和力材料。其他工业用特殊设计的离心机或水力旋流器。

而另一种CTour工艺也正在北海油田应用，使用水力旋流器和额外的液化天然气（LNG）（平台上必须有），同时从采出水中提取包括分散和溶解态多环芳香烃及苯、甲苯和二甲苯等在内的原油。通常首先使用絮凝剂来降低水中的原油含量，再通过CTour工艺进一步将原油含量降至5mg/L以下。

包括油滴凝聚器、旋流阀和有机物吸附剂等在内的非化学除油工艺，已有发明专利和文献报道。还在现场流程中测试了表面活性剂改性的沸石吸附床和膜生物反应器，以从采出水中去除极性和非极性有机物。有机改性黏土（有机黏土）也可用作水澄清剂。

从水中分离油类的传统技术都是基于斯托克斯定律的物理方法，该定律预测球体在流体中的沉降速度，其中主要依靠油水密度差、重力、黏度和分散的烃类化合物大小4个参数。使用絮凝剂的常见设备是浮选池。有时使用初级絮凝剂，去除油性絮凝物，然后用二级絮凝剂进行第二次处理。这些技术对溶解态的烃类化合物没有作用。在亚洲的一个大型油田，澄清的水在第二阶段进行生物氧化，以去除氨和有机碳（生物需氧量）。

13.2 絮凝理论

水中的油和颗粒，在电荷排斥等因素下处于稳定状态。疏水性颗粒表面的水分子取向会给颗粒带来阴离子表面电位，与其他类似阴离子表面颗粒相互排斥。这些颗粒的絮凝必须克服电荷排斥作用。随着产水矿化度的增加，电荷排斥力变弱，有效作用范围缩小。原油中的胶质和有机酸等极性分子，以及黏土、水垢、铁锈和其他极性产物，将出现在水包油乳状液的油水界面。这些极性部位会被水化层所包围，防止油滴聚集。这是一种短程效应。

两个油滴或固体颗粒聚并的方法是加入絮凝剂，并进行良好的混合。絮凝剂的作用是中和分散在油和颗粒上的电荷，使其聚集（絮凝），也可以为颗粒搭桥而絮凝。这通常是在如高分子量聚合物或原位生成的聚合物等带电大分子间发生。聚合物一旦桥接了两个或更多的颗粒，电荷就会更加中和，导致聚合物的构象从开放、线型转变为更为盘绕或球状结构。这样聚合物塌陷，将絮凝的颗粒包裹成絮状物。

13.3 絮凝剂的种类及性能

石油工业中使用的絮凝剂种类有高价金属盐、阳离子聚合物、二硫代氨基甲酸盐（DTCs，原位形成的阳离子"聚合物"）、阴离子聚合物、非离子聚合物、两性聚合物。

表面活性剂或其他小分子也作为一些配方的增效剂。但添加表面活性剂时需谨慎，因其很容易稳定乳液，也可能会造成乳液不稳定。水处理行业已经开发了生物絮凝剂，但在石油行业中似乎尚未有应用。

高价金属盐由于具有较高的正电荷密度而具有絮凝或凝聚特性。在油田应用中通常不单独使用，而是添加到聚合物絮凝剂中以提高性能。最常见的是铁（Ⅲ）、锌（Ⅱ）和铝

(Ⅲ)盐。由于水解，其在溶液中呈酸性，会引起腐蚀问题。水解后的产物为金属氢氧化物，会形成额外的淤泥。在一些地区，对某些金属离子的排放水平也有限制。在水中产生H_3O^+的酸，通常比混凝剂盐更有效地打破水包油乳状液，由此产生的腐蚀性酸性废水必须在油水分离后进行中和。有人提出用Fe^{3+}的电凝聚方法，从油田采出水中去除溶解的烃类。

离子聚合物（聚电解质）是石油工业中使用的初级絮凝剂的主体。一家服务公司销售能在原位形成阳离子"聚合物"的二硫代氨基甲酸盐化学品。阳离子聚合物比阴离子聚合物使用得更多，说明分散在水中的油性颗粒通常带负电，需要阳离子聚合物中和。而阴离子絮凝剂会使具有负表面电荷的黏土结块。在初级水包油破乳过程中，通常首先使用阳离子聚合物。如果水相的pH值很低，如在酸压处理之后，或者残留在水包油乳状液中的油包水破乳剂是阳离子，那么分散的油和颗粒表面可能是正电荷，需要使用阴离子聚合物。此外，可能还需要第二种甚至第三种絮凝剂，以使水充分澄清，满足排放水的环境要求。已经研究了如聚环氧乙烷-b-环氧丙烷、聚乙烯醇（PVA）和疏水改性的PVA等非离子聚合物类絮凝剂，但通常其性能比聚电解质更差。

离子聚合物的分子量应相当高（$>1\times10^6$），以有效促进架桥絮凝机制发生。但如果分子量太高，聚合物流动性差，阻碍凝聚。此外，高分子量可能意味着聚合物水溶液的黏度可能难以控制，为此通常将这类水溶液的絮凝剂质量分数限制在5%～10%。

为了降低黏度以达到注入的目的，可以使用反相乳液或乳胶分散体中的聚合物水溶液。但反相乳液或乳胶聚合物会给待处理流体增加更多的油，因为这些聚合物通常包括20%～30%（质量分数）的烃类连续相。另一个缺点是，这些聚合物产品在使用前必须反相，这使得将聚合物送入系统的过程更加复杂。与使用干聚合物相同，油田中通常没有预配反相乳液的设备，此时只能将聚合物直接加入系统。与这种进料方法相关的许多问题，使许多用户避免使用乳胶聚合物。此外，乳胶的处理范围通常很窄，常常导致在较高剂量下的过度处理。已经有很多人尝试在水基组合物中提供水溶性、相对高分子量的聚合物絮凝剂（从而避免溶解粉末或使用油类连续相的缺点），其中所产生的组合物具有可接受的黏度，但比溶解在水中的高分子量聚合物的浓度高得多。这些尝试包括通过改性分散在其中的连续水相或改性聚合物来抑制高分子量聚合物的溶胀或溶解。这种产品通常被称为水包水乳剂，尽管高分子量材料的物理状态不一定是真正的乳液。有专利称，通过将高分子量聚合物分散在含有溶解无机盐的低分子量聚合物溶液中，可以制成高浓度（至少15%）的高分子量聚合物。另一项相关专利使用第二种分散聚合物絮凝剂，制成高浓度溶液。稀释后，絮凝剂将完全溶于水。在多价阴离子盐（如磷酸盐或硫酸盐）溶液中进行单体聚合，可以改善阳离子絮凝剂的可操作性。在聚合开始之前，为提高分散性，可加入一种不溶于多价阴离子盐水溶液的水溶性阳离子聚合物，起到种子聚合物的作用。

离子聚合物的电荷密度也是一个重要因素。过高的电荷密度会使颗粒上的电荷发生逆转，使其重新稳定。因此，大多数离子聚合物絮凝剂是包含离子单体和中性单体的共聚物。两种单体的比例可以变化，在一个结构类别中可以得到几种具有不同电荷密度的潜在絮凝剂。据报道，少量的分支或交联可以提高某些聚合物絮凝剂的性能。水处理行业的其他技术人员报道了乙烯基聚合物型交联水溶性聚合物1与乙烯基聚合物型线型水溶性聚合物2的混合物。其中，聚合物1的电荷包容率为20%或以上，聚合物2的电荷包容率为5%~20%。

13.3.1 絮凝剂的性能测试

测试絮凝剂通常采用摇瓶试验（类似于水包油乳状液的标准瓶试法）。絮凝剂添加到"油性"水样中，并摇动盖紧的瓶子，静置一段时间后，检查水的透明度。通过这种测试方法，确定的絮凝剂浓度通常超过现场实际所需的浓度。测试时，加量一般从10mg/L开始，然后在1~50mg/L范围内增加或减少。与解决油包水乳状液问题的破乳剂相比，絮凝剂的加量要求通常较低。

加入大罐中的絮凝剂通常采用低剪切力的试杯试验进行评价。专门的"试杯试验"仪器或微型池模拟絮凝剂投加的工况，配套温和的搅拌。根据特定的应用情况，可以调整搅拌时长和速度。为了模拟化学剂投加的工况，刚开始是高速搅动，之后变为低速搅拌，以模拟工厂正常工况。

许多系统利用气泡的作用，将油和固体污染物从水相中漂浮出来。这些系统通常有大量的搅拌，并在水相上方产生泡沫或浮沫。气浮测试是用浮选试验池（有时称为"台式Wemco"），以模拟工厂实际情况。

13.3.2 阳离子聚合物

许多类型的阳离子聚合物（聚季铵盐）是有效的絮凝剂。这类聚合物的共同点都是通过加入叠氮原子而实现阳离子化，包括：二烯丙基二甲基氯化铵（DADMAC）聚合物，丙烯酰胺或丙烯酸酯类阳离子聚合物，聚亚烷基亚胺，聚烷醇胺，聚乙烯氯化铵，聚烯丙基氯化铵，支链型聚乙烯咪唑啉酸盐，阳离子多糖和壳聚糖，凝结的单宁酸，其中前两类最常用。

13.3.2.1 二烯丙基二甲基氯化铵聚合物

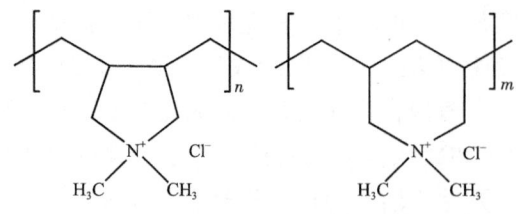

图13.1 DADMAC聚合物（其中以五环吡啶单体为主要成分，六环哌啶单体为次要成分）

均聚物聚二烯丙基二甲基氯化铵（聚DADMAC）等基本的DADMAC聚合物，长期以来一直被认为是性能良好的絮凝剂（图13.1）。通常使用分子量大于1×10^6的高分子聚合物，主要是五元环阳离子基团，但也有些六元环。仲胺和叔胺聚二烯丙基铵盐显示出高的抗菌活性。

已经提出了如水溶液、分散体或乳液等制备聚合物的方法。对聚DADMAC均聚物的一种改进是在配方中加入质量分数为5%~10%的残余DADMAC单体作为增强剂。

DADMAC 单体可以与丙烯酰胺或其他廉价的非离子亲水单体共聚，改变聚合物中的电荷密度。一些聚合物上少量的支链可以增加絮凝性能。典型的交联剂是乙烯基三甲氧基硅烷和亚甲基双丙烯酰胺。疏水改性的 DADMAC 聚合物具有较好的絮凝性能。首选的疏水单体是乙基己基丙烯酸酯，疏水基团被吸引到疏水油滴，可以增强其絮凝性能。氯酸铝和聚胺（如聚 DADMAC）的组合也可以提供优良的絮凝剂性能。

13.3.2.2　丙烯酰胺或丙烯酸酯类阳离子聚合物

曼尼希丙烯酰胺聚合物是众所周知的阳离子絮凝剂（图 13.2）。一般来说，这些聚合物是丙烯酰胺的均聚物或其与丙烯腈、甲基丙烯酰胺或丙烯酸等共聚单体的共聚物，其含量最高约为合成共聚物的 50%。这些聚合物的分子量为 $1\times10^4 \sim 3\times10^6$，通过与甲醛和二甲胺（或其他次级 C_1—C_8 烷基胺）的曼尼希反应进行化学改性，以提供二甲胺甲基。这些基团通过季铵化改性，如用硫酸二甲酯，制成水性阳离子聚合物。以反相微乳剂的形式，曼尼希丙烯酰胺聚合物可以得到更好的性能。这种方法可以制备出高固含量、低黏度的聚合物。

各种丙烯酰胺或丙烯酸酯类阳离子单体用于合成阳离子聚合物絮凝剂（图 13.3）。最常见的阳离子单体是甲基丙烯酸二甲胺基乙酯和甲基丙烯酰胺丙基三甲基氯化铵的季铵盐。另一种阳离子聚合物是基于 1-丙烯酰-4-甲基哌嗪的季铵盐。与 DADMAC 聚合物一样，使用交联剂获得的少量支链可以提高絮凝剂性能。

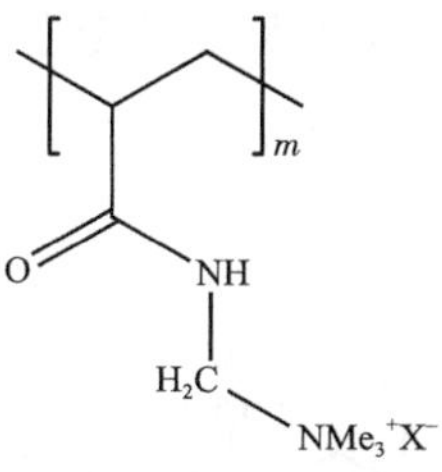

图 13.2　曼尼希丙烯酰胺聚合物的活性单体结构

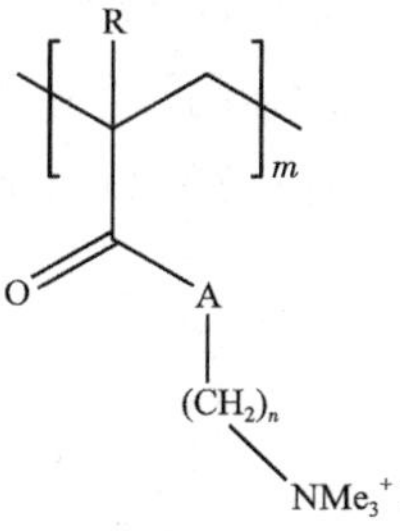

图 13.3　阳离子絮凝剂中丙烯酰胺或丙烯酸酯基单体（其中 A=NH 或 O，R=H、CH_3，$n=2\sim3$）

阳离子密度和作为絮凝剂的有效性可以通过与中性亲水单体（如丙烯酰胺）的共聚来改变。例如，物质的量比为 20∶80 的丙烯酰氧乙基三甲基氯化铵和丙烯酰胺的共聚物，其分子量高于约 2×10^6。其他关于该单体的使用见相关文献。20%~30% 的低分子量聚合物与此类高分子量聚合物混合使用，絮凝效果更佳。

在这些高分子量的情况下，浓缩聚合物溶液的黏度非常高。如前所述，反相乳液或乳胶分散体可用于克服高黏度。一种降低黏度的新方法是在水外相乳状液中，聚合如二甲基氨基乙基甲基丙烯酸酯和丙烯酸乙酯等中性疏水性单体。加入盐溶液后，聚合物变成阳离子和水溶性。丙烯酰胺的亲水阳离子共聚物在盐介质中的分散也被认为是处理这些高黏度聚合物溶液的更简单方法。用于制备阳离子聚合物的二季铵丙烯酸单体可由乙烯基叔胺

（如二甲胺丙基甲基丙烯酰胺），通过与（3-氯-2-羟基丙基）三烷基氯化铵反应制备。

与均聚物相比，含有一定比例疏水单体的阳离子聚合物具有更多的絮凝性。例如，DADMAC 或丙烯酸酯阳离子聚合物，与乙烯基三甲氧基硅烷或二甲基氨基乙基丙烯酸酯苄基氯化物季盐的共聚物。有专利将亲油性烷基丙烯酸酯共聚单体与阳离子丙烯酸酯或丙烯酸单体结合使用。由丙烯酰胺单体、水溶性阳离子单体和不溶于水的疏水性单体［如烷基（甲基）丙烯酰胺或烷基（甲基）丙烯酸酯］聚合而成的水分散型三元共聚物是一种优异的絮凝剂。作者描述了通过引入疏水单体来提高阳离子聚合物性能的一种可能机制。因此，传统聚合物可以通过库仑引力、氢键及其他未定义或不清楚的机制而附着在油滴上，但这些新型三元共聚物的疏水基团也可以通过疏水基团—疏水油滴的关联而附着。此外，也可能是不同聚合物分子上的疏水基团相互作用，形成桥梁或网络，从而有助于絮凝物的形成和油的浮选。虽然库仑引力似乎仍然是最强的吸引力类型，但疏水缔合或疏水效应似乎对这种吸引力有明显的加强作用，这一点从改善破乳和废水清洁中得到证明。

13.3.2.3 其他阳离子聚合物

二乙醇胺、三乙醇胺和异丙醇胺是廉价的起始材料，可以缩合成带支链的聚醇胺，起到水包油破乳剂的作用（图 13.4）。三乙醇胺聚合物含有周边的—OH 基团，以及被取代的二氧杂环己烷基团。这种聚合物也可以被季铵化，可以添加尿素以赋予分支或交联。

另一类阳离子聚合物絮凝剂是基于 1,2-二氯乙烷与小分子多胺（如 1,2-二氨基乙烷或二乙烯三胺）的反应（图 13.5）。这些聚合物单独使用需要很高的浓度才能起效。这些聚合物和上述乙醇胺聚合物的阳离子电荷是在主链上，而不是在侧链上，这种情况不常见。这些聚合物在与聚 DADMAC 混合时表现良好。

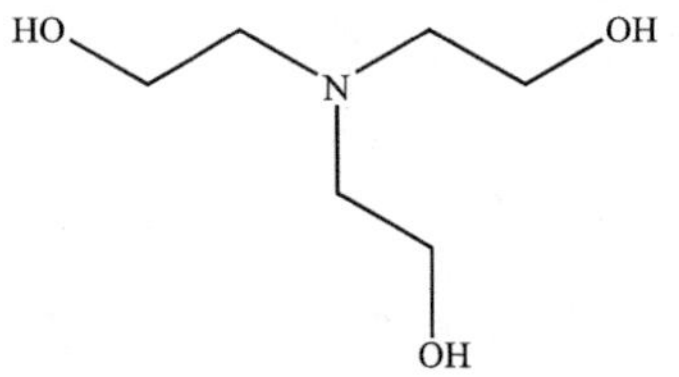

图 13.4　三乙醇胺（含这类分子的聚合物有—OH，以及取代的二氧己环基团）

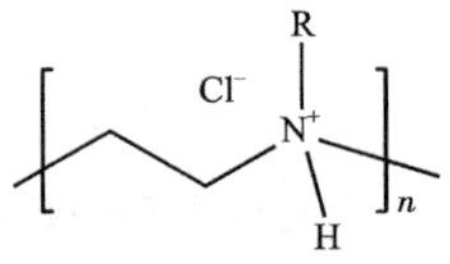

图 13.5　1,2-二氯乙烷与 1,2-二氨基乙烷的聚合反应产物

通过环氧氯丙烷和二甲胺或更大分子的多胺反应，可以得到类似的结构，其主干是季氮（图 13.6）。季铵化的聚亚烷基多胺（如 2-羟基-3-氯丙基三甲基氯化铵和聚乙烯多胺的加合物），已作为絮凝剂。这些聚亚烷基多胺中，优选更高分子量的聚乙烯多胺和聚丙烯多胺（如数均分子量为 100~15000）。特别值得关注的是用 1,2-二氯乙烷或类似物质交联的聚亚烷基多胺，以及这种交联的多胺与其他聚亚烷基多胺的混合物。

聚合或缩合的单宁也可作为絮凝剂。其中一类是由单宁、氨基化合物（单乙醇胺）和甲醛制成。单宁主要含有通过糖苷键与葡萄糖相连接的没食子酸残基。缩合单宁还能与阳离子单体（如丙烯酸二甲氨基乙酯的甲基氯化物季铵盐）反应。

聚乙烯胺盐也是一种阳离子聚合物型絮凝剂。市售的聚乙烯胺通过聚乙烯甲酰胺水解生成。聚烯丙基氯化铵也被认为是一种阳离子絮凝剂。已经开发出了相应的聚脒基絮凝剂产品（图 13.7）。

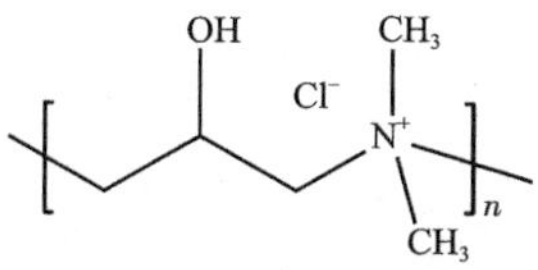

图 13.6　环氧氯丙烷和二甲胺的聚合反应产物

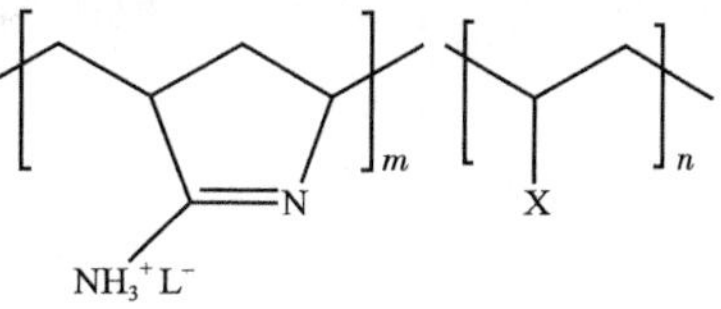

图 13.7　聚脒基絮凝剂

还有研究者研究了其他阳离子聚合物型絮凝剂，包括季铵化聚合吡啶和喹啉（也可用作缓蚀剂和杀菌剂）、N-二烯丙基-3-羟基氮杂环丁烷盐聚合物、聚环氧乙烷的噻嗪类季铵盐（也可用作杀菌剂或油包水破乳剂），以及支链聚乙烯咪唑啉酸盐（已验证絮凝效果优于线型聚合物）。枝状阳离子聚合物（如枝状大分子、超支化聚合物），也可用作絮凝剂，具体包括枝状多胺、枝状聚酰氨基胺和超支化聚乙烯亚胺，以及其与葡萄糖酸内酯、亚烷基氧化物、3-氯-2-羟基丙磺酸盐、烷基卤化物、苄基卤化物和二烷基硫酸盐的反应产物。有研究报道了 3 个系列的分子量分布集中、骨架上带有疏水取代基的阳离子聚合物［聚（N-乙烯基苄基-N, N, N-三甲基氯化铵）、聚（N-乙烯基苄基-N, N-二甲基-N-丁基氯化铵）和聚（N-乙烯基苄基氯化吡啶）］的絮凝行为。絮凝过程中，基材具有低电荷密度时，疏水相互作用就会更为重要。

13.3.2.4　环境友好型阳离子聚合物絮凝剂

海水环境中有关阳离子聚合物絮凝剂的数据非常少（DTCs 将在后面讨论）。由于其分子量和水溶性特点，预计毒性和生物积累量都较低。但配方中可能残留的单体和低聚物会增加毒性。例如，残留的阳离子单体有生物急毒性，甲醛和丙烯酰胺单体具有致癌性（可能出现在曼尼希聚合物产品中）。DADMAC 和乙烯基聚合物的生物降解性很差，这是含全碳主链的聚合物的典型特征。由哌嗪衍生物和胺制成的聚酰胺絮凝剂，因为有酰胺键可能更容易生物降解，但这一点未见报道。例如，N, N'-双（甲氧基羰基乙基）-哌嗪和 N, N-双（3-氨基丙基）-甲胺或二乙烯三胺或更高级聚亚烷基胺的反应产物。与 DADMAC 和某些阳离子乙烯基单体相比，哌嗪衍生物的较高成本可能会限制其应用。将四乙烯戊胺添加到乙二醇二丙烯酸酯中，通过迈克尔加成反应合成了一种水溶、可降解的新型聚（β-氨基酯）阳离子聚合物。

聚亚烷基二醇接枝水溶性乙烯基不饱和单体的接枝聚合物也是一种絮凝剂。带有聚乙二醇主链的聚合物，其生物降解性取决于接枝程度。例如，聚乙二醇接枝阳离子单体 2-（丙烯酰氧基）乙基三甲基氯化铵的聚合物。这类接枝聚合物的平均分子量应在 1×10^5 以上。

还有一类生物降解性良好的阳离子聚合物是阳离子多糖，其还没有作为石油工业絮凝剂的应用报道，但广泛应用于纸浆和造纸工业的废水处理。淀粉、糖原、葡甘露聚糖和黄胞胶等多糖可以通过季铵盐分子的羟基衍生，如 N-（3-氯-2-羟基丙基）三甲基氯化铵，或通过环氧等效等进行衍生，生成阳离子多糖（图13.8）。多糖也可以接枝丙烯酰胺或其他乙烯基阳离子单体，生成阳离子多糖。

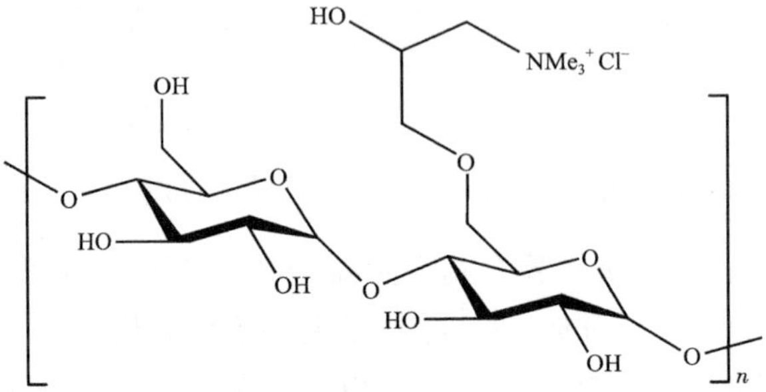

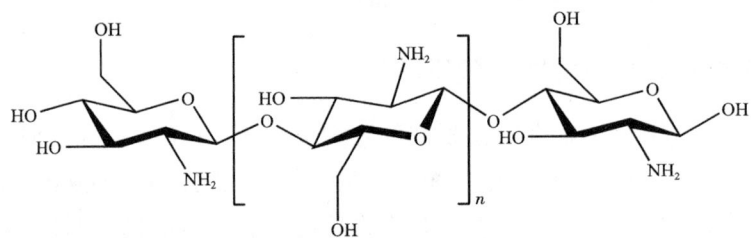

图 13.8　阳离子淀粉结构

壳聚糖有侧伯氨基，由于其中氮的季铵化，而在酸化时起到阳离子聚合物作用（图13.9）。通过 N-（3-氯-2-羟基丙基）三甲基氯化铵，将阳离子基团接入壳聚糖。有报道称疏水改性的壳聚糖衍生物可作为水包油乳状液的改进型絮凝剂，但其成本高，难以广泛应用。甲壳类动物（螃蟹、虾等）外骨骼中获得的市售甲壳素，通过去乙酰化可得到壳聚糖。还有报道将壳聚糖接枝聚丙烯酰胺的共聚物，作为絮凝剂使用。

图 13.9　壳聚糖结构（完全脱乙酰化）

13.3.2.5　二硫代氨基甲酸盐类（DTCs）

比高分子量阳离子聚合物更有优势的一类絮凝剂是二硫代氨基甲酸盐类，是由聚合/低聚的伯胺或仲胺与 CS_2、碱的水或醇溶液反应得到。DTCs生物降解性很好，据称比阳离子聚合物的急毒性低，有良好环境特性。由于其是亲油性，不会出现在排放水中。DTCs是低分子量的水溶性阴离子聚合物，通过与 Fe^{2+} 络合，原位形成高分子量的阳离子型、亲油性聚合物。油井产出水中通常存在足够浓度的 Fe^{2+}，或者可以人工添加。可以添加一种终止剂，以增加聚合物基体水溶性。终止剂选自含有磺酸盐或硫酸盐基团的非乳化助水溶剂。图13.10显示了DTCs简单结构式，以说明聚合物如何与 Fe^{2+} 络合而变成阳离子结构。由于此类聚合物的分子量低，故加注使用时可以配制成相当高的浓度，并且不用担心过于黏稠。

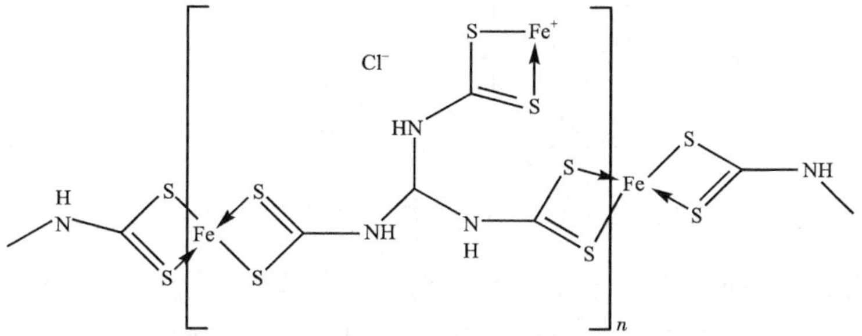

图 13.10 DTCs 简单结构（与 Fe^{2+} 交联生成阳离子）

多年来，DTCs 絮凝剂技术在持续改进。DTCs 可以由任何合适的胺制成，包括但不限于双（六亚甲基）三胺、六亚甲基二胺五亚甲基六胺、聚氧乙烯三胺、氨基乙基乙醇胺，以及主要由三亚甲基四胺和氨基乙基哌嗪组成的混合物。环氧改性的 DTCs 絮凝剂也被认为是对早期 DTCs 絮凝剂的改进。在该专利中，与 CS_2 反应所需的多胺是由双酚 A/环氧氯丙烷基环氧树脂与乙二胺反应而成。多胺反应产物的结构如图 13.11 所示。由这些或其他多胺制成的二硫代氨基甲酸盐也可用作防垢剂、缓蚀剂或杀菌剂。

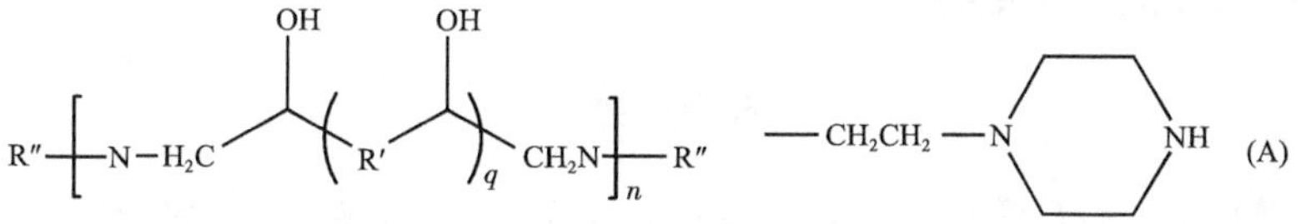

图 13.11 制备 DTCs 絮凝剂的环氧改性多胺前体［R″ 从结构—R—NH_2 和（A）组成的基团中选择，其中 R 为烃基，R′ 为—$(CH_2)_m$—O—R—O—$(CH_2)_m$—，其中 n 和 m 分别从 1～5，q 为 0 或 1］

已经发现诸如胺、醇、氨基醇、醚及其混合物（包括其卤化加合物）等材料是水澄清剂中有用的絮状物改性剂，以形成有效的整体水澄清剂成分。改进后的絮凝物容易脱脂且不会在系统中堆积——基本上更容易处理，如乙醇胺。

13.3.3 阴离子聚合物

如果颗粒和油滴带有正电荷，则需要阴离子聚合物絮凝剂。这可能是在酸化操作之后，或在用初级阳离子絮凝剂处理之后，以促进液滴之间的聚合物架桥并加速絮凝物的形成。市场上的阴离子聚合物主要是高分子量丙烯酸聚合物的盐类（图 13.12）。其通常是用碱中和的丙烯酸与丙烯酰胺的共聚物，也可以使用部分水解的聚丙烯酰胺（PAM）。如果需要纯净的非离子聚合物，可以使用未水解的 PAM。据称，丙烯酰胺共聚物与离子型亲水表面活性剂结合使用，对水包油乳状液具有优异的絮凝剂特性。具有良好环境特性的阴离子和中性聚丙烯酰胺可用于井口清理作业的絮凝剂。聚丙烯酰胺可以用亚磷酸衍生得到含膦聚合型絮凝剂。当完全去质子化时，膦酰侧基被转化为膦酸侧基，每个基团有两个负电位，即每个侧基的阴离子电荷数是羧酸化絮凝剂的两倍。此类聚合物也可作为防垢剂。

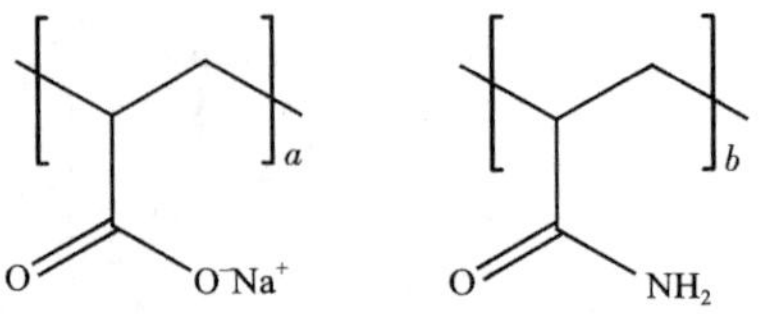

图 13.12 丙烯酸钠－丙烯酰胺共聚物结构（两个单体基团通常见于市售阴离子絮凝剂中）

聚 γ-谷氨酸已被研究为可生物降解的阴离子絮凝剂，但没有关于其在石油工业中使用的报道。与聚天冬氨酸不同，聚 γ-谷氨酸可以通过生物合成，制成分子量可高达 2×10^6。磷酸改性的蒟蒻（葡甘露聚糖，一种多糖）是可生物降解的阴离子絮凝剂。其他潜在的可生物降解的阴离子絮凝剂是部分水解海藻酸钠接枝的 PAM 和水解 PAM 接枝的黄胞胶。其他的接枝多糖也可以使用。

13.3.4 两性聚合物

尽管市面上有若干种两性聚合物絮凝剂，但其性能少有报道。曾经有两性淀粉－接枝－聚丙烯酰胺（S-g-PAM）的报道。与阳离子、水解和两性 PAM 处理油田废水的效果相比，S-g-PAM 的效果更好。

参 考 文 献

[1] E. Garland, "Discharge of Produced Water in the North Sea: Where We Are, Where We Go," SPE 97048（paper presented at the Offshore Europe, Aberdeen, UK, 6–9 September 2005）.

[2] B. Holland, T. Cooksley, and C. A. Malbrel, "Polymeric Flocculants Prove Most Effective Chemicals for Wellbore Cleanup Operations," SPE 27408（paper presented at the SPE Formation Damage Control Symposium, Lafayette, LA, 7–10 February 1994）.

[3]（a）J. Robinson and J. Veil, "An Overview of Offshore and Other Onshore Produced Water Treatment Technologies"（paper presented at the 14th Annual International Petroleum Environmental Conference, Houston, TX, 8 November 2007）.（b）M. J. Plebon, X. Chen, and M. A. Saad, "Adapting a De-Oiling Solution to the Unique Characteristics of Produced Water"（paper presented at the 14th Annual International Petroleum Environmental Conference, Houston, TX, 8 November 2007）.（c）M. Davies and P. J. B. Scott, *Oilfield Water Technology*, Houston: National Association of Corrosion Engineers（NACE）, 2006.

[4] S. E. Oseroed, International Patent Application WO/2002/041965.

[5]（a）D. Th. Meijer and C. A. T. Kuijvenhoven, "Field-Proven Removal of Dissolved Hydrocarbons from Offshore Produced Water by the Macro Porous Polymer-Extraction Technology," OTC 13217（paper presented at the 2001 Offshore Technology Conference, Houston, TX, 30 April–3 May 2001）.（b）D. Meijer, "The Removal of Toxic Dissolved and Dispersed Hydrocarbons from Oil and Gas Produced Water with the Macro Porous Polymer Extraction Technology"（paper presented at the Offshore Mediterranean Conference and Exhibition, Ravenna, Italy, 28–30 March 2007）.

[6]（a）Z. I. Khatib, M. S. Faucher, and E. L. Sellman, "Field Evaluation of Disc-Stack Centrifuges for Separating Oil/Water Emulsions on Offshore Platforms," SPE 30674（paper presented at the SPE Annual Technical Conference and Exhibition, Dallas, TX, 22–25 October 1995）.（b）S. E. Rye, "A New

Method for Removal of Oil in Produced Water," SPE 26775 (paper presented at the Offshore Europe, Aberdeen, UK, 7-10 September 1993).

[7] (a) J. C. Ditria and M. E. Hoyack, "The Separation of Solids and Liquids with Hydrocyclone-Based Technology for Water Treatment and Crude Processing," SPE 28815 (paper presented at the SPE Asia Pacific Oil and Gas Conference, Melbourne, Australia, 7-10 November 1994). (b) L. Nnabuihe, "Novel Compact Oil/Water Separator Tested in Nimr," SPE 68150 (paper presented at the SPE Middle East Oil Show, Bahrain, 17-20 March 2001). (c) A. Sinker, "Produced Water Treatment Using Hydrocyclones: Theory and Practical Application" (paper presented at the 14th Annual International Petroleum Environmental Conference, Houston, TX, 8 November 2007).

[8] I. B. Henriksen, International Patent Application WO/2005/123213.

[9] H. Goksoeyr and N. Henriksen, International Patent Application WO/2004/069753.

[10] (a) T. Husveg, International Patent Application WO/2007/024138. (b) B. L. Knudsen, M. Hjelsvold, T. K. Frost, M. B. E. Svarstad, P. G. Grini, C. F. Willumsen, and H. Torvik, "Meeting the Zero Discharge Challenge for Produced Water," SPE 86671 (paper presented at the SPE International Conference on Health, Safety, and Environment in Oil and Gas Exploration and Production, Calgary, Alberta, Canada, 29-31 March 2004). (c) T. Arato, H. Iizuka, A. Mochizuki, T. Suzuki, H. Honji, S. Komatsu, H. Isogami, and H. Sasaki, U.S. Patent Application 20080023401.

[11] S. Kwon, E. J. Sullivan, L. Katz, K. Kinney, and R. Bowman, "Pilot Scale Test of a Produced Water-Treatment System for Initial Removal of Organic Compounds," SPE 116209 (paper presented at the SPE Annual Technical Conference and Exhibition, Denver, CO, 21-24 September 2008).

[12] G. Alther and T. Wilkinson, "Organoclays Can Cut the Cost of Cleanup of Produced Water, Wastewater and Groundwater by 50%" (paper presented at the 13th Annual International Petroleum Environmental Conference, San Antonio, TX, 17-20 October 2006).

[13] E. S. Madian and R. J. Jan, "Treating of Produced Water at the Giant Arun Field," SPE 27130 (paper presented at the SPE Health, Safety and Environment in Oil and Gas Exploration and Production Conference, Jakarta, Indonesia, 25-27 January 1994).

[14] P. R. Hart, "The Development and Application of Dithiocarbamate (DTC) Chemistries for Use as Flocculants by North Sea Operators," *Proceedings of the Chemistry in the Oil Industry VII Symposium*, Royal Society of Chemistry, Manchester, UK, 2002, p.149.

[15] R. S. Fernandes, G. Gonzalez, and E. F. Lucas, *Colloid and Polymer Science* 283 (2005): 375.

[16] A. L. Feder, G. Gonzalez, C. L. S. Teixeira, and E. F. Lucas, *Journal of Applied Polymer Science* 94 (4) (2004): 1473.

[17] M. S. Ghafoor, M. Skinner, and I. M. Johnson, U.S. Patent 6001920, 1999.

[18] J. W. Sparapany and J. R. Hurlock, U.S. Patent 5938937, 1999.

[19] J. R. Hurlock, U.S. Patent 6025426, 2000.

[20] N. E. Byrne, R. A. Marble, and M. Ramesh, U.S. Patent 5330650, 1994.

[21] T. Higashino and S. Shimosato, U.S. Patent 6890996, 2005.

[22] W. E. Hunter and T. P. Sieder, U.S. Patent 4151202, 1979.

[23] M. Hofinger, M. Hille, and R. Bohm, U.S. Patent 4686066, 1987.

[24] A. Sivakumar and P. G. Murray, U.S. Patent 6036868, 2000.

[25] W. L. Whipple, C. Maltesh, C. C. Johnson, A. Sivakumar, T. M. Guddendorf, and A. P. Zagala, U.S. Patent 6753388, 2004.

[26] M. Ramesh and A. Sivakumar, U.S. Patent 5635112, 1997.

[27] P. R. Hart, U.S. Patent 5607574, 1997.

[28] A. T. Coscia and M. N. D. O'Connor, U.S. Patent 4137164, 1979.

[29] T. V. Vyshkina, U.S. Patent 5744563, 1998.

[30] J. J. Kozakiewicz and S.-Y. Huang, U.S. Patent 5723548, 1998.

[31] M. S. Raman, U.S. Patent 4160742, 1979.

[32] P. Flesher, D. Farrar, M. Hawe, and J. Langley, U.S. Patent 4702844, 1987.

[33] D. W. Fong and A. M. Halverson, U.S. Patent 4802992, 1989.

[34] R. S. Buriks, A. R. Fauke, and D. W. Griffiths, U.S. Patent 4224150, 1980.

[35] P. R. Hart, J. M. Brown, and E. J. Connors, U.S. Patent 5730905, 1995.

[36] M. L. Braden and S. J. Allenson, U.S. Patent 4931191, 1990.

[37] (a) J. R. Hurlock, U.S. Patent 6025426, 2000. (b) S.-Y. Huang, L. Rosati, and J. J. Kozakiewicz, U.S. Patent 6702946, 2004.

[38] L. Z. Liu, J. D. Kiplinger, and D. Radtke, International Patent Application WO/2008/118315.

[39] A. Sivakumar and M. Ramesh, U.S. Patent 5560832, 1996.

[40] P. R. Hart, F. Chen, W. P. Liao, and W. J. Burgess, U.S. Patent 5921912, 1999.

[41] J. Bock, P. L. Valint, T. J. Pacansky, and H. W. H. Yang, U.S. Patent 5362827, 1994.

[42] T. J. Bellos, U.S. Patent 4459220, 1984.

[43] R. Fikentscher, K. Oppenlaender, J. P. Dix, W. Sager, H.-H. Vogel, and G. Elfers, U.S. Patent 5234626, 1993.

[44] C. W. Burkhardt, U.S. Patent 4411814, 1983.

[45] B. Lehmann and U. Litzinger, U.S. Patent 5707531, 1998.

[46] B. S. Fee, U.S. Patent 4387028, 1983.

[47] J. E. Quamme and A. H. Kemp, U.S. Patent 4558080, 1985.

[48] P. R. Hart, J.-C. Chen, F. Chen, and T. H. Duong, U.S. Patent 5851433, 1998.

[49] P. L. Dubin, U.S. Patent 4217214, 1980.

[50] A. G. Sommese and A. Sivakumar, U.S. Patent 5702613, 1997.

[51] D. N. Roark, U.S. Patent 4614593, 1986.

[52] P. M. Quinlan, U.S. Patent 4339347, 1982.

[53] R. S. Buriks and E. G. Lovett, U.S. Patent 4383926, 1983.

[54] P. M. Quinlan, U.S. Patent 4331554, 1982.

[55] M. B. Manek, M. J. Howdeshell, K. E. Wells, H. A. Clever, and W. K. Stephenson, U.S. Patent Application 20060289359, 2006.

[56] S. Schwarz, W. Jaeger, B.-R. Paulke, S. Bratskaya, N. Smolka, and J. Bohrisch, *Journal of Physical Chemistry B* 111 (29) (2007): 8649.

[57] J. M. Rice, "The Carcinogenicity of Acrylamide," *Mutation Research/Genetic Toxicology and Environmental Mutagenesis* 580 (2005): 3-20.

[58] U.-W. Hendricks, B. Lehmann, and U. Litzinger, European Patent Application EP691150, 1996.

[59] M. Singh, B. Dymond, A. Hooley, and K. Symes, International Patent Application WO/2006/050811.

[60] S. Pal, D. Mal, and R. P. Singh, *Colloids and Surfaces A: Physicochemical and Engineering Aspects* 289 (2006): 193.

[61] S. Bratskaya, S. Schwarz, T. Liebert, and T. Heinze, "Starch Derivatives of High Degree of Functionalization: 10. Flocculation of Kaolin Dispersions," *Colloids and Surfaces A: Physicochemical and Engineering Aspects* 254 (2005): 75-80.

[62] S. Pal, D. Mal, and R. P. Singh, "Cationic Starch: An Effective Flocculating Agent," *Carbohydrate Polymers* 59 (2005): 417.

[63] L. Järnström, L. Lason, and M. Rigdahl, "Flocculation in Kaolin Suspensions Induced by Modified Starches 1. Cationically Modified Starch—Effects of Temperature and Ionic Strength," *Colloids and Surfaces A: Physicochemical and Engineering Aspects* 104 (1995): 191.

[64] E. Gunn, A. Gabbianelli, R. Crooks, and K. Shanmuganandamurthy, International Patent Application WO/2006/055877.

[65] D. W. Fong and A. M. Halverson, U.S. Patent 4568721, 1986.

[66] A. Pinotti, A. Bevilacqua, and N. Zaritzky, *Journal of Surfactants and Detergents* 4 (2001): 57.

[67] S. Bratskaya, V. Avramenko, S. Schwarz, and I. Philippova, *Colloids and Surfaces A: Physicochemical and Engineering Aspects* 275 (2006): 168.

[68] N. E. S. Thompson and R. G. Asperger, U.S. Patent 4689177, 1987.

[69] T. J. Bellos, U.S. Patent 6019912, 2000.

[70] T. J. Bellos, U.S. Patent 6130258, 2000.

[71] D. K. Durham, "Advances in Water Clarifier Chemistry for Treatment of Produced Water on Gulf of Mexico and North Sea Offshore Production Facilities," SPE 26008 (paper presented at the SPE/EPA Exploration and Production Environmental Conference, San Antonio, TX, 7–10 March 1993).

[72] D. K. Durham, U. C. Conkle, and H. H. Downs, U.S. Patent 5006274, 1991.

[73] G. T. Rivers, U.S. Patent 5247087, 1993.

[74] N. E. S. Thompson and R. G. Asperger, U.S. Patent 5089619, 1992.

[75] E. J. Evain, H. H. Downs, and D. K. Durham, U.S. Patent 5302296, 1994.

[76] P. R. Hart, J. M. Brown, and E. J. Connors, Canadian Patent CA2156444, 1996.

[77] M. N. M. Yunus, A. D. Procyk, C. A. Malbrel, and K. L. C. Ling, "Environmental Impact of a Flocculant Used to Enhance Solids Transport during Well Bore Clean-Up Operations," SPR 29736 (paper presented at the SPE/EPA Exploration and Production Environmental Conference, Houston, TX, 27–29 March 1995).

[78] L. E. Nagan, U.S. Patent 5393436, 1995.

[79] H. Yokoi, T. Arima, J. Hirose, S. Hayashi, and Y. Takasaki, *Journal of Fermentation and Bioengineering* 82 (1996): 84.

[80] I.-L. Shih and Y.-T. Van, "The Production of Poly-(Gamma-Glutamic Acid) from Microorganisms and Its Various Applications," *Bioresource Technology* 79 (2001): 207–225.

[81] C. Xie, Y. Feng, W. Cao, Y. Xia, and Z. Lu, "Novel Biodegradable Flocculating Agents Prepared by Phosphate Modification of Konjac," *Carbohydrate Polymers* 67 (2007): 566.

[82] T. Tripathy and R. P. Singh, *European Polymer Journal* 36 (2000): 1471.

[83] P. Adhikary and R. P. Singh, *Journal of Applied Polymer Science* 94 (2004): 1411.

[84] J. M. Walsh and W. J. Georgie, "Produced Water Treating Systems—Comparison between North Sea and Deepwater Gulf of Mexico," SPE 159713 (paper presented at the SPE Annual Technical Conference and Exhibition, San Antonio, TX, 8–10 October 2012).

[85] K. Johansen, personal communication.

[86] W. Lei, D. Portehault, D. Liu, S. Qin, and Y. Chen, *Nature Communications* 4 (2013): 1777.

[87] L. Moore, C. Cardoso, M. Costa, and A. Mahmoudkhani, "Removal of Total Organics and Grease from Oil Production Effluents by an Adsorption Process," SPE 163272 (paper presented at the SPE Kuwait International Petroleum Conference and Exhibition, Kuwait City, Kuwait, 10–12 December 2012).

[88] I. B. Henriksen, K. Voldum, and E. Garpestad, "The Ctour Process, an Option to Comply with The 'Zero Discharge-Legislation' in Norwegian Waters," Paper 118012 (paper presented at the Abu Dhabi International Petroleum Exhibition and Conference, Abu Dhabi, UAE, 3–6 November 2008).

[89] Y.-S. Wong, S.-A. Ong, T.-T. Teng, L. N. Aminah, and K. Kumaran, *Water Air and Soil Pollution* 223 (7) (2012): 3775.

[90] M. Fujita, M. Ike, S. Tachibana, G. Kitada, S. M. Kim, and Z. Inoue, *Journal of Bioscience and Bioengineering* 89 (1) (2000): 40–46.

[91] W. X. Gong, S. G. Wang, X. F. Sun, X. W. Liu, Q. Y. Yue, and B. Y. Gao, *Bioresource Technology* 99 (11) (2008): 4668–4674.

[92] N. He, Y. Li, and J. Chen, *Bioresource Technology* 94 (2004): 99–105.

[93] J. Younker, S. Y. Lee, G. A. Gagnon, and M. E. Walsh, "Removal of Dissolved Hydrocarbons from Oilfield Produced Water by Chemical Coagulation and Electro-Coagulation," OTC 22003 (paper presented at the Offshore Technology Conference, Houston, TX, 2–5 May 2011).

[94] R. Yonemoto and S. Wakatuki, U.S. Patent Application 20090137720.

[95] Z. Amjad, ed., *The Science & Technology of Industrial Water Treatment*. Boca Raton, FL: CRC Press, 2010.

[96] X. Jia and Y. Zhang, *Journal of Applied Polymer Science* 118 (2010): 1152–1159.

[97] L. M. Timofeeva, N. A. Kleshcheva, A. F. Moroz, and L. V. Didenko, *Biomacromolecules* 10 (2009): 2976.

[98] L.-J. Wang, J.-P. Wang, S.-J. Zhang, Y.-Z. Chen, S.-J. Yuan, G.-P. Sheng, and H.-Q. Yu, *Separation and Purification Technology* 67 (2009): 331–335.

[99] G. González, J. C. de la Cal, and J. M. Asua, *Colloids and Surfaces A: Physicochemical and Engineering Aspects* 385 (2011): 166–170.

[100] Available at http://www.mrc.co.jp/english/corporate/envsafe/pdf/10_e_all.pdf.

[101] C.-B. Wu, J.-Y. Hao, and X.-M. Deng, *Polymer* 48 (2007): 6272.

[102] S. Krishnamoorthi, P. Adhikary, D. Mal, and R. P. Singh, *Journal of Applied Polymer Science* 118 (2010): 3593.

[103] S. Pal, G. Sen, S. Ghosh, and R. P. Singh, *Carbohydrate Polymers* 87 (2012): 336–342.

[104] S. Akbar Ali, S. Pal, and R. P. Singh, *Journal of Applied Polymer Science* 118 (2010): 2592.

[105] W. Zhang, Y. Shang, B. Yuan, Y. Jiang, Y. Lu, Z. Qin, A. Chen, X. Qian, H. Yang, and R. Cheng, *Journal of Applied Polymer Science* 117 (2010): 2016.

[106] Y. Lu, Y. Shang, X. Huang, A. Chen, Z. Yang, Y. Jiang, J. Cai, W. Gu, X. Qian, H. Yang, and R. Cheng, *Industrial & Engineering Chemistry Research* 50 (2011): 7141.

[107] H. Song, "Preparation of Novel Amphoteric Flocculant and Its Application in Oilfield Water Treatment," SPE 140965 (paper presented at the SPE International Symposium on Oilfield Chemistry, The Woodlands, TX, 11–13 April 2011).

14 杀 菌 剂

14.1 概述

在石油和天然气领域广泛应用的杀菌剂，也称抑菌剂或抗菌剂。油气生产中加注杀菌剂的目的是杀死微生物（尤其是细菌）或干扰其生物活动，减轻微生物腐蚀（MIC）及相关的生物污染，减少硫化物的产生。油田地层或注入水中的微生物通常按其作用类别进行分类。在油田待处理水体中可能存在的细菌有硫酸盐还原菌（SRB）、硝酸盐还原菌（hNRB）、黏液形成菌、铁氧化细菌及其他微生物，如藻类、硝酸盐还原硫化物氧化细菌（NR-SOB）、酵母菌、霉菌及原生动物。甚至有报道称在中东某油田发现了可形成碳酸盐垢的细菌。细菌可在溶液中以分散（浮游）的菌落形式或固定的沉积物（固着细菌及其废物）形式存在于水体中，利用多种氮、磷和碳化合物（如有机酸）来维持其生长。细菌通常可依靠地层水中的氮和磷维持生长，但外加有机含氮、含磷化合物会有效激发其生长潜力。

根据对氧气的需求情况，可以将细菌分为好氧菌和厌氧菌，如油田生产运行中所有水体都存在的厌氧菌 SRB（为脱硫弧菌属）。SRB 可将硫酸盐转化为酸性的 H_2S，从而导致储层酸化并生成硫化物垢。细菌菌落的固体沉积物又被称为生物膜或生物污垢。MIC 的重要指征是管路中的硫化亚铁或水体中水溶性硫化物浓度增加。因此控制 MIC 时，首要是防止在管道和容器表面形成生物膜，同时要对浮游和固着细菌采取有效的处理措施。油田回注水可能是产出水和海水的混合物，这时 SRB 的生物活性可能更强。具有适宜 SRB 生长的条件包括硫酸根离子、有机碳、氨氮化合物等营养物，以及低温环境。尽管最适宜 SRB 生存的温度为 5~80℃，有些种类可以在极端的温度、压力、矿化度和 pH 值下存活。

细菌虽体型微小（约 1.5μm³），比表面积却在所有生命体中最大。因此，细菌具有吸附极低浓度金属阳离子的界面，能将金属从周围的水环境中富集。这主要是由于细菌表面形成的大分子结构，使其整体带阴离子电荷。一旦金属阳离子与细菌分子上电负性位点相互作用，就会利用外部环境的阴离子作为反离子形成微粒矿物的核，发生进一步的金属络合。油田的生物膜中常发现含有碳酸盐和硫酸盐/硫化物等多种类型矿物，在中东油田注水系统的生物膜中甚至发现了与硫化亚铁共存的放射性铀盐。

油气生产过程中，分离器的上游和下游都会用到杀菌剂。其他包括钻井、压裂作业及油井措施的上游过程，使用杀菌剂以减轻硫化氢和硫化物垢、控制生物污垢和腐蚀，从而提高油井产能。

杀菌剂在油田用量最大的用途是注入水处理，注水的目的是保持压力和提高采收率。但水中的 SRB 及其生成的硫化氢，会造成注水系统的点蚀、管壁穿孔等重大损害。还需要去除水中的氧气，以防止氧腐蚀。首先将注入水流经除氧器，再加入除氧剂将氧含量降至痕量。在除氧器的上游，向水中加入氯等氧化型生物杀菌剂。氯通常由海水的电化学反应产生，在 pH 值为 8~9 时，形成的氯化物主要是次氯酸盐离子。电解氯发生器控制除氧上游的细菌生长，但残存的细菌还会继续污染除氧器下游的系统。此外，除氧剂（通常是亚硫酸氢盐）会与余氯反应，将水中杀菌剂消耗殆尽，因此需要间歇加注非氧化型有机杀菌剂。如果有机杀菌剂会引起注入水的发泡问题，就在除氧塔的下游加注，但这样也会使除氧塔失去杀菌保护。

人们曾认为硫酸盐还原菌只能代谢硫酸盐离子、有机酸和乙醇，但最新的研究表明，某些 SRB 菌株可以代谢饱和烃类，甚至是甲苯。因此，SRB 产生量和储层酸化程度可能大于最初根据有机酸/乙醇等 SRB 营养物质所预测的量值。系统中出现不同的 SRB 菌株，也意味着在用的杀菌剂并非对所有的菌株都能有效。

为最大限度地降低储层酸化，有五种经过实践检验的基本方法：

（1）添加杀菌剂以杀灭 SRB。
（2）采用控制 SRB 生长的生物抑制剂（生物杀伤剂或代谢抑制剂）。
（3）添加如硝酸根离子等营养物质，刺激生成 NR-SOB。这将耗尽碳基营养物，形成亚硝酸盐生物杀伤剂，从而抑制 SRB 的生长。
（4）注水井中使用不含硫的地层水或脱硫酸根的海水。
（5）使用 H_2S 脱除剂（见第 15 章）。

膜技术可以大幅度降低海水中硫酸根离子的浓度，并减少生产井中潜在的硫酸盐结垢。磷酸盐等含磷化合物也可作为 SRB 生长的营养物质，但不能被 SRB 利用生成硫化氢。利用膜技术脱除注入海水中的水溶性含磷化合物，也可以减弱 SRB 生长。优选的分离膜可以是反渗透膜或纳滤膜。

还有研究采用了紫外线辐射杀菌方法。因无法证明其能杀死全部的 SRB，还需要进行二次化学杀菌处理。上述方法（1）至方法（4）通常用于注入井，同时将硫化氢脱除剂注入井下或平台顶部的产出流体中。方法（2）和方法（3）可以同时使用硝酸盐处理，后文将一并讨论。

14.2　控制细菌的化学剂

油田行业中用于控制细菌的化学品可分为杀菌剂（氧化型和非氧化型/有机）和生物抑制剂（生物杀伤剂或代谢抑制剂）两大类。

杀菌剂在正常使用浓度下可以杀死细菌。而生物抑制剂并不会杀死细菌，只是通过干扰细菌的新陈代谢活动，使硫化物的形成量最小。正如本节所述，某些不同类别的有机杀菌剂复配时可以发挥协同作用，比单独使用任何一种的效果更佳；杀菌剂和生物抑制剂的组合也可能比单一产品的效果更有优势。此外，多数杀菌剂似乎对浮游生物效果更好，仅有少许种类的杀菌剂能够在相同剂量下有效减少固着菌群量（生物污垢）。因为没有普适性的杀菌方案，作业方与供应商或服务公司需要在措施前有效协调，以确保加注正确杀菌剂的正确时机和合理剂量。

杀菌剂的评价是一个系统工程，其性能除受浓度影响外，还有诸多其他因素。通常可以利用浮游细菌粗筛杀菌剂，但研究人员普遍认为最终确定合适的杀菌剂，必须测试固着细菌的清除效果。杀菌剂效果的试验评价可以在实验室或在现场水系统开展，监测固着SRB种群生长（生物污损）、次生硫化物、微生物腐蚀情况和硫化亚铁的产生。其中，监测SRB生长的方法包括观测物理外观、微生物数量、显微镜分析、压力波动和热传递。荧光显微镜也可有效用于分析现场样品的细菌。实验室内通常采用循环回路或旋转的生物膜圆筒高压釜，测试杀菌剂减少生物膜（固着细菌）的效果。实验室内SRB培养测试的操作难度较大，有研究人员选择用荧光假单胞菌等其他细菌与SRB混合培养，但在有氧环境的生物膜下SRB仍然可能生长。评价过程中需要明确：监测重点应在于固着菌群（生物膜）的减少，而非浮游生物。生物污损的循环管路研究表明，杀菌剂段塞剂量消耗完后，生物膜种群就会迅速恢复。此外，测试过程中可选用石英晶体微天平监测微生物沉积物。

氯酸盐或次氯酸盐等氧化型杀菌剂通常需要比有机杀菌剂更长的保留时间（长达30min），以达到完全杀死细菌的目的。有机杀菌剂的特点是"杀灭速度"或"击溃"细菌的性能强，但使用浓度相对较高（400～1000mg/L）。有机杀菌剂通常以段塞形式加注到注水系统，这样可以在相对短周期内加注。典型的有机杀菌剂加注通常是每隔几天一次，每次持续若干小时。使用杀菌剂后，生物膜细菌的高种群水平会迅速降低。而杀菌剂一旦加注完成，生物膜内的细菌又会开始再生并快速增殖，达到加注杀菌剂之前的同等水平；细菌正是在生长期间活性最强，造成的危害也最大。上述方法不包括加注氯酸盐/次氯酸盐等氧化型杀菌剂。

尽管水系统已添加杀菌剂，许多注水井还是会出现MIC增加的情况。出现该问题的原因，除可能使用无效的杀菌剂外，还有接触并杀灭目标细菌的杀菌剂浓度、保留时间或频次不满足要求。另有研究表明，一些水处理添加剂是可以促进细菌生长的营养物，因此在选用前应进行详细评估。MIC问题更常见的原因是杀菌剂加量不足。

生物抑制剂不会杀灭微生物，但可以抑制其进一步生长，这就意味着微生物暴露于生物抑制剂中时，无法增殖，而一旦与生物抑制剂脱离接触，微生物即可继续增殖。生物抑制剂在油气生产中也常使用，通过保持SRB的低含量水平并抑制其代谢活动，高效地防止生成H_2S。生物抑制剂的需用添加浓度（通常为2～10mg/L）远低于有机杀菌剂。有些有机杀菌剂兼具生物抑制剂的功能，但并非所有的生物抑制剂都具有与杀菌剂同等的功效。通常在定期段塞式添加有机杀菌剂的基础上，同时以连续或段塞式的剂量添加生物抑制剂。有研究对比了杀菌剂和生物抑制剂的测试方法。

14.3 杀菌剂的种类及性能

杀菌剂根据其化学性质分为氧化型杀菌剂和非氧化型有机杀菌剂两种。

注水系统的处理手段通常是以氧化型杀菌剂为主，辅以非氧化型有机杀菌剂。海水提升系统的氧化型杀菌剂通常采用氯酸盐或次氯酸盐，下文将对其及其他氧化型杀菌剂进行介绍。

可采用肉桂醛或肉桂基添加剂，来防止或减缓生物污损，以增强常规氧化型或非氧化型杀菌剂的效果。

14.3.1 氧化型杀菌剂

氧化型杀菌剂能够作用于微生物的蛋白质基团，以及微生物与设备表面结合处的多糖，使其发生不可逆转的氧化/水解作用。这一过程导致细菌的酶活性降低和细胞死亡。因此，有些 SRB 可能对部分非氧化型有机杀菌剂有耐受性，但氧化型杀菌剂对所有的 SRB 菌株均有杀灭作用。

氧化型杀菌剂主要类别如下：电化学过程产生的氯/次氯酸盐（以及溴/次溴酸盐），次氯酸盐和次溴酸盐类，稳定的氯化溴，羟基自由基，氯胺酮，二氧化氯，氯异氰脲酸酯，含卤素的海因类，过氧化氢和过氧乙酸。

有文献详述了现场使用氧化型杀菌剂时需要考虑的因素。目前只有氯酸盐和次氯酸盐主要在石油生产的注水系统中应用。

氯气和液溴均属于剧毒、强腐蚀性和难以处理的化学物质，但可以通过直流电电解氧化海水中的氯离子，在现场直接制备氯气（Cl_2）：

$$2Cl^- \longrightarrow Cl_2 + 2e^- \text{（阳极反应）}$$

同时，H^+ 在阴极被还原成 H_2，产生更多的碱性溶液。氯气与水反应产生次氯酸和盐酸。

$$Cl_2 + H_2O \longrightarrow HClO + HCl$$

由于海水呈碱性，酸会进一步与 OH^- 反应，生成的次氯酸根离子（ClO^-）也是一种杀菌剂。

$$OH^- + HClO \longrightarrow ClO^- + H_2O$$

次氯酸的杀菌效果比次氯酸根离子更强，有文献报道了其对微生物膜外蛋白质氧化的作用机理。因此，处理的效果随 pH 值降低而增大，特别是 pH 值低于 6。但实际上海水的典型 pH 值通常为 8~9，此时的游离态次氯酸量极其有限，所以杀菌效果并不理想。海水中还存在溴离子（Br^-），会被电解产生的 Cl_2 氧化成溴单质（Br_2），并进一步反应生成比 ClO^- 氧化性更强的杀菌剂次溴酸根离子（BrO^-）。因此目前的共识是次生的 BrO^- 可能在防止细菌生长方面也起到重要作用。使用电解氯化法杀菌的问题是氯可能与水中的有机物反应，使控制细菌的有效氯浓度大幅降低。除氧器下游有时不是使用非氧化型杀菌剂，而是采用电解氯化法作为控制细菌的次要手段。但氯气对钢铁有腐蚀性，并且会除去水中防止氧腐蚀的亚硫酸氢盐除氧剂。

通过现场电催化产生的羟基自由基也是一种强效杀菌剂，但目前还未在油气生产领域应用。

市售的次氯酸钠（NaClO）碱性溶液或固体次氯酸钙[$Ca(ClO)_2$，漂白粉]也可用作氧化型杀菌剂。但在海水注入系统中，受到物流、储罐和运输成本高的限制，这些盐类实用性不强，主要用于密闭或半密闭的水系统。目前应用次氯酸盐溶液作为杀菌剂的新方法是连续调节注入水的pH值恒定为7，以使游离氯能存在于井内水体中。有报道称，稳定态的次氯酸盐或溴组分性能优于次氯酸盐。次氯酸盐和Fe（Ⅲ）在碱性溶液中反应，会生成Fe^{3+}（铁酸根离子，FeO_3^{3-}），是处理油田被细菌和有机化合物污染水体的一种强力氧化剂。

氯化溴（BrCl）与次氯酸盐有相同的缺点，但利用氨基磺酸盐对其进行稳定，可以有效降低处理难度。在近中性的pH值条件下，氯化溴可以生成强力杀菌剂亚溴酸HOBr。在氯化溴—氨基磺酸盐混合物中加入C_8—C_{14}烷基胺，可起到提升杀菌效果的增效剂作用，而烷基胺本身作为非氧化型杀菌剂，杀菌效果不佳。

在氯胺（N—Cl键）中可以找到高价氯的来源。氯胺的杀菌性能优于次氯酸盐，在造纸工业中应用较为广泛，其缺点是后续处理困难。尽管有研究报道了如通过次氯酸钠活化溴化铵等原位生成氯胺的方法，但截至目前石油行业还未开展过验证试验。

亚氯酸根离子形式的二氧化氯（ClO_2^-，ClO_2在空气中以高浓度存在时是一种爆炸性气体）并非采用氧化杀菌机制，但通常也归类为氧化型杀菌剂。相较于氯气，ClO_2在较高pH值、含氮气或有机物污染体系中的杀菌效果更强。ClO_2可以通过亚氯酸盐和酸原位反应制备。还可以通过亚氯酸盐溶液（如$NaClO_2$溶液）、氯生成剂（如NaClO）和碱（如NaOH）混合体系，解决亚氯酸盐溶液的稳定性问题，得到高活性含量的活性ClO_2。ClO_2溶液具有腐蚀性。工业使用时，有时会添加重铬酸钠（$Na_2Cr_2O_7$）以抑制ClO_2溶液的腐蚀性。考虑到$Na_2Cr_2O_7$的高毒性，有专利研究了如乙二醇、乙酸、脂肪咪唑啉和乙氧基化脂肪二胺等混合制成的缓蚀剂，其毒性较低，但缓蚀效果非常有限。

溴氯海因（1-溴-3-氯-5,5-二甲基海因，BCDMH）是白色粉末（图14.1），在水中可缓慢水解生成次溴酸（HBrO）和次氯酸（HClO），因此是氯和溴的最佳来源。相较于次氯酸钠，BCDMH在相同剂量下对微生物细胞膜的杀灭活性更强，且腐蚀性更低。二氯海因、二溴海因也归类为氧化型杀菌剂。海因类杀菌剂已广泛应用于各工业领域，也开发出了可泵送、液态BCDMH产品。非卤化的羟甲基海因（如1-羟甲基-5,5-二甲基海因）还具有脱除H_2S的功能（见第15章）。

氯代异氰尿酸盐类是一类易于处理的粉末化合物，可在水中缓慢水解释放氯和氰尿酸。最典型的产品是二氯异氰尿酸钠（图14.2）。氯代异氰尿酸盐类产品与其他含氯杀菌剂有类似的缺点，即有效pH值的范围较窄和潜在的腐蚀性问题。

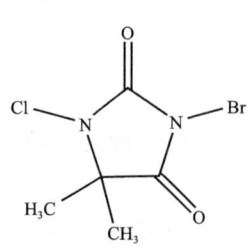

图 14.1　溴氯海因　　　　　　　图 14.2　二氯异氰尿酸钠的结构式

过氧乙酸（CH_3COOOH）与过氧化氢（H_2O_2）复配也起到杀菌剂作用，特别是用于油井措施液。注水系统采用过氧乙酸的腐蚀性过强。但与传统的非氧化型杀菌剂相比，过氧乙酸因其环保性和杀菌效果更优，建议应用于水力压裂的水处理环节。

14.3.2　非氧化型有机杀菌剂

非氧化型有机杀菌剂的作用机理一般是其可以改变微生物细胞膜的渗透性，并干扰微生物的生物过程。与氧化型杀菌剂相比，非氧化型有机杀菌剂的腐蚀性较弱，部分种类的杀菌剂还可抑制腐蚀。非氧化型有机杀菌剂主要包括醛类、季磷盐类化合物、季铵盐表面活性剂、阳离子聚合物、有机溴化物、甲硝唑、异噻唑酮（或异噻唑啉酮）和硫酮类、有机硫氰酸酯、酚类化合物、烷基胺、二胺类和三胺类、二硫代氨基甲酸盐类、2-（癸硫基）乙胺（DTEA）及其盐酸盐、三嗪类衍生物、恶唑烷类、其他特定类表面活性剂类。

科研人员曾开展试验以考察常用的非氧化型杀菌剂在浓度为 $1\times10^4 \sim 1\times10^5$ mg/L 时，是否会引起油田腐蚀问题。试验结果证实，腐蚀速率随杀菌剂浓度变化很大，特别是在高浓度时。在同样的高浓度下，有些杀菌剂仅产生轻微腐蚀性，而有的腐蚀性则要强得多。通常杀菌剂的腐蚀性与其使用剂量呈直接正相关性，即加量越大，腐蚀速率越高。不过统计数据表明，杀菌剂处理在防止 MIC 和生物污损方面带来的裨益，远超过其潜在腐蚀性带来的不利影响。

上面所列举的杀菌剂中，只有几种因成本低廉或环境友好而用于油田生产，其他的则更多用于循环水或冷却水系统。石油行业中最常用的非氧化型有机杀菌剂是戊二醛、四（羟甲基）硫酸磷（THPS）及低剂量的甲醛和丙烯醛（后两种醛有疑似致癌性）。这些杀菌剂有时也会与季铵盐表面活性剂或其他增效剂复配使用。季铵盐表面活性剂在其自身的杀菌行为外，还有助于穿透生物膜并增强杀菌作用。下文将详细讨论这几类有机杀菌剂。

14.3.2.1　醛类杀菌剂

醛类杀菌剂（图 14.3）主要包括：C_3—C_7 二醛，尤其是戊二醛（1,5-戊二醛）；甲醛；丙烯醛；邻苯二甲醛。

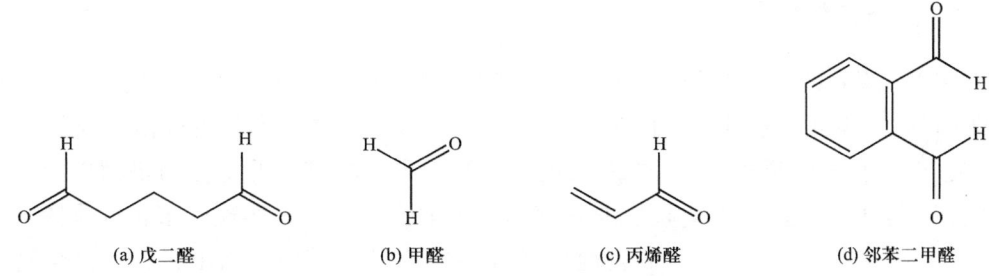

图 14.3　主要醛类杀菌剂的结构式

在水中可以释放醛的半缩醛也有一定规模的应用。半缩醛末端的碳原子中含有—OR 和—OH 基团，如甲醛的半缩醛分子式为 $H_2C(OCH_3)(OH)$。半缩醛最初的发明是基于医学需求，二醛及二醛的单/双缩醛衍生物组成的混合物具有协同杀菌作用。

醛类杀菌剂除了可以杀死 SRB，还具有脱除硫化氢的作用，因此在油田生产中有很大的使用潜力（见第 15 章）。1，5-戊二醛是迄今用量最大、最常见的醛类杀菌剂。其杀菌机理是与细菌细胞外部蛋白质发生交联及阻碍细胞渗透性。戊二醛如果储存不当，会不稳定而分解。有室内研究发现，单独以戊二醛作为杀菌剂就可使细菌数量减少、代谢活动降低，生物膜的增长速度也有所下降，且整体效果与加量呈一定正相关性。而经常将戊二醛与其他表面活性剂、季铵或季鏻化合物等杀菌剂复配使用，可以提高其杀灭细菌的速度并减少戊二醛的用量（单独使用戊二醛时加量会很高），同时复配物也缩短了杀菌所必需的接触时间。戊二醛也能与二硫氰基甲烷协同起效。戊二醛对 pH 值比较敏感，其在中性到碱性的水中有效，因此在注入水中可能有效，而对采出水的处理效果通常不理想。

甲醛作为一种低成本醛，能有效杀灭生物膜中的固着微生物。有研究表明油田现场的杀菌剂需求加量往往远高于室内实验结果。因甲醛的致癌特性，2007 年欧盟已禁止在欧洲使用甲醛。甲醛被美国环境保护局（EPA）列为可能的人类致癌物。因此，相较于甲醛水溶液和固体多聚甲醛，以能够分解产生甲醛的聚甲醛作为杀菌剂更具优势。聚甲醛不溶于水，其分解产生甲醛的速度可以通过 pH 值、温度和某些特定的促分解催化剂来控制。虽然在北海地区禁止使用甲醛，但有油田通过油服公司选购能释放甲醛的化合物（如乙二醇半缩甲醛）作为杀菌剂。

丙烯醛可以从以下三个方面解决 SRB 活动引起的油田生产问题：（1）丙烯醛是一种有效的杀菌剂；（2）与其他小分子醛类相同，丙烯醛也可以清除硫化氢；（3）丙烯醛可以溶解硫化亚铁。但即使丙烯醛具有一定生物降解性，基于其本身很强的急性毒性及疑似致癌性，使用时必须特别小心。应用时，使用密闭输送系统的流程，可以最大限度地降低潜在的化学品暴露概率。

邻苯二甲醛作为戊二醛的有效改进产物，主要应用于医学领域，可在较低剂量和较短接触时间内清除固着菌群。

14.3.2.2 季鏻盐类化合物

四羟甲基鏻（THP）离子具有非氧化和非发泡性，是广泛用于石油天然气工业的性能优异的杀菌剂，但最近被挪威禁止使用。TPHS 对外排水水质的影响也有相关报道。

稳定 THP 盐（包括 THPS）的方法也有专利提及。季鏻盐类化合物杀菌机制比较多样，但主要是通过与蛋白质交联使细胞膜崩塌（细胞裂解）来实现。市售的季鏻盐类杀菌剂通常为四羟甲基硫酸鏻（THPS），该化合物在储层岩石上无明显吸附（图 14.4）。THP 盐只能缓慢生物降解。THPS 可在 pH 值为 3~10 的范围内杀死所有类型的 SRB 厌氧菌。THPS 处理浮游菌时，通常浓度为 50~100mg/L，而对于已形成生物膜的情况，则需要更高浓度。生产商针对 THPS 处理所给出的结论如下：THPS 处理后，在系统排出阶段就会即刻失去抗菌性，并降解为无毒的三羟甲基氧化膦；此外的碱性条件下，THPS 降解速度会加快。在某些特定条件下，THPS 可以有效抑制腐蚀，而有些条件下却会加剧腐蚀，因此有时使用 THPS 时需注入成膜型缓蚀剂。THPS 也会与某些表面活性剂协同起效，杀灭生物膜。不含表面活性剂的新一代更高效 THPS 配方已有报道。这些新的配方也能尽量减少标准 THPS 配方会引起的硫酸钙或硫酸钡/硫酸锶结垢。THPS 也能作为杀菌剂与某些醛（如甲醛）协同起效。

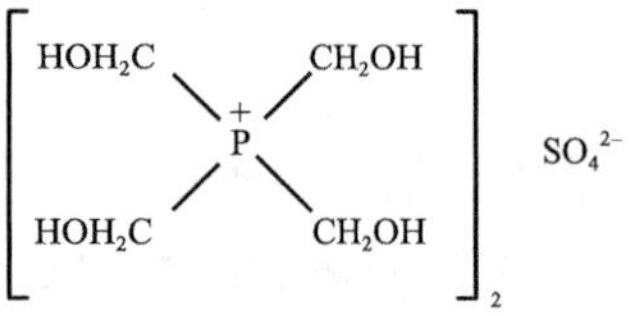

图 14.4 四羟甲基硫酸鏻的结构式

目前市售的 THPS 通常为液态，虽然实验室内已制备了可吸附于己二酸的固态 THPS，但尚未实现商业化生产。将 THPS 吸附于二氧化硅所制备的杀菌剂，可用于压裂作业。同多数杀菌剂一样，THPS 等液态季鏻化合物与常见的除氧剂（如亚硫酸盐类、异抗坏血酸等）会发生反应或间接干扰除氧性能。针对此问题，有研究人员提出了解决办法并已获得授权专利，即将鏻化合物嵌入基质材料中后再进行应用，其中鏻化合物优选三羟有机磷、THP^+ 盐（四羟基有机鏻盐）或 THP 与含氮化合物（最好是尿素）的缩合物等。

当 THPS 与铵盐共注时可以去除井底和平台顶部的硫化亚铁垢。许多油田的采出水中都含有足量的铵根离子，只需加入适量的 THPS 即可达到预期的除垢效果。研究表明，$Fe(Ⅲ)$ 可与氮—磷配体发生螯合形成配合物，并使水体最终呈现红色。在某油田利用 THPS 和表面活性剂在杀灭 SRB 的同时，脱除硫化亚铁垢，有几口油井取得了极好的增产效果，最高增产达到了 300%。除去采出水中的颗粒状硫化亚铁垢，还有助于后续的顺利破乳。

THPS 和相关盐类也可用于控制或抑制聚合物成分的解聚速度，从而在特定酸性条件下，降低溶液黏度，使杀菌处理过程顺利进行。

THPS 的杀菌活性可以通过与润湿剂（不同类型的表面活性剂）、助水溶剂（互溶剂）或生物渗透剂，如聚[氧乙烯（二甲基亚氨基）乙烯（二甲基亚氨基）二氯化物]一起使用而得到协同增强。

另一种增效剂混合物包括阴离子防垢剂（如 1-羟乙基-1,1-二膦酸和二乙烯三胺五亚甲基膦酸的混合物），与 THPS 或聚［氧乙烯（二甲基亚氨基）乙烯（二甲基亚氨基）二氯化物］等带阳离子的杀菌剂混合。

丙烯酸和 THPS 的反应产物已证明是性能良好的杀菌剂，也可以溶解硫化亚铁，但目前还未在市场上销售（图 14.5）。

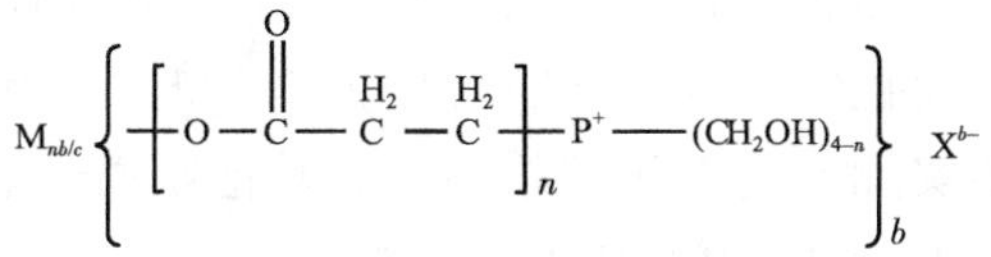

$$M_{nb/c}\left\{\left[O-\overset{O}{\underset{\|}{C}}-\overset{H_2}{C}-\overset{H_2}{C}\right]_n P^+ —(CH_2OH)_{4-n}\right\}_b X^{b-}$$

图 14.5　丙烯酸和 THPS 的反应产物

有报道介绍了基于三正丁基十四烷基氯化膦（TTPC）的一种新型季鏻盐杀菌表面活性剂，并进行了毒理测试。室内和现场数据表明，TTPC 在低浓度下有效，作用迅速，对产酸菌和硫酸盐还原菌都有效。在杀菌效果对比试验中，TTPC 的效果优于戊二醛和 THPS。TTPC 与氧化型杀菌剂、硫化氢脱除剂和除氧剂兼容，并且热稳定性和缓蚀性优良。但与 THPS 不同的是，TTPC 会产生少量泡沫，易吸附在管壁或容器内表面且不容易失去活性，也不能溶解硫化亚铁垢。

有文章介绍了非油田环境应用的含不同长度单或双烷基链（C_{10}—C_{18}）的其他膦酸盐表面活性剂杀菌剂。其杀菌性能表现优于对应的铵盐。对于烷基三甲基鏻盐，杀菌活性随着烷基链长度的增加而增强。相反，二烷基二甲基鏻盐的杀菌活性随着取代物的链长增加而降低。由一级或二级膦酸盐制成的四烷基膦酸盐的膦酸酯或硫酸酯，也是一种杀菌剂。

14.3.2.3　季铵盐表面活性剂

季铵盐表面活性剂杀菌剂具有很强的表面活性，有时会与其他杀菌剂混合使用以提高性能。这一类中最常见的是长链正烷基二甲基苄基氯化铵，其中的烷基是 12 个碳或更多（图 14.6）。通常在碱性 pH 值范围内，季铵盐对藻类和细菌最有效，但在油污、垢和其他细菌残骸所污染的系统中会失去活性。低浓度（<250mg/L）的烷基二甲基苄基铵表面活性剂已证实可以抑制包括 SRB 在内的许多细菌株生长。杀菌机制是由于其阳离子性质，即与细胞膜形成静电键，从而影响渗透性和蛋白质变性。单独使用时，季铵盐类杀菌剂可能需要 10min 才能杀死细菌。有人建议在石油工业以外的杀菌剂应用中，使用如抗坏血酸或乙醇酸等细胞膜破坏剂，以提高杀灭率。

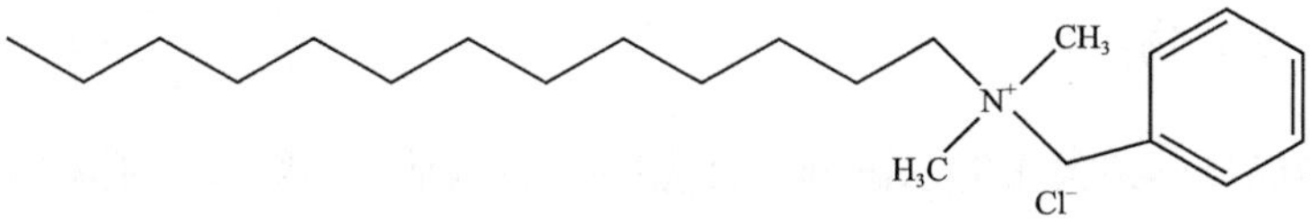

图 14.6　长链正烷基二甲基苄基氯化铵表面活性杀菌剂

二烷基二甲基季铵盐也用作杀菌剂。有人提出用二癸基二甲基氯化铵和锌盐的混合物来有效控制储层酸化和产生 H_2S。

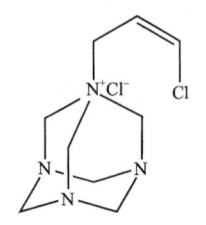

图14.7 1-(3-氯丙烯)-3,5,7-三氮杂-1-氮鎓结构式

由于季铵盐是表面活性剂,具有成膜型缓蚀剂的特性。季铵盐表面活性剂经常与可可二胺等胺类物质一起配制。季铵盐表面活性剂通常是起泡剂,但是也发现有非起泡的季铵盐表面活性剂杀菌剂。

如 N,N-二甲基-N-乙基-N-丙基三溴化铵等非聚合季铵多卤化物,也已证明可作为杀菌剂。三溴离子是溴源,所以其可能实际上是氧化型杀菌剂。其他阳离子杀菌剂包括长链烷基胍盐和1-(3-氯丙烯)-3,5,7-三氮杂-1-氮鎓(图14.7)。

14.3.2.4 阳离子聚合物

文献报道称聚表卤代醇与叔胺反应,可得到一种兼具缓蚀和杀菌作用且可生物降解的季铵盐聚合物。

如聚六亚甲基双胍盐酸盐(图14.8)等双胍类化合物,通常是短链聚合物或低聚物,属于另一类比较典型的阳离子聚合物杀菌剂,主要用作消毒剂。近期有研究称二次和三次聚二烯丙基铵盐也具有抗菌活性。这类聚合物还可用作絮凝剂(见第13章)。

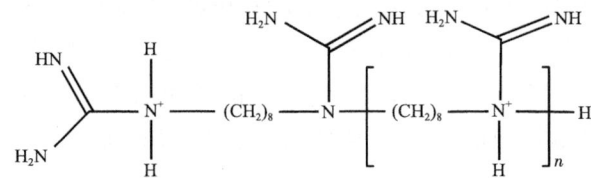

图14.8 聚六亚甲基双胍盐酸盐结构式

由环氧氯丙烷、二胺和三胺制得的季铵聚合物既是非表面活性剂型杀菌剂,还兼具缓蚀效果。其他相关的季铵类聚合物,如 N,N,N',N'-四甲基-1,2-二氨基乙烷与双(2-氯乙基)醚通过共聚反应,得到的聚[氧乙烯(二甲基亚氨基)乙烯(二甲基亚氨基)二氯化物],也都是杀菌剂。聚[氧乙烯(二甲基亚氨基)乙烯(二甲基亚氨基)二氯化物]与阴离子防垢剂[如1-羟基乙烷-1,1-二膦酸和二乙烯三胺五(亚甲基膦酸)的混合物]的混合物,据称也是一种优良的杀菌剂。

杀菌剂聚合物的主链或重复基团中,接入聚氧化亚烷基和季铵基团,可以制备得到新的杀菌剂,但尚未在油田应用。

14.3.2.5 有机溴化物

有机溴化物也是一类实用的杀菌剂,常见的产品包括2-溴-2-硝基丙二醇(BNPD)、2,2-二溴-3-腈基丙酰胺(DBNPA)、1,2-二溴-2,4-二氰基丁烷(DBDBC)和2,2-二溴-2-硝基乙醇等(图14.9),其中二溴丙二酰胺与氧化型杀菌剂复配得到的体系,是适用于多个领域的多用途杀菌剂。BNPD发明于20世纪60年代,已用作油田作业的杀菌剂。DBNPA对pH值敏感,在酸性和碱性条件下都能迅速水解。得益于其在水中的不稳定性，能快速杀菌并迅速降解为氨和溴离子。有研究提出了解决其低水溶性的方法。DBNPA 建议与一些氧化型（如次氯酸盐）和非氧化型杀菌剂协同使用。还有专利提出在浓缩岩盐溶液等现场环境下，原位生成活性杀菌剂的配方。

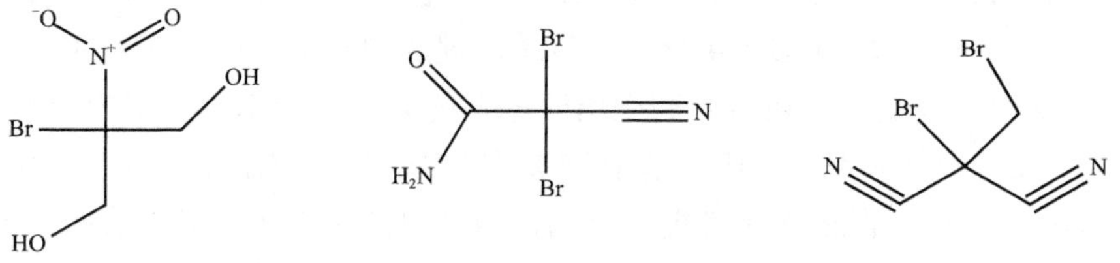

(a) 2-溴-2-硝基丙二醇　　(b) 2,2-二溴-3-腈基丙酰胺　　(c) 1,2-二溴-2,4-二氰基丁烷

图 14.9　2-溴-2-硝基丙二醇、2,2-二溴-3-腈基丙酰胺、1,2-二溴-2,4-二氰基丁烷结构式

14.3.2.6　甲硝唑

甲硝唑是咪唑的一种衍生物，具有一定的缓蚀特性（图 14.10）。甲硝唑可以选择性地被厌氧细菌吸收，进入细胞后，甲硝唑的硝基会被递质毒素（或与递质毒素相关的代谢过程）还原，生成的氧化产物可以破坏 DNA 螺旋结构，从而抑制核酸的合成。研究表明，甲硝唑可以有效抑制油田注水设施中生物性硫化物的产生。

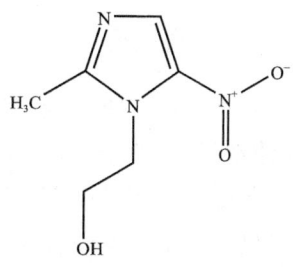

图 14.10　甲硝唑结构式

通过衍生甲硝唑制备得到的季铵盐表面活性剂，是表面活性更强的杀菌剂，还兼具良好的缓蚀性能（结构见第 8 章）。

14.3.2.7　异噻唑酮（或异噻唑啉酮）和硫酮类

这类产品有氯-2-甲基-4-异噻唑啉-3-酮（CMIT）、5-和2-甲基-4-异噻唑啉-3-酮及4,5-二氯-2-（正辛基）-4-异噻唑啉-3-酮（图 14.11）。水溶性较差的苯并异噻唑啉酮也可在市场上购买。异噻唑啉酮类是广谱抗菌剂，具有良好的生物降解性，对固着菌（生物膜）有效。异噻唑啉酮类杀菌的机理是抑制微生物的呼吸和通过细胞膜的食物运输。CMIT 和这类的其他生物杀伤剂被广泛用于各种工业水处理的微生物控制。由于异噻唑啉酮作为生物杀伤剂的性能有限，而且有其他更具成本效益的处理方法，目前没有在油田生产中得到应用。

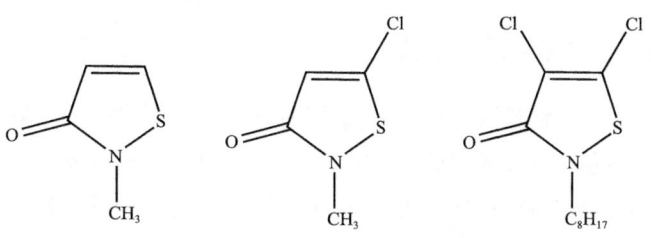

图 14.11　常见的异噻唑啉酮杀菌剂

尽管异噻唑啉酮可用作杀菌剂，但其效果会因现有 SRB 形成的硫化氢而降低。异噻唑啉酮是非常有效的生物抑制剂，通过保持 SRB 的低水平和抑制其代谢活动来防止硫化氢的形成。因此其最好与其他有机杀菌剂结合使用。例如，减少固着 SRB 的方法包括定期段塞加入诸如戊二醛之类的烷基二醛作为杀菌剂，同时连续加入异噻唑啉酮作为生物抑制剂。改进的方法是，在段塞加入生物有效量杀菌剂的同时或之后，间歇性地加入有效量的异噻唑啉酮等生物抑制剂。还有专利提出改进型的方法，即基于异噻唑啉酮的组合物中，将 3-碘-2-丙炔-N-丁基氨基甲酸酯作为第二种生物杀伤剂。另有专利提出，使用锌盐来提高异噻唑啉酮的杀菌性能。抗菌剂组合物还可以包含如吡硫鎓（包括吡硫鎓锌或吡硫鎓铜在内）的共杀菌剂。六环硫酮也是有效的油田杀菌剂。优选的硫酮是 3，5-二甲基-1，3，5-噻二唑啉-2-硫酮（图 14.12）。2-甲基异噻唑啉-3-酮和一个或多个卤代烷基砜据称具有协同杀菌特性。

14.3.2.8 有机硫氰酸酯

这类产品中唯一被广泛使用的是亚甲基双硫氰酸酯（图 14.13）。亚甲基双硫氰酸酯在 pH 值超过 8.0 时会迅速分解，最终释放出有毒的氰化氢。其价格相对昂贵，并且通常需要一种分散剂来使杀菌剂有效地渗透到生物污损处。如果单独使用，加注浓度相当高。在低浓度下，亚甲基双硫氰酸酯抗菌谱狭窄，不能完全阻止微生物的生长。其已证实能与如戊二醛、DTEA 和有机溴化物等其他几种杀菌剂协同工作。亚甲基双硫氰酸酯杀伤机制是阻止微生物中的电子转移，防止氧化/还原机制。

图 14.12　3，5-二甲基-1，3，5-噻二唑啉-2-硫酮　　图 14.13　亚甲基双硫氰酸酯结构式

14.3.2.9 酚类化合物

如五氯酚钠等酚类杀菌剂是一类成本低、效果强的杀菌剂，但因对环境有隐患，这类杀菌剂逐渐退出应用。如 4-羟基苯甲酸等对羟基苯甲酸酯类也是酚类杀菌剂，未在石油和天然气工业中规模应用。目前，市售的酚类工业杀菌剂包括苯基苯酚钠、正苯基苯酚、二氯-M-二甲苯酚、对羟基苯甲酸酯类（如 4-羟基苯甲酸）等。

14.3.2.10 烷基胺、二胺类和三胺类

椰油二胺表面活性剂长期以来一直用作石油工业中的杀菌剂（图 14.14）。如 N，N-双（3-氨基丙基）十二烷基胺或双（3-氨基丙基）辛胺等三胺类化合物，也可用作杀菌剂。此外研究表明某些胺类表面活性剂也可能表现出缓蚀特性。

如 N，N′-二甲氨基丙基丙烯酰胺或 N，N′-二甲氨基丙基甲基丙烯酰胺等非表面活

性剂的叔胺类丙烯酰胺单体,可以抑制微生物生长(图14.15)。此外,研究人员还发现缓蚀剂与杀菌剂之间存在协同作用,缓蚀剂的存在与否,对于杀菌剂的需求量有重要影响,即体系中存在缓蚀剂时,用少量的杀菌剂即可实现100%的杀菌效果;反过来杀菌剂的存在又能提高缓蚀效果。

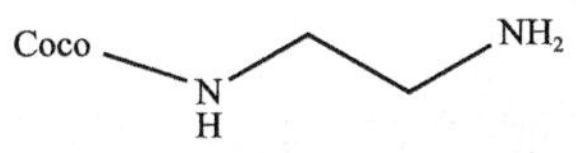

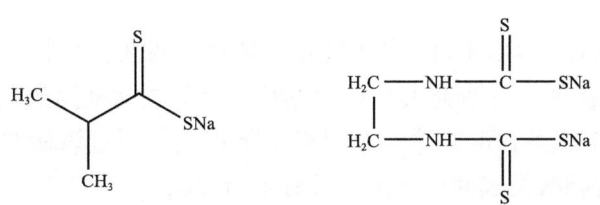

图14.14　椰油二胺结构式 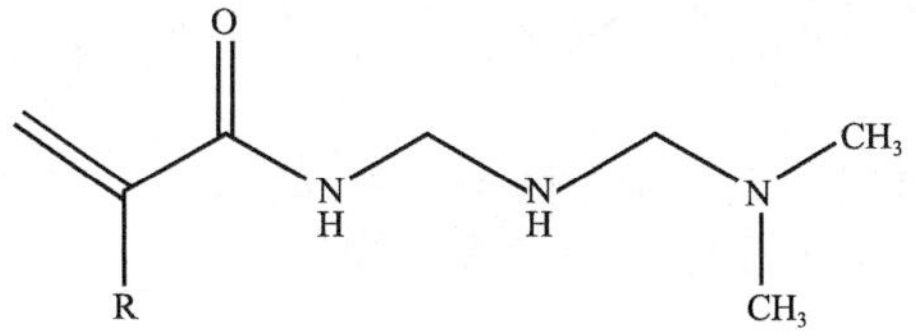　　　　图14.15　N,N'-二甲氨基丙基甲基丙烯酰胺结构式

14.3.2.11　二硫代氨基甲酸盐类

二硫代氨基甲酸盐杀菌剂的实例是二甲基二硫代氨基甲酸钠和1,2-亚乙基双二硫代氨基甲酸二钠(代森钠)(图14.16)。其更常用于造纸行业,有时与醛类混用,但也已经在北美很多油田注入水中评价,但在欧盟不允许使用。二硫代氨基甲酸盐具有稳定性好、无腐蚀性、对人体低毒的特性。研究表明,即使存在亚硫酸氢盐除氧剂时,其依然对SRB有较好的清除作用。此外,含有硫代氨基甲酸酯基团的聚合物也可以用于油田杀菌处理。

(a) 二甲基二硫代氨基甲酸钠　　(b) 1,2-亚乙基双二硫代氨基甲酸二钠(代森钠)

图14.16　二甲基二硫代氨基甲酸钠和1,2-亚乙基双二硫代氨基甲酸二钠(代森钠)

14.3.2.12　2-(癸巯基)乙胺(DTEA)及其盐酸盐

DTEA在低pH值环境下以盐酸盐形式存在(图14.17),在高pH值时则以游离胺形式存在。DTEA及其盐酸盐都具有生物杀伤性。DTEA的盐酸盐是一种季铵盐类表面活性剂,因而还具有缓蚀特性。有研究人员建议将戊二醛与DTEA结合使用,在美国的一些油田也开始试验这种复配体系。DTEA的盐酸盐与亚甲基双硫氰酸酯有协同作用。

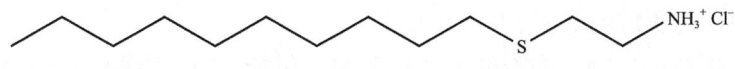

图14.17　2-(癸巯基)乙胺盐酸盐结构式

14.3.2.13 三嗪类衍生物

如含硫三嗪和 2-（叔丁氨基）-4-氯-6-（氨乙基）-S-三嗪（特丁津）（图 14.18）等三嗪类衍生物，已证实具有抗微生物特性。H_2S 清除剂的三嗪类化学品也已作为市售杀菌剂（详见第 15 章）。

图 14.18　2-（叔丁氨基）-4-氯-6-（氨乙基）-S-三嗪结构式

14.3.2.14 恶唑烷类

如 7-乙基双环恶唑烷、4,4-二甲基恶唑烷和亚甲基双恶唑烷，以及一些卤代恶唑烷酮等恶唑烷类，均是广谱市售杀菌剂（图 14.19）。

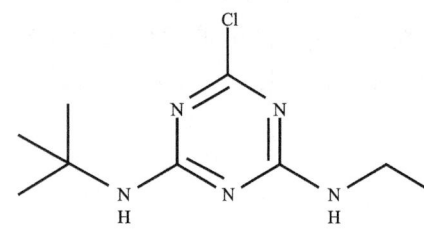

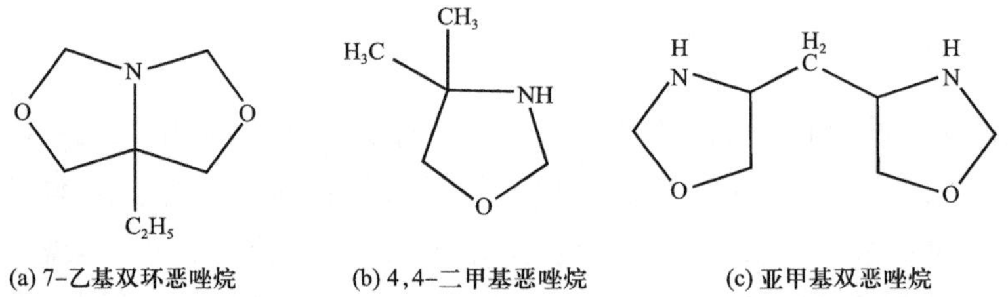

(a) 7-乙基双环恶唑烷　　(b) 4,4-二甲基恶唑烷　　(c) 亚甲基双恶唑烷

图 14.19　7-乙基双环恶唑烷、4,4-二甲基恶唑烷及亚甲基双恶唑烷的结构式

14.3.2.15 其他特定类表面活性剂类

通式为 $R'R_2NCH_2P(O)(OH)_2$ 的烷基氨亚甲基膦酸两亲化合物（如辛基氨甲基膦酸），已建议用作生物杀伤剂。有研究人员认为具有多个膦酸酯基团的疏水改性膦酸盐类防垢剂也可能具有生物杀伤性，但对这一结果尚未下定论。不过该研究小组已明确了如十二烷基氨基磺酸等氨基磺酸类表面活性剂，具有杀菌性能。

14.4 生物抑制剂（"控制性生物杀伤剂"或代谢抑制剂）

生物抑制剂不一定能直接杀死细菌，但会干扰其代谢过程，从而抑制生长。正如在有机杀菌剂一节中所讨论的内容，异噻唑啉酮是生物抑制剂，可以使 SRB 保持低含量水平并抑制其代谢活动，从而防止形成硫化亚铁垢（通过 H_2S 形成）。代谢抑制剂接触 SRB 时，通过剥夺 SRB 产生 ATP 的能力，使细胞无法生长或分裂。不能生长或分裂可能最终导致部分 SRB 死亡，但与杀菌剂不同，细胞的死亡不是接触代谢抑制剂的直接结果。烷基苄基二甲基铵盐也是市售的重要生物抑制剂。实验室研究表明，将生物杀伤剂和生物抑制剂结合使用时，可有效抑制生物硫化物的产生，且所需药剂浓度远低于生物杀伤剂或生物抑制剂单独使用时所需的浓度。

本节讨论了不属于生物杀伤剂的生物抑制剂，包括：(1) 蒽醌；(2) 罗丹酮；(3) 叠氮离子；(4) 硝酸盐和亚硝酸盐；(5) 钼酸盐或钨酸盐；(6) 硒酸盐离子。此外，也有报道称，氯酸盐的还原作用可以减轻储层酸化问题。

14.4.1 蒽醌类"控制性生物杀伤剂"

自 20 世纪 90 年代后期以来，蒽醌已被用作诸多项目中的生物抑制剂（图 14.20）。蒽醌不溶于水，而具有醇式结构的 9,10-蒽二醇二钠盐则是水溶性，且其具有与蒽醌相同的抑制机理。研究证实蒽醌可抑制 SRB 在硫酸盐中的呼吸作用，从而有效阻断硫化物生成，并对其他类别的细菌影响较小。在一项注水项目中，试验者成功地以更易处理的段塞剂量蒽醌产品代替了连续注入季铵杀菌剂。此外，还使用了段塞剂量的丙烯醛作为协同杀菌剂。蒽醌与 THPS 杀菌剂的间歇处理，也已证明可有效阻止污水处理系统采出水池中生物硫化物的产生。

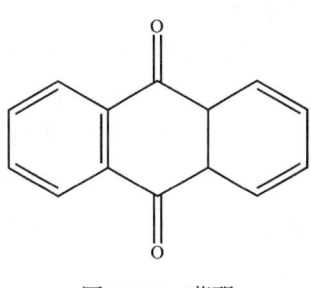

图 14.20 蒽醌

14.4.2 硝酸盐和亚硝酸盐处理

硝酸/亚硝酸钙、钠等硝酸盐（NO_3^-）及亚硝酸盐（NO_2^-）都是价格低廉、处理简单、环境可接受的无机化学品，越来越多地被用于油气生产中抑制 SRB。但使用硝酸盐作为 SRB 抑制剂仍是一种优缺参半的策略，因为其造成的注入管线腐蚀速率超出可接受的范围。目前已有研究评估了油田中使用硝酸盐、亚硝酸盐的腐蚀后果。针对腐蚀问题，有小组试验通过连续加注高剂量的亚硝酸盐，可有效防止酸化和腐蚀问题。

亚硝酸盐可直接抑制 SRB 还原硫酸盐，相较于亚硫酸盐被硫酸盐还原过程中最后阶段，亚硫酸盐还原酶对亚硫酸盐的还原作用，比对亚硝酸盐的还原过程更慢，所以亚硝酸盐可以直接抑制 SRB 还原硫酸盐。因此采出水或注入海水中，加注足够高浓度的亚硝酸盐离子，就可以有效抑制生物硫化物的生成［结果取决于 pH 值、亚硝酸盐离子（非 NO_3^-）也会直接与已存在的 H_2S 反应，形成硫和还原氮化合物］。有现场试验表明，脉冲式加入比连续加入相同量亚硝酸盐，更为有效。有研究表明，与戊二醛相比，亚硝酸盐在抑制酸化和延长 SRB 恢复时间方面的效果都更佳。亚硝酸盐注入也成功用于生产井中消除 H_2S 和防止 SRB 活性。在亚硝酸盐处理的典型案例中，措施后原油产量短时间内显著增长，推测是近井筒区域沉淀的硫化亚铁被溶解所致。

大多数油田都存在硝酸盐还原菌（NRB）和硝酸盐还原硫化物氧化细菌（NR-SOB），均可将硝酸盐还原为亚硝酸盐。石油上游产业已成功引入一种基于硝酸盐的微生物处理技术，对于预防和清除油藏、采出水、地面设施、管线和储气库中的生物硫化物及提高原油采收率均有较好效果。这种储层处理措施的工作原理是通过注入无机硝酸盐体系配方，增强天然有益微生物的生物活性，并最终取代 SRB。如储层是 SRB 的不良碳源（小分子有机酸含量低），注入的硝酸盐将刺激"生命力更顽强"的反硝化细菌快速生长（营养增加），最终占据支配地位并抑制 SRB 的生长。此外，NRB 和 NR-SOB 以硝酸盐为基底物，分解过程中会产生亚硝酸根离子（NO_2^-），有时还有一氧化氮（NO），NO_2^- 和 N_2O 也会通过毒性直接抑制 SRB 的生长；除了毒素，NRB 和 NR-SOB 代谢活动中还可能产生化合物，通过提高环境的氧化还原电位达到抑制 SRB 生长的作用。

醋酸盐等挥发性有机酸离子向来被认为是 SRB 最青睐的营养来源，但近期一些现场经验证实，如苯、甲苯、乙苯和二甲苯（BTEX）的可溶性有机碳源，在一定环境条件下也可以作为 SRB 代谢活动的营养基底物，此时只加入硝酸盐并不能完全抑制 H_2S 的生成。

硝酸钙、硝酸钠的环保性较好，如果将其作为储层中天然存在有机酸的补充，可以选择性地刺激目标硝酸盐还原菌的生长。北美的大多数及北海的若干气田已成功应用了此项处理方案，即通过被动和主动策略来降低储层酸化程度，但不能达到完全消除的目的。这无疑是一种防止生物成因硫化物形成的简易方法，可以在很大程度上减少有机杀菌剂的用量。

基于上述原因，研究人员认为，使用硝酸盐和亚硝酸盐混合物可能比单纯的硝酸盐处理效果更佳。但也有研究表明在只存在 SRB 的储层中，单纯使用亚硝酸盐处理效果反而会更好。此外有研究表明，钼是 SRB 中一种已知氢化酶的酶抑制剂，因此通过注入含钼化合物（如钼酸盐）也可以提升处理效果。值得注意的是，注入硝酸盐/亚硝酸盐的时机对处理效果也有重要影响，如果加注时间较晚，则可能达不到彻底杀菌的效果。例如，尼日利亚的深海油田在用硝酸钙处理后，仍然加注了 THPS 来防止注入设施中形成生物膜。此外，利用硝酸盐处理注入水还有一个潜在优势——微生物提高原油采收率（MEOR）。研究发现，储层中形成的 NR-SOB 生物膜有助于将岩石表面黏附的原油释放出来，从而使原来无法流动的原油随水驱迁移至生产井。基于此，有人建议将硝酸盐与其他合适的营养物质（如维生素和磷酸盐）一同注入，作为 MEOR 的一种改进方法。

除了依靠原生细菌，还研究了注入非 SRB 细菌和硝酸根离子等营养物质的方法，以避免使用有害的丙烯醛杀菌剂。有研究报道了以氧化硫化物和含硝酸盐油类有机物为营养物质的新型细菌。目前人们对于亚硝酸盐/硝酸盐处理还有很多未知，如硝酸盐和亚硝酸盐的相对有效性和减少生物硫化物产生的机制中哪一种为决定性机制。

14.4.3 其他生物抑制剂

以叠氮化钠（NaN_3）为代表的叠氮盐长期以来被称为生物抑制剂，并已建议用于防止油井的生物污损。叠氮化钠通过抑制革兰氏阴性菌中的细胞色素氧化酶起生物抑制剂的作用。不过截至目前尚无有关 NaN_3 对 SRB 清除效果方面的报道。

硒酸盐和一些过渡金属氧离子等（如钒酸盐、钼酸盐、高锰酸盐和钨酸盐）也均具有抑制 SRB 生长的特性，但这些化学品成本昂贵，应用规模相对有限。钼酸根离子也可用于一些缓蚀剂混合物。金属氧酸盐 SRB 抑制剂可以消耗 SRB 的 ATP 池，从而导致 SRB 死亡。

鱼藤酮是一种广谱杀菌剂、杀鱼剂（对鱼类有毒）和杀虫剂，近期有研究称鱼藤酮可以通过干扰微生物线粒体中的电子传递链，起到杀菌作用（图 14.21）。

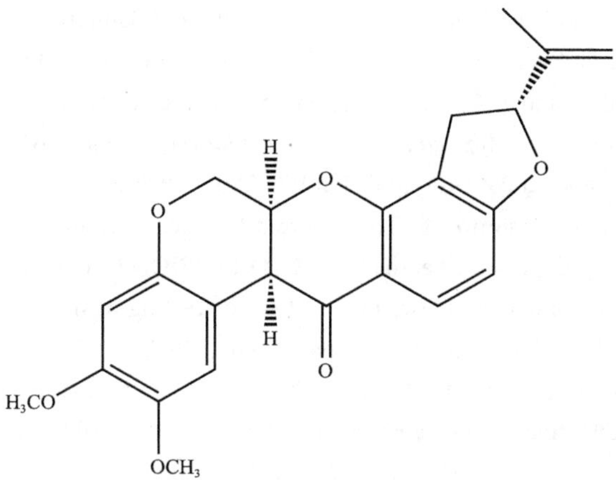

图 14.21　鱼藤酮结构式

总而言之，使用生物杀伤剂和生物抑制剂是防止生物成因硫化物形成（储层酸化）和 MIC 的最常见方法。如果前期已经产生了有毒的 H_2S，则还可以使用 H_2S 清除剂将其转化为无害的化学物质，这部分内容将在第 15 章讨论。

参 考 文 献

[1] M. Magot, "Indigenous Microbial Communities in Oil Fields," in *Petroleum Microbiology*, eds. B. Ollivier and M. Magot, Washington, DC: ASM Press, 2005, 21.

[2] P. A. Lapointe, M. A. Muhsin, and A. F. Maurin, "Microbial Corrosion and Biologically Induced New Products Example from a Seawater Injection System, Umm Shaif Field, U.A.E.," SPE 21367 (paper presented at the SPE Middle East Oil Show, Bahrain, 16-19 November 1991).

[3] J. L. Lynch and R. G. J. Edyvean, "Biofouling in Oilfield Water Systems—A Review," *Biofouling* 1(1988): 147-162.

[4] R. Cord-Ruwisch, W. Kleintz, and F. Widdel, "Sulfate-Reducing Bacteria and Their Activities in Oil Production," SPE 13554, *Journal of Petroleum Technology* 39 (1987): 97.

[5] P. F. Sanders and P. J. Sturman, "Biofouling in the Oil Industry," in *Petroleum Microbiology*, eds. B. Ollivier and M. Magot, Washington DC: ASM Press, 2005, 171.

[6] J.-L. Crolet, "Microbial Corrosion in the Oil Industry: A Corrosionist's View," in *Petroleum Microbiology*, eds. B. Ollivier and M. Magot Washington, DC: ASM Press, 2005, 143.

[7] S. Schultze-Lam, D. Fortin, B. S. Davis, and T. J. Beveridge, "Mineralization of Bacterial Surfaces," *Chemical Geology* 132 (1996): 171.

[8] A. F. Bird, H. R. Rosser, M. E. Worrall, K. A. Mously, and O. I. Fageeha, "Sulfate Reducing Bacteria Biofilms in a Large Seawater Injection System," SPE 73959 (paper presented at the International Conference on Health, Safety and Environment in Oil and Gas Exploration and Production, Kuala Lumpur, Malaysia, 20-22 March 2002).

[9] (a) M. Davies and P. J. B. Scott, Oilfield Water Technology, Houston, TX: National Association of Corrosion Engineers (NACE), 2006. (b) E. A. Morris, R. Gomez, and R. Peterson, "Application of Chemical and Microbiological Data for Sulfide Control," SPE 52705 (paper presented at the SPE/EPA Exploration and Production Environmental Conference, Austin, TX, 28 February-3 March 1999).

[10] F. Akersburg, F. Bak, and F. Widdel, "Anaerobic Oxidation of Saturated Hydrocarbons to CO_2 by a New Type of Sulfate-Reducing Bacterium," *Archives of Microbiology*, 156 (Feb 1991): 5.

[11] H. Beller, P. Spormann, and J. Cole, "Isolation and Characterization of a Novel Toluene-Degrading Sulfate-Reducing Bacterium," *Applied and Environmental Microbiology* 62 (1996): 1188.

[12] J. E. McElhiney, International Patent Application WO/2007/106691.

[13] (*a*) L. Watkins and J. W. Costerton, "The inherent biocide resistance of corrosion-causing biofilm bacteria," Paper 246 (paper presented at the NACE CORROSION Conference, Anaheim, CA, 1983). (*b*) I. Ruseska, J. Robbins, and J. W. Costerton, "Biocide Testing against Corrosion-Causing Oil-Field Bacteria Helps Control Plugging," *Oil & Gas Journal* 80 (1982): 253. (*c*) V. Keasler, B. Bennett, R. Diaz, P. Lindmuth, D. Kasowski, C. Adelizzi, Nalco, and L. Santiago-Vazquez, "Identification and Analysis of Biocides Effective Against Sessile Organisms," SPE 121082, SPE International Symposium on Oilfield Chemistry, The Woodlands, TX, 20-22 April 2009.

[14] T. Thorstenson, G. Boedtker, B.-L. P. Lilleboe, T. Torsvik, E. Sunde, and J. Beeder, "Biocide Replacement by Nitrate in Sea Water Injection," Paper 02033 (paper presented at the NACE CORROSION Conference, 2002).

[15] D. B. McIlwaine, J. Diemer, and L. Grab, "Determining the Biofilm Penetrating Ability of Various Biocides Utilizing an Artificial Biofilm Matrix," Paper 97400 (paper presented at the NACE CORROSION Conference, 1997).

[16] C. J. Nalepa, H. Ceri, and C. A. Stremick, "A Novel Technique for Evaluating the Activity of Biocides against Biofilm Bacteria," Paper 00347 (paper presented at the NACE CORROSION Conference, 2000).

[17] B. Yin, J. Yang, U. Bertheas, and J. Adams, "A High Throughput Evaluation of Biocides for Biofouling Control in Oilfields," *Chemistry in the Oil Industry X: Oilfield Chemistry*, Royal Society of Chemistry, Manchester, UK, 5-7 November 2007.

[18] R. G. Eager, J. Leder, J. P Stanley, and A. B. Theis, "The Use of Glutaraldehyde for Microbiological Control in Waterflood Systems," *Materials Performance* 27 (1988): 40.

[19] N. T. Macchiarolo, B. McGuire, and J. M. Scalise, U.S. Patent 4297224, 1981.

[20] T. K. Haack, E. S. Lashen, and D. E. Greenly, *Developments in Industrial Microbiology* (*Journal of Industrial Microbiololgy* Suppl. 3) (1988): 247.

[21] T. K. Haack, D. E. Greenley, U.S. Patent 5026491, 1991.

[22] S. Maxwell, "Controlling Corrosive Biofilms by the Application of Biocides," SPE 93172 (paper presented at the SPE International Symposium on Oilfield Corrosion, Aberdeen, UK, 13 May 2005).

[23] E. Sunde, T. Thorstenson, and T. Torsvik, "Growth of Bacteria on Water Injection Additives," SPE 20690 (paper presented at the SPE Annual Technical Conference and Exhibition, New Orleans, LA, 23-26 September 1990).

[24] C. J. Nalepa, J. N. Howarth, and F. D. Azarnia, "Factors to Consider when Applying Oxidizing Biocides in the Field," Paper 02223 (paper presented at the NACE CORROSION Conference, 2002).

[25] K. B. Flatval, S. Sathyamorrthy, C. Kuijvenhoven, and D. Ligthelm, "Building the Case for Raw Seawater Injection Scheme in Barton," SPE 88568 (paper presented at the SPE Asia Pacific Oil and Gas Conference and Exhibition, Perth, Australia, 18-20 October 2004).

[26] (*a*) D. M. Clementz, D. E. Patterson, R. J. Aseltine, and R. E. Young, "Stimulation of Water Injection Wells in the Los Angeles Basin by Using Sodium Hypochlorite and Mineral Acids," SPE 10624, *Journal of Petroleum Technology* 34 (1982): 2087. (*b*) J. Winter, M. Ilbert, P. C. F. Graf, D. Özcelik, and U.

Jakob, *Cell*, 135（2008）: 691.
[27] S. Yang, World Patent Application WO/2004/026770.
[28] R. M. Moore Jr. and C. J. Nalepa, U.S. Patent 6322822, 2001.
[29] C. J. Nalepa, U.S. Patent 7087251 2006.
[30] J. F. Carpenter and C. J. Nalepa, "Bromine-Based Biocides for Effective Microbiological Control in the Oil Field," SPE 92702（paper presented at the SPE International Symposium on Oilfield Chemistry, Houston, TX, 2-4 February 2005）.
[31] C. J. Nalepa, U.S. Patent 6419838, 2002.
[32] J. A. Findlay, "The Potential of Alkyl Amines as Antifouling Biocides I: Toxicity and Structure-Activity Relationships," *Biofouling* 9（4）（1996）: 257.
[33] M. J. Mayer and F. L. Singleton, U.S. Patent Application 20070045199.
[34] H. N. Cheng, D. Sharoyan, M. J. Mayer, and F. L. Singleton, International Patent Application, WO/2008/091678.
[35] A. Gupta, M. Ramesh, and R. Elliott, International Patent Application WO/2008/083182.
[36] G. H. Zaid and D. W. Sanders, U.S. Patent 6431279, 2002.
[37] T. J. Parkinson and A. T. Harris, U.S. Patent 6325970, 2001.
[38] J. R. Ohlsen, J. M. Brown, G. F. Brock, and V. K. Mandlay, U.S. Patent 5459125, 1995.
[39] M. L. Ludyanakiy and F. J. Himpler, "The Effect of Halogenated Hydantoins on Biofilms," Paper 97405（paper presented at the NACE CORROSION Conference, 1997）.
[40] B. R. Sook, T. F. Ling, and A. D. Harrison, "A New Thixotropic Form of Bromochlorodimethylhydantoin: A Case Study," Paper 03715（paper presented at the NACE CORROSION Conference, 2003）.
[41]（a）D. M. Brandon, J. P. Fillo, A. E. Morris, and J. M. Evans, "Biocide and Corrosion Inhibition Use in the Oil and Gas Industry: Effectiveness and Potential Environmental Impacts," SPE 29735（paper presented at the SPE/EPA Exploration and Production Environmental Conference, Houston, TX, 27-29 March 1995）.（b）H. R. Mcginley, M. Enzien, G. Hancock, S. Gonisor, and M. Mikoztal, "Glutaraldehyde: An Understanding of its Ecotoxicity Profile and Environmental Chemistry," Paper 09405（paper presented at the NACE, Corrision 2009 Conference and Exposition, Atlanta, GA, 22-26 March 2009）.
[42] R. G. Eager, J. Leder, J. P. Stanley, and A. B. Theis, "Glutaraldehyde: Impact of Corrosion Causing Biofilms," Papers 86-125（papers presented at the NACE CORROSION Conference, 1986）.
[43] T. M. LaMarre and C. H. Martin, U.S. Patent 4616037, 1986.
[44] W. G. McLelland, "Results of Using Formaldehyde in a Large North Slope Water Treatment System," SPE 35675（paper presented at the SPE Computer Applications, April 1997, 55）.
[45] B. G. Kriel, A. B. Crews, E. D. Burger, E. Vanderwende, and D. O. Hitzman, "The Efficacy of Formaldehyde for the Control of Biogenic Sulfide Production in Porous Media," SPE 25196（paper presented at the SPE International Symposium on Oilfield Chemistry, New Orleans, LA, 2-5 March 1993）.
[46] J. T. Fenton and J. F. Miller, U.S. Patent 4911923, 1990.
[47] R. M. Jorda, "Aqualin Biocide in Injection Waters," SPE 280（paper presented at the SPE Production Research Symposium, Tulsa, OK, 12-13 April 1962）.
[48] C. Reed, J. Foshee, J. E. Penkala, and M. Roberson, "Acrolein Application to Mitigate Biogenic Sulfides and Remediate Injection Well Damage in a Gas Plant Water Disposal System," SPE 93602

[49] J. Penkala, M. D. Law, D. Horaska, and A. L. Dickinson, "Baker Petrolite, Acrolein 2-Propenal: A Versatile Microbiocide for Control of Bacteria in Oilfield Systems," Paper 04749 (paper presented at the NACE CORROSION Conference, 2004).

(paper presented at the SPE International Symposium on Oilfield Chemistry, The Woodlands, TX, 2-4 February 2005).

[50] A. B. Theis and J. Leder, U.S. Patent 5128051, 1992.

[51] K. P. Davis and R. E. Talbot, K Patent Application GB2145798, 1984.

[52] N. Macleod, T. Bryan, A. J. Buckley, R. E. Talbot, and M. A. Veale, SPE 30171, Society of Petroleum Engineers eLibrary, 1994.

[53] B. L. Downward, R. E. Talbot, and T. K. Haack, "Tetrakishydroxymethylphosphonium sulfate (THPS), A New Industrial Biocide with Low Environmental Toxicity," Paper 97401 (paper presented at the NACE CORROSION Conference, 1997).

[54] R. L. Martin, "Unusual Oilfield Corrosion Inhibitors," SPE 80219 (paper presented at the SPE International Symposium on Oilfield Chemistry, Houston, TX, 5-7 February 2003).

[55] K. G. Cooper, R. E. Talbot, and M. J. Turvey, U.K. Patent Application GB2178960, 1986.

[56] C. R. Jones, G. Collins, B. L. Downward, and K. Hernandez, "THPS: A Holistic Approach to Treating Sour Systems," Paper 08659 (paper presented at the NACE CORROSION Conference & Exposition, New Orleans, LA, 16-20 March 2008).

[57] E. Bryan, M. A. Veale, R. E. Talbot, K. G. Cooper, and N. S. Matthews, European Patent EP0385801, 1990.

[58] C. R. Jones and R. Diaz, International Patent Application WO/2005/079578.

[59] P. D. Gilbert, J. M. Grech, R. E. Talbot, M. A. Veale, and K. A. Hernandez, "Tetrakishydroxymethylphosphonium Sulfate (THPS), for Dissolving Iron Sulfides Downhole and Topside—A Study of the Chemistry Influencing Dissolution," Paper 02030 (paper presented at the NACE CORROSION Conference, 2002).

[60] H. A. Nasr-El-Din, A. M. Al-Mohammad, M. A. Al-Hajri, and J. B. Chesson, "A New Chemical Treatment to Remove Multiple Damages in a Water Supply Well," SPE 95001 (paper presented at the SPE European Formation Damage Conference, Sheveningen, The Netherlands, 25-27 May 2005).

[61] J. C. Jeffery, B. Odell, N. Stevens, and R. E. Talbot, "Self-Assembly of a Novel Water-Soluble Iron (II) Macrocylic Phosphine Complex from Tetrakis (Hydroxymethyl) Phosphonium Sulfate and Iron (II) Ammonium Sulfate: Single Crystal X-ray Structure of the Complex $[Fe(H_2O)_2\{RP(CH_2N(CH_2PR_2)CH_2)_2PR\}]SO_4 \times 4H_2O$ ($R=CH_2OH$)," *Chemical Communications* (2000): 101.

[62] P. R. Rincon, J. P. McKee, C. E. Tarazon, L. A. Guevara, and B. Vinccler, "Biocide Stimulation in Oilwells for Downhole Corrosion Control and Increasing Production," SPE 87562 (paper presented at the SPE International Symposium on Oilfield Corrosion, Aberdeen, United Kingdom, 28 May 2004).

[63] K. G. Cooper, R. E. Talbot, and M. J. Turvey, U.S. Patent 5741757, 1998.

[64] C. R. Jones and R. E. Talbot, U.S. Patent 6784168, 2004.

[65] T. F. McNeel, D. L. Comstock, M. Z. Anstead, and R. A. Clark, U.S. Patent 6180056, 2001.

[66] G. P. Otter, S. G. Breen, G. Woodward, R. E. Talbot, R. S. Padda, K. P. Davis, S. D'Arbeloff-Wilson, and C. R. Jones, International Patent Application WO/2003/021031.

[67] J. F. Kramer, F. O'Brien, and S. F. Strba, "A New High Performance Quaternary Phosphonium Biocide for Microbiological Control in Oilfield Water Systems," Paper 08660 (paper presented at the NACE CORROSION Conference & Exposition, New Orleans, LA, 16-20 March 2008).

[68] V. Vlasaty and D. Q. Cao, International Patent Application WO/2008/019320

[69] D. T. Murray, "A New Quat Demonstrates High Biocidal Efficacy with Low Foam," Paper 97406 (paper presented at the NACE CORROSION Conference, 1997).

[70] J. E. Gannon and S. Thornburgh, World Patent Application WO/1988/002351.

[71] A. Naraghi and N. Obeyesekere, International Patent Application WO/2006/034101.

[72] V. L. Colclough, U.S. Patent 6303557, 2001.

[73] R. F. Stockel, U.S. Patent 4891423, 1990.

[74] J. G. Fenyes and J. D. Pera, U.S. Patent 4778813, 1988.

[75] L. G. Kleina, M. H. Czechowski, J. S. Clavin, W. K. Whitekettle, and C. R. Ascolese, "Performance and Monitoring of a New Non-Oxidizing Biocide—The Study of BNPD/ISO and ATP," Paper 97403 (paper presented at the NACE CORROSION Conference, 1997).

[76] D. M. Bryce, B. Croshaw, J. E. Hall, V. R. Holland, and B. Lessel, "The Activity and Safety of the Antimicrobial Agent Bronopol (2-Bromo-Nitropan-1, 3-Diol)," *Journal of the Society of Cosmetic Chemists* 29 (1978): 3.

[77] C. D. Gartner, U.S. Patent 5627135, 1997.

[78] J. M. Cronan Jr. and M. J. Myer, U.S. Patent 7008545, 2006.

[79] E. S. Littmann and T. L. McLean, "Chemical Control of Biogenic H_2S in Producing Formations," SPE 16218 (paper presented at the SPE Production Operations Symposium, Olahoma City, OK, 8-10 March 1987).

[80] H. Jin-Ying, Z. Jia-Shen, F. Chao-Yang, Q. Jun-e, and L. Jian-Guo, "The Inhibition Effects of a New Heterocyclic Bisquaternary Ammonium Salt in Simulated Oilfield Water," *Anti-Corrosion Methods and Materials* 51 (2004): 272.

[81] T. M. Williams, R. Levy, and B. Hegarty, "Control of SRB Biofouling and MIC by Chloromethyl-Methyl-Isothiazolone," Paper 01273 (paper presented at the NACE CORROSION Conference, 2001).

[82] S. N. Lewis, G. A. Miller, and A. B. Law, U.S. Patent 4322475.

[83] R. P. Clifford and G. A. Birchall, U.S. Patent 4539071, 1985.

[84] T. M. Williams, B. M. Hegarty, and R. Levy, "Control of Oilfield Biofouling," Paper 01273 (paper presented at the NACE CORROSION Conference, 2001).

[85] T. K. Haack, European Patent P 337624A, 1989.

[86] B. M. Hegarty and R. Levy, U.S. Patent 5827433, 1998.

[87] D. Antoni-Zimmermann, R. Baum, T. Wunder, and H.-J. Schmidt, U.S. Patent Application 20050124674.

[88] N. E. Thompson and M. Greenhalgh, International Patent Application WO/2007/139950.

[89] R. J. Starkey, G. A. Monteith, and C. W. Aften, International Patent Application WO/2008/016662.

[90] J. H. Payton, U.S. Patent 3996378, 1976.

[91] A. J. Telang, S. Ebert, J. M. Foght, D. W. S. Westlake, and G. Voordouw, "Effects of Two Diamine Biocides on the Microbial Community from an Oil Field," *Canadian Journal of Microbiology/Review* 44 (11) (1998): 1060.

[92] M. Ludensky, C. Hill, and F. C. A. Lichtenberg, U.S. Patent Application 20030228373.

[93] F. W. Valone, U.S. Patent 4647589, 1987.

[94] J. G. La Zonby, U.S. Patent 5209824, 1993.

[95] R. W. Walter, A. G. Relenyi, and R. L. Johnson, U.S. Patent 4816061, 1989.

[96] R. W. Walter and L. M. Cooke, Paper 410 (paper presented at the NACE CORROSION Conference,

1997).

[97] W. K. Whitekettle and D. K. Donofrio, U.S. Patent 4916158, 1990.

[98] B. Heer, G. Tiedtke, and B. M. Hegarty, U.S. Patent Application 20040198713.

[99] J. S. Gill and A. Gupta, U.S. Patent Application 20030200997.

[100] H. Eggensperger and K.-H. Diehl, U.S. Patent 4148905, 1979.

[101] T. E. McNeel, M. S. Whittlemore, S. D. Bryant, and G. H. Vunk, World Patent Application WO/1999/006325.

[102] T. E. McNeel, M. S. Whittlemore, S. D. Bryant, and G. H. Vunk, World Patent Application WO/1999/005912.

[103] G. E. Jenneman, A. Greene, and G. Voordouw, U.S. Patent Application 20050238729.

[104] E. D. Burger, A. B. Crews, and H. W. Ikerd II, Paper 01274 (paper presented at the NACE CORROSION Conference, 1999).

[105] E. D. Burger and J. M. Odom, "Mechanisms of Anthraquinone Inhibition of Sulfate-Reducing Bacteria," SPE 50764 (paper presented at the SPE International Symposium on Oilfield Chemistry, Richardson, TX, 1999).

[106] F. B. Cooling, C. L. Maloney, E. Nagel, J. Tabinowski, and J. D. Odom, "Inhibition of Sulfaterespiration by 1, 8-Dihydroxyanthraquinone and Other Anthaquinone Derivatives," *Applied and Environmental Microbiology* 62 (1996): 2999.

[107] M. D. Law, M. B. Kretsinger, E. D. Burger, J. G. Schoenenberger, and M. A. Ulman, "A Field Case History: Chemical Treatment of a Produced-Water Injection System Using Anthraquinone Improves Water Quality and Reduces Costs," SPE 65023 (paper presented at the SPE International Symposium on Oilfield Chemistry, Houston, TX, 13-16 February 2001).

[108] M. D. Johnson, M. L. Harless, A. L. Dickinson, and E. D. Burger, "A New Chemical Approach to Mitigate Sulfide Production in Oilfield Water Injection Systems," SPE 50741 (paper presented at the SPE International Symposium on Oilfield Chemistry, Houston, TX, 16-19 February 1999).

[109] A. Dickinson, G. Peck, and B. Arnold, "Effective Chemistries to Control SRB, H_2S, and FeS Problems," SPE 93007 (paper presented at the SPE Western Regional Meeting, Irvine, CA, 30 March-1 April 2005).

[110] E. D. Burger, C. A. Andrade, M. Rebelo, and R. Ribeiro, "Flexible Treatment Program for Controlling H_2S in FPSO Produced-Water Tanks," SPE 106106 (paper presented at the SPE International Symposium on Oilfield Chemistry, Houston, TX, 28 February-2 March 2007).

[111] R. L. Martin, "Corrosion Consequences of Nitrate/Nitrite Additions to Oilfield Brines," SPE 114923 (paper presented at the Annual Technical Conference and Exhibition, Denver, CO, 22-24 September 2008).

[112] M. A. Reinsel, J. T. Sears, P. S. Stewart, and M. J. McInerney, "Isolation and Characterization of Strains CVO and FWKO B, Two Novel Nitrate-Reducing, Sufide-Oxidizing Bacteria Isolated from Oil Field Brine," *Journal of Industrial Microbiology* 17 (2) (1996): 128.

[113] D. O. Hitzman and D. M. Dennis, "Sulfide Removal and Prevention in Gas Wells," SPE 50980, *SPE Reservoir Evaluation & Engineering* 1 (4) (1998): 367.

[114] P. J. Sturman and D. M. Goeres, "Control of Hydrogen Sulfide in Oil and Gas Wells with Nitrite Injection," SPE 56772 (paper presented at the SPE Annual Technical Conference and Exhibition, Houston, TX, 3-6 October 1999).

[115] C. Hubert, M. Nemati, G. Jenneman, and G. Voordouw, "Containment of Biogenic Sulfide Priduction

in Continuous Up-Flow Packed-Bed Bioreactors with Nitrate or Nitrite," *Biotechnology Progress* 19(2)(2003): 338.

[116] G. Voordouw, M. Nemati, and G. E. Jenneman, "Use of Nitrate-Reducing, Sulfide-Oxidizing Bacteria to Reduce Souring in Oil Fields," Paper 02034 (paper presented at the NACE CORROSION Conference, 2002).

[117] E. A. Morris, R. M. Derr, T. M. Kenney, and D. H. Pope, "Field and Laboratory Tests on Nitrate Treatment for Potential Use in Natural Gas Operations—Stimulate Non-SRB Bacteria," SPE 29738 (paper presented at the SPE/EPA Exploration and Production Environmental Conference, Houston, TX, 27-29 March 1995).

[118] J. F. D. Scott, "Modern Concepts of Chemical Treatment for the Control of Microbially-Induced Corrosion in Oilfield Water Systems," Proceedings of Chemistry in the Oil Industry IX Symposium, Royal Society of Chemistry Publications, Manchester, UK, 31 October-2 November 2005.

[119] C. Hubert, G. Voordouw, M. Nemati, and G. E. Jenneman, "Is Souring and Corrosion by Sulfate-Reducing Bacteria in Oil Fields Reduced More Efficiently by Nitrate or by Nitrate?," Paper 04762 (paper presented at the NACE CORROSION Conference, 2004).

[120] E. A. Greene, C. Hubert, M. Nemati, G. E. Jenneman, and G. Voordouw, "Nitrite Reductase Activity of Sulfate-Reducing Bacteria Prevents Their Inhibition by Nitrate-Reducing, Sulfide-Oxidizing Bacteria," *Environmental Microbiology* 5 (2003): 607.

[121] D. O. Hitzman and G. T. Sperle, "A New Microbial Technology for Enhanced Oil Recovery and Sulfide Prevention and Reduction," SPE 27752 (paper presented at the SPE/DOE Enhanced Oil Recovery Symposium, Tulsa, OK, 17-20 April 1994).

[122] K. A. Sandbeck and D. O. Hitzman, "Biocompetitive Exclusion Technology: A Field System to Control Reservoir Souring and Increase Production" (paper presented at the 5th Int. Conf. on Microbial Enhanced Oil Recovery and Related Biotechnology for Solving Environmental Problems, Sponsored by U.S. DOE, 1995).

[123] D. O. Hitzman and D. M. Dennis, "New Technology for Prevention of Sour Oil and Gas," SPE 37908 (paper presented at the SPE/EPA Exploration and Production Environmental Conference, Dallas, TX, 3-5 March 1997).

[124] D. O. Hitzman, G. T. Sperl, and K. A. Sandbeck, U.S. Patent 5750392, 1998.

[125] M. A. Reinsel, J. T. Sears, P. S. Stewart, and M. J. McInerney, "Control of Microbial Souring by Nitrate, Nitrite or Glutaraldehyde Injection in a Sandstone Column," *Journal of Industrial Microbiology* 17 (1996): 128.

[126] D. O. Hitzman, M. Dennis, and D. C. Hitzman, "Recent Successes: MEOR Using Synergistic H_2S Prevention and Increased Oil Recovery Systems," SPE 89453 (paper presented at the SPE/DOE Symposium on Improved Oil Recovery, Tulsa, OK, 17-21 April 2004).

[127] M. Collison, "Biological H_2S Removal Is Gaining a Toehold in the Sour Gas Fields of Western Canada," *Oilweek Magazine*, April 2006.

[128] A. Anchliya, "New Nitrate-Based Treatments—A Novel Approach to Control Hydrogen Sulfide in Reservoir and to Increase Oil Recovery," SPE 100337 (paper presented at the SPE Europec/EAGE Annual Conference and Exhibition, Vienna, Austria, 12-15 June 2006).

[129] J. Larsen, "Downhole Nitrate Applications to Control Sulfate Reducing Bacteria Activity and Reservoir Souring," Paper 02025 (paper presented at the NACE CORROSION Conference, 2002).

[130] E. Sunde and T. Torsvik, "Microbial Control of Hydrogen Sulfide Production in Oil

Reservoirs," *Petroleum Microbiology*, eds. B. Ollivier and M. Magot, Washington, DC: ASM Press, 2005, 201.

[131] D. O. Hitzman, G. T. Sperl, and K. A. Sandbeck, U.S. Patent 5405531, 1995.

[132] G. Voordouw, B. Buziak, S. Lin, A. Grigoryan, K. M. Kaster, G. E. Jenneman, and J. J. Arensdorf, "Use of Nitrate or Nitrite for the Management of the Sulfur Cycle in Oil and Gas Fields," SPE 106288 (paper presented at the SPE International Symposium on Oilfield Chemistry, Houston, TX, 28 February-2 March 2007).

[133] D. M. Dennis and D. O. Hitzman, "Advanced Nitrate-Based Technology for Sulfide Control and Improved Oil Recovery," SPE 106154 (paper presented at the SPE International Symposium on Oilfield Chemistry, Houston, TX, 28 February-2 March 2007).

[134] C. Kuijvenhoven, A. Bostock, D. Chappell, J. C. Noirot, and A. Khan, "Use of Nitrate to Mitigate Reservoir Souring in Bonga Deepwater Development Offshore Nigeria," SPE 92795 (paper presented at the SPE International Symposium on Oilfield Chemistry, The Woodlands, TX, 2-4 February 2005).

[135] C. Koijvenhoven, J. C. Noirot, P. Hubbard, and L. Oduola, "One Year Experience with the Injection of Nitrate to Control Souring in Bonga Deepwater Development Offshore Nigeria," SPE 105784 (paper presented at the SPE International Symposium on Oilfield Chemistry, Houston, TX, 28 February-2 March, 2007).

[136] E. Sunde and T. Torsvik, U.S. Patent 6758270, 2004.

[137] R. B. Cassinis, W. A. Farrone, and J. H. Portwood, "Microbial Water Treatment: An Alternative Treatment to Manage Sulfate Reducing Bacteria (SRB) Activity, Corrosion, Scale, Oxygen, and Oil Carry-Over at Wilmington Oil Field, Wilmington, CA," SPE 49152 (paper presented at the SPE Annual Technical Conference and Exhibition, New Orleans, LA, 27-30 September 1998).

[138] C. Hubert, G. Voordouw, J. Arensdorf, and G. E. Jenneman, "Control of Souring Through a Novel Class of Bacteria That Oxidize Sulfide as Well as Oil Organics with Nitrate," Paper 06669 (paper presented at the NACE CORROSION Conference, 2006).

[139] D. R. Grimshaw, U.S. Patent Application, 20060185851.

[140] S. L. Percival, "The Effect of Molybdenum on Biofilm Development," *Journal of Industrial Microbiology and Biotechnology* 23 (1999): 112.

[141] P. Angell and D. C. White, "Is Metabolic Activity by Biofilms with Sulfate-Reducing Bacterial Consortia Essential for Long-Term Propagation of Pitting Corrosion of Stainless Steel?," *Journal of Industrial Microbiology and Biotechnology* 15 (4) (1995): 329.

[142] D. Stepan and D. Ye, International Patent Application WO/2008/076928.

[143] B. Ollivier and M. Magot, *Petroleum Microbiology*, ASM Press, 2005.

[144] C. Whitby and T. L. Skovhus, eds., "Applied Microbiology and Molecular Biology in Oilfield Systems," Proceedings from the International Symposium on Applied Microbiology and Molecular Biology in Oil Systems (ISMOS-2), Springer, 2011.

[145] M. D. Lohithesh, A. K. Agnihotri, and B. Lal, "Control of Sulfate Reducing Bacteria in Oil and Gas Pipelines," Paper 118410 (paper presented at the Abu Dhabi International Petroleum Exhibition and Conference, Abu Dhabi, UAE, 3-6 November 2008).

[146] L. Abney, U.S. Patent Application 20090140133.

[147] Z. Amjad, ed., *The Science & Technology of Industrial Water Treatment*, Chapter 19. Boca Raton, FL: CRC Press, 2010.

[148] S. M. Shevchenko, M. J. Murcia, and P. J. Macuch, International Patent Application WO/2009/155390.

[149] K. G. Wunch and J. E. Penkala, U.S. Patent Application 20110195938.

[150] J. M. Wilson, J. D. Weaver, and B. F. Slabaugh, U.S. Patent Application 20090062156.

[151] A. Gupta and E. H. K. Zeiher, International Patent Application WO/2010/083070.

[152] J. R. Rovison, M. John, S. Huang, and H. A. Pfeffer, U.S. Patent Application 20100160449.

[153] R. M. De Paula, V. Keasler, J. Li, D. McSherry, and R. Staub, "Development of Peracetic Acid (PAA) as an Environmentally Safe Biocide for Water Treatment During Hydraulic Fracturing Applications," SPE 164088 (paper presented at the 2013 SPE International Symposium on Oilfield Chemistry, The Woodlands, TX, 8–10 April 2013).

[154] M. Wiencek and J. S. Chapman, "Water Treatment Biocides: How Do They Work and Why Should You Care?," Paper 99308 (paper presented at the CORROSION 99, San Antonio, TX, 25–30 April 1999).

[155] V. V. Keasler, H. R. McGinley, and B. M. Bennett, "Analysis of Bacterial Kill versus Corrosion from Use of Common Oilfield Biocides," IPC2010-31593 (paper presented at the 8th International Pipeline Conference, Alberta, Canada, 27 September–1 October 2010).

[156] T. O. Glasbey, G. D. Probert, G. S. Whiteley, and R. K. Whiteley, International Patent Application WO/2011/134005.

[157] D. D. Horaska, J. E. Penkala, C. A. Reed, M. D. Law, S. H. Gaffney, and M. M. Srour, SPE, Baker Hughes Incorporated; and A. S. Al-Harthy, "Field Experiences Detailing Acrolein (2-Propenal) Treatment of a Produced Water Injection System in the Sultanate of Oman," SPE 120238 (paper presented at the SPE Middle East Oil and Gas Show and Conference, Bahrain, Bahrain, 15–18 March 2009).

[158] D. D. Horaska, C. San Juan, A. L. Dickinson, S. L. Lear, and A. Colquhoun, "Acrolein Provides Benefits and Solutions to Offshore Oilfield Production Problems," SPE 146080 (paper presented at the SPE Annual Technical Conference and Exhibition, Denver, CO, 30 October–2 November 2011).

[159] K. Annadorai and A. Darwin, "Effect of THPS on Discharge Water Quality: Lessons Learned Study," SPE 125785 (paper presented at the SPE International Conference on Health, Safety and Environment in Oil and Gas Exploration and Production, Rio de Janeiro, Brazil, 12–14 April 2010).

[160] C. Jones, S. Edmunds, and A. Fellows, U.S. Patent Application 20120046248.

[161] K. Zhao, J. Wen, T. Gu, A. Kopliku, and I. Cruz, "Mechanistic Modeling of Anaerobic THPS Biocide Degradation under Alkaline Conditions," *Materials Performance* 62 (2009).

[162] S. Kesavan, G. Woodward, A. Adedeji, T. Curtis, and F. Smith, International Patent Application WO/2010/093473.

[163] S. M. Tawfik, A. Sayed, and I. Aiad, *Journal of Surfactants and Detergents* 15 (2012): 577.

[164] M. Petkovic, D. O. Hartmann, G. Adamová, K. R. Seddon, L. P. N. Rebelo, and C. S. Pereira, *New Journal of Chemistry* 36 (2012): 56–63.

[165] A. Kanawaza, T. Ikeda, and T. Endo, *Antimicrobial Agents and Chemotherapy* 38 (5) (1994): 945–952.

[166] J. C. Bradaric-Baus and Y. Zhou, International Patent Application WO/2004/094438.

[167] C. Riverol and V. Pilipovik, *Industrial & Engineering Chemistry Research*, submitted 2013.

[168] L. M. Timofeeva, N. A. Kleshcheva, A. F. Moroz, and L. V. Didenko, *Biomacromolecules* 10 (2009): 2976–2986.

[169] R. DeSousa, N. L. Dassanayake, and H. A. Ketelson, U.S. Patent Application 20100158853.

[170] D. Feldman, M. Adda, and R. J. Roccon, U.S. Patent Application 20090117202.

[171] T. M. Williams, "Efficacy of Isothiazolone Biocide versus Sulfate Reducing," Paper 09059 (paper presented at the CORROSION 2009, Atlanta, GA, 22-26 March 2009).

[172] M. A. Diehl and D. A. Shaw, U.S. Patent Application 20110224270.

[173] R. Baum, H.-J. Schmidt, and T. Wunder, Savides, European Patent Application EP2213166.

[174] Available at http://sc.akzonobel.com/en/watertreatment/Documents/AkzoNobel_tb_Aquatreat_biocides.pdf.

[175] O. M. Musa, International Patent Application WO/2011/163317.

[176] C. Hubert, "Microbial Ecology of Oil Reservoir Souring Control by Nitrate Injection," in *Handbook of Hydrocarbon and Lipid Microbiology*, ed. K.N. Timmis. Berlin: Springer, (2010): 2753-2766.

[177] J. F. D. Stott, "Implementation of Nitrate Treatment for Reservoir Souring Control: Complexities and Pitfalls," SPE 155155 (paper presented at the SPE International Conference and Exhibition on Oilfield Corrosion, Aberdeen, UK, 28-29 May 2012).

[178] C. Hubert, M. Nemati, G. Jenneman, and G. Voordouw, *Applied Microbiology and Biotechnology* 68 (2005): 272-282.

[179] J. Wallace, U.S. Patent Application 20110056693.

[180] J. Waage, A.-M. B. Hårvik, K. Kroknes, and A. F. Mitchell, "Reservoir Souring: How to Minimize the Impact Based on Current Knowledge?," (paper presented at the Tekna 23rd International Oil Field Chemistry Symposium, Geilo, Norway, 18-21 March 2012).

[181] E. D. Burger, G. E. Jenneman, and X. Gao, "The Impact of Dissolved Organic-Carbon Type on the Extent of Reservoir Souring," SPE 164068 (paper presented at the 2013 SPE International Symposium on Oilfield Chemistry, The Woodlands, TX, 8-10 April 2013).

[182] J. B. Harris, A. Stepp, T. Pierce, R. Webb, G. E. Jenneman, and E. D. Burger, "Laboratory Evaluation of H_2S Bioscavenging in Produced Water at 60℃," SPE 164129 (paper presented at the 2013 SPE International Symposium on Oilfield Chemistry, The Woodlands, TX, 8-10 April 2013).

[183] P. Evans, "Reservoir Souring Challenges and Solutions from the Operators Perspective," (paper presented at the 4th International Symposium on Applied Microbiology and Molecular Biology in Oil Systems (ISMOS4), Rio de Janeiro, Brazil, 25-28 August 2013).

[184] B. Lomans, "Potentials of Chlorate Reduction for Souring Mitigation," (paper presented at the 4th International Symposium on Applied Microbiology and Molecular Biology in Oil Systems (ISMOS4), Rio de Janeiro, Brazil, 25-28 August 2013).

[185] G. Neal, K. Kleinwolterink, L. Abney, and L. Gloe, "Nonchemical Bacteria-Control Process," SPE 133368 (paper presented at the SPE Asia Pacific Oil and Gas Conference and Exhibition, Brisbane, Queensland, Australia, 18-20 October 2010).

[186] J. A. Haggstrom, J. E. Bryant, J. Holtsclaw, and J. D. Weaver, U.S. Patent Application 20120103919.

[187] J. Holtsclaw, J. D. Weaver, L. Gloe, and M. A. McCabe, U.S. Patent 8276663, 2012.

15 硫化氢清除剂

15.1 概述

硫化氢（H_2S）是一种剧毒和刺激性气体，对石油和天然气工业的上游和下游都会造成危害。即使在较低的浓度下，H_2S 也会对暴露的人员造成严重的伤害。通常要求商品天然气中 H_2S 的浓度约低于 4mg/L。H_2S 通常伴有少量的硫醇（RSH 或 R_2S），如甲硫醇（CH_3SH）、芳香族硫化物、多硫化物和羰基硫化物（COS）。

H_2S 是一种易溶于水的酸性气体，形成的弱酸能部分解离成 HS^- 和 S^{2-}。

$$H_2S + H_2O \Longleftrightarrow H_3O^+ + HS^- \qquad pK_a = 6.9$$

$$HS^- + H_2O \Longleftrightarrow H_3O^+ + S^{2-} \qquad pK_a = 19$$

阴离子的浓度由 pH 值控制，特别是由另一种酸性气体 CO_2 决定。关于 H_2S 在油田系统的溶解分散关系已有资料介绍。H_2S 具有腐蚀性，可与油井和管道的钢材发生反应，造成点蚀和应力腐蚀开裂，并造成硫化亚铁垢的沉积。其他的硫化物垢还包括硫化锌和硫化铅（见第 3 章）。

储层中有若干个自然过程可产生 H_2S，包括本地硫酸盐还原菌（SRB）造成的细菌性硫酸盐还原、热裂解和硫酸盐热化学还原（TSR）作用。一般而言，对于储层温度在 140~150℃ 以上的油田，H_2S 问题可以忽略不计，因为该温度下 SRB 不会存活。有人认为 TSR 产生了大多数的 H_2S，但是其他研究表明，烃类的 TSR 作用只发生在储层中温度远高于 140℃ 的部位。这种现象涉及烃类的氧化和硫酸盐（来自天然的或由于注入海水引入硫酸根离子所形成的石膏）的还原，并产生 H_2S、CO_2、碳酸盐矿物和重质有机硫化合物等副产物。

在为二次采油而注入海水的油藏中，产生的 H_2S 通常会显著增加。海水含有浓度约为 2800mg/L 的硫酸根离子，会被储层中的本地 SRB 和 TSR 过程还原成 H_2S，最终到达生产井。SRB 的生长还需要有如有机酸等容易代谢的碳源。通常其在储层流体中的量足以维持 SRB 的生长。挥发性脂肪酸离子（如乙酸根）历来被认为是 SRB 生长的有利营养源，但最近的现场经验表明，其他溶解的有机碳源也能促进 SRB 生长。减少储层 H_2S 含量的一种方法是用足够的杀菌剂处理注入水，其他预防方法包括控制 SRB 的代谢或促进非本地 SRB 的生长，详细内容可参考第 14 章。最后，防止生物硫化物形成的另一种方法是注入不含硫酸根的地层水，或者注入利用膜技术脱除硫酸根的海水。在后一种情况下，并非能脱除海水中的所有硫酸根离子，但是这一方法足以大幅降低储层 H_2S 含量及硫酸盐的结垢。

生产的天然气中必须去除 H_2S，以满足商品天然气的销售要求，即 H_2S 最大含量为几毫克每升。吸附 H_2S 和其他硫化合物的间歇处理方法可用于脱除极少量的 H_2S，即要求低气体流速或低浓度的 H_2S。间歇处理可以使用固体或液体泥浆、亚硝酸盐、ClO_2 水溶液、氧化锌或氧化铁泥浆、甲醛/甲醇/水、分子筛、海绵铁及其他基于金属的专有工艺。在大型生产设施中，去除生产气流中 H_2S 的最经济解决方案是安装再生系统来处理含 H_2S 的天然气。这些系统通常通过吸收塔中的化合物来接触产出流体，选择性地吸收 H_2S 和可能的硫醇等有毒物质，这个过程被称为气体脱硫。通常通过加热使吸收化合物和 H_2S 随后再生。吸收材料得以在系统中重新使用，分离出的 H_2S 则通过改良的克劳斯脱硫处理工艺（Claus Process），形成单质硫。

目前从产出天然气中脱除 H_2S 的最常用化学品是与活化剂混合的浓缩胺类水溶液（实际上是烷醇胺）。这些溶液与酸性气体发生可逆反应，并可在需要时循环再生（通常通过加热），去除绝大多数的硫和 CO_2。根据对含 H_2S 气体的要求，可以使用若干种类型的胺溶液，典型的胺液包括单乙醇胺（MEA）、二乙醇胺（DEA）、N- 甲基二乙醇胺（MDEA）、2-（2- 氨基乙氧基）乙醇胺（DGA）或 2-（2- 氨基乙氧基）乙醇胺（图 15.1）。

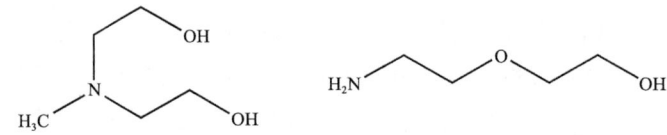

图 15.1　N- 甲基二乙醇胺和 2-（2- 氨基乙氧基）乙醇胺（DGA）

单乙醇胺和 DEA 早前曾被用于气体脱硫。这些醇胺同时吸收 H_2S 和 CO_2。DEA 是一种二级胺，能与 CO_2 快速反应生成氨基甲酸盐，再部分水解为碳酸氢盐。MDEA 吸收 H_2S 有选择性，大多数现代的胺类气体脱硫工艺都是基于 MDEA。MDEA 是不能与 CO_2 反应的叔胺（因为氮原子上没有游离的氢原子），只能通过形成碳酸氢盐/碳酸盐来吸收 CO_2。与醇胺和 H_2S 之间的反应相比，该反应速率慢。但可以根据需要添加活化剂（如聚亚烷基胺、含二级胺的聚胺、烷氧基丙胺、哌嗪、氨基哌嗪、取代哌嗪的聚胺、氨基哌啶、丁基二乙醇胺和氨基乙基乙醇胺）来改善 CO_2 的吸收。DGA｛也称［2-（2- 氨基乙氧基）］乙醇胺｝水溶液已用于天然气或炼厂气的处理中。DGA 是一种一级胺，不仅能去除 H_2S 和 CO_2，还能去除气流和液流中的 COS 和硫醇。据报道，一种复合溶剂配方既可对烃类有限吸收，又提高了能量效率，对 H_2S 的选择性优于 CO_2。三乙醇胺和二异丙醇胺也用于吸收 H_2S，但不常用。有研究利用空间位阻胺（如 2- 氨基 -2- 甲基 -1- 丙醇）进行 CO_2 捕集和 H_2S 吸收。

其他脱除 H_2S 的商用溶剂工艺包括使用聚乙二醇、N- 甲基吡咯烷酮（NMP）、丙烯基碳酸酯和甲醇。碳酸钾单独使用或与其他活化剂的胺类溶液混合使用，多年来一直用于脱除石化行业气流中的 H_2S 和 CO_2。砜（可选择与胺一起配制）也是一种商用的脱除 H_2S

溶剂。铁（或钒）络合脱硫工艺属于液相氧化还原法脱硫工艺，反应效率高，最适于处理规模为 10~12t/d 的低硫量。而对 H_2S 和 CO_2 的"分子篮"纳米多孔吸附剂，只在实验室内进行过研究。

在 H_2S 浓度只有百万分之几甚至更低的情况下，用非再生型脱硫剂处理含 H_2S 的烃类产出流体最经济。这种方法常见于湿气输送管线，在原油生产输送管线也有应用。

15.2 非再生型脱硫剂

已经过研究的非再生型脱硫剂可分为几类：（1）固体、碱性金属化合物；（2）氧化性化学品；（3）醛和醛相关产品；（4）醛和胺的反应产物（包括三嗪类）；（5）金属羧酸盐和螯合物（其中部分为再生型）；（6）其他氨基产品。

目前最常用的是三嗪类脱硫剂，但其不能脱除硫醇。这类产品的主要缺点是会提高采出水的 pH 值，从而加剧碳酸盐结垢。最近开发的新型脱硫剂避免了这种问题（见 15.2.6）。部分过渡金属络合物和螯合物（如铁螯合物）是再生型、选择性脱硫剂，将在下文讨论。

据称二醛和含氮脱硫剂（如三嗪）的混合物能产生协同作用，提高反应速率和整体 H_2S 脱除效率。脱硫剂与 H_2S 反应生成无毒的化合物，可以从油相中移除或排放到水相中。脱硫剂分类的第 2 至第 6 类产品大多是水溶性配方。测试脱硫剂的室内设备可见相关资料。液体脱硫剂可以在生产管线的任何部位注入，有时是在井口，但通常是在更下游的平台上部，通常通过雾化头加入湿气流或静态管内混合器。北海某油田每天使用 20000L 质量分数为 40% 的液体脱硫剂。一种脱硫剂雾化注入的专利设备据称可以将脱硫剂的消耗量减少 30%~35%。仅仅在设备弯头前注入液体脱硫剂，就可能将其功效减半。这是因为气溶胶产物受其密度影响，会在管壁上形成液滴，其中大部分不会进入气相。脱硫剂液滴应尽可能小，以产生更大的表面积和良好的扩散效果，但也要避免因液滴过小导致的脱水问题。将脱硫剂注入气相也可避免对采出水进行处理。

水基脱硫剂通常要求被处理气体可以水饱和，以防止生成固体反应产物。影响脱硫剂效率的其他条件包括：

（1）气体流速，影响混合效果。
（2）保留时间，以达到最佳反应时间。
（3）压力，影响 H_2S 分压。
（4）温度。
（5）与液体和其他生产化学品的相容性。

15.2.1 固体脱硫剂

固体脱硫剂无法以液体形式注入，只适用于处理工艺设备内的含 H_2S 气体。固体脱硫剂一般是锌或铁基材料。某北海运营商曾成功地使用氧化锌床除去产出气体中的 H_2S。目前工业界使用的改进型固体氧化铁脱硫剂会在反应后生成无害的黄铁矿（FeS_2）。催化剂可浸渍在惰性陶瓷材料上，解决了早期的铁海绵型脱硫剂自燃问题。

15.2.2 氧化性脱硫剂

具有氧化性阴离子的水溶性盐类脱硫剂有亚氯酸盐（如 $NaClO_2$）、溴酸盐/碘酸盐（如 $NaBrO_3$）、亚硝酸盐（如 $NaNO_2$）和过酸盐（如过氧化物、过硫酸盐、二氧化硫脲、过碳酸盐、过硼酸盐、二乙基羟胺、过氧乙酸、超氧化物）。这些氧化剂（除过氧化物外）与 H_2S 的反应很复杂，但产物通常含有单质硫。

氯酸盐氧化剂可与 H_2S 迅速反应，但其会产生如硫黄沉积和腐蚀等处理和操作问题，使用受到限制。这种脱硫剂可有效地用于脱除废弃储罐中的 H_2S，此时的腐蚀就不再是主要问题。

含 H_2S 油气井的井下挤注可采用亚硝酸钠溶液，以有效地去除水相和气相中的 H_2S，还可减少 H_2S 造成的腐蚀，并去除近井眼的硫化亚铁垢。平台上部的油水分离设备中也采用亚硝酸盐以加速脱除 H_2S。

常用的三嗪类脱硫剂（见 15.2.4）会使外排水含有未耗尽的三嗪或三嗪与 H_2S 反应形成的有毒胺，给北海某区域造成环境排放问题。现在已经开发出一种基于过氧化物的环境友好型 H_2S 清除剂来解决这一问题。虽然最简单的产品是过氧化氢，而过硼酸盐或过硫酸盐等同样含有 O—O 过氧化物连接的过酸盐可能同样有效。有机过氧化物也已经进行了现场测试，但导致了腐蚀问题。过氧化物将硫化物转化为硫酸盐，如过氧化氢与 H_2S 的反应产生硫酸根离子和质子。

$$H_2S + 4H_2O_2\,(aq) \longrightarrow SO_4^{2-}\,(aq) + 4H_2O\,(l) + 2H^+\,(aq)$$

海上经验表明，H_2S 必须以离子形式存在（HS^- 或 S^{2-}）才能快速反应，而且过氧化物产品必须有高 pH 值。但高 pH 值通常会破坏过氧化氢的稳定性，特别是在有污染物（如微量过渡金属）存在时。在与运营商的合作中，有服务公司配制出稳定、高 pH 值的过氧化物产品，并在实验室和现场试验中成功使用（稳定剂可能是咪唑或三唑，可防止过氧化物与金属位点结合后催化分解）。但该稳定剂不符合 OSPAR 对北海的环保要求，故而又找到了新的稳定剂。强氧化性过氧化物造成的腐蚀问题也没有得到解决。

许多文章中介绍的液体脱硫剂评价方法将在下文讨论。有资料介绍了评价脱硫剂的新方法和设备。

15.2.3 醛类

醛类与 H_2S 反应会形成多种硫化物。已经使用的典型醛类包括甲醛、戊二醛、丙烯醛和乙二醛（图 15.2）。作为 H_2S 清除剂的醛可能对如钢、铁和铝等金属有腐蚀作用，但通过选用二基可溶性磷酸盐、三基可溶性磷酸盐、磷酸酯、硫代磷酸酯、硫胺、曼尼希反应产物及其组合物的缓蚀剂，可以减轻醛的腐蚀作用。

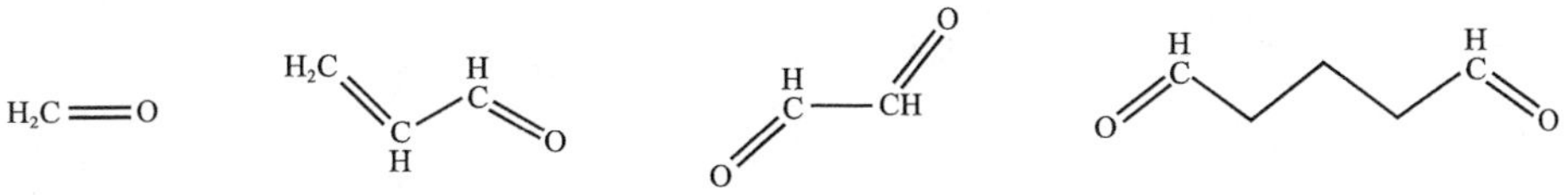

（a）甲醛　（b）丙烯醛　（c）乙二醛　（d）戊二醛

图 15.2　甲醛、丙烯醛、乙二醛和戊二醛结构式

甲醛主要与 H_2S 反应形成 1，2，3-三噻烷环状化合物（图 15.3）。甲醛是疑似的致癌物，在欧洲已被禁止使用，所以现在并不常用。更安全的水溶性化合物六亚甲基四胺（HMTA）与强酸反应后会在原位分解出甲醛，可用于酸化增产液。但醛和酸反应形成的油性聚合物有可能造成储层伤害，同时，H_2S 和甲醛反应生成不溶于水的固体，会带来处理问题。

$$3HCHO + 3H_2S \longrightarrow \text{（1,3,5-三噻烷）} + 3H_2O$$

图 15.3　甲醛与 H_2S 反应

乙二醛可单独使用，也可与季铵盐（如二甲基椰苯扎氯铵）协同使用。据称使用乙二醛的中性水溶液（pH 值为 6~8.5）比单独使用乙二醛或单独使用碱能更好地脱除天然气和石油中的 H_2S。这些组合脱硫剂大幅提高了反应速率和整体脱除效率，比单独使用乙二醛的处理量更高。可以选择使用缓冲剂、不含氮的表面活性剂，使反应速率和总体脱除效率显著提高。乙二醛与过量的 H_2S 反应，产生部分水溶性的反式 -4,4,5,5′-四羟基 -2,2′-双（1,3-硫杂环戊烷）。将乙二醛与至少有两个伯胺或仲胺基团的化合物（如聚胺）反应所制备的脱硫剂效果好，且产生的腐蚀轻微。据称乙醛酸与聚乙二醇的协同组合也是一种脱硫剂，与 H_2S 反应后只生成水溶性产物。还有一种由乙二醛和环氧丙烷制成的聚合物（如聚丙二醇）组成的组合脱硫添加剂。pH 值为 2.5~6.0 的乙二醛溶液能分散到液态烃类介质中以脱除 H_2S。

醛基越大，分散到液态烃相的比例就越高。甲醛、戊二醛和乙二醛主要溶于水（98%~99% 分散于盐水相，1%~2% 分散于烃相），丙烯醛有 50% 以上分散到液态烃相。丙烯醛虽是有效的硫化物脱除剂，但因高毒性和处理问题而未广泛使用。

醛类的缺点是其与 H_2S 反应的大多数产物难溶于水。但醛类的额外优势是能作为杀菌剂，杀灭 SRB，防止其产生 H_2S（见第 14 章）。因为醛类既能控制 SRB 的形成，又能脱除既有的 H_2S，这就使其成为注海水驱油、近井或挤注作业中的理想产品。

丙烯醛实际上有三种功能：作为脱硫剂；同时还是杀菌剂；溶解硫化亚铁垢。丙烯醛与 H_2S 反应，首先形成硫醇醛，而硫醇醛又与丙烯醛进一步反应，形成水溶性的噻喃（图 15.4）。

丙烯醛还被用于多相流产出液的脱硫。由丙烯醛和甲醛缩合产生的水溶性低分子量聚缩合产品已被建议用于脱除水系统中的 H_2S 和硫化亚铁。丙烯醛是一类共轭的不饱和醛。据称不饱和醛的酯类也是一种脱硫剂（图 15.5）。例如，反式 -4-氧代 -2-丁烯酸乙酯，也称为 3-醛基丙烯酸乙酯（图 15.5 中左图 R_1= 乙基）。

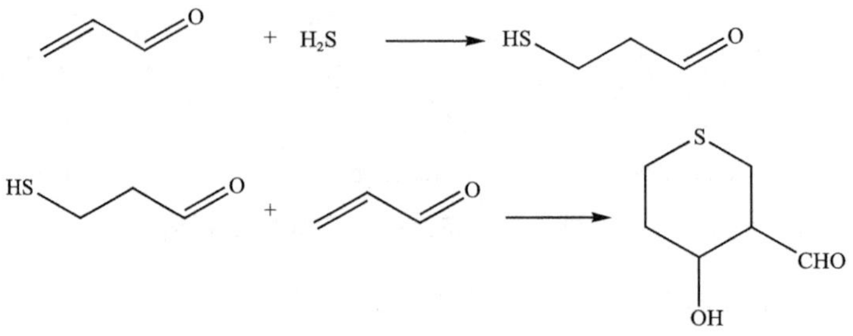

图 15.4 丙烯醛与 H₂S 的反应

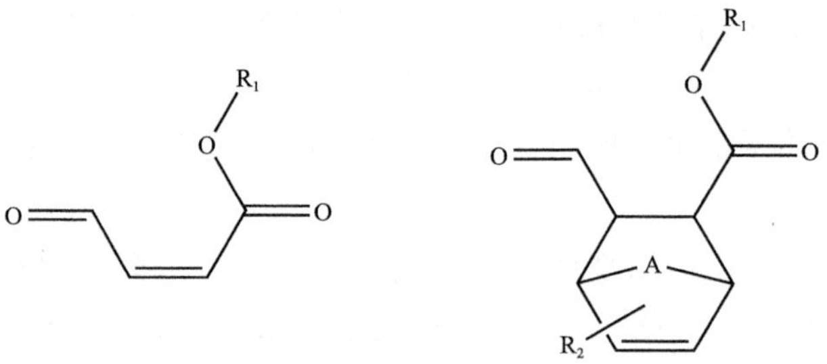

图 15.5 不饱和醛的酯类

醛类的另一个缺点是与 H₂S 的反应较慢，特别是在低温时。相比之下，基于三嗪的脱硫剂（见下文）与 H₂S 的反应明显更快（但不是在低温下）。而醛类不会像胺和三嗪产品会提高产出水的 pH 值，避免了可能导致的碳酸盐结垢并加剧乳化倾向。

研究发现，醛、多羟基或尿素基化合物的反应产物是优良的脱硫剂。这项技术不含胺，但发明人指出最好是与烷醇胺或三嗪混合使用。使用这种或任何醛基脱硫剂可以避免三嗪类脱硫剂提高水 pH 值而导致的碳酸盐结垢（见 15.2.4）。典型的产品是通过甲醛与乙二醇或甘油反应制成。由乙二醇制备的产品是乙二醇半缩甲醛，也被称为 1，2- 乙烷二基双（氧）- 双甲醇或 1，6- 二羟基 -2，5- 二氧六环（图 15.6）。这种产品可以与尿素和甲醛反应形成的二甲基尿素［也称为 N, N- 双（羟甲基）尿素］结合使用。也可以使用由乙二醇和甲醛按 1∶1 反应制成的杂环缩醛 1，3- 二氧戊环，与甲醛同 H₂S 反应的效果相似，该反应产生 1，3，5- 三噻烷并再生乙二醇。而二氧杂环己烷和三氧杂环己烷不与 HS⁻ 反应。丁基甲醛（丁氧基甲醇）也可用于脱除 H₂S。

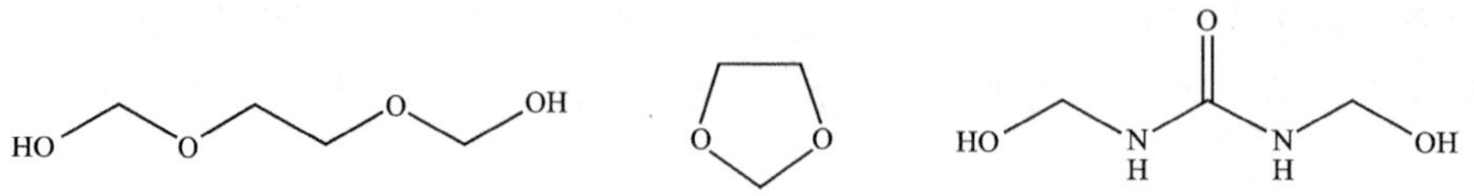

(a) 1,6-二羟基-2,5-二氧六环　　(b) 1,3-二氧戊环　　(c) 二甲戊环

图 15.6　1，6- 二羟基 -2，5- 二氧六环、1，3- 二氧戊环和二甲戊环

醛基脱硫剂可用于含 H₂S 碳酸盐岩油藏的近井酸化增产。中东地区的一项早期研究将醛类作为注水井酸化增产的脱硫剂，但发现高脱硫剂浓度下，形成的聚合物影响酸与硫化

亚铁反应效果，并产生地层伤害。后来的一项研究使用了 HMTA，其在低 pH 值溶液（如 7.5% 的盐酸）中会产生甲醛，但发现 HTMA 脱硫效果比甲醛差。在实验室测试中能促进硫化亚铁良好溶解，并在现场使用的最好的脱硫剂是一种脂肪族和芳香族醛的混合物。由于担心醛对地层的伤害，在后来的现场酸化增产应用中，成功使用了一种羟烷基三嗪（关于三嗪的详细介绍见 15.2.4）。

15.2.4 醛类和胺类（特别是三嗪类）反应产物

醛类与胺类反应形成亚胺，但只要有一个当量的甲醛就会形成环状的 1，3，5-六氢三嗪产品。其中的一个副产物是 N,N'-亚甲基双恶唑烷，如果调整胺和甲醛的比例，也可以使其成为主要产物。据称双恶唑烷本身也是一种脱硫剂；在石油和天然气工业中最广泛使用的是 1，3，5-六氢三嗪产品。脱硫的副产品 5-羟基乙基二噻嗪（5-HEDT）是 FFCI 的良好增效剂，2-羟基丙基衍生物效果更好。因为甲氧基效果差，羟基组的作用很重要。三羟乙基三嗪的增效效果较差。含有 5-HEDT 的未分离废液效果更好，这表明当含有少量的其他脱硫产物时，脱硫的增效作用更活跃。尽管 IUPAC 正确名称是 1，3，5-六氢三嗪，但生产化学家通常称其为三嗪。实验室实验表明，在较宽的 CO_2 分压范围内和 H_2S/CO_2 值范围内，CO_2 对三嗪类的 H_2S 脱除性能影响甚微。

第一批被研究的三嗪是卤代三嗪，如三氯-S-三嗪三酮。如今最常用的三嗪是由烷醇胺或甲胺与甲醛反应制成的水溶性产物。例如，乙醇胺等反应生成的产物主要是 1，3，5-三（2-羟乙基）-六氢-S-三嗪（图 15.7）。作为副产品（如果改变反应比例则为主要产品）形成的 N,N'-亚甲基双恶唑烷也会与 H_2S 反应。1，3，5-三甲基六氢三嗪是另一种常见的脱硫剂。

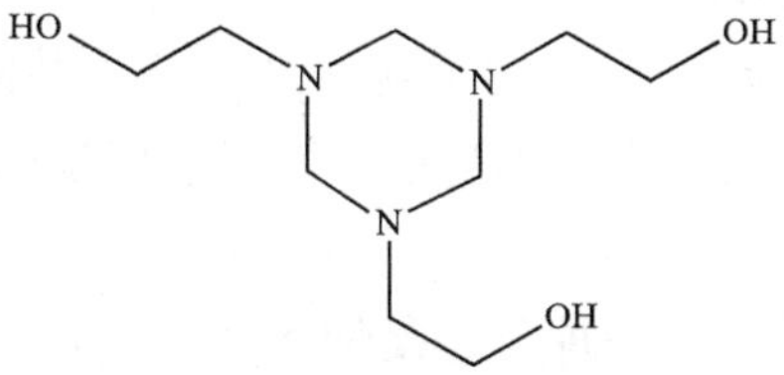

(a) 1,3,5-三(2-羟乙基)-六氢-S-三嗪 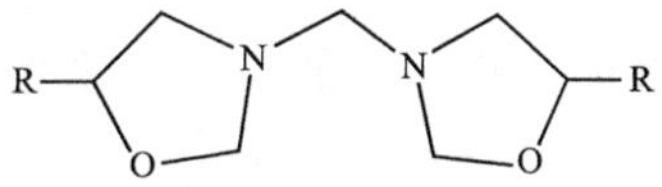　　　　(b) N,N'-亚甲基双恶唑烷

图 15.7　1，3，5-三（2-羟乙基）-六氢-S-三嗪和 N,N'-亚甲基双恶唑烷（R= 烷基）

还有由其他胺类制成的三嗪，如 3-甲氧基丙胺（MOPA）、甲胺和甲醛的反应产物（据称是比羟乙基三嗪更好的脱硫剂）。也可以由醇胺与甲醛在上游足够远的地方，原位形成三嗪类物质并与 H_2S 反应。

三嗪类与 H_2S 的反应速度比醛类快，而且还能与 HS^- 反应。使用三嗪类液体脱硫剂的经济性上限是每天最多可以清除约 50kg H_2S，在 H_2S 浓度相对较低的流体中，可以将 H_2S 浓度降低到约 5mg/L 的水平。三嗪类的效率相当低，如果正确注入，12～20kg 典型的质量分数为 40%～50% 的三嗪类产品可以去除 1kg H_2S。

三嗪类具有低毒性特征，通常可生物降解，但可能含有少量游离的疑似致癌物——甲醛，其通常以水或甲醇—水溶液的形式注入。最初的三嗪可以通过热碱水处理而再生，但在实践中通常不经济。三嗪类脱硫剂通常在温度为 120～150℃时分解。

1，3，5-三（2-羟乙基）-六氢 -S- 三嗪与 H_2S 反应的主要产物取决于 pH 值，主要是 5-（2-羟乙基）六氢 -1,3,5- 二噻嗪、3,5- 双（2-羟乙基）六氢 -1,3,5- 噻嗪，以及 1，3，5- 三噻嗪（图 15.8），发现还有三噻烷和四噻烷。双羟乙基化合物的水溶性相当好，不会出现问题；但是三噻烷和其他聚合产品的水溶性较差，尤其是在低温的寒冷管道或处理设施中可能造成沉积问题。二噻烷在气体处理设备中形成一个独立的液相或层。据称，在大约 20℃ 或更低的温度下，此层中会形成固体二噻嗪晶体并从溶液中沉淀析出。最近的研究结果证实，这种材料并不是单体二噻嗪或三噻嗪的不同物理形式，而是二噻嗪环开环后形成的聚合结构。同一报告提到了一种新型脱硫剂，其巯基化学沉积较少，但没有给出结构细节。

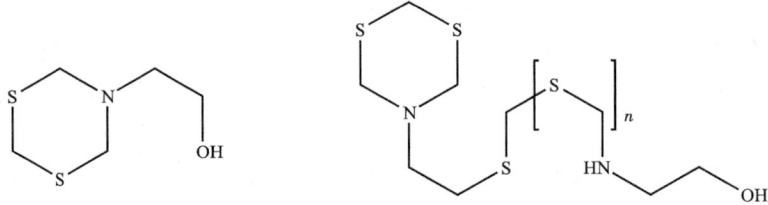

图 15.8 1，3，5-三（2-羟乙基）-六氢 -S- 三嗪与 H_2S 反应的含硫产物

还有研究使用一种由三甘醇和胺与醛（三嗪）的反应产物组成的配方，以缓解与二噻嗪和其他相关固体的沉积问题。另一种方法是泵送初级胺（如乙醇胺）和可选的氨，作为碱与任何不溶性的三噻嗪反应，得到水溶性更好的产品。

1，3，5-三（2-羟乙基）-六氢 -S- 三嗪已成功用于含 H_2S 水注入井的酸化作业，以去除 H_2S 并防止进一步的硫化亚铁垢。该反应产物为水溶性，而且该脱硫剂也有一定的缓蚀作用。

已经有许多尝试来改善三嗪类脱硫剂的性能。一种改进是使用如 MEA、DGA 和甲醛的反应产物等更亲水的三嗪，在与 H_2S 反应后得到更多水溶性产物。得到的混合物包括 1，3，5-三（2-羟乙基）-六氢 -S- 三嗪和 1，3，5-三（2-乙氧基乙醇）六氢三嗪。二甲胺基丙胺与甲醛的反应可以得到亲水性更强的三嗪，分子中有五个水溶性的二甲胺基。具有羟烷基和烷基胺的三嗪脱硫剂被认为比只含有羟烷基的产品性能更好。这些脱硫剂是由至少一个烷醇胺和至少一个烷基胺与醛（如甲醛）反应而成。

还有一种脱硫剂组合物包括甲醛与非常亲水的氨基乙基哌嗪的三嗪反应产物和第二种或"增强型"胺（如正丁胺或 MEA），据称可以避免有机硫固体沉积问题，并有若干案例报道。将三嗪与甲醛或胺类混合，可起到增效作用。已经报道了一种选定优化混合物的方法。

另一种改进是使用季铵化合物来加速三嗪与 H_2S 的反应。首选的六氢三嗪是 1，3，5-三甲氧基丙基六氢 -1，3，5- 三嗪（MOPA 六氢三嗪），季铵化合物可以是苄基焦烷基二甲基季铵氯，其同时也是一种成膜型缓蚀剂。据称，三烷基或羟烷基胺的氧化物（硝基化物）也是脱硫剂（如烷基三嗪）与 H_2S 反应的促进剂。硝基氧化物能有效地加速脱硫剂的活性，可避免将促进腐蚀的卤离子引入烃类流体。

还有一种改进是使用由—CH_2 基团连接的两个或多个三嗪环的低聚物。这些化合物的优点是不容易产生游离甲醛，而且与 H_2S 的反应产物刺激性小。双三嗪是通过乙二胺与甲醛反应制成。其他复杂的多三嗪是亚烷基多胺如二亚乙基三胺与甲醛反应制成。二甲胺基丙胺与过量甲醛反应的三嗪产物也是有用的脱硫剂。三嗪和醛类混合后也用于泡沫环境的处理，兼具杀菌和脱硫功能。

也可以通过疏水性的烷基胺与甲醛反应制成油溶性三嗪，如据称制备了油溶性的 1，3，5- 三六氢 -1，3，5- 叔丁基三嗪。但在为烃类液体选择脱硫剂时，需要注意对凝析油或原油下游加工的影响，往往需要保持低氮含量以避免重整催化剂中毒。水溶性脱硫剂对去除凝析油中的 H_2S 同样有效，因为溶解度和密度的不同，脱硫剂不会进入凝析油相。

三嗪类脱硫剂的一个问题是含有叔氨基，在水溶液中呈碱性，会提高水的 pH 值。三嗪与 H_2S 反应后形成的胺也呈碱性。三嗪类脱硫剂造成的 pH 值增高会导致新的或加重碳酸钙结垢问题。

$$Ca^{2+}(aq) + 2HCO_3^-(aq) \rightleftharpoons CaCO_3(s) + H_2O(l) + CO_2(g)$$

$$HCO_3^-(aq) + R_2R'NH(aq) \longrightarrow CO_3^{2-}(aq) + R_2R'NH^+(aq)$$

$$CO_3^{2-}(aq) + Ca^{2+}(aq) \longrightarrow CaCO_3(s)$$

有报道称三嗪类药剂造成的较高 pH 值也导致硫化钙结垢。为了克服这些问题，可以使用非碱性的醛脱硫，或配套碳酸盐防垢剂。防垢剂—脱硫剂的混合配方在现场应用取得了良好效果。在一个综合处理方案中，发现磷酸盐防垢剂的性能最优。有人建议将不含游离甲醛的芳香族亚胺化合物作为三嗪类化合物的改进型、低毒性替代品。

在某油田，使用传统的防垢剂处理日益增多的碳酸盐垢并不可行。替代性的解决方案是在与采出的水混合之前，向脱硫剂中加入降低 pH 值的化学剂，如甲酸。

15.2.5　金属羧酸盐和螯合剂

水溶性和油溶性高价金属螯合物都可作为脱硫剂，已用于钻井液和受污染的水和油流处理。金属离子通常是 Zn^{2+} 或 Fe^{3+}，螯合物是含有羧基的氨基三乙酸、EDTA、聚氨基二琥珀酸或生物降解性更好的葡萄糖酸盐。通常，反应会生成金属硫化物；但如果金属处于高氧化价态，可以将 H_2S 氧化成硫。例如，$N-$（2- 羟乙基）EDTA 的铁螯合物，其与 H_2S 反应后被还原成亚铁螯合物。必要时，铁螯合物可以通过亚铁螯合物与 O_2 在高温下的反应来再生。其中，需要抗氧化稳定剂，如破坏羟基自由基的硫代硫酸根离子或过氧化氢酶，以防止螯合物的降解。这种螯合物和其他更常见的螯合物如 EDTA 都可降解。

还开发出了基于锌和铁羧酸盐的快速起效、油溶性脱硫剂。通过现场处理含 1200mg/L H_2S 原油的试验，验证了该技术的有效性。典型产品是由长链脂肪酸制成的油溶性羧酸锌，与 H_2S 反应后，生成的 ZnS 微粒可以分散到集输的原油相。

$$Zn（OOCR）_2 + H_2S \rightleftharpoons ZnS + 2HOOCR$$

15.2.6 其他氨基产品

除了非选择性的再生烷醇胺，也有研究涉及其他几类选择性更强的氨基产品。小分子的水溶性脒[RC（=NH）NH_2]或低聚物脒已证明具有良好的脱硫特性。胺氧化物和特定酶的混合物已获得专利授权，但未见到现场试验数据。哌嗪酮或其烷基取代衍生物如 1,4-二甲基哌嗪酮也是良好的脱硫剂。具有杀菌性的马来酰亚胺也被证明是良好的脱硫剂。另一项专利申请声称，一种脱硫剂由 1,3,5-三烷胺六氢-1,3,5-三嗪衍生物、吗啉或哌嗪衍生物、氧化胺、烷醇胺、脂肪族或芳香族多胺组成。

烃类中的 H_2S 和小分子硫醇可使用季铵醇盐或氢氧化物以及含有高氧化态金属如钴、铁、铬或镍的配方来脱除。高氧化态的金属是一种氧化剂，可能作为催化剂改进对 H_2S 和硫醇的脱除效果。该方法的确切运作机制尚不清楚，但硫醇与脱硫剂反应的产物是二硫化物。此外，引入氧气（如用空气喷入处理过的液体）会大幅提高脱硫活性。还发现在 50～70℃的更高温度下，脱硫剂的性能得到改善。针对含硫原油研究的一种典型脱硫剂是由二甲基大豆胺和环氧乙烷制备的季铵氢氧化物，其中含有 Co^{3+}。

有文章报道一种环境友好的非三嗪类脱硫剂不会提高 pH 值和增加碳酸钙的结垢倾向，但未详细提及化学结构式。由醛类和相关化学剂与氨基酸的反应产物、羟甲基膦或羟甲基鏻盐，以及可选的季铵化合物或一个或多个 N-氢化合物（如 5,5-二烷基海因或胺）组成的配方也是一种快速脱硫剂。据称这些无腐蚀性的配方具有高 H_2S 脱除能力、良好的可生物降解性，对 pH 值相对不敏感。具体配方实例是将甘氨酸与四（羟甲基）硫酸磷结合，然后加入图 15.9 所示的两种甲基海因与二甲基十二烷基氯化铵的混合物。

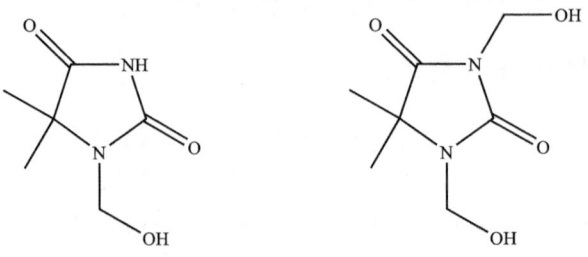

(a) 1-羟甲基-5,5-二甲基海因　　(b) 1,3-二羟甲基-5,5-二甲基海因

图 15.9　1-羟甲基-5,5-二甲基海因和 1,3-二羟甲基-5,5-二甲基海因

还有资料提出了一种脱除 H_2S 或硫醇的方法，涉及使用二级胺和甲醛反应的混合物或单一成分。反应产物主要是 α-氨基醚及少量的 N, N, N', N'-四烷基甲二胺，结构示例如图 15.10 所示。氨基醚据称也能去除硫醇。当烷基优先为丁基时，脱硫反应将硫醇转化为油溶性的化合物。

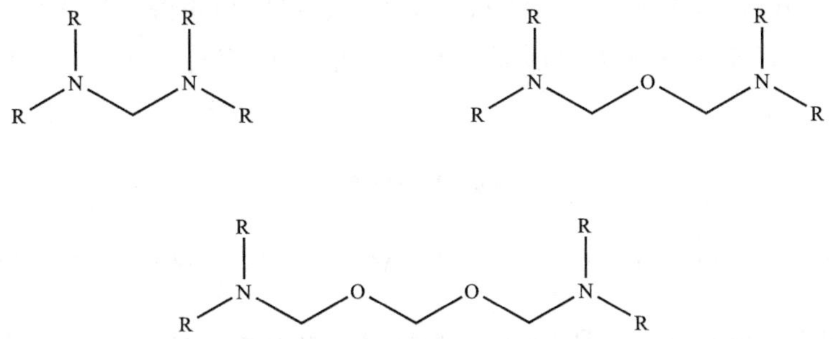

图 15.10 N, N, N', N'-四烷基甲二胺和 α-氨基醚（H_2S 和硫醇清理剂）

一些仲胺—甲醛加合物（N-甲基仲胺）对 H_2S 是惰性的。这些胺缺少醚基或聚醚基。可以通过优化反应物的物质的量比，控制（减少）N-甲基仲胺的产生。

偶氮二甲酰胺类据称可以减少产出液中的 H_2S、硫化亚铁或硫醇（图 15.11）。相对常见的是偶氮二甲酰胺（所有的 R=H）。

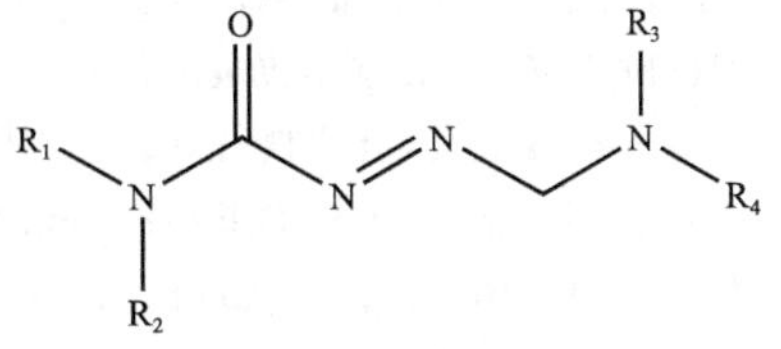

苯醌（如对苯醌）也是有用的脱硫剂，最好在添加碱或表面活性剂的碱性环境使用。苯醌类的氧化胺衍生物，是多效脱硫剂（图 15.12）。这些化合物可用于清除硫醇、硫化物、氰化物和一级或二级胺。

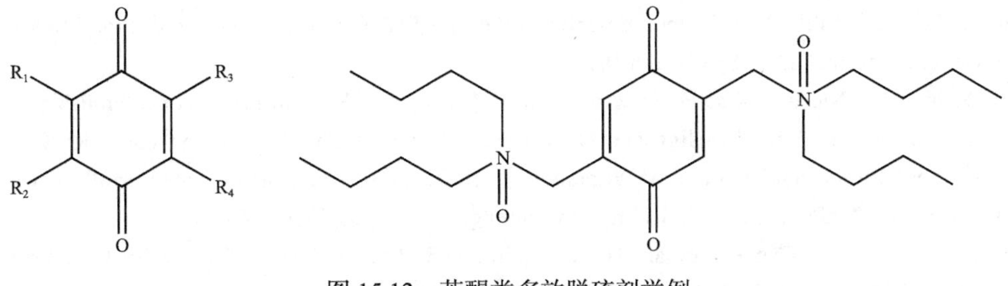

图 15.11 偶氮二甲酰胺

图 15.12 苯醌类多效脱硫剂举例

H_2S 控制策略的选择取决于 H_2S 浓度、温度、产出液化学成分、环境要求及系统中的处理位置等多种因素。总的来说，有五种基本方法来控制储层产生 H_2S：

（1）加入兼具脱硫作用的杀菌剂来杀灭 SRB。

（2）用代谢抑制剂处理 SRB，防止硫酸盐被还原成硫化物。

（3）通过添加营养物质促进非 SRB 菌类形成，如用于促进反硝化细菌生长的硝酸根离子。

（4）在注水井中使用无硫酸根的地层水或脱硫酸根的海水。

（5）使用脱硫剂。

方法（6）是用微生物将硫化物氧化成硫酸根离子，在北美已成功使用。现已开发出一种抗 H_2S 的热塑性聚合物层，用于柔性管道的环空。

方法（1）至方法（4）通常通过注水井注入，也可以将杀菌剂注入采出流体。脱硫剂可以在井下注入，最常见的是平台顶部注入。第 14 章中有对方法（1）至方法（4）的详细讨论。

参 考 文 献

[1] *Hydrogen Sulfide in Production Operations* (*Petroleum*), 2nd ed., Austin, TX: University of Texas at Austin, 1996.

[2] N. P. Tung, P. V. Hung, and H. D. Tien, "Study of Corrosion Control Effect of H_2S Scavengers in Multiphase System," SPE 65399 (paper presented at the SPE International Symposium on Oilfield Chemistry, Houston, TX, 13–16 February 2001).

[3] (*a*) P. Mougin, V. Lamoureux-Var, A. Bariteau, and A. Y. Huc, *Journal of Petroleum Science and Technology* 58 (2007): 413. (*b*) R. H. Worden, P. C. Smalley, and M. M. Cross, "The Influence of Rock Fabric and Mineralogy on Thermochemical Sulfate Reduction: Khuff Formation, Abu Dhabi," *Journal of Sediment Research* 70 (2000): 1210–1211.

[4] I. Vance and D. R. Thrasher, "Reservoir Souring: Mechanism and Prevention," in *Petroleum Microbiology*, eds. B. Ollivier, and M. Magot. Washington, DC: ASM Press, 2005, 123.

[5] (*a*) E. D. Berger, I. Vance, G. F. Gammack, and S. E. Duncan, The 5th International Conference on Microbial Enhanced Oil Recovery and Related Biotechnology for Solving Environmental Problems, sponsored by U.S. DOE, 1995. (*b*) D. O. Hitzman and D. M. Dennis, "New Technology for Prevention of Sour Oil and Gas," SPE 37908 (paper presented at the SPE/EPA Exploration and Production Environmental Conference, Dallas, TX, 3–5 March 1997).

[6] E. A. Morris, R. Gomez, and R. Peterson, "Application of Chemical and Microbiological Data for Sulfide Control," SPE 52705 (paper presented at the SPE/EPA Exploration and Production Environmental Conference, Austin, TX, 1–3 March 1999).

[7] A. L. Kohl and R. Nielsen, *Gas Purification*, 5th ed. Houston, TX: Gulf Professional Publishing, 1997.

[8] M. Abedinzadegan Abdi, "Design and Operations of Natural Gas Sweetening Facilities Course for the National Iranian Gas Company Workshop," 2nd Iranian Gas Forum, Memorial University of Newfoundland (MUN), June 2008 (http://www.engr.mun.ca/people/mabdi.php).

[9] (*a*) J.-L. Peytavy, S. Capdeville, and H. Lacamoire, U.S. Patent 6290754, 2001. (*b*) J.-L. Peytavy, P. Le Coz, and O. Oliveau, U.S. Patents 5348714 and 5209914, 1993.

[10] (*a*) C. J. N. Buisman, A. J. H. Janssen, and R. J. Van Bodegraven, U.S. Patent 6656249, 2003. (*b*) C. Cline, A. Hoksberg, R. Abry, and A. Janssen, "Biological Process for H_2S Removal from Gas Streams: The Shell-Paques/THIOPAQ Gas Desulfurization Process" (paper presented at the Laurance Reid Gas Conditioning Conference, LRGCC, Norman, OK, 23–26 February 2003).

[11] (*a*) H. L. Smith, A. F. Johnsen, and B. Knudsen, U.S. Patent 7078005, 2006. (*b*) J. van Dijk and A. Bos, "An Experimental Study of the Reactivity and Selectivity of Novel Polymeric 'Triazine-Type' H_2S Scavengers," *Proceedings of the Chemicals in the Oil Industry VI*, 14–17 April 1997, 170.

[12] H. A. Nasr-El-Din, M. Zabihi, S. K. Kelkar, and M. Samuel, "Development and Field Application of a New Hydrogen Sulfide Scavenger for Acidizing Sour-Water Injectors," SPE 106442 (paper presented at the SPE International Symposium on Oilfield Chemistry, Houston, TX, 28 February-2 March 2007).

[13] H. Linga, F. P. Nilsen, R. Abiven, and B. H. Kalgraff, International Patent Application WO/2006/038810.

[14] D. R. Wilson, "Hydrogen Sulfide Scavengers: Recent Experience in a Major North Sea Field," SPE 36943 (paper presented at the European Petroleum Conference, Milan, Italy, 22-24 October 1996).

[15] E. E. Burnes and K. Bhatia, U.S. Patent 4515759, 1985.

[16] D. Geverte and G. E. Jenneman, U.S. Patent 5820766, 1998.

[17] A. G. Hunton, P. A. Read, and R. D. Wilson, "Evaluation and Field Application of a New Hydrogen Sulfide Scavenger" (paper presented at the 10th International Oil Field Chemical Symposium, Fagernes, Norway, 1-3 March 1999).

[18] P. J. Sturman and D. M. Goeres, Center for Biofilm Engineering, Montana State University; and M. A. Winters, "Control of Hydrogen Sulfide in Oil and Gas Wells with Nitrite Injection," SPE 56772 (paper presented at the SPE Annual Technical Conference and Exhibition, Houston, TX, 3-6 October 1999).

[19] B. Knudsen, S. Tjelle, and H. Linga, "A New Approach Towards Environmentally Friendly Desulfurization," SPE 73957 (paper presented at the SPE International Conference on Health, Safety and Environment in Oil and Gas Exploration and Production, Kuala Lumpur, Malaysia, 20-22 March 2002).

[20] J. G. Edmondson, U.S. Patent 4680127, 1987.

[21] R. Roehm, U.S. Patent 3459852, 1969.

[22] J. A. Hardy and J. W. Georgie, U.S. Patent Application 20040074813.

[23] J. R. Elliott, M. B. Raymond, B. Kalpakci, and N. F. Magri, "Theory and Measurements of Fates of H_2S Scavengers," SPE 28949 (paper presented at the SPE International Symposium on Oilfield Chemistry, San Antonio, TX, 14-17 February 1995).

[24] C. T. Bedford, A. Fallah, E. Mentzer, and F. A. Williamson, "The First Characterisation of a Glyoxal-Hydrogen Sulfide Adduct," *Journal of the Chemical Society*, *Chemical Communications* 1035, 1992.

[25] B. G. Kriel, A. B. Crews, E. D. Burger, E. Vanderwende, and D. O. Hitzman, "The Efficacy of Formaldehyde for the Control of Biogenic Sulfide Production in Porous Media," SPE 25196 (paper presented at the SPE International Symposium on Oilfield Chemistry, New Orleans, LA, 2-5 March 1993).

[26] C. L. Kissel, J. L. Brady, H. Gottry, H. N. Clifton, M. J. Meshishnek, and M. W. Preus, "Factors Contributing to the Ability of Acrolein to Scavenge Corrosive Hydrogen Sulfide," SPE 11749, *SPE Journal* 25 (1985): 647-655.

[27] J. E. Penkala, C. Reed, and J. Foshee, "Acrolein Application to Mitigate Biogenic Sulfides and Remediate Injection-Well Damage in a Gas-Plant Water-Disposal System," SPE 98067 (paper presented at the International Symposium and Exhibition on Formation Damage Control, Lafayette, LA, 15-17 February 2006).

[28] T. Salma, "Cost Effective Removal of Iron Sulfide and Hydrogen Sulfide from Water Using Acrolein," SPE 59708 (paper presented at the SPE Permian Basin Oil and Gas Recovery Conference, Midland, TX, 21-23 March 2000).

[29] J. Penkala, M. D. Law, D. D. Horaska, and A. L. Dickinson, "Baker Petrolite, Acrolein 2-Propenal: A Versatile Microbiocide for Control of Bacteria in Oilfield Systems," Paper 04749 (paper presented at

the NACE CORROSION 2004).

[30] C. Reed, J. Foshee, J. E. Penkala, and M. Roberson, "Acrolein Application to Mitigate Biogenic Sulfides and Remediate Injection Well Damage in a Gas Plant Water Disposal System," SPE 93602 (paper presented at the SPE International Symposium on Oilfield Chemistry, The Woodlands, TX, 2–4 February 2005).

[31] J. J. Howell and M. B. Ward, "The Use of Acrolein as an H_2S Scavenger in Multiphase Production," SPE 21712 (paper presented at the SPE Production Operations Symposium, Oklahoma City, OK, 7–9 April 1991).

[32] W. Merk and K.-H. Rink, U.S. Patent 4501668, 1985.

[33] A. Y. Al-Humaidan and H. A. Nasr-El-Din, "Optimization of Hydrogen Sulfide Scavengers Used during Well Stimulation," SPE 50765 (paper presented at the SPE International Symposium on Oilfield Chemistry, Houston, TX, 15–19 February 1999).

[34] H. A. Nasr-El-Din, A. Y. Al-Humaidan, B. A. Fadhel, W. W. Frenler, and D. G. Hill, "Investigation of Sulfide Scavengers in Well-Acidizing Fluids," SPE 80289, *SPE Production and Facilities* 17 (4) (2002): 229.

[35] (a) J. M. Bakke and J. B. Buhaug, "Hydrogen Sulfide Scavenging by 1, 3, 5-Triazinanes. Comparison of the Rates of Reaction," *Industrial & Engineering Chemistry Research* 43 (9) (2004): 1962–1965. (b) L. Zea, P. Jepson, and R. Kumar, "Role of Pressure and Reaction Time on Corrosion Control of H_2S Scavenger," SPE 114175 (paper presented at the SPE International Oilfield Corrosion Conference, Aberdeen, UK, 27 May 2008).

[36] (a) G. T. Rivers, U.S. Patent 6117310, 2000. (b) G. T. Rivers and J. T. Hackerott, U.S. Patent 6339153, 2002.

[37] T. Salma, "Effect of Carbon Dioxide on Hydrogen Sulfide Scavenging," SPE 59765 (paper presented at the SPE/CERI Gas Technology Symposium, Calgary, Alberta, 3–5 April 2000).

[38] J. D. Allison and J. W. Wimberley, U.S. Patent 4710305, 1987.

[39] E. T. Dillon, World Patent Application WO90/07467.

[40] E. T. Dillon, *Hydrocarbon Process, International Edition* 70 (12) (1991): 65.

[41] E. T. Dillon, U.S. Patent 4978512, 1990.

[42] G. T. Rivers and R. L. Rybacki, U.S. Patent 5347004, 1994.

[43] G. T. Rivers and R. L. Rybacki, U.S. Patent 5554349, 1996.

[44] A. J. Galloway, U.S. Patent 5405591, 1995.

[45] G. J. Nagl, "Removing Hydrogen Sulfide," *Hydrocarbon Engineering* 6 (2) (2001): 35.

[46] T. Salma, M. L. Briggs, D. T. Herrmann, and E. K. Yelverton, "Hydrogen Sulfide Removal from Sour Condensate Using Non-regenerable Liquid Sulfide Scavengers: A Case Study," SPE 71078 (paper presented at the SPE Rocky Mountain Petroleum Technology Conference, Keystone, CO, 21–23 May 2001).

[47] E. A. Trauffer and R. D. Evans, U.S. Patent 5347003, 1994.

[48] J. Bakke, J. Buhaug, and J. Riha, "Hydrolysis of 1, 3, 5-Tris (2-Hydroxyethyl) Hexahydro-s-Triazine and its Reaction with H_2S," *Industrial & Engineering Chemistry Research* 40 (2001): 6051–6054.

[49] T. R. Owens, International Patent Application WO/2008/049188, 1990.

[50] C. W. Titley and P. H. Wieninger, U.S. Patent 7115215, 2006.

[51] L. W. Gatlin, World Patent Application WO/2004/043938.

[52] T. Salma, A. A. Lambert III, and G. T. Rivers, International Patent Application WO/2008/027721.

[53] J. C. Warrender, World Patent Application WO98/19774, 1998.

[54] S. R. Schieman, "Solids-Free H_2S Scavenger Improves Performance and Operational Flexibility," SPE 50788 (paper presented at the SPE International Symposium on Oilfield Chemistry, Houston, TX, 16–19 February 1999).

[55] D. Sullivan III, A. R. Thomas, J. M. Garcia, and P. Yon-Hin, U.S. Patent 5744024, 1998.

[56] J. F. Vasil, U.S. Patent 5314672, 1994.

[57] J. J. Weers and T. J. O'Brien, U.S. Patent 6024866, 2000.

[58] L. W. Gatlin, U.S. Patent Application 20030089641.

[59] T. Arnold, W. R. Graham, and J. D. Cranmer, U.S. Patent 6942037, 2005.

[60] D. S. Sullivan III, A. R. Thomas, M. A. Edwards, G. N. Taylor, P. Yon-Hin, and J. M. Garcia III, U.S. Patent 5674377, 1997.

[61] J. C. Millan, S. Dubey, and W. Koot, "Accelerated Mechanism of Scale Deposition in UW Production Operation," SPE 87446 (paper presented at the SPE International Symposium on Oilfield Scale, Aberdeen, UK, 26–27 May 2004).

[62] M. M. Jordan, K. Mackin, C. J. Johnston, and N. D. Feasey, "Control of Hydrogen Sulfide Scavenger Induced Scale (Due to Raised pH) and the Associated Challenge of Sulfide Scale Formation within a North Sea High Temperature/High Salinity Field Production Wells: Laboratory Evaluation to Field Application," SPE 87433 (paper presented at the SPE International Symposium on Oilfield Scale, Aberdeen, UK, 26–27 May 2004).

[63] C. D. Sitz, D. K. Barbin, and B. J. Hampton, "Scale Control in a Hydrogen Sulfide Treatment Program," SPE 80235 (paper presented at the International Symposium on Oilfield Chemistry, Houston, TX, 5–7 February 2003).

[64] J. G. R. Eylander, H. A. Holtman, T. Salma, M. Yuan, M. Callaway, and J. R. Johnstone, "The Development of Low-Sour Gas Reserves Utilizing Direct-Injection Liquid Hydrogen Sulfide Scavengers," SPE 71541 (paper presented at the SPE Annual Technical Conference and Exhibition, New Orleans, LA, 30 September–3 October 2001).

[65] E. Davidson, J. Hall, and C. Temple, "A New Iron-Based, Environmentally Friendly Hydrogen Sulfide Scavenger for Drilling Fluids," SPE 84313, *SPE Drilling & Completion* 19 (4) (2004): 229–234.

[66] J. Buller and J. F. Carpenter, "H_2S Scavengers for Non-aqueous Systems," SPE 93353 (paper presented at the SPE International Symposium on Oilfield Chemistry, The Woodlands, TX, 2–4 February 2005).

[67] A. S. Deshpande, N. V. Sankpal, and B. D. Kulkarni, U.S. Patent Application 20040192995.

[68] E. Davidson, U.S. Patent Application 20040167037, 2004.

[69] D. A. Wilson and D. K. Crump, U.S. Patent 5569443, 1996.

[70] S. Piché and F. Larachi, "Dynamics of pH on the Oxidation of HS^- with Iron (III) Chelates in Anoxic Conditions," *Chemical Engineering Science* 61 (23) (2006): 7673–7683.

[71] D. C. Olson, U.S. Patent 4443423, 1984.

[72] D. McManus and A. E. Martell, "The Development, Chemistry and Application of a Chelated Iron, Hydrogen Sulphide Removal Process," *Recent Advances in Oilfield Chemistry* (1994): 207.

[73] J. J. Weers and C. E. Thomasson, U.S. Patent 5223127, 1993.

[74] B. C. Collins, P. A. Mestetsky, and N. J. Savaiano, U.S. Patent 5807476, 1998.

[75] J. W. Bozzelli, G. D. Shier, R. L. Pearce, and C. W. Martin, U.S. Patent 4112049, 1978.

[76] E. T. Kool and C. E. Uebele, U.S. Patent 4569766, 1986.

[77] M. K. Pakulski, P. Logan, and R. Matherly, U.S. Patent Application 20050238556.

[78] T. J. O'Brien and J. J. Weers, International Patent Application WO/2008/115704.

[79] G. Westlund and D. Weller, International Patent Application WO/2009/035570.

[80] E. D. Burger, G. E. Jenneman, and J. J. Carroll, "On the Partitioning of Hydrogen Sulfide in Oilfield Systems," SPE 164067 (paper presented at the SPE International Symposium on Oilfield Chemistry, The Woodlands, TX, 8-10 April 2013).

[81] K. Robinson, J. Tuck, W. R. Ginty, E. Samuelsen, T. Lundgaard, and D. L. Roberts, "Reservoir Souring in a Field with Sulphate Removal: A Case Study," SPE 132697 (paper presented at the SPE Annual Technical Conference and Exhibition, Florence, Italy, 19-22 September 2010).

[82] E. D. Burger, G. E. Jenneman, and X. Gao, "The Impact of Dissolved Organic-Carbon Type on the Extent of Reservoir Souring," SPE 164068 (paper presented at the SPE International Symposium on Oilfield Chemistry, The Woodlands, TX, 8-10 April 2013).

[83] J. Waage, A.-M. Brurås Hårvik, K. Kroknes, and A. F. Mitchell, "Reservoir Souring: How to Minimize the Impact Based on Current Knowledge?" Tekna 23rd International Oil Field Chemistry Symposium, Geilo, Norway, 18-21 March 2012.

[84] T. C. Klaver and F. Guezebroek, "Development of Contaminated Gas & Oil Fields using Existing and Breakthrough CO_2/H_2S Separation Technologies," IPTC 12912 (paper presented at the SPE International Petroleum Technology Conference, Kuala Lumpur, Malaysia, 3-5 December 2008).

[85] P.-E. Just, M. A. Ouimet, and L. E. Hakka, International Patent Application WO/2011/009195.

[86] A. M. Blair, K. N. Garside, W. J. Andrews, and K. B. Sawant, International Patent Application WO/2012/068327.

[87] J. Zhao, C. Weiss, R. Cadours, and V. Shah, "Hysweet Technology: A Major Progress In Sour Gas Processing," SPE161457 (paper presented at the Abu Dhabi International Petroleum Conference and Exhibition, Abu Dhabi, UAE, 11-14 November 2012).

[88] X. Ma, X. Wang, and C. Song, *Journal of the American Chemical Society* 131 (16) (2009): 5777-5783.

[89] C. W. Aften and G. Roberts, "New Compounds for Hydrogen Sulfide Scavenging and Iron Sulfide Control," SPE 141286 (paper presented at the SPE International Symposium on Oilfield Chemistry, The Woodlands, TX, 11-13 April 2011).

[90] C. M. Menendez, V. Jovancicevic, and S. Ramachandran, U.S. Patent Application 20130004393.

[91] J. G. Frost, J. R. Snyder, and J. Kenneth, U.S. Patent Application 20100243578.

[92] C. B. Talley, U.S. Patent Application 20100056404.

[93] M. H. Al-Khaldi and Y. Duailej, "Triazine-Based Scavengers: Can They Be a Potential for Formation Damage?" SPE 157109 (paper presented at the SPE International Production and Operations Conference & Exhibition, Doha, Qatar, 14-16 May 2012).

[94] C. Mendez, J. Vera, I. A. O. Magalh, P. Alto, Ferreira, and E. C. Bastos, "Development of a Novel Testing Protocol and Equipment for the Evaluation of H_2S Scavengers, 2012-1466," CORROSION 2012, Salt Lake City, UT, 11-15 March 2012.

[95] L. J. Karas and A. E. Goliaszewski, International Patent Application WO/2010/027353.

[96] S. Ramachandran, V. Jovancicevic, K. C. Cattanach, and M. P. Squicciarini, International Patent Application WO/2012/128935.

[97] J. L. Stark, R. A. Steele, K. Babic-Samardzija, J. A. Schield, W. J. Cappel, and M. T. Barnes, U.S. Patent Application 20120329930.

[98] M. Subramaniyam, International Patent Application WO/2010/128523.

[99] G. Kaplan and L. J. Karas, International Patent Application WO/2013/049027.

[100] J. Yang, M. T. Barnes, and J. L. Stark, European Patent EP2465975, 2012.

[101] G. N. Taylor, "The Isolation and Formulation of Highly Effective Corrosion Inhibitors from the Waste Product of Hexahydrotriazine Based Hydrogen Sulphide Scavengers," *Chemistry in the Oil Industry XIII—New Frontiers*, Manchester, UK, November 2013.

[102] E. G. Sørgaard, H. T. Madsen, C. V. Jensen, M. Streek, and M. Hentz, "A Conceptual Model for the Understanding of Fouling Phenomena When Using Triazine Based H_2S Scavengers," *Chemistry in the Oil Industry XIII—New Frontiers*, Manchester, UK, November 2013.

[103] G. N. Taylor, P. Prince, R. Matherly, R. Ponnapati, R. Tompkins, V. Panchalingam, V. Jovancicevic, and S. Ramachandran, "The Formation and Chemical Nature of Amorphous Dithiazine Produced from the Use of Hexahydrotriazine Based Hydrogen Sulfide Scavengers and the Use of an Alternative Scavenger to Minimize Solid Formation in Sour Gas Applications," SPE 164134 (paper presented at the SPE International Symposium on Oilfield Chemistry, The Woodlands, TX, 8-10 April 2013).

[104] R. L. Horton, R. Stoker, and P. Davis, International Patent Application WO/2008/124404.

[105] R. Rodriguez Gonzalez and A. Grinrod, International Patent Application WO/2009/127604.

[106] D. R. Compton and B. J. Strickland, International Patent Application WO/2012/009391.

[107] J. Rosnes and A. Grinrod, International Patent Application WO/2010/150107.

[108] N. Goodwin, J. M. Walsh, R. Wright, S. Dyer, and G. M. Graham, "Modeling the Effect of Triazine Based Sulphide Scavengers on the *In Situ* pH, and Scaling Tendency," SPE 141583 (paper presented at the International Symposium on Oilfield Chemistry, The Woodlands, TX, 11-13 April 2011).

[109] G. Graham, H. Williams, and N. Goodwin, "Sulphide Scavengers, Performance, Deployment and Impact on Carbonate Scaling," Oilfield Chemical Symposium, Geilo, Norway, 17-20 March 2013.

[110] J. J. Weers and D. R. Gentry, U.S. Patent 5840177, 1998.

[111] F. Chaudhry, G. Mobley, Y. H. Tsang, S. Ramachandran, V. Jovancicevic, S. C. Braman, E. Rowton, A. McDonald, and J. Davis, "Laboratory Development of A Novel, Non-Triazine-Based Hydrogen Sulfide Scavenger and Field Implementation in the Haynesville Shale," SPE 164077 (paper presented at the SPE International Symposium on Oilfield Chemistry, The Woodlands, TX, 8-10 April 2013).

[112] K. Janak, International Patent Application WO/2013/041654.

[113] K. E. Janak, "Beyond Triazines: Development of a Novel Chemistry for Hydrogen Sulfide Scavenging," Paper, 2012-1520, CORROSION 2012, Salt Lake City, UT, 11-15 March 2012.

[114] D. R. Compton, B. J. Strickland, and J. M. Garcia III, International Patent Application WO/2012/009396.

[115] D. R. Compton, International Patent Application WO/2012/009390.

[116] P. Lue and G. Kaplan, International Patent Application WO/2013/077949.

[117] P. Lue and G. Kaplan, International Patent Application WO/2013/077965.

[118] D. Smith, C. Smith, and D. Watson, U.S. Patent Application 20090314720.

[119] S. Ramachandran, V. Jovancicevic, Y. H. Tsang, M. P. Squicciarini, P. Prince, J. Yang, and K. C. Cattanach, International Patent Application WO/2012/003267.

[120] J. Yang, T. Salma, J. A. Schield, J. J. Weers, and J. L. Stark, International Patent Application WO/2009/052127.

[121] G. Voordouw, S. M. Armstrong, M. F. Reimer, B. Fouts, A. J. Telang, Y. Shen, and D. Gevertz,

Applied and Environmental Microbiology 62（5）（1996）: 1623.

[122] T. Epsztein, F. Demanze, and X. Lefebvre, "Toward a H_2S-Free Environment in Flexible Pipe Annulus: The Use of a New Anti-H_2S Polymer Layer," OTC 21371（paper presented Offshore Technology Conference, Houston, TX, 2-5 May 2011）.

[123] S. R. Keenan, J. Collins, S. Ramachandran, V. Jovancicevic, R. Tompkins, G. N. Taylor, R. L. Martin, and M. L. Walker, U.S. Patent Application 20130224092.

[124] M. Subramaniyam, U.S. Patent Application 20130240409.

16 除 氧 剂

16.1 概述

水中的溶解氧会对金属管道和工艺设备造成破坏性的氧腐蚀，而且腐蚀的副产物还会堵塞地层造成伤害，因此需要从油田水中去除氧气。与仅含 CO_2 的体系相比，含 CO_2/O_2 体系的缓蚀抑制难度更大，此时也需要良好的除氧。

在油气生产中，除氧剂最常见的用途是注海水驱油生产系统、静水压测试（在压力下，用水测试管道和容器等的完整性）和酸化增产作业。对于注水系统，通常通过真空塔或气提塔进行初级除氧，将水中氧含量从约 9mg/L 降至 50μg/L 以下。自 20 世纪 90 年代初以来，已在部分注海水系统中应用了氮气脱氧的专利技术，可将氧含量降低至 5~15μg/L。这种工艺具有无须添加化学品、重量轻的优点。最近公开了一种专利方法，使用气提工艺将海水进水端中的溶解氧含量从 8mg/L 脱除到约 10μg/L 及以下。也可以通过在除氧容器的下游加注除氧剂，将氧含量降至 10μg/L 以下（最好小于 5μg/L，以避免显著的氧腐蚀）。

16.2 除氧剂的种类

已经使用的除氧剂有许多种，包括：（1）亚硫酸氢盐（$MHSO_3$）、焦亚硫酸盐（$M_2S_2O_5$）和亚硫酸盐（M_2SO_3）；（2）连二亚硫酸盐和 $Na_2S_2O_4$；（3）肼类，包括 1-氨基吡咯烷；（4）胍类；（5）氨基脲和碳酰肼；（6）羟胺类；（7）肟类；（8）活性醛类；（9）多羟基化合物；（10）有活性贵金属催化剂的氢气；（11）对基质和氧的反应起到催化作用的酶；（12）固体硫化亚铁。

由于过渡金属离子具有不同的氧化状态和配体络合能力，可以通过添加过渡金属离子进一步改善很多上述除氧剂的性能。

目前，石油工业用于注水和静水压测试的除氧剂主要是亚硫酸盐、亚硫酸氢盐和焦亚硫酸盐，有时会配套添加催化剂。其他类别的除氧剂主要是用于钻井液或锅炉水的有机氮化合物。以下先简要讨论这些除氧剂，再讨论亚硫酸氢盐、焦亚硫酸盐和亚硫酸盐等。

16.2.1 连二亚硫酸盐

在钻井和完井作业中的除氧，建议使用如连二亚硫酸钠（$Na_2S_2O_4$）等连二亚硫酸盐。

这些盐中的硫为 +3 价氧化态，在与氧反应时，先被氧化为 +4 价（亚硫酸盐），最终为 +6 价（硫酸盐）。

16.2.2 肼和胍类盐

肼类（RNH_2NH_2）呈碱性并会提高水的 pH 值。如果水中含有 Ca^{2+}、Mg^{2+} 和 HCO_3^-，就有可能新生成或次生更多的碳酸钙/碳酸镁垢。肼（NH_2NH_2）是一种疑似致癌物，需要采取特殊的处置预防措施。肼与氧气的反应相当缓慢，可以加入如铜（Ⅱ）和锰（Ⅱ）等过渡金属离子催化剂及提高温度来加快反应速率。使用过渡金属而非其他金属的原因是其可能有两个或更多的氧化态，并且可以通过金属 d 轨道与分子氧 p 轨道的重叠而与氧分子较强配合。与亚硫酸氢盐不同，肼在高温下（这种情况很少遇到）不会分解。肼可以以肼盐的形式作为酸化增产作业中的除氧剂，也可以使用如苯肼的其他肼，但其成本比亚硫酸氢盐、焦亚硫酸盐高。1-氨基吡咯烷是一种环状的 1,1-二烷基肼，同样存在过渡金属离子催化剂时，应用于锅炉可产生良好的除氧效果。肼类也用作控制油气井套管外腐蚀的除氧剂，完井后的防腐保护效果可达到 12～18 个月。已经证明，钴与水杨酸 [N, N'-双（水杨酸）乙二胺] 或 3,4-二氨基甲苯配体组成的络合物可以催化锅炉水中的肼除氧效果。

如乙酸胍等胍盐 [$H_2NC(=NH)NH_3^+X^-$] 主要用作注入海水的除氧剂。氨基脲（$H_2NNHCONH_2$）和碳酰肼（$RCONHNH_2$）也有使用，因其不像巯基除氧剂会增加溶解性无机固体，特别适用于锅炉水除氧。除氧过程需要对苯二酚或钴离子等催化剂。碳酰肼/肼等有机化合物对乙二醇系统的除氧也很有效。

16.2.3 羟胺和氧化剂

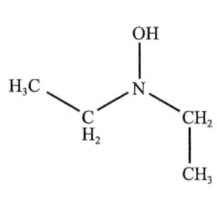

图 16.1 二乙基羟胺

如二乙基羟胺（DEHA）（图 16.1）或 N, N-双（2-羟乙基）羟胺等羟胺类是挥发性液体，目前主要用于密闭循环水系统的除氧。其优点是可将氧导致的点蚀和潜在腐蚀降至最低，不受温度影响，并且不增加溶解的无机固体。过渡金属催化剂、胺、丹宁、特丁基邻苯二酚、对苯二酚或邻苯三酚都可以作为催化剂。DEHA 还原氧气的速率很慢，单位质量只能吸收少量氧气，因此除氧时的相对加量较高。

酸性气体的气提设施中，如添加甲基乙基酮肟和乙醛肟等肟类（R=NOH 和 RR'NOH）等除氧剂，还可减少胺或乙二醇的降解产物量。

16.2.4 活性醛和多羟基化合物

甲醛等醛类因为氧化为羧酸的速率很慢，除氧性能较差。水杨醛和没食子酸因为其芳香环上的羟基具有活性，除氧效果相对较好。对苯二酚能催化醛类与氧气的反应。异抗坏血酸的分子结构含有一个酮和四个羟基的互变异构体，也可以归为这一类别（图 16.2）。

该分子经常用于锅炉水的除氧，也有专利将其与烷基羟胺结合使用于油田完井液的除氧。配套 Mg^{2+} 或 Cu^{2+} 作为催化剂，或使用 pH 值控制剂将 pH 值保持在 7 以上，可以提高除氧性能。含有过渡金属离子的聚合物已用于去除锅炉水中的溶解氧。这些聚合物基于氢醌—醌氧化还原体系，以 4-（2，5-二甲氧基苄基）肉桂酸甲酯的聚合和该单体与 4-（4'-乙烯基苯乙基）-1,10-菲罗啉的共聚来实现。聚 -4-（2，5-二羟基苄基）肉桂酸可在 70s 内将氧含量降低到 0.1mg/L 以下，其氧化还原能力达到 69.7mg 氧 /g 聚合物。

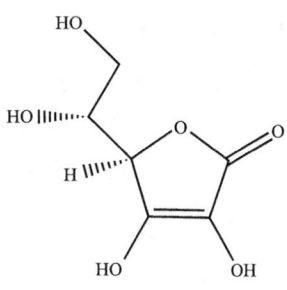

图 16.2　异抗坏血酸
（也称赤藻糖酸）

如葡萄糖酸盐、山梨醇、甘露醇、木糖醇、抗坏血酸和焦糖醇等其他多羟基化合物，与过渡金属的催化剂结合，也用作除氧剂。例如，有建议将酮基葡萄糖酸盐或酮基葡萄糖酸的立体异构体盐用于锅炉水体系或油田注入水或盐水的除氧。一种典型的多羟基除氧混合物是通过等物质的量的 1mol/L Mn^{2+} 和 1mol/L 葡萄糖酸钠相混合，并用氢氧化钠将 pH 值提高到约 9 而制成。这类混合物在弱碱性环境下的除氧效率最高。

16.2.5　催化加氢除氧

在贵金属催化剂（如钯树脂）的激活下，氢气可以与氧反应，在 60s 内将海水中的氧气浓度降低到可忽略的水平，甚至低于发生氧点蚀的浓度。元素周期表的Ⅷ族贵金属是氢和氧的强吸附剂，可以促进二者反应生成水。有文献针对这项技术在安哥拉、巴西和墨西哥湾及北海等海上油田的应用情况，进行了效果评价。其潜在优势是反应步骤少，设备重量轻、体积小。

16.2.6　酶

据称一种能催化基质和氧反应的酶可降低氧含量及其引起的腐蚀。这种酶可以是醇氧化酶，而基质是一种小分子醇。水中还添加了原油或其他烃类材料。

16.2.7　硫化亚铁试剂

使用硫化亚铁试剂可以去除氧气。这种方法使用细粉状的碳酸亚铁（优选为菱铁矿），可以从含有硫化氢和硫醇等硫化物的烃类化合物或二氧化碳流体中去除氧气。这些硫化物将碳酸亚铁转化为硫化亚铁，而硫化亚铁是活性除氧剂。

16.2.8　亚硫酸氢盐、偏亚硫酸氢盐和亚硫酸盐

油田生产中最常见的除氧剂是亚硫酸盐（M_2SO_3）、亚硫酸氢盐（$MHSO_3$）和焦亚硫酸盐（$M_2S_2O_5$）。目前主要使用的是亚硫酸氢盐离子（HSO_3^-）的盐类。过去也曾使用亚硫酸盐，但不能像亚硫酸氢盐那样制备为高浓度溶液。最常见的亚硫酸氢盐是亚硫酸氢铵

（NH_4HSO_3）和亚硫酸氢钠（$NaHSO_3$），亚硫酸氢铵比亚硫酸氢钠更易溶于水，在4℃时可制成质量分数高达65%的溶液。虽然亚硫酸氢钠是弱酸强碱盐，其水溶液的碱性较强，但亚硫酸氢铵碱性较弱，pH值只在8左右。经验表明，将水中氧浓度从9mg/L降到50μg/L时，需要添加浓度为60mg/L的质量分数为65%的亚硫酸氢铵。

在没有催化剂、温度低于约93℃的条件下，亚硫酸或亚硫酸氢盐与氧气的反应会比较慢（速率差异经常是2~3倍），形成的硫酸氢根离子如下：

$$2HSO_3^- + O_2 \longrightarrow 2HSO_4^-$$

加速反应过程的最佳催化剂是过渡金属离子，有研究通过公式量化了各种过渡金属离子的除氧率。因海水中含有少量的过渡金属离子，使用亚硫酸氢铵对海水进行除氧时，无须再额外添加过渡金属离子。但当使用亚硫酸氢钠时，需要在配方中额外添加过渡金属离子。Co^{2+}已证明是最好的金属离子催化剂，至今仍在一些地区使用，但在环境敏感地区则被铁（Ⅲ）盐取代。在pH值为5时，Mn^{2+}与Fe^{3+}协同作用可提高亚硫酸氢盐的除氧率。特别是加入浓度为5mg/L的高锰酸钾（$KMnO_4$）氧化剂，可以显著提高亚硫酸氢铵的除氧效率，此时亚硫酸氢铵浓度也只需要为5mg/L。可能的原因是高锰酸盐被亚硫酸氢盐还原为一种或多种含锰催化性物质。有证据表明，亚硫酸氢盐水溶液的铁催化氧化是通过硫酸根的自由基链机制起作用。天然或人为添加的过渡金属离子催化剂会因螯合物作用而失去活性。因此，如果注入水中存在如聚磷酸盐缓蚀剂、聚羧酸盐和聚磷酸盐防垢剂及聚氨基羧酸盐溶垢剂等螯合剂，会与过渡金属离子催化剂络合，而大大降低亚硫酸氢盐的性能。二氧化氯或其水溶性的盐（如亚氯酸钠或亚氯酸铵）也会催化亚硫酸氢盐的除氧过程。

尽管亚硫酸盐除氧剂广泛使用，但仍有以下缺点：

（1）亚硫酸氢铵中的氮为高温环境中的硫酸盐还原菌（SRB）等细菌提供了食物，但适当使用杀菌剂可以控制这种情况。

（2）戊二醛等醛类杀菌剂会降低亚硫酸氢盐的除氧效率。

（3）次氯酸盐和氯杀菌剂会与亚硫酸氢盐发生反应，使其失去活性，在实践中，要计算去除氧气所需的亚硫酸盐量及氯/次氯酸盐的残留量。

（4）锡离子、甘露醇、乙醇和有机酸会降低亚硫酸氢盐的除氧能力。

（5）亚硫酸盐和亚硫酸氢盐的除氧能力对水溶液pH值变化很敏感，有研究表明，在pH值大于7时，亚硫酸盐离子能迅速清除溶解氧，而在pH值小于6.0时，反应速率太慢，无法实际使用。在其他研究中，亚硫酸氢铵在pH值为6.5时，也能很好地发挥作用，在pH值为7.5~9时表现最佳。海水pH值为7.8。亚硫酸盐和亚硫酸氢盐在强酸性溶液中会发生分解，因此不能用于常规酸化增产作业。

（6）在接近0℃的低温海水中，除氧的速率会明显变慢。

在静水压测试中，因为水相可能在设备系统中留存多年，必须除氧以防止氧腐蚀，还应加入杀菌剂以消除微生物引起的腐蚀。但并非所有的杀菌剂都能与亚硫酸盐和亚硫酸氢盐兼容。其次，杀菌剂有毒性，而法规可能对排放水的毒性有规定。在静水压测试中，比较环保的防腐方法是首先在作业用水中加入亚硫酸氢盐除氧剂，当氧被清除后，再加入适量的杀菌剂，用氢氧化钠等碱调整 pH 值至 9.5 左右以限制细菌生长，并加入阻垢剂。

参 考 文 献

[1] M. Davies and P. J. B. Scott, *Oilfield Water Technology*, National Association of Corrosion Engineers (NACE), Houston, TX: 2006.

[2] H. G. Byars and B. R. Gallop, "Injection Water + Oxygen = Corrosion and/or Well Plugging Solids," SPE 4253 (paper presented at the SPE Symposium on Handling of Oilfield Water, Los Angeles, CA, 4-5 December 1972).

[3] R. L. Martin, "Corrosion Consequences of Oxygen Entry into Sweet Oilfield Fluids," SPE 71470 (paper presented at the SPE Annual Technical Conference and Exhibition, New Orleans, LA, 30 September-3 October 2001).

[4] K. B. Flatval, S. Sathyamoorthy, C. Kuijvenhoven, and D. Ligthelm, "Building the Case for Raw Seawater Injection Scheme in Barton," SPE 88568 (paper presented at the SPE Asia Pacific Oil and Gas Conference and Exhibition, Perth, Australia, 18-20 October 2004).

[5] S. Yntema, P. De Boer, R. A. Trompert, R. M. de Jonge, and B. J. Gellekom, "Oxygen-Free Acid Stimulation in an Underground Gas Storage Well Completed with Pre-Packed Screens," SPE 99846 (paper presented at SPE/IcoTA Coiled Tubing and Well Intervention Conference, Woodlands, TX, 4-5 April 2006).

[6] S. L. Wellington, "Biopolymer Solution Viscosity Stabilization—Polymer Degradation and Antioxidant Use," SPE 9296, *SPE Journal* 14 (1974): 643.

[7] N. Henriksen, European Patent Application EP0234771, 1986.

[8] J. L. Watson and L. L. Carney, U.S. Patent 4059533, 1977.

[9] C. J. Philips, European Patent 106666, 1984.

[10] Y. Shimura, S. Taya, and T. Shiro, U.S. Patent Application 20030141483, 2003.

[11] Y. Shimura, S. Taya, K. Uchida, and T. Sato, "The Performance of New Volatile Oxygen Scavenger and Its Field Application in Boiler Systems," Paper 00327 (paper presented at the Corrosion 2000, NACE International Conference).

[12] F. W. Schremp, J. F. Chittum, and T. S. Arczynski, "Use of Oxygen Scavengers to Control External Corrosion of Oil-String Casing," SPE 1606, *Journal of Petroleum Technology* 13 (7) (1961): 703.

[13] F. Dawans, D. Binet, N. Kohler, and D. V. Quang, U.S. Patent 4454620, 1984.

[14] M. Slovinsky, U.S. Patent 4269717, 1981.

[15] B. Greaves, S. C. Poole, C. M. Hwa, and J. C.-J. Fan, U.S. Patent 5830383, 1998.

[16] A. M. Rossi and P. R. Burgmayer, U.S. Patent 5256311, 1993.

[17] Y. Shimura and J. Takahashi, U.S. Patent 7112284, 2006.

[18] R. R. Veldman and D. Trahan, U.S. Patent 5686016, 1997.

[19] E. J. Burcik and G. C. Thankur, "Reaction of Polyacrylamide with Commonly Used Additives," SPE 4164, *Journal of Petroleum Technology* 24 (1972): 1137-1139.

[20] J. A. Muccitelli, U.S. Patent 4569783, 1986.

[21] C. A. Soderquist, J. A. Kelly, and F. S. Mandel, U.S. Patent 4968438, 1990.

[22] M. Slovinsky, Canadian Patent CA1186425, 1985.

[23] H. L. Gewanter and R. D. May, U.S. Patent 5114618, 1992.

[24] Article in *Offshore* 59 (11) (1 November 1999).

[25] E. Gobina, International Patent Application, WO/2001/085622.

[26] L. A. Cantu and L. D. Harrison, "Field Evaluation of Catalytic Deoxygenation Process for Oxygen Scavenging in Oilfield Waters," SPE 14284, *SPE Production Engineering* 3 (4) (1988): 619−624.

[27] Y. Wu, U.S. Patent 4501674, 1985.

[28] N. Matsuka, Y. Nakagawa, M. Kurihara, and T. Tonomura, "Reaction Kinetics of Sodium Bisulfite and Dissolved Oxygen in Seawater and Their Applications to Seawater Reverse Osmosis," *Desalination* 51 (2) (1984): 163−171.

[29] R. K. Ulrich, G. T. Rochelle, and R. E. Prada, "Enhanced Oxygen Absorption into Bisulphite Solutions Containing Transition Metal Ion Catalysts," *Chemical Engineering Science* 41 (8) (1986): 2183−2191.

[30] T. Chen and C. H. Barron, "Some Aspects of the Homogeneous Kinetics of Sulfite Oxidation," *Industrial & Engineering Chemistry Fundamentals* 11 (4) (1972): 466.

[31] J. Nakajima, M. Yamashita, and K. Kimura, U.S. Patent 6402984, 2002.

[32] J. Ziajka, F. Beer, and P. Warneck, "Iron-Catalysed Oxidation of Bisulphite Aqueous Solution: Evidence for a Free Radical Chain Mechanism," *Atmospheric Environment* 28 (1994): 2549−2552.

[33] A. J. McMahon, A. Chalmers, and H. Macdonald, "Optimising Oilfield Oxygen Scavengers," *Proceedings of the Chemistry in the Oil Industry VII*, Royal Society of Chemistry, Manchester, UK, 2002, 163.

[34] J. R. Stanford, J. H. Martin, and G. D. Chappell, U.S. Patents 3996135, 1976, and 4098716, 1978.

[35] T. G. Braga, "Effects of Commonly Used Oilfield Chemicals on the Rate of Oxygen Scavenging by Sulfite/Bisulfite," SPE 13556, *SPE Production Engineering*, 2 (2) (1987): 137.

[36] E. S. Snavely, "Chemical Removal of Oxygen from Natural Waters," SPE 3262, *Journal of Petroleum Technology* 23 (1971): 443−446.

[37] R. W. Mitchell, "The Forties Field Seawater Injection System," SPE 6677, *SPE Journal* (1978): 877.

[38] R. Prasad, U.S. Patent 6815208, 2002.

[39] K.-Y. Wu, A. T. Lee, L. Vuong, E. K. Liu, and K. T. Chuang, International Patent Application WO/2012/065243.

[40] K. Sanlo Rane and S. Bandodkar, "Metal Complexes as Catalysts in Enhancing the Oxygen Scavenging Action of Hydrazine in the Boiler Feed Water," (paper presented at the 2013 AIChE Annual Meeting, San Francisco, 3−8 November 2013).

[41] G. M. Noack, U.S. Patent 4026664, 1976.

[42] G. M. Noack, "Catalyzed Hydrazine Compound Corrosion Inhibiting Composition and Use," U.S. Patent 4012195, 1975.

[43] T. Tsumaki, *Bulletin of the Chemical Society of Japan* 13 (1938): 252.

[44] K. Sanlo Rane and S. S. Bandodkar, "Metal Complexes for Enhancing the Oxygen Scavenging Action of Hydrazine" India Patent-IP15596 (provisional Filing, January 2011) [Sifali Bandodkar, PhD thesis, Goa University, Goa, India, November 2011].

[45] D. J. Cookson, T. D. Smith, J. F. Boas, P. R. Hicks, and J. R. Pilbrow, "Electron Spin Resonance Study of the Autoxidation of Hydrazine, Hydroxylamine and Cysteine Catalysed by the Cobalt (II) Chelate Complex of 3, 10, 17, 24-Tetrasulpho Phthalocyanine" *Journal of Chemical Society Dalton Transactions.* (1977): 109−114.

[46] J. P. Deville, U.S. Patent Application 20120118569.
[47] C. Waterlot and D. Couturier, *Journal of Applied Polymer Science* 7-16 (2010): 118.
[48] I. C. Callaghan and R. Shaunak, European Patent EP0272887, 1988.
[49] F. E. Farha and J. A. Kane, U.S. Patent Application 20100126346.
[50] R. Barr, 2013, personal communication.

17 减 阻 剂

17.1 概述

　　液体在导管（如管道）内的流动，会产生摩擦能量损失，导致管内液体压力沿着流动方向下降。导管直径固定时，压降随着流速的增加而增大。减阻剂添加到液体中的作用是减少液体在湍流流场（雷诺数大于2100）下的摩擦阻力（相较于纯液体）。这个效应有时以此现象发明者的名字命名，称为汤姆斯效应。减阻剂（DRA）有时也被称为减摩擦剂或增流（流动增强）剂，后一个术语可能会与防蜡剂/倾点抑制剂（如聚甲基丙烯酸酯）相混淆。减阻剂在湍流过程中与流体相互作用，减少摩擦压力损失，从而降低定流速下的压降，或增加定压降下的流量。在多数石油管道中，液体是以湍流状态流动，因此减阻剂在其中都能展现良好的效果。减阻剂可以减少摩擦能量损失，从而能够增加管道、软管和其他导管中的液体流动能力；还能降低泵运行成本和设备成本；并能在给定的流量下使用较小的管道直径。当原油冷却到接近倾点时，减阻剂的效果可能会变差。油溶性减阻剂的最大用途是精炼成品油（而非原油）的管道运输。

　　有室内研究使用流变仪和流动管路揭示了某些特种聚合物降低稠油集输阻力的机理。结果显示在较低的剪切应力下，具有卷曲—拉伸过渡的聚合物应该更有效。除了17.3节所讨论的更为标准的油溶性减阻剂，还有改善稠油与采出水混合物流动性的其他聚合物（也称为乳化降黏剂）。与第4章和第10章所述的沥青质分散剂或倾点抑制剂相似，也有专门针对沥青质原油设计的减阻聚合物。

　　超高分子量（UHMW）聚合物通常是最有效的减阻剂；但是表面活性剂也能表现出良好的减阻效果，不过通常需要较高剂量。浓度20~30mg/L的聚合物实现高达70%~80%减阻率的情况也并不罕见。通常管径越大，需要的减阻剂浓度越高。纤维也可表现出减阻剂特性。减阻剂在水或烃类流体中的性能，取决于诸多参数：流体黏度、管道直径、液体和气体速度、油的成分、管道表面粗糙度、含水率、管道倾斜度、减阻剂浓度和类型、减阻剂的剪切降解和温度，甚至是水溶性减阻剂的pH值。在注水过程中，使用UHMW聚合物减阻剂的问题是会造成储层内天然气或石油流速降低，在17.4.1中讨论。

　　石油工业中最早使用减阻剂的现场报道之一是1965年采用瓜尔胶降低水基压裂液的泵送成本。此后减阻剂在石油和天然气行业的钻井液、压裂、酸化增产、注水、连续油管作业和原油集输等环节得到了广泛使用。在原油集输中的应用将在后文详细讨论。减阻剂通常设计为油溶性（用于集油管道）或水溶性（用于输水管道），还有研究用于多相流（油和水一起流动，有时还有气体）体系。最外端的流动相（油相或水相）将决定油溶性减阻剂或水溶性减阻剂是更加合适的减阻剂。油溶性减阻剂的有效性会随着含水率上升而降低。

用于气体集输的减阻剂已有现场应用。例如，注入成膜型两亲缓蚀剂或简单的脂肪酸胺，据称可以减少湍流中气体的摩擦或阻力。缓蚀气体减阻剂也有在油田成功应用的报道。减阻剂的工作原理是带正电的胺和酰胺官能团与金属表面紧密结合，同时长链烃部分作为顺滑或润滑的表面来减弱气相边界的湍流。

减阻剂性能的初步筛选常采用流变仪。有报道称基于保持样品在湍流中旋转所施加的扭矩，研究了一种流变仪的新实验技术。在有/无添加剂的两种油样中，水动力阻力的减少与施加扭矩的差异成正比。还有利用湍流流变仪评价减阻聚合物性能的报道。

在实验室中测试减阻剂性能的常见方法是使用流动环路并测量压降。本章引用的许多参考文献（特别是专利）描述了相关设备和测试方法。环路测试的结果可以直接与现场观察结果进行比较。但使用与现场操作环境相当规格的全尺寸管路流动系统评价减阻剂，其成本很高且需要大量液体，往往难以实现，这导致文献中关于管径对减阻剂性能的影响存在一些争议。试验环路有时会高估减阻剂效果，特别是在低浓度条件时。有研究人员试图解决实验室表征和现场操作之间的差距。该研究表明，选取正确的实验测试参数，对利用实验结果预测现场条件下减阻剂的有效性和性能至关重要。该研究提出了一种评估气液两相流中减阻效果的新方法。

还有资料介绍了使用带有浸没式喷射池的环路及快速筛选的多重测试仪器。常用来测量减阻剂性能的更简化、更小型仪器是转盘或筛网拉伸流变仪，还有一种特别设计的毛细管黏度计。如果一种市售减阻剂已有良好的现场应用经验，在将其用于其他类似应用之前，可能不需要进行实验室测试。

17.2 减阻剂机制

尽管减阻剂领域已有大量研究，但对于聚合物或表面活性剂在湍流中的减少摩擦机制尚没有普遍接受的模型。一种早期理论认为，随机卷曲聚合物的拉伸增加了有效黏度，这样导致小涡流减弱，进而导致黏性层增厚和阻力减小。最近又有理论认为阻力减小是由弹性而非黏度引起。这一结论是基于实验中发现当聚合物在管道中心活跃时阻力减少，而此处黏性力并不发挥作用。另有研究团队观察到，阻力的减小量受经验渐近线（称为 Virk 渐近线）的限制，也有其他研究者发现了矛盾的结果。现场应用时，可参考并放大实验室内测得的聚合物弹性参数。当聚合物以不同喷射速度流经细直管时，通过测量压降、速度和弛豫时间，就可以确定减阻剂的摩擦压降梯度。

随后的一个理论分三部分定性讨论管道中的湍流（图17.1）。在管道最中心是湍流核心区，其中可以存在涡流。该部分包括管道中的大部分流体，所占区域体积最大。最贴近管壁的是层流亚层。该区域的流体以片状形式横向移动。在层流亚层和湍流核心区之间是缓冲区，湍流在这里首次形成。层流亚层的一部分称为"条纹"，偶尔会移动到缓冲区。

在该区域的条纹开始涡旋和振荡，随着其越接近湍流核心区，移动速度越快。最后条纹变得不稳定，当其把流体抛入流动的核心时就会破裂。这种将流体抛入湍流核心区的现象称为湍流爆裂。这种爆裂运动和在湍流核心区中的爆裂增加导致了能量的浪费。减阻聚合物会干扰爆裂过程，并减少核的湍流。聚合物像减震器一样吸收条纹中的能量，减少随后的湍流爆裂。因此，减阻聚合物在缓冲区内最为活跃。总体效果可能是增加层流亚层的厚度，从而减少对流传热。

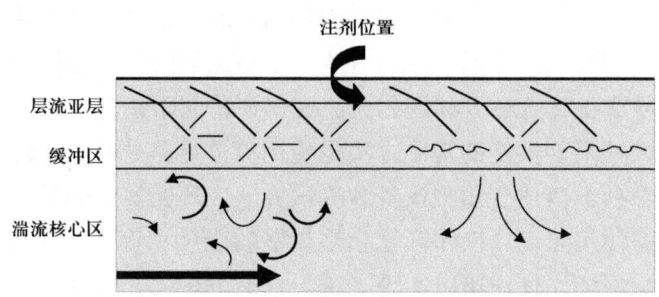

图 17.1 减阻聚合物注入湍流内，抑制能量爆裂

另一项用水溶性聚氧乙烯（PEO）进行的流动环路研究，使用不同的术语表明阻力减少伴随着"剪切层"（即高空间速度梯度的细丝状区域）的出现，这些剪切层作为界面，分开靠近管壁的低动量流区和靠近中心线的高动量流区。剪切层不是静态的，其厚度与测量的阻力减小量有关。

显然，分子量、分子间聚集和分子链柔性（相对于聚合物刚性）都是影响聚合物减阻剂性能的重要因素。关于分子量的理论认为，分子链较长的聚合物将最适合打破流动中的湍流爆裂或涡流。一些研究者根据实验室研究提出，聚合物的流体力学体积（卷曲体积）是比分子量更重要的关键因素，许多油溶性减阻剂就是有长的烷基侧链以增加聚合物体积。聚合物体积随着溶剂的变化而变化，如果是水溶液，有时也会随着 pH 值和电离度的变化而变化。最近的一项研究表明，在低矿化度水的流动环路中，相对于分子量相似的非离子丙烯酰胺均聚物，95% 的丙烯酰胺和 5% 的丙烯酸钠组成的阴离子聚合物具有更好的减阻性能。离子聚合物由于电斥力导致摩尔质量增大，从而可能是更好的减阻剂。

向多相流中加入聚合物的显著作用是不仅可以降低可测量的阻力，还会改变相的构成或流动模式。例如有研究发现，将聚丙烯酰胺（PAM）和丙烯酸钠的浓缩溶液注入水平管道的空气—水流体中，通过破坏液膜中的干扰波，将环形模式变为分层模式。在平均浓度为 10~15mg/L 时，测得的阻力降低了 48%。

长链聚合物对管道内输送过程中的机械降解很敏感，这种现象会逐渐降低减阻剂的整体效率。流变仪和湍流环路的实验结果凸显出降解动力学和流动耗散能量之间的明确联系。由此产生的模型能从设计开发阶段就考虑，在原油出口管道中更好地使用减阻剂，而不是仅将其用于解决流动的瓶颈段问题。

17.3 油溶性减阻剂

原油集输使用油溶性减阻剂始于跨阿拉斯加管道的成功应用。从 1979 年 7 月开始首个商用减阻剂在该管道中应用，到 1980 年该管道的输量已经增加到 9940m^3/h 的水平，其中大约 1300m^3/h 是加注减阻剂的直接结果。此后，许多项目都使用减阻剂以提高管道的输量或减少泵站。原油集输中，相比于表面活性剂型减阻剂，聚合物减阻剂尽管单位成本较高，但给定浓度下性能更好，也更受欢迎。

17.3.1 油溶性聚合物减阻剂

在油溶性减阻剂的增效型分子结构中，聚合物的尺寸小于湍流的最小尺度。众所周知的影响是聚合物会使流体的剪切黏度增加，这让人怀疑聚合物可以影响微观尺度上的湍流。但超高分子量聚合物在湍流的微观尺度和宏观尺度上都很活跃。因此，减阻剂的一个关键特征是其分子链越长越好，分子量应尽可能高。

在流变仪和流动环路中对各种流体（包括石蜡基原油）进行了测试。实验结果表明，减阻剂的效率会被蜡状晶体或乳化水滴改变，但不受蜡质沉积物或低温的影响。

市售原油集输用减阻剂多数是基于烯烃的齐格勒－纳塔有机金属聚合制备得到的超高分子量聚合物。可以使用的低成本单体有异丁烯、异戊二烯、苯乙烯、己烯、辛烯、癸烯、十四烯。

只有齐格勒－纳塔聚合才能够产生重均分子量为 $(1\sim3)\times10^7$ 的超高烯烃聚合物，这是良好减阻剂性能所必备的条件。术语"超高分子量"对应的是至少约 10$mPa\cdot s$ 的固有聚合物黏度。由于减阻剂聚合物的分子量极高，很难可靠和准确地测量实际分子量；但可以用固有黏度估计分子量。甲基丙烯酸酯也可以通过聚合得到超高分子量聚合物减阻剂。

17.3.1.1 聚乙烯（聚烯烃）减阻剂

多年来，油溶性聚合物减阻剂是以异丁烯等小分子单体为基础，合成聚异丁烯（PIB）（图 17.2）。减阻所需的典型浓度为 10～30mg/L。近期研究人员发现，含有较大单体（如己烯、辛烯、癸烯和十四烯）的共聚物具有更好的减阻性能。这可能是因为两个因素：首先，长链聚合物容易被管道或泵中的湍流（剪切力）所降解。增加分子量可以提高减阻性能，但同时也增加了聚合物的降解可能性。少量或没有侧链的聚合物（如聚乙烯或 PIB）几乎没有抗剪切降解的能力；而具有较大支链或含有较大的烯类结构的聚合物，具有更强的抗剪切降解能力。其次，对于给定链长度的聚合物，较大侧链的聚合物有较大的流体力学体积（卷曲体积），将更可能破坏湍流。两种烯烃的共聚物而不是单一烯烃的均聚物似乎是优选。由于使用不同的单体，共聚物的结晶度似乎比均聚物低。较低结晶度的共聚物非常有利于其在烃类流体中的溶解，进而增加了减阻效果。

图17.2 聚异丁烯（PIB）的片段

大多数合成的较大烯烃都是1-异构体类型（α-烯烃）。事实上，多年来只有壳牌公司生产的α-烯烃产品适用于高性能聚合物减阻剂。因此，有研究提出1-己烯和1-辛烯、1-辛烯和1-癸烯、1-癸烯和1-十四烯的共聚物作为改进的油溶性减阻剂（图17.3）。随着单体尺寸的增加，由于齐格勒-纳塔金属催化剂中心（通常是钛）的空间拥堵，超高分子量材料的聚合变得越来越困难。可以合成最多16个碳原子的1-烯烃单体，但优选范围是6～10个碳原子。

另一种具有相当大侧链的聚烯烃聚合物是聚苯乙烯（图17.4）。用分子量为$7.1×10^6$的苯聚合物开展管道实验，阻力显著降低。对于湍流中降解导致的减阻效率下降，聚苯乙烯样品有良好的抵抗力。研究人员得出结论，阻力减少和降解在很大程度上取决于分子量的分布。具有超高分子量（$>10^7$）的聚苯乙烯可能效果会更好。苯乙烯和1-烯烃的超高分子量共聚物也是减阻剂，还有如t-丁基苯乙烯-己烯-十二烯三元共聚物的烷基苯乙烯共聚物。

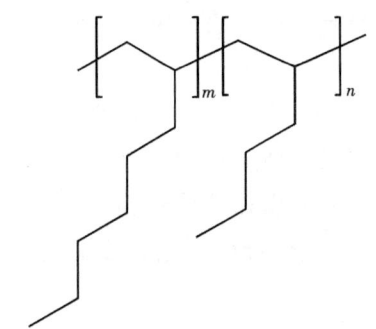

图17.3 1-己烯和1-辛烯的共聚物

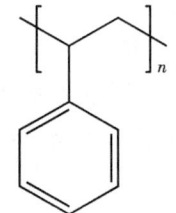

图17.4 聚苯乙烯

聚烯烃减阻剂的另一项改进是在共聚物中使用1-烯烃（α-烯烃）的一种或多种异构体，如1-己烯和1-十二烯的共聚物或1-辛烯和1-十四烯的共聚物，在聚合前将一种或两种单体进行异构化。研究表明，使用1-烯烃的异构体能减少大约50%的催化剂用量。此外，通过在聚合物减阻剂中加入支链，可以改善其在烃类流动流体中的溶解度。支链的平均链长应至少为4～8个碳原子。相同分子量的支化聚合物由于回旋半径（R_g）减小，其整体尺寸也会变小，因此更容易溶解。

17.3.1.2 聚（甲基）丙烯酸酯减阻剂

除了聚烯烃接枝长侧链可以改善聚合物减阻剂的性能外，甲基丙烯酸烷基酯、丙烯酸烷基酯和苯乙烯烷基酯的乳液聚合也可以得到具有长侧链的聚乙烯聚合物。（甲基）丙烯酸烷基酯可以单独用于合成聚合物减阻剂，这种产物通常含有至少 6~8 个碳原子的侧基。例如，聚甲基丙烯酸异癸酯的减阻性能优于市售的 PIB，特别是在抗剪切稳定性方面（图 17.5）。在给定聚合物浓度和管道尺寸的情况下，将分子量从 10×10^6 提高到 26×10^6，阻力减少的幅度更大。该聚合物已商业化用于压裂液，但尚未用于石油集输。一般来说，与聚烯烃相比，超高分子量聚烷基（甲基）丙烯酸酯的合成更为困难。已有方法将极性或离子型单体短链置入聚合物的末端，提高聚甲基丙烯酸酯减阻剂的分子量（通过聚合物网络）。所用的极性单体是在甲基丙烯酸月桂酯基聚合物上的 2-乙烯基吡啶和甲基丙烯酸甲酯。此外使用长侧链（如月桂基），可使聚合物具有高的流体力学体积。这些和其他特有因素使商品化（甲基）丙烯酸酯聚合物的性能与聚烯烃减阻剂同样出色。

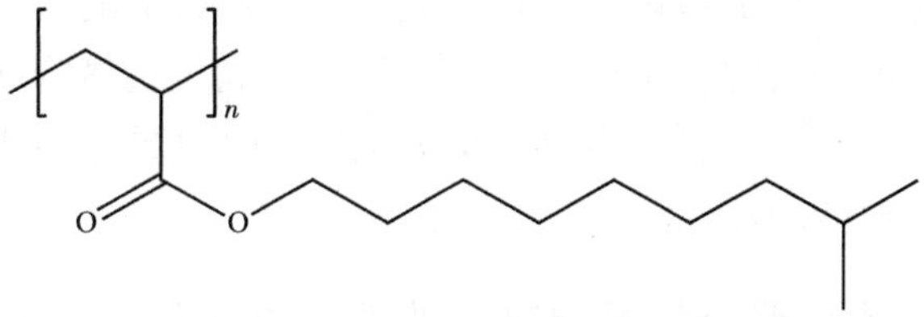

图 17.5 聚甲基丙烯酸异癸酯

分子量大于 10^7 的聚甲基丙烯酸十二烷基酯是良好的减阻剂，同样具有良好的剪切稳定性（图 17.6）。这些丙烯酸烷基酯聚合物的大规模生产价格因单体的成本较高通常比较昂贵。另一个例子是二十二烷基丙烯酸苯酯共聚物（C_{22}）。如 1-辛烯和 10-十一烯酸甲酯的共聚物等烯烃和乙烯基酯的共聚物也有报道。在由水、THI（MEG）、表面活性剂、引发剂和缓冲剂组成的乳液中，聚合烷基（甲基）丙烯酸酯（如甲基丙烯酸 2-乙基己酯）可以制备水合抑制减阻胶乳。

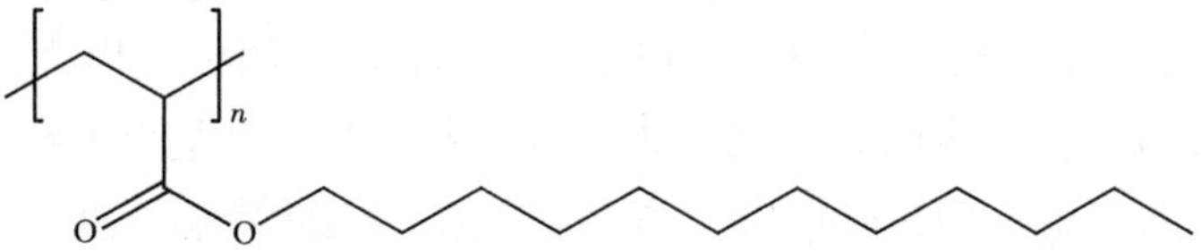

图 17.6 聚甲基丙烯酸十二烷基酯

与聚烯烃相比，丙烯酸烷基酯聚合物更适用于低 API 度或高沥青质含量的原油。例如，甲基丙烯酸 2-乙基己酯的超高分子量聚合物，可选择与其他丙烯酸烷基酯（如丙烯酸正丁酯）共聚。据称，丁基（甲基）丙烯酸酯或分子量稍小的烷基丙烯酸聚合物也是减阻剂。

17.3.1.3 其他油溶性减阻剂聚合物

对有低比例极性缔合基团的油溶性聚合物减阻剂，研究表明在存在聚合物分子间相互作用时，产物的减阻效果比均聚物好；在存在聚合物分子内相互作用时，产物的减阻效果比均聚物差。同一研究团队的相关文章还指出，分别与烯烃共聚而具有油溶性的一种阳离子聚合物和一种阴离子聚合物的混合物，在减阻及抗剪切降解方面表现出色。该聚合物的分子量不需要达到超高分子量的范围，这样更容易加注和处理。例如，苯乙烯/乙烯基吡啶共聚物（阳离子）和中和的磺化共聚物（阴离子）的混合物。其他研究报道氢键结合的聚合物体系有更好的减阻性能。

酸碱相互作用的聚合物混合物也有作为减阻剂的报道，如 α-烯烃（如 1-辛烯和 10-十一烯）共聚物与苯乙烯和乙烯基吡啶的共聚物的混合物。这种酸碱相互作用的聚合物可以通过聚合物网络而不是通过高分子量来更大幅度减少阻力，从而这种聚合物网络结构对流动剪切降解的敏感性更低。对氢键结合聚合物混合物的研究也有后续报道。

关于油溶性聚合物减阻剂的可生物降解类产品研究报道较少。秋葵黏液—丙烯腈接枝聚合物可生物降解，而且在烃类液体中非常有效，可用作柴油的可溶性减阻剂。但受到泵和管道内湍流涡流的高剪切力时，由于接枝聚合物分子发生机械降解，随时间延长，其减阻效果会逐步降低。水溶性秋葵黏液由分子量为 1.7×10^6 的糖蛋白组成，会对溶液起到增黏、剪切变稀和改变黏弹性的作用。

17.3.1.4 克服超高分子量聚合物减阻剂在处理、泵送和注入方面的困难

高分子量聚合物减阻剂的合成只是工艺问题的一部分，其主要困难在于如何获得低黏度、自由流动、不结块和高浓度的最终产物，并使减阻剂能顺利和容易地注入管道。聚合物减阻剂的分子量极大，溶液非常黏稠，因此常在烃类溶剂中稀释到活性组分占比小于 10%，以将黏度降低到可控水平。有许多专利详细介绍了改善聚合物减阻剂的黏度、处理和在管道流体中溶解的方法。还提出了在水、含氧溶剂及烃类溶液中，聚合物以乳液、分散或悬浮液等形式的许多变化。一种优选方法是通过研磨将聚合物减阻剂制成固体颗粒，并将其分散在无溶剂液体或乳液中。与溶解于溶剂相比，这样可以得到更高浓度（25%~30%）的减阻剂产品，还能提高性能。聚合物减阻剂也可以微囊化，使其易于使用并具有定时释放的特性。使用这种聚合物减阻剂固体微粒也避免了在注入阀位置的剪切降解问题。研究表明，由于高拉伸应变，聚合物在注入阀处降解程度最大，而不是在湍流的管道内。据称，胶乳悬浮液减阻剂可以更好地避免运输和处理危险溶剂，并避免产品的现场储存和高温安全性等问题。

另一项旨在克服聚合物减阻剂早期剪切降解的改进，是使用两种减阻剂产品的混合物。其中一种产品溶解较快，而另一种不直接溶解，只是在管道中逐步溶解，这样当前者因剪切降解而失去作用时，后者仍可保持整体的减阻性能。还可以是快速溶解的沉淀浆液产品与溶解较慢的研磨聚合物浆液产品的混合物。两种聚合物减阻剂的混合颗粒也能提高产品的性能。

固体颗粒减阻剂的一项专利改进是使用双峰或多峰的粒度分布。粒度大的减阻剂溶解速度远慢于粒度小的。使用至少是双峰的粒度分布，较小的颗粒在管道内溶解得更早或更近距离内，较大的颗粒溶解得更晚或更远距离外，从而减阻效果可以在管道全长度内更均匀地分布。还有一种方法是将聚合物与蜡晶体改进剂研磨后制成悬浮液，以改善超高分

量聚合物减阻剂的处理和注入性能。通过使用脂肪酸悬浮介质（如大豆油）和由含氧极性有机化合物（如乙醇）组成的改性剂，可以制成低黏度聚烯烃减阻剂颗粒的冷冻保护浓缩悬浮液并长期保持稳定。

有报道称制备了用于原油管道运输的微胶囊化 α-烯烃减阻剂聚合物。微胶囊化并不影响 α-烯烃聚合物的减阻性能。

还报道了一种制备有机金属油溶性聚合物减阻剂的新方法。在毛细管流变仪中，正己烷（雷诺数 25000）加入浓度为 0.1% 的三正丁基氟化锡，阻力可减少 75%，与 PIB 的结果相似。在相同条件下，三正丁基氯化锡却没有减阻效果。此外，氟化物产品在高剪切力下的减阻能力也没有损失。三正丁基氟化锡的效果可解释为形成了—Sn—F—Sn—F—线型聚合物，其中的五价锡通过氟桥接。

17.3.1.5　多相流中的油溶性聚合物减阻剂

聚合物减阻剂已经在两相流中进行了测试。超高分子量聚烯烃减阻剂在凝析气两相流环境进行的性能测试，减阻率可达到 65%。这些减阻剂一般是用于油流环境。在油—CO_2 气的多相流环境进行了油溶性聚合物减阻剂测试，与预期形成鲜明对比的是，阻力降低主要来自加速部分，这表明减阻剂不但在测试流体的缓冲区起作用，在段塞流的混合区也起作用。加速阻力的减少占总阻力减少的 88%。

17.3.2　油溶性表面活性剂减阻剂

在水或多相流中，表面活性剂减阻剂的应用得到了更广泛的探索。但是，非聚合物油溶性减阻剂研究的报告十分有限。

4 种减阻剂的单相正己烷流动测试中，当浓度大于 200mg/L 和摩擦速度小于 0.3ft/s 时，一种烷基膦酸酯的性能优于三种超高分子量聚合物（20~100mg/L）。该条件下膦酸酯使阻力降低高达 85%，并且剪切降解可忽略。对于浓度小于 100mg/L 和摩擦速度大于 0.3ft/s 的情况，超高分子量聚合物产品在没有降解时减阻性能优越。但在正己烷和天然气的两相流中，聚合物产品更胜一等，而且浓度更低。

有资料介绍了羧酸铝减阻剂。这类减阻剂不会发生永久性的剪切降解，也不会导致被处理液体的乳化或液体质量变差，或产生不良的泡沫，另外其可注入性好。例如，二辛酸铝、二硬脂酸铝和各种混合物。这种产品的另一种变化是在现场将单羧酸铝与至少一种羧酸混合，产生二羧酸铝减阻剂。这就避免了非常黏稠溶液的处理、运输和加注等难题。一些用于减少阻力的水溶性表面活性剂（见下文）具有足够的疏水性，可用作烃类液体或多相流集输的减阻剂。

17.4　水溶性减阻剂

水溶性减阻剂可分为高分子量线型聚合物和表面活性剂两类。下文将讨论这两类物质，以及其在减少阻力和缓蚀之间的关系。

17.4.1 水溶性聚合物减阻剂

许多类别的水溶性聚合物都具有减阻特性，包括阳离子、阴离子或非离子聚合物。已用作减阻剂的水溶性聚合物包括：（1）PAM 和部分水解聚丙烯酰胺（PHPA）；（2）丙烯酰胺、丙烯酰胺衍生物和丙烯酸酯的其他共聚物；（3）PEO；（4）聚乙烯醇；（5）多糖和衍生物（如瓜尔胶、羟丙基瓜尔胶）、黄胞胶、羧甲基纤维素（CMC）；（6）羟乙基纤维素（HEC）。

以上类别中，丙烯酰胺超高分子量聚合物和共聚物是石油工业注水中使用最多的减阻剂。另一类已被充分研究的水溶性合成聚合物减阻剂是超高分子量 PEO。

在注水中使用聚合物减阻剂会出现储层吸附的问题，导致进入油井的天然气或原油流量减少。这在极低渗透率的页岩气藏中尤其不可取。此外，与非缔合聚合物的同系物相比，使用缔合聚合物时，有可能存在更高的聚合物吸附风险。目前正在研究的潜在解决方案是使用对刺激敏感的缔合聚合物，可以在缔合和分离的状态之间切换。17.4.1.3 中介绍了一个热敏聚合物的实例。

中国有人提出通过减少水在储层岩石微孔道中的层流阻力，来大幅降低注水压力的减阻方法。具体是将含有 SiO_2 疏水纳米颗粒（HNPs）的溶液注入储层微孔道，HNPs 被吸附于岩壁，形成一个强或超强的疏水层，达到形成滑移的边界条件，减少对流体的阻力。

17.4.1.1 多糖和衍生物

通常合成的水溶性聚合物可能由于较高分子量和较低的剪切力，性能要好于生物可降解性较强的多糖。还发现多糖和 PAM 溶液之间具有协同效应。石油工业中多糖及衍生物可用作如压裂作业等的减阻剂。多糖比乙烯基聚合物更容易受到剪切力和生物降解的影响。通常认为性能最好的天然多糖类减阻剂是瓜尔胶及其衍生物（图 17.7）。HEC（图 17.8）和 CMC 也是油井作业中增黏、减阻的典型多糖衍生物。多糖的线型聚合物是能起到减阻剂作用的关键因素，这可以解释为什么葡聚糖和阿拉伯胶等高度支化的多糖减阻性能很差。瓜尔胶和其他多糖还可以通过接枝丙烯酰胺，来提高抗剪切和生物降解的能力。

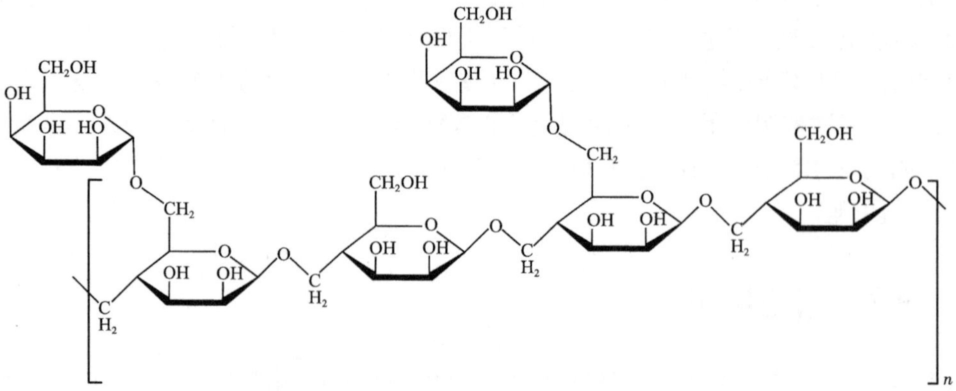

图 17.7　瓜尔胶

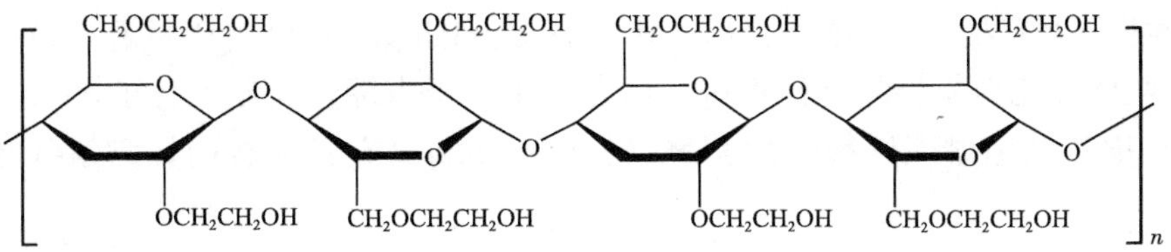

图 17.8 HEC

17.4.1.2 聚氧化乙烯减阻剂

聚氧化乙烯由金属催化环氧乙烷聚合而成，可以得到非常高的分子量（高达 8×10^6）（图 17.9）。碱基催化的环氧乙烷聚合得到低分子量的聚乙二醇（PEO）（分子量 $<1\times10^5$）。PEO 有成熟产品可用作增稠剂，但在注入或在湍流下容易剪切降解，很少用于石油工业。

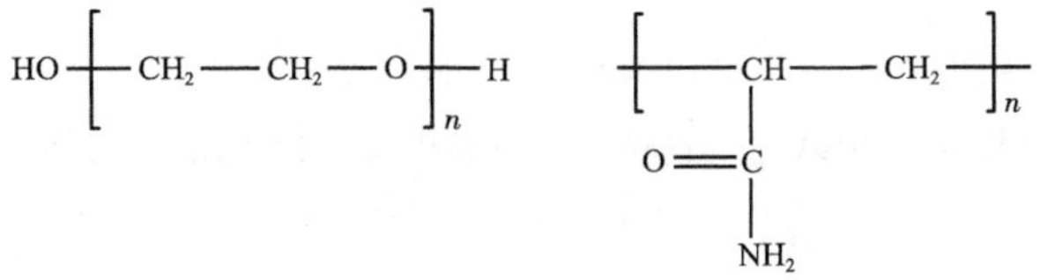

(a) 聚乙烯氧化物　　(b) 聚丙烯酰胺

图 17.9 聚乙烯氧化物和聚丙烯酰胺

已经对减阻型非离子 PEO 和减阻型阳离子表面活性剂（十八烷基三甲基氯化铵）之间的相互作用进行了研究。在流动环路测试中，将这种表面活性剂添加到聚合物中总是能增强减阻效果。在混合体系中，低浓度聚合物和高浓度表面活性剂的协同效应更强。这与下文所述的阳离子表面活性剂与阴离子聚合物的拮抗减阻剂效应形成鲜明对比。

17.4.1.3 丙烯酰氨基减阻剂

聚丙烯酰胺、PHPA 及相关的丙烯酰胺或丙烯酸酯聚合物和共聚物作为减阻剂比 PEO 有更多的实际用途，因为其有侧链，不易发生剪切降解（图 17.9）。因此，PAM 和相关衍生物通常是石油工业中用于注水的首选水溶性聚合物减阻剂。在水相中的典型加量为 20~30mg/L。丙烯酰胺共聚物的分子量可高达 2×10^7。PHPA 主要是丙烯酰胺和丙烯酸酯单体的共聚物，已广泛用于注聚合物驱油项目的增注。在单相流室内实验中，使用 PHPA 阻力可减少 70%，在现场降低的数值通常为 50%~60%。在多相流中，阻力减少值通常较低。PHPA 是通过部分水解 PAM 制成。另外，具有类似减阻剂性能的聚合物可以通过丙烯酰胺与少量丙烯酸酯单体的共聚而制成。

一些具有较大单体的共聚物，如丙烯酰胺与双丙酮丙烯酰胺、丙烯酰氨基-2-甲基丙烷磺酸钠或丙烯酰氨基-3-甲基丁烯酸钠共聚（图 17.10），显示出比 PAM 更好的性能。获得的聚合物分子量高达 2.8×10^7。这些聚合物能够进行分子间缔合。而另一项流动环路研究表明，PHPA 比含磺酸盐的 PAM 具有更好的减阻效果。在溶液中带有缔合基团的疏水改性 PAM 比聚氨酯的性能更好，后者由于分子内缔合而导致分子塌缩。据称丙烯酰胺和烷基聚氧乙烯醚丙烯酸酯的共聚物，也可作为具有长侧链丙烯酰氨基减阻剂的改进型。

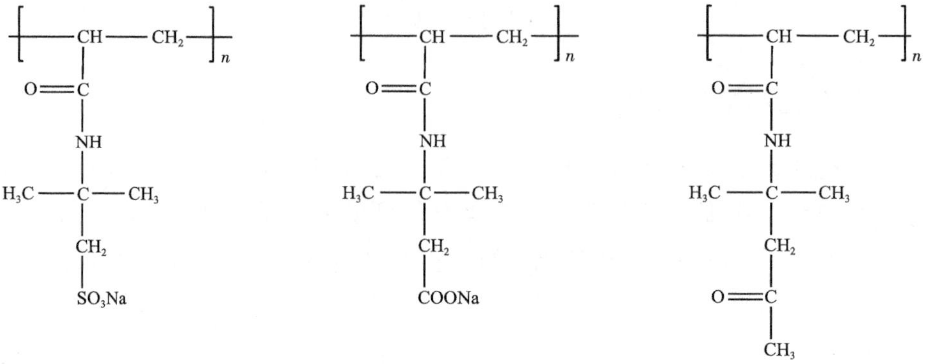

(a) 丙烯酰氨基-2-甲基丙烷磺酸钠　　(b) 丙烯酰氨基-3-甲基丁烯酸钠　　(c) 双丙酮丙烯酰胺

图 17.10　丙烯酰氨基 -2- 甲基丙烷磺酸钠（AMPS 钠盐）、丙烯酰氨基 -3- 甲基丁烯酸钠、双丙酮丙烯酰胺

PHPA 或其他含有丙烯酸酯单体聚合物的一个缺点是其与高钙盐水的兼容性降低。如磺酸盐（AMPS）（图 17.10）等其他离子基团，兼容性更强，可以加至丙烯酰胺共聚物。另一个例子是丙烯酰胺和（N-3- 磺丙基）-N- 甲基丙烯醇 - 氧乙基 -N，N- 二甲基甜菜碱铵共聚物，其中的甜菜碱基团确保了良好的钙兼容性。在存在铁离子时，PAMs 性能会降低。

水溶液中溶解低分子量添加剂（酸、碱或盐）后，多数聚合物减阻剂的性能会降低。这些添加剂会屏蔽沿分子链骨架固定的电荷，导致聚合物分子的尺寸减小。只要分子链继续收缩，减阻效果就会变弱。但丙烯酰胺/金属苯乙烯磺酸盐/甲基丙烯酰胺丙基三甲基氯化铵的两性离子三元共聚物显示，随着水溶液的离子强度增加，黏度和减阻剂性能也在增加。

另有研究表明，在水相中加入尿素（氢键破坏剂）会大幅降低丙烯酰胺共聚物的减阻效果，而通过加入氯化钠增加离子强度，会产生相反的效果。溶液中聚合物分子链聚集不是减阻效果所必需，但会产生更大的体积破坏湍流，在 PAM 和 PEO 的稀溶液中发现了这种聚集。高度的分子链柔性似乎能带来良好的减阻效果。因此，相同链长时，聚甲基丙烯酰胺的性能应该比 PAM 差。

虽然早期的理论认为聚合物的分支不利于减阻效果，但一些研究表明情况并非如此。例如，高度支化的高分子量 PAM 显示出更好的减阻性能和剪切稳定性。可能与线型 PAM 相比，此时的更高流体力学体积才是关键。这就是为什么具有长侧链的线型聚合物与具有短侧链或无侧链的聚合物相比是更好的减阻剂。但很难得到具有高比例长侧链的超高分子量 PAM。有可能通过齐格勒 - 纳塔催化的烯烃聚合，在类似丙烯酰胺的极性单体存在时，这些有机金属钛或锆催化剂并不牢靠。

与油溶性超高分子量聚合物减阻剂一样，处理这种高分子量的水溶性聚合物也有困难，因为其在极性溶剂中的黏度很高。对于丙烯酰氨基减阻剂，通常的解决方法是使用聚合物的油包水乳状液，以大幅降低产品黏度。为避免高黏度溶液，一种新的选择是使用阴离子丙烯酰胺共聚物和分子量相对较低的阳离子丙烯酰胺共聚物的混合物。当注入水介质中时，其将结合并形成复合物，使减阻性能比单独使用共聚物高得多。

在提高石油采收率（EOR）水驱项目中，使用水溶性减阻剂，聚合物在储层上的吸附作用降低了从页岩等致密低渗透储层中采油、采气的效果。此时分子骨架中含弱键的水溶性聚合物减阻剂，在经历如温度、pH 值或还原剂等某种触发因素后可被降解，可能是非常有利的。在合成的 PAM 主干上放置偶氮基团，可以得到和纯 PAM 同样性能良好的减阻剂，额外好处是一旦受到高温影响，该聚合物就会失去减阻性能。

简单单体的表面活性剂会影响减阻剂（如 PEO 或 PAM）的性能。表面活性剂浓度低时，PEO 性能较差，但在高浓度下，性能相对于纯 PEO 有所提高。这可能是由于高浓度的胶束结构有助于减阻。已研究了阴离子聚合物（丙烯酰胺和丙烯酸钠的共聚物）和阳离子表面活性剂（十八烷基三甲基氯化铵）之间的相互作用。在聚合物溶液中加入表面活性剂，会导致溶液黏度和减阻剂效果大幅下降。

水溶性聚合物减阻剂也用于多相流动，尽管目前似乎还没有任何现场应用的报道。例如，具有烃类外相和基于 PAM 减阻剂水溶液内相的聚合物纳米乳液，据称可以减少多相流管道的阻力和摩擦。

在油—水流动管道室内实验中，水溶性聚合物减阻剂扩张了层状流的区域，并推迟了向段塞流动的过渡。还有研究声称，与阳离子或中性修饰的同系物相比，聚合物骨架中含有阴离子的 PAM 的乳化倾向大幅降低。还应提及的是，通过在井口注入破乳剂，可以减少容易出现黏性乳化问题的湍流多相流体的阻力。

17.4.2 水溶性表面活性剂减阻剂

表面活性剂能够减少湍流液体的阻力已久为人知。表面活性剂添加剂对摩擦阻力有双重影响。首先，其引入黏弹性剪切应力，从而增加摩擦阻力；其次，其抑制湍流涡流结构，减少剪切应力，然后降低摩擦阻力。由于第二种效应大于第一种效应，因此会出现阻力降低。

有人建议注水时，将表面活性剂减阻剂作为聚合物减阻剂的替代品。表面活性剂减阻剂只在比超高分子量聚合物减阻剂更高的浓度下才有活性。例如，为达到与 20mg/L 超高分子量聚合物减阻剂相同的减阻性能，可能需要浓度为 200mg/L 的表面活性剂减阻剂。使用表面活性剂减阻剂已观察到高达 80% 的阻力降低。表面活性剂减阻剂的高浓度原因是，表面活性剂或混合的表面活性剂需要高于临界胶束浓度，才能结合成足够大的胶束。有些表面活性剂的胶束呈棒状。只有这些相关的表面活性剂棒状胶束，而不是单个的表面活性剂分子或球状胶束，才能减少湍流中的涡流和爆裂。一些水溶性表面活性剂在环境温度下没有减阻特性，但在温度升高到刚好低于其浊点时，显示出减阻剂效果。这可能是由于表面活性剂的聚集，但表面活性剂与水的相互作用并未完全破坏。

自20世纪90年代末以来，石油工业对表面活性剂减阻剂重新产生了兴趣，特别是用于注入海水及集输管道的水循环。可能的原因如下：首先，尽管与超高分子量聚合物减阻剂相比，需要更高浓度的表面活性剂减阻剂，但超高分子量聚合物减阻剂的成本相当高。因此更便宜的表面活性剂可以在成本效益的基础上进行竞争。其次，表面活性剂减阻剂不像超高分子量聚合物减阻剂那样容易受到永久性剪切降解的影响，因此可以应用于有长期湍流的地方。如果线状胶束结构被损坏，其可以沿着管道进一步自我修复。对于注入海水，这意味着表面活性剂减阻剂可以在泵之前注入。最后，一些表面活性剂减阻剂可生物降解，更容易被环保部门接受。

许多类别的表面活性剂都显示出减阻效果。由于良好的减阻剂性能需要较高的浓度，制备表面活性剂需要相当便宜的原料。许多研究是针对如十六烷基三甲基氯化铵等简单的阳离子表面活性剂，但如十二烷基硫酸钠（SDS）等阴离子表面活性剂也证明可以减少阻力。例如，蒸馏水中400mg/L的SDS可以减少压降损失25%~40%。

对于阳离子表面活性剂，相应阴离子的性质和大小会显著影响胶束形状和减阻性能。在一项有关阳离子表面活性剂的研究中，阴离子在2-氯苯甲酸根、3-氯苯甲酸根或4-氯苯甲酸根三种异构体中变化。每种异构体都显示出不同类型的流变学和减阻行为及不同的胶束结构。4-氯体系有良好的减阻效果和棒状的胶束网络，而2-氯体系没有显示减阻效果，且表观拉伸黏度低，只有球形胶束。3-氯体系的行为更为复杂。

具有大尺寸阴离子的阳离子表面活性剂似乎有利于形成棒状胶束。例如，十六烷基三甲基水杨酸铵和乙酰吡啶水杨酸铵是特别首选的减阻剂，尽管阴离子也可以是硫代水杨酸盐、磺酸盐或羟基萘酸盐（图17.11）。芥子气甲基双（2-羟乙基）氯化铵不需要大的反离子来形成黏弹胶束。一种商业化的氧化胺黏弹性表面活性剂是水性流体中的胶凝剂，用于辅助注水提高采收率，还研究了其在盘管中应用时的特性。

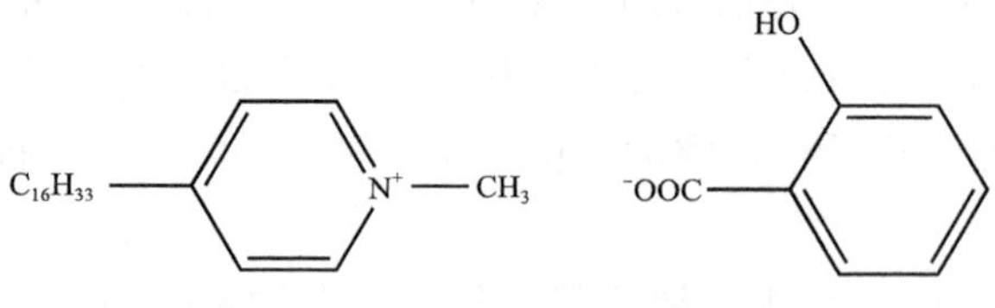

图17.11 十六烷基水杨酸吡啶

其他可作为减阻剂的廉价表面活性剂包括脂肪酸、脂肪酸的烷氧基化衍生物、脂肪酸的有机盐和无机盐及其烷氧基化衍生物（图17.12）。此类中更亲油的表面活性剂还可用于油相减阻。这些减阻剂和前面提到的阳离子表面活性剂还具有额外的缓蚀功能，可以降低化学剂的总成本。其他低成本的表面活性剂减阻剂还有顺丁烯二酸及其酯类和其有机盐、无机盐或胺盐。特别优选的盐类是咪唑啉盐，可以起到缓蚀作用。

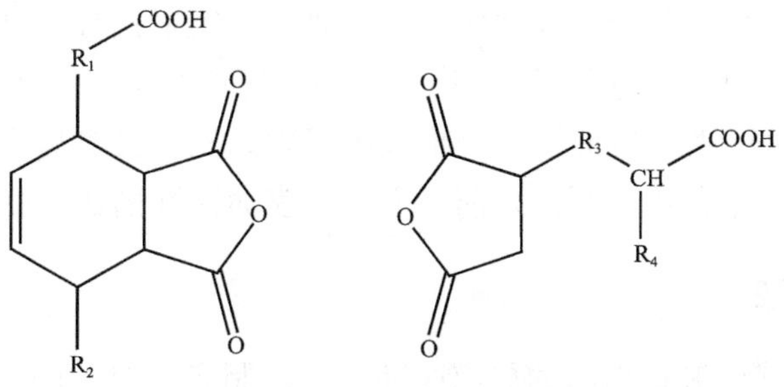

图 17.12　马来酸脂肪酸表面活性剂减阻剂（这些产品的单酯也属于咪唑啉盐）

有研究报道了聚合物和表面活性剂协同的减阻效果。例如，PEO 与同系列的羧酸皂盐混合结果表明，二者形成了协同胶束。此外，其降阻能力的增强与表面活性剂分子及聚合物链疏水结合的模型一致，其中相邻的极性表面活性剂基团之间的排斥力促进了聚合物线团的膨胀。

一家公司在烷醇酰胺和两性离子表面活性剂水溶性减阻剂及与其他表面活性剂的混合方面做了大量工作。这些表面活性剂能够形成长圆柱形胶束，是减少阻力的理想选择。该公司最早开发的表面活性剂减阻剂是基于烷氧基化的烷醇酰胺，但只在有限的温度范围内和在低矿化度水中表现良好。烷氧基化的烷醇酰胺与烷氧基化的醇或离子表面活性剂，磺化的、两性的或两性离子表面活性剂的混合被认为是改进。最近的研究集中于双重和三重混合物中的两性离子表面活性剂。例如，将一种两性离子表面活性剂与一种醚基硫酸盐或醚基羧酸盐表面活性剂混合（图 17.13），如 $N-$ 苯基甜菜碱与十二烷基醚硫酸钠混合。这种混合物对高矿化度水的敏感性很低。但为实现本质性降低阻力，所需的表面活性剂浓度高于 500mg/L。此外，胶束的形成及随之而来的阻力降低，预计会受到大量电解质的负面影响。因此这种类型的减阻剂不适用于注海水条件。进一步的改进是一种由两个两性离子表面活性剂组成的三重混合物，每个两性离子表面活性剂都包括一个酰基和一种阴离子表面活性剂（其中亲水基团是硫酸根、磺酸根或醚硫酸根）。这种混合物在浓度为 50~400mg/L（最佳浓度 60~300mg/L）时，2~70℃的较大温度区间内，甚至在电解质质量分数高达 7% 的水溶液中都有非常好的减阻效果。

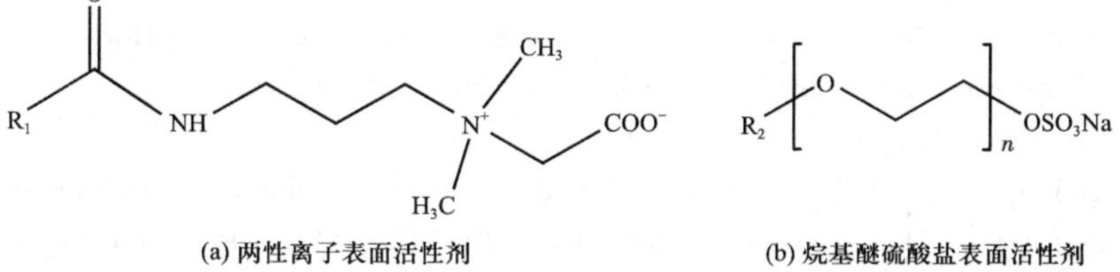

(a) 两性离子表面活性剂　　(b) 烷基醚硫酸盐表面活性剂

图 17.13　优选的两性表面活性剂和烷基醚硫酸盐表面活性剂组成的表面活性剂减阻剂混合物

两性离子 $N-$ 烷基甜菜碱/阴离子表面活性剂减阻剂混合物已经在北海油田现场应用，以提高用乙二醇水溶液对 10km 长管束的加热能力。使用聚合物减阻剂的试验失败，因为在加热液循环过程中，泵内聚合物发生了降解。为了实现在注入海水时使用表面活性剂减阻剂的目标，在实验室实验的合成海水中使用一种两性表面活性剂和一种阴离子表面活性剂，平均速度为 1.9m/s 时，使用该混合物 200mg/L 可降低阻力 75%～80%，在 2.9m/s 时可降低阻力 50%～55%。由于表面活性剂形成的减阻结构与会被剪切力永久破坏的聚合物不同，具有自我修复特性，可在泵注之前加入，该表面活性剂还可生物降解。

17.4.3 减阻和缓蚀

应用减阻剂可以减少流动引起的局部腐蚀，其机制有两个方面：首先，如果减阻剂是表面活性剂，还可以作为成膜型缓蚀剂。例如，十六烷基三甲基水杨酸铵和乙酰吡啶水杨酸铵可发挥两种功效。许多成膜型缓蚀剂在超过临界胶束浓度后，显示出部分减阻特性。其次，减阻剂会减少管壁附近的湍流。这样就可以减缓侵蚀，或有助于防止成膜型缓蚀剂从管壁脱落。一家制造商声称其水溶性聚合物减阻剂的缓蚀率可达到 40%。

参 考 文 献

[1] B. A. Jubran, Y. H. Zurigat, M. S. Al-Shukri, and H. H. Al-Busaidi, *Polymer-Plastics Technology and Engineering* 45（4）（2006）：553.

[2] A. Gyr and H. W. Bewersdorff, *Drag Reduction of Turbulent Flows by Additives（Fluid Mechanics and Its Applications）*, Dordrecht, The Netherlands: Kluwer Academic Publishers, 1995.

[3] G. T. Pruitt, C. M. Simmons, G. H. Neil, and H. R. Crawford, "A Method to Minimize the Cost of Pumping Fluids Containing Friction Reducing Additives," SPE 997, *Journal of Petroleum Technology* 17（6）（1965）：641.

[4] R. A. Woodroof and R. W. Anderson, "Synthetic Polymer Friction Reducers Can Cause Formation Damage," SPE 6812（paper presented at the 52nd Annual Fall Technical Conference and Exhibition, Denver, CO, 9–12 October 1977）.

[5] H. A. Al-Anazi, M. G. Al-Faifi, F. Tulbah, and J. Gillespie, "Evaluation of Drag Reducing Agent（DRA）for Seawater Injection System: Lab and Field Cases," SPE 100844（paper presented at the SPE Asia Pacific Oil & Gas Conference and Exhibition, Adelaide, Australia, 11–13 September 2006）.

[6] M. Ke, Q. Qu, R. F. Stevens, N. Bracksieck, C. Price, and D. Copeland, "Evaluation of Friction Reducers for High-Density Brines and Their Application in Coiled Tubing at High Temperature," SPE 103037（paper presented at the SPE Annual Technical Conference and Exhibition, San Antonio, TX, 24–27 September 2006）.

[7] R. L. J. Fernandes, B. M. Jutte, and M. G. Rodriguez, "Drag Reduction in Horizontal Annular Two-Phase Flow," *International Journal of Multiphase Flow* 30（2004）：1051.

[8] Y.-H. Li, U.S. Patent 5020561, 1991.

[9] Y.-H. Li, G. R. Chesnut, R. D. Richmond, G. L. Beer, and V. P. Caldarera, "Laboratory Tests and Field Implementation of Gas-Drag-Reduction Chemicals," SPE 37256, *SPE Production & Facilities* 13（1998）：53.

[10] A. A. Hamouda, "Drag Reduction-Performance in Laboratory Compared to Pipelines," SPE 80258（paper presented at the SPE International Symposium on Oilfield Chemistry, Houston, TX, 5–7 February 2003）.

[11] G. Schmitt, C. Bosch, H. Bauer, and M. Mueller, "Modelling the Drag Reducing Effect of CO_2

Corrosion Inhibitors," Paper 00002 (paper presented at the NACE CORROSION Conference, 2000).

[12] K. Slater and M. Zamora, U.S. Patent Application 20080289435.

[13] (*a*) M. E. Cowan, R. D. Hester, and C. L. McCormick, "Water-Soluble Polymers: LXXXII. Shear Degradation Effects on Drag Reduction Behavior or Dilute Polymer Solutions," *Journal of Applied Polymer Science* 82 (2001): 1211. (*b*) M. E. Cowan, C. Garner, R. D. Hester, and C. L. McCormick, "Water-Soluble Polymers: LXXXII. Correlation of Drag Experimentally Determined Drag Reduction Efficiency and Extensional Viscosity of High Molecular Weight Polymers in Dilute Aqueous Solution," *Journal of Applied Polymer Science* 82 (2001): 1222.

[14] K. Lee, C. A. Kim, S. T. Lim, D. H. Kwon, H. J. Choi, and M. S. Jhon, "Mechanical Degradation of Polyisobutylene in Kersosene," *Colloid and Polymer Science* 280 (2002): 779.

[15] C. A. Kim, D. S. Jo, H. J. Choi, C. B. Kim, and M. S. Jhon, "A High-Precision Disk Apparatus for Drag Reduction Characterization," *Polymer Testing* 20 (2000): 43.

[16] M. S. Figueiredo, L. C. Almeida, F. G. Costa, M. D. Clarisse, L. Lopes, R. Leal, A. L. Martins, and E. F. Lucas, "Development of a Method to Evaluate the Performance of Aqueous Polymer Solutions as Drag Reduction Agents in Bench Scale," *Macromolecular Symposia* 245-246 (2006): 260.

[17] J. L. Lumley, "Drag Reduction by Additives," *Annual Review of Fluid Mechanics* 1 (1969): 367.

[18] P. G. De Gennes, *Introduction to Polymer Dynamics*, Cambridge: Cambridge University Press, 1990.

[19] P. S. Virk, H. S. Mickley, and K. A. Smith, "The Ultimate Asymptote and Mean Flow Structure in Tom's Phenomenon," *ASME, Journal of Applied Mechanics* 37 (1970): 480.

[20] P. Peyser and R. C. Little, "The Drag Reduction of Dilute Polymer Solutions as a Function of Solvent Power, Viscosity, and Temperature," *Journal of Applied Polymer Science* 15 (1971): 2623.

[21] B. A. Jubran, Y. H. Zurigat, and M. F. A. Goosen, "Drag Reducing Agents in Multiphase Flow Pipelines: Recent Trends and Future Needs," *Petroleum Science & Technology* 23 (2005): 1403.

[22] A. A. Hamouda and F. S. Evensen, "Possible Mechanism of the Drag Reduction Phenomenon in Light of the Associated Heat Transfer Reduction," SPE 93405 (paper presented at the SPE International Symposium on Oilfield Chemistry, Houston, TX, 2-4 February 2005).

[23] A. P. Matjukhatov, B. P. Mironov, and I. A. Animisov, *The Influence of Polymer Additives on Velocity and Temperature Fields*, ed. B. Gampert, Berlin: Springer-Verlag, 1985, 107.

[24] A. Al-Sarkhi and T. J. Hanratty, "Effect of Drag-Reducing Polymers on Pseudo-Slugs Interfacial Drag and Transition to Slug Flow," *International Journal of Multiphase Flow* 28 (2002): 1911-1927.

[25] E. D. Burger, W. R. Munk, and H. A. Wahl, "Flow Increase in the Trans Alaska Pipeline Through Use of a Polymeric Drag-Reducing Additive," SPE 9419, *Journal of Petroleum Technology* 34 (1982): 377.

[26] H. A. Wahl, W. R. Beatty, J. G. Dopper, and G. R. Hass, "Drag Reducer Increases Oil Pipeline Flow Rates," SPE 10446 (paper presented at the Offshore South East Asia 82 Conference, Singapore, 9-12 February 1982).

[27] B. K. Berge and O. Solsvik, "Increased Pipeline Throughput Using Drag Reducing Additives (DRA): Field Experiences," SPE 36835 (paper presented at the SPE European Petroleum Conference, Milan, Italy, 22-24 October 1996).

[28] J. U. Ibrahim and L. A. Braimoh, "Drag Reducing Agent Test Result for ChevronTexaco, Eastern Operations, Nigeria," SPE 98819 (paper presented at the 29th Annual SPE International Technical Conference and Exhibition, Abuja, Nigeria, 1-3 August 2003).

[29] P. K. Ptasinski, B. J. Boersma, F. T. M. Nieuwstadt, M. A. Hulsen, B. H. A. A. Van den Brule, and J. C. R. Hunt, "Turbulent Flow of Polymer Solutions Near Maximum Drag Reduction; Experiments, Simulations

and Mechanisms," *Journal of Fluid Mechanics* 490 (2003): 251.

[30] M. P. Mack, U.S. Patent 4493903, 1985.

[31] J. A. Lescarboura, J. D. Culter, and H. A. Wahl, "Drag Reduction with a Polymeric Additive in Crude Oil Pipelines," SPE 3087 (paper presented at the SPE 45th Annual Fall Meeting, Houston, TX, 4–7 October 1970).

[32] H. J. Choi and M. S. Jhon, "Polymer-Induced Turbulent Drag Reduction," *Industrial & Engineering Chemistry Research* 35 (9) (1996): 2993.

[33] K. Lee, C. A. Kim, S. T. Lim, D. H. Kwon, H. J. Choi, and M. S. Jhon, "Mechanical Degradation of Polyisobutylene under Turbulent Flow," *Colloid & Polymer Science* 280 (2002): 779.

[34] R. L. Johnston and S. N. Milligan, U.S. Patent 6596832, 2003.

[35] E. Karhu, M. Karhu, L. Rockas, Leif, and H. Harjuhahto, U.S. Patent Application, 20020173569, 2002.

[36] W. Brostow, "Drag Reduction and Mechanical Degradation in Polymer Solutions in Flow," *Polymer* 24 (1983): 631.

[37] K. W. Smith, L. V. Haynes, and D. F. Massouda, U.S. Patent 5449732, 1995.

[38] D. L. Hunston, "Effects of Molecular Weight Distribution in Drag Reduction and Shear Degradation," *Journal of Polymer Science: Polymer Chemistry Edition* 14 (1976): 713.

[39] B. Liu, Bing, X. Bao, Y. Gao, C. Li, and G. Li, U.S. Patent Application 20070004837.

[40] S. N. Milligan and K. W. Smith, U.S. Patent 6576732, 2003.

[41] G. B. Eaton, M. J. Monahan, A. K. Ebert, R. J. Tipton, and E. Baralt, U.S. Patent 6730750, 2004.

[42] J. R. Harris, U.S. Patent Application 20060281832, 2006.

[43] S. Malik, S. N. Shintre, and R. A. Mashelkar, U.S. Patent 5080121, 1992.

[44] D. E. Farley, "Drag Reduction in Non-Aqueous Solutions: Structure-Property Correlations for Poly (isodecylmethacrylate)," SPE 5308 (paper presented at the SPE International Symposium on Oilfield Chemistry, Dallas, TX, 16–17 January 1975).

[45] M. D. Holtmyer and J. Chatterji, *Polymer Engineering & Science* 20 (1980): 473.

[46] Y. Ma, X. Zheng, F. Shi, Y. Li, and S. Sun, "Synthesis of Poly (Dodecyl Methacrylate) s and Their Drag-Reducing Properties," *Journal of Applied Polymer Science* 88 (2003): 1622.

[47] W. Ritter and C. P. Herold, International Patent Application WO9002766, 1990.

[48] D. N. Schulz, K. Kitano, T. J. Burkhardt, and A. W. Langer, U.S. Patent 4518757, 1985.

[49] (a) K. W. Smith, W. R. Dreher, and T. I. Burden, International Patent Application WO/2008/014190. (b) S. N. Milligan, R. L. Johnston, T. L. Burden, W. R. Dreher, K. W. Smith, and W. F. Harris, International Patent Application WO/2008/079642.

[50] R. M. Kowalik, I. Duvdevani, D. G. Pfeiffer, R. D. Lundberg, K. Kitang, and D. N. Schulz, "Enhanced Drag Reduction via Interpolymer Associations," *Journal of Non-Newtonian Fluid Mechanics* 24 (1987): 1.

[51] R. M. Kowalik, I. Duvdevani, D. G. Peiffer, and R. D. Lundberg, U.S. Patent 4508128, 1985.

[52] S. Malik and R. A. Mashelkar, "Hydrogen Bonding Mediated Shear Stable Clusters as Drag Reducers," *Chemical Engineering Science* 50 (1995): 105.

[53] R. M. Kowalik, I. Duvdevani, K. Kitano, and D. N. Schulz, U.S. Patent 4625745, 1986.

[54] S. Malik and R. A. Mashelkar, *Chemical Engineering Science* 50 (1995): 195.

[55] R. L. Johnston and L. G. Fry, U.S. Patent 5376697, 1994.

[56] K. Fairchild, R. Tipton, J. F. Motier, and N. Kommareddi, U.S. Patent 5733953 1998.

[57] G. B. Eaton and M. J. Monahan, U.S. Patent 5869570, 1999.

[58] (a) K. M. Labude, K. W. Smith, and T. L. Burden, U.S. Patent 6399676, 2002. (b) T. Mathew and N. S. Kommareddi, U.S. Patent Application 20080287568.

[59] J. R. Harris, J. F. Motier, L.-C. Chou, and T. J. Martin, U.S. Patent 7119132, 2006.

[60] N. S. Kommareddi, R. Dinius, N. Vasishtha, and D. E. Barlow, U.S. Patent 6841593, 2005.

[61] T. Moussa and C. Tiu, "Factors Affecting Polymer Degradation in Turbulent Pipe Flow," *Chemical Engineering Science* 49 (1994): 1681.

[62] W. F. Harris, F. William, K. W. Smith, S. N. Milligan, R. L. Johnston, and V. S. Anderson, International Patent Application WO/2006/081010.

[63] J. F. Motier. J. R. Harris, L. C. Chou, and N. S. Kommareddi, U.S. Patent Application 20070021531.

[64] J. R. Harris, L. C. Chou, G. G. Ramsay, J. F. Motier, N. S. Kommareddi, and T. Mathew, U.S. Patent Application 20060293196.

[65] (a) K. M. Labude, K. W. Smith, and R. L. Johnston, U.S. Patent Application 20030187123. (b) B. A. Bucher, M. J. Monahan, and S. B. Erikson, International Patent Application WO/2008/073293.

[66] A. P. Evans, "A New Drag-Reducing Polymer with Improved Shear Stability for Nonaqueous Systems," *Journal of Applied Polymer Science* 18 (1974): 1919.

[67] D. Mowla and A. Naderi, "Experimental Study of Drag Reduction by a Polymeric Additive in Slug Two-Phase Flow of Crude Oil and Air in Horizontal Pipes," *Chemical Engineering Science* 61 (2006): 1549.

[68] D. Mowla, M. Moshfeghian, and M. S. Hatamipour, "A Simple Model for Prediction of Pressure Drop in Horizontal Two-Phase Flow," *Iranian Journal of Science and Technology* 15 (1991): 177.

[69] R. L. Fernandes, "Multiphase Drag Reduction. Part I: Proof-of-Concept Experiments," Internal Shell Report EP 2003-5028, Shell Rijswijk, 2003.

[70] C. Kang and W. P. Jepson, "Effect of Drag-Reducing Agents in Multiphase, Oil/Gas Horizontal Flow," SPE 58976 (paper presented at the SPE International Petroleum Conference and Exhibition, Villahermosa, Mexico, 1-3 February 2000).

[71] M. Daas, C. Kang, and W. P. Jepson, "Quantitative Analysis of Drag Reduction in Horizontal Slug Flow," SPE 62944 (paper presented at the SPE Annual Technical Conference and Exhibition, Dallas, TX, 1-4 October 2000).

[72] N. D. Sylvester, R. H. Dowling, H. Paz-y-Mino, and J. Brill, "Drag Reduction in Two-Phase Gas-Liquid Flow," Prepared for the Materials Committee Pipeline Research Committee of Pipeline Research Council International, Inc., 1977.

[73] V. Jovancicevic, S. Campbell, S. Ramachandran, P. Hammonds, and S. J. Weghorn, U.S. Patent Application 20040142825.

[74] P. Hammonds, V. Jovancicevic, C. M. Means, C. Mitch, and D. Green, U.S. Patent Application 20040216780.

[75] K. Oh-Kil and C. Ling Siu, *Drag Reducing Polymers: The Polymeric Materials Encyclopedia*, Boca Raton, FL: CRC Press Inc., 1996.

[76] S. E. Morgan and C. L. McCormick, "Water-Soluble Copolymers: XXXII. Macromolecular Drag Reduction. A Review of Predictive Theories and the Effects of Polymer Structure," *Progress in Polymer Science* 15 (1990): 507.

[77] J. W. Hoyt, "Drag Reduction in Polysaccharide Solutions," *Trends Biotechnology* 3 (1985): 17.

[78] W. Interthal and H. Wilski, "Drag Reduction Experiments with Very Large Pipes," *Colloid & Polymer Science* 263 (1985): 217.

[79] J. P. Malhotra, P. N. Chaturvedi, and R. P. Singh, "Drag Reduction by Polymer-Polymer Mistures," *Journal*

of *Applied Polymer Science* 36（1988）：837.

[80] S. R. Deshmukh and R. P. Singh, "Drag Reduction Characteristics of Graft Copolymers of Xanthangum and Polyacrylamide," *Journal of Applied Polymer Science* 32（1986）：6163.

[81] R. P. Singh, G. P. Karmakar, S. K. Rath, N. C. Karmakar, S. R. Pandey, T. Tripathy, J. Panda, K. Kanan, S. K. Jain, and N. T. Lan, "Biodegradable Drag Reducing Agents and Flocculants Based on Polysaccharides: Materials and Applications," *Polymer Engineering & Science* 40（2000）：46.

[82] R. C. Little and M. Wiegard, "Drag Reduction and Structural Turbulence in Flowing Polyox Solutions," *Journal of Applied Polymer Science* 14（2）（1969）：409.

[83] S. A. Nosier, Y. A. Alhamed, A. A. Bakry, and I. S. Mansour, "Forced Convection Solid-Liquid Mass Transfer at a Surface of Tube Bundles under Single Phase Flow," *Chemical and Biochemical Engineering Quarterly* 21（3）（2007）：213.

[84] S. U. S. Choi, Y. I. Cho, and K. E. Kasza, "Degradation Effects of Dilute Polymer Solutions on Turbulent Friction and Heat Transfer Behavior," *Journal of Non-Newtonian Fluid Mechanics* 41（1992）：289.

[85] D. H. Fisher and F. Rodriguez, "Degradation of Drag-Reducing Polymers," *Journal of Applied Polymer Science* 15（1971）：2975.

[86] P. K. Ptasinski, F. T. M. Nieuwstadt, B. H. A. A. van den Brule, and M. A. Hulsen, "Experiments in Turbulent Pipe Flow with Polymer Additives at Maximum Drag Reduction," *Flow Turbulence and Combustion* 66（2001）：159.

[87] B. L. Knight, J. S. Rhudy, and W. B. Gogarty, U.S. Patent 4236545, 1980.

[88] C. L. McCormick, R. D. Hester, S. E. Morgan, and A. M. Safieddine, "Water-Soluble Copolymers: 30. Effects of Molecular Structure on Drag Reduction Efficiency," *Macromolecules* 23（8）（1990）：2124.

[89] C. L. McCormick, R. D. Hester, S. E. Morgan, and A. M. Safieddine, "Water-Soluble Copolymers: 31. Effects of Molecular Parameters, Solvation, and Polymer Associations on Drag Reduction Performance," *Macromolecules* 23（8）（1990）：2139.

[90] P. S. Mumick, R. D. Hester, and C. L. McCormick, "Water-Soluble Copolymers: 55. N-Isopropylacrylamide Copolymers in Drag Reduction: Effect of Molecular Structure, Hydration, and Flow Geometry on Drag Reduction Performance," *Polymer Engineering & Science* 34（1994）：1429.

[91] D. N. Schulz, R. M. Kowalik, J. Bock, and J. J. Maurer, U.S. Patent 4546784, 1985.

[92] D. N. Schulz, D. G. Peiffer, R. M. Kowalik, and J. J. Kaladas, U.S. Patent 4560710, 1985.

[93] J. W. Hoyt, "Effect of Ferric Ions on Drag Reduction Effectiveness of Polyacrylamide," *Polymer Engineering & Science* 20（1980）：493.

[94] D. G. Peiffer, R. D. Lundberg, R. M. Kowalik, and S. R. Turner, U.S. Patent 4460758, 1984.

[95] C. L. McCormick, B. H. Hutchinson, and S. E. Morgan, "Water-Soluble Copolymers: 16. Studies of the Behavior of Acrylamide/N-（1,1-Dimethyl-3-Oxobutyl）Acrylamide Copolymers in Aqueous Salt Solution," *Makromolekulare Chemie* 188（1987）：357.

[96] O. K. Kim, R. C. Little, R. L. Patterson, and R. Y. Ting, "Polymer Structure," *Nature* 250（1974）：408.

[97] R. D. Lundberg, D. G. Peiffer, I. Duvdevani, and R. M. Kowalik, U.S. Patent 4489180, 1984.

[98] R. L. Patterson and R. C. Little, *Journal of Colloid Interface Science* 53（1975）：110.

[99] R. N. Grabois and Y. N. Lee, U.S. Patent 5027843, 1991.

[100] C. Kang and W. P. Jepson, "Multiphase Flow Conditioning Using Drag-Reducing Agents," SPE 56569

(paper presented at the SPE Annual Technical Conference and Exhibition, Houston, TX, October 1999).
[101] J. Yang, Jiang, and S. J. Weghorn, U.S. Patent Application 20050209368.
[102] T. Al-Wahaibi, M. Smith, and P. Angeli, *Journal of Petroleum Science and Engineering* 57 (2007): 334.
[103] V. Jovancicevic, S. J. Weghorn, and P. R. Hart, U.S. Patent Application 20050049327.
[104] H. JianZhong, "Reducing the Drag Force of the Multiphase Flow in Gathering Lines by Injecting Demulsifiers at the Wellhead," SPE 10576 (paper presented at the International Meeting on Petroleum Engineering).
[105] A. V. Shenoy, *Colloids and Polymer Science* 262 (1984): 319.
[106] B. Yu, F. Li, and Y. Kawaguchi, *International Journal of Heat & Flow* 25 (2004): 961.
[107] J. Myska, P. Stepanek, and J. L. Zakin, *Colloids and Polymer Science* 275 (1997): 254.
[108] A. V. Shenoy, *Rheology Acta* 15 (1976): 658.
[109] J. Myska and J. L. Zakin, *Industrial & Engineering Chemistry Research* 36 (12) (1997): 5483.
[110] H. Inaba and N. Haruki, *Heat Transfer—Japanese Research* 27 (1998): 1.
[111] R. J. Wilkiens and D. K. Thomas, "Influence of Gravity and Lift on Particle Velocity Statistics and Transfer Rates in Turbulent Vertical Channel Flow," *International Journal of Multiphase Flow* 33(2007): 134.
[112] B. Lu, X. Li, L. E. Scriven, H. T. Davis, Y. Talmon, and J. L. Zakin, *Langmuir* 14 (1998): 8.
[113] H. W. Bewersdoff and D. Ohlendorf, *Colloids and Polymer Science* 266 (1988): 941.
[114] B. A. M. O. Alink and V. Jovancicevic, U.S. Patent Application 20040206937.
[115] L. Chaal, C. Deslouis, A. Pailleret, and B. Saidani, "On the Mitigation of Erosion-Corrosion of Copper by a Drag-Reducing Cationic Surfactant in Turbulent Flow Conditions Using a Rotating Cage," *Electrochimica Acta* 52 (2007): 7786.
[116] V. Jovancicevic and K. A. Bartrip, U.S. Patent 6774094, 2004.
[117] V. Jovancicevic and Y. S. Ahn, U.S. Patent 7137401, 2006.
[118] M. Hellsten and I. Harwigsson, U.S. Patent 5339855, 1994.
[119] M. Hellsten and I. Harwigsson, U.S. Patent 5979479, 1999.
[120] M. Hellsten and I. Harwigsson, U.S. Patent 5911236 1999.
[121] M. Hellsten and I. Harwigsson, U.S. Patent, 5902784, 1999.
[122] M. Hellsten and H. Oskarsson, U.S. Patent Application 20040077734.
[123] M. Hellsten and H. Oskarsson, International Patent Application WO/2004/007630.
[124] E. Sletfjerding, A. Gladsø, Statoil, S. Elsborg, and H. Oskarsson, "Boosting the Heating Capacity of Oil-Production Bundles Using Drag-Reducing Surfactants," SPE 80238 (paper presented at the International Symposium on Oilfield Chemistry, Houston, TX, 5-7 February 2003).
[125] M. Hellsten, *Journal of Surfactants and Detergents* 4 (2002): 65.
[126] H. Oskarsson, I. Uneback, and M. Hellsten, "Surfactants as Flow Improvers in Water Injection," SPE 93116 (paper presented at the SPE International Symposium on Oilfield Chemistry, The Woodlands, TX, 2-4 February 2005).
[127] G. H. Sedahmed, M. S. E. Abdo, M. A. Amer, and G. Abd El-Latif, *Intenational Communications in Heat and Transfer* 26 (4) (1999): 531.
[128] S. E. Campbell and V. Jovancicevic, "Performance Improvements from Chemical Drag Reducers," SPE 65021 (paper presented at the SPE International Symposium on Oilfield Chemistry, Houston, TX,

13–16 February 2001).

[129] G. Schmitt, "Drag Reduction by Corrosion Inhibitors: A Neglected Option for Mitigation of Flow Induced Localized Corrosion," *Materials and Corrosion* 52 (2001): 329.

[130] G. Schmitt, M. Bakalli, and M. Hoerstmeier, "Contribution of Drag Reduction to the Performance of Corrosion Inhibitors in One and Two Phase Flow," Paper 07615 (paper presented at the NACE CORROSION Conference, 2007).

[131] B. A. Toms, First International Congress on Rheology, Amsterdam, North Holland, 1948.

[132] P. Glénat, F. Dang, I. Hénaut, and M. Darbouret, "Experimental Study of Drag Reducers Suitable for Heavy Oil Cold Production," Oilfield Chemical Symposium, Geilo, Norway, 14–17 March 2010.

[133] (a) S. J. Allenson, A. T. Yen, and F. Lang, "Application of Emulsion Viscosity Reducers to Lower Produced Fluid Viscosity," OTC 22443, OTC Brasil 2011 Preliminary Technical Program. (b) J. R. Faust, A. K. Flatt, J. R. Weathers, M. Thomas, and D. T. Nguyen, U.S. Patent Application 20110009556.

[134] D. T. Nguyen, International Patent Application WO/2010/053969.

[135] S. N. Milligan, R. L. Johnston, T. L. Burden, W. R. Dreher, K. W. Smith, and W. F. Harris, U.S. Patent Application 20120298209.

[136] (a) S. Asomaning and S. E. Lehrer, U.S. Patent Application 20130096043. (b) S. Asomaning and S. E. Lehrer, International Patent Application WO/2010/027862.

[137] J. Guo, H. Wang, C. Chen, Y. Chen, and X. Xie, *Petroleum Science* 7 (4) (2010): 536–540.

[138] B. Wu and J. Gao, *Petroleum Science and Technology* 28 (18) (2010): 1919–1935.

[139] M. Ashworth, "DRA for Gas Pipelining Successful in Gulf of Mexico Trial," *Oil & Gas Journal* 06/05/2000.

[140] M. A. da Silva, N. de O. Rocha, C. H. Carvalho, and E. Sabadini, *Energy & Fuels* 23 (9) (2009): 4529–4532.

[141] J. Zhou, H. Sun, R. Stevens, and Q. Qu, "Bridging the Gap between Laboratory Characterization, and Field Applications of Friction Reducers," SPE 140942 (paper presented at the SPE Production and Operations Symposium, Oklahoma City, OK, 27–29 March 2011).

[142] M. J. Zhou, H. Sun, Q. Qu, and B. Bai, "An Effective Model of Pipe Friction Prediction from Laboratory Characterization to Field Applications for Friction Reducers," SPE 146674 (paper presented at the SPE Annual Technical Conference and Exhibition, Denver, CO, 30 October–2 November 2011).

[143] A. M. M. Gomaa, J. Zhou, H. Sun, and Q. Qu, U.S. Patent Application 20130041587.

[144] I. Zadrazil, A. Bismarck, G. F. Hewitt, and C. N. Markides, *Chemical Engineering Science* 72 (2012): 142–154.

[145] J. J. Wu, U.S. Patent Application 20100324166.

[146] P. Glénat, I. Hénaut, C. Cassar, and P. Pagnier, "Mechanical Degradation Kinetics of Polymeric DRAs," (paper presented at the Tekna 23rd International Oil Field Chemistry Symposium, Geilo, Norway, 18–21 March 2012).

[147] I. Henaut, M. Darbouret, T. Palermo, P. Glenat, and C. Hurtevent, "Experimental Methodology to Evaluate DRA: Effect of Water Content and Waxes on their Efficiency," SPE 121544 (paper presented at the SPE International Symposium on Oilfield Chemistry, The Woodlands, TX, 20–22 April 2009).

[148] M. Darbouret, I. Hénaut, T. Palermo, C. Hurtevent, and P. Glénat, "Experimental Methodology to Evaluate DRA Efficiency and Mutual Effects of Wax Deposit and DRA Action," (paper presented at the

BHR 12th International Conference on Multiphase, Cannes, 17–19 June 2009).
[149] P. Yu, C. Li, C. Zhang, S. Chen, S. Fang, and H. Sun, *Petroleum Science* 8 (3) (2011): 357–364.
[150] S. N. Milligan, R. L. Johnston, T. L. Burden, J. R. Dreher, R. Wayne, K. W. Smith, and W. F. Harris, U.S. Patent Application 20130041094.
[151] S. N. Milligan, U.S. Patent Application 20100029843.
[152] Z. Bao, S. M. Milligan, and M. Olechnowicz, U.S. Patent Application 20130037118.
[153] H. A. A. Siti and N. Nour, *Chemical Industry & Chemical Engineering Quarterly* 18 (2012): 361–371.
[154] S. Ganji, K. M. Chinnala, and J. Aukunuru, *Pharmacognosy Magazine* 4 (15) (2008): 73–77.
[155] T. L. Burden, R. L. Johnston, W. F. Harris, K. W. Smith, W. R. Dreher, and S. N. Milligan, U.S. Patent Application 20090111714.
[156] S. N. Milligan, W. F. Harris, and T. L. Burden, U.S. Patent Application 20090107554.
[157] B. Li, W. Xing, G. Dong, X. Chen, N. Zhou, Z. Qin, and C. Zhang, *Petroleum Science* 8 (1) (2011): 99–107.
[158] Q. Di, C. Gu, Y. Qian, C. Shen, Z. Wang, B. Jing, C. Gu, and Y. Qian, "Innovative Drag Reduction of Flow in Rock's Microchannels Using Nano Particles Adsorbing Method," SPE 130994 (paper presented at the International Oil and Gas Conference and Exhibition in China, Beijing, China, 8–10 June 2010).
[159] H. K. Gurchran Singh and A. Bt. Jaafar, "Study of Drag Reduction Ability of Naturally Produced Polymers from a Local Plant Source," IPTC 17207 (paper presented at the 6th International Petroleum Technology Conference, Beijing, China, 26–28 March 2013).
[160] A. M. Shetty and M. J. Solomon, *Polymer* 50 (1) (2009): 261–270.
[161] A. Mohsenipour and R. Pal, *Canadian Journal of Chemical Engineering* 91 (2013): 181–189.
[162] A. Kamel and S. N. Shah, *Journal of Petroleum Science and Engineering* 67 (2009): 23–33.
[163] A. Patel, H. J. Zhang, M. Ke, and B. Panamarathupalayam, "Lubricants and Drag Reducers for Oilfield Applications—Chemistry, Performance, and Environmental Impact," SPE 164049 (paper presented at the 2013 SPE International Symposium on Oilfield Chemistry, The Woodlands, TX, 8–10 April 2013).
[164] E. Kot, R. K. Saini, L. R. Norman, and A. Bismarck, *SPE Journal* 17 (3) (2012): 924–930.
[165] E. Kot and A. Bismarck, *Macromolecules* 43 (15) (2010): 6469–6475.
[166] A. Mohsenipour and R. Pal, *Canadian Journal of Chemical Engineering* 91 (2013): 190–201.
[167] A. H. Ahmed Kamel and S. N. Shah, "Friction Pressure Losses of Surfactant-Based Fluids Flowing in Coiled Tubing," SPE 135826 (paper presented at the SPE Production and Operations Conference and Exhibition, Tunis, Tunisia, 8–10 June 2010).
[168] L. Liu, *Chemical Engineering Science* 95 (2013): 54–64.
[169] J. W. Veneman, H.-J. Oschmann, and J. A. J. Brunink, "Performance Assessment of Drag Reducing Polymers Utilizing a Turbulent Flow Rheometer," *Chemistry in the Oil Industry XIII: Oilfield Chemistry—New Frontiers*', Manchester Conference Centre, UK, 4–6 November 2013.

18 静水压测试用化学剂

18.1 概述

静水压试验是在设备、管道等的调试运行前,将其充满水并加压到最大许可(或可接受)运行压力(MAOP)的125%,以检查或检测是否有泄漏(图18.1)的过程。这样做是为了确保管道、设备不会对周边环境和居民造成重大危害。从油田集输管线到大口径长输管道的静水压检测,要特别加以重视。对整个或部分管道的静水压试验是管道结构完整性测试的首个阶段。例如,设计运行压力7.1MPa(1000psi)的管道应在测试时,加压到8.9MPa(1250psi)。海底环境中的泄漏检测通常需要通过远程操作的车辆进行紫外线检测,具有特别的挑战性。与此相关的特殊化学染料见下文。

图18.1 管道静水压测试用的清管器收发装置

管道设计是参照如美国机械工程师协会ASME B31.8、美国石油学会API 5L等标准,管道材质选用基于最低屈服值等级的管材,常见钢级有X60或X80等。这些设计参数明确了管道或容器的荷载,提供了额定最小屈服应力和环向应力,得出其最大许可运行工作压力(MAOP)。

静水压测试的目的如下:(1)检查是否泄漏;(2)验证并提高容器或管道的强度;(3)识别和消除结构中的缺陷;(4)减少缺陷的影响,从而提高抗疲劳寿命;(5)明确管道的预应力。

静水压试验通常使用水或油等不可压缩流体,常用的技术规范是API-RP 14E、API-RP 1110和API-RP 1111。在多数情况下,要使用如罗丹明红或荧光染料等显色染料,以提供一种简便的视觉检漏手段。其他泄漏检测方法包括液体流入与流出的流量平衡评价法、检测泄漏的振动频率超过20Hz的声学方法和动态管道建模。

当使用水试压时,推荐使用饮用水,但多数情况会使用海水、地下水或河水等易获得的水源，此类水需要化学处理。主要是为了防止腐蚀和因氧气存在而带来的加速腐蚀等影响，如某些特定水体里的细菌［特别是但不限于硫酸盐还原菌（SRB）］、藻类和真菌。因此需要对水进行缓蚀和杀菌处理，以控制静滞水环境中的细菌繁殖，并确定构筑物是否存在任何泄漏。针对静水压试验，特别是海底环境下，使用化学剂复配混合物的做法越来越普遍。

18.2 静水压测试用化学剂配方

静水压测试用化学剂配方有多种功能：
（1）除去氧气，通常是使用亚硫酸盐除氧剂，将氧含量降低到 10μg/L。
（2）控制腐蚀，包括延缓使用海水时的盐水效应，以及细菌繁殖造成的潜在微生物诱导腐蚀（MIC）。
（3）防止微生物生长和活动导致的生物污损和相关的 MIC。
（4）当被测设施处于海底环境中时，通过视觉或紫外光检测的染料显色。

因此，复配药剂包含除氧剂、缓蚀剂和杀菌剂等成分，还可能包含显色染料。供应商有时为使用方便，将上述化学单剂复配为含多种成分的单一产品，这种方式与多个单剂独立包装使用的方式并没有差别。采用复配产品时，对不同化学剂的相容性有特别高的要求。

当使用河水或不太理想的水源，需要特别考虑水质处理，而在热带气候环境作业时，要重点做好细菌控制，需要更高浓度的杀菌剂。另一个需要考虑的因素是静水压测试的时间跨度，长周期就需要更高浓度的缓蚀剂、杀菌剂。

在典型的静水压试验液体体系中，通常使用超过规定浓度 20mg/L 的除氧剂、50～100mg/L 的缓蚀剂及 300mg/L 的杀菌剂。杀菌剂浓度将根据前面提到的因素而变化，如果试验跨度在 3 个月以上，建议使用浓度 500mg/L；在管线中的试验跨度在 6 个月及以上，建议使用浓度 1000mg/L。特别是长管道的分段静水压试验中，液体体系重复使用，长时间跨度很常见。

杀菌剂和缓蚀剂等化学剂的补充添加非常关键。多年来，输气管道出现了若干起"黑粉"事件，其来源归因以下：
（1）轧制锈层——来自制造过程。
（2）闪锈——来自静水压测试的水腐蚀。
（3）内腐蚀——微生物、H_2S 与钢材的反应。

这种现象在含 CO_2、H_2S 的系统中都有发现，其后果是导致严重下游问题（如压缩机的侵蚀磨损、过滤器和控制装置的堵塞等）；"黑粉"被水或冷凝水浸湿后，还会形成坚硬段塞，从管内清除并暴露于空气中会发生自燃。

所涉及的反应可以表示如下：
（1）SRB 活动。

$$SO_4^{2-} + 2H^+ + 营养物 \longrightarrow H_2S + CO_2 + 2H_2O$$

（2）静水压测试活动：静水压测试水中 SRB 活动产生 H_2S 的作用影响。

锈层：$\quad Fe_2O_3 \cdot H_2O + 3H_2S \longrightarrow 2FeS + S + 4H_2O$

轧制层：$\quad Fe_3O_4 + 4H_2S \longrightarrow 3FeS + S + 4H_2O$

钢铁：$\quad Fe + H_2S \longrightarrow FeS + H_2$

（3）运行活动：湿态下的 FeS 氧化影响。

$$3FeS + S + 5O_2 \longrightarrow Fe_3O_4 + 3SO_2 + S$$

$$3FeS + S + 6O_2 \longrightarrow Fe_3O_4 + 4SO_2$$

$$4FeS + 7O_2 + 2H_2O \longrightarrow 2Fe_2O_3 \cdot H_2O + 4SO_2$$

多数对"黑粉"的分析表明存在铁的氧化物和硫化物。虽然在通常情况下粉末量有限，而且可以被清管流程去除，但大型长输管道可能包含数吨黑粉。

18.2.1 杀菌剂

杀菌剂是通过杀灭细菌或形成细菌的不耐受环境，以控制水中细菌活动的方法。油田典型的杀菌剂测试是针对浮游性细菌，而在静水压试验时，更重要的是针对固着性细菌。这与微生物膜的形成及其潜在腐蚀问题特别相关。好氧菌有腐蚀性，易杀灭，而且半衰期相对短，对静水压试验的水处理影响有限。厌氧菌的作用时间较长，通常腐蚀性较低。油田最常见的杀菌剂是季铵盐表面活性剂、戊二醛、稳定态的甲醛和四羟甲基硫酸磷（THPS）（见第 14 章）。对于海底立管和管道内的化学处理海水，THPS 降解问题已有相应的调查报告。

在化学复配药剂中，当戊二醛和 THPS 等典型杀菌剂与常规亚硫酸氢盐除氧剂或类似产品混合时，会出现迅速降解问题，会使杀菌剂失去效力。而 n-正烷基二甲基苄基氯化铵等季铵盐则不会出现类似问题，已经成为静水压试验介质的标准杀菌剂。季铵盐类产品的半衰期较长，作为表面活性剂对固体表面的润湿性良好，兼具缓蚀剂的特性。而 THPS 等其他杀菌剂则需要添加表面活性剂成分，通过辅助金属表面的润湿作用对附着细菌膜进行生物渗透。

还有与杀菌剂持久性相关的问题，因为季铵盐类比其他杀菌剂的降解速度慢，其生态毒理学特征差。

18.2.2 除氧剂

主要是亚硫酸铵或亚硫酸钠。静水压试验期间的氧腐蚀与 MIC 是腐蚀问题的最重要来源。需要将氧含量降至低于 $50\mu g/L$，通常低于 $10\mu g/L$，可以通过溶解氧与亚硫酸盐比例 1:（8~10）来有效实现。因此，要确保复配药剂中的除氧剂可用并且不与杀菌剂发生反应。其他种类的除氧剂也可以达到同样效果，但其成本更高、反应速率慢，而且通常需要某种形式的催化作用。

针对淡水或地层水作为静水压试验介质，使用亚硫酸钠时，也可能需要配套催化剂。使用海水则情况并非如此，因为海水中少量的过渡金属就足以有效促进氧的消耗反应。

18.2.3 缓蚀剂

静水压试验的标准液体体系通常不包含额外的缓蚀剂；如果组分是单独添加的，且杀菌剂不是季铵盐时，则可能需要配套缓蚀剂。当添加杀菌剂导致静水压试验用水的pH值较低，或者不能接受任何先前的腐蚀，此时也需要配套缓蚀剂。具有缓蚀作用的化学品很多，季铵类化合物仍然是首选。

在水中加入如氨基羧酸基化学品的低浓度气相缓蚀剂（VCI），能够在测试期间、储存期间和测试后都起到防腐作用。气相缓蚀剂的机理是在溶液内、水线处和水线以上三种相态下起到防腐作用。基于气相缓蚀剂的化合物可对其他缓蚀剂无法达到的区域起到防腐效果。

18.2.4 染料

静水压试验过程中采用的标准染料是荧光素，因为其使用浓度低，10mg/L下依然可见，并且在更低浓度下，仍可由紫外线检测到（图18.2）。因荧光素的持久性和毒性，已在北海油田被禁止使用（见18.2.6）。

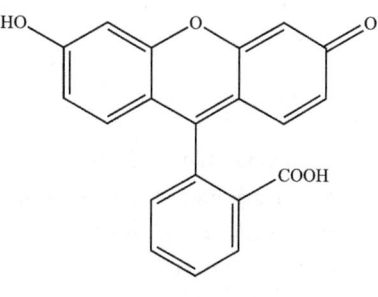

图18.2 荧光素

18.2.5 其他静水压测试化学品

对于深海管道，必须消除试验后的残留水，以防止开井生产后管道生成水合物。为此，使用如甲醇、乙醇或乙二醇等热力学水合物抑制剂。有关集输管道的研究表明，仅靠甲醇置换不能完全抑制水合物，还需要进行复杂的氮气置换脱水。相反，管道测试采用高矿化度的完井盐水有对环境影响小等若干优点。

18.2.6 环境友好型的发展

近年来，针对海底管道和相关的关键工艺过程，开发和应用环境可接受的静水压试验复配药剂成为一项特别挑战，特别是在北海盆地应用时。北海盆地的化学品使用和排放受2002年OSPAR条约中海上法规管辖，其中规定了化学品的环境危害和风险等级、使用量及允许排放的内容。

在海底环境的静水压试验中，使用组合复配药剂来提供缓蚀、杀菌及其他防垢、控制氧气侵入等策略。此外，某些复配药剂也配套专门的染料，肉眼可见或通过"照相机"检测。通常情况下，染料单独施用。

颇具挑战性的工作是开发一种环境可接受的复配药剂，确保在危险评级中达到最高的可接受等级（英国和荷兰部门为金带和无替代标志，挪威和丹麦部门为黄色评级）。这种产品应有最佳的允许使用和排放情况，同时最大限度地减少对环境的风险。

近年来，已经开发了与亚硫酸盐除氧剂兼容的杀菌剂，特别是二癸基二甲基氯化铵（图18.3）。基于罗丹明盐类和其等价物的环境友好型染料可作为荧光素的替代品。但某些情况下，仍然不能采用以上染料时，可以考虑采用含光学增白剂的紫外线指示性染料。

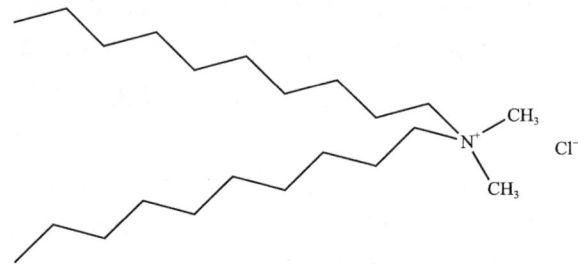

图18.3 二癸基二甲基氯化铵

参 考 文 献

［1］B. E. Mappus and A. G. Torstrick, "Pipeline Hydrotesting, Dewatering, and Commissioning," OTC 19062（paper presented at the Offshore Technology Conference, Houston, TX, 30 April–3 May 2007）.

［2］E. W. McAllister, *Hydrostatic Testing in Pipeline Rules of Thumb Handbook*, Houston：Gulf Publishing CO, 1998, 121–132.

［3］*Protection of Pipelines during Hydrotesting*. Baker Petrolite Publication.

［4］W. Boulton, "Pipeline Hydrotest Water Management," SPE 111781-MS（paper presented at the SPE International Conference on Health, Safety, and Environment in Oil and Gas Exploration and Production, Nice, France, 15–17 April 2008）.

［5］G. Ruschau, W. Huang, E. Sullivan, and M. Surkein, "Hydrotesting of LNG Tanks Using Untreated Brackish Water," Paper 11216（paper presented at the CORROSION 2011, Houston, TX, 13–17 March 2011）.

［6］A. Darwin, K. Annadorai, and K. Heidersbach, "Prevention of Corrosion in Carbon Steel Pipelines Containing Hydrotest Water—An Overview," Paper 10401（paper presented at the CORROSION 2010, San Antonio, TX, 14–18 March 2010）.

［7］R. Prasad, "Chemical Treatment Options for Hydrotest Water to Control Corrosion and Bacterial Growth," Paper 03572（paper presented at the CORROSION 2003, San Diego, CA, March 2003）.

［8］K. J. Grobe and P. S. Stewart, "Characterisation of Glutaraldehyde Efficacy Against Bacterial Biofilm," NACE, Corrosion 2000, Paper 00124, Orlando, FL, 26–31 March 2000.

［9］C. J. Nalepa, J. N. Howarth, and F. D. Azarnia, "Factors to Consider When Applying Oxidizing Biocides in the Field," Paper 02223（paper presented at the CORROSION 2002, Denver, CO, 7–11 April 2002）.

［10］*Selection, Application and Evaluation of Biocides in the Oil and Gas Industry*. NACE International Publication, Houston, TX, 2006, 31205.

［11］R. Prasad, "Chemical Treatment for Hydrostatic Test," U.S. Patent 6815208, 2004.

［12］B. L. Downward, R. E. Talbot, and T. K. Hack, "Tetrakishyddroxymethylphosphonium Sulphate（THPS），

A New Industrial Biocide with Low Environmental Toxicity," Paper 401, CORROSION 1997.

[13] J. Willmon, "THPS Degradation in the Long-Term Preservation of Sub-sea Flow-Lines and Risers," Paper 10402 (paper presented at the CORROSION 2010, San Antonio, TX, 14-18 March 2010).

[14] T. M. Williams and H. R. McGinley, "Deactivation of Industrial Water Treatment Biocides," Paper 10049, NACE CORROSION 2010.

[15] *Spills, Deactivation, and Disposal of Glutaraldehyde*. Dow Biocides, 2008.

[16] API RP1110-2007, Pressure Testing of Steel Pipelines for the Transportation of Gas, Petroleum Gas, Hazardous Liquids, Highly Volatile Liquids or Carbon Dioxide. American Petroleum Institute, 2007.

[17] A. J. McMahon, A. Chalmers, and H. MacDonald, "Optimising Oilfield Oxygen Scavengers," Chemistry in the Oil Industry VII, Royal Society of Chemistry, 2002.

[18] J. Holden, A. Hansen, A. Furman, R. Kharshan, and E. Austin, "Vapor Corrosion Inhibitors in Hydro-Testing and Long Term Storage Applications," Paper 10405 (paper presented at the CORROSION 2010, San Antonio, TX, 14-18 March 2010).

[19] K. Zhao, T. Gu, I. Cruz, and A. Kopliku, "Laboratory Investigation of MIC in Hydrotesting Using Seawater 10406," (paper presented at the CORROSION 2010, San Antonio, TX, 14-18 March 2010).

[20] M. Erdogmus, L. Cowie, R. Chapman, G. Fung, and P. Bollavaram, "A Novel Approach to Green, Safe and Economic Subsea Hydrotesting," OTC 17326 (paper presented at the Offshore Technology Conference, Houston, TX, 2-5 May 2005).

19 排水采气用泡排剂

19.1 概述

气井中的液体聚集会造成额外的回压,降低天然气产量,严重时会使气井完全停产。气井排水采气是欧洲年度会议的主题,读者可以浏览相关报告,以了解详情。除了机械排水技术,最重要的化学方法是注入发泡表面活性剂来泡沫排水,即将表面活性剂注入井底,与水、天然气混合,降低表面张力,形成密度低于水的泡沫,从井中产出。可以将浓缩的表面活性剂连续或间歇注入,也可以制成泡排棒使用,后者在低产气井中更常用(表面活性剂被复合加工成蜡质柱状体,然后将其投入气井内)。使用发泡的表面活性剂来举升排出气井中的液体非常普遍。全球大约有40%的气井存在积液问题,造成生产速度低于最佳水平。多数泡排剂设计用于天然气井和凝析气井,很少有产品可以发泡原油。泡排剂的通用测试方法需要使用模型化的表面活性剂,这通常涉及测量动态表面张力及用定制设备测试发泡性能。实验室内对泡排剂在多种原油和不同含水率情况下的性能进行了测试,开发的新型泡排剂已在现场试验成功。

19.2 泡排剂的特性和类别

包括阴离子、阳离子、两性离子和非离子型的各种表面活性剂都可以作为泡排剂,其性质各不相同。如果有液态烃时,泡排剂不应造成储层伤害或乳化问题。需要考虑的其他因素包括:

(1)高温或高矿化度下的浊点问题。
(2)对液态烃类(凝析油)的起泡耐受性。
(3)高温下的降解。
(4)受pH值的影响程度。
(5)防冻能力。

有些表面活性剂也具有缓蚀性能,可以按需求添加特定的缓蚀剂和防垢剂等其他化学剂。典型的发泡表面活性剂有α-烯烃磺酸盐、醇醚硫酸盐和甜菜碱,后两类的生物降解性更好。有机溶剂经常用以改善泡排剂配方的倾点、黏度或泵送性等辅助性能。随着凝析油/水比率的增加,传统泡排剂往往会失效。有报道称开发了一种专用于携带井内凝析油的新型泡排剂。

有报道称为了提高耐高温、适当含量的原油、凝析油、缓蚀剂和黏土微粒等的能力,开发了无腐蚀性的发泡剂。这类泡排剂不受盐水矿化度的影响，同时与地层水相容，已成功用于现场。还有报道称提高油气井产量的咪唑啉基泡排剂和缓蚀剂，已用于现场凝析油含量大于90%井的携液生产。有发明还提出通过将发泡剂引入油气井而使流体发泡的方法，如椰油基羟乙基咪唑啉和丙烯酸在醇或乙二醇溶剂中的中和反应产物。

据称带有酰胺间隔基团的季铵盐可作为泡排剂，用于气井排液和提高气举油井的产量（图19.1）。表面活性剂的尾部含有萘基等芳香基团，头端烷基最好是甲基、乙基或苄基。

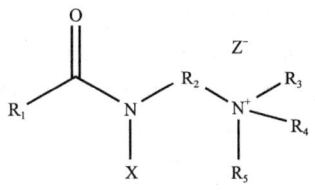

图19.1 季酰胺类气井泡排剂

通过对泡排剂基础表面活性剂特性的测量，将这些特性与携液能力有效关联，为加深泡排剂工作原理认识和快速评价技术提供了手段。该技术可用于优化含有两种不同类型的发泡表面活性剂、缓蚀剂和抗冻剂乙二醇甲醚的泡排剂配方。使用统计设计方法评价不同配方的协同效应。在测试两性表面活性剂和阴离子表面活性剂的发泡性能时，通过测定二元表面活性剂混合液的表面张力和临界胶束浓度，探索二者之间协同作用。还讨论了基于流速和其他因素的泡排剂排液优化过程。对环空坐封、有井下安全阀的气井可以提供泡沫排液的定时、定制方法和处理系统。

一种新型的泡排方法通过化学反应在井下产生气体和泡沫。该体系包括起泡剂、泡沫增强剂和产生气体的添加剂。起泡剂、泡沫增强剂被吸收或吸附在两类不同基质上。产生气体的添加剂优选由包含酸性成分、碳酸盐或碳酸氢盐的两种释放机制载体组成。药剂制成胶囊类型，可以确保分阶段和有针对性地将化学剂注入近井裂缝或井眼中。

还有专利报道了一种高温稳定的发泡剂制备工艺，该组合物具有阴离子表面活性剂、中和作用的胺、其他生产化学剂和溶剂，具体举例是通过结合 α-烯烃磺酸盐、三乙醇胺及防垢剂、缓蚀剂和互溶剂来制备。

参 考 文 献

[1] See, for example, articles from the 6th（2011）or 7th（2012）European Conference on Gas Well Deliquification, Hampshire Hotel, Plaza Groningen, Netherlands.

[2] J. F. Lea, H. V. Nickens, and M. R. Wells, *Use of Foam to Deliquify Gas Wells*, *Gas Well Deliquification*, Chap. 8. Gulf Publishing, Burlington, MA 2008, 193-240.

[3] See also Annual European Gas Well Deliquification Workshops such as http: //www.alrdc.com/workshops/2012_2012EuropeanGasWell/.

[4] B. P. Price and B. Gothard, "Foam Assisted Lift—Importance of Selection and Application," SPE 106465（paper presented at the Production and Operations Symposium, Oklahoma City, OK, 31 March-3 April 2007）.

[5] S. Ramachandran, J. Bigler, and D. Orta, "Surfactant Dewatering of Production and Gas Storage Wells," SPE 84823（paper presented at the SPE Eastern Regional Meeting, Pittsburgh, PA, 6-10 September 2003）.

[6] M. J. Willis, D. Horsup, and D. Nguyen, "Chemical Foamers for Gas Well Deliquification," SPE 115633

（paper presented at the SPE Asia Pacific Oil and Gas Conference and Exhibition, Perth, Australia, 20−22 October 2008）.

[7] M. J. Willis, "Chemical Foamers for Gas Well Deliquification," *Exploration & Production-Oil & Gas Review* 8（1）（2010）: 1.

[8] J. Visser and M. Willis, "Selection and Application of Chemical Foamers for Offshore North Sea," Gas Well Deliquification Conference, Denver, CO, 2009.

[9] S. Ramachandran, J. Collins, C. Gamble, P. Schorling, J. G. R. Eylander, and C. Wittfield, *Proceedings of the Chemistry in the Oil Industry IX*, Royal Society of Chemistry, UK, 31 October−2 November 2005, 205.

[10] D. Van Nimwegen, L. Portela, and R. Henkes, "The Effect of Surfactants on Vertical Air/Water Flow for Prevention of Liquid Loading," SPE 164095（paper presented at the SPE International Symposium on Oilfield Chemistry, The Woodlands, TX, 8−10 April 2013）.

[11] S. Lehrer, S. B. Debord, C. Means, and S. Crosby, "Development and Application of Foamers to Enhance Crude Oil Production," SPE 141026（paper presented at the SPE Production and Operations Symposium, Oklahoma City, OK, 27−29 March 2011）.

[12] W. Jelinek and L. L. Schramm, "Improved Production from Mature Gas Wells by Introducing Surfactants into Wells," SPE 11028（paper presented at the International Petroleum Technology Conference, Doha, Qatar, 21−23 November 2005）.

[13] S. Campbell, S. Ramachandran, and K. Bartrip, "Corrosion Inhibition/Foamer Combination Treatment to Enhance Gas Production," SPE 67325（paper presented at the SPE Production and Operations Symposium, Oklahoma City, OK, 24−27 March 2001）.

[14] J. J. Wylde and E. Welmer, "Chemical Deliquification—Multifunctional Product Development and Application for Low Pressure, Low Rate Gas Wells," 6th European Gas Well Deliquification Conference at the Hampshire Hotel—Plaza in Groningen, Netherlands, 27−29 September 2011.

[15] S. T. Davis and V. Panchalingam, U.S. Patent Application 20060128990.

[16] J. Yang, V. Jovancicevic, and S. Ramachandran, *Colloids and Surfaces A: Physicochemical and Engineering Aspects* 309（2007）: 177.

[17] D. Orta, S. Ramanchandran, J. Yang, M. Fosdick, T. Salma, J. Long, J. Blanchard, A. Allcorn, C. Atkins, and O. Salinas, "A Novel Foamer for Deliquification of Condensate-Loaded Wells," SPE 107980（paper presented at the Rocky Mountain Oil & Gas Technology Symposium, Denver, CO, 16−18 April 2007）.

[18] S. Heuvel, "Non-Corrosive Foamers," *Innovoil*（5）（2012）: 4.

[19] D. T. Nguyen and G. R. Meyer, International Patent Application WO/2009/064719.

[20] N. Nguyen and F. Huang, International Patent Application WO/2012/154521.

[21] D. T. Nguyen, *Petroleum Science and Technology* 27（2009）: 733.

[22] F. F. Huang and D. Nguyen, *Fuel* 97（2012）: 523.

[23] J. Cabanilla and B. D. Dotson, US Patent Application 20110155378.

[24] D. K. Durham, J. Archer, and E. J. Gothard, US Patent Application 20110155378.

[25] K. Dismuke, P. B. Kaufman, H. Juppe, G. S. Penny, R. N. Fox, and B. Zhou, International Patent Application WO2013019188.

[26] T. J. Sanner, G. S. Penny, and R. Padgham, International Patent Application WO/2005/100534.

附录　油田化学品的 OSPAR 环境条例

东北大西洋的 OSPAR 指南 2001 年开始实施统一的强制控制计划，对拟在北海海上油田使用的油田化学品，其所有成分都要按规定进行生态毒理学测试。OSPAR 要求对油田化学品进行急性毒性、生物积累性、海水中的生物降解性三类生态毒理学测试。必须确定化学品配方中每种单独化学成分（而不是成品）的环境特性。

完整的 OSPAR 海洋急性毒性数据集包括：

（1）中肋骨条藻（海洋藻类；ISO 10253）。

（2）纺锤水蚤（海洋桡足类；ISP 14669，OSPARCOM 提出的建议）。

（3）滚卷螺赢蛰［海洋两栖动物；《巴黎委员会准则》（1995 年），OSPAR 关于海上工业中使用的化学品测试方法的议定书 A 部分］。这种测试被称为沉积物再加工测试，只有在化学品具有某些特性的情况下，如具有很强的生物积累性或表面活性，或已知会吸附在颗粒上或沉积在沉积物中时，才需要进行这种测试。

（4）大菱鲆［海洋鱼类幼虫；《巴黎委员会准则》（1995 年），《关于近海使用的化学品测试方法的议定书》的 B 部分：鱼类急性毒性试验的议定书］，如果该化学品对中肋骨条藻或纺锤水蚤有很强的毒性，则不需要进行这种测试。

中肋骨条藻是对毒性最敏感的物种。因此，为衡量新型生产化学品的毒性水平，毒性测试通常首先采用中肋骨条藻。表面活性剂（特别是阳离子表面活性剂）往往是一些毒性最大的生产化学品。除了杀菌剂，如季铵化合物和咪唑啉等成膜型表面活性剂缓蚀剂（在酸性产出水中质子化而成为阳离子）往往是毒性最大的生产化学品。而现在许多服务公司和化学品供应商也提供一系列毒性较低的产品，有时会在性能上有所折中。

四种毒性试验的试验期长度不同。对于中肋骨条藻，结果以 EC 50 表示，即导致 50% 生长抑制的试验物质的浓度。对于其他海洋生物，结果以 LC 50 表示，即测试物质的浓度能使其有 50% 的死亡（固定化）。毒性数据的评估通常分为五类（<1mg/L，>1~10mg/L，>10~100mg/L，>100~1000mg/L 和 >1000mg/L），即决定一种化学品毒性的剂量水平。对于含有多种成分的配方（如缓蚀剂或破乳剂），全面的毒性测试可能相当昂贵。测试少量可溶化合物的方法已有报道。

针对产出水中毒性化合物的新型筛选技术已有报道，还讨论了矿化度对外排水毒性的影响。评估生态毒理学数据的系统性方法也已经发表。对一种由 CAS 通用编号定义的常见物质，调查了其相关的毒性测试数据的差异性。该物质存在于 100 多种数据库中注册的产品，形成了分布于 10 个不同物种的 460 个生态毒理学端点库。该研究表明毒性结果的明显差异，可归因于以下因素：

（1）实验室样品应用方法。
（2）使用的测试实验室。
（3）多成分物质的化学批次的组成。
（4）依靠 CAS 号作为唯一标识符。

测定生物积累潜力时，对于水溶性化学品，通常采用高效液相色谱法（经合组织，OECD 117）；对于亲油性较强的化学品，采用摇瓶法（OECD 107）。因为表面活性剂会在界面上累积，而且会形成乳液，所以评估生物积累非常困难。

已报告了一种利用对线型醇氧乙烯醚的研究，确定表面活性剂的生物富集系数（BCF）的新方法。计算机建模和定量构效关系（QSAR）软件建立在大量的非表面活性剂物质的平均片段贡献上，最初用来计算拟 $\lg P_{ow}$。这个疏水参数与流水式生物测试中确定的 BCF 值相关，具有很好的一致性（$R^2 > 0.99$）。该方法不涉及动物试验，设置和执行成本低，所有人都可以使用，快速、便捷和易于验证，并能提供清晰的输出。

生物积累性被记录为 $\lg P_{ow}$（正辛醇与水的比值）。英国的化学危害评估和风险管理（CHARM）模型中也使用了 $\lg P_{ow}$ 值，以估计物质在油和水之间的分配情况，目的是预测环境浓度（PEC）。如果计算或实验确定的 $\lg P_{ow} \geq 3$，就假定存在生物积累性（除非实验性 BCF 测试表明情况相反）。一种测量非离子表面活性剂的分配系数的新方法已经公布。

那些 $\lg P_{ow}$ 值大于 4.5~5.0 的物质被认为具有潜在的高生物积累性。油溶性聚合物被认为具有较高的 $\lg P_{ow}$ 值；但分子量超过约 700，就被认为不太可能通过亲脂性细胞膜并造成累积性损害。

生物降解性测试是新化学品环境风险评估的一个基本部分，经合组织的《有机化学品测试指南》被广泛接受为欧洲的共识方法。采用的生物降解策略包括 0 级筛选测试在有氧水环境中的生物降解性，然后是固有的生物降解性测试（1 级），最后是模拟测试（2 级）。如果筛选（0 级）测试结果阳性，则不必进行进一步的测试（更高级别）。在 OECD 306 方法中提出了海洋生物降解性的 0 级测试，可以根据摇瓶溶解有机碳（DOC）衰减法或闭瓶生物需氧量（BOD）测试方法［OECD，1992；在某些情况下，淡水测试方法 OECD 301 可以被英国和荷兰的环境、渔业和水产养殖科学中心（CEFAS）接受。由于在淡水中获得的数值通常高于海水中的数值，因此应声明使用该方法］。通过呼吸测定法进行的闭瓶测试是基于矿化测试化合物和呼吸氧之间的化学计量关系。生物降解计算为理论需氧量（ThOD）和测量的降解过程中耗氧量（BOD）之间的比率，如果消耗的氧气等于或超过 ThOD 的 60%，该化合物是（容易）生物降解。这种方法的一个问题是没有考虑到部分测试化学品被同化为新的生物质。因此，在计算生物降解程度时，必须对提供高细胞产量（Y_{xs}）的测试化合物进行补偿。更正确的数据分析基于测量的 BOD 和计算的测试化合物的理论生物需氧量（ThBOD）之间的比率，给定为 $1-Y_{xs}$。葡萄糖是推荐用于阳性对照的化合物之一，其 ThBOD 约为 ThOD 的 33%，这会导致错误的结论，即根据 OECD 306 方法，葡萄糖在海水中永远不会被轻易地生物降解。在实验室测试中仍然（成功）使用葡萄糖作为阳性对照化合物；但是高度的生物降解性只能通过在总的 BOD 估算中包括生长后的内源性呼吸阶段来确定。

OECD 306 海洋生物降解试验通常要进行 28 天。如果 28 天内生物降解率小于 20%，则该化学品被认为具有持久性，如果生物降解率大于 60%～70%，则被认为是易生物降解。OECD 306 方法指出，"由于与大多数自然系统相比，试验测试浓度相对较高，试验物质和其他碳源物质的浓度之间存在不利比例，因此该方法应被视为一种初步测试，可用于表明某种物质是否易于生物降解。而低的结果并不一定意味着测试物质在海洋环境中不能被生物降解，而是表明需要做更多的工作来确定该点认识"。因此，鼓励化学品供应商提交更多的数据以证明一种物质是否可以生物降解。对于在 OECD 306 测试中显示生物降解性小于 60% 或小于 70%（取决于测试终点）的物质，可进行延长的 OECD 306 测试，如超过 60 天。在英国，OECD 301 测试也可以接受。如果 OECD 306 测试显示该物质生物降解性小于 20%，则推荐进行适当的固定测试。可以进行专门用于溶解性差物质的海洋 BODIS 测试。目前法规规定，生物降解测试必须对配方中包含的每一种有机溶剂成分进行单独测试，而不是对整个配方。

因为生物降解率随着温度的升高而增加，OECD 306 测试规定了进行生物降解试验的标准温度。而对于海水样品中存在的生物体的浓度和类型没有给出标准。这可能会因地点和海水的深度不同而有很大差异，并会影响生物降解的速度。测试物质的浓度也会影响生物降解速度。OECD 306 闭瓶法建议使用 2～10mg/L 的测试化合物，也可以使用更高的浓度，特别是如果选择在瓶子里有一个空气顶空并测量压降时。摇瓶法（OECD 306 的替代方法）建议使用浓度为 5～40mg/L 的 DOC，相当于 10～80mg/L（根据化合物的碳含量）的测试化合物。一般来说，在较低的浓度下，生物降解会更快。水溶性物质，特别是非聚合性物质，通常比不溶性物质降解得快，因为后者倾向于强烈吸附于固相。此外，如果该物质具有很强的毒性或部分降解为有毒化合物，就可能会杀死进行生物降解的生物体。

如果化学品沉入海底，就会受阳光的影响较少；而在海面附近有光照，阳光的存在可以增加生物降解的速度。例如，二乙烯醚在 OECD 306 试验中，二乙烯三胺五乙酸的 1 类盐降解较差，部分原因是试验是在黑暗中进行。然而，这些盐类在阳光下的降解速度更快。

在接下来的内容中，对英国（与荷兰非常相似）和挪威的海上环境法规进行介绍，以说明两国在评估生产化学品方面的差异。北海有关国家的法规还在不断修订，读者可通过相关的污染管理部门查询最新信息。

附录 1　英国和荷兰的北海生态毒理学条例

英国环境、食品和农村事务部（DEFRA）的下属机构 CEFAS 管理着海上化学品通报计划（OCNS）。自 2007 年以来，CEFAS（英国）一直在管理荷兰的 OCNS 计划。OCNS 对海上使用的化学产品进行危险评估。CHARM 模型计算预测效应浓度与无效应浓度的比率（PEC/NEC），并表示为危险商数（HQ），然后以此来对产品进行排名。在 CHARM 评估中使用的数据包括产品中各成分的百分比、预期产品剂量率、毒性、生物降解和生物积累。HQ 被转换为一个色带，然后被公布在批准产品的明确排名名单上，有金色、银色、白色、蓝色、橙色和紫色六个色带（按危险程度递增）。

不适用于 CHARM 模型的产品（无机物、液压油或仅用于管道的化学品）被分配到 OCNS 组别 A—E，其中 A 的潜在环境危害最大，E 代表最小。挪威/丹麦的黄 1 和黄 2 相当于英国/荷兰的金色。英国/荷兰的银色基于 CHARM 及生态毒理学数据，不同于挪威/荷兰的红色。

替代物是 OSPAR 统一强制控制计划的一个重要组成部分。当化学品被确定为候选替代物或含有被确定为候选替代物的成分，英国就有义务执行替代该化学品的策略。符合以下条件的海上化学品应被替代：

（1）被列入 OSPAR 战略附件 2 中的危险物质。

（2）受理申请的当局认为该物质对海洋环境的影响等同于前款规定的物质。

（3）LC_{50} 或 EC_{50} 小于 1mg/L 的无机物。

（4）在 28 天内的生物降解率小于 20%，或符合以下三个标准中的两个：

① 28 天内的生物降解率小于 60%（OECD 306，英国和荷兰现在不接受 OECD 301 作为生物降解试验，只接受 OECD 306）。

② 生物积累性 $\lg P_{ow}$ 大于 3 且分子量小于 600，或 BCF 大于 100 的物质。

③ LC_{50} 或 EC_{50} 毒性小于 10mg/L。

附录 2　挪威海上生态毒理学条例

在挪威，化学品被列为绿色、黄色、红色和黑色四种颜色。环境可接受性依次递减。黄色化学品进一步细分为黄 1、黄 2 和黄 3，第一类环境可接受度最高。"绿色"化学品，即 OSPAR 的 PLONOR（风险小或无风险）清单中的化学品，允许在挪威近海使用。而"黑色"化学品是不允许的。有几类"黑色"化学品，包括内分泌干扰剂和致癌物，以及以下：

（1）生物降解率小于 20% 且 $\lg P_{ow}$ 不小于 5 的化学品。

（2）生物降解率小于 20% 且毒性 EC_{50} 或 LC_{50} 不大于 10mg/L 的化学品。

甲醇在 PLONOR 名单中，但因为员工有接触的风险，在健康、环境和安全（HES，挪威语的字母缩写为 HMS）评级中被评为"黑色"。

"红色"化学品被归类为对环境有害的化学品，如果正在使用，应优先替换。除了 OSPAR 污点清单上的化学品，"红色"化学品包括：

（1）毒性 EC_{50} 或 LC_{50} 不大于 1mg/L 的无机化学品。

（2）生物降解率小于 20% 的有机化学品。

符合以下三个标准中的两个的有机化学品或混合物：

（1）28 天内生物降解率小于 60% 的有机化学品。

（2）生物积累性 $\lg P_{ow}$ 不小于 3。

（3）急性毒性 EC_{50} 或 LC_{50} 不大于 10mg/L。

挪威环境署（KLIF，挪威语为 Miljødirektoratet）规定如果一个化学品的 $\lg P_{ow}$ 大于 3，毒性 EC_{50} 小于 10mg/L，即使其很容易被生物降解，也属于"红色"类别。而在公海上作业时，可以使用这样的化学品。该问题已经引起监管机构的注意，但是到目前为止还没有对法规进行修改。

与"绿色"化学品一样，"黄色"化学品通常被允许在挪威近海使用，不需要特别批准。"黄色"化学品指不属于其他颜色类别，或必须满足以下要求的化学品：

（1）生物降解率大于 20%。

（2）$\lg P_{ow}$ 小于 3。

（3）EC_{50} 或 LC_{50} 大于 10mg/L。

理论上，$\lg P_{ow}$ 为 2.99 可以认为是理想值，因为刚好在批准的极限值之下，意味着进入水相排放的化学品数量最少。如果一种化学品的分子量超过 700，就认为不具有生物积累的潜力（其分子太大，无法穿过细胞膜），也不需要进行这方面的测试。这也引起了一个潜在的困境，即如果化学品因为分子量大而不能轻易通过膜，其在某些降解机制就不容易被生物降解（高分子量并不保证低毒性，如阳离子聚丙烯酰胺和聚合季铵化合物等某些聚合物对浮游生物和鱼类等海洋生物有相当大的毒性。其毒性机制与小分子化学品不同，后者可以穿过细胞膜，如阻碍氧气运输到生物体内进行呼吸的过程）。对于生物降解率在 20%~60% 之间的化学品，还需要关注其生物降解的产物，以防其具有持久性和毒性。挪威当局要求对与测试化学品的生物降解产物有关的危险性进行文献评估。化学品的生物降解百分比也应在 28 天后继续增加。

参 考 文 献

[1] OSPAR Guidelines for Completing the Harmonized Offshore Chemical Notification Format (2005-13) Available at http://www.ospar.org.

[2] S. Glover and I. Still, "HMCS (Harmonized Mandatory Control Scheme) and the Issue of Substitution" (paper presented at the Ninth Annual International Petroleum Environmental Conference, 22-25 October 2002).

[3] M. Thatcher and G. Payne, "Impact of the OSPAR Decision on the Harmonized Mandatory Control System on the Offshore Chemical Supply Industry," Proceedings of the Chemistry in the Oil Industry VII Symposium, Royal Society of Chemistry, Manchester, UK, 13-14 November 2001.

[4] G. M. Rand, *Fundamentals of Aquatic Toxicology: Effects, Environmental Fate and Risk Assessment*, 2nd ed., Philadelphia: Taylor & Francis, 1995.

[5] G.-G. Ying, "Fate, Behavior and Effects of Surfactants and Their Degradation Products in the Environment," *Environment International* 32 (2006): 417.

[6] M. A. Ottoboni, *The Dose Makes the Poison*. New York: Wiley-Blackwell, 1996.

[7] Organisation for Economic Cooperation and Development, *OECD Guideline for the Testing of Chemicals 117—Partition Coefficient (n-Octanol/Water), High Performance Liquid Chromatography (HPLC)*

Method, 1989.

[8] Organisation for Economic Cooperation and Development, *OECD Guideline for the Testing of Chemicals 107—Partition Coefficient (n-Octanol/Water), Shake Flask Method*, 1995.

[9] A. J. Millais, R. J. Rycroft, M. A. Tolhurst, and D. A. Sheahan, "Bioconcentration: Comparison of Methods for Assessing Potential Hazards of Offshore Chemicals," Proceedings of the Chemistry in the Oil Industry X Symposium RSC/EOSCA, Manchester, UK, 5–7 November 2007.

[10] A. Karcher, H. Wiggins, I. Robb, and J. M. Wilson, "A Method for Measuring n-Octanol/Water Partition Coefficients for Non-ionic Surfactants," Proceedings of the Chemistry in the Oil Industry X Symposium RSC/EOSCA, Manchester, UK, 5–7 November 2007.

[11] N. Nyholm, "The European System of Standardized Legal Tests for Assessing the Biodegradability of Chemicals," *Environmental Toxicology and Chemistry* 10 (1991): 1237.

[12] L. Del Villano, R. Kommedal, and M. A. Kelland, "Class of Kinetic Hydrate Inhibitors with Good Biodegradability," *Energy Fuels* 22 (2008): 3143.

[13] M. M. Jordan, N. Feasey, C. Johnston, D. Marlow, and M. Elrick, "Biodegradable Scale Inhibitors: Laboratory and Field Evaluation of 'Green' Carbonate and Sulfate Scale Inhibitors with Deployment Histories in the North Sea," Proceedings of the Chemistry in the Oil Industry X Symposium RSC/EOSCA, Manchester, UK, 5–7 November 2007.

[14] ECETOC Technical Report No. 20 (1986) Annex III of OECD 1992 301 and ISO Guidance Document ISO 10634.

[15] R. S. Boethling, E. Sommer, and D. DiFiore, "Designing Small Molecules for Biodegradability," *Chemical Reviews* 107 (2007): 2207.

[16] Available at http://www.cefas.co.uk/offshore-chemical-notification-scheme-(ocns)/hazard-assessment.aspx.

[17] D. Sheahan, J. Girling, P. Neall, R. Rycroft, S. Thompson, M. Tolhurst, L. Weiss, M. Kirby, and A. Millais, "Evaluating and Forecasting Trends in Chemical Use and Impacts by the UK Offshore Oil and Gas Industry; Management and Reduction of Use of Those Substances Considered of Greatest Environmental Concern, CEFAS," Proceedings of the Chemistry in the Oil Industry X: Oilfield Chemistry, Royal Society of Chemistry, Manchester, UK, 5–7 November 2007.

[18] "Supplementary Guidance for the Completing of Harmonized Offshore Chemical Notification Format (HOCNF) 2000 for Norwegian Sector," *Harmonized Offshore Chemical Notification Format OSPAR Recommendation 2000/5*.

[19] J. Beyer, A. Skadsheim, M. A. Kelland, K. Alfsnes, and S. Sanni, "Ecotoxicology of Oilfield Chemicals: The Relevance of Evaluating Low-Dose and Long-Term Impact on Fish and Invertebrates in Marine Recipients," SPE 65039 (paper presented at the SPE International Symposium on Oilfield Chemistry, Houston, TX, 13–16 February 2001).

[20] J. M. Getliff and S. G. James, "The Replacement of Alkylphenol Ethoxylates to Improve the Environmental Acceptability of Drilling Fluid Additives," SPE 35982 (paper presented at the International Conference on Health, Safety & Environment, New Orleans, LA, June 1996).

[21] H. Rufli, P. R. Fisk, A. E. Girling, J. M. H. King, R. Lange, X. Lejeuned, N. Stelter, C. Stevens, P. Suteau, J. Tapp, J. Thus, D. J. Versteeg, and H. J. Niessend, "Aquatic Toxicity Testing of Sparingly Soluble, Volatile, and Unstable Substances and Interpretation and Use of Data," *Ecotoxicology and Environmental Safety* 39 (1998): 72–77.

[22] W. Kirit, "Whole Effluent Toxicity of Offshore Discharge Water—Methodology and Testing of Produced

Water in the UK," Tekna Produced Water Management Conference, Stavanger, Norway, 23-25 January 2012.

[23] H. J. Klimisch, M. Andreae, and U. Tillmann, "A Systematic Approach for Evaluating the Quality of Experimental Toxicological and Ecotoxicological Data," *Regulatory Toxicology and Pharmacology* 25 (1997): 1-5.

[24] M. A. La Vedrinea, E. J. Smitha, A. Millaisa, D. Dorana, L. Jonesa, R. Rowlesa, S. Supplea, and J. Corton, "The Impact of Active Compound Application Methods and Other Factors on the Validity of Ecotoxicity Study Data," RSC Chemistry in Oil Industry XI, Manchester, UK, 2-4 November 2009.

[25] D. J. Miller, "Octanol-Water Distribution Coefficients for Surfactants," Proceedings of the Chemistry in the Oil Industry X Symposium RSC/EOSCA, Manchester, UK, 5-7 November 2007.

[26] G. Fraser, "Method for Determining the Bioconcentration Factor of Linear Alcohol Ethoxylates," SPE 123846, Offshore Europe, Aberdeen, UK, 8-11 September 2009.